A B S T R A C T
ALGEBRA

A B S T R A C T
ALGEBRA

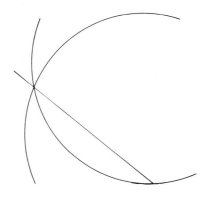

D E N N I S K L E T Z I N G
Stetson University

HBJ

HARCOURT BRACE JOVANOVICH, PUBLISHERS
and its subsidiary, Academic Press
San Diego New York Chicago Austin Washington, D.C.
London Sydney Tokyo Toronto

To my mother and memory of my father.

ISBN: 0-15-500391-7

Library of Congress Catalog Card Number: 90-83664

Printed in the United States of America

PREFACE

Abstract Algebra is an undergraduate textbook designed to introduce the junior/senior student to the basic concepts of abstract algebra. The text begins with a review chapter and is then divided into three parts: groups, rings, and fields. Each part contains several chapters of core material as well as additional selected topics that may be used to supplement the basic course.

Part One consists of Chapters 2, 3, and 4 and addresses the basic concepts of group theory. These chapters introduce students to the language of groups, subgroups, cyclic groups, group-homomorphism, group-isomorphism, and the construction of quotient groups, as well as to specific groups, such as the symmetric groups, the group of integers modulo n, and various groups of geometric transformations, such as the group of the triangle. There are four additional chapters of selected topics, including a more complete discussion of the representation of groups by means of permutations begun in Chapter 4, the theory of conjugation, p-groups and the Sylow theory, and the geometric groups. The last selected topic, Chapter 8, The Geometric Groups, is designed to show students how familiar groups of geometric transformations of the Euclidean plane, such as the group of the triangle, extend to higher dimensions and how the geometry and algebra of Euclidean space interact to give a richer and more meaningful understanding of both the geometry and algebra of these transformations and their groups.

Part Two consists of Chapters 9 and 10 and concerns the basic concepts of ring theory, including subrings, ideals, ring-homomorphism, ring-isomorphism, and the construction of quotient rings. These concepts are illustrated via the ring of integers, the ring of integers modulo n, the ring of Gaussian integers, rings of matrices and functions, and the ring of quaternions, among others. Chapter 10 then concentrates on commutative rings with identity, and includes a discussion of integral domains and their quotient fields; ideals, including prime and maximal ideals and their quotient rings; and concludes with a discussion of polynomial rings and their properties. In addition, there is a supplementary chapter on unique factorization domains that introduces students to basic factorization theory, including Euclidean domains and principal ideal domains.

Part Three, consisting of Chapters 12, 13, and 14, introduces students to the basic concepts of field theory and the Galois theory of fields. The Galois theory begins with a discussion and characterization of normal field extensions, both in terms of splitting fields as well as in terms of the degree of the extension, and concludes with a discussion and proof of the fundamental theorem of Galois theory. Applications to finite fields, cyclotomic fields and roots of unity, symmetric polynomials, and a proof of the fundamental theorem of algebra illustrate the basic ideas and significance of Galois theory. In this part two additional topics are included: the solvability of algebraic equations by means of radicals—a topic not only of mathematical importance but also of historical significance since it led to the evolution of group theory—and Euclidean constructibility.

In addition to the mathematical content of the book, three appendices are included which deal with the history and evolution of groups, rings, and fields. Virtually all of the basic concepts of abstract algebra were conceived of and formalized during the eighteenth century, a time that was phenomenally rich in mathematical as well as general intellectual activity. The evolution of the fundamental concepts of abstract algebra is traced in these appendices through the life and times of prominent mathematicians, providing a basis for understanding how and why the concepts emerged from the existing mathematical climate.

The book provides an extensive number of examples. Examples are an important part of any mathematics course; examples that are worked out in explicit detail, not just described in general terms. Some of the examples are routine and are designed to illustrate a particular result; others are more extended and expository in nature. In choosing examples, two goals were kept in mind: first, a specific example, worked out in detail, should show students how the theory works in practice; and second, an example should provide a model for students to use in doing their own problems. This last point—using an example as a guide for doing problems—is especially important in abstract mathematics since most students find it difficult to make the transition from a theoretical discussion to computing a specific problem. For these reasons the text contains a large number of examples, most of which are worked out and discussed with a great deal of care and detail.

It is assumed that students are familiar with elementary set theory, functions, equivalence relations, natural numbers, and mathematical induction, although these topics are reviewed at the beginning of the book. It also is assumed that students are familiar with the basic concepts and results of linear algebra. Linear algebra is an essential component of the mathematics curriculum today and most students are introduced to it at an early stage in their undergraduate training. It is important because it provides both a technical language as well as a geometric framework in which to visualize many different mathematical concepts. Linear algebra provides a way to interpret algebraic concepts within the framework of geometry and thus to explore some of the important relationships between algebra and geometry.

By the time students come to the study of abstract mathematics at the junior/senior level, they have a rich heritage of mathematical knowledge and experience. I believe that a course in abstract algebra should build on this knowledge and experience and should evolve naturally as an extension of material with which students already are familiar. For example, they are familiar with various types of number systems, such as the integers and the real numbers; with vectors, matrices, and functions; and with techniques for manipulating and dealing with these objects. In most cases they have studied these objects from a single perspective, concentrating on the properties of individual objects rather than on general properties common to many different types of objects. A primary goal in the study of abstract algebra is the construction of algebraic models, or structures, called groups, rings, and fields that reflect or model the properties common to these different types of mathematical objects. By studying the abstract properties of these structures, unclouded by the particular details of any one system, results can be obtained that apply to any and all such systems. Studying mathematics at the abstract level therefore serves not only to unify material that students already are familiar with from previous courses, but also to demonstrate the power of the abstract approach. By letting the basic concepts of abstract algebra evolve naturally from everyday mathematical experiences, it is hoped that students will gain a better understanding of abstract algebra and its role within mathematics itself.

ACKNOWLEDGMENTS

For their guidance and suggestions, I would like to thank the following reviewers: Vasily C. Cateforis, State University College at Potsdam; Morton L. Curtis, Rice University; Stephen L. Davis, Davidson College; Robert Etter, California State University, Sacramento; Frederick Hoffman, Florida Atlantic University; Ellen Kirkman, Wake Forest University; Ancel C. Mewborn, University of North Carolina, Chapel Hill; Hiram Paley, University of Illinois, Urbana; George D. Poole, East Tennessee State.

I would like to express a word of thanks to the professionals at Harcourt Brace Jovanovich: Richard Wallis for his patience and encouragement over the years; Michael Johnson, senior acquisitions editor, for his encouragement and continued support; Pamela Whiting, associate editor; Zanae Rodrigo, manuscript editor; Michael Biskup, production editor; Kay Faust, designer; Elizabeth Banks, art editor; and Lynne Bush, production management.

Finally, I must add a special word of appreciation to my wife for all her patience and understanding throughout the writing of this textbook.

Dennis Kletzing

CONTENTS

A B S T R A C T
ALGEBRA

1

A Review of Set Theory and Linear Algebra

As noted in the Preface, our goal in the study of abstract algebra is the construction of general algebraic structures that reflect, or model, the algebraic properties common to familiar mathematical objects such as numbers, functions, and matrices. Let us begin by briefly reviewing these topics and, at the same time, indicate our basic notation and terminology.

1. SET THEORY

By a *set* we mean, informally, any well-defined collection of objects. We denote sets by using uppercase italic letters, although certain sets used frequently have special symbols:

$\mathbb{N}$ the set of natural numbers
$\mathbb{Z}$ the set of integers
$\mathbb{Q}$ the set of rational numbers
$\mathbb{R}$ the set of real numbers
$\mathbb{C}$ the set of complex numbers

The objects in a set are called its *elements*, and we indicate that an object x is or is not an element of a set S by writing $x \in S$ or $x \notin S$, respectively. Sometimes it is convenient to talk about a set that has no elements; we call this set the *null set* and denote it by the symbol $\emptyset$. Sets may be described either by indicating the elements of the set explicitly or by indicating a property that identifies certain objects as belonging to the set. For example, $A = \{1, 2\}$

means that A is the set containing two elements, the numbers 1 and 2, while $B = \{n \in \mathbb{N} \mid n \text{ is even}\}$ means that B is the set of even natural numbers.

If A and B are sets such that every element of A is also an element of B, we say that A is a *subset* of B and write $A \subseteq B$, or equivalently, $B \supseteq A$. By convention, the null set is a subset of every set. If there is at least one element in B that is not an element of A, we say that A is a *proper subset* of B and write $A \subsetneq B$. If A is a set, the *power set* of A is the set $\mathscr{P}(A)$ consisting of all subsets of A. For example, if $A = \{1, 2\}$, then $\mathscr{P}(A) = \{\varnothing, \{1\}, \{2\}, \{1, 2\}\}$. Finally, two sets are *equal* if and only if they contain precisely the same elements; that is, $A = B$ if and only if $A \subseteq B$ and $B \subseteq A$. In this case, $x \in A$ if and only if $x \in B$.

There are three important ways of combining sets; union, intersection, and complementation. To describe these operations, let S be a set and let A and B be subsets of S. The *union* of A and B is the subset of S consisting of those elements belonging to either A or B and is denoted by $A \cup B$; thus, $A \cup B = \{x \in S \mid x \in A \text{ or } x \in B\}$. Recall that the connective "or" is used mathematically in the inclusive sense; that is, to say that $x \in A$ or $x \in B$ means that x is an element of either set A or B, *or possibly both sets*. The *intersection* of A and B is the subset of S consisting of those elements belonging to both A and B and is denoted by $A \cap B$; thus, $A \cap B = \{x \in S \mid x \in A \text{ and } x \in B\}$. If A and B have no elements in common, then $A \cap B = \varnothing$, and we say that A and B are *disjoint*. Finally, the *complement* of A in S is the subset of S consisting of those elements not in A and is denoted by $S - A$; thus, $S - A = \{x \in S \mid x \notin A\}$. Figures 1–3 illustrate the union, intersection, and complement of subsets. The following result summarizes the basic properties of these operations.

PROPOSITION 1. Let A, B, and C be subsets of a set S. Then:

(1) $A \cup (B \cup C) = (A \cup B) \cup C$

$\quad A \cap (B \cap C) = (A \cap B) \cap C$

(2) $A \cup (B \cap C) = (A \cup B) \cap (A \cup C)$

$\quad A \cap (B \cup C) = (A \cap B) \cup (A \cap C)$

(3) $S - (S - A) = A$

$\quad A \cup (S - A) = S$

$\quad$ if $A \subseteq B$, then $S - B \subseteq S - A$

(4) $S - (A \cup B) = (S - A) \cap (S - B)$

$\quad S - (A \cap B) = (S - A) \cup (S - B)$

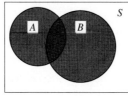

FIGURE 1. $A \cup B$

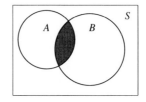

FIGURE 2. $A \cap B$

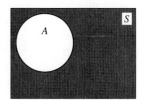

FIGURE 3. $S - A$

Proof.   Let us prove the first equality in statement (2) and leave the remaining proofs as an exercise for the reader. Let $x \in A \cup (B \cap C)$. Then x is either an element of A, in which case it belongs to both $A \cup B$ and $A \cup C$ and therefore to their intersection, or x belongs to $B \cap C$, in which case it belongs to both B and C and hence also to the intersection $(A \cup B) \cap (A \cup C)$. In each case x lies in the intersection of $A \cup B$ and $A \cup C$ and it follows, therefore, that $A \cup (B \cap C) \subseteq (A \cup B) \cap (A \cup C)$. Conversely, if $x \in (A \cup B) \cap (A \cup C)$, then x belongs to both of the sets $A \cup B$ and $A \cup C$. Therefore x is either an element of A or, if not, both B and C; in either case, $x \in A \cup (B \cap C)$. Hence, $(A \cup B) \cap (A \cup C) \subseteq A \cup (B \cap C)$, and therefore $A \cup (B \cap C) = (A \cup B) \cap (A \cup C)$. ∎

The equations in statement (4) of Proposition 1 are usually referred to as De Morgan's laws for set complementation, named after the British mathematician August De Morgan (1806–1871); they describe the complement of the union and intersection of two sets in terms of the complements of each of the sets. Also note that the operations of set union and intersection, as well as their properties, are easily extended to any finite number of sets. If $A_1, \ldots, A_n$ are a finite number of subsets of a set S, then the union $A_1 \cup \cdots \cup A_n$ is the set of all elements in S that belong to any one or more of the subsets $A_1, \ldots, A_n$ and is written as $\bigcup_{i=1}^{n} A_i$, while the intersection $A_1 \cap \cdots \cap A_n$ is the set of all elements in S that belong to each of the subsets $A_1, \ldots, A_n$ and is written as $\bigcap_{i=1}^{n} A_i$.

PROPOSITION 2. Let A be a set containing n elements. Then A has 2^n subsets.

Proof.   We prove this result by enumerating the subsets of A according to the number of elements they contain; first the null set, which contains no elements, then the subsets containing one element, followed by those containing two elements, until we arrive at the set A itself which contains n elements. To this end, let j be any number between 0 and n. Then the number of subsets containing exactly j elements is equal to the binomial coefficient $\binom{n}{j}$, which represents the number of ways of choosing j objects from n objects. The number of subsets of A is therefore equal to the sum $\binom{n}{0} + \binom{n}{1} + \cdots + \binom{n}{n}$. But $\binom{n}{0} + \binom{n}{1} + \cdots + \binom{n}{n} = (1 + 1)^n = 2^n$. Hence, A has 2^n subsets. ∎

If A is a finite set, we let $|A|$ stand for the number of elements in A. If A is a finite set containing n elements, then by Proposition 2, A has 2^n subsets and hence $|\mathscr{P}(A)| = 2^n$. For example, if $A = \{1, 3, 5\}$, then $\mathscr{P}(A) = \{\varnothing, \{1\}, \{3\}, \{5\}, \{1, 3\}, \{1, 5\}, \{3, 5\}, \{1, 3, 5\}\}$. In this case, $|A| = 3$ and $|\mathscr{P}(A)| = 8 = 2^3$.

Finally, let us recall that if A and B are sets, the *Cartesian product* of A and B is the set $A \times B = \{(a, b) \mid a \in A, b \in B\}$ consisting of all ordered pairs (a, b), where $a \in A$ and $b \in B$. By definition, two ordered pairs are equal if and only if their corresponding components are equal; that is, if $a, a' \in A$

and b, $b' \in B$, then $(a, b) = (a', b')$ if and only if $a = a'$ and $b = b'$. For example, if $A = \{1, 2\}$ and $B = \{2, 3\}$, then $A \times B = \{(1, 2), (1, 3), (2, 2), (2, 3)\}$. In general, if A and B are finite sets, then $A \times B$ is a finite set and $|A \times B| = |A||B|$. Note that for any set A, $A \times \varnothing = \varnothing \times A = \varnothing$.

Exercises

1. Let $S = \{1, 2, 3, 4, 5, 6\}$ and let $A = \{1, 2\}$, $B = \{2, 3\}$, $C = \{3, 4\}$, $D = \{4, 5\}$ and $E = \{5, 6\}$.
    (a) Find $A \cup B \cup C \cup D \cup E$.
    (b) Find $A \cap B \cap C \cap D \cap E$.
    (c) Write $S - A$ as a union of some of the five subsets A, B, C, D, and E.
    (d) Find $A \cup (B \cap C)$, $A \cup B$, and $A \cup C$, and verify that $A \cup (B \cap C) = (A \cup B) \cap (A \cup C)$.
    (e) Verify that $S - (A \cup B) = (S - A) \cap (S - B)$.
2. Let A and B be finite sets.
    (a) Show that $|A \cup B| = |A| + |B| - |A \cap B|$.
    (b) Show that $|A \cup B| = |A| + |B|$ if and only if A and B are disjoint.
3. Let $A = \{n \in \mathbb{N} \mid n$ is a multiple of $6\}$ and $B = \{n \in \mathbb{N} \mid n$ is a multiple of $10\}$.
    (a) Find the set $A \cap B$.
    (b) Find the set $A \cup B$.
    (c) Find the set $\mathbb{N} - B$.
    (d) Find the set $(A \cup B) - B$. Does $(A \cup B) - B = A$?
4. Let A and B be subsets of a set S such that $S = A \cup B$. If X is any subset of S, show that $X = (X \cap A) \cup (X \cap B)$.
5. Let S be a set and let $\mathscr{S} = \{S_\alpha \mid \alpha \in I\}$ be a collection of subsets of S indexed by some set I. Define the union of the sets in $\mathscr{S}$ to be the set $\bigcup_{\alpha \in I} S_\alpha = \{x \in S \mid x \in S_\alpha$ for some element $\alpha \in I\}$, consisting of those elements in S that belong to any one of the subsets S_α, and the intersection of the sets in $\mathscr{S}$ to be the set $\bigcap_{\alpha \in I} S_\alpha = \{x \in S \mid x \in S_\alpha$ for every element $\alpha \in I\}$, consisting of those elements in S that belong to every one of the sets S_α.
    (a) Show that $S - \bigcap_{\alpha \in I} S_\alpha = \bigcup_{\alpha \in I} (S - S_\alpha)$.
    (b) Show that $S - \bigcup_{\alpha \in I} S_\alpha = \bigcap_{\alpha \in I} (S - S_\alpha)$.
    (c) For each real number $\varepsilon > 0$, let $S_\varepsilon = (1 - 1/\varepsilon, 1 + 1/\varepsilon)$ stand for the open interval from $1 - 1/\varepsilon$ to $1 + 1/\varepsilon$. Find $\bigcup_{\varepsilon > 0} S_\varepsilon$ and $\bigcap_{\varepsilon > 0} S_\varepsilon$.
6. Let $A = \{0, 1\}$.
    (a) Find $\mathscr{P}(A)$. Is $A \subseteq \mathscr{P}(A)$?
    (b) Find $A \times A$. Is $A \subseteq A \times A$?
7. Let S be a finite set containing n elements and let $x \in S$.
    (a) How many subsets of S contain the element x?
    (b) How many subsets of S do not contain x?

8. Let A and B be subsets of a set S. The *symmetric difference* of A and B is the set $A \triangle B = (A \cup B) - (A \cap B)$.
   (a) Show that $A \triangle A = \varnothing$ and $A \triangle \varnothing = A$.
   (b) Show that $A \triangle B = B \triangle A$.
   (c) Show that $A \triangle (B \triangle C) = (A \triangle B) \triangle C$.

2.   FUNCTIONS

Functions, like sets, occur everywhere in mathematics and, like sets, have their own special terminology and properties. Since we will be using these terms and properties throughout the book, let us take a moment to review them. Recall that if X and Y are nonempty sets, a *function*, or *mapping*, from X into Y is a rule $f : X \to Y$ that associates with each element in X a uniquely determined element in Y. If $x \in X$, the uniquely determined element in Y that f associates to x is denoted by $f(x)$ and is called the *image* of x; in this case we write $x \mapsto f(x)$ and say that x maps to $f(x)$ under f. The set X is called the *domain* of f, while the set of images of elements in X is called the *image* of f and is denoted by $\operatorname{Im} f$. Thus, $\operatorname{Im} f = \{ f(x) \in Y \mid x \in X \}$.

For example, let $f : \mathbb{R} \to \mathbb{R}$ stand for the function defined by setting $f(x) = x^2 + 1$ for all real numbers x, as illustrated in Figure 1. Then

$$\operatorname{Im} f = \{ x^2 + 1 \in \mathbb{R} \mid x \in \mathbb{R} \}$$
$$= \{ a \in \mathbb{R} \mid a \geq 1 \}$$
$$= [1, \infty);$$

that is, the image of f consists of the half-line of real numbers that are greater than or equal to 1. Let us make two observations about the function f. First, not every real number is in its image; the number 0, for example, cannot be written in the form $x^2 + 1$ for some $x \in \mathbb{R}$ and hence is not in the image. Second, observe that there are distinct real numbers that have the same image under f; for example, $f(-1) = f(1) = 2$. The function g illustrated in Figure 2 is similar to f in that it is defined by means of an arithmetic rule: to find $g(x)$, simply calculate $2x + 3$. But unlike f, we claim that every real number is in the image of g. For if y is any real number, we can easily solve the equation $y = g(x) = 2x + 3$ for x and find that $x = \dfrac{y - 3}{2}$. Thus, $g\left(\dfrac{y - 3}{2}\right)$

$$= 2\left(\frac{y - 3}{2}\right) + 3 = y$$ and therefore every real number is in the image of g.

Figures 3 and 4, on the other hand, illustrate two functions that are not defined by an arithmetic rule. In Figure 3, the function p maps the line segment L to the line segment M; the image of a point x on L is the projection of x to its image $p(x)$ on M, as illustrated in the figure. In Figure 4, the function C maps an arbitrary subset A of a given set S to its complement $S - A$; for example, $C(\varnothing) = S$ and $C(S) = \varnothing$.

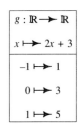

FIGURE 1.

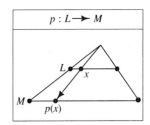

FIGURE 2.

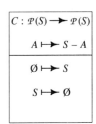

FIGURE 3.

FIGURE 4.

As the above examples show, a function may or may not map distinct elements of the domain to distinct elements of the image. A function $f : X \to Y$ is called a *1–1 function* if, whenever x and x' are distinct elements in X, $f(x)$ and $f(x')$ are distinct elements in Y; that is, if $x \neq x'$, then $f(x) \neq f(x')$. Thus, 1–1 functions preserve the distinctness of elements in X. To show that a given function f is 1–1, it is usually more convenient in practice to verify the contrapositive; that is, if $f(x) = f(x')$ for some $x, x' \in X$, then $x = x'$. For example, let us show that the function g in Figure 2 is 1–1. Suppose that x and x' are real numbers such that $g(x) = g(x')$. Then $2x + 3 = 2x' + 3$, from which it follows easily that $x = x'$. Hence, g is 1–1. The function C in Figure 4 is also a 1–1 function; for if $C(A) = C(A')$ for subsets A, A' of S, then $S - A = S - A'$ and therefore $S - (S - A) = S - (S - A')$, or equivalently, $A = A'$. And finally, we leave the reader to verify that the projection function $p : L \to M$ in Figure 3 is also a 1–1 function. But the function f in Figure 1 that maps a real number x to $x^2 + 1$ is not 1–1 since, for example, 1 and -1 are distinct real numbers that have the same image under f.

The above examples also show that if $f : X \to Y$ is a function, then not every element in Y is necessarily in the image of f. We say that f is an *onto* function, or that f maps X onto Y, if every element in Y is the image under f of some element in X. The function g discussed previously, for example, is an onto function since, if y is any real number, $g\left(\dfrac{y - 3}{2}\right) = y$ and hence y is in the image of g. Similarly, we leave the reader to verify that the projection function p and the complementation function C are both onto functions. When a function $f : X \to Y$ is both 1–1 and onto, every element in Y is the image of exactly one element in X, and consequently, we may pair-off the elements in X with those in $Y : x \leftrightarrow f(x)$. In this case we say that the function is a *1–1 correspondence*. For example, the function g, the projection mapping p, and the complementation function C just discussed are all 1–1 correspondences. The function f, however, is not a 1–1 correspondence.

One-to-one correspondences are important because they may be reversed, or inverted. For if $f : X \to Y$ is a 1–1 correspondence and $y \in Y$, there is a uniquely determined element $x \in X$ such that $f(x) = y$. Consequently, we may define a new function $g : Y \to X$ that maps y to x; that is, $g(y) = x$. Clearly, $g(f(x)) = g(y) = x$ for every element $x \in X$, and $f(g(y)) = f(x) = y$ for every element $y \in Y$. The function $g : Y \to X$ is called the *inverse* of f. For example, the function mapping $n \mapsto n + 1$ defines a 1–1 correspondence between the set $\mathbb{N}_E$ of even natural numbers, including zero, and the set $\mathbb{N}_O$ of odd natural numbers. In this case, the inverse function maps a typical number n to $n - 1$.

Let us now show that 1–1 correspondences are in fact the only functions that have an inverse. Recall that if $f : X \to Y$ is an arbitrary function, then a function $g : Y \to X$ is an inverse for f if $g(f(x)) = x$ for all $x \in X$ and $f(g(y)) = y$ for all $y \in Y$.

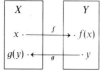

$\mathbb{N}_E$	$\leftrightarrow$	$\mathbb{N}_O$
n	$\leftrightarrow$	$n + 1$
0	$\leftrightarrow$	1
2	$\leftrightarrow$	3
4	$\leftrightarrow$	5
6	$\leftrightarrow$	7

PROPOSITION 1. Let $f:X \to Y$ be a function. Then f has an inverse if and only if f is a 1–1 correspondence. Moreover, if f has an inverse, it has only one inverse.

Proof. We showed above that if f is a 1–1 correspondence, then f has an inverse. Now suppose, conversely, that f has an inverse, say $g:Y \to X$. Then f is 1–1; for if x and x' are elements in X such that $f(x) = f(x')$, then $g(f(x)) = g(f(x'))$ and hence $x = x'$. Moreover, f is an onto mapping; for if $y \in Y$ and $x = g(y)$, then $f(x) = f(g(y)) = y$ and hence f maps x onto y. It follows, therefore, that f is 1–1 and onto and hence is a 1–1 correspondence. Finally, suppose that the function f has two inverses, say $g:Y \to X$ and $g':Y \to X$. Then $g = g'$; for if y is any element in Y, then $f(g(y)) = y = f(g'(y))$ and hence, since f is 1–1, $g(y) = g'(y)$. Therefore, $g = g'$ and the proof is complete. ∎

The nineteenth century German mathematician Georg Cantor (1845–1918) used the concept of 1–1 correspondence as the basis for comparing the size of two sets, infinite sets, in particular. We say that two sets X and Y have the same *cardinality* and write $|X| = |Y|$ if there is a 1–1 correspondence $f:X \to Y$. Intuitively, if two sets have the same cardinality, then their elements may be paired-off with one another and thus the sets have the same number of elements, although this statement must be interpreted cautiously when dealing with infinite sets. If a set is finite, it contains n elements for some integer n, and two finite sets, one containing n elements, the other m elements, have the same cardinality if and only if $n = m$. But for infinite sets, cardinality is deceptive. The projection mapping p illustrated in Figure 3, for example, is a 1–1 correspondence between line segment L and line segment M, and consequently, the sets L and M have the same cardinality. In other words, the two line segments contain the same number of points in the sense that the points on L may be paired off with those on M, even though M "looks bigger" than L. As another example, we previously showed that the mapping $n \mapsto n + 1$ is a 1–1 correspondence between the sets $\mathbb{N}_E$ and $\mathbb{N}_O$ of even and odd numbers; these sets therefore have the same cardinality, or said informally, there are as many even natural numbers as there are odd natural numbers.

A set X is said to be *countable* if it is either finite or if it is in 1–1 correspondence with the set $\mathbb{N}$ of natural numbers. In the later case there is a 1–1 correspondence $f:\mathbb{N} \to X$, which means that the elements in X may be enumerated, $a_1, a_2, \ldots$, where $a_i = f(i)$ for $i = 1, 2, \ldots$. For example, the set $\mathbb{Z}$ of integers is countable since the function $f:\mathbb{N} \to \mathbb{Z}$ defined by setting $f(2n) = n$ and $f(2n + 1) = -n$ for $n \in \mathbb{N}$ is a 1–1 correspondence.

Now recall that if $f:X \to Y$ and $g:Y \to Z$ are functions, then the *composition* of f and g is the function $g \circ f:X \to Z$ defined by setting $(g \circ f)(x) = g(f(x))$ for every element $x \in X$; in other words, begin with the element $x \in X$, map it to $f(x)$ under f, then map $f(x)$ to $g(f(x))$ under g. For

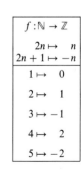

$f:\mathbb{N} \to \mathbb{Z}$
$2n \mapsto \quad n$
$2n + 1 \mapsto -n$
$1 \mapsto \quad 0$
$2 \mapsto \quad 1$
$3 \mapsto -1$
$4 \mapsto \quad 2$
$5 \mapsto -2$

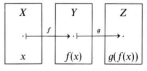

X	Y	Z
x	$f(x)$	$g(f(x))$

example, if $f: \mathbb{R} \to \mathbb{R}$ and $g: \mathbb{R} \to \mathbb{R}$ are defined by setting $f(x) = x^2 + 1$ and $g(x) = 2x + 3$ for all $x \in \mathbb{R}$, then

$$(g \circ f)(x) = g(f(x)) = g(x^2 + 1) = 2(x^2 + 1) + 3 = 2x^2 + 5$$

and

$$(f \circ g)(x) = f(g(x)) = f(2x + 3) = (2x + 3)^2 + 1 = 4x^2 + 12x + 10.$$

Note, in particular, that $g \circ f \neq f \circ g$. Thus, even though both composites $f \circ g$ and $g \circ f$ exist, they are not necessarily equal.

In conclusion, let us mention a function that plays an especially important role in algebra: the identity function. The *identity function* on a nonempty set X is the function $1_X: X \to X$ defined by setting $1_X(x) = x$ for all elements $x \in X$. In other words, the identity function maps every element of the set X to itself. For example, if $f: X \to Y$ is a 1–1 correspondence between two sets X and Y, then f has an inverse, say $g: Y \to X$, and $(g \circ f)(x) = g(f(x)) = x$ for all $x \in X$ and $(f \circ g)(y) = f(g(y)) = y$ for all $y \in Y$. Therefore, $g \circ f = 1_X$, the identity function on X, and $f \circ g = 1_Y$, the identity function on Y.

Exercises

1. Listed here are several functions. In each case, find the image of the function and determine if the function is 1–1.
   (a) $f: \mathbb{R} \to \mathbb{R}$, $f(x) = 5x + 3$
   (b) $f: \mathbb{R} \to \mathbb{R}$, $f(x) = x^2 + 4x - 1$
   (c) $f: [0, +\infty) \to \mathbb{R}$, $f(x) = x^3 - 2x^2 + x + 2$
   (d) $f: (0, +\infty) \to \mathbb{R}$, $f(x) = \ln x$, the natural logarithm of x
   (e) $f: \mathbb{R} \to \mathbb{R}$, $f(x) = 2^{x+1} - 3$
   (f) $f: \mathbb{R} \to \mathbb{R}$, $f(x) = \sin x$
   (g) $f: \mathbb{Z} \to \mathbb{Z}$, $f(n) = 3n + 2$
2. Let f stand for the rule that associates to each rational number n/m the number $(n + 1)/(m + 1)$. Is f a function from $\mathbb{Q}$ to $\mathbb{Q}$?
3. Let $f: \mathbb{R} \to \mathbb{R}$ stand for the function defined by setting $f(x) = 4x - 3$ for all $x \in \mathbb{R}$.
   (a) Show that f is a 1–1 correspondence.
   (b) Let $g: \mathbb{R} \to \mathbb{R}$ stand for the inverse of f. Find $g(x)$ in terms of x.
4. Let $f: [2, +\infty) \to [-3, +\infty)$ stand for the function defined by setting $f(x) = x^2 - 4x + 1$ for all real numbers $x \in [2, +\infty)$.
   (a) Show that f is a 1–1 correspondence.
   (b) Let $g: [-3, +\infty) \to [2, +\infty)$ stand for the inverse of f. Find $g(x)$ in terms of x.
5. Let X and Y be sets of real numbers and let $f: X \to Y$ be a function.
   (a) Show that f is a 1–1 correspondence if and only if every horizontal line $y = b$ intersects the graph of $y = f(x)$ in exactly one point for each number $b \in Y$.

(b) Suppose that f is a 1–1 correspondence and let $g:Y \to X$ stand for the inverse of f. Show that a point (a, b) is on the graph of $y = f(x)$ if and only if the point (b, a) is on the graph of $y = g(x)$. Conclude, therefore, that the graph of $y = g(x)$ may be obtained by reflecting the graph of $y = f(x)$ about the line $y = x$.

6. Let $S = \{1, 2, 3\}$.
   (a) Find $\mathscr{P}(S)$.
   (b) Let $C:\mathscr{P}(S) \to \mathscr{P}(S)$ stand for the complementation function defined by setting $C(A) = S - A$ for every subset $A \in \mathscr{P}(S)$. Find the image of each element in $\mathscr{P}(S)$.

7. Let $C:\mathscr{P}(S) \to \mathscr{P}(S)$ stand for the complementation function on a set S. Show that C is a 1–1 correspondence and is its own inverse.

8. Show that the set of even natural numbers is countable.

9. Show that the set of even integers is countable.

10. Let $f:X \to Y$ and $g:Y \to Z$ be 1–1 correspondences. Show that the composition $g \circ f:X \to Z$ is a 1–1 correspondence.

11. Let $f:X \to Y$ be a function. If A is a subset of X, let $f(A) = \{f(x) \in Y \mid x \in A\}$ stand for the set of images under f of the elements in A. If A and A' are subsets of X, show that
    (a) $f(A) \subseteq f(A')$ if $A \subseteq A'$
    (b) $f(A \cup A') = f(A) \cup f(A')$
    (c) $f(A \cap A') \subseteq f(A) \cap f(A')$

12. Let $f:X \to Y$ be a function. If B is a subset of Y, let $f^{-1}(B) = \{x \in X \mid f(x) \in B\}$ stand for the subset of elements in X whose image under f lies in B. If B and B' are subsets of Y, show that
    (a) $f^{-1}(B) \subseteq f^{-1}(B')$ if $B \subseteq B'$
    (b) $f^{-1}(B \cup B') = f^{-1}(B) \cup f^{-1}(B')$
    (c) $f^{-1}(B \cap B') = f^{-1}(B) \cap f^{-1}(B')$
    (d) $f^{-1}(Y - B) = X - f^{-1}(B)$

3.   EQUIVALENCE RELATIONS

Sometimes the elements of a set may be separated, or partitioned, into disjoint subsets according to some property that the elements have in common. Equivalence relations provide the mathematical tools for carrying out this partitioning. Recall that if X is a nonempty set, a *relation* on X is any subset R of the Cartesian product $X \times X$. If R is a relation on X and $(x, y) \in R$ for some elements $x, y \in X$, we write $x \, R \, y$ and say that x is related to y; if x is not related to y, we write $x \, \cancel{R} \, y$. For example, "less than or equal to" is a familiar relation on the set of real numbers; in this case the relation is denoted by the symbol $\leq$, and $x \leq y$ means that $y = x + r$ for some nonnegative real number r. Or, we may define a relation R on the set of integers as follows: if $n, m \in \mathbb{Z}$, define $n \, R \, m$ to mean that $n + 2m$ is even. In this case we find, for example, that $0 \, R \, 2$, $6 \, R \, -4$, and $2 \, R \, 1$, but $1 \, \cancel{R} \, 1$.

A relation R on a set X is called an *equivalence relation* on X if the following three conditions are satisfied:

(1) *reflexivity*: $x \, R \, x$ for all elements $x \in X$
(2) *symmetry*: if $x \, R \, y$ for some elements $x, y \in X$, then $y \, R \, x$
(3) *transitivity*: if $x \, R \, y$ and $y \, R \, z$ for some elements $x, y, z \in X$, then $x \, R \, z$

We denote equivalence relations by the symbol $\sim$. Thus, to say that $\sim$ is an equivalence relation on a set X means that $x \sim x$ for all $x \in X$; if $x \sim y$ for some $x, y \in X$, then $y \sim x$; and if $x \sim y$ and $y \sim z$ for some $x, y, z \in X$, then $x \sim z$.

For example, let $\mathscr{L}$ stand for the set of lines in the plane. Define a relation $\sim$ on $\mathscr{L}$ as follows: if $l, l' \in \mathscr{L}$, let $l \sim l'$ mean that l is parallel to l'. Then $\sim$ is an equivalence relation on $\mathscr{L}$. For if l is any line, then l is parallel to itself and hence $l \sim l$. Moreover, if l and l' are lines such that $l \sim l'$, then l is parallel to l' and therefore l' is parallel to l, which shows that $l' \sim l$. Finally, if $l \sim l'$ and $l' \sim l''$, then l is parallel to l', which is parallel to l'', and hence l is parallel to l'', which shows that $l \sim l''$. Thus, the relation of parallelism is an equivalence relation on the set of lines in the plane.

Now, let $\sim$ be an equivalence relation on a set X and let $x \in X$. The set $[x] = \{ y \in X \, | \, y \sim x \}$ of elements in X equivalent to x is called the *equivalence class* of x.

PROPOSITION 1. Let $\sim$ be an equivalence relation on a set X. Then:

(1) $x \in [x]$ for every element $x \in X$; that is, every element in X belongs to its own equivalence class;

(2) if $x, y \in X$, then $[x] = [y]$ if and only if $x \sim y$; that is, two elements determine the same equivalence class if and only if they are equivalent.

Proof. If $x \in X$, then $x \sim x$ and therefore $x \in [x]$, which proves statement (1). To prove the second statement, suppose that $[x] = [y]$ for some elements $x, y \in X$. Then, since $x \in [x]$, it follows that $x \in [y]$ and therefore $x \sim y$. On the other hand, if $x \sim y$ for some $x, y \in X$, then $[x] = [y]$. To show this, let $z \in [x]$. Then $z \sim x$. But $x \sim y$. Therefore, $z \sim y$ and hence $z \in [y]$. Thus, $[x] \subseteq [y]$. Similarly, $[y] \subseteq [x]$. Therefore $[x] = [y]$, as required. ∎

COROLLARY.   Equivalence classes are either disjoint or identical.

Proof. Let $\sim$ be an equivalence relation on a set X and let $[x]$ and $[y]$ be typical equivalence classes of $\sim$. If $[x]$ and $[y]$ are disjoint, we are done. Otherwise, there is an element $z \in [x] \cap [y]$ and hence $x \sim z$ and $z \sim y$. Therefore, $x \sim y$ and it follows that $[x] = [y]$. Thus, $[x]$ and $[y]$ are either disjoint or identical. ∎

If $\sim$ is an equivalence relation on a set X, we let $X/\sim$ stand for the set of equivalence classes of $\sim$. In view of the preceding corollary, equivalence classes separate, or partition, the set X into disjoint subsets. In general, a *partition* of X is any collection $\mathscr{P}$ of disjoint subsets of X such that every element in X lies in some subset of $\mathscr{P}$. Thus, an equivalence relation on a set X partitions the set into equivalence classes, as illustrated in the figure, and we write $X = \bigcup_{x \in X} [x]$ to indicate this partitioning.

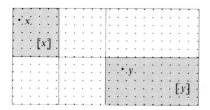

For example, consider the equivalence relation on the set $\mathscr{L}$ of lines in the plane just discussed. In this case, two lines are equivalent if and only if they are parallel. Consequently, the equivalence class $[l]$ of a line l consists of all lines parallel to l, as shown below.

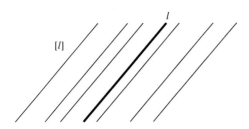

As a second example of an equivalence relation and its partitioning of the underlying set, let us discuss the congruence relation modulo 3 on the set $\mathbb{Z}$ of integers. If n and m are integers, we say that n is congruent to m modulo 3 and write $n \equiv m \pmod{3}$ if $n - m$ is a multiple of 3. Congruence mod 3 is clearly a reflexive relation since $n \equiv n \pmod{3}$ for all $n \in \mathbb{Z}$. It is also symmetric; for if $n \equiv m \pmod{3}$, then $n - m = 3s$ for some integer s. Therefore, $m - n = 3(-s)$ and hence $m \equiv n \pmod{3}$. And finally, if $n \equiv m \pmod{3}$ and $m \equiv r \pmod{3}$, then $n - m = 3s$ and $m - r = 3t$ for some integers s, t, and hence $n - r = (n - m) + (m - r) = 3(s + t)$. Therefore, $n \equiv r \pmod{3}$, which shows that congruence mod 3 is a transitive relation. Thus, congruence modulo 3 is an equivalence relation on the set $\mathbb{Z}$ of integers. The equivalence class of a typical integer n is $[n] = \{m \in \mathbb{Z} \mid m \equiv n \pmod{3}\} = \{m \in \mathbb{Z} \mid m - n = 3s, s \in \mathbb{Z}\} = \{n + 3s \in \mathbb{Z} \mid s \in \mathbb{Z}\}$. For example,

$$[0] = \{3s \in \mathbb{Z} \mid s \in \mathbb{Z}\} = \{\ldots, -3, 0, 3, \ldots\}$$

$$[1] = \{1 + 3s \in \mathbb{Z} \mid s \in \mathbb{Z}\} = \{\ldots, -2, 1, 4, \ldots\}$$

$$[2] = \{2 + 3s \in \mathbb{Z} \mid s \in \mathbb{Z}\} = \{\ldots, -1, 2, 5, \ldots\}$$

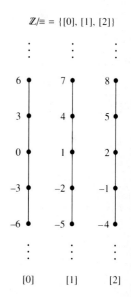

We find, for example, that $[11] = [2]$ since $11 \equiv 2 \pmod 3$. In fact, if n is any integer and r is the remainder of n upon division by 3, then $n \equiv r \pmod 3$ since $n = 3q + r$ for some $q \in \mathbb{Z}$; that is, every integer is congruent mod 3 to its remainder upon division by 3. It follows, therefore, that $[n] = [r]$. Moreover, since the remainder of n upon division by 3 is either 0, 1, or 2, $[n]$ is either $[0]$, $[1]$, or $[2]$. Thus, the three equivalence classes just listed represent all equivalence classes of the congruence mod 3 relation, and hence $\mathbb{Z}/\!\equiv \, = \{[0], [1], [2]\}$.

We have shown that every equivalence relation on a set partitions the set into disjoint subsets called equivalence classes. Let us conclude by showing that every partitioning of a set is in fact obtained from an equivalence relation in this manner. Suppose that $\mathscr{P}$ is a partition of a set X. Then $\mathscr{P}$ consists of disjoint subsets of X, and every element of X lies in exactly one subset in $\mathscr{P}$. Now, define a relation $\sim$ on X as follows: if $x, y \in X$, let $x \sim y$ mean that x and y lie in the same subset of $\mathscr{P}$. Then $x \sim x$ for every element $x \in X$. Moreover, if $x \sim y$ for some $x, y \in X$, then x and y lie in the same subset of $\mathscr{P}$. Hence, y and x lie in the same subset and therefore $y \sim x$. Finally, if $x \sim y$ and $y \sim z$ for some $x, y, z \in X$, then it is clear that x, y, and z all lie in the same subset of $\mathscr{P}$ and hence $x \sim z$. Thus, $\sim$ is an equivalence relation on X. If A is one of the subsets in the partition $\mathscr{P}$ and $x \in A$, then $[x] = A$ since the equivalence class $[x]$ consists of those elements in X that lie in the same subset as x, namely A. Thus, the equivalence classes of $\sim$ are just the subsets in the partition $\mathscr{P}$. We conclude, therefore, that equivalence relations on a set partition the set, and conversely, partitions of the set define equivalence relations on the set.

Exercises

1. Let $\mathscr{S}$ stand for the set of squares in the plane. Define a relation $\sim$ on $\mathscr{S}$ as follows: if S and S' are squares in the plane, $S \sim S'$ means that S is congruent to S'.
   (a) Show that $\sim$ is an equivalence relation on $\mathscr{S}$.
   (b) Describe the equivalence class of a typical square.
2. Let $\mathscr{P}$ stand for the set of points in the Cartesian plane. Define a relation $\sim$ on $\mathscr{P}$ as follows: if P and Q are points in the plane, $P \sim Q$ means that P and Q lie at the same distance from the origin.
   (a) Show that $\sim$ is an equivalence relation on $\mathscr{P}$.
   (b) Describe geometrically the equivalence class of a typical point.
3. Let $\leq$ stand for the usual "less than or equal to" relation on the set $\mathbb{Z}$ of integers. Is $\leq$ an equivalence relation on $\mathbb{Z}$?
4. Define the congruence relation modulo 6 on the set of integers as follows: if $n, m \in \mathbb{Z}$, $n \equiv m \pmod 6$ means that $n - m$ is a multiple of 6.
   (a) Show that congruence mod 6 is an equivalence relation on $\mathbb{Z}$.
   (b) Find the equivalence classes of the congruence mod 6 relation.
   (c) Illustrate the equivalence classes by means of a diagram.

5. Let S be a nonempty set. Define a relation $\sim$ on the power set $\mathcal{P}(S)$ as follows: if $A, B \in \mathcal{P}(S)$, $A \sim B$ means that A and B have the same cardinality.
   (a) Show that $\sim$ is an equivalence relation on $\mathcal{P}(S)$.
   (b) If S is a finite set, describe the equivalence class of a typical subset.
   (c) If S is an infinite set, show by means of an example that S may contain a proper subset that is equivalent to S.
6. Let $f : X \to Y$ be a function. Define a relation $\sim$ on X as follows: if x, $x' \in X$, $x \sim x'$ means that $f(x) = f(x')$.
   (a) Show that $\sim$ is an equivalence relation on X.
   (b) Describe the equivalence class of an element $x \in X$.
   (c) Let $f : \mathbb{R} \to \mathbb{R}$ stand for the function defined by setting $f(x) = x^2$ for all real numbers x. Find the equivalence class of a typical real number x.
   (d) Let $f : \mathbb{R} \to \mathbb{R}$ stand for the function defined by setting $f(x) = \sin x$ for all real numbers x. Find the equivalence class of a typical real number x.

4. THE NATURAL NUMBERS

By the natural numbers we mean the ordinary counting numbers $0, 1, 2, \ldots$. Do we take the natural numbers for granted, or do we attempt to construct them axiomatically from more primitive concepts such as sets? The nineteenth century German mathematician Leopold Kronecker (1823–1891) once remarked, for example, that "God made the integers, all the rest is the work of man." He was strongly opposed to the axiomatic developments of the various number systems that were being formulated during the later part of the nineteenth century, especially when the concept of infinity came into the picture. Perhaps the most well-known axiomatic treatments of the number system that evolved during this time are the ones due to Richard Dedekind (1831–1916) and Giuseppe Peano (1858–1932). Dedekind's work, *The Nature and Meaning of Numbers*, was published in 1887 and is still in print today, while Peano's construction of the natural numbers was published as a pamphlet, *Principles of Arithmetic*, in 1889. Peano chose set, set membership, and successor as part of his basic concepts, and he then laid out five axioms from which he began the systematic development of the number system which today bear his name:

(1) Zero is a natural number.

(2) If n is a natural number, then its successor n' is a uniquely determined natural number.

(3) If n and m are natural numbers such that $n' = m'$, then $n = m$.

(4) Zero is not the successor of any natural number.

(5) If S is a set of natural numbers that contains zero and contains the successor n' whenever it contains n, then S is the set of all natural numbers.

The arithmetic of natural numbers is then defined as follows. If n is a natural number, the sum $n + 1$ is defined to be the successor n', while the sum $n + m$ is defined recursively, in general, by setting $n + (m + 1) = (n + m) + 1$. For products, we set $0n = 0$ for all numbers n, and $n(m + 1) = nm + n$ for all n and m. Finally, since the natural numbers are defined sequentially, we may define an order relation on the numbers by setting $n \leq m$ whenever $m = n + s$ for some number s.

The fifth Peano axiom is referred to as the *principle of mathematical induction*. It states that if a mathematical statement about the natural numbers is true for the number zero, and if its truth for any number n implies its truth for $n + 1$, then the statement is true for all natural numbers. Mathematical induction is an especially useful method for proving certain statements about the natural numbers. Let us now use it to prove two important properties of the natural number system.

PROPOSITION 1.   (THE   WELL-ORDERING   PRINCIPLE)
The set of natural numbers is well-ordered by the relation $\leq$; that is, every nonempty set of natural numbers contains a least number.

Proof.   Let S be a nonempty set of natural numbers and let $S^* = \{n \in \mathbb{N} \mid n \leq s$ for all $s \in S\}$ stand for the set of numbers that are less than or equal to every number in S. Clearly, $0 \in S^*$. On the other hand, S^* is not the entire set of natural numbers since there is at least one number in S. Thus, by the principle of induction, there is some number m such that $m \in S^*$ but $m + 1 \notin S^*$. Then $m \leq s$ for all $s \in S$. If $m < s$ for all $s \in S$, then $m + 1 \leq s$ for all $s \in S$ and hence $m + 1 \in S^*$, which is a contradiction. Therefore, $m = s$ for some $s \in S$ and hence $m \in S$. Thus, m is the least element in S and the proof is complete.   ■

PROPOSITION 2.   (THE PRINCIPLE OF STRONG INDUCTION)   If S is a set of natural numbers containing zero, and if S contains a number n whenever it contains all numbers less than n, then S is the set of all natural numbers.

Proof.   Let $S^* = \{n \in \mathbb{N} \mid n \notin S\}$. If the set S is not the entire set of natural numbers, then the set S^* is a nonempty set of numbers and hence, by the Well-Ordering Principle, contains a least number. Let $n \in S^*$ be the smallest number in S^*. Then every number smaller than n must belong to S and hence, by assumption, $n \in S$, which is a contradiction. Therefore, S is the entire set of natural numbers.   ■

The principle of induction is useful in proving statements about the natural numbers when the truth of the statement for $n + 1$ is easily verified from the truth of the statement for n. For example, let us prove by induction that $1 + 2 + \cdots + n = n(n + 1)/2$ for all natural numbers n. This statement is clearly true when $n = 1$. Now, if the equation is true for n, then its truth for $n + 1$ follows easily since

$$1 + 2 + \cdots + (n + 1) = (1 + 2 + \cdots + n) + (n + 1)$$
$$= \frac{n(n + 1)}{2} + (n + 1)$$
$$= \frac{n^2 + 3n + 2}{2}$$
$$= \frac{(n + 1)(n + 2)}{2},$$

which is the formula $n(n + 1)/2$ with n replaced by $n + 1$. Thus, the formula $1 + 2 + \cdots + n = n(n + 1)/2$ is valid for all natural numbers n. But in many situations, especially those dealing with algebraic structures, the inference from n to $n + 1$ is not so easy. In these cases, strong induction is useful; that is, first verify the statement for $n = 1$ and then infer its truth for n assuming it is true for all numbers $<n$. Let us illustrate the principle of strong induction by proving some basic arithmetic properties of the natural numbers. Recall that if n and m are natural numbers and $m \neq 0$, then m *divides* n if $n = ms$ for some number s. If a number $n \neq 1$ has no divisors other than itself and 1, n is said to be *prime*.

PROPOSITION 3. Every natural number >1 is divisible by some prime.

Proof. Let $n > 1$ be a natural number. If $n = 2$, then n is prime and we are done. Now assume as induction hypothesis that $n > 2$ and that every number less than n has some prime divisor. If n is prime, it is its own prime divisor and we are done. Otherwise $n = st$ for some numbers s and t, where $1 < s, t < n$, and hence, by the induction assumption, there is some prime p that divides s. Therefore, p divides n and hence n is divisible by some prime. It now follows by the principle of strong induction that every natural number >1 is divisible by some prime, and the proof is complete. ■

COROLLARY. There are infinitely many primes.

Proof. Suppose, to the contrary, that there are only a finite number of primes, say $p_1, \ldots, p_m$, and let $n = p_1 \cdots p_m + 1$. Then, by Proposition 3, n is divisible by some prime p which is clearly not one of the primes $p_1, \ldots, p_m$. Thus, we have a contradiction and conclude, therefore, that the number of primes is infinite. ■

Let us now turn our attention to the integers. The integers, we recall, include both the natural numbers and their negatives. We let $\mathbb{Z}$ stand for the set of integers. Thus, $\mathbb{Z} = \{\ldots, -2, -1, 0, 1, 2, \ldots\}$. If n and m are integers with $m \neq 0$, we say that m *divides* n and write $m|n$, if $n = sm$ for some integer s. A nonzero integer n other than ± 1 is said to be *prime* if its only divisors are ± 1 and $\pm n$; otherwise, it is *composite*.

PROPOSITION 4. Let a, b, and c be integers. Then:

(1) If $a|b$ and $b|c$, then $a|c$.

(2) If $a|b$, then $ca|cb$, provided $c \neq 0$.

(3) If $a|b$ and $b|a$, then $b = \pm a$.

(4) If $c|a$ and $c|b$, then $c|(sa + tb)$ for all integers s and t.

Proof. We leave this proof as an exercise for the reader. ∎

PROPOSITION 5. (THE DIVISION ALGORITHM FOR IN-TEGERS) Let n and m be integers, with $m > 0$. Then there are unique integers q and r such that $n = qm + r$, where $0 \leq r < m$.

Proof. Let $S = \{n - qm \in \mathbb{Z} | q \in \mathbb{Z}\}$ stand for the set of all integers of the form $n - qm$, where q is an arbitrary integer. Since q may be chosen to be negative, zero, or positive, the set S contains both positive as well as negative integers. Consequently, it contains a smallest nonnegative integer r. Then $r = n - qm$ for some integers q and m, and $r \geq 0$. We claim that $r < m$. For suppose that $r \geq m$. Then $r = m + s$ for some integer s, where $0 \leq s < r$. It follows that $n - (q + 1)m = (n - qm) - m = r - m = s < r$, and hence $n - (q + 1)m$ is a nonnegative integer in S that is smaller than the least integer in S, which is a contradiction. Therefore, $n = qm + r$, where $0 \leq r < m$, as required. To show that q and r are unique, suppose that $n = qm + r = q'm + r'$ for some integers q, q', r, r', where $0 \leq r \leq r' < m$ and where we have assumed that r is the smaller of r and r'. Then $r' - r = (q - q')m$. But $r' - r < m$, while $(q - q')m \geq m$ if $q - q' \neq 0$. Therefore, $q - q' = 0$ and hence $r' - r = 0$. Thus, $q' = q$ and $r' = r$, which shows that the integers q and r are uniquely determined. ∎

COROLLARY. If p is a prime integer and $p|nm$ for some integers n and m, then $p|n$ or $p|m$.

Proof. If p divides n, we are done. Assume, therefore, that p does not divide n and consider the set $S = \{ap + bn \in \mathbb{Z} | a, b \in \mathbb{Z}, ap + bn \geq 0\}$. S is nonempty since it contains the number p, for example, and hence contains a smallest positive integer. Let d be the smallest positive integer in S. Then $d = ap + bn$ for some integers a and b. We claim first that $d|p$. For, by the division algorithm, integers q and r exist such that $p = qd + r$, where $0 \leq r < d$. Therefore, $r = p - qd = p - q(ap + bn) = (1 - qa)p + (-qb)n$. Hence, r is a nonnegative number in S. If $0 < r < d$, then r is a positive number in S

smaller than the smallest positive number d in S, which is a contradiction. Therefore, $r = 0$, which shows that $p = qd$ and hence that $d \mid p$. Similarly, $d \mid n$. Now, since $d \mid p$ and p is prime, d is either 1 or p. But if $d = p$, then $p \mid n$, which is a contradiction. Hence, $d = 1$ and therefore $1 = ap + bs$. It follows that $m = (ma)p + b(nm)$, and therefore p divides m since p divides both p and nm. Thus, the prime p divides either n or m, as required, and the proof is complete. ■

Let us note that this result extends by induction to any finite number of factors; that is, if p is prime and $p \mid n_1 \cdots n_s$ for some integers $n_1, \ldots, n_s$, then p divides at least one of the integers $n_1, \ldots, n_s$. We leave the details as an exercise for the reader.

COROLLARY. (THE FUNDAMENTAL THEOREM OF ARITHMETIC) Every positive integer may be written as the product of a finite number of uniquely determined prime numbers.

Proof. Let n be a positive integer, $n \neq 1$. If $n = 2$, then n is prime and we are done. Otherwise, $n > 2$ and we assume, as induction hypothesis, that all numbers $< n$ may be written as the product of a finite number of uniquely determined primes. By Proposition 3, $n = pm$ for some prime p and some number m. Since $m < n$, it follows from the induction hypothesis that there are primes $p_1, \ldots, p_s$ such that $m = p_1 \cdots p_s$. Therefore, $n = pp_1 \cdots p_s$, and hence n is a product of finitely many primes. It now follows from the principle of strong induction that every positive integer > 1 may be written as the product of primes. To show that the primes in such a factorization are uniquely determined, suppose that $n = p_1 \cdots p_s = q_1 \cdots q_t$ are two factorizations of n as a product of primes. Then $p_1 \mid q_1 \cdots q_t$ and hence $p_1 \mid q_i$ for some i, $1 \leq i \leq t$. Since q_i is prime, $q_i = p_1$. Thus, $p_1 \cdots p_s = p_1 q_1, \ldots, q_{i-1} q_{i+1} \cdots q_t$, and therefore $p_2 \cdots p_s = q_1 \cdots q_{i-1} q_{i+1} \cdots q_t$. But this last equation represents two factorizations of a number $< n$ into a product of primes. Hence, by the induction hypothesis, the primes $p_2, \ldots, p_s$ are the same as the primes $q_1, \ldots, q_{i-1}, q_{i+1}, \ldots, q_t$, and therefore the primes $p_1, \ldots, p_s$ are the same as the primes $q_1, \ldots, q_t$. Thus, the prime factors of n are uniquely determined. ■

By collecting the common factors that occur in the prime factorization of a number, it follows that we may write every nonzero integer other than ± 1 uniquely in the form $\pm p_1^{m_1} \cdots p_s^{m_s}$, where $p_1, \ldots, p_s$ are the distinct prime divisors of the number and $m_1, \ldots, m_s$ are positive integers.

Exercises

1. Prove the following statements by mathematical induction.

(a) $1^2 + 2^2 + \cdots + n^2 = \dfrac{n(n + 1)(2n + 1)}{6}$

(b) $1^3 + 2^3 + \cdots + n^3 = \left(\dfrac{n(n+1)}{2}\right)^2$

(c) If $n \geq 3$, the sum of the interior angles of a convex polygon in the plane is $180n$ degrees.

(d) If $n \geq 5$, $2^n > n^2$.

2. For any positive number n, show that 3 divides at least one of the numbers n, $n + 1$, or $n + 2$.

3. If p and n are positive integers, p a prime, and $p \mid n^s$ for some integer $s > 0$, show that $p \mid n$.

4. If p is prime and $p \mid n_1 \ldots n_s$ for some integers $n_1, \ldots, n_s$, show that p divides at least one of the integers n_i, $1 \leq i \leq s$.

5. Using mathematical induction, show that every set containing n elements has 2^n subsets.

5. VECTOR SPACES AND LINEAR TRANSFORMATIONS

We conclude our review chapter by discussing some of the basic ideas in linear algebra. Let $\mathbb{R}$ stand for the set of real numbers and let n be a positive integer. Then *real n-dimensional space*, or simply *real n-space*, is the set $\mathbb{R}^n = \mathbb{R} \times \cdots \times \mathbb{R}$ of ordered n-tuples with entries in the set $\mathbb{R}$, together with addition and scalar multiplication. The elements in $\mathbb{R}^n$ are called *vectors*, elements in $\mathbb{R}$ are called *scalars*, and addition in $\mathbb{R}^n$ is called *vector addition*. Vector addition, we recall, is defined by setting $(a_1, \ldots, a_n) + (b_1, \ldots, b_n) = (a_1 + b_1, \ldots, a_n + b_n)$ for all vectors $(a_1, \ldots, a_n)$, $(b_1, \ldots, b_n) \in \mathbb{R}^n$, while scalar multiplication is defined by setting $c(a_1, \ldots, a_n) = (ca_1, \ldots, ca_n)$ for all scalars $c \in \mathbb{R}$. The vector $(0, \ldots, 0)$ all of whose entries are zero is called the *zero vector* and is denoted by 0. Vector addition and scalar multiplication then satisfy the following easily verified properties: for all vectors $A, B, C \in \mathbb{R}^n$ and all scalars $c, d \in \mathbb{R}$,

(1) $A + (B + C) = (A + B) + C$

(2) $A + B = B + A$

(3) $A + 0 = A$

(4) $A + -A = 0$, where $-A = (-1)A$

(5) $c(A + B) = cA + cB$

(6) $(c + d)A = cA + dA$

(7) $c(dA) = (cd)A$

(8) $1A = A;\ 0A = 0$

For example, real 2-space is the vector space $\mathbb{R}^2$ consisting of all ordered pairs (a, b) of real numbers together with vector addition and scalar multiplication. In this case, $(a, b) + (c, d) = (a + c, b + d)$ and $c(a, b) = (ca, cb)$ for all vectors (a, b), $(c, d) \in \mathbb{R}^2$ and all scalars $c \in \mathbb{R}$. We represent vectors in $\mathbb{R}^2$ geometrically as points in the Cartesian plane; the vector $A = (a, b)$ is represented by the point (a, b), as illustrated in the figure in the margin. The sum of two vectors may then be obtained geometrically by using the *parallelogram rule*: if A and B are vectors not lying on a line, then $A + B$ is the diagonal of the parallelogram whose sides are A and B. Similarly, vectors

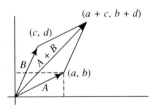

in real 3-space $\mathbb{R}^3$ are represented by means of the 3-dimensional Cartesian coordinate system.

Now, let $A_1, \ldots, A_m$ be a finite number of vectors in $\mathbb{R}^n$. A *linear combination* of $A_1, \ldots, A_m$ is any vector of the form $c_1 A_1 + \cdots + c_m A_m$, where $c_1, \ldots, c_m$ are scalars. Clearly, the zero vector 0 is a linear combination of any collection of vectors since $0 = 0A_1 + \cdots + 0A_m$. The vectors $A_1, \ldots, A_m$ are said to be *linearly independent* if the only scalars $c_1, \ldots, c_m$ for which $c_1 A_1 + \cdots + c_m A_m = 0$ are $c_1 = \cdots = c_m = 0$; otherwise they are said to be *linearly dependent*. Thus, $A_1, \ldots, A_m$ are linearly dependent if there are scalars $c_1, \ldots, c_m$, not all of which are zero, such that $c_1 A_1 + \cdots + c_m A_m = 0$. For example, the vectors $A = (6, -9)$ and $B = (-4, 6)$ are linearly dependent since $2A + 3B = (12, -18) + (-12, 18) = (0, 0)$. The vectors $A_1, \ldots, A_m$ are said to *span* the space $\mathbb{R}^n$ if every vector in $\mathbb{R}^n$ may be written as a linear combination of $A_1, \ldots, A_m$ and said to form a *basis* for $\mathbb{R}^n$ if they span $\mathbb{R}^n$ and are linearly independent. For example, the vectors $e_1 = (1, 0, \ldots, 0)$, $e_2 = (0, 1, 0, \ldots, 0), \ldots, e_n = (0, \ldots, 0, 1)$ span $\mathbb{R}^n$ since $(a_1, \ldots, a_n) = a_1 e_1 + \cdots + a_n e_n$ for all vectors $(a_1, \ldots, a_n) \in \mathbb{R}^n$, and they are linearly independent since, if $a_1 e_1 + \cdots + a_n e_n = 0$, then $a_1 = \cdots = a_n = 0$. Therefore, $\{e_1, \ldots, e_n\}$ is a basis for $\mathbb{R}^n$. We refer to this basis as the *standard basis* for $\mathbb{R}^n$. For example, the vectors $e_1 = (1, 0)$ and $e_2 = (0, 1)$ form a basis for real 2-space $\mathbb{R}^2$.

Let us emphasize, however, that there are many other bases for $\mathbb{R}^n$ in addition to the standard basis; for example, the vectors $A = (1, -2)$ and $B = (2, 3)$ are linearly independent and span $\mathbb{R}^2$ and hence form a basis for real 2-space. But regardless of what basis one uses for $\mathbb{R}^n$, every basis contains exactly n vectors. A *coordinate system* for $\mathbb{R}^n$ is any ordered basis; that is, an ordered collection of vectors $(A_1, \ldots, A_n)$ in which the order of the base vectors $A_1, \ldots, A_n$ is taken into account. The coordinate system (e_1, e_2), for example, is the standard coordinate system for the real plane.

Finally, a nonempty subset U of $\mathbb{R}^n$ is a *subspace* of $\mathbb{R}^n$ if the following two conditions are satisfied:

(1) if $A, B \in U$, then $A + B \in U$
(2) if $A \in U$ and $c \in \mathbb{R}$, then $cA \in U$

If condition (1) is satisfied, we say that U is closed under vector addition; if condition (2) is satisfied, we say that U is closed under scalar multiplication. Thus, a subspace of $\mathbb{R}^n$ is a nonempty subset of vectors that is closed under vector addition and scalar multiplication. An easy way to generate subspaces is by taking all possible linear combinations of a finite number of vectors: if $A_1, \ldots, A_m$ are a finite number of vectors in $\mathbb{R}^n$, then the subset

$$\langle A_1, \ldots, A_m \rangle = \{c_1 A_1 + \cdots + c_m A_m \in \mathbb{R}^n \,|\, c_1, \ldots, c_m \in \mathbb{R}\}$$

consisting of all linear combinations of $A_1, \ldots, A_m$ is closed under vector addition and scalar multiplication and is therefore a subspace of $\mathbb{R}^n$. We call this subspace the subspace *spanned* or *generated* by the vectors $A_1, \ldots, A_m$.

For example, let $A_1 = (2, -1, 3)$, $A_2 = (1, -1, 1)$, and $A_3 = (1, 1, 3)$, and let $U = \langle A_1, A_2, A_3 \rangle$ stand for the subspace spanned by these vectors in $\mathbb{R}^3$. Then a typical vector in U has the form

$$c_1 A_1 + c_2 A_2 + c_3 A_3 = (2c_1 + c_2 + c_3, -c_1 - c_2 + c_3, 3c_1 + c_2 + 3c_3),$$

where c_1, c_2, and c_3 are arbitrary scalars. To determine if these vectors are linearly independent, we set the typical vector equal to the zero vector and solve the resulting homogeneous system of three equations in three unknowns; thus,

$$2c_1 + c_2 + c_3 = 0$$

$$-c_1 - c_2 + c_3 = 0$$

$$3c_1 + c_2 + 3c_3 = 0.$$

Solving this system, we find that $c_1 = -2c_3$ and $c_2 = 3c_3$, where c_3 is arbitrary. The vectors A_1, A_2, and A_3 are therefore linearly dependent. For example, if $c_3 = 1$, then $c_1 = -2$ and $c_2 = 3$, and hence $(-2)A_1 + 3A_2 + A_3 = (0, 0, 0)$ is a linearly dependent relation among the three vectors A_1, A_2, and A_3. Moreover, this dependence relation shows that $A_3 = 2A_1 - 3A_2$, and therefore every linear combination of A_1, A_2, and A_3 may in fact be written as a linear combination of just A_1 and A_2. Thus, A_1 and A_2 span U and hence $U = \langle A_1, A_2 \rangle$. We leave the reader to verify that A_1 and A_2 are in fact linearly independent and conclude that U is a real 2-dimensional vector space having A_1 and A_2 as basis.

Let us now turn our attention to linear mappings between vector spaces. If V_1 and V_2 are vector spaces, a *linear transformation* from V_1 to V_2 is a function $L : V_1 \to V_2$ that satisfies the following two conditions:

(1) $L(A + B) = L(A) + L(B)$ for all vectors $A, B \in V_1$
(2) $L(cA) = cL(A)$ for all scalars $c \in \mathbb{R}$ and all vectors $A \in V_1$

If $L : V_1 \to V_2$ is a linear transformation, it follows easily by induction that $L(c_1 A_1 + \cdots + c_n A_n) = c_1 L(A_1) + \cdots + c_n L(A_n)$ for all vectors $A_1, \ldots, A_n \in V_1$ and all scalars $c_1, \ldots, c_n \in \mathbb{R}$. Moreover, $L(0) = 0'$, where 0 stands for the zero vector in V_1 and $0'$ the zero vector in V_2, and $L(-A) = -L(A)$ for all vectors $A \in V_1$.

For example, if V is any real vector space, the identity map $1_V : V \to V$ is a linear transformation mapping V to itself. As another example of a linear transformation, consider the function $L : \mathbb{R}^2 \to \mathbb{R}^2$ defined by setting $L(a, b) = (2a - 6b, -a + 3b)$ for all vectors $(a, b) \in \mathbb{R}^2$. To show that L is a linear transformation, we verify the two conditions listed above: if $(a, b), (a', b') \in \mathbb{R}^2$ and $c \in \mathbb{R}$, then

$$
\begin{aligned}
L((a, b) + (a', b')) &= L(a + a', b + b')\\
&= (2(a + a') - 6(b + b'), -(a + a') + 3(b + b'))\\
&= ((2a - 6b) + (2a' - 6b'), (-a + 3b) + (-a' + 3b'))\\
&= (2a - 6b, -a + 3b) + (2a' - 6b', -a' + 3b')\\
&= L(a, b) + L(a', b'),
\end{aligned}
$$

and

$$L(c(a, b)) = L(ca, cb) = (2(ca) - 6(cb), -(ca) + 3(cb))$$
$$= c(2a - 6b, -a + 3b) = cL(a, b).$$

Therefore L is a linear transformation.

Observe that in this example, the image $L(a, b)$ of a typical vector (a, b) was obtained by taking linear combinations of the coordinates a and b. In general, any such linear combination of the coordinates defines a linear transformation. That is, if n and m are positive integers and $a_{11}, \ldots, a_{1n}, \ldots, a_{m1}, \ldots, a_{mn}$ are scalars, then it is easily verified, as above, that the function $L:\mathbb{R}^n \to \mathbb{R}^m$ defined by setting $L(c_1, \ldots, c_n) = (a_{11}c_1 + \cdots + a_{1n}c_n, \ldots, a_{m1}c_1 + \cdots + a_{mn}c_n)$ for every vector $(c_1, \ldots, c_n) \in \mathbb{R}^n$ is a linear transformation. We represent this equation in terms of matrices by writing

$$L\begin{pmatrix} c_1 \\ \vdots \\ c_n \end{pmatrix} = \begin{pmatrix} a_{11}c_1 + \cdots + a_{1n}c_n \\ \vdots \\ a_{m1}c_1 + \cdots + a_{mn}c_n \end{pmatrix} = \begin{pmatrix} a_{11} & a_{12} & \cdots & a_{1n} \\ a_{21} & a_{22} & \cdots & a_{2n} \\ & & \vdots & \\ a_{m1} & a_{m2} & \cdots & a_{mn} \end{pmatrix} \begin{pmatrix} c_1 \\ \vdots \\ c_n \end{pmatrix}.$$

When doing so, however, it must be emphasized that we are using the column matrix representations for the vectors $(c_1, \ldots, c_n)$ and $L(c_1, \ldots, c_n)$. For example, the matrix equation of the linear transformation previously discussed is $L\binom{a}{b} = \left(\begin{smallmatrix} 2 & -6 \\ -1 & 3 \end{smallmatrix}\right)\binom{a}{b}$.

Now, let $L:V_1 \to V_2$ be a linear transformation between vector spaces V_1 and V_2. Then the *image* of L is the set $\text{Im } L = \{L(A) \in V_2 \mid A \in V_1\}$ consisting of the images of all vectors in V_1, and the *kernel* of L is the set $\text{Ker } L = \{A \in V_1 \mid L(A) = 0\}$ consisting of all vectors in V_1 that L maps to the zero vector in V_2.

PROPOSITION 1. Let $L:V_1 \to V_2$ be a linear transformation. Then $\text{Ker } L$ is a subspace of V_1 and $\text{Im } L$ is a subspace of V_2. Moreover, the mapping L is 1–1 if and only if $\text{Ker } L = \{0\}$.

Proof. To show that $\text{Ker } L$ is a subspace of V_1, we verify the two conditions just listed. Let $A, B \in \text{Ker } L$ and let $c \in \mathbb{R}$ be an arbitrary scalar. Then $L(A) = L(B) = 0$, and hence $L(A + B) = L(A) + L(B) = 0$ and $L(cA) = cL(A) = c0 = 0$. Therefore, $A + B \in \text{Ker } L$ and $cA \in \text{Ker } L$, and hence $\text{Ker } L$ is a subspace of V_1. Similarly, $\text{Im } L$ is a subspace of V_2, Finally, suppose that L is a 1–1 mapping and let $A \in \text{Ker } L$. Then $L(A) = 0 = L(0)$ and hence $A = 0$. Thus, if L is 1–1, $\text{Ker } L = \{0\}$. Conversely, if $\text{Ker } L = \{0\}$, then L is 1–1. For suppose that $L(A) = L(B)$ for some vectors $A, B \in V_1$. Then $L(A - B) = L(A) - L(B) = 0$, and hence $A - B \in \text{Ker } L$. Since $\text{Ker } L = \{0\}$, $A - B = 0$ and hence $A = B$. Thus, L is a 1–1 mapping. ■

We say that a linear transformation is *nonsingular* if it is a 1–1 mapping, and a *linear isomorphism* if it is both 1–1 and onto. In view of Proposition 1, a linear transformation is nonsingular if and only if its kernel is trivial, that is, it consists of only the zero vector. In this case, the mapping is a linear isomorphism from the domain space to the image space. For example, the identity mapping $1_V : V \rightarrow V$ from a vector space V to itself is clearly non-singular and onto, and hence is a linear isomorphism. But the linear transformation $L: \mathbb{R}^2 \rightarrow \mathbb{R}^2$ previously discussed is neither 1–1 nor onto. Here we have $L(a, b) = (2a - 6b, -a + 3b)$ for all vectors $(a, b) \in \mathbb{R}^2$ and hence, in particular, $L(3, 1) = (0, 0)$. Thus, $(3, 1) \in \text{Ker } L$. To find Ker L, we set $L(a, b) = (0, 0)$ and find that $2a - 6b = 0$ and $-a + 3b = 0$. Therefore, $a = 3b$, with b arbitrary, and hence,

$$\text{Ker } L = \{(3b, b) \in \mathbb{R}^2 \,|\, b \in \mathbb{R}\} = \{b(3, 1) \in \mathbb{R}^2 \,|\, b \in \mathbb{R}\} = \langle(3, 1)\rangle.$$

Thus, the kernel of L is the 1-dimensional subspace of $\mathbb{R}^2$ having the vector $(3, 1)$ as basis. For the image of L, we find that

$$
\begin{aligned}
\text{Im } L &= \{L(a, b) \in \mathbb{R}^2 \,|\, (a, b) \in \mathbb{R}^2\} \\
&= \{(2a - 6b, -a + 3b) \in \mathbb{R}^2 \,|\, a, b \in \mathbb{R}\} \\
&= \{a(2, -1) - 3b(2, -1) \in \mathbb{R}^2 \,|\, a, b \in \mathbb{R}\} \\
&= \{(a - 3b)(2, -1) \in \mathbb{R}^2 \,|\, a, b \in \mathbb{R}\} = \langle(2, -1)\rangle.
\end{aligned}
$$

Thus, the image of L is the 1-dimensional subspace of $\mathbb{R}^2$ having the vector $(2, -1)$ as basis. The linear transformation L therefore collapses real 2-space onto real 1-space in such a way that the line spanned by $(3, 1)$ is mapped to the zero vector, as illustrated in Figure 1.

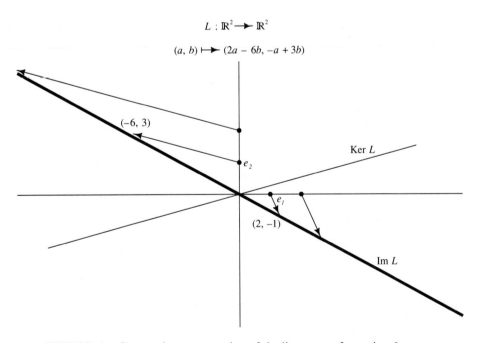

$$L : \mathbb{R}^2 \longrightarrow \mathbb{R}^2$$

$$(a, b) \longmapsto (2a - 6b, -a + 3b)$$

FIGURE 1.   Geometric representation of the linear transformation L.

We showed earlier that an $m \times n$ matrix $M = (a_{ij})$ defines a linear transformation $L: \mathbb{R}^n \to \mathbb{R}^m$ by setting $L(c) = Mc$ for every vector $c = (c_1, \ldots, c_n)$ $\in \mathbb{R}^n$. Let us now conclude our review of linear algebra by showing that every linear transformation may in fact be represented in this form. Let $L: \mathbb{R}^n \to \mathbb{R}^m$ be a linear transformation and let $(A_1, \ldots, A_n)$ and $(B_1, \ldots, B_m)$ be coordinate systems for $\mathbb{R}^n$ and $\mathbb{R}^m$, respectively. Then every vector in $\mathbb{R}^n$ may be written uniquely in the form $c_1 A_1 + \cdots + c_n A_n$ for some scalars $c_1, \ldots, c_n$, and $L(c_1 A_1 + \cdots + c_n A_n) = c_1 L(A_1) + \cdots + c_n L(A_n)$. It follows that the image of every vector in $\mathbb{R}^n$ is completely determined by the images $L(A_1), \ldots,$ $L(A_n)$ of the base vectors. But each of these images may be expressed uniquely in terms of the coordinate system $(B_1, \ldots, B_m)$ for $\mathbb{R}^m$. Thus, if

$$L(A_1) = a_{11} B_1 + a_{21} B_2 + \cdots + a_{m1} B_m$$

$$L(A_2) = a_{12} B_1 + a_{22} B_2 + \cdots + a_{m2} B_m$$

$$\vdots$$

$$L(A_n) = a_{1n} B_1 + a_{2n} B_2 + \cdots + a_{mn} B_m,$$

where $a_{11}, \ldots, a_{mn}$ are scalars, then

$$L(c_1 A_1 + \cdots + c_n A_n) = c_1 L(A_1) + \cdots + c_n L(A_n)$$
$$= c_1 (a_{11} B_1 + \cdots + a_{m1} B_m) + \cdots + c_n (a_{1n} B_1 + \cdots + a_{mn} B_m)$$
$$= (a_{11} c_1 + \cdots + a_{1n} c_n) B_1 + \cdots + (a_{m1} c_1 + \cdots + a_{mn} c_n) B_m.$$

In terms of matrices, we write this equation as

$$L \begin{pmatrix} c_1 \\ \vdots \\ c_n \end{pmatrix} = \begin{pmatrix} a_{11} & a_{12} & \cdots & a_{1n} \\ a_{21} & a_{22} & \cdots & a_{2n} \\ & & \vdots & \\ a_{m1} & a_{m2} & \cdots & a_{mn} \end{pmatrix} \begin{pmatrix} c_1 \\ \vdots \\ c_n \end{pmatrix}.$$

The $m \times n$ matrix (a_{ij}) whose columns are the coefficients of the images $L(A_1), \ldots, L(A_n)$ in the $(B_1, \ldots, B_m)$ coordinate system is called the *matrix representation* of L relative to the coordinate systems $(A_1, \ldots, A_n)$ and $(B_1, \ldots, B_m)$. Thus, by first choosing coordinate systems for $\mathbb{R}^n$ and $\mathbb{R}^m$, every linear transformation $L: \mathbb{R}^n \to \mathbb{R}^m$ has associated with it an $m \times n$ matrix M, and $L(A) = MA$ for every vector $A \in \mathbb{R}^n$ provided we use the column matrix representation for A in terms of the given coordinate system. Let us emphasize, once again, that the matrix representation of a linear transformation depends upon the choice of coordinate systems and, in general, changes when the coordinate systems change.

For example, let $L: \mathbb{R}^2 \to \mathbb{R}^3$ be the linear transformation defined by setting $L(a, b) = (a + 2b, 2a + b, a + b)$ for all $(a, b) \in \mathbb{R}^2$. Let (e_1, e_2) be the standard coordinate system for $\mathbb{R}^2$ and (e_1, e_2, e_3) be the standard coordinate system for $\mathbb{R}^3$. Then $L(e_1) = L(1, 0) = (1, 2, 1) = e_1 + 2e_2 + e_3$ and $L(e_2) = L(0, 1)$ $= (2, 1, 1) = 2e_1 + e_2 + e_3$. Therefore, the matrix of L relative to the standard coordinate systems is

$$\begin{pmatrix} 1 & 2 \\ 2 & 1 \\ 1 & 1 \end{pmatrix}.$$

In terms of matrices,

$$L\begin{pmatrix} a \\ b \end{pmatrix} = \begin{pmatrix} 1 & 2 \\ 2 & 1 \\ 1 & 1 \end{pmatrix}\begin{pmatrix} a \\ b \end{pmatrix} = \begin{pmatrix} a + 2b \\ 2a + b \\ a + b \end{pmatrix}.$$

PROPOSITION 2. Let $L_1: \mathbb{R}^n \to \mathbb{R}^m$ and $L_2: \mathbb{R}^m \to \mathbb{R}^s$ be linear transformations, and let M_1 and M_2 be the matrix representations of L_1 and L_2, respectively, relative to given coordinate systems. Then the composition $L_2 \circ L_1: \mathbb{R}^n \to \mathbb{R}^s$ is a linear transformation, and the matrix representation of $L_2 \circ L_1$ relative to the given coordinate systems is the matrix product $M_2 M_1$.

Proof. Let $A_1, \ldots, A_n, B_1, \ldots, B_m$, and $C_1, \ldots, C_s$ stand for the given coordinate systems for $\mathbb{R}^n$, $\mathbb{R}^m$, and $\mathbb{R}^s$, respectively, and let M stand for the matrix of $L_2 \circ L_1$ relative to these coordinate systems. Then, for $1 \le i \le n$, the ith column of M is $(L_2 \circ L_1)(A_i)$. But $(L_2 \circ L_1)(A_i) = L_2(L_1(A_i))$ $= L_2(M_1 A_i) = M_2(M_1 A_i) = (M_2 M_1)A_i$, which is the ith column of the product matrix $M_2 M_1$. It follows that $M = M_2 M_1$, as required. ∎

PROPOSITION 3. Let $L: \mathbb{R}^n \to \mathbb{R}^n$ be a linear transformation and let M stand for the matrix of L relative to any given coordinate system for $\mathbb{R}^n$. Then L is nonsingular if and only if $\det(M) \ne 0$, that is, if and only if the determinant of M is nonzero. In this case, L is a linear isomorphism.

Proof. Let M^* stand for the reduced echelon form of the matrix M. Then

$$M^* = \begin{pmatrix} I_{n-s} & O \\ O & O_s \end{pmatrix},$$

where s is the dimension of Ker L and I_{n-s} stands for the $(n-s) \times (n-s)$ identity matrix. Moreover, L is nonsingular if and only if $s = 0$ or, equivalently, if and only if $\det(M^*) \ne 0$. But M^* is obtained from M through a sequence of elementary row operations and hence $\det(M^*)$ is a scalar multiple of $\det(M)$. Therefore, L is nonsingular if and only if $\det(M) \ne 0$. Finally, if L is nonsingular, then the mapping L is 1–1. In this case, Im L is an n-dimensional subspace of $\mathbb{R}^n$ and hence Im $L = \mathbb{R}^n$, which shows that L is an onto mapping and hence a linear isomorphism. ∎

For example, the matrix of the linear transformation $L:\mathbb{R}^2 \to \mathbb{R}^2$ defined by $L(a, b) = (2a - 6b, -a + 3b)$ is $M = \left(\begin{smallmatrix} 2 & -6 \\ -1 & 3 \end{smallmatrix}\right)$. In this case $\det(M) = 0$ and hence L is singular, that is, L is not 1–1; indeed, we found the kernel of L to be the 1-dimensional subspace spanned by the vector $(3, 1)$. On the other hand, the linear transformation $L:\mathbb{R}^2 \to \mathbb{R}^2$ defined by the matrix equation

$$L\begin{pmatrix} a \\ b \end{pmatrix} = \begin{pmatrix} 3 & 1 \\ 2 & 4 \end{pmatrix}\begin{pmatrix} a \\ b \end{pmatrix} = \begin{pmatrix} 3a + b \\ 2a + 4b \end{pmatrix}$$

is nonsingular since $\det\left(\begin{smallmatrix} 3 & 1 \\ 2 & 4 \end{smallmatrix}\right) = 10 \neq 0$.

Exercises

1. Let $U = \{(a + c, a + 2b - c, b - c) \in \mathbb{R}^3 \,|\, a, b, c \in \mathbb{R}\}$.
   (a) Show that U is a subspace of $\mathbb{R}^3$.
   (b) Find a basis for U and the dimension of U.
   (c) Is the vector $(1, -1, 2)$ in the subspace U?
2. Let $U = \{(a, b, c) \in \mathbb{R}^3 \,|\, 2a - b + 3c = 0\}$.
   (a) Show that U is a subspace of $\mathbb{R}^3$.
   (b) Find a basis for U and the dimension of U.
3. Let $A_1 = (2, 1)$ and $A_2 = (-1, 3)$.
   (a) Show that A_1 and A_2 form a basis for $\mathbb{R}^2$.
   (b) Find the coordinates of the vector $A = (-5, 1)$ relative to the (A_1, A_2) coordinate system.
4. Define the function $L:\mathbb{R}^2 \to \mathbb{R}^3$ by setting $L(a, b) = (a, a + b, a - b)$ for all vectors $(a, b) \in \mathbb{R}^2$.
   (a) Show that L is a linear transformation.
   (b) Find the matrix representation of L relative to the standard coordinate systems for $\mathbb{R}^2$ and $\mathbb{R}^3$.
   (c) Show that L is nonsingular and find a basis for Ker L.
   (d) Find a basis for Im L.
   (e) Is the vector $(3, 5, 1)$ in the image of L?
5. Let $L:\mathbb{R}^3 \to \mathbb{R}^3$ be the linear transformation defined by the matrix equation

$$L\begin{pmatrix} a \\ b \\ c \end{pmatrix} = \begin{pmatrix} 1 & 0 & 1 \\ 0 & 2 & 1 \\ 2 & 1 & 1 \end{pmatrix}\begin{pmatrix} a \\ b \\ c \end{pmatrix}$$

relative to the standard coordinate system for $\mathbb{R}^3$.
   (a) Show that L is a linear isomorphism.
   (b) Find the image of the vector $(4, 2, 3)$.
   (c) Find the unique vector $A \in \mathbb{R}^3$ such that $L(A) = (1, 1, 1)$.

(d) Let $U = \langle(2, 0, 1), (1, 1, 0)\rangle$ stand for the subspace of $\mathbb{R}^3$ spanned by the vectors $(2, 0, 1)$ and $(1, 1, 0)$. Find a basis for the image $L(U)$.

6. Let $L:V_1 \to V_2$ be a linear transformation from a vector space V_1 to a vector space V_2.
   (a) If U_1 is a subspace of V_1, show that the image $L(U_1)$ is a subspace of V_2.
   (b) If U_2 is a subspace of V_2, show that $L^{-1}(U_2)$ is a subspace of V_1, where $L^{-1}(U_2) = \{A \in V_1 \,|\, L(A) \in U_2\}$.

7. Let $L:\mathbb{R}^3 \to \mathbb{R}^3$ stand for the linear transformation discussed in Exercise 5 and let $U = \langle(1, 0, 2), (0, 1, 1)\rangle$ stand for the subspace of $\mathbb{R}^3$ spanned by the vectors $(1, 0, 2)$ and $(0, 1, 1)$.
   (a) Find a basis for the image $L(U)$.
   (b) Find a basis for $L^{-1}(U)$.

8. Let $L:\mathbb{R}^3 \to \mathbb{R}^3$ be the linear transformation defined by setting $L(a, b, c) = (a + b, a + b + c, b + c)$ for all vectors $(a, b, c) \in \mathbb{R}^3$.
   (a) Find the matrix representation of L relative to the standard coordinate system for $\mathbb{R}^3$.
   (b) Let $A_1 = (2, 1, 0)$, $A_2 = (0, 1, 2)$, and $A_3 = (1, 1, 0)$. Show that (A_1, A_2, A_3) is a coordinate system for $\mathbb{R}^3$, and find the matrix representation of L relative to (A_1, A_2, A_3).
   (c) Show that L is a linear isomorphism.

9. **Transition matrices.** Let $L:\mathbb{R}^n \to \mathbb{R}^m$ be a linear transformation, $(A_1, \ldots, A_n)$ and $(B_1, \ldots, B_m)$ coordinate systems for $\mathbb{R}^n$ and $\mathbb{R}^m$, respectively, and let M be the matrix of L relative to these coordinate systems. If $(A'_1, \ldots, A'_n)$ is another coordinate system for $\mathbb{R}^n$, then there are scalars $s_{11}, \ldots, s_{nn}$ such that

$$A'_1 = s_{11}A_1 + \cdots + s_{n1}A_n$$
$$\vdots$$
$$A'_n = s_{1n}A_1 + \cdots + s_{nn}A_n.$$

The matrix $S = (s_{ij})$ is called the *transition matrix* from the coordinate system $(A_1, \ldots, A_n)$ to the coordinate system $(A'_1, \ldots, A'_n)$ and is an invertible matrix. In terms of matrices, $(A'_1 \cdots A'_n) = (A_1 \cdots A_n)S$.
   (a) If $(c_1, \ldots, c_n)$ are the coordinates of a vector relative to the $(A_1, \ldots, A_n)$ coordinate system and $(c'_1, \ldots, c'_n)$ are its coordinates relative to $(A'_1, \ldots, A'_n)$, show that

$$\begin{pmatrix} c'_1 \\ \vdots \\ c'_n \end{pmatrix} = S^{-1} \begin{pmatrix} c_1 \\ \vdots \\ c_n \end{pmatrix}.$$

   (b) Let $(B'_1, \ldots, B'_m)$ be another coordinate system for $\mathbb{R}^m$ and let T be the transition matrix from $(B_1, \ldots, B_m)$ to $(B'_1, \ldots, B'_m)$. If $A \in \mathbb{R}^n$, show that $L(A) = (T^{-1}MS)A$ relative to the coordinate systems $(A'_1, \ldots, A'_n)$ and $(B'_1, \ldots, B'_m)$. Thus, if M' stands for the matrix of L relative to the new coordinate systems, then $M' = T^{-1}MS$.

(c) Find the transition matrix S that changes the standard coordinate system (e_1, e_2, e_3) for $\mathbb{R}^3$ to the new coordinate system (A_1, A_2, A_3) discussed in Exercise 8b. Verify that the matrix obtained in Exercise 9b for L relative to the new coordinate system is $S^{-1}MS$, where M is its matrix relative to the standard coordinate system.

ONE

BASIC GROUP THEORY

2

Groups

As noted in the Preface, abstract algebra deals with the construction and study of algebraic structures that reflect, or model, algebraic properties common to many different types of mathematical objects. Objects such as numbers, functions, and matrices, for example, may be added or multiplied, and these operations—addition and multiplication—satisfy certain basic properties such as associativity. Our goal in this book is to create algebraic structures, called groups, rings, and fields, that model these operations and their properties. By studying the abstract properties of these operations, unclouded by the details of any one particular system, we obtain results that apply to all such systems. In the end, it is the algebraic operations and their properties that are the central theme in the study of abstract algebra and not the objects themselves.

In the first part of this book we concentrate on groups, an algebraic structure in which elements are combined according to a single rule that satisfies certain standard arithmetic properties. Groups provide the basic algebraic model for many types of mathematical objects, from groups of real and complex numbers, to groups of permutations, matrices, and linear transformations. The concept itself evolved out of early nineteenth century attempts to solve algebraic equations by means of radicals. By the beginning of the twentieth century, however, groups were no longer associated exclusively with algebraic equations but were becoming an important part of number theory and geometry. Today, the generality and simplicity of the basic concepts of group theory permeate virtually all areas of mathematics. Our goal in the first part of this book is to introduce the reader to these basic concepts, illustrating them with examples drawn from number theory and geometry. By doing so, we hope to convince the reader that group theory is not just an abstract theory but an integral part of our everyday mathematical experience.

Before we begin, let us take a moment to discuss the basic ideas of the chapter more informally. As we previously noted, a group is a set together with a rule, called a binary operation, for combining the elements of the set and which satisfies certain standard properties such as associativity. Integers, real numbers, and complex numbers, for example, all form groups under the operation of addition. We will discuss these groups, as well as groups of functions under the operation of function composition, groups of matrices under matrix addition, and various groups of geometric transformations.

An especially important class of groups arises when all the elements in a group are obtained from a single element by repeatedly combining that element with itself. These groups are called cyclic groups and include the group of integers, for example, as well as the group of integers modulo n, which are the familiar groups of modular arithmetic. Cyclic groups are especially useful in number theory because they provide information about the Euler phi-function, for example, and about complex roots of unity. These applications to number theory and the geometry of complex numbers illustrate how the algebraic properties of a group can provide important mathematical information about the objects in the group.

We conclude the chapter by turning our attention to the concept of group-isomorphism. Isomorphism is, in general, the mathematician's way of saying that two mathematical structures are essentially the same. In set theory, for example, two sets are said to be isomorphic, or equivalent, if there is a 1–1 correspondence between them, while in linear algebra two vector spaces are isomorphic if there is a linear transformation from the one to the other that is a 1–1 correspondence. Similarly, two groups are considered to be the same, or isomorphic, if there is a 1–1 correspondence between their elements such that corresponding elements combine in the same way. From the point of view of isomorphism, the particular elements in a group are not important. It does not matter, for example, whether the elements of the group are numbers, functions, or matrices—the only thing that is important is how the elements combine, that is, the arithmetic of the group. By focusing on the arithmetic of the group and not on its particular elements, group-isomorphism gives us the freedom to think of a group not in terms of a particular set of elements, but rather as an abstract model.

1. BINARY OPERATIONS

As we noted in the introduction, the basic concepts of group theory stem from the idea of a binary operation. Let us begin therefore by discussing this concept in more detail.

Definition

Let X be a nonempty set. A *binary operation* on X is a function from the Cartesian product $X \times X$ into X. That is, it is a rule that associates to each ordered pair (a, b) of elements in X a well-defined element $a * b$ in X.

Addition of integers, for example, is a binary operation on the set $\mathbb{Z}$ of integers; if n and m are integers, we write $n + m$ to indicate the sum of n and m instead of the abstract symbol $n * m$. Subtraction and multiplication of integers are also binary operations on $\mathbb{Z}$. Division, however, is not a binary operation on the set $\mathbb{Z}$ since, if n is an integer, $n \div 0$ is not even defined let alone an element of $\mathbb{Z}$. In fact, even if we restrict our attention to the set $\mathbb{Z}^*$ of nonzero integers, we still do not have a binary operation since $1 \div 2$, for example, is not in $\mathbb{Z}^*$. Here are some other examples of familiar binary operations.

EXAMPLE 1

Let $\mathbb{Q}$ stand for the set of rational numbers and $\mathbb{Q}^*$ for the set of nonzero rational numbers. Then addition, subtraction, and multiplication of fractions are binary operations on $\mathbb{Q}$. We indicate the sum, difference, and product of two rational numbers $r, s \in \mathbb{Q}$ by writing $r + s, r - s$, and rs, respectively. Division, however, is not a binary operation on $\mathbb{Q}$, although it is a binary operation on $\mathbb{Q}^*$; for if $r, s \in \mathbb{Q}^*$, then $r \div s = r/s$ and r/s is a number in $\mathbb{Q}^*$. Note, however, that subtraction is not a binary operation on $\mathbb{Q}^*$ since, for example, $1 \in \mathbb{Q}^*$, but $1 - 1 = 0 \notin \mathbb{Q}^*$.

EXAMPLE 2

Let X be a nonempty set and let $F(X)$ stand for the set of all functions mapping X into X. Then function composition is a binary operation on $F(X)$; for if $f, g \in F(X)$, the composition $f \circ g$ is defined by setting $(f \circ g)(x) = f(g(x))$ for every element $x \in X$ and hence is a well-defined element in $F(X)$. Thus, the formula $f * g = f \circ g$ defines a binary operation on $F(X)$.

EXAMPLE 3

Let X be any set and let $\mathscr{P}(X)$ stand for the power set of X. Recall that $\mathscr{P}(X)$ is the set of all subsets of X. Then union and intersection are binary operations on $\mathscr{P}(X)$; for if $A, B \in \mathscr{P}(X)$, both $A \cup B$ and $A \cap B$ are well-defined subsets of X. Thus, union and intersection are binary operations on $\mathscr{P}(X)$.

EXAMPLE 4

Let V be a real vector space. Then vector addition is a binary operation on V since, if $A, B \in V$, their sum $A + B$ is a well-defined vector in V.

EXAMPLE 5

Let V be a real vector space and let $L(V)$ stand for the set of all linear transformations mapping V into V. Then function composition is a binary operation on $L(V)$: if f, $g \in L(V)$, then their composition $f \circ g$ is a linear transformation from V into V. To show this, let A, $B \in V$ and let c and d be real numbers. Then

$$(f \circ g)(cA + dB) = f(g(cA + dB)) = f(cg(A) + dg(B)) = cf(g(A)) + df(g(B))$$
$$= c(f \circ g)(A) + d(f \circ g)(B)$$

since both f and g are linear transformations. Therefore, $f \circ g$ is a linear transformation and hence an element of $L(V)$. Thus, function composition is a binary operation on $L(V)$.

EXAMPLE 6

Let $\text{Mat}_2(\mathbb{R})$ stand for the set of 2×2 matrices with real number entries. Then addition and multiplication of matrices are two binary operations on $\text{Mat}_2(\mathbb{R})$ since the sum and product of 2×2 real matrices are 2×2 real matrices:

$$\begin{pmatrix} a & b \\ c & d \end{pmatrix} + \begin{pmatrix} e & f \\ g & h \end{pmatrix} = \begin{pmatrix} a+e & b+f \\ c+g & d+h \end{pmatrix}$$

$$\begin{pmatrix} a & b \\ c & d \end{pmatrix}\begin{pmatrix} e & f \\ g & h \end{pmatrix} = \begin{pmatrix} ae+bg & af+bh \\ ce+dg & cf+dh \end{pmatrix},$$

where $a, b, c, d, e, f, g,$ and h are real numbers.

The above examples illustrate some familiar binary operations. Observe that in many of these examples, the binary operation satisfies additional properties. Addition of integers, for example, is a commutative operation; that is, the order in which we add two integers does not matter. Matrix multiplication, on the other hand, does not have this property since the product of two matrices depends upon the order in which we multiply them. In general, there are four basic properties that a binary operation on a set may or may not satisfy. To describe these properties, let $*$ be a binary operation on a nonempty set X.

(1) *Associativity.* We say that $*$ is *associative* if $x * (y * z) = (x * y) * z$ for all $x, y, z \in X$. One way to think of associativity is the following. If $x, y,$ and z are elements in X, then the expression $x * y * z$ is meaningless because $*$ is a "binary" operation on X; that is, it is defined only for pairs of elements in X. Now, there are two ways to insert parentheses into the expression $x * y * z$ that will make it meaningful; either $x * (y * z)$ or $(x * y) * z$. Associativity of $*$ simply says that these two meaningful expressions are equal.

(2) *Commutativity.* We say that $*$ is *commutative* if $x * y = y * x$ for all x, $y \in X$; in other words, the order in which elements are combined does not matter. One may think of this property as expressing a type of symmetry of the binary operation.

(3) *Identity element.* We say that an element $e \in X$ is an *identity element* for $*$ if $e * x = x * e = x$ for every element $x \in X$.

(4) *Inverse element.* Let $e \in X$ be an identity element for $*$ and let $x \in X$. We say that an element $y \in X$ is an *inverse for x* if $x * y = y * x = e$.

Observe that in order for an element $e \in X$ to be an identity element for the operation $*$, it is not enough to show that $e * x = x$ or $x * e = x$ for every element $x \in X$. Both of these equations must be verified. Similarly, for an element y to be an inverse for some element x, we must show that both $x * y = e$ and $y * x = e$. Of course, if the operation $*$ is commutative, then only one equation need be verified in each case. Note also that the concept of an inverse element first requires the existence of an identity element for the operation $*$. For example, consider the set $\mathbb{Z}$ of integers under addition. Clearly, addition of integers is both an associative and commutative binary operation since $n + (m + s) = (n + m) + s$ and $n + m = m + n$ for all integers n, m, and s. Moreover, the integer 0 is an identity element for addition since $n + 0 = n$ for every integer n, and the integer $-n$ is an additive inverse for n since $n + (-n) = 0$ for every integer n.

EXAMPLE 7

Let X be a nonempty set and let $F(X)$ stand for the set of functions from X into X. Then, as we noted earlier, function composition is a binary operation on $F(X)$. Let's discuss the properties of this operation.

(A) Function composition is associative. To show this, let f, g, and h be functions from X into X, and let $x \in X$. Then

$$[f \circ (g \circ h)](x) = f((g \circ h)(x)) = f(g(h(x))) = (f \circ g)(h(x)) = [(f \circ g) \circ h](x),$$

and therefore $f \circ (g \circ h) = (f \circ g) \circ h$. Hence, function composition is an associative binary operation on $F(X)$.

(B) We claim, however, that function composition is not a commutative operation in general. For example, choose $X = \mathbb{R}$, the set of real numbers, and define $f, g : \mathbb{R} \to \mathbb{R}$ by setting $f(x) = 2x$ and $g(x) = x + 1$ for all $x \in \mathbb{R}$. Then $(f \circ g)(x) = 2x + 2$, but $(g \circ f)(x) = 2x + 1$. Therefore, $f \circ g \neq g \circ f$, and hence function composition is not, in general, a commutative operation.

(C) Now, let $1_X : X \to X$ stand for the identity function on X; recall that 1_X is the function defined by setting $1_X(x) = x$ for all $x \in X$. Then 1_X is an identity element for function composition since $1_X \circ f = f \circ 1_X = f$ for every function $f \in F(X)$.

(D) Finally, we ask about inverses. Recall that if f is a function from X into X, an inverse for f is any function $g : X \to X$ such that $f \circ g = g \circ f = 1_X$, or equivalently, $f(g(x)) = g(f(x)) = x$ for all $x \in X$. Recall also that such a function g exists if and only if f is both 1–1 and onto; that is, if and only if f is a 1–1 correspondence. Thus, the only elements in $F(X)$ that have an inverse under function composition are the 1–1 correspondences. For example, the function $f:\mathbb{R} \to \mathbb{R}$ defined by $f(x) = 2x + 3$ is a 1–1 correspondence whose inverse $g:\mathbb{R} \to \mathbb{R}$ is given by the formula $g(x) = 1/2(x) - 3/2$.

EXAMPLE 8

Matrix addition is an associative and commutative binary operation on the set $\text{Mat}_2(\mathbb{R})$ and has the 2×2 zero matrix $\left(\begin{smallmatrix} 0 & 0 \\ 0 & 0 \end{smallmatrix}\right)$ as identity matrix. Moreover, every matrix has an additive inverse; the additive inverse of $\left(\begin{smallmatrix} a & b \\ c & d \end{smallmatrix}\right)$ is $\left(\begin{smallmatrix} -a & -b \\ -c & -d \end{smallmatrix}\right)$. Thus, matrix addition is an associative and commutative binary operation on the set $\text{Mat}_2(\mathbb{R})$ and has an identity element, and every matrix has an additive inverse. Matrix multiplication is also an associative binary operation on the set $\text{Mat}_2(\mathbb{R})$, but it is well known that this operation is not commutative; that is, not every pair of 2×2 real matrices commute under multiplication. For example, $\left(\begin{smallmatrix} 1 & 1 \\ 0 & 0 \end{smallmatrix}\right)\left(\begin{smallmatrix} 0 & 0 \\ 1 & 1 \end{smallmatrix}\right) = \left(\begin{smallmatrix} 1 & 1 \\ 0 & 0 \end{smallmatrix}\right)$, but $\left(\begin{smallmatrix} 0 & 0 \\ 1 & 1 \end{smallmatrix}\right)\left(\begin{smallmatrix} 1 & 1 \\ 0 & 0 \end{smallmatrix}\right) = \left(\begin{smallmatrix} 0 & 0 \\ 1 & 1 \end{smallmatrix}\right)$. Nevertheless, there are certain pairs of 2×2 matrices that do commute; for example, every 2×2 matrix commutes with itself. Matrix multiplication also has an identity element, namely, the 2×2 identity matrix $\left(\begin{smallmatrix} 1 & 0 \\ 0 & 1 \end{smallmatrix}\right)$; for if $\left(\begin{smallmatrix} a & b \\ c & d \end{smallmatrix}\right)$ is a typical 2×2 real matrix, then

$$\begin{pmatrix} a & b \\ c & d \end{pmatrix}\begin{pmatrix} 1 & 0 \\ 0 & 1 \end{pmatrix} = \begin{pmatrix} a & b \\ c & d \end{pmatrix} = \begin{pmatrix} 1 & 0 \\ 0 & 1 \end{pmatrix}\begin{pmatrix} a & b \\ c & d \end{pmatrix}.$$

However, not every matrix has an inverse under matrix multiplication. Those matrices that have an inverse are called *invertible matrices* and are, we recall, the 2×2 matrices that have a nonzero determinant; that is, a matrix M is invertible if and only if $\det(M) \neq 0$. Thus, matrix multiplication is an associative binary operation on the set $\text{Mat}_2(\mathbb{R})$ and has an identity element, but it is not commutative and not every element has an inverse.

We have discussed function composition and matrix multiplication in considerable detail because many of the groups we will introduce in the next section consist of functions or matrices. It is especially important, therefore, to understand the basic ideas discussed in these examples.

Let $*$ be an associative binary operation on a nonempty set X and let x, y, $z \in X$. Then associativity of $*$ means that if parentheses are inserted into the expression $x * y * z$ in either one of the two meaningful ways, the same element of X is obtained. But suppose we have four factors, say $x * y * z * w$. Then there are five meaningful ways to insert parentheses: $(x * y) * (z * w)$, $x * ((y * z) * w)$, $x * (y * (z * w))$, $(x * (y * z)) * w$, and $((x * y) * z) * w$, and each of these expressions indicates a different way of calculating $x * y * z * w$. We claim, however, that they are all equal. Consider the expression $(x * (y * z)) * w$, for example. By repeated use of associativity, we find that

$$(x * (y * z)) * w = x * ((y * z) * w) = x * (y * (z * w)).$$

Similarly, each of the other expressions may be written in the form $x * (y * (z * w))$, and hence all five expressions are equal.

More generally, let $x_1, \ldots, x_n$ be any finite number of elements in X. We say that the operation $*$ satisfies *general associativity* if, whenever the n-fold product $x_1 * \cdots * x_n$ is calculated in any meaningful way by the insertion of parentheses, the same result is always obtained. Similarly, $*$ satisfies *general commutativity* if, whenever the factors in any meaningful n-fold product are interchanged in any manner, the same result is always obtained.

PROPOSITION 1. Let $*$ be an associative binary operation on a nonempty set X. Then

(1) $*$ satisfies general associativity.

(2) If $*$ is commutative, then $*$ satisfies general commutativity.

(3) There is at most one identity element for $*$ in X.

(4) If e is an identity element for $*$, then every element in X has at most one inverse.

Proof.   We prove statement (1) by induction on the number n of factors. If $n = 3$, the statement is true because $*$ is associative by definition. Now, let $n > 3$ and suppose that general associativity is satisfied for any collection of fewer than n factors. Assume that we have a meaningful product of n factors and that this product is equal to some element y. Let the factors be $x_1, \ldots, x_n$, where x_i is understood to be the ith factor, $1 \le i \le n$. We now write $y = A * B$, where each of A and B contain at least one of the factors x_i; this is possible since, when parentheses are inserted in a meaningful way, it must lead to the calculation of y as a product of two elements. But A and B involve fewer than n of the factors $x_1, \ldots, x_n$. Hence, by the induction hypothesis, we may write $A = x_1 * (x_2 * \cdots * (x_{i-1} * x_i) \ldots)$, where $x_1, \ldots, x_i$ are the factors in A, and therefore

$$y = A * B = \left[x_1 * \underbrace{(x_2 * \cdots * (x_{i-1} * x_i) \ldots)}_{C} \right] * B = (x_1 * C) * B = x_1 * (C * B).$$

But the expression $C * B$ contains fewer than n factors, and hence by the induction hypothesis, it may be written in the form $x_2 * (x_3 * \cdots * (x_{n-1} * x_n) \ldots)$. Therefore, $y = x_1 * (x_2 * \cdots * (x_{n-1} * x_n) \ldots)$. It now follows that if the expression $x_1 * \cdots * x_n$ is calculated in any meaningful way, the same result is always obtained, namely, the element $x_1 * (x_2 * \cdots * (x_{n-1} * x_n) \ldots)$. Hence, by mathematical induction, $*$ satisfies general associativity. The proof of statement (2) is similar and is left as an exercise for the reader. To show that there is at most one identity element for $*$ in X, we assume that there are two such elements and show that they are in fact the same element. Let e and e' be identity elements for $*$ in X. Then $e * e' = e$ because e' is an identity element. Likewise, $e * e' = e'$ since e is an identity element. Since $*$

is well defined, we conclude that $e = e'$, which proves statement (3). Finally, to prove statement (4), suppose that e is an identity element for $*$ in X and let $x \in X$. To show that x has at most one inverse, assume that y and y' are inverses for x. Then $x * y = y * x = e = x * y' = y' * x$. Hence, $y = y * e = y * (x * y') = (y * x) * y' = e * y' = y'$. Thus, y and y' are the same element and therefore x has at most one inverse, as required. ∎

If $*$ is an associative binary operation on a nonempty set X and $x_1, \ldots, x_n$ are a finite number of elements in X, we may now refer to "the" product $x_1 * \cdots * x_n$; how we choose to calculate this product is irrelevant. In particular, if n is a positive integer and x is an element in X, we let x^n stand for the n-fold product $x * \cdots * x$.

COROLLARY. Let $*$ be an associative binary operation on a nonempty set X and let x be an element in X. Then $x^n * x^m = x^{n+m}$ and $(x^n)^m = x^{nm}$ for all positive integers n and m.

Proof. For any positive integer t, x^t is the product of t factors all of which equal x and, by general associativity, may be calculated in any meaningful way. In particular, since $x^n * x^m$ and x^{n+m} are two ways of calculating the product of $n + m$ equal factors, they are equal. Similarly, $(x^n)^m$ and x^{nm} are the same since they represent two ways of calculating the product of nm equal factors. ∎

We conclude this section by discussing the closure of a subset under a binary operation. Let $*$ be a binary operation on a nonempty set X and let Y be a subset of X. If y and y' are elements of Y, then $y * y'$ exists as an element in X, but may or may not lie in the subset Y.

> **Definition**
>
> The subset Y is said to be *closed* with respect to the binary operation $*$ on X if, for all elements $y, y' \in Y$, the product $y * y'$ is also an element in Y. In this case we say that Y *inherits* the binary operation on X.

For example, addition is a binary operation on the set of integers, and the subset of even integers inherits this operation since the sum of two even integers is an even integer. But the subset of odd integers does not inherit addition since the sum of two odd numbers is not an odd number. Division is a binary operation on the set of nonzero rational numbers, but the subset of nonzero integers does not inherit division. If V is a real vector space and U is a subspace of V, then vector addition is a binary operation on V and U is closed under vector addition. And, as a final example, matrix addition is a binary operation on the set of $n \times n$ real matrices, but the subset of invertible matrices is not closed under matrix addition since the sum of two invertible matrices may not be invertible. For example, both $\left(\begin{smallmatrix} 1 & 0 \\ 0 & 1 \end{smallmatrix}\right)$ and $\left(\begin{smallmatrix} -1 & 0 \\ 0 & -1 \end{smallmatrix}\right)$

are invertible matrices since their determinants are nonzero, but

$$\begin{pmatrix} 1 & 0 \\ 0 & 1 \end{pmatrix} + \begin{pmatrix} -1 & 0 \\ 0 & -1 \end{pmatrix} = \begin{pmatrix} 0 & 0 \\ 0 & 0 \end{pmatrix},$$

which is not invertible.

If $*$ is a binary operation on a set X and if Y is a subset of X that is closed under $*$, then Y not only inherits $*$, it also inherits many of the properties of $*$. For example, if $*$ is associative and $x, y, z \in Y$, then $x * (y * z)$ and $(x * y) * z$ are calculated in Y the same way they are calculated in X and therefore are equal as elements in Y. Hence, $*$ is an associative binary operation on Y; that is, Y inherits associativity. The following proposition summarizes some of the properties of a binary operation that a closed subset inherits.

PROPOSITION 2. Let $*$ be a binary operation on a nonempty set X and let Y be a subset of X that is closed under $*$.

(1) If $*$ is associative, then $*$ on Y is associative.

(2) If $*$ is commutative, then $*$ on Y is commutative.

(3) If e is an identity element for $*$ in X, and if $e \in Y$, then e is an identity element for $*$ in Y.

Proof.   We leave the details as an exercise for the reader.   ■

EXERCISES

1. Determine which of the following formulas defines binary operations on the indicated set. If the formula is a binary operation, determine whether it is associative or commutative, whether it has an identity element, and if so, which elements have an inverse.
   (a) $x * y = x^2 + y^2$, on the set $\mathbb{R}$ of real numbers
   (b) $x * y = x^y$, on the set $\mathbb{R}$
   (c) $x * y = x$, on the set $\mathbb{R}$
   (d) $n * m = nm$, on the set $\mathbb{N}$ of natural numbers
   (e) $n * m = $ any number that divides both n and m, on the set $\mathbb{N}$
   (f) $n * m = n^m$, on the set $\mathbb{N}$
2. Let $\mathrm{Mat}_2(\mathbb{R})$ stand for the set of 2×2 matrices with real number entries. Define a binary operation $*$ on $\mathrm{Mat}_2(\mathbb{R})$ as follows:

$$\begin{pmatrix} a & b \\ c & d \end{pmatrix} * \begin{pmatrix} e & f \\ g & h \end{pmatrix} = \begin{pmatrix} ae & bf \\ cg & dh \end{pmatrix},$$

   where $a, b, c, d, e, f, g, h \in \mathbb{R}$.
   (a) Show that $*$ is an associative and commutative binary operation on $\mathrm{Mat}_2(\mathbb{R})$.
   (b) Show that $*$ has an identity element.

(c) Describe the 2×2 matrices that have an inverse under the operation $*$.

3. Let $F(\mathbb{R})$ stand for the set of functions from $\mathbb{R}$ into $\mathbb{R}$, and let C stand for the subset consisting of the constant functions. Does C inherit function composition?

4. Let f and g be functions in $F(\mathbb{R})$. Define the function $f + g : \mathbb{R} \to \mathbb{R}$ by setting $(f + g)(x) = f(x) + g(x)$ for every $x \in \mathbb{R}$. This operation is called *function addition* on $F(\mathbb{R})$.
   (a) Show that function addition is an associative and commutative binary operation on $F(\mathbb{R})$.
   (b) Show that function addition has an identity element.
   (c) Show that every element in $F(\mathbb{R})$ has an inverse under function addition.
   (d) Is the subset C of constant functions closed under function addition?

5. We know that set union is a binary operation on the power set $\mathscr{P}(X)$ of a set X.
   (a) Show that set union is an associative and commutative binary operation on $\mathscr{P}(X)$.
   (b) Show that this operation has an identity element.
   (c) Which elements in $\mathscr{P}(X)$ have an inverse?

6. Let X be a set and let $\mathscr{P}(X)$ stand for the power set of X.
   (a) Show that set intersection is an associative and commutative binary operation on $\mathscr{P}(X)$.
   (b) Show that this operation has an identity element.
   (c) Which elements in $\mathscr{P}(X)$ have an inverse?

7. Let X be a set and let $\varnothing$ stand for the empty set.
   (a) Does $\mathscr{P}(X) - \{\varnothing\}$ inherit the operation of set union from $\mathscr{P}(X)$?
   (b) Does $\mathscr{P}(X) - \{\varnothing\}$ inherit the operation of set intersection from $\mathscr{P}(X)$?

8. Let X be a set. If A and B are subsets of X, let $A \triangle B = (A \cup B) - (A \cap B)$ stand for the symmetric difference of A and B (see Chapter 1, Section 1, Exercise 8).
   (a) Show that $\triangle$ is an associative binary operation on $\mathscr{P}(X)$.
   (b) Show that $\triangle$ is a commutative operation.
   (c) Show that $\triangle$ has an identity element.
   (d) Show that every element in $\mathscr{P}(X)$ has an inverse under the operation $\triangle$.
   (e) Suppose that X is a finite set. Let $E(X)$ stand for the collection of all subsets of X that contain an even number of elements. Show that $E(X)$ is closed under the operation $\triangle$. Does $E(X)$ contain the identity element of $\mathscr{P}(X)$ under $\triangle$? If A is an element of $\mathscr{P}(X)$, then A has an inverse in $\mathscr{P}(X)$ by Exercise 8(d); does the inverse of A belong to $E(X)$?
   (f) Suppose that X is a finite set. Let $O(X)$ stand for the collection of all subsets of X that contain an odd number of elements. Is $O(X)$ closed under the operation $\triangle$?

9. Let $V = \mathbb{R}^3$ stand for real 3-space. If A and B are vectors in V, let $A \times B$ stand for the vector cross-product of A and B; that is, if $A = (a_1, a_2, a_3)$ and $B = (b_1, b_2, b_3)$, then $A \times B = (a_2 b_3 - a_3 b_2, a_3 b_1 - a_1 b_3, a_1 b_2 - a_2 b_1)$.

(a) Show that cross-product is a binary operation on V.

(b) Show that cross-product is not commutative.

(c) Let $A = (1, 0, 0)$ and $B = (0, 1, 0)$. Calculate $A \times (A \times B)$ and $(A \times A) \times B$. Is cross-product an associative binary operation?

2. GROUPS AND THEIR ARITHMETIC PROPERTIES

In the previous section we discussed several binary operations such as addition and multiplication of integers, addition and multiplication of matrices, and function composition, and we showed that in each case these operations satisfied additional properties such as associativity and the existence of an identity element. In this section our goal is to construct a general algebraic model that reflects these properties—the concept of a group. We begin by defining the group concept and establishing some of the basic arithmetic properties common to all groups. Following this, we look more closely at several important examples of groups, including groups of complex numbers, groups of transformations, and groups of matrices. We then conclude the section by discussing the order of an element in a group.

Definition

A *group* is a set G together with a binary operation $*$ that satisfies the following three conditions:

(1) $*$ is associative,

(2) $*$ has an identity element in G, and

(3) every element in G has an inverse with respect to $*$.

If the binary operation on a group is commutative, the group is called an *abelian*, or *commutative*, group. If G is a group and the set G is finite, we say that G is a *finite group* and use the symbol $|G|$ to stand for the number of elements in G; in this case, $|G|$ is called the *order* of G.

Before we look at examples of groups, let us first summarize some of the basic properties common to all groups.

PROPOSITION 1. Let G be a group. Then

(1) G has a unique identity element.

(2) Every element in G has a unique inverse.

(3) The binary operation on G satisfies general associativity.

(4) If G is commutative, then there is general commutativity.

Proof. All of these statements follow immediately from Proposition 1 in Section 1. We leave the details as an exercise for the reader. ■

Consider the set $\mathbb{Z}$ of integers under the operation of addition. Addition of integers is an associative binary operation and has the integer 0 as an identity element. Moreover, if n is an integer, then the integer $-n$ is an additive inverse for n. The set $\mathbb{Z}$ of integers is therefore a group under addition; we refer to it as the *additive group of integers*. Observe that it is an abelian group since addition of integers is commutative. Similarly, the set $\mathbb{Q}$ of rational numbers and the set $\mathbb{R}$ of real numbers are both abelian groups under addition. We refer to these groups as the *additive group of rational numbers* and the *additive group of real numbers*, respectively.

Now consider the set $\mathbb{Z}^*$ of nonzero integers. Multiplication of integers is an associative binary operation on $\mathbb{Z}^*$ and has the integer 1 as identity element. But $\mathbb{Z}^*$ is not a group under multiplication since the only integers that have an inverse are ± 1. On the other hand, the set $\mathbb{Q}^*$ of nonzero rational numbers and the set $\mathbb{R}^*$ of nonzero real numbers are groups under multiplication. In both of these groups the identity element is the number 1. As for inverses, if $n/m \in \mathbb{Q}^*$, then the multiplicative inverse of n/m is $(n/m)^{-1} = (m/n)$, while $r^{-1} = 1/r$ for any nonzero real number $r \in \mathbb{R}^*$. We refer to $\mathbb{Q}^*$ as the *multiplicative group of nonzero rational numbers* and $\mathbb{R}^*$ as the *multiplicative group of nonzero real numbers*. Finally, observe that the additive groups $\mathbb{Z}$, $\mathbb{Q}$, and $\mathbb{R}$, as well as the multiplicative groups $\mathbb{Q}^*$ and $\mathbb{R}^*$, are all infinite abelian groups. Let us now look at some other examples of groups in more detail.

EXAMPLE 1. Groups of complex numbers

A *complex number* is any expression of the form $a + bi$, where a and b are real numbers and i is a fixed symbol. Two such numbers, $a + bi$ and $c + di$, are equal if and only if $a = c$ and $b = d$. Complex numbers are added and multiplied using the formulas

$$(a + bi) + (c + di) = (a + c) + (b + d)i$$

$$(a + bi)(c + di) = (ac - bd) + (bc + ad)i.$$

It follows that addition and multiplication of complex numbers are binary operations on the set of complex numbers, and these operations are easily seen to be associative and commutative. Moreover, $i^2 = -1$ and hence we think of i as $\sqrt{-1}$. Let's discuss some of the algebraic properties of the complex numbers. Let $\mathbb{C}$ stand for the set of complex numbers.

(A) $\mathbb{C}$ is an abelian group under addition of complex numbers. For, as we noted earlier, addition is both associative and commutative. The identity element is the number $0 = 0 + 0i$, and, if $z = a + bi$ is a typical complex number, the additive inverse of z is the number $-z = (-a) + (-b)i$.

(B) Let $\mathbb{C}^*$ stand for the set of nonzero complex numbers; that is, all complex numbers of the form $a + bi$, where not both a and b are zero. Then $\mathbb{C}^*$ is an abelian group under multiplication of complex numbers. To show this, we must first verify that multiplication is, in fact, a binary operation on the set $\mathbb{C}^*$; it is obviously a binary operation on $\mathbb{C}$, but not so obviously a binary operation on $\mathbb{C}^*$. Let $a + bi$ and $c + di$ be nonzero complex numbers. Then not both a and b are zero and not both c and d are zero. Therefore, $a^2 + b^2 \neq 0$ and $c^2 + d^2 \neq 0$. Now suppose that the product $(a + bi)(c + di) = 0$. Then $(a - bi)(a + bi)(c + di) = (a^2 + b^2)c + (a^2 + b^2)di = 0$. Hence, $(a^2 + b^2)c = 0$ and $(a^2 + b^2)d = 0$. But since $a^2 + b^2 \neq 0$, it follows that $c = d = 0$, which is a contradiction since $c^2 + d^2 \neq 0$. Therefore, $(a + bi)(c + di) \neq 0$. Thus, the product of two nonzero complex numbers is a nonzero complex number, and therefore complex multiplication is a binary operation on $\mathbb{C}^*$. It now follows that $\mathbb{C}^*$ inherits complex multiplication from $\mathbb{C}$ and, along with it, associativity and commutativity. The identity element is the number $1 = 1 + 0i$. Finally, we must show that every element in $\mathbb{C}^*$ has a multiplicative inverse. To this end, let $z = a + bi$ be an element in $\mathbb{C}^*$. Then $a^2 + b^2 \neq 0$. Set

$$w = \frac{a}{a^2 + b^2} - \frac{b}{a^2 + b^2}\, i.$$

Then $w \in \mathbb{C}^*$, and it is easily verified that $zw = wz = 1$. Therefore, $z^{-1} = w$. Thus, all elements in $\mathbb{C}^*$ have an inverse, and we therefore conclude that $\mathbb{C}^*$ is an abelian group under complex multiplication.

(C) Complex numbers may be represented geometrically as follows. The set of complex numbers is a 2-dimensional real vector space with vector addition defined as above and scalar multiplication defined, for $c \in \mathbb{R}$ and $z = a + bi \in \mathbb{C}$, by setting $cz = ca + cbi$. This 2-dimensional space is called the *complex plane* and is visualized as shown in Figure 1. The number $z = a + bi$ corresponds to the point (a, b) in the plane; a is called

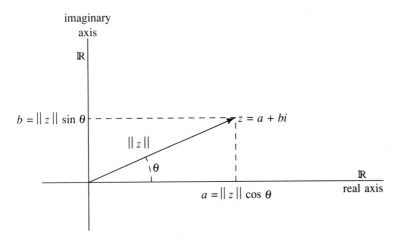

FIGURE 1.   The complex number plane.

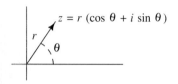

the *real part* of z and b the *imaginary part*. The *norm* of z is the real number $\|z\| = \sqrt{a^2 + b^2}$ and is the magnitude of z as a vector. By using the complex number plane, we may interpret addition of complex numbers geometrically as follows. Let z, $z' \in \mathbb{C}$. Then, by vector addition, $z + z'$ is the complex number represented by the diagonal of the parallelogram having z and z' as sides, as illustrated in Figure 2. To interpret complex multiplication geometrically, we first represent complex numbers trigonometrically. Recall that if z is a complex number, then $z = r(\cos \theta + i \sin \theta)$, where $r = \|z\|$, as shown in the margin. If $z' = r'(\cos \theta' + i \sin \theta')$ is another complex number, then

$$zz' = rr'[(\cos \theta \cos \theta' - \sin \theta \sin \theta') + (\sin \theta \cos \theta' + \cos \theta \sin \theta')i]$$
$$= rr'[\cos(\theta + \theta') + i \sin(\theta + \theta')].$$

In other words, zz' is obtained by rotating z through an angle of θ' radians and multiplying by the scalar factor r', as illustrated in Figure 3. For example, $z^2 = r^2(\cos 2\theta + i \sin 2\theta)$ and is obtained by rotating z counterclockwise through an angle equal to its own argument θ. More generally, for any integer n, $z^n = r^n(\cos n\theta + i \sin n\theta)$, which shows that z^n is obtained by rotating z counterclockwise (if $n > 0$) or clockwise (if $n < 0$) through a total of $n - 1$ multiples of its angle. This result, called *DeMoivre's theorem*, expresses an important relationship between complex numbers and their trigonometric representation in the complex number plane. Finally, let us note that if z and z' lie on the unit complex circle, then $\|z\| = \|z'\| = 1$, and therefore zz' also lies on the unit complex circle since $\|zz'\| = \|z\|\,\|z'\| = 1$. Thus, we may interpret multiplication of complex numbers on the unit circle as rotation of the circle about the origin. This concludes the example.

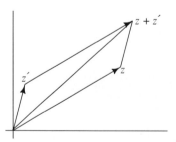

FIGURE 2. Geometric interpretation of addition of complex numbers.

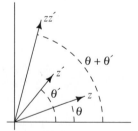

FIGURE 3. Geometric interpretation of multiplication of complex numbers.

EXAMPLE 2. The group of the triangle

A *rigid motion* of the Euclidean plane is any mapping of the plane into itself that leaves the distance between any two points unchanged. Rotations, reflections, and translations of the plane are typical examples of rigid motions. In this example we show that the set of rigid motions of the plane that map an equilateral triangle to itself is a group of order 6 under function composition called the group of the triangle, find its multiplication table, and then discuss a matrix representation for the elements in the group.

(A) Let T stand for the set of all rigid motions of the plane that bring an equilateral triangle into coincidence with itself. Referring to Figure 4, we see that there are six such rigid motions of the plane: three rotations, $1, R_1, R_2$, about the center of the triangle, and three reflections, S_1, S_2, S_3, about axes passing through the center of the triangle and each of the three vertices. The elements in T are functions from the plane into itself, and the binary operation is function composition; for clearly, if f and g are rigid motions of the plane that map the triangle to itself, then the composition $f \circ g$ is also a rigid motion of the plane that maps the triangle to itself. Thus, function composition is a binary operation on T and, as such, is associative. Moreover, the identity function 1_P, or simply

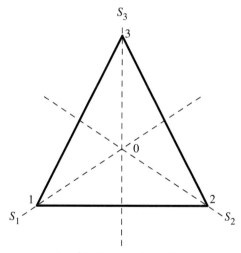

1 = identity transformation
R_1 = 120° counterclockwise rotation about 0
R_2 = 240° counterclockwise rotation about 0
S_1, S_2, S_3 = reflections about lines passing through 0 and a vertex of the triangle

FIGURE 4. Rigid motions of the plane that map an equilateral triangle to itself.

1, on the plane P is the identity element for T under function composition. Finally, each of the six transformations in T has an inverse in T:

$$1^{-1} = 1 \qquad S_1^{-1} = S_1$$
$$R_1^{-1} = R_2 \qquad S_2^{-1} = S_2$$
$$R_2^{-1} = R_1 \qquad S_3^{-1} = S_3.$$

Therefore, T is a finite group of order 6. We refer to T as the *group of the triangle*.

(B) To calculate the product, that is, the composition of two elements in T, first note that any transformation in T is uniquely determined by where it maps the three vertices of the triangle. Thus, to calculate the product $R_1 S_1$, for example, we reason that $R_1 S_1(1) = R_1(1) = 2$, $R_1 S_1(2) = R_1(3) = 1$, and $R_1 S_1(3) = R_1(2) = 3$. But the transformation that maps $1 \mapsto 2$, $2 \mapsto 1$, and $3 \mapsto 3$ is S_3. Hence, $R_1 S_1 = S_3$. Similarly, we find that $R_1^2 = R_1 R_1 = R_2$ and $R_1^2 S_1 = S_2$. By setting $R = R_1$ and $S = S_1$, it now follows that every element in T may be written as the product of R's and S's: $R_1 = R$, $R_2 = R^2$, $S_1 = S$, $S_2 = R^2 S$, and $S_3 = RS$. Hence, $T = \{1, R, R^2, S, RS, R^2 S\}$. We express this fact by saying that R and S *generate* T. The multiplication table for T, called its *Cayley table* and shown below, is the square array whose rows and columns are indexed by the elements in T; the entry in row x and column y is the product xy.

	1	R	R^2	S	RS	R^2S
1	1	R	R^2	S	RS	R^2S
R	R	R^2	1	RS	R^2S	S
R^2	R^2	1	R	R^2S	S	RS
S	S	R^2S	RS	1	R^2	R
RS	RS	S	R^2S	R	1	R^2
R^2S	R^2S	RS	S	R^2	R	1

It is clear from this table that T is a nonabelian group since, for example, $RS \neq SR$. Notice that an abelian group will have a multiplication table that is symmetric about the diagonal going from the upper left corner to the lower right corner.

(C) Up to this point we have worked with the elements in T as functions from the plane to itself. But these functions are also linear transformations of real 2-space $\mathbb{R}^2$ into itself if we choose the center of the triangle as the origin of coordinates. As such, we may find a matrix representation for the six elements in T relative to a given coordinate system and thus obtain a group of matrices. To do this, first choose a coordinate system

(A, B) for the real plane, with origin at 0, as illustrated in the marginal figure. To find the matrix of R, first calculate the image of the base vectors A and B under R, and write each image as a linear combination of the vectors A and B: $R(A) = B = 0A + 1B$, $R(B) = -A - B = -1A + -1B$. The matrix of R in the (A, B) coordinate system is therefore $\begin{pmatrix} 0 & -1 \\ 1 & -1 \end{pmatrix}$. Similarly, we find the matrix representation for each of the remaining elements in T and have listed them here.

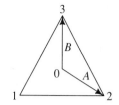

$$1 \leftrightarrow \begin{pmatrix} 1 & 0 \\ 0 & 1 \end{pmatrix} \qquad 1 \leftrightarrow \begin{pmatrix} 0 & 1 \\ 1 & 0 \end{pmatrix}$$

$$R \leftrightarrow \begin{pmatrix} 0 & -1 \\ 1 & -1 \end{pmatrix} \qquad S_2 \leftrightarrow \begin{pmatrix} 1 & -1 \\ 0 & -1 \end{pmatrix}$$

$$R^2 \leftrightarrow \begin{pmatrix} -1 & 1 \\ -1 & 0 \end{pmatrix} \qquad S_3 \leftrightarrow \begin{pmatrix} -1 & 0 \\ -1 & 1 \end{pmatrix}$$

Observe that since matrix multiplication corresponds to function composition, the matrix of R^2, for example, may be found by squaring the matrix of R. The matrices for S_2 and S_3 are found similarly by using the fact that $S_2 = R^2S$ and $S_3 = RS$. By representing the elements of T as matrices, we obtain a new group of order 6 whose elements are matrices. In general, it is not a simple matter to write down a set of matrices and have them form a group under matrix multiplication. This example illustrates one way to do this by using the matrix representation of familiar geometric transformations.

EXAMPLE 3

Let $\text{Mat}_n^*(\mathbb{R})$ stand for the set of $n \times n$ invertible matrices with real number entries. Then $\text{Mat}_n^*(\mathbb{R})$ is a group under matrix multiplication. To show this, observe first that matrix multiplication is in fact a binary operation on $\text{Mat}_n^*(\mathbb{R})$ since the product of two invertible matrices is invertible: if M and N are invertible with inverses M^{-1} and N^{-1}, respectively, then the inverse of MN is $N^{-1}M^{-1}$. We recall that matrix multiplication is associative and has the $n \times n$ identity matrix I_n as identity element. Finally, every element in $\text{Mat}_n^*(\mathbb{R})$ has an inverse, by definition, and the inverse is an $n \times n$ invertible matrix since $(M^{-1})^{-1} = M$ for every invertible matrix M. Therefore, $\text{Mat}_n^*(\mathbb{R})$ is a group under matrix multiplication. Let us note that since there are infinitely many invertible matrices, as well as invertible matrices that do not commute, $\text{Mat}_n^*(\mathbb{R})$ is an infinite nonabelian group.

EXAMPLE 4

Let V be a real vector space. Then V is an abelian group under vector addition. To show this, we need only recall that vector addition is associative and commutative and has the zero vector as identity element; the inverse of an arbitrary vector $A \in V$ is the vector $-A$. Thus, V is an abelian group under vector addition. Note that the additive group of real numbers and the additive group of complex numbers are particular examples of this type of group since they are both real vector spaces.

EXAMPLE 5. The general linear group

Let V be a real vector space and let $GL(V)$ stand for the set of nonsingular linear transformations from V into V. A nonsingular linear transformation, we recall, is a linear transformation $\sigma: V \to V$ that is 1–1; it is necessarily onto and consequently is a linear isomorphism mapping V onto V. We claim that the set $GL(V)$ of all such linear transformations is a group under function composition.

(A) To show that $GL(V)$ is a group, let us first recall a few facts from linear algebra. First, recall that if σ and τ are linear transformations from V into V, then the composition $\sigma\tau$ is also a linear transformation. If, in addition, both σ and τ are nonsingular, then $\sigma\tau$ is also nonsingular. It follows, therefore, that composition is a binary operation on the set $GL(V)$. The identity function 1_V on V is a nonsingular linear transformation and is, clearly, the identity element for $GL(V)$ under function composition. The second fact that we need to recall is that if σ is nonsingular, then not only does σ^{-1} exist, but it is, in fact, a nonsingular linear transformation whose inverse is σ. Hence, if σ is an element of $GL(V)$, then σ^{-1} is also an element of $GL(V)$. It now follows that $GL(V)$ is a group under function composition. We call $GL(V)$ the *general linear group of* V.

(B) To gain more perspective on the group $GL(V)$, observe that the elements in $GL(V)$ are, first and foremost, functions from V into V, and consequently $GL(V)$ is a subset of $F(V)$, the set of all functions on V. What we have shown in part (A) of this example is that $GL(V)$ inherits the binary operation on $F(V)$, namely, function composition, and thus also inherits associativity of the operation. Since $1_V \in GL(V)$, it also inherits the identity element.

(C) Let us consider in more detail the case when V is a 1-dimensional real vector space. Let $V = \langle A \rangle$ stand for the line spanned by a nonzero vector A. In this case, the linear transformations from V into V are just the scalar mappings $\sigma_c: V \to V$ defined by setting $\sigma_c(A) = cA$, where c is any real number. σ_c is nonsingular if and only if $c \neq 0$. Hence, $GL(V) = \{\sigma_c \mid c \in \mathbb{R}^*\}$. It is clear from this description of $GL(V)$ that there is a 1–1 correspondence $\sigma_c \leftrightarrow c$ between $GL(V)$ and $\mathbb{R}^*$. More importantly, we find that $\sigma_c \sigma_d(A) = cdA = \sigma_{cd}(A)$ for all $c, d \in \mathbb{R}^*$ and hence $\sigma_c \sigma_d = \sigma_{cd}$, which means that the product $\sigma_c \sigma_d$ calculated in $GL(V)$ corresponds to the product of c and d calculated in the group $\mathbb{R}^*$. In other words, the correspondence $\sigma_c \leftrightarrow c$ preserves the multiplication on $GL(V)$. For this reason we make no distinction between the groups $GL(V)$ and $\mathbb{R}^*$. More generally, we will show in Section 5 that if V is an n-dimensional real vector space, there is essentially no difference between the general linear group $GL(V)$ and the multiplicative group $\mathrm{Mat}_n^*(\mathbb{R})$ discussed in Example 3.

These examples have illustrated particular kinds of groups. In the next example we discuss an important method for combining any two groups to form a new group.

EXAMPLE 6. Direct product of groups

Let G and H be groups, and let $G \times H$ stand for the Cartesian product of G and H. Define a binary operation on $G \times H$ as follows: if (g, h), (g', h') $\in G \times H$, set $(g, h)(g', h') = (gg', hh')$, where gg' is calculated in G and hh' is calculated in H. We refer to this operation as *componentwise multiplication*.

(A) We claim that $G \times H$ is a group under componentwise multiplication. To show this, let (g, h), (g', h'), and (g'', h'') be typical elements in $G \times H$, where $g, g', g'' \in G$ and $h, h', h'' \in H$. Then

$$(g, h)[(g', h')(g'', h'')] = (g, h)(g'g'', h'h'')$$
$$= (g(g'g''), h(h'h'')) = ((gg')g'', (hh')h'')$$
$$= (gg', hh')(g'', h'') = [(g, h)(g', h')](g'', h'').$$

Therefore, componentwise multiplication is associative. Moreover, if 1_G and 1_H stand for the identity elements of G and H, respectively, then $(1_G, 1_H)$ is, clearly, an identity element for componentwise multiplication. And finally, it is clear that $(g, h)^{-1} = (g^{-1}, h^{-1})$ for all $(g, h) \in G \times H$. Hence, $G \times H$ is a group under componentwise multiplication. $G \times H$ is called the *direct product* of G and H. Note that if G and H are finite groups, then $G \times H$ is a finite group whose order is equal to $|G||H|$.

(B) For example, consider the direct product $T \times T$, where T is the group of the triangle. $T \times T$ is a group of order 36; Figure 5 displays these

R^2S	$(1, R^2S)$	(R, R^2S)	(R, R^2S)	(S, R^2S)	(RS, R^2S)	(R^2S, R^2S)
RS	$(1, RS)$	(R, RS)	(R^2, RS)	(S, RS)	(RS, RS)	(R^2S, RS)
S	$(1, S)$	(R, S)	(R^2, S)	(S, S)	(RS, S)	(R^2S, S)
R^2	$(1, R^2)$	(R, R^2)	(R^2, R^2)	(S, R^2)	(RS, R^2)	(R^2S, R^2)
R	$(1, R)$	(R, R)	(R^2, R)	(S, R)	(RS, R)	(R^2S, R)
1	$(1, 1)$	$(R, 1)$	$(R^2, 1)$	$(S, 1)$	$(RS, 1)$	$(R^2S, 1)$
	1	R	R^2	S	RS	R^2S

FIGURE 5.   The 36 elements in the direct product $T \times T$.

elements. The group $T \times T$ is nonabelian since, for example, $(R, 1)(S, 1) = (RS, 1)$, but $(S, 1)(R, 1) = (SR, 1)$. Furthermore, the four elements $(R, 1)$, $(S, 1)$, $(1, R)$, and $(1, S)$ generate $T \times T$. To show this, observe that if $(x, y) \in T \times T$, then $(x, y) = (x, 1)(1, y)$. Since both x and y may be written as products of R's and S's, it follows that (x, y) may be written as a product of the four elements listed above. For example, $(RS, S) = (R, 1)(S, 1)(1, S)$.

(C) As a second example of direct products, consider the group $\mathbb{R} \times \mathbb{R}$, where $\mathbb{R}$ stands for the additive group of real numbers. The componentwise operation on $\mathbb{R} \times \mathbb{R}$ is now written additively as $(a, b) + (c, d) = (a + c, b + d)$, where $a, b, c, d \in \mathbb{R}$. $\mathbb{R} \times \mathbb{R}$ is just the additive group of 2-dimensional real space $\mathbb{R}^2$, the ordered pair $(a, b) \in \mathbb{R} \times \mathbb{R}$ corresponding to the vector $A \in \mathbb{R}^2$ whose Cartesian coordinates are a and b. This concludes the example.

As Examples 1 through 6 have shown, a group is not just a set of elements, but a set of elements together with an associative binary operation that has an identity element and in which every element has an inverse. It is the binary operation and its properties that make, or do not make, the set into a group. Throughout these examples, we have used two different notations for the binary operation on a group. When the *multiplicative notation* is used, the unique identity element is denoted by the symbol 1 and the unique inverse of an element x is denoted by x^{-1}. Using the multiplicative notation, we define the symbol x^n for any integer n as follows. If n is positive, x^n stands for the product of x with itself n times; this makes sense since the binary operation on a group satisfies general associativity. If n is negative, set $x^n = (x^{-1})^{-n}$. Finally, set $x^0 = 1$. Observe that $(x^{-1})^{-1} = x$ for all elements x in a group since $xx^{-1} = x^{-1}x = 1$.

Sometimes the symbol $+$ is used to denote the binary operation on a group, especially if the group is abelian. This is the *additive notation*. When the additive notation is used, the symbols 0 and $-x$ are used for the identity element and the inverse of an element, respectively. In this case, if n is a positive integer, $nx = x + \cdots + x$ stands for the sum of x with itself n times, while, for $n < 0$, $nx = (-n)(-x) = -x + \cdots + -x$. Let us emphasize, however, that regardless of which notational system is used, the symbols 1 and x^{-1}, or 0 and $-x$, are abstract symbols that stand for certain uniquely determined elements in a group. When we deal with a particular group, these symbols must be interpreted within the context of the group. Let us now show that in any group, these symbols satisfy the usual exponential formulas.

PROPOSITION 2. Let G be a group and let x be an element in G. Then $(x^n)^m = x^{nm}$ and $x^n x^m = x^{n+m}$ for all integers n and m.

Proof. We prove the first statement by considering different cases and leave the proof of the second statement as an exercise for the reader.

(A) Suppose that both n and m are positive. Then $(x^n)^m = x^{nm}$ by the corollary to Proposition 1 in Section 1.

(B) Suppose that $m > 0$ and $n < 0$. Then $n = -s$ for some integer $s > 0$. Therefore, $x^n = x^{-s} = (x^{-1})^s$ by definition, and hence $(x^n)^m = [(x^{-1})^s]^m = (x^{-1})^{sm}$ by part (A). But, by definition, $(x^{-1})^{sm} = x^{-sm} = x^{nm}$. Hence, $(x^n)^m = x^{nm}$.

(C) Suppose that $m < 0$ and $n > 0$. Then $m = -s$ for some integer $s > 0$. Therefore, $(x^n)^m = (x^n)^{-s} = [(x^n)^{-1}]^s$. Now consider the element $(x^n)^{-1}$. Using induction on n, let us show that $(x^n)^{-1} = (x^{-1})^n$ for every positive integer n. If $n = 1$, this is obvious. Next, suppose that $n > 1$ and $(x^n)^{-1} = (x^{-1})^n$, and consider the element $(x^{n+1})^{-1}$. Since

$$x^{n+1}(x^{-1})^{n+1} = (xx^n)[(x^{-1})^n x^{-1}]$$
$$= x[x^n(x^n)^{-1}]x^{-1} = xx^{-1} = 1,$$

and similarly, $(x^{-1})^{n+1}x^{n+1} = 1$, it follows that $(x^{n+1})^{-1} = (x^{-1})^{n+1}$. Hence, by induction, $(x^n)^{-1} = (x^{-1})^n$ for all positive integers n. Returning to $(x^n)^m$, it now follows that $(x^n)^m = [(x^n)^{-1}]^s = [(x^{-1})^n]^s = x^{-ns} = x^{nm}$.

(D) Suppose, finally, that $m < 0$ and $n < 0$. Then $m = -s$ and $n = -t$ for some integers $s, t > 0$. Therefore, $(x^n)^m = (x^{-t})^{-s} = [(x^{-1})^t]^{-s} = (x^{-1})^{-st}$ by part (C). Now, by definition this last expression is equal to $[(x^{-1})^{-1}]^{st}$. But $(x^{-1})^{-1} = x$. Thus, $(x^n)^m = [(x^{-1})^{-1}]^{st} = x^{st} = x^{nm}$, and the proof is complete. ∎

It follows from Proposition 2 that multiplication in a group satisfies the usual properties of exponents. When stated in terms of the additive notation, this result shows that $m(nx) = (nm)x$ and $nx + mx = (n + m)x$ for all integers n and m.

Let us now conclude this section by discussing the order of an element.

Definition

Let G be a group and let x be an element in G. The *order of* x, denoted by $|x|$, is the smallest positive integer n such that $x^n = 1$, where 1 stands for the identity of G. If there is no such integer, we say that x has *infinite order*.

For example, using the multiplication table for the group T of the triangle, we find that $R \neq 1$, $R^2 \neq 1$, but $R^3 = 1$. Hence, $|R| = 3$. We find, similarly, that $|R^2| = 3$, $|S| = |RS| = |R^2S| = 2$ and that $|1| = 1$. Note that the order of the identity element in any group is always equal to 1. Now, 3 is not the only exponent that reduces R to the identity; for example, $R^6 = (R^3)^2 = 1$. Let us show that the only way this may happen is when the exponent is a multiple of the order.

PROPOSITION 3. Let G be a group and let x be an element in G that has finite order. Then $x^n = 1$ if and only if $|x|$ divides n.

Proof. Observe that we need only prove this statement when n is positive since $x^n = 1$ if and only if $x^{-n} = 1$. To this end, suppose that n is a positive integer and that $|x|$ divides n. Then $n = k|x|$ for some integer k and hence $x^n = (x^{|x|})^k = 1^k = 1$, as required. Now suppose, conversely, that $x^n = 1$ for some positive integer n. By the division algorithm, there are integers q and r such that $n = q|x| + r$, where $0 \le r < |x|$. Therefore, $1 = x^n = (x^{|x|})^q x^r = x^r$. But $r < |x|$ and $|x|$ is the least positive integer that reduces x to the identity. Therefore, $r = 0$ and hence $n = q|x|$. Thus, $|x|$ divides n and the proof is complete. ∎

PROPOSITION 4. Every element of a finite group has finite order.

Proof. Let G be a finite group and let x be an element in G. Since G is finite, the subset $\{1, x, x^2, \ldots\}$ is also finite. It follows that there are distinct integers n and m such that $x^n = x^m$, and we choose the notation so that $n < m$. Then $x^{m-n} = 1$ and hence x must have finite order. ∎

Since the elements of a finite group must have finite order, we may find the order of any such element by calculating successively the powers 1, x, $x^2, \ldots$. Eventually we must obtain the identity element, and when we do, we will have the order of x. For example, in the group $T \times T$, we find that $(R, S)^2 = (R^2, 1)$, $(R, S)^3 = (1, S)$, $(R, S)^4 = (R, 1)$, $(R, S)^5 = (R^2, S)$, $(R, S)^6 = (1, 1)$. Hence, $|(R, S)| = 6$.

As we just showed, every element in a finite group must have finite order. But what happens if the group is infinite? Do all the elements have infinite order except for the identity element? Consider the multiplicative group $\mathbb{R}^*$ of nonzero real numbers. Clearly, $\mathbb{R}^*$ is an infinite group. But nevertheless, the number -1 has order 2 since $(-1)^2 = 1$. In fact, for a number $x \in \mathbb{R}^*$ to have finite order, it must be a solution of the equation $x^n = 1$ for some positive integer n. But the only possible real solutions of this equation are ± 1. Hence, ± 1 are the only elements in $\mathbb{R}^*$ that have finite order. As a second example, consider the multiplicative group $\text{Mat}_2^*(\mathbb{R})$ of invertible 2×2 real matrices. If $M = \left(\begin{smallmatrix} 0 & 1 \\ 1 & 0 \end{smallmatrix}\right)$, we find that $M^2 = I_2$, the identity element, and therefore $|M| = 2$. These examples show that infinite groups may contain elements of finite order other than the identity element. As our final example, let us now construct an infinite group in which, rather surprisingly, every element has finite order.

EXAMPLE 7. An infinite group all of whose elements have finite order

Let G stand for the set of all complex numbers z such that $z^n = 1$ for some integer n. The numbers 1 and i, for example, belong to G. We claim that G is an infinite abelian group under multiplication of complex numbers and that every element in G has finite order.

(A) We begin by showing that multiplication of complex numbers is a binary operation on G. Let $z, w \in G$. Then $z^n = 1$ and $w^m = 1$ for some integers n and m. Then $(zw)^{nm} = z^{nm}w^{nm} = 1$ and hence $zw \in G$. Therefore, complex multiplication is a binary operation on G.

(B) Let us now show that G is an abelian group under complex multiplication. Since G is a subset of the multiplicative group $\mathbb{C}^*$ of nonzero complex numbers and since it inherits complex multiplication by part (A), it follows that G also inherits associativity and commutativity of multiplication from $\mathbb{C}^*$. Moreover, the number 1 is clearly an element in G and is, therefore, an identity element for G. As for inverses, if $z \in G$, with $z^n = 1$ for some integer n, then $(z^{-1})^n = (1/z)^n = 1$, and hence $z^{-1} \in G$. Thus, every element in G has an inverse in G and therefore G is an abelian group under multiplication of complex numbers.

(C) To show that every element in G has finite order, let $z \in G$. Then $z^n = 1$ for some integer n. If n is positive, then z has finite order. If n is negative, then $z^{-n} = 1$ and $-n$ is positive. Thus, in either case z has finite order.

(D) Finally, let us show that G is infinite. We do this by constructing an infinite subset of G as follows. For each positive integer n, let $z_n = \cos(2\pi/n) + i\sin(2\pi/n)$. Then by DeMoivre's theorem, $z_n^n = \cos n(2\pi/n) + i\sin n(2\pi/n) = 1$. Therefore, z_n is an element in G for every positive integer n. To see why the numbers z_n form an infinite subset of G for $n = 1, 2, \ldots$, observe that z_n is the point on the unit complex circle having coordinates $(\cos(2\pi/n), \sin(2\pi/n))$. It is clear that the numbers $z_1, z_2, z_3, \ldots$ form an infinite subset of G, and hence G itself must be infinite. We conclude therefore that G is an infinite abelian group all of whose elements have finite order.

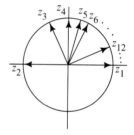

Exercises

1. **The group of the square.** Let D_4 stand for the set of rigid motions of the Euclidean plane that bring a square into coincidence with itself. Find the elements in D_4 and show that it is a group under function composition. D_4 is called the *group of the square*. Construct the multiplication table for D_4 and find the inverse and order of each element. Is D_4 commutative?

2. Let D_4 stand for the group of the square discussed in Exercise 1.
   (a) Find the matrix representation for each transformation in D_4 relative to the coordinate system (A, B) indicated in the top marginal figure.
   (b) Change the coordinate system to the one shown in the bottom marginal figure. Find the matrix representation for the elements in D_4 relative to this new coordinate system.
   (c) Compare the two groups of matrices obtained in parts (a) and (b). Which transformations in D_4 have the same matrix representation in both coordinate systems?

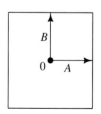

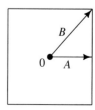

3. Let K stand for the set of rigid motions of the plane that bring a rectangle into coincidence with itself. Show that K is a group under function composition of order 4 and that every nonidentity element in K has order 2. K is the *group of the rectangle*, but it is usually referred to as the *Klein 4-group*.

4. Show that the set $\text{Mat}_n(\mathbb{R})$ of $n \times n$ matrices over the real numbers is an abelian group under matrix addition.

5. Let G stand for the set of 2×2 matrices of the form $\left(\begin{smallmatrix} 1 & a \\ 0 & 1 \end{smallmatrix}\right)$, where a is any real number. Show that G is a group under matrix multiplication. Is G commutative? If $M \in G$, find an explicit description of M^{-1}; that is, describe the entries in M^{-1} in terms of the entries in M.

6. Let G stand for the set of complex numbers z such that $\|z\| = 1$.
   (a) Show that G is an abelian group under complex multiplication.
   (b) If $z \in G$, find an explicit description for z^{-1}; that is, describe the real and imaginary parts of z^{-1} in terms of those of z. Illustrate the relationship between z and z^{-1} on the unit complex circle.
   (c) If $z = a + bi$ is a complex number, the number $\bar{z} = a - bi$ is called the *complex conjugate* of z. Show that $\|\bar{z}\| = \|z\|$ and that $z\bar{z} = \|z\|^2$. Illustrate geometrically the relationship between z and $\bar{z}$ in the complex plane.
   (d) If $z \in G$, show that $z^{-1} = \bar{z}$.

7. Let G stand for the set of complex numbers z such that $\|z\| = 2$. Is G a group under complex multiplication?

8. Show that the following real-valued functions form a group under function composition and find $|f_3|$ and $|f_5|$:

$$f_1(x) = x \qquad f_4(x) = 1 - \frac{1}{x}$$

$$f_2(x) = 1 - x \qquad f_5(x) = \frac{1}{1 - x}$$

$$f_3(x) = \frac{1}{x} \qquad f_6(x) = \frac{x}{x - 1}.$$

9. Let x and y be elements of a group G.
   (a) Show that $(xy)^{-1} = y^{-1}x^{-1}$.
   (b) Give an example of a group and elements x, y such that $(xy)^{-1} \neq x^{-1}y^{-1}$.

10. Let G be an abelian group. Show that $(xy)^n = x^n y^n$ for all elements $x, y \in G$ and all integers n.

11. Let G be a group in which every nonidentity element has order 2. Show that G is abelian.

12. Let $F(\mathbb{R})$ stand for the set of functions from $\mathbb{R}$ into $\mathbb{R}$.
   (a) Show that $F(\mathbb{R})$ is a group under addition of functions.
   (b) If $f, g \in F(\mathbb{R})$, define the product function $fg : \mathbb{R} \to \mathbb{R}$ by setting $(fg)(x) = f(x)g(x)$ for all $x \in \mathbb{R}$. Is $F(\mathbb{R})$ a group under this operation?

3. SUBGROUPS

In the previous section we introduced the concept of a group and discussed several examples of groups. In many of these examples we showed that the given set of elements formed a group under the binary operation by showing that it was in fact part of a larger group and inherited the binary operation on the larger group. By doing so, we avoided having to verify certain properties of the operation on the given subset since, if the subset inherits the operation, it also inherits some of the properties of the operation. For example, the set G of complex numbers z such that $z^n = 1$ for some integer n discussed in Example 7 is part of the larger multiplicative group $\mathbb{C}^*$ of nonzero complex numbers and is closed under multiplication. G therefore inherits complex multiplication from $\mathbb{C}^*$ and thus also inherits the associativity and commutativity of complex multiplication.

In this section our goal is to extend this method to arbitrary groups by defining the concept of a subgroup. We will show that only two requirements need to be verified for a nonempty subset to be a subgroup: closure under the binary operation on the group and closure under inverses. If these conditions are satisfied, then all the axioms for a group are satisfied. We then conclude the section by showing how to generate subgroups using any collection of elements in a group.

> **Definition**
>
> Let G be a group. A *subgroup* of G is a nonempty subset H that inherits the binary operation on G and forms a group under this operation. We indicate that H is a subgroup of G by writing $H \leq G$.

PROPOSITION 1. Let H be a nonempty subset of a group G. Then H is a subgroup of G if and only if the following two conditions are satisfied:

(1) H is closed under multiplication; that is, if $x, y \in H$, then $xy \in H$.
(2) H is closed under inverses; that is, if $x \in H$, then $x^{-1} \in H$.

Proof. Suppose that H is a subgroup of G. Then, by definition, H inherits the binary operation on G and is a group with respect to this operation. Condition (1) is therefore satisfied since this is what is meant by saying that H inherits the binary operation on G. Condition (2), however, must be approached more carefully. By assumption, H is a group and therefore has an identity element, say 1_H. But this means that 1_H is an identity element only for the elements in H, not necessarily for all elements in G. Let us first show that 1_H is in fact the identity element 1 of G. To this end, observe that $1_H 1_H = 1_H$ in H. If we regard this equation as an equation among elements in G, which we may do since the operations on H and G are the same, then it follows that $1 = 1_H 1_H^{-1} = (1_H 1_H) 1_H^{-1} = 1_H (1_H 1_H^{-1}) = 1_H 1 = 1_H$ and hence $1_H = 1$. Therefore, the identity for H is the same as the identity for G. We can now verify that condition (2) is satisfied. Let $x \in H$. Then x has an inverse $h \in H$ and hence $xh = hx = 1_H = 1$. But h is then an inverse for x in G. Since inverses in a group are unique, it follows that $x^{-1} = h$ and therefore $x^{-1} \in H$. Thus, condition (2) is satisfied.

Conversely, suppose that H is a nonempty subset of G that satisfies conditions (1) and (2). To show that H is a subgroup of G, we must show that H inherits the binary operation on G and forms a group under this operation. Now, condition (1) shows that H inherits the binary operation on G. Moreover, since the operation is associative on G, the operation on H inherits associativity. To show that H contains an identity element for the operation, recall that H is nonempty. Let $h \in H$. Then $h^{-1} \in H$ by condition (2), and hence, by condition (1), $hh^{-1} \in H$. But $hh^{-1} = 1$. Therefore, H contains the identity element of G, which is clearly an identity element for H. Now that we know that G and H have the same identity element, it follows from condition (2) that every element in H has an inverse in H. Therefore, H is a group under the binary operation on G and is therefore a subgroup of G. ∎

Let us emphasize that to apply Proposition 1 to a subset of a group, we must know that the subset is nonempty, a requirement that is easily overlooked. By way of analogy, note that conditions (1) and (2) are similar to what happens in a vector space. For if V is a vector space, then a subspace of V is a nonempty subset U that forms a vector space under the vector operations on V, and U is a subspace of V if and only if it is closed with respect to vector addition and scalar multiplication.

COROLLARY.   Let H be a nonempty subset of a group G and suppose that H is finite. Then H is a subgroup of G if and only if the product xy is in H whenever x and y are in H.

Proof.   If H is a subgroup of G and $x, y \in H$, then $xy \in H$ by condition (1) of Proposition 1. Conversely, if H is a nonempty finite subset of G and H is closed under products, then condition (1) is immediately satisfied and it remains only to verify condition (2). To this end, let $x \in H$. Since H is closed under products, the set $\{x, x^2, \ldots\}$ is a subset of H and hence is finite. Therefore, $x^n = x^m$ for some positive integers n and m, and we assume that $n < m$. Then $x^{m-n} = 1$. Now, if $m - n > 1$, then $x^{-1} = x^{m-n-1}$ and hence $x^{-1} \in H$. Otherwise $m - n = 1$, in which case $x = 1$ and hence $x^{-1} \in H$ since $x^{-1} = 1$. Thus, in all cases $x^{-1} \in H$ and hence condition (2) of Proposition 1 is satisfied. Therefore, H is a subgroup of G. ∎

COROLLARY.   Let H and K be subgroups of a group G. Then $H \cap K$ is a subgroup of G.

Proof.   We leave the proof as an exercise for the reader. ∎

The first corollary is especially useful when the group is finite since, in this case, every subset is also finite. The subgroups of a finite group are therefore those subsets that are closed under multiplication. For example, in the group T of the triangle, the subset $\{1, R\}$ is not a subgroup since it contains R but does not contain R^2. The subset $\{1, R, R^2\}$ is a subgroup, however, since the product of any two powers of R is a power of R. Notice also that the proof of the first corollary brings out an important point about the inverse of an element in a finite group: in a finite group, the inverse of an element may always be expressed as a positive integral power of the element itself. In the group T of the triangle, for example, we find that $R^{-1} = R^2$.

As a second example of subgroups, consider a vector space V over the real numbers. If U is a subspace of V, we claim that U is a subgroup of the additive group of V. For, as we noted earlier, U is closed under vector addition and, if $A \in U$, the inverse of A is the vector $-A = -1A$, which lies in U since U is closed under scalar multiplication. Therefore, U is a subgroup of the additive group of V. Thus, one way to obtain subgroups of V is to choose any subspace of V. But V also has subgroups that are not subspaces; in the exercises we give an example of this type of subgroup.

Here are some more examples of subgroups. In each case we show that the particular subset is a subgroup by verifying that the two conditions in Proposition 1 are satisfied, unless the group is finite, in which case we need only check for closure under multiplication.

EXAMPLE 1

Let G be any group. Then both of the subsets $\{1\}$ and G are closed under multiplication and inverses, and both are therefore subgroups of G. We call these the *trivial subgroups* of G.

EXAMPLE 2. Subgroups of $\mathbb{Z}$

Let n be an integer and let $n\mathbb{Z} = \{nk \in \mathbb{Z} \mid k \in \mathbb{Z}\}$ stand for the set of integral multiples of n. We claim that $n\mathbb{Z}$ is a subgroup of the additive group $\mathbb{Z}$ of integers and that every subgroup has this form for some n.

(A) Let us first show that $n\mathbb{Z}$ is a subgroup of $\mathbb{Z}$ for any integer n. Clearly, $n\mathbb{Z}$ is a nonempty subset of $\mathbb{Z}$. Moreover, if $ns, nt \in n\mathbb{Z}$, then $ns + nt$ $= n(s + t) \in n\mathbb{Z}$ and $-(ns) = n(-s) \in n\mathbb{Z}$. Hence, $n\mathbb{Z}$ is closed under addition and inverses, and it is therefore a subgroup of $\mathbb{Z}$. Note that $0\mathbb{Z}$ $= \{0\}$ and $1\mathbb{Z} = \mathbb{Z}$ are the trivial subgroups of $\mathbb{Z}$, while $2\mathbb{Z}$ is the subgroup of even integers.

(B) To show that every subgroup of $\mathbb{Z}$ has the form $n\mathbb{Z}$ for some integer n, let H be a subgroup of $\mathbb{Z}$. If $H = \{0\}$, then $H = 0\mathbb{Z}$ and we are done. Otherwise, H contains some integer $h \neq 0$. If $h < 0$, then $-h > 0$, and

$-h \in H$ since $-h$ is the additive inverse of h and H is closed under inverses. It follows that H contains at least one positive integer and hence a smallest positive integer, say n. We claim that $H = n\mathbb{Z}$. To show this, we must verify that both $n\mathbb{Z} \subseteq H$ and $H \subseteq n\mathbb{Z}$. Let us first show that $n\mathbb{Z} \subseteq H$. Choose $nk \in n\mathbb{Z}$. If $k = 0$, then $nk = 0$, and this is an element in H. If $k > 0$, then $nk = n + \cdots + n$, the sum of n with itself k times, and this is an element in H since H is a subgroup and hence closed under addition. Finally, if $k < 0$, then $nk = -n + \cdots + -n$, the sum of $-n$ with itself $-k$ times, and this, too, is an element in H since H contains the inverse $-n$ and is closed under addition. It follows, therefore, that $nk \in H$ for all integers k and hence $n\mathbb{Z} \subseteq H$. Conversely, to show that $H \subseteq n\mathbb{Z}$, let $h \in H$. By the division algorithm, there are integers q and r such that $h = qn + r$, where $0 \le r < n$. Then $r = h - qn = h + (-q)n$, and hence $r \in H$ since $h \in H$ and $(-q)n \in n\mathbb{Z} \subseteq H$. Thus, if $r \neq 0$, then r is a positive integer in H that is smaller than n, the smallest positive integer in H. Since this is impossible, we conclude that $r = 0$. Therefore, $h = qn$ and hence $h \in n\mathbb{Z}$. Thus, $H \subseteq n\mathbb{Z}$. It now follows that $H = n\mathbb{Z}$. Thus, every subgroup of $\mathbb{Z}$ has the form $n\mathbb{Z}$ for some integer n.

(C) Let us note that in this example we have not only illustrated how to find subgroups of $\mathbb{Z}$, but we have also classified the subgroups of $\mathbb{Z}$; they are the subsets $n\mathbb{Z}$ consisting of all integral multiples of a fixed integer n.

EXAMPLE 3

Let $\mathbb{C}^*$ stand for the multiplicative group of nonzero complex numbers, let n stand for a positive integer, and let $H_n = \{z \in \mathbb{C}^* | z^n = 1\}$. Let us show that H_n is a subgroup of $\mathbb{C}^*$. Clearly, $1 \in H_n$ and hence H_n is nonempty. Furthermore, if $z, w \in H_n$, then $z^n = w^n = 1$ and therefore $(zw)^n = z^n w^n = 1$. Hence, $zw \in H_n$, which shows that H_n is closed under products. Finally, if $z \in H_n$, then $(z^{-1})^n = (z^n)^{-1} = 1$ and hence $z^{-1} \in H_n$. Thus, H_n is closed under inverses. Therefore, $H_n \le \mathbb{C}^*$. Now, let $U = \{z \in \mathbb{C}^* | \|z\| = 1\}$. Then U is also a subgroup of $\mathbb{C}^*$ since $1 \in U$, and if $z, w \in U$, then $\|zw\| = \|z\| \|w\| = 1$, which shows that $zw \in U$, and $\|z^{-1}\| = \|z\|^{-1} = 1$, which shows that $z^{-1} \in U$. The subgroups H_n and U are related to each other as follows. If z is a complex number such that $z^n = 1$, then $\|z\|^n = 1$. But $\|z\|$ is a positive real number. Therefore, $\|z\| = 1$. Consequently, $H_n \subseteq U$ for every positive integer n. Since H_n is already a group under multiplication, it follows that $H_n \le U$.

EXAMPLE 4. Subgroups of the group of the triangle

Recall that the group T of the triangle consists of the six rigid motions of the real plane that bring an equilateral triangle into coincidence with itself: $T = \{1, R, R^2, S, RS, R^2S\}$, where R stands for the $120°$ counterclockwise rotation of the triangle about its center, and S stands for a reflection of the

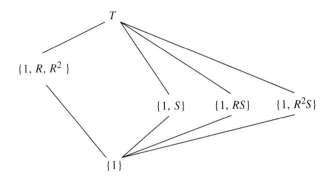

triangle about a line passing through the center and vertex of the triangle. Let us find all subgroups of T. Note that since T is finite, we need only look for subsets that are closed under multiplication. Now, it is clear that the three rotations 1, R, and R^2 form a subgroup of T, as well as the two-element subsets consisting of the identity and any reflection. This gives us the four subgroups $\{1, R, R^2\}$, $\{1, S\}$, $\{1, RS\}$, and $\{1, R^2S\}$. There are also the trivial subgroups $\{1\}$ and T. It is not difficult to verify that any other combination of elements must result in the entire group T. For example, if a subgroup contains R^2 and RS, then it must also contain R and S since $R = (R^2)^2$ and $S = R^2(RS)$; hence it must be T. The group T therefore has six subgroups, $\{1\}$, $\{1, R, R^2\}$, $\{1, S\}$, $\{1, RS\}$, $\{1, R^2S\}$, and T. Finally, observe that since every element in T is also a nonsingular linear transformation of $\mathbb{R}^2$, T is a subgroup of the general linear group $GL(\mathbb{R}^2)$. The group of matrices obtained in Example 2 of Section 2, which are the matrix representations for the elements in T, is then a subgroup of the group of invertible 2×2 matrices:

$$\left\{ \begin{pmatrix} 1 & 0 \\ 0 & 1 \end{pmatrix}, \begin{pmatrix} 0 & -1 \\ 1 & -1 \end{pmatrix}, \begin{pmatrix} -1 & 1 \\ -1 & 0 \end{pmatrix}, \begin{pmatrix} 0 & 1 \\ 1 & 0 \end{pmatrix}, \begin{pmatrix} 1 & -1 \\ 0 & -1 \end{pmatrix}, \begin{pmatrix} -1 & 0 \\ -1 & 1 \end{pmatrix} \right\} \leq \text{Mat}_2^*(\mathbb{R}).$$

EXAMPLE 5

Let $V = \mathbb{R}^2$ stand for the 2-dimensional real plane.

(A) For each real number θ, let $\rho_\theta : V \to V$ stand for the rotation of the plane through an angle of θ radians about the origin, counterclockwise if $\theta > 0$, clockwise if $\theta < 0$, and let $\text{Rot}(V) = \{\rho_\theta | \theta \in \mathbb{R}\}$ stand for the set of all rotations of V about the origin. We claim that $\text{Rot}(V) \leq GL(V)$. Observe first that a rotation of V is in fact a nonsingular linear transformation of V and hence is an element of $GL(V)$. Now, $1 \in \text{Rot}(V)$ since $1 = \rho_0$, the rotation of the plane through an angle of 0 radians about the origin. Moreover, if ρ_θ and $\rho_{\theta'}$ are rotations of V through θ and θ' radians, respectively, then $\rho_\theta \rho_{\theta'}$ is a rotation of V through an angle of $\theta + \theta'$ radians. Hence, $\rho_\theta \rho_{\theta'} = \rho_{\theta + \theta'} \in \text{Rot}(V)$. Finally, ρ_θ^{-1} is the rotation

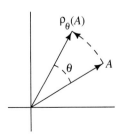

through $-\theta$ radians; hence $\rho_\theta^{-1} = \rho_{-\theta} \in \text{Rot}(V)$. It now follows that $\text{Rot}(V) \le GL(V)$.

(B) A *reflection* of V is a transformation of the form $W_l: V \to V$, where l is any line in the plane passing through the origin, and where $W_l(A)$ is the reflection of the vector A through the line l. Let $\text{Refl}(V)$ stand for the set of reflections of V. It is easily verified that a reflection is a nonsingular linear transformation of V and hence $\text{Refl}(V) \subseteq GL(V)$. But we claim that $\text{Refl}(V)$ is not a subgroup of $GL(V)$. For the only vectors that a reflection W_l fixes are those lying on the line l. The identity map 1, on the other hand, fixes all vectors in V. Hence, $1 \notin \text{Refl}(V)$ and therefore $\text{Refl}(V)$ is not a subgroup of $GL(V)$.

The above examples illustrate several types of subgroups. In each case, the subgroup consists of those elements in a group that satisfy some special property. Let us now show how to generate a subgroup using any collection of elements in a group. Let G be a group and let X be a nonempty subset of G. Let (X) stand for the set of all elements in G of the form $x_1^{s_1} \cdots x_n^{s_n}$, where n is any positive integer, $s_1, \ldots, s_n$ are arbitrary integers, and every x_i is an element of X. We think of (X) as consisting of all finite products in which each factor is either an element in X or the inverse of an element in X. For example, if $X = \{x_1, x_2\}$, then $x_1 x_2^{-1} x_1$, $x_1^3 x_2 x_1^{-1} x_2^{-3}$, and $x_1^{-5} x_2^{-2}$ are typical elements in (X).

PROPOSITION 2. Let X be a nonempty subset of a group G. Then (X) is a subgroup of G.

Proof. Clearly, (X) is nonempty since it contains the set X, and is closed under multiplication since the product of any two finite products of the form $x_1^{s_1} \cdots x_n^{s_n}$ is again a finite product of this form. Finally, if $x = x_1^{s_1} \cdots x_n^{s_n}$ is a typical element in (X), then $x^{-1} = x_n^{-s_n} \cdots x_1^{-s_1}$, and this is an element in (X) since it has the required form. Therefore, (X) is a subgroup of G. $\blacksquare$

Definition

Let X be a nonempty subset of a group G. The subgroup (X) is called the subgroup of G *generated by* X.

This terminology is appropriate since (X) contains those elements and only those elements that are needed to form a subgroup; that is, closure under products and closure under inverses. In the exercises we ask the reader to show that (X) is also equal to the intersection of all subgroups of G that contain X. Thus, in this sense (X) is the smallest subgroup of G that contains X. For example, the group T of the triangle is generated, as we have seen, by the elements R and S. Hence, $T = (R, S)$. The subgroup of T generated by the single element R, for example, is $(R) = \{1, R, R^2\}$, while the subgroup generated by S is $(S) = \{1, S\}$. Let us mention that when the additive notation is used, a typical element in (X) is written in the form $s_1 x_1 + \cdots + s_n x_n$, where $x_1, \ldots, x_n$ are any finite number of elements in X and $s_1, \ldots, s_n$ are arbitrary integers. It is helpful to think of this expression as an integral linear combination of elements in X.

EXAMPLE 6

Let us determine the subgroup $(-6, 4, 10)$ of the additive group $\mathbb{Z}$ of integers generated by the numbers -6, 4, and 10. Let $H = (-6, 4, 10)$. Then every number in H has the form $-6s_1 + 4s_2 + 10s_3$, where s_1, s_2, and s_3 are arbitrary integers, and hence $2 \in H$ since $2 = -6(-1) + 4(-1) + 10(0)$. It follows, therefore, that $2\mathbb{Z} \leq H$. On the other hand, $-6s_1 + 4s_2 + 10s_3 = 2(-3s_1 + 2s_2 + 5s_3)$ for all integers s_1, s_2, and s_3, and hence $H \leq 2\mathbb{Z}$. Therefore, $H = 2\mathbb{Z}$. Thus, the subgroup of $\mathbb{Z}$ generated by the three numbers -6, 4, and 10 is in fact generated by the single number 2 and is therefore the subgroup of even integers. This is not surprising since we showed in Example 2 that every subgroup of $\mathbb{Z}$ is generated by a single element; it is just a matter of finding which element.

EXAMPLE 7

Let W_X and W_Y stand for the reflections of the real plane $\mathbb{R}^2$ through the X- and Y-axes, respectively. Then W_X and W_Y are nonsingular linear transformations of $\mathbb{R}^2$ and hence $W_X, W_Y \in GL(\mathbb{R}^2)$. Let us determine the subgroup (W_X, W_Y) of $GL(\mathbb{R}^2)$ generated by W_X and W_Y. We begin by calculating the product $W_X W_Y$. Since $W_X W_Y$ is a linear transformation, it is completely determined by its effect on a basis for $\mathbb{R}^2$. Let $e_X = (1, 0)$ and $e_Y = (0, 1)$ be the standard basis for $\mathbb{R}^2$. Then

$$W_X W_Y(e_X) = W_X(-e_X) = -e_X$$

$$W_X W_Y(e_Y) = W_X(e_Y) = -e_Y.$$

Therefore, $W_X W_Y(A) = -A$ for every vector $A \in \mathbb{R}^2$. In other words, $W_X W_Y = \rho_\pi$, the $180°$ rotation of the plane about the origin. It now follows that $\{1, W_X, W_Y, \rho_\pi\} \subseteq (W_X, W_Y)$. To verify that this set of four elements is closed under function composition, we construct the multiplication table for the four elements, as shown in the margin. As before, products are calculated by determining their effect on the standard basis e_1, e_2. The multiplication table shows that the set of four elements is closed under function composition and hence is a group. We therefore conclude that $(W_X, W_Y) = \{1, W_X, W_Y, \rho_\pi\}$.

	1	W_X	W_Y	ρ_π
1	1	W_X	W_Y	ρ_π
W_X	W_X	1	ρ_π	W_Y
W_Y	W_Y	ρ_π	1	W_X
ρ_π	ρ_π	W_Y	W_X	1

This concludes our discussion of subgroups. We have characterized subgroups as those nonempty subsets of a group that are closed with respect to both products and inverses, and we have illustrated several ways to obtain them, either by choosing elements of a group that satisfy some special property or by forming the subgroup generated by a given collection of elements in the group. In the next section we will concentrate on groups that are generated by a single element.

Exercises

1. Let $H = \{(\begin{smallmatrix} a & 0 \\ 0 & b \end{smallmatrix}) \in \text{Mat}_2(\mathbb{R}) | a, b \in \mathbb{R}\}$. Show that H is a subgroup of the additive group $\text{Mat}_2(\mathbb{R})$.

2. Let $H = \{(\begin{smallmatrix} a & 0 \\ 0 & b \end{smallmatrix}) \in \text{Mat}_2(\mathbb{R}) | a, b \in \mathbb{R}^*\}$. Show that H is a subgroup of the multiplicative group $\text{Mat}_2^*(\mathbb{R})$.

3. Find all subgroups of the group D_4 of the square discussed in Exercise 1, Section 2. Draw a lattice of subgroups to indicate the relationship among these subgroups.

4. Let $V = \mathbb{R}^2$ stand for the real plane and let A be a nonzero vector in V. Let $H = \{nA \in V | n \in \mathbb{Z}\}$. Show that H is a subgroup of the additive group of V, but that H is not a subspace of V.

5. Let $H = \{z \in \mathbb{C}^* | \|z\| = 2\}$. Is H a subgroup of $\mathbb{C}^*$?

6. Let n and m be integers. Show that $n\mathbb{Z} = m\mathbb{Z}$ if and only if $n = \pm m$.

7. Find a single generator for each of the following subgroups of $\mathbb{Z}$. In each case, express the single generator in terms of the given generators.
   (a) $(6, 20)$ (b) $(12, 15, 21)$ (c) $(-10, 12)$ (d) $(12, 35, 21)$

8. Let V be a real vector space and let A be a nonzero vector in V.
   (a) Let $H = \{\sigma \in GL(V) | \sigma(A) = A\}$. Show that $H \leq GL(V)$.
   (b) Let $K = \{\sigma \in GL(V) | \sigma(\langle A \rangle) = \langle A \rangle\}$, where $\langle A \rangle$ stands for the subspace spanned by the vector A. Show that $K \leq GL(V)$.
   (c) Show that $H \leq K$.
   (d) Is it true that $H = K$?

9. Let G be a group and let $\{H_\alpha | \alpha \in I\}$ be a collection of subgroups of G, where I stands for an index set that may or may not be finite.
   (a) Let $H = \bigcap H_\alpha$ be the intersection of the subgroups in the family $\{H_\alpha | \alpha \in I\}$. Show that $H \leq G$.
   (b) Let X be a nonempty subset of G and let $\mathscr{S} = \{H | H \leq G, X \subseteq H\}$ stand for the family of subgroups that contain X. Show that $(X) = \bigcap H$, where the intersection is over the subgroups in $\mathscr{S}$. Thus, the subgroup of G generated by X is the intersection of those subgroups that contain X.
   (c) If X and Y are subsets of G with $Y \subseteq X$, show that $(Y) \leq (X)$.

10. Let G, H, and K be groups with $H \leq K$ and $K \leq G$. Show that $H \leq G$.

11. Is it true that the union of two subgroups is a subgroup?

12. Let U and H_n stand for the subgroups of complex numbers discussed in Example 3 in this section.
    (a) Let $H = \bigcup_{n=0}^{\infty} H_n$. Show that $H \leq U$.
    (b) Show that H is the group used in Example 7, Section 2, to illustrate an infinite group all of whose elements have finite order.
    (c) Using DeMoivre's theorem, show that the numbers $\cos(2\pi/n)k + i \sin(2\pi/n)k$, for $k = 1, \ldots, n$ are the complex solutions of the equation $X^n = 1$. From this fact, show that $H = \{\cos 2\pi r + i \sin 2\pi r \in \mathbb{C} | r \in \mathbb{Q}\}$. Illustrate the numbers in H on the unit complex circle.
    (d) Does $H = U$?

13. Let G be a group. For each positive integer d, let $G_d = \{x \in G | x^d = 1\}$.

(a) If G is abelian, show that $G_d \leq G$ for every d.

(b) Find G_2 and G_3 when $G = T$, the group of the triangle.

(c) If G is nonabelian, is every subset G_d necessarily a subgroup of G?

14. Let G be an abelian group and let $G_T = \{x \in G \,|\, x$ has finite order$\}$. Show that $G_T \leq G$. If G is nonabelian, is it still true that G_T is a subgroup of G?

15. **The center of a group.** Let G be a group and let $Z(G) = \{x \in G \,|\, xy = yx$ for every element $y \in G\}$ stand for the set of elements in G that commute with every element in G.

(a) Show that $Z(G)$ is a subgroup of G. $Z(G)$ is called the *center of G*.

(b) Find $Z(D_4)$ and $Z(T)$.

(c) Show that $Z(G)$ is an abelian group.

(d) Show that $Z(G) = G$ if and only if G is abelian.

16. **The center of the matrix groups.** The purpose of this exercise is to determine the center of the matrix group $\mathrm{Mat}_2^*(\mathbb{R})$.

(a) Let $M = \left(\begin{smallmatrix} a & b \\ c & d \end{smallmatrix}\right) \in \mathrm{Mat}_2^*(\mathbb{R})$. Show that

$$\begin{pmatrix} 1 & 1 \\ 1 & 0 \end{pmatrix} M \begin{pmatrix} 1 & 1 \\ 1 & 0 \end{pmatrix}^{-1} = \begin{pmatrix} b + d & a + c - b - d \\ b & a - b \end{pmatrix}$$

and

$$\begin{pmatrix} 1 & 1 \\ 0 & 1 \end{pmatrix} M \begin{pmatrix} 1 & 1 \\ 0 & 1 \end{pmatrix}^{-1} = \begin{pmatrix} a + c & -a - c + b + d \\ c & -c + d \end{pmatrix}.$$

(b) Using the calculations in part (a), show that $M \in Z(\mathrm{Mat}_2^*(\mathbb{R}))$ if and only if M is a scalar matrix; that is, if and only if $M = aI_2 = \left(\begin{smallmatrix} a & 0 \\ 0 & a \end{smallmatrix}\right)$ for some nonzero real number a.

17. **The special linear subgroup.** Let $SL_2(\mathbb{R}) = \{M \in \mathrm{Mat}_2^*(\mathbb{R}) \,|\, \det(M) = 1\}$ stand for the subset of 2×2 matrices over $\mathbb{R}$ having determinant equal to 1.

(a) Show that $SL_2(\mathbb{R}) \leq \mathrm{Mat}_2^*(\mathbb{R})$. $SL_2(\mathbb{R})$ is called the *special linear group over R in dimension 2*.

(b) Using the ideas in Exercise 16, show that $Z(SL_2(\mathbb{R})) = \{I_2, -I_2\}$.

(c) A geometric interpretation of the group $SL_2(\mathbb{R})$. Let $V = \mathbb{R}^2$ stand for the real plane. Every mapping $\sigma \in GL(V)$ is a nonsingular linear transformation of V onto itself and, as such, maps the unit square S into a parallelogram $\sigma(S)$, as illustrated here. But the area of $\sigma(S)$ need not equal 1. Show that the area of $\sigma(S)$ is equal to 1 for every $\sigma \in SL_2(\mathbb{R})$. Thus, the elements in $SL_2(\mathbb{R})$ are area-preserving transformations.

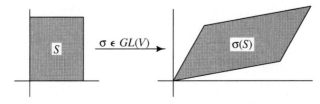

18. Let n be a nonzero integer and let H be the subset of all rational numbers of the form n^s, where $s \in \mathbb{Z}$. Show that $H \leq \mathbb{Q}^*$.

19. Let X stand for the set of prime numbers. Describe the subgroup of $\mathbb{Q}^*$ generated by X.

20. Let $D = \{(x, x) \in \mathbb{R}^2 \,|\, x \in \mathbb{R}\}$ stand for the graph of the equation $y = x$ in the real plane; D is the $45°$ line through the origin. Show that the subgroup (W_D, W_X) of $GL(\mathbb{R}^2)$ generated by reflections through D and the X-axis is the group D_4 of the square.

4. CYCLIC GROUPS

In this section we turn our attention to groups that are generated by a single element. These groups are called cyclic groups. After establishing the basic properties of cyclic groups, we will construct both an arithmetic and geometric model of a cyclic group of order n for every positive integer n, thus showing that finite cyclic groups of every possible order exist. We then study in detail the generators of finite cyclic groups, determining the number of such generators and how the generators are related to each other. The answers to these questions, it turns out, have important applications in number theory. We conclude the section by discussing some of these applications.

PROPOSITION 1. Let G be a group, $x \in G$, and let $(x) = \{x^n \in G \,|\, n \in \mathbb{Z}\}$. Then (x) is a subgroup of G.

Proof. If x^n and x^m are typical elements in (x), then $x^n x^m = x^{n+m}$ and $(x^n)^{-1} = x^{-n}$. Hence, (x) is closed under products and inverses and is therefore a subgroup of G. ∎

> **Definition**
>
> The subgroup (x) is called the *cyclic subgroup of G generated by x*. If $G = (x)$ for some element $x \in G$, we say that G is a *cyclic group generated by x*.

For example, in the group T of the triangle the cyclic subgroup generated by the rotation R is $(R) = \{1, R, R^2\}$, while the cyclic subgroup generated by R^2 is $(R^2) = \{1, R^2, R\}$, which is equal to (R). Thus, both R and R^2 are generators of the cyclic group (R). The group T itself is not cyclic, however, since there is no element $x \in T$ such that $T = (x)$. If the additive notation is used, then the cyclic subgroup generated by an element x is written $\{nx \in G \,|\, n \in \mathbb{Z}\}$. The additive group $\mathbb{Z}$ of integers, for example, is a cyclic group; for, using additive notation, $n = n1 = (-n)(-1)$ for every integer n and hence $(1) = (-1) = \mathbb{Z}$. Thus, $\mathbb{Z}$ is a cyclic group having both 1 and -1 as generators. More generally, if n is any integer, the cyclic subgroup generated by n is $(n) = \{nk \in \mathbb{Z} \,|\, k \in \mathbb{Z}\} = n\mathbb{Z}$. In particular, the subgroup of even integers is just the cyclic subgroup generated by the number 2.

PROPOSITION 2. Let G be a cyclic group. Then

(1) G is commutative.
(2) Every subgroup of G is cyclic.
(3) If G is finite and has x as a generator, then $|G| = |x|$.

Proof. Let x be a generator for G.

(A) To prove statement (1), let $a, b \in G$. Then $a = x^n$ and $b = x^m$ for some integers n, m, and therefore $ab = x^n x^m = x^{n+m} = x^m x^n = ba$. Hence, G is commutative.

(B) To prove statement (2), let H be a subgroup of G. If $H = \{1\}$, then H is cyclic since $H = (1)$, and we are done. Assume therefore that H contains some element $x^n \neq 1$; we may assume that $n > 0$ since, if $n < 0$, $x^{-n} = (x^n)^{-1} \in H$ and $-n > 0$. Now, let s be the smallest positive integer such that $x^s \in H$. We claim that $H = (x^s)$. For clearly, $(x^s) \leq H$. To show the reverse inclusion, choose $h \in H$. Then $h = x^n$ for some integer n and, by the division algorithm, there are integers q and r such that $n = qs + r$, where $0 \leq r < s$. Then $h = x^n = (x^s)^q x^r$. Hence, $x^r = (x^s)^{-q} h$, and since both of these factors lie in H, $x^r \in H$. But if $r > 0$, then r is a positive integer smaller than s such that $x^r \in H$, which contradicts the definition of s. Therefore, $r = 0$ and hence $n = qs$. It follows that $h = x^n = (x^s)^q$ and hence that $h \in (x^s)$. Therefore, $H = (x^s)$ and hence H is a cyclic group.

(C) To prove statement (3), assume that G is finite and let $|x| = n$. We will show that $G = \{1, x, \ldots, x^{n-1}\}$ and that these powers of x are distinct, from which it follows that $|G| = n$. Let x^m be an arbitrary element in G. Then $m = qn + r$ for some integers q and r, and r must be one of the numbers $0, 1, \ldots, n - 1$. It follows that $x^m = x^r$, and therefore $x^m \in \{1, x, \ldots, x^{n-1}\}$. Hence, $G \subseteq \{1, x, \ldots, x^{n-1}\}$. Since the reverse inclusion is obvious, we conclude that $G = \{1, x, \ldots, x^{n-1}\}$. Finally, to show that these powers of x are distinct, suppose that $x^i = x^j$ for some integers i and j, with $0 \leq i \leq j \leq n - 1$. Then $x^{j-i} = 1$ and therefore n divides $j - i$. But $0 \leq j - i < n$. Hence, $j - i = 0$ and therefore $i = j$. Thus, G consists of the n distinct elements $1, x, \ldots, x^{n-1}$, and therefore $|G| = n = |x|$, which completes the proof. ∎

It follows from Proposition 2 that cyclic groups are commutative, that cyclic groups have only cyclic subgroups, and that the number of elements in a finite cyclic group is the same as the order of any generator of the group. Let us also note that the proof of statement (2)—that every subgroup of a cyclic group is cyclic—is very similar to the proof that every subgroup of $\mathbb{Z}$ has the form $n\mathbb{Z}$ for some integer n discussed in Example 2, Section 3. Indeed, since $\mathbb{Z}$ is a cyclic group, it is now clear that Example 2 is just a special case of this more general result. It should also be observed that the proof of statement (2) in Proposition 2 reveals how to find a generator for a subgroup H of a cyclic group: select the least positive integer s such that $x^s \in H$.

Let us now turn our attention to the construction of finite cyclic groups. We showed earlier that there is a cyclic group of order 3, namely, the subgroup $\{1, R, R^2\}$ of the group of the triangle generated by the rotation R. But is there a cyclic group of order 23? To answer this question, and every other such question, we will construct a finite cyclic group of order n, called the group of integers modulo n, for every positive integer n. By doing so we not only obtain a new and important family of groups, we also show that there is, in fact, a finite cyclic group of every possible order. Before we begin, let us take a moment to explain the basic idea behind the construction. We begin by using the integer n to define a relation on the set $\mathbb{Z}$ of integers, called the congruence relation modulo n, and show that this relation is an equivalence relation on $\mathbb{Z}$. This results in a partitioning of $\mathbb{Z}$ into equivalence classes which are called congruence classes modulo n. Using the addition operation on $\mathbb{Z}$, we then define addition of congruence classes, and show, finally, that these classes form a cyclic group of order n under addition.

To begin the construction, let n be a positive integer. Define a relation on the set $\mathbb{Z}$ of integers as follows: if $s, t \in \mathbb{Z}$, then $s \equiv t \pmod{n}$ (pronounced "s is congruent to t mod n") if $s - t$ is a multiple of n. The relation $\equiv \pmod{n}$ is called *congruence modulo n*.

PROPOSITION 3. Congruence modulo n is an equivalence relation on $\mathbb{Z}$. If s is any integer, the equivalence class of s is the set $[s] = \{s + kn \in \mathbb{Z} \mid k \in \mathbb{Z}\}$.

Proof.   To show that congruence is an equivalence relation, we must verify that it is reflexive, symmetric, and transitive.

(A) Reflexivity. This means that $s \equiv s \pmod{n}$ for every integer $s \in \mathbb{Z}$, and this is true since $s - s = 0n$ for all integers s.

(B) Symmetry. This means that if $s \equiv t \pmod{n}$ for some integers $s, t \in \mathbb{Z}$, then $t \equiv s \pmod{n}$. This is true because, if $s, t \in \mathbb{Z}$ and $s - t = kn$ for some integer k, then $t - s = (-k)n$.

(C) Transitivity. This means that if $s \equiv t \pmod{n}$ and $t \equiv r \pmod{n}$ for some integers s, t, and r, then $s \equiv r \pmod{n}$. It is true because, if $s, t, r \in \mathbb{Z}$ and $s - t = kn$ and $t - r = pn$ for some integers k and p, then $s - r = (s - t) + (t - r) = kn + pn = (k + p)n$.

It follows from statements (A), (B), and (C) that congruence modulo n is an equivalence relation on the set $\mathbb{Z}$ of integers. To find the equivalence class $[s]$ of a typical integer s, recall that, by definition, $[s] = \{t \in \mathbb{Z} \mid t \equiv s \pmod{n}\}$. Since $t - s = kn$ for some integer k if and only if $t = s + kn$, it follows that $[s] = \{s + kn \in \mathbb{Z} \mid k \in \mathbb{Z}\}$. ■

The equivalence class $[s]$ of an integer s under the congruence relation modulo n is called the *congruence class of s*. Thus, by Proposition 3,

$[s] = \{s + kn \in \mathbb{Z} \mid k \in \mathbb{Z}\}$; that is, to find the congruence class of an integer modulo n, simply add all integral multiples of n to s. For example, the congruence class of 3 modulo 5 is $[3] = \{3 + 5k \in \mathbb{Z} \mid k \in \mathbb{Z}\} = \{\ldots, -2, 3, 8, 13, \ldots\}$. Moreover, since congruence modulo n is an equivalence relation on $\mathbb{Z}$, it follows that $[s] = [t]$ if and only if $s \equiv t \pmod{n}$. For example, $[8] = [3]$ since $8 \equiv 3 \pmod{5}$.

PROPOSITION 4. If s is an integer and r is the remainder of s upon division by n, then $[s] = [r]$. In particular, every integer is congruent modulo n to exactly one of the numbers $0, 1, \ldots, n-1$, and hence there are exactly n distinct congruence classes modulo n, namely $[0], [1], \ldots, [n-1]$.

Proof. If r is the remainder of s upon division by n, then $s = qn + r$ for some integer q. Therefore, $s - r$ is a multiple of n and hence $s \equiv r \pmod{n}$, or equivalently, $[s] = [r]$. Since the possible remainders upon division by n are $0, 1, \ldots, n-1$, it follows that $[0], [1], \ldots, [n-1]$ are the only possible congruence classes modulo n. Finally, these classes are distinct since no two distinct integers lying between 0 and $n-1$ can have a difference that is a multiple of n. Thus, there are exactly n distinct congruence classes modulo n, namely, $[0], [1], \ldots, [n-1]$. ∎

Let $\mathbb{Z}_n$ stand for the set of congruence classes modulo n; that is, $\mathbb{Z}_n = \{[s] \mid s \in \mathbb{Z}\}$. By Proposition 4, $\mathbb{Z}_n = \{[0], [1], \ldots, [n-1]\}$. For example, $\mathbb{Z}_2 = \{[0], [1]\}$, where $[0] = \{0 + 2k \in \mathbb{Z} \mid k \in \mathbb{Z}\} = \{\text{even integers}\}$ and $[1] = \{1 + 2k \in \mathbb{Z} \mid k \in \mathbb{Z}\} = \{\text{odd integers}\}$. We find, for example, that $[35] = [1]$ since $35 \equiv 1 \pmod{2}$, and $[26] = [0]$ since $26 \equiv 0 \pmod{2}$. Similarly, $\mathbb{Z}_3 = \{[0], [1], [2]\}$, where

$$[0] = \{0 + 3k \in \mathbb{Z} \mid k \in \mathbb{Z}\} = \{\ldots, -3, 0, 3, \ldots\} = \{\text{multiples of 3}\},$$

$$[1] = \{1 + 3k \in \mathbb{Z} \mid k \in \mathbb{Z}\} = \{\ldots, -2, 1, 4, \ldots\}$$
$$= \{\text{integers having remainder 1 upon division by 3}\}, \text{ and}$$

$$[2] = \{2 + 3k \in \mathbb{Z} \mid k \in \mathbb{Z}\} = \{\ldots, -1, 2, 5, \ldots\}$$
$$= \{\text{integers having remainder 2 upon division by 3}\}.$$

Let us now use addition of integers to define addition of congruence classes and then show that the classes form a cyclic group under addition. Congruence classes, we recall, partition the set $\mathbb{Z}$ of integers into disjoint subsets. Thus, we may use addition of integers to add the integers in one class to those in any other class as follows: if $[s]$ and $[t]$ are typical congruence classes, we define their sum to be the subset

$$[s] + [t] = \{x + y \in \mathbb{Z} \mid x \in [s], y \in [t]\}.$$

But for this operation to be a binary operation on the set $\mathbb{Z}_n$ we must show that the subset $[s] + [t]$ is in fact the congruence class of some integer. We do this by using the description of the classes given in Proposition 3:

$$[s] + [t] = \{x + y \in \mathbb{Z} \mid x \in [s], y \in [t]\}$$
$$= \{s + kn + t + pn \in \mathbb{Z} \mid k, p \in \mathbb{Z}\}$$
$$= \{(s + t) + (k + p)n \in \mathbb{Z} \mid k, p \in \mathbb{Z}\}$$
$$= \{(s + t) + qn \in \mathbb{Z} \mid q \in \mathbb{Z}\}$$
$$= [s + t].$$

Therefore, $[s] + [t]$ is a congruence class, and consequently, addition of classes is a binary operation on the set $\mathbb{Z}_n$. Informally, we say that the sum of two classes is the class of the sum. Let us now show that the set of congruence classes modulo n is a finite cyclic group of order n under addition of classes.

PROPOSITION 5. Let n be a positive integer. Then $\mathbb{Z}_n$ is a finite cyclic group of order n under addition of classes.

Proof.   To show that $\mathbb{Z}_n$ is a group under addition of classes, we must show that class addition is an associative binary operation on $\mathbb{Z}_n$ that has an identity element and for which every class has an inverse.

(A) Associativity. Let $[s]$, $[t]$, and $[r]$ be typical classes in $\mathbb{Z}_n$. Then

$$[s] + ([t] + [r]) = [s] + [t + r] = [s + (t + r)]$$
$$= [(s + t) + r] = [s + t] + [r] = ([s] + [t]) + [r].$$

Hence, addition of classes is associative.

(B) Existence of an identity element. The class $[0]$ is an identity element for class addition since $[0] + [s] = [0 + s] = [s] = [s + 0] = [s] + [0]$ for every class $[s] \in \mathbb{Z}_n$.

(C) Existence of inverses. The inverse of a typical class $[s]$ is $[-s]$ since

$$[s] + [-s] = [s + (-s)] = [0] = [(-s) + s] = [-s] + [s].$$

That is, $-[s] = [-s]$ for every class $[s] \in \mathbb{Z}_n$.

It follows from statements (A), (B), and (C) that $\mathbb{Z}_n$ is a group under addition of classes. Since we showed earlier that there are exactly n distinct classes, $[0], [1], \ldots, [n - 1]$, it follows that $|\mathbb{Z}_n| = n$. Finally, to complete the proof we must show that the group $\mathbb{Z}_n$ is cyclic. Let $[s]$ be a typical element in $\mathbb{Z}_n$, where s is one of the integers $0, 1, \ldots, n - 1$. Then $[s] = [1] + \cdots + [1]$, the sum of $[1]$ with itself s times. Therefore, $\mathbb{Z}_n$ is a cyclic group with generator $[1]$, and the proof is complete.   ∎

The group $\mathbb{Z}_n$ is called the *group of integers modulo n*. For example, $\mathbb{Z}_2 = \{[0], [1]\}$ is the group of integers modulo 2. In this case $[1] + [1] = [0]$ and hence, in particular, $-[1] = [1]$ and $|[1]| = 2$. For $\mathbb{Z}_3$, we find that $-[1] = [2]$, $-[2] = [1]$, and $|[1]| = |[2]| = 3$; both $[1]$ and $[2]$ are generators for the group $\mathbb{Z}_3$. Similarly, in $\mathbb{Z}_4$ we find that $-[1] = [3]$, $-[2]$

$\mathbb{Z}_2$	[0]	[1]
[0]	[0]	[1]
[1]	[1]	[0]

$\mathbb{Z}_3$	[0]	[1]	[2]
[0]	[0]	[1]	[2]
[1]	[1]	[2]	[0]
[2]	[2]	[0]	[1]

$\mathbb{Z}_4$	[0]	[1]	[2]	[3]
[0]	[0]	[1]	[2]	[3]
[1]	[1]	[2]	[3]	[0]
[2]	[2]	[3]	[0]	[1]
[3]	[3]	[0]	[1]	[2]

FIGURE 1. Addition tables for the cyclic groups $\mathbb{Z}_2$, $\mathbb{Z}_3$, and $\mathbb{Z}_4$.

$= [2]$, and $-[3] = [1]$, while $\|[1]\| = \|[3]\| = 4$, but $\|[2]\| = 2$; thus $\mathbb{Z}_4$ has two generators, namely [1] and [3]. Figure 1 illustrates the addition tables for these groups.

EXAMPLE 1

The group $\mathbb{Z}_{12}$ is a finite cyclic group of order 12, and consequently, every element has finite order and every subgroup is cyclic. Let us determine the order of each element and the cyclic subgroup it generates. We begin with [1]. Since [1] generates $\mathbb{Z}_{12}$, $\|[1]\| = 12$. Now consider the element [2]. We find by direct calculation that $1[2] = [2]$, $2[2] = [4]$, $3[2] = [6]$, $4[2] = [8]$, $5[2] = [10]$, and $6[2] = [12] = [0]$. Therefore, $\|[2]\| = 6$ and the cyclic subgroup generated by [2] is $([2]) = \{[0], [2], [4], [6], [8], [10]\}$. Similarly, for the remaining elements in $\mathbb{Z}_{12}$ we find the following:

$\|[0]\| = 1,\quad ([0]) = \{[0]\}$

$\|[1]\| = 12,\quad ([1]) = \mathbb{Z}_{12}$

$\|[2]\| = 6,\quad ([2]) = \{[0], [2], [4], [6], [8], [10]\}$

$\|[3]\| = 4,\quad ([3]) = \{[0], [3], [6], [9]\}$

$\|[4]\| = 3,\quad ([4]) = \{[0], [4], [8]\}$

$\|[5]\| = 12,\quad ([5]) = \mathbb{Z}_{12}$

$\|[6]\| = 2,\quad ([6]) = \{[0], [6]\}$

$\|[7]\| = 12,\quad ([7]) = \mathbb{Z}_{12}$

$\|[8]\| = 3,\quad ([8]) = \{[0], [8], [4]\}$

$\|[9]\| = 4,\quad ([9]) = \{[0], [9], [6], [3]\}$

$\|[10]\| = 6,\quad ([10]) = \{[0], [10], [8], [6], [4], [2]\}$

$\|[11]\| = 12,\quad ([11]) = \mathbb{Z}_{12}$

$\mathbb{Z}_{12} = ([1]) = ([5]) = ([7]) = ([11])$

$([3]) = ([9])$
$= \{[0], [3], [6], [9]\}$

$([2]) = ([10])$
$= \{[0], [2], [4], [6], [8], [10]\}$

$([6]) = \{[0], [6]\}$

$([4]) = \{[0], [4], [8]\} = ([8])$

$([0]) = \{[0]\}$

It follows from these calculations that $\mathbb{Z}_{12}$ has four generators, namely, [1], [5], [7], and [11], and six distinct subgroups, namely, ([0]), ([1]), ([2]), ([3]), ([4]), and ([6]), as shown in the diagram.

The group of integers modulo n is a finite cyclic group of order n and, as such, provides an arithmetic model for cyclic groups. In the following example we discuss a geometric model for cyclic groups.

EXAMPLE 2

Let n be a positive integer and let P_n stand for a regular n-gon in the real plane $\mathbb{R}^2$ with its center at the origin. Let $\text{Rot}_n(\mathbb{R}^2)$ stand for the set of rotations of the plane that bring the polygon P_n into coincidence with itself. Let us show that $\text{Rot}_n(\mathbb{R}^2)$ is a cyclic group of order n under function composition, and then discuss a matrix representation for the transformations in the group.

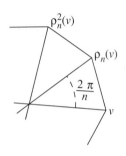

(A) To show that $\text{Rot}_n(\mathbb{R}^2)$ is a cyclic group under function composition, observe that the rotations in $\text{Rot}_n(\mathbb{R}^2)$ are precisely those rotations of the plane about the origin that map an arbitrary vertex v to any vertex on the n-gon. There are n such rotations of the plane. To describe them, let ρ_n stand for the counterclockwise rotation of the plane through an angle of $2\pi/n$ radians about the origin. Then the n rotations that bring the n-gon into coincidence with itself are $1, \rho_n, \rho_n^2, \ldots, \rho_n^{n-1}$, and therefore $\text{Rot}_n(\mathbb{R}^2) = \{1, \rho_n, \ldots, \rho_n^{n-1}\}$, which is a cyclic group of order n generated by ρ_n. Thus, the rotation group $\text{Rot}_n(\mathbb{R}^2)$ gives us a geometric interpretation of a cyclic group of order n.

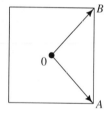

(B) Let us illustrate the results in part (A) by discussing the group $\text{Rot}_4(\mathbb{R}^2)$. In this case $n = 4$, hence we are dealing with a square, and $\text{Rot}_4(\mathbb{R}^2) = \{1, \rho_4, \rho_4^2, \rho_4^3\}$. The mapping ρ_4 rotates the square 90° counterclockwise about its center, ρ_4^2, 180°, and ρ_4^3, 270°. Now, every element in $\text{Rot}_4(\mathbb{R}^2)$ is a linear transformation mapping $\mathbb{R}^2$ onto itself and hence has a 2×2 matrix representation. Let us find a matrix representation for the transformations in $\text{Rot}_4(\mathbb{R}^2)$ and thereby obtain a cyclic group of matrices of order 4. We begin by choosing a coordinate system (A, B) for $\mathbb{R}^2$ as shown in the figure. Since ρ_4 rotates the square 90° counterclockwise, it follows that $\rho_4(A) = B$ and $\rho_4(B) = -A$. Hence, the matrix of ρ_4 in the (A, B) coordinate system is $\left(\begin{smallmatrix} 0 & -1 \\ 1 & 0 \end{smallmatrix}\right)$. Since matrix multiplication corresponds to function composition, which is the operation on $\text{Rot}_4(\mathbb{R}^2)$, we may find the matrices for the remaining powers of ρ_4 by calculating the powers of the matrix for ρ_4, as shown here.

$$\rho_4^2 \leftrightarrow \begin{pmatrix} 0 & -1 \\ 1 & 0 \end{pmatrix}\begin{pmatrix} 0 & -1 \\ 1 & 0 \end{pmatrix} = \begin{pmatrix} -1 & 0 \\ 0 & -1 \end{pmatrix}$$

$$\rho_4^3 \leftrightarrow \begin{pmatrix} -1 & 0 \\ 0 & -1 \end{pmatrix}\begin{pmatrix} 0 & -1 \\ 1 & 0 \end{pmatrix} = \begin{pmatrix} 0 & 1 \\ -1 & 0 \end{pmatrix}$$

$$\rho_4^4 \leftrightarrow \begin{pmatrix} -1 & 0 \\ 0 & -1 \end{pmatrix}\begin{pmatrix} -1 & 0 \\ 0 & -1 \end{pmatrix} = \begin{pmatrix} 1 & 0 \\ 0 & 1 \end{pmatrix}$$

It follows that the set $\{(\begin{smallmatrix}1&0\\0&1\end{smallmatrix}),(\begin{smallmatrix}0&-1\\1&0\end{smallmatrix}),(\begin{smallmatrix}-1&0\\0&-1\end{smallmatrix}),(\begin{smallmatrix}0&1\\-1&0\end{smallmatrix})\}$ is a cyclic group of matrices under matrix multiplication. In the exercises we indicate a general method for finding a matrix representation of $\mathrm{Rot}_n(\mathbb{R}^2)$ for any positive integer n.

Let us now turn our attention to the generators of a finite cyclic group. As we have seen, finite cyclic groups usually have more than one generator. For example, the group $\mathbb{Z}_{12}$ discussed in Example 1 has four generators, namely, [1], [5], [7], and [11]. Observe that the numbers 1, 5, 7, and 11 are precisely the numbers between 1 and 12 that have no factor in common with 12. Our goal for the rest of this section is to show that the generators of a finite cyclic group always have this form. Since greatest common divisors play an important role throughout this discussion, we begin by first discussing this concept.

Let n and m be nonzero integers and let (n, m) stand for the subgroup of $\mathbb{Z}$ generated by n and m. Since the group $\mathbb{Z}$ is cyclic, every subgroup of $\mathbb{Z}$ is cyclic. Therefore, there is some integer d such that $(n, m) = (d) = d\mathbb{Z}$. Moreover, since $(-d) = (d)$, we may assume that d is positive. The unique positive integer d such that $(n, m) = (d)$ is called the *greatest common divisor* of n and m and is denoted by $\gcd(n, m)$. The name is appropriate since d is not only a common divisor of n and m, but is, in fact, the greatest of all common divisors in the sense that if d' is any divisor of both n and m, then d' divides d. To show that d in fact divides both n and m, observe that $n \in (n, m) = (d)$ and hence $n = dk$ for some integer k. Therefore, d divides n and, similarly, d divides m. To show that d is the greatest of the common divisors of n and m, first recall that $(n, m) = \{s_1 n + s_2 m \in \mathbb{Z} \mid s_1, s_2 \in \mathbb{Z}\}$. Since $d \in (n, m)$, $d = s_1 n + s_2 m$ for some integers s_1, s_2, and consequently if d' divides both n and m, d' also divides d. Thus, we conclude that $d = \gcd(n, m)$ is a common divisor of n and m and is the greatest of all such common divisors. For example, $\gcd(6, 15) = 3$, and we find that $3 = (-2)6 + (1)15$, while $\gcd(2, 3) = 1$, where $1 = (-1)2 + (1)3$.

If $\gcd(n, m) = 1$, we say that n and m are *relatively prime*. In this case n and m have no nontrivial factor in common, and there are integers s_1 and s_2 such that $s_1 n + s_2 m = 1$. For each positive integer n, let $\varphi(n)$ stand for the number of positive integers between 1 and n that are relatively prime to n; that is, $\varphi(n) = |\{s \in \mathbb{N}^* \mid 1 \le s \le n, \gcd(s, n) = 1\}|$. For example, $\varphi(1) = 1$, $\varphi(2) = 1$, $\varphi(3) = 2$, $\varphi(10) = 4$, and $\varphi(12) = 4$. The function φ is called the *Euler phi-function*.

PROPOSITION 6. Let G be a group and let x be an element of G that has finite order. Then $|x^s| = |x|/\gcd(s, |x|)$ for every nonzero integer s.

Proof. Let $n = |x|$, $m = |x^s|$, and $d = \gcd(s, |x|)$. Since d divides both n and s, it follows that $(x^s)^{n/d} = (x^n)^{s/d} = 1$. Therefore, m divides n/d since $m = |x^s|$. Conversely, to show that n/d divides m, recall that there are integers s_1 and s_2 such that $d = s_1 s + s_2 n$. Therefore,

$$x^{dm} = x^{s_1 sm + s_2 nm} = [(x^s)^m]^{s_1}(x^n)^{s_2 m} = 1.$$

Since $n = |x|$, it follows that n divides dm, or equivalently, that n/d divides m. Therefore, $m = n/d$, as required.   ∎

Proposition 6 is an extremely important result since it tells us exactly how to find the order of any power of x if we know the order of x itself. If x has order 6, for example, then x^2 has order 3; if x has order 12, then x^9 has order 4. Or, using additive notation, if x has order 20, then $12x$ has order 5. Using this result, let us now describe the generators and subgroups of a finite cyclic group.

COROLLARY.   Let G be a finite cyclic group of order n and let x be a generator of G. Then the element x^s generates G if and only if $\gcd(s, n) = 1$; that is, if and only if s and n are relatively prime. In particular, G has $\varphi(n)$ generators.

Proof.   Since the order of a cyclic group is the same as the order of any generator, it follows that x^s generates G if and only if $|x^s| = |x| = n$. But $|x^s| = n/\gcd(n, s)$ and this is equal to n if and only if $\gcd(n, s) = 1$. Thus, the generators of G are precisely those elements of the form x^s, where s is a positive integer between 1 and n that is relatively prime to n. Since these elements are necessarily distinct, it follows that G has $\varphi(n)$ generators.   ∎

COROLLARY.   Let G be a finite cyclic group of order n. Then the order of every subgroup of G is a divisor of n, and conversely, G has a unique subgroup of order d for every divisor d of n. Thus, there is a 1–1 correspondence between the subgroups of G and the divisors of n.

Proof.   Let $G = (x)$, where $|x| = n$, and let H be a subgroup of G. Then H is cyclic and hence $H = (x^s)$ for some nonzero integer s. It follows that $|H| = |x^s| = n/\gcd(s, n)$, and this number clearly divides n, the order of G. Thus, the order of every subgroup divides the order of G. Now, let d be any positive divisor of n. To construct a subgroup of order d, we first set $n = qd$ for some integer q and observe that $|x^q| = n/\gcd(q, n) = qd/\gcd(q, qd) = d$. Therefore, (x^q) is a subgroup of order d. To show that (x^q) is the only subgroup of order d, let H be a subgroup of G such that $|H| = d$. Since H is cyclic, $H = (x^s)$ for some positive integer s, and hence $d = |H| = |x^s| = n/\gcd(s, n) = qd/\gcd(s, n)$, from which it follows that $\gcd(s, n) = q$. Hence, q divides s, say $s = kq$. Then $x^s = (x^q)^k \in (x^q)$, and it follows that $H \leq (x^q)$. Since $|H| = d = |x^q|$, we conclude that $H = (x^q)$. Thus, G has a unique subgroup of order d, and the proof is complete.   ∎

These results give us a complete description of the generators and subgroups of a finite cyclic group: if G is a cyclic group of order n generated by an element x, then G has $\varphi(n)$ generators that have the form x^s, where s and n are relatively prime, and has a unique subgroup of order d for each divisor d of n, namely, $(x^{n/d})$. For example, the cyclic group $\mathbb{Z}_{12}$ has four generators,

[1], [5], [7], and [11], which we found by direct calculation in Example 1. This result, on the other hand, tells us immediately that the generators have the form $s[1] = [s]$, where $s = 1, 5, 7, 11$, since these are the only numbers between 1 and 12 that are relatively prime to 12. This is obviously a much better way to find the generators than by checking each element individually. As for the subgroups of $\mathbb{Z}_{12}$, here again we found them by direct calculation. Using these results, we see at once that there is one and only one subgroup of order d for $d = 1, 2, 3, 6, 12$, since these numbers are the only divisors of 12. The subgroup of order d has the form $([12/d])$. For example, the subgroup of order 4 is $([12/4]) = ([3]) = \{[0], [3], [6], [9]\}$. The marginal diagram illustrates the 1–1 correspondence between the subgroups of $\mathbb{Z}_{12}$ and the divisors of 12. Lines in the subgroup lattice indicate set inclusion, while lines in the divisor lattice indicate divisibility. Finally, notice that $([3])$ is itself a cyclic group of order 4 and hence has $\varphi(4)$, or 2, generators; they are $1[3] = [3]$ and $3[3] = [9]$, since 1 and 3 are the only numbers between 1 and 4 that are relatively prime to 4.

EXAMPLE 3

Consider the cyclic group $\mathbb{Z}_{16}$. Since $16 = 2^4$, the only subgroups of $\mathbb{Z}_{16}$ are those of order 16, 8, 4, 2, and 1. The subgroup of order 8, for example, is generated by $[2]$ since $16/8 = 2$ and $2[1] = [2]$. Similarly, $[4]$ generates the subgroup of order 4, and $[8]$ generates the subgroup of order 2. Moreover, $([0]) \le ([8]) \le ([4]) \le ([2]) \le ([1]) = \mathbb{Z}_{16}$, which shows that the subgroups of $\mathbb{Z}_{16}$ form an ascending chain.

16 $\mathbb{Z}_{12}$
|
8 $([2]) = \{[0], [2], [4], [6], [8], [10], [12], [14]\}$
|
4 $([4]) = \{[0], [4], [8], [12]\}$
|
2 $([8]) = \{[0], [8]\}$
|
1 $([0]) = \{[0]\}$

EXAMPLE 4. Roots of unity

Let n be a positive integer and let $H_n = \{z \in \mathbb{C}^* \mid z^n = 1\}$ stand for the set of complex numbers whose nth power is 1. Recall from Example 3, Section 3, that H_n is a subgroup of the multiplicative group $\mathbb{C}^*$ of nonzero complex numbers. Let us show that H_n is in fact a cyclic group of order n, and then discuss its generators.

(A) To show that H_n is a cyclic group, let $z_n = \cos(2\pi/n) + i \sin(2\pi/n)$. Then, by DeMoivre's theorem, $z_n^n = \cos(2\pi/n)n + i \sin(2\pi/n)n = 1$, and hence

$z_n \in H_n$. Therefore, $(z_n) \leq H_n$. We claim that $(z_n) = H_n$. To show this, let z be an arbitrary element in H_n and set $z = \cos \theta + i \sin \theta$. Since $z^n = \cos n\theta + i \sin n\theta = 1$, it follows that $n\theta = 2\pi k$ for some integer k. Therefore, $\theta = (2\pi/n)k$ and hence $z = \cos(2\pi/n)k + i \sin(2\pi/n)k = z_n^k$. It follows that $z \in (z_n)$ and hence $H_n = (z_n)$. Since the smallest positive integer k such that $z_n^k = 1$ is $k = n$, we conclude that H_n is a cyclic group of order n.

(B) The complex numbers $1, z_n, \ldots, z_n^{n-1}$ are the complex roots of the equation $X^n = 1$ and are called the *nth roots of unity*. It follows from part (A) that the nth roots of unity form a cyclic group H_n of order n. Any generator of this group is called a *primitive nth root of unity*. From our general discussion of cyclic groups, we know that there are exactly $\varphi(n)$ primitive nth roots of unity, that they have the form z_n^s, where s is relatively prime to n, and that any one of these primitive roots generates all the roots of the equation $X^n = 1$. We may illustrate the roots of unity geometrically by observing that they all lie on the unit complex circle and divide the circle into n equal parts of length $(2\pi/n)$ units. In the marginal figures we have indicated these roots for $n = 3$ and $n = 4$; the double arrows ($\Longrightarrow$) denote primitive roots. Let us determine the third and fourth roots of unity explicitly. For $n = 3$, we find that

$$z_3 = \cos \frac{2\pi}{3} + i \sin \frac{2\pi}{3} = -\frac{1}{2} + \frac{1}{2}\sqrt{3}i$$

$$z_3^2 = \cos \frac{4\pi}{3} + i \sin \frac{4\pi}{3} = -\frac{1}{2} - \frac{1}{2}\sqrt{3}i$$

$$z_3^3 = \cos 2\pi + i \sin 2\pi = 1,$$

while for $n = 4$,

$$z_4 = \cos \frac{2\pi}{4} + i \sin \frac{2\pi}{4} = i$$

$$z_4^2 = \cos \frac{4\pi}{4} + i \sin \frac{4\pi}{4} = -1$$

$$z_4^3 = \cos \frac{6\pi}{4} + i \sin \frac{2\pi}{4} = -i$$

$$z_4^4 = \cos \frac{8\pi}{4} + i \sin \frac{8\pi}{4} = 1.$$

(C) Let us now bring together some of the ideas that we have discussed up to this point. The set $\mathbb{C}$ of complex numbers is a 2-dimensional vector space over the real numbers $\mathbb{R}$ and has the set $\{1, i\}$ as a basis. As such, we call it the complex plane. Define the mapping $\sigma : \mathbb{C} \to \mathbb{C}$ by setting $\sigma(z) = z_n z$, where $z_n = \cos(2\pi/n) + i \sin(2\pi/n)$, the primitive nth root of unity whose argument is $2\pi/n$. Then σ is a linear transformation on the complex plane since $\sigma(az + bz') = z_n(az + bz') = az_nz + bz_nz' = a\sigma(z)$ $+ b\sigma(z')$ for every $a, b \in \mathbb{R}$ and $z, z' \in \mathbb{C}$, and σ is nonsingular since $\sigma(z)$ $= 0$ if and only if $z = 0$. The mapping σ is just complex multiplication by the fixed number z_n. We recall from Example 1, Section 2, however, that complex multiplication corresponds to rotation of the complex plane. Specifically, $\sigma(z) = z_nz$ is obtained from z by a counterclockwise rotation of the complex plane through an angle of $2\pi/n$ radians about the origin. Thus, the transformation σ is rotation of the complex plane through $2\pi/n$ radians, and therefore the cyclic subgroup (σ) of $GL(\mathbb{C})$ is just the cyclic rotation group $\text{Rot}_n(\mathbb{R}^2)$ discussed in Example 2 of this section. To find the matrix representation of σ, we first calculate the image of the base vectors 1 and i under $\sigma:\sigma(1) = z_n = (\cos(2\pi/n))1$ $+ (\sin(2\pi/n))i$, $\sigma(i) = iz_n = (-\sin(2\pi/n))1 + (\cos(2\pi/n))i$. Hence, the matrix of σ relative to the coordinate system $(1, i)$ is

$$M_\sigma = \begin{pmatrix} \cos\dfrac{2\pi}{n} & -\sin\dfrac{2\pi}{n} \\ \sin\dfrac{2\pi}{n} & \cos\dfrac{2\pi}{n} \end{pmatrix}.$$

The reader may recall that this is the standard rotation matrix corresponding to an angle of $2\pi/n$ radians. The order of M_σ as an element of the group $\text{Mat}_2^*(\mathbb{R})$ is n, and M_σ generates the cyclic subgroup (M_σ) $= \{I_2, M_\sigma, \ldots, M_\sigma^{n-1}\}$, where

$$M_\sigma^k = \begin{pmatrix} \cos\dfrac{2\pi}{n}k & -\sin\dfrac{2\pi}{n}k \\ \sin\dfrac{2\pi}{n}k & \cos\dfrac{2\pi}{n}k \end{pmatrix}.$$

This concludes the example.

Earlier in this section we introduced the Euler phi-function to count the number of generators of a finite cyclic group. But the function φ also has several important arithmetic properties. Let us conclude this section by using our results on the generators of finite cyclic groups to derive some of these properties.

COROLLARY.   Let n be a positive integer. Then $\Sigma\ \varphi(d) = n$, where the summation is over all divisors d of n.

Proof.   Let $G = (x)$ be a cyclic group of order n and let $d_1, \ldots, d_s$ stand for all divisors of n. Then there are unique cyclic subgroups $(x_1), \ldots, (x_s)$ of G such that $|(x_i)| = d_i$ for $1 \le i \le s$, and these are all of the subgroups of G. Hence, since (x_i) has $\varphi(d_i)$ generators, the total number of elements in G that may generate some subgroup is $\varphi(d_1) + \cdots + \varphi(d_s)$. But every element in G generates a subgroup. Therefore, $\varphi(d_1) + \cdots + \varphi(d_s) = n$.   ∎

COROLLARY. Let n and m be positive integers that are relatively prime. Then $\varphi(nm) = \varphi(n)\varphi(m)$. Thus, the function φ is multiplicative on relatively prime pairs.

Proof. Let G_n and G_m be cyclic groups of order n and m, respectively, and let $G = G_n \times G_m$ stand for their direct product.

(A) We claim that G is a cyclic group of order nm. To show this, let x be a generator for G_n and y a generator for G_m. Since the group G has order nm, the element $(x, y) \in G$ has order at most nm. Let $s = |(x, y)| \le nm$. Then $(x, y)^s = (x^s, y^s) = (1, 1)$ and hence $x^s = 1$ and $y^s = 1$. Therefore, both n and m divide s. Since n and m are relatively prime, it follows that the product nm divides s. But $s \le nm$. Thus, $s = nm$. Hence, (x, y) is an element of order nm in G and therefore generates G since $|G| = nm$. In particular, G is a cyclic group of order nm.

(B) Now, if (x, y) is an arbitrary element in G, we claim that (x, y) generates G if and only if x generates G_n and y generates G_m. For, by part (A), if x generates G_n and y generates G_m, then (x, y) generates G. Conversely, if (x, y) generates G and g is any element in G_n, then $(g, 1) \in G$, and hence $(g, 1) = (x, y)^s$ for some integer s. Therefore, $(g, 1) = (x^s, y^s)$, and hence $g = x^s$. Thus, x generates G_n. Similarly, y generates G_m.

It follows from parts (A) and (B) that G is a cyclic group whose generators are precisely those ordered pairs whose first entry is any generator of G_n and whose second entry is any generator of G_m. It follows that the number of generators of G is equal to the product of the number of generators G_n and number of generators of G_m. That is, $\varphi(nm) = \varphi(n)\varphi(m)$. ∎

COROLLARY. Let n be a positive integer and let $p_1^{s_1} \cdots p_t^{s_t}$ be the prime factorization of n, where $p_1, \ldots, p_t$ are distinct primes. Then

$$\varphi(n) = p_1^{s_1-1} \cdots p_t^{s_t-1}(p_1 - 1) \cdots (p_t - 1) = n \prod_{i=1}^{t}\left(1 - \frac{1}{p_i}\right).$$

Proof. Since $n = p_1^{s_1} \cdots p_t^{s_t}$ and the factors $p_1^{s_1}, \ldots, p_t^{s_t}$ are relatively prime, it follows that $\varphi(n) = \varphi(p_1^{s_1}) \cdots \varphi(p_t^{s_t})$. To calculate $\varphi(p^s)$, where p is prime and s is a positive integer, we use the summation formula $\Sigma \varphi(d) = p^s$. In this case the only divisors of p^s are $1, p, \ldots, p^s$, and therefore $\varphi(1) + \varphi(p) + \cdots + \varphi(p^s) = p^s$. Hence, $\varphi(p^s) = p^s - (\varphi(1) + \cdots + \varphi(p^{s-1}))$. But $\Sigma \varphi(d) = p^{s-1}$ when summed over all divisors of p^{s-1}. Hence, $\varphi(p^s) = p^s - p^{s-1}$, and therefore

$$\varphi(n) = \varphi(p_1^{s_1}) \cdots \varphi(p_t^{s_t})$$

$$= p_1^{s_1-1}(p_1 - 1) \cdots p_t^{s_t-1}(p_t - 1)$$

$$= p_1^{s_1}\left(1 - \frac{1}{p_1}\right) \cdots p_t^{s_t}\left(1 - \frac{1}{p_t}\right)$$

$$= n \prod_{i=1}^{t}\left(1 - \frac{1}{p_i}\right). \quad ∎$$

These results now give an effective computational procedure for calculating $\varphi(n)$ for any positive integer n: factor n into a product of distinct prime powers p^s, find $\varphi(p^s)$ using the fact that $\varphi(p^s) = p^{s-1}(p-1)$, and then multiply the $\varphi(p^s)$ for each factor p^s. For example, $\varphi(16) = \varphi(2^4) = 2^3(2-1) = 8$ and $\varphi(504) = \varphi(2^3 \, 3^2 \, 7) = \varphi(2^3)\varphi(3^2)\varphi(7) = 2^2(2-1)3(3-1)(7-1) = 144$.

Exercises

1. Let $\mathbb{Q}$ stand for the additive group of rational numbers.
   (a) Determine the cyclic subgroup generated by 1.
   (b) Determine the cyclic subgroup generated by $1/2$.
   (c) Is 8 an element of the cyclic subgroup generated by $2/3$?
2. Let $\mathbb{Q}^*$ stand for the multiplicative group of nonzero rational numbers.
   (a) Determine the cyclic subgroup generated by -1.
   (b) Determine the cyclic subgroup generated by $1/2$.
   (c) Is $8/27$ an element of the cyclic subgroup generated by $3/2$?
3. Write out the addition table for the group $\mathbb{Z}_5$ and find the inverse and order of each element.
4. Write out the addition table for the group $\mathbb{Z}_8$.
   (a) Find the inverse and order of each element in $\mathbb{Z}_8$.
   (b) Find all generators of $\mathbb{Z}_8$.
   (c) Find the cyclic subgroup $([2])$ generated by $[2]$.
   (d) Find the cyclic subgroup $([4])$ generated by $[4]$.
   (e) Draw the lattice of subgroups of $\mathbb{Z}_8$.
5. In each of the following groups determine the order of the element x and find all generators of the cyclic subgroup (x).
   (a) $\mathbb{Z}_{20}; x = [6]$ (d) $\mathbb{Z}_9; x = [6]$
   (b) $\mathbb{Z}_{20}; x = [5]$ (e) $\mathbb{C}^*; x = \cos(6\pi/8) + i \sin(6\pi/8)$
   (c) $\mathbb{Z}_9; x = [2]$ (f) $\mathrm{Mat}_2^*(\mathbb{R}); x = \left(\begin{smallmatrix} 0 & -1 \\ 1 & 0 \end{smallmatrix}\right)$.
6. Find all subgroups of $\mathbb{Z}_{30}$ and all generators of each subgroup. Draw the lattice of subgroups of $\mathbb{Z}_{30}$ and the corresponding lattice of divisors of 30.
7. Let $G = \mathbb{Z}_3 \times \mathbb{Z}_5$.
   (a) Find the inverse and order of each element in G.
   (b) Show that G is a cyclic group of order 15 and find all its generators.
   (c) Find all subgroups of G and draw the lattice of subgroups.
8. Is the group $\mathbb{Z}_6 \times \mathbb{Z}_9$ a cyclic group?
9. Find all of the sixth roots of unity and sketch them on the unit complex circle, indicating which ones are primitive. How many primitive sixth roots of unity are there?
10. In this exercise we give a matrix description of the transformations in the cyclic group $\mathrm{Rot}_n(\mathbb{R}^2)$. Let $n \geq 3$ and let A and B be adjacent vertices of a regular n-gon P_n labeled so that $\rho(A) = B$, where ρ is the counterclockwise rotation of the plane through $2\pi/n$ radians. Then (A, B) is a coordinate system for the real plane with its origin at the center of P_n.

(a) Show that $\rho(B) + A = \lambda B$ for some scalar λ. The cases where $n = 3$, 4 require separate attention, while all cases follow from one argument when $n \geq 5$.

(b) Using the formula $X \cdot Y = |X| |Y| \cos(X, Y)$ for the dot product of two vectors, where (X, Y) is the angle between vectors X and Y, show that $\lambda = 2 \cos(2\pi/n)$.

(c) Using parts (a) and (b), find the matrix M_ρ of ρ relative to the (A, B) coordinate system.

(d) Write out the matrices for ρ, ρ^2, ρ^3, ρ^4, and ρ^5 for the case $n = 5$.

(e) We may use the fact that the group $\text{Rot}_n(\mathbb{R}^2) = (\rho)$ has order n to show that the number λ is the root of a polynomial having integer coefficients. Begin with the fact that $M_\rho^n = I_2$. The entries in M_ρ^n are polynomials in λ with integer coefficients. Equating these polynomials to the corresponding entries in I_2 yields several equations having λ as a root. Carry out these calculations for $n = 5$, where $\lambda = 2 \cos(2\pi/5)$, and thus find a polynomial with integer coefficients that has λ as a root.

11. Let G be a cyclic group of order n, d a positive integer that divides n, and $G_d = \{x \in G \,|\, x^d = 1\}$. Show that G_d is a subgroup of G of order d.

12. Let G be an arbitrary finite group and d a positive integer. Let N_d stand for the number of elements in G that have order d; this number may be zero. Show that $\varphi(d)$ divides N_d.

13. Let $N_1, \ldots, N_s$ be nonzero integers.

(a) Show that there is a unique positive integer d such that $(N_1, \ldots, N_s) = (d)$ as subgroups of the additive group $\mathbb{Z}$ of integers. The integer d is called the *greatest common divisor* of the numbers $N_1, \ldots, N_s$, and is denoted by the symbol $\gcd(N_1, \ldots, N_s)$.

(b) Show that $\gcd(N_1, \ldots, N_s)$ divides each integer N_i.

(c) If d' is any integer that divides $N_1, \ldots, N_s$, show that d' divides $\gcd(N_1, \ldots, N_s)$.

(d) Find $\gcd(12, 18, 24)$, $\gcd(6, 8, -12, 15)$, $\gcd(-10, 12)$, and $\gcd(12, 15, 35)$.

(e) Show that there are integers $t_1, \ldots, t_s$ such that $\gcd(N_1, \ldots, N_s) = t_1 N_1 + \cdots + t_s N_s$.

(f) Is it true that $\gcd(N_1, \ldots, N_s) = (\gcd, \ldots, (\gcd(N_1, N_2), N_3), \ldots, N_s)$?

14. Let n and m be positive integers.

(a) Show that there is a unique positive integer r such that $(n) \cap (m) = (r)$ as subgroups of the additive group $\mathbb{Z}$ of integers. The integer r is called the *least common multiple* of n and m and is denoted by $\text{lcm}(n, m)$.

(b) Show that both n and m divide $\text{lcm}(n, m)$.

(c) Show that if an integer t is divisible by both n and m, then t is divisible by $\text{lcm}(n, m)$.

(d) Show that $\dfrac{nm}{\gcd(n, m)} = \text{lcm}(n, m)$.

5.   GROUP-ISOMORPHISM

The reader will have noticed by now that many of the groups we have discussed up to this point are really quite similar. For example, we have worked with four different cyclic groups of order 3: $\mathbb{Z}_3 = \{[0], [1], [2]\}$ $= ([1])$, the additive group of integers modulo 3; $(R) = \{1, R, R^2\}$, the rotation subgroup of the group of the triangle; $(z_3) = \{1, z_3, z_3^2\}$, the group of cube roots of unity; and $\text{Rot}_3(\mathbb{R}^2) = (\rho) = \{1, \rho, \rho^2\}$. To what extent do these groups differ? On the one hand, they contain different kinds of elements; the group $\mathbb{Z}_3$, for example, consists of congruence classes of integers, while (R) consists of linear transformations of the plane. But aside from this, the elements in these groups combine in exactly the same way; that is, there is a 1–1 correspondence between any two of these groups such that the product of any two elements in the one group corresponds to the product of the corresponding elements in the other group. This is the basic idea behind the concept of group-isomorphism. The purpose of this section is to discuss and illustrate this concept and then use it to classify cyclic groups according to their order.

Definition

Let G and G' be groups. A *group-isomorphism* from G to G' is a function $f: G \to G'$ that satisfies the following two properties:

(1) f is a 1–1 correspondence; that is, f is both 1–1 and onto.
(2) $f(xy) = f(x)f(y)$ for every $x, y \in G$.

If G and G' are groups, we say that G is isomorphic to G' if there is a group-isomorphism $f: G \to G'$ and indicate this by writing $G \cong G'$. In the exercises we ask the reader to show that the isomorphism relation $\cong$ is an equivalence relation on the class of all groups. We express statement (2) in this definition informally by saying that f "preserves" the arithmetic on the groups. Thus, a group-isomorphism is a 1–1 correspondence that preserves the arithmetic on the groups. For example, define the function $f:(z_3) \to (R)$ by setting $f(1) = 1$, $f(z_3) = R$, and $f(z_3^2) = R^2$. Then f is clearly a 1–1 correspondence. To verify that $f(xy) = f(x)f(y)$ for all $x, y \in (z_3)$, we use the multiplication table for (z_3) and calculate its image under f. Since the "image" table is the multiplication table for the group (R), it follows that $f(xy)$ $= f(x)f(y)$ for all $x, y \in (z_3)$ and hence f is a group-isomorphism from (z_3) to (R).

(z_3)	1	z_3	z_3^2
1	1	z_3	z_3^2
z_3	z_3	z_3^2	1
z_3^2	z_3^2	1	z_3

$\xrightarrow{f}$

(R)	1	R	R^2
1	1	R	R^2
R	R	R^2	1
R^2	R^2	1	R

Let us emphasize that in the definition of group-isomorphism, the equation $f(xy) = f(x)f(y)$ should be clearly understood. As written, it is expressed in multiplicative notation. But if the operation on either group is written additively, the equation will be different. It may be $f(xy) = f(x) + f(y)$ if the operation on G' is written additively, or $f(x + y) = f(x)f(y)$ if the operation on G is written additively, or $f(x + y) = f(x) + f(y)$ if the operations on both groups are written additively. For example, the function $f: \mathbb{Z}_3 \to (R)$ defined by setting $f([0]) = 1$, $f([1]) = R$, and $f([2]) = R^2$ is a 1–1 correspondence such that $f([s] + [t]) = f([s])f([t])$ for all $[s], [t] \in \mathbb{Z}_3$, and hence is an isomorphism.

EXAMPLE 1

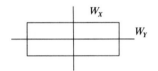

Let G be the Klein 4-group, that is, the group of the rectangle discussed in Exercise 3, Section 2. Then $G = \{1, W_X, W_Y, \rho_\pi\}$, where W_X and W_Y are the reflections of the rectangle about the X- and Y-axes, as shown in the marginal figure, and ρ_π is the 180° rotation of the rectangle about its center. Let us show that $G \cong \mathbb{Z}_2 \times \mathbb{Z}_2$. Define the function $f: G \to \mathbb{Z}_2 \times \mathbb{Z}_2$ by setting $f(1) = ([0], [0])$, $f(W_Y) = ([0], [1])$, $f(W_X) = ([1], [0])$, $f(\rho_\pi) = ([1], [1])$. Clearly, f is a 1–1 correspondence. Moreover, the image under f of the multiplication table for G is easily seen to be the addition table for the group $\mathbb{Z}_2 \times \mathbb{Z}_2$. Therefore, f is a group-isomorphism and hence $G \cong \mathbb{Z}_2 \times \mathbb{Z}_2$.

G	1	W_X	W_Y	ρ_π
1	1	W_X	W_Y	ρ_π
W_X	W_X	1	ρ_π	W_Y
W_Y	W_Y	ρ_π	1	W_X
ρ_π	ρ_π	W_Y	W_X	1

$\xrightarrow{f}$

$\mathbb{Z}_2 \times \mathbb{Z}_2$	$(0, 0)$	$(1, 0)$	$(0, 1)$	$(1, 1)$
$(0, 0)$	$(0, 0)$	$(1, 0)$	$(0, 1)$	$(1, 1)$
$(1, 0)$	$(1, 0)$	$(0, 0)$	$(1, 1)$	$(0, 1)$
$(0, 1)$	$(0, 1)$	$(1, 1)$	$(0, 0)$	$(1, 0)$
$(1, 1)$	$(1, 1)$	$(0, 1)$	$(1, 0)$	$(0, 0)$

EXAMPLE 2

We have used matrix representations many times to obtain a group of matrices from a group of linear transformations. Let us now show that this process is actually a group-isomorphism. Let $\mathbb{R}^n$ stand for real n-dimensional space.

(A) Let X be any coordinate system for $\mathbb{R}^n$. Then X defines a function $f_X: GL(\mathbb{R}^n) \to \text{Mat}_n^*(\mathbb{R})$ by assigning to every transformation $\sigma \in GL(\mathbb{R}^n)$ its matrix representation $f_X(\sigma)$. Note that since σ is nonsingular, $f_X(\sigma)$ is, in fact, invertible and hence an element of the group $\text{Mat}_n^*(\mathbb{R})$. We claim that f_X is a group-isomorphism. To show that f_X is 1–1, recall that the matrix of a linear transformation completely determines the transformation. Hence, if σ and σ' are two transformations that have the same matrix, that is, if $f_X(\sigma) = f_X(\sigma')$, then σ and σ' must be the same transformation. It follows, therefore, that f_X is a 1–1 mapping. To show that f_X maps onto $\text{Mat}_n^*(\mathbb{R})$, recall that every $n \times n$ invertible matrix M defines a nonsingular linear transformation $\sigma_M : \mathbb{R}^n \to \mathbb{R}^n$ by setting

$$\sigma_M \begin{pmatrix} x_1 \\ \vdots \\ x_n \end{pmatrix} = M \begin{pmatrix} x_1 \\ \vdots \\ x_n \end{pmatrix},$$

and that the matrix of σ_M is M. It follows, therefore, that f_X is an onto mapping. Finally, $f_X(\sigma\sigma') = f_X(\sigma)f_X(\sigma')$ for all $\sigma, \sigma' \in GL(\mathbb{R}^n)$ since, relative to X, the matrix of the composition $\sigma\sigma'$ is equal to the product of the matrix of σ and the matrix of σ'. It now follows that f_X is a group-isomorphism.

(B) For example, let G be a subgroup of $GL(\mathbb{R}^2)$ consisting of the rigid motions of the plane that bring some geometric figure into coincidence with itself. When we choose a coordinate system X for the plane and write the matrices of the transformations in G relative to X, we are simply using the isomorphism f_X to transfer the group G over to a group G' of matrices in $\text{Mat}_2^*(\mathbb{R})$, as illustrated in the marginal diagram. Since there are many different coordinate systems for $\mathbb{R}^2$, there are many different ways to map $GL(\mathbb{R}^2)$ into $\text{Mat}_2^*(\mathbb{R})$; consequently, there are many different subgroups of $\text{Mat}_2^*(\mathbb{R})$ that are isomorphic to G. In Chapter 6 we discuss how these different groups are related to each other.

(C) Let us illustrate the ideas in parts (A) and (B) by using the group T of the triangle. Recall that $T = \{1, R, R^2, S, RS, R^2S\}$. Let us choose two different coordinate systems for $\mathbb{R}^2$, say $X = (A, B)$ and $Y = (A, C)$, as illustrated in Figure 1. Then the matrix representations $f_X(T)$ and $f_Y(T)$ of the group T determined by these coordinate systems are isomorphic groups of matrices, and they are both isomorphic to the group T itself. This concludes the example.

Group-isomorphism is an extremely important concept in group theory because it allows us to classify groups solely on the basis of algebraic structure. It permits us to ignore differences in the nature of elements and to concentrate, instead, on the way in which elements combine—that is, on the arithmetic of the groups. As we have seen, the fundamental question underlying group-isomorphism is whether or not the elements of the groups can be paired-off in such a way that they have exactly the same multiplication table. Determining when this is possible is usually a difficult task. If the groups are cyclic, however, there is an easy way to decide the question: cyclic groups are isomorphic if and only if they have the same order. Let us now prove this result and, in doing so, establish our first classification theorem.

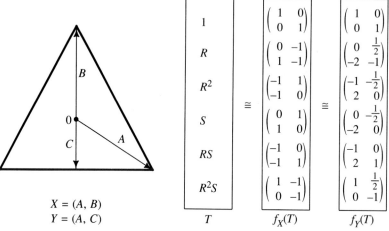

$$GL(\mathbb{R}^2) \xrightarrow{\ f_X,\, f_Y\ } Mat_2^*(\mathbb{R})$$

$$\sigma \longmapsto f_X(\sigma) \qquad\qquad f_Y(\sigma)$$

T	$f_X(T)$	$f_Y(T)$
1	$\begin{pmatrix} 1 & 0 \\ 0 & 1 \end{pmatrix}$	$\begin{pmatrix} 1 & 0 \\ 0 & 1 \end{pmatrix}$
R	$\begin{pmatrix} 0 & -1 \\ 1 & -1 \end{pmatrix}$	$\begin{pmatrix} 0 & \frac{1}{2} \\ -2 & -1 \end{pmatrix}$
R^2	$\begin{pmatrix} -1 & 1 \\ -1 & 0 \end{pmatrix}$	$\begin{pmatrix} -1 & -\frac{1}{2} \\ 2 & 0 \end{pmatrix}$
S	$\begin{pmatrix} 0 & 1 \\ 1 & 0 \end{pmatrix}$	$\begin{pmatrix} 0 & -\frac{1}{2} \\ -2 & 0 \end{pmatrix}$
RS	$\begin{pmatrix} -1 & 0 \\ -1 & 1 \end{pmatrix}$	$\begin{pmatrix} -1 & 0 \\ 2 & 1 \end{pmatrix}$
R^2S	$\begin{pmatrix} 1 & -1 \\ 0 & -1 \end{pmatrix}$	$\begin{pmatrix} 1 & \frac{1}{2} \\ 0 & -1 \end{pmatrix}$

$X = (A, B)$
$Y = (A, C)$

FIGURE 1. Two coordinate systems X, Y for $\mathbb{R}^2$ and the corresponding isomorphisms f_X and f_Y that represent the group T of the triangle as a group of matrices.

PROPOSITION 1. Let G and G' be cyclic groups. Then G and G' are isomorphic if and only if they have the same order.

Proof. If G and G' are isomorphic, then the isomorphism between them is a 1–1 correspondence and shows that they have the same order. Now suppose, conversely, that G and G' are cyclic groups that have the same order, and let $G = (x)$ and $G' = (y)$. Since every element in G may be written in the form x^s for some integer s, we define $f : G \to G'$ by setting $f(x^s) = y^s$ for all elements $x^s \in G$. We claim that f is a group-isomorphism. To show this, we must first verify that f is well defined since the integer s used to represent the element x^s is not necessarily unique. The details of verifying this, as well as showing that f is an isomorphism, depend upon the order of G. Thus, we separate the proof into two cases.

(A) Suppose that $|G|$ is finite, and let $|G| = |G'| = n$. Then $|x| = |y| = n$. To show that f is well defined, we must show that if an element in G is represented in two different ways, say $x^s = x^t$, then the images y^s and y^t are also the same. Now, if $x^s = x^t$, then $x^{s-t} = 1$ and therefore n divides $s - t$. Therefore, $y^{s-t} = 1$, or equivalently, $y^s = y^t$, since $|y| = n$, and hence f is well defined. Now, to show that f is a group-isomorphism, we must verify that it is 1–1, onto, and preserves the arithmetic on the groups. To this end, suppose that $f(x^s) = f(x^t)$ for some elements $x^s, x^t \in G$. Then

$y^s = y^t$ and it follows, as before, that $x^s = x^t$. Hence, f is 1–1. Moreover, since every element in G' has the form y^s, and since $y^s = f(x^s)$, it follows that f is an onto mapping. Finally, since $f(x^s x^t) = f(x^{s+t}) = y^{s+t} = y^s y^t = f(x^s) f(x^t)$ for all integers s and t, it follows that f preserves the arithmetic on G and G'. Therefore, f is a group-isomorphism and hence $G \cong G'$.

(B) Suppose that $|G|$ is infinite. In this case $|x|$ and $|y|$ are infinite and hence $x^n = 1$ and $y^n = 1$ if and only if $n = 0$. It follows that the integer s used to represent the element x^s is uniquely determined, and consequently, the mapping $f : x^s \mapsto y^s$ is well defined. To show that f is 1–1, suppose that $f(x^s) = f(x^t)$ for some integers s, t. Then $y^s = y^t$ and hence $s = t$. Therefore, $x^s = x^t$, and hence f is 1–1. Finally, the proof that f is onto and satisfies the equation $f(x^s x^t) = f(x^s) f(x^t)$ is the same as in part (A). Therefore, f is a group-isomorphism. Hence, $G \cong G'$ and the proof is complete. ∎

Let us take a moment to discuss the result in Proposition 1. If n is a positive integer, the group $\mathbb{Z}_n$ of integers modulo n is a finite cyclic group of order n. Hence, by Proposition 1, every finite cyclic group of order n is isomorphic to $\mathbb{Z}_n$. On the other hand, the additive group $\mathbb{Z}$ of integers is an infinite cyclic group and hence every infinite cyclic group is isomorphic to $\mathbb{Z}$. Thus, Proposition 1 tells us that every cyclic group is isomorphic to either $\mathbb{Z}_n$ or $\mathbb{Z}$, depending upon whether it is finite or infinite, and consequently these groups, $\mathbb{Z}_n$ and $\mathbb{Z}$, provide the basic algebraic models for all cyclic groups. It is also important to realize that the function f constructed in the proof of Proposition 1 tells us explicitly how to find an isomorphism from one cyclic group to another cyclic group of the same order. If G and G' are cyclic groups having the same order, then f maps any fixed generator x of G to any fixed generator y of G' and is extended to all of G by mapping x^s to y^s for every integer s. Thus, the prescription *generator of $G \mapsto$ generator of G'* allows us to construct many isomorphisms from G to G'. That this map will always be a group-isomorphism is exactly what the proof in Proposition 1 proved.

EXAMPLE 3. Isomorphisms between (z_3) and $\mathrm{Rot}_3(\mathbb{R}^2)$

The group $\mathrm{Rot}_3(\mathbb{R}^2)$ is a cyclic group of order 3 having two generators, ρ_3 and ρ_3^2. Since $(z_3) = \{1, z_3, z_3^2\}$ is also a cyclic group of order 3, it follows that there are two isomorphisms from (z_3) to $\mathrm{Rot}_3(\mathbb{R}^2)$, namely, $f : (z_3) \to \mathrm{Rot}_3(\mathbb{R}^2)$, defined by $f(1) = 1$, $f(z_3) = \rho_3$, $f(z_3^2) = \rho_3^2$, and $g : (z_3) \to \mathrm{Rot}_3(\mathbb{R}^2)$, defined by $g(1) = 1$, $g(z_3) = \rho_3^2$, $g(z_3^2) = \rho_3$.

EXAMPLE 4

Consider the cyclic group $\mathbb{Z}_{12}$. Since the element $[3]$ has order 4, it generates a cyclic subgroup $([3])$ of order 4. Therefore, $([3]) \cong \mathbb{Z}_4$. Let us construct an isomorphism $f : \mathbb{Z}_4 \to ([3])$. Using the generator $[1]$ for $\mathbb{Z}_4$ and the generator $[3]$ for $([3])$, we set $f([1]) = [3]$ and then extend f to all of $\mathbb{Z}_4$ by setting $f([0]) = [0]$, $f([1]) = [3]$, $f([2]) = [3] + [3] = [6]$, and $f([3]) = [3] + [3] + [3] = [9]$. Thus, f is a group-isomorphism mapping $\mathbb{Z}_4$ onto $([3])$.

EXAMPLE 5. A special case of the Chinese Remainder Theorem

Consider the group $\mathbb{Z}_2 \times \mathbb{Z}_3$. This is a cyclic group of order 6 generated by the element $([1], [1])$ since

$$2([1], [1]) = ([0], [2]), \ 3([1], [1]) = ([1], [0]), \ 4([1], [1]) = ([0], [1]), \ 5([1], [1])$$
$$= ([1], [2]), \text{ and } 6([1], [1]) = ([0], [0]).$$

Therefore, $\mathbb{Z}_2 \times \mathbb{Z}_3 \cong \mathbb{Z}_6$. Since $\mathbb{Z}_2 \times \mathbb{Z}_3$ is a cyclic group of order 6 and $\varphi(6) = 2$, $\mathbb{Z}_2 \times \mathbb{Z}_3$ has 2 generators: $([1], [1])$ and $5([1], [1]) = ([1], [2])$. Hence, there are two isomorphisms $f, g : \mathbb{Z}_6 \to \mathbb{Z}_2 \times \mathbb{Z}_3$, f defined by setting $f([1]) = ([1], [1])$, and g defined by setting $g([1]) = ([1], [2])$.

$$\mathbb{Z}_6 \overset{f}{\to} \mathbb{Z}_2 \times \mathbb{Z}_3 \qquad \mathbb{Z}_6 \overset{g}{\to} \mathbb{Z}_2 \times \mathbb{Z}_3$$

$\mathbb{Z}_6 \overset{f}{\to} \mathbb{Z}_2 \times \mathbb{Z}_3$	$\mathbb{Z}_6 \overset{g}{\to} \mathbb{Z}_2 \times \mathbb{Z}_3$
$[0] \to ([0], [0])$	$[0] \to ([0], [0])$
$[1] \to ([1], [1])$	$[1] \to ([1], [2])$
$[2] \to ([0], [2])$	$[2] \to ([0], [1])$
$[3] \to ([1], [0])$	$[3] \to ([1], [0])$
$[4] \to ([0], [1])$	$[4] \to ([0], [2])$
$[5] \to ([1], [2])$	$[5] \to ([1], [1])$

Let us use the isomorphism f to illustrate an important result in number theory called the *Chinese Remainder Theorem*, which states that if M and N are integers, there is an integer s uniquely determined modulo 6 such that $s \equiv M \pmod 2$ and $s \equiv N \pmod 3$. We begin by observing that if $[s] \in \mathbb{Z}_6$, then $f([s]) = ([s], [s])$; that is, the function f assigns to $[s]$ the ordered pair $([s], [s])$, where the first component is the class of s mod 2 and the second component is the class of s mod 3. Since f is an isomorphism, it follows that every element in $\mathbb{Z}_2 \times \mathbb{Z}_3$ has the form $([s], [s])$ for some integer s:

$$f([0]) = ([0], [0]) = ([0], [0]) \qquad f([3]) = ([1], [0]) = ([3], [3])$$
$$f([1]) = ([1], [1]) = ([1], [1]) \qquad f([4]) = ([0], [1]) = ([4], [4])$$
$$f([2]) = ([0], [2]) = ([2], [2]) \qquad f([5]) = ([1], [2]) = ([5], [5]).$$

Now, let M and N be arbitrary integers. Then the pair $([M], [N])$ is an element of the group $\mathbb{Z}_2 \times \mathbb{Z}_3$ and hence there is an integer s such that $([M], [N]) = ([s], [s])$; in other words, $s \equiv M \pmod 2$ and $s \equiv N \pmod 3$.

Moreover, if t is another integer such that $([M]), [N]) = ([t], [t])$, then $(s - t)([1], [1]) = ([0], [0])$ and therefore 6 divides $s - t$ since $|([1], [1])| = 6$. Thus, $s \equiv t \pmod 6$ and hence s is uniquely determined modulo 6. What we have shown, therefore, is that the system of linear congruences $X \equiv M$ (mod 2), $X \equiv N$ (mod 3) has a simultaneous solution $X = s$ and that this solution is uniquely determined modulo 6. This is an example of the Chinese Remainder Theorem in two variables.

This concludes our discussion of group-isomorphism and with it our introduction to the basic concepts of group theory, from the group concept itself, through subgroups, cyclic groups, and group-isomorphism. Using examples drawn from geometry and number theory, we illustrated how the algebraic properties of a group may be used to obtain information about the underlying objects in the group. We discussed a number of important groups, including the matrix groups and the general linear group, and we constructed the cyclic groups $\mathbb{Z}_n$ of integers modulo n, a new and important family of groups that serve as the basic algebraic model for all finite cyclic groups. In the next chapter we turn our attention to another important family of groups, the symmetric groups, and show that, in one sense, all groups are represented by the symmetric groups.

Exercises

1. Find all isomorphisms from $\mathbb{Z}_3$ to (z_3), where z_3 is a primitive cube root of unity.
2. Find all isomorphisms from (z_6) to $\mathrm{Rot}_6(\mathbb{R}^2)$, where z_6 is a primitive sixth root of unity.
3. Let G and G' be finite cyclic groups of order n. How many isomorphisms are there from G to G'?
4. Let $f : G \to G'$ be a group-isomorphism.
   (a) Show that $f(1) = 1$; thus, a group-isomorphism must map the identity element to the identity element.
   (b) Show that $f(x^{-1}) = f(x)^{-1}$; thus, a group-isomorphism must map the inverse of an element to the inverse of the image of the element.
5. Let G and G' be isomorphic cyclic groups and let $f : G \to G'$ be an isomorphism. If x is a generator of G, show that $f(x)$ is a generator of G'.
6. Let G, G' and G'' be groups.
   (a) Show that $G \cong G$.
   (b) If $G \cong G'$, show that $G' \cong G$.
   (c) If $G \cong G'$ and $G' \cong G''$, show that $G \cong G''$.
   (d) Using parts (a), (b), and (c), show that the relation "is isomorphic to" is an equivalence relation on the class of all groups.
7. Let G and G' be isomorphic groups. Show that G contains an element of order n if and only if G' contains an element of order n. This result

is frequently a useful method for showing that two groups are not isomorphic. Using this approach, answer the following questions:
(a) Is the group D_4 of the square cyclic?
(b) Is the Klein 4-group cyclic?
(c) Is the group $\mathbb{Z}_4 \times \mathbb{Z}_6$ cyclic?
(d) Is the multiplicative group $\mathbb{Q}^*$ of nonzero rational numbers cyclic?
(e) Is the multiplicative group $\mathbb{R}^*$ of nonzero real numbers cyclic?

8. Is the additive group of real numbers cyclic?

9. Is the additive group of rational numbers cyclic?

10. Show that $\mathbb{Z} \cong n\mathbb{Z}$ for every nonzero integer n. Thus, it is possible for a group to be isomorphic to a proper subgroup of itself. Can this happen if the group is finite?

11. Below we have listed several cyclic groups. In each case, decide if the group is isomorphic to $\mathbb{Z}$ or $\mathbb{Z}_n$; in the latter case, find n.
(a) $G = (\rho)$, where ρ is the counterclockwise rotation of $\mathbb{R}^2$ through an angle of $\pi/5$ radians about the origin.
(b) $G = (\rho)$, where ρ is the counterclockwise rotation of $\mathbb{R}^2$ through an angle of $7\pi/16$ radians about the origin.
(c) $G = (\rho)$, where ρ is the counterclockwise rotation of $\mathbb{R}^2$ through an angle of 2 radians about the origin.
(d) $G = (M)$, where

$$M = \begin{pmatrix} \dfrac{1}{2} & -\dfrac{1}{2}\sqrt{3} \\ \dfrac{1}{2}\sqrt{3} & \dfrac{1}{2} \end{pmatrix}.$$

(e) $G = (M)$, where $M = \begin{pmatrix} 1 & 1 \\ 1 & 0 \end{pmatrix}$.

12. Find a 2×2 invertible real matrix $M \neq I_2$ such that $M^7 = I_2$.

13. If G, G', H, and H' are groups such that $G \cong G'$ and $H \cong H'$, show that $G \times H \cong G' \times H'$.

14. Let G and H be groups and let $G^* = \{(g, 1) \in G \times H \mid g \in G\}$ and $H^* = \{(1, h) \in G \times H \mid h \in H\}$.
(a) Show that $G^* \leq G \times H$ and $H^* \leq G \times H$.
(b) Show that $G^* \cong G$ and $H^* \cong H$.
(c) Show that $G^* \times H^* \cong G \times H$.

15. Let $G = \mathbb{Z}_4 \times \mathbb{Z}_5$.
(a) Show that G is a cyclic group of order 20.
(b) Construct all isomorphisms $f : \mathbb{Z}_{20} \to G$.
(c) Find an integer s such that $s \equiv 8 \pmod 4$ and $s \equiv 36 \pmod 5$.

16. **The Chinese Remainder Theorem.** Let n and m be positive integers that are relatively prime.
(a) Show that $\mathbb{Z}_{nm} \cong \mathbb{Z}_n \times \mathbb{Z}_m$.
(b) Let r_1 and r_2 be any integers. Show that there is an integer s such that $s \equiv r_1 \pmod n$ and $s \equiv r_2 \pmod m$ and that s is uniquely determined modulo nm.

17. **The generalized Chinese Remainder Theorem.** Let $n_1, \ldots, n_t$ be positive integers that are pairwise relatively prime and let $r_1, \ldots, r_t$ be any integers. Show that there is an integer s such that $s \equiv r_1 \pmod{n_1}, \ldots, s \equiv r_t \pmod{n_t}$, and that s is uniquely determined modulo $n_1 \cdots n_t$.

18. Let n be a positive integer and d a positive divisor of n such that $\gcd(d, \frac{n}{d}) = 1$. Show that $\mathbb{Z}_n \cong \mathbb{Z}_d \times \mathbb{Z}_{n/d}$.

19. Let n be a positive integer and let $n = p_1^{s_1} \cdots p_t^{s_t}$ be the prime factorization of n. Show that $\mathbb{Z}_n \cong \mathbb{Z}_{p_1^{s_1}} \times \cdots \times \mathbb{Z}_{p_t^{s_t}}$

20. Let G be a group. A *factorization* of G is any representation of G in the form $G \cong H \times K$, where H and K are subgroups of G. If H and K are proper subgroups of G, then $G \cong H \times K$ is called a *proper factorization* of G.

    (a) Find a proper factorization of the Klein 4-group.
    (b) Find a proper factorization of $\mathbb{Z}_6$.
    (c) Find a proper factorization of $\mathbb{Z}_{12}$.
    (d) Show that the group of the triangle does not have a proper factorization.

21. Let A and B be distinct nonzero vectors in $\mathbb{R}^2$ that are not collinear, that is, do not lie on the same line, and let $U = \langle A \rangle$ and $W = \langle B \rangle$ be the lines spanned by A and B, respectively. For any vector $X \in \mathbb{R}^2$, let X_U stand for the projection of X onto U in the direction parallel to W, and let X_W stand for the projection of X onto W in the direction parallel to U, as illustrated in the diagram. Define the function $f : \mathbb{R}^2 \to U \times W$ by setting $f(X) = (X_U, X_W)$. Show that f is a group-isomorphism. Thus, $\mathbb{R}^2 \cong U \times W$. This factorization of the real plane corresponds to parallel projection onto the lines U and W. By choosing different lines for U and W, we obtain different factorizations. For example, if $A = e_1$ and $B = e_2$, then U and W are the X- and Y-axes, respectively, and f is the usual Cartesian representation of vectors in $\mathbb{R}^2$.

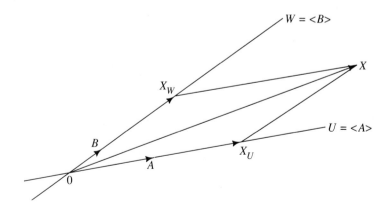

3

The Symmetric Groups

In this chapter we turn our attention to a new and especially important family of groups called the symmetric groups. The symmetric groups consist of all permutations of a given set, that is, 1–1 correspondences of the set with itself. In the first section we discuss the basic properties of permutations and show that the set of all permutations of a given set X form a group, the symmetric group Sym(X). We then turn our attention to permutations of a finite set. Permutations of finite sets have their own special notation and terminology. Our goals are first to identify a special type of permutation called a cycle and then to show that every permutation can be written as a product of cycles in an essentially unique way. This factorization, or cycle decomposition of a permutation, shows that cycles are the basic building blocks for all permutations, and is similar, from an algebraic point of view, to the factorization of integers into primes. Finally, in the last section we indicate how to represent the elements of any group as permutations of some set. This representation, called the Cayley representation of the group, is important because it allows us to use the language and properties of permutations when we discuss arbitrary groups.

1. PERMUTATIONS AND THE SYMMETRIC GROUPS

Let us begin with an observation. If a collection of functions is to form a group under function composition, then the functions must map a given set to itself and each function must be a 1–1 correspondence since 1–1 correspondences are the only functions that have an inverse under function composition. With this observation in mind, we now give a special name to 1–1 correspondences of a set with itself.

Definition

Let X be a nonempty set. A 1–1 correspondence of the set X with itself is called a *permutation of* X. Thus, a permutation of X is any mapping $\sigma:X \to X$ that is both 1–1 and onto. We denote permutations by lowercase Greek letters such as σ and τ.

PROPOSITION 1. Let σ and τ be permutations of a set X. Then the composition $\sigma\tau$ and the inverse σ^{-1} are also permutations of X.

Proof. The composition of two functions both of which are 1–1 and onto is also 1–1 and onto. Hence, $\sigma\tau$ is a permutation of X. Moreover, since $\sigma:X \to X$ is 1–1 and onto, its inverse $\sigma^{-1}:X \to X$ exists and is also 1–1 and onto and hence is a permutation of X. ∎

Let $\mathrm{Sym}(X)$ stand for the set of all permutations of X. Then it follows from Proposition 1 and the fact that function composition is an associative operation that $\mathrm{Sym}(X)$ is a group under function composition. The identity element is the identity function $1_X:X \to X$, which is clearly a permutation of X. The group $\mathrm{Sym}(X)$ is called *symmetric group on* X. For convenience, we let $\mathrm{Sym}(n)$ stand for the symmetric group on the set $\{1, \ldots, n\}$ when n is a positive integer; an element in $\mathrm{Sym}(n)$ is any permutation of the set $\{1, \ldots, n\}$, that is, a 1–1 correspondence of this set with itself, and hence may be thought of simply as a rearrangement of the numbers $1, \ldots, n$.

PROPOSITION 2. $|\mathrm{Sym}(n)| = n!$

Proof. The permutations of the set $\{1, \ldots, n\}$ are just the rearrangements of the n symbols $1, \ldots, n$. Since these n symbols may be rearranged in $n!$ ways, it follows that $|\mathrm{Sym}(n)| = n!$ ∎

The symmetric groups $\mathrm{Sym}(n)$ have their own system of notation. If σ is a permutation of the set $\{1, \ldots, n\}$ and $\sigma(1) = i_1, \sigma(2) = i_2, \ldots, \sigma(n) = i_n$, then we display σ by means of the matrix

$$\begin{pmatrix} 1 & 2 & \cdots & n \\ i_1 & i_2 & \cdots & i_n \end{pmatrix}.$$

For example, $\sigma = \begin{pmatrix} 1 & 2 & 3 \\ 2 & 1 & 3 \end{pmatrix}$ is a typical element in $\mathrm{Sym}(3)$; σ is the permutation of the set $\{1, 2, 3\}$ defined by $\sigma(1) = 2$, $\sigma(2) = 1$, and $\sigma(3) = 3$.

Now, the operation on the group $\mathrm{Sym}(n)$ is function composition. Thus, if σ and τ are elements in $\mathrm{Sym}(n)$ and i is a number in the set $\{1, \ldots, n\}$, then $(\sigma\tau)(i) = \sigma(\tau(i))$. In matrix notation, this means that the matrix for τ acts first on i to give $\tau(i)$, and then the matrix for σ acts on $\tau(i)$ to give $\sigma(\tau(i))$. For example,

$$\begin{pmatrix} 1 & 2 & 3 \\ 2 & 1 & 3 \end{pmatrix}\begin{pmatrix} 1 & 2 & 3 \\ 3 & 1 & 2 \end{pmatrix} = \begin{matrix} 1 \xrightarrow{\text{right matrix}} 3 \xrightarrow{\text{left matrix}} 3 \\ 2 \longmapsto 1 \longmapsto 2 \\ 3 \longmapsto 2 \longmapsto 1 \end{matrix} = \begin{pmatrix} 1 & 2 & 3 \\ 3 & 2 & 1 \end{pmatrix}$$

and

$$\begin{pmatrix} 1 & 2 & 3 & 4 & 5 \\ 1 & 4 & 5 & 2 & 3 \end{pmatrix}\begin{pmatrix} 1 & 2 & 3 & 4 & 5 \\ 3 & 1 & 2 & 5 & 4 \end{pmatrix} = \begin{pmatrix} 1 & 2 & 3 & 4 & 5 \\ 5 & 1 & 4 & 3 & 2 \end{pmatrix}.$$

This notation makes it easy to find the inverse of a permutation; simply read from the bottom row to the top row. For example, $\begin{pmatrix} 1 & 2 & 3 & 4 & 5 \\ 3 & 1 & 2 & 5 & 4 \end{pmatrix}^{-1} = \begin{pmatrix} 1 & 2 & 3 & 4 & 5 \\ 2 & 3 & 1 & 5 & 4 \end{pmatrix}$.

EXAMPLE 1. The symmetric group Sym(3)

The symmetric group Sym(3) consists of all permutations of the set $\{1, 2, 3\}$. Since $|\text{Sym}(3)| = 3! = 6$, there are six such permutations:

$$1 = \begin{pmatrix} 1 & 2 & 3 \\ 1 & 2 & 3 \end{pmatrix}, \qquad \sigma = \begin{pmatrix} 1 & 2 & 3 \\ 2 & 3 & 1 \end{pmatrix}, \qquad \sigma^2 = \begin{pmatrix} 1 & 2 & 3 \\ 3 & 1 & 2 \end{pmatrix},$$

$$\tau = \begin{pmatrix} 1 & 2 & 3 \\ 1 & 3 & 2 \end{pmatrix}, \qquad \sigma\tau = \begin{pmatrix} 1 & 2 & 3 \\ 2 & 1 & 3 \end{pmatrix}, \qquad \tau\sigma = \begin{pmatrix} 1 & 2 & 3 \\ 3 & 2 & 1 \end{pmatrix}.$$

It is easily verified by direct calculation that $\sigma^3 = \tau^2 = 1$ and $\tau\sigma\tau^{-1} = \sigma^2$, and hence $|\sigma| = 3$ and $|\tau| = 2$. Thus, Sym(3) is a group of order 6 generated by two elements, σ and τ, of orders 3 and 2, respectively, and related to each other by the equations $\sigma^3 = \tau^2 = 1$ and $\tau\sigma\tau^{-1} = \sigma^2$. The similarity between the group Sym(3) and the group T of the triangle should be apparent at this point. Indeed, if we label the vertices of the triangle as indicated to the upper right, then each congruence motion of the triangle permutes the vertices of the triangle and is uniquely determined by its permutation of the vertices, and every permutation of the vertices in fact corresponds to some congruence motion of the triangle. Moreover, the mapping $f : T \to$ Sym(3) that assigns to each motion in T its permutation on the vertices is an isomorphism of these two groups. Hence, Sym(3) $\cong T$.

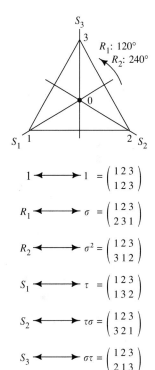

$$1 \longleftrightarrow 1 = \begin{pmatrix} 1 & 2 & 3 \\ 1 & 2 & 3 \end{pmatrix}$$

$$R_1 \longleftrightarrow \sigma = \begin{pmatrix} 1 & 2 & 3 \\ 2 & 3 & 1 \end{pmatrix}$$

$$R_2 \longleftrightarrow \sigma^2 = \begin{pmatrix} 1 & 2 & 3 \\ 3 & 1 & 2 \end{pmatrix}$$

$$S_1 \longleftrightarrow \tau = \begin{pmatrix} 1 & 2 & 3 \\ 1 & 3 & 2 \end{pmatrix}$$

$$S_2 \longleftrightarrow \tau\sigma = \begin{pmatrix} 1 & 2 & 3 \\ 3 & 2 & 1 \end{pmatrix}$$

$$S_3 \longleftrightarrow \sigma\tau = \begin{pmatrix} 1 & 2 & 3 \\ 2 & 1 & 3 \end{pmatrix}$$

EXAMPLE 2

Let V be a real vector space. Then every element in the general linear group $GL(V)$ is a nonsingular linear transformation of V and, as such, is a 1–1 mapping of V onto V and hence is a permutation of V. Thus, $GL(V)$ is a subset of Sym(V). Since $GL(V)$ is already a group under function composition, it follows that it is a subgroup of Sym(V). We may think of $GL(V)$ as the subgroup of Sym(V) consisting of those permutations of V that are linear transformations.

Exercises

1. Write out the elements of the group Sym(4) and find the order of each element.

2. Calculate the following products and inverses:

(a) $\begin{pmatrix} 1 & 2 & 3 & 4 & 5 \\ 3 & 1 & 4 & 2 & 5 \end{pmatrix}\begin{pmatrix} 1 & 2 & 3 & 4 & 5 \\ 1 & 4 & 2 & 3 & 5 \end{pmatrix}$

(b) $\begin{pmatrix} 1 & 2 & 3 & 4 & 5 & 6 & 7 \\ 4 & 7 & 2 & 1 & 3 & 5 & 6 \end{pmatrix}\begin{pmatrix} 1 & 2 & 3 & 4 & 5 & 6 & 7 \\ 4 & 3 & 5 & 1 & 6 & 7 & 2 \end{pmatrix}$

(c) $\begin{pmatrix} 1 & 2 & 3 & 4 & 5 & 6 \\ 6 & 3 & 2 & 4 & 1 & 5 \end{pmatrix}^{-1}$

(d) $\begin{pmatrix} 1 & 2 & 3 & 4 & 5 & 6 \\ 4 & 2 & 6 & 5 & 1 & 3 \end{pmatrix}^{-1}$

(e) $\left[\begin{pmatrix} 1 & 2 & 3 & 4 \\ 3 & 4 & 1 & 2 \end{pmatrix}\begin{pmatrix} 1 & 2 & 3 & 4 \\ 2 & 4 & 3 & 1 \end{pmatrix}\right]^{-1}$

(f) $\begin{pmatrix} 1 & 2 & 3 & 4 \\ 3 & 4 & 2 & 1 \end{pmatrix}^{-1}$

3. Find the order of the following permutations:

(a) $\begin{pmatrix} 1 & 2 & 3 & 4 & 5 \\ 3 & 1 & 4 & 2 & 5 \end{pmatrix}$ (b) $\begin{pmatrix} 1 & 2 & 3 & 4 & 5 & 6 \\ 4 & 6 & 5 & 1 & 3 & 2 \end{pmatrix}$

(c) $\begin{pmatrix} 1 & 2 & 3 & 4 \\ 4 & 2 & 1 & 3 \end{pmatrix}\begin{pmatrix} 1 & 2 & 3 & 4 \\ 1 & 4 & 2 & 3 \end{pmatrix}$

4. Let D_4 stand for the group of the square. Then every congruence motion in D_4 defines a permutation of the four vertices of the square.
   (a) Find the permutations of the vertices as labeled in the marginal figure corresponding to each element in D_4. If H stands for this set of permutations, show that $H \leq \text{Sym}(4)$. Is $H \cong D_4$, the group of the square? Is there a permutation of the vertices that does not correspond to a congruence motion of the square?
   (b) Find the permutations of the vertices as labeled in the marginal figure corresponding to each element in D_4. Let K stand for this set of permutations. Show that $K \leq \text{Sym}(4)$. Is $K \cong D_4$? Does $K = H$?
   (c) Let $\sigma = \begin{pmatrix} 1 & 2 & 3 & 4 \\ 2 & 1 & 4 & 3 \end{pmatrix}$. The permutation σ changes the labeling of the square in part (a) to the labeling in part (b). For each element $\tau \in H$, calculate $\sigma\tau\sigma^{-1}$ and show that $\sigma\tau\sigma^{-1} \in K$.
   (d) Define the function $f: H \to K$ by setting $f(\tau) = \sigma\tau\sigma^{-1}$ for every element $\tau \in H$. Show that f is a group-isomorphism.

5. (a) Show that Sym(1) and Sym(2) are abelian groups.
   (b) If $n \geq 3$, construct two permutations σ and τ in Sym(n) such that $\sigma\tau \neq \tau\sigma$. Thus, for $n \geq 3$, Sym(n) is a nonabelian group.
   (c) If $n \geq 3$, show that the center of Sym(n) contains only the identity permutation (see Exercise 15, Section 3, Chapter 2). Thus, for $n \geq 3$, $Z(\text{Sym}(n)) = \{1\}$.

6. Which of the following functions are permutations? If the function is a permutation, determine whether or not it has finite order.

(a) $f: \mathbb{Z} \to \mathbb{Z}, f(n) = n + 1$ (e) $f: \mathbb{R} \to \mathbb{R}, f(x) = x^4 + x + 1$
(b) $f: \mathbb{N} \to \mathbb{N}, f(n) = n + 1$ (f) $f: \mathbb{R}^2 \to \mathbb{R}^2, f(x, y) = (x + y, x - y)$
(c) $f: \mathbb{Z} \to \mathbb{Z}, f(n) = 2n$ (g) $f: \mathbb{Z}_4 \to \mathbb{Z}_4, f([s]) = 2[s]$
(d) $f: \mathbb{R} \to \mathbb{R}, f(x) = x^3$ (h) $f: \mathbb{Z}_4 \to \mathbb{Z}_4, f([s]) = 3[s]$

7. Let $f: X \to X$ be a function on a set X. If n is a positive integer, let f^n stand for the composition of f with itself n times. Suppose that $f^n = 1_X$ for some n. Show that f is a permutation of X.

8. Let X be a set and let $\mathscr{P}(X)$ stand for the power set of X. Define the function $f: \mathscr{P}(X) \to \mathscr{P}(X)$ by setting $f(A) = X - A$, the complement of A in X, for each subset A of X. Show that f is a permutation of $\mathscr{P}(X)$. Does f have finite order?

9. Let $f: X \to Y$ be a 1–1 correspondence between sets X and Y. Show that $\mathrm{Sym}(X) \cong \mathrm{Sym}(Y)$.

10. Let i be any number in the set $\{1, \ldots, n\}$ and let $\mathrm{Sym}(n)_i = \{\sigma \in \mathrm{Sym}(n) \mid \sigma(i) = i\}$. Show that $\mathrm{Sym}(n)_i \cong \mathrm{Sym}(n - 1)$.

11. Let G be a group. We say that a group K may be *embedded* in G if there is a subgroup H of G such that $K \cong H$. In this case, any isomorphism $f: K \to H$ is called an *embedding* of K in G.
   (a) Show that the group T of the triangle may be embedded in $\mathrm{Sym}(3)$. How are the embeddings obtained?
   (b) Show that the group T of the triangle may be embedded in $\mathrm{Mat}_2^*(\mathbb{R})$. How are the embeddings obtained?
   (c) Show that the group D_4 of the square may be embedded in both $\mathrm{Sym}(4)$ and $\mathrm{Mat}_2^*(\mathbb{R})$. How are these embeddings obtained?
   (d) Is it possible to embed the cyclic group $\mathbb{Z}_4$ in the group $\mathrm{Sym}(3)$?
   (e) Is it possible to embed the cyclic group $\mathbb{Z}_4$ in the group $\mathrm{Sym}(4)$?

12. We have represented congruence motions of a triangle and a square by means of permutations of the vertices by first labeling the vertices in some manner. The purpose of this exercise is to formalize the concept of labeling and describe how the representation of a given congruence motion changes when the labeling changes. The concept of labeling is similar to the idea of a coordinate system for a vector space; indeed, a coordinate system may be thought of as a labeling of the vectors in the vector space. Let X be a finite set containing n elements. A *labeling* of X is any 1–1 correspondence $f: X \to \{1, \ldots, n\}$.
   (a) Let f be a labeling of the set X. Define $f^*: \mathrm{Sym}(X) \to \mathrm{Sym}(n)$ by setting $f^*(\pi) = f\pi f^{-1}$ for every element $\pi \in \mathrm{Sym}(X)$, as illustrated in the diagram here. Show that f^* is a group-isomorphism.

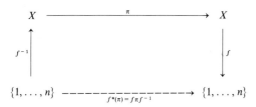

(b) Using part (a), show that if X and Y are finite sets containing the same number of elements, then $\text{Sym}(X) \cong \text{Sym}(Y)$.

(c) Let f and g be labelings of X and let $\sigma = gf^{-1}$. Show that $\sigma \in \text{Sym}(n)$ and that $\sigma f^*(\pi)\sigma^{-1} = g^*(\pi)$ for every element $\pi \in \text{Sym}(X)$.

(d) Let f be a labeling of X. If $\sigma \in \text{Sym}(n)$, show that the composition σf is also a labeling of X. Conversely, if g is any labeling of X, show that $g = \sigma f$ for some element $\sigma \in \text{Sym}(n)$.

(e) We say that two permutations σ and σ' in $\text{Sym}(n)$ *represent* the same permutation π of X if there are labelings f and g of X such that $\sigma = f^*(\pi)$ and $\sigma' = g^*(\pi)$. Show that σ and σ' represent the same permutation of X if and only if there is some element $\tau \in \text{Sym}(n)$ such that $\sigma' = \tau\sigma\tau^{-1}$.

(f) Interpret Exercise 4 of this section from the point of view of labelings. In particular, explain how the permutation σ in part (c) in Exercise 4 is obtained.

2. CYCLE DECOMPOSITION AND CYCLE STRUCTURE

In this section we turn our attention to permutations of finite sets. Our goal is to identify a special type of permutation called a cycle and show that every permutation of a finite set may be factored as a product of cycles. This factorization, or cycle decomposition, shows that cycles are the basic building blocks from which all permutations are constructed, and it is important because the cycles of a given permutation are essentially unique.

Definition

Let σ be a permutation in $\text{Sym}(n)$. For each number i in the set $\{1, \ldots, n\}$, the *σ-orbit* of i is the set $O_\sigma(i) = \{i, \sigma(i), \sigma^2(i), \ldots\}$ of images of i under σ. The sets $O_\sigma(1), O_\sigma(2), \ldots, O_\sigma(n)$ are called the *orbits* of σ.

For example, if $\sigma = \left(\begin{smallmatrix} 1 & 2 & 3 & 4 & 5 \\ 4 & 5 & 1 & 3 & 2 \end{smallmatrix}\right)$, then the orbits of σ are $O_\sigma(1) = \{1, 4, 3\}$ $= O_\sigma(4) = O_\sigma(3)$ and $O_\sigma(2) = \{2, 5\} = O_\sigma(5)$, while for $\tau = \left(\begin{smallmatrix} 1 & 2 & 3 & 4 & 5 \\ 3 & 2 & 4 & 1 & 5 \end{smallmatrix}\right)$ we find that $O_\sigma(1) = \{1, 3, 4\} = O_\sigma(3) = O_\sigma(4)$, $O_\sigma(2) = \{2\}$, and $O_\sigma(5) = \{5\}$.

PROPOSITION 1. Let σ be a permutation in $\text{Sym}(n)$. Then the orbits of σ partition the set $\{1, \ldots, n\}$ into disjoint subsets.

Proof.   Define a relation $\sim$ on the set $\{1, \ldots, n\}$ as follows: if $i, j \in \{1, \ldots, n\}$, then $i \sim j$ means that $j = \sigma^s(i)$ for some integer $s \geq 0$. Then $\sim$ is an equivalence relation on the set I. To show this, we must verify that $\sim$ is reflexive, symmetric, and transitive.

(A) Reflexivity; that is, $i \sim i$ for every $i \in I$. This is true since we may choose $s = 0$ to get $\sigma^0(i) = 1(i) = i$.

(B) Symmetry; that is, if $i \sim j$ for some $i, j \in I$, then $j \sim i$. This is true since, if $\sigma^s(i) = j$ for some integer s, then $\sigma^{-s}(j) = i$. Since σ has finite order, $\sigma^{-s} = \sigma^t$ for some integer $t \geq 0$. Hence, $j \sim i$.

(C) Transitivity; that is, if $i \sim j$ and $j \sim k$ for some $i, j, k \in I$, then $i \sim k$. This is true since, if $\sigma^s(i) = j$ and $\sigma^t(j) = k$ for some integers $s \geq 0$ and $t \geq 0$, then $\sigma^{s+t}(i) = \sigma^t(\sigma^s(i)) = \sigma^t(j) = k$.

Thus, the relation $\sim$ satisfies the three requirements for an equivalence relation. Since the equivalence class of a number i is the orbit $O_\sigma(i)$, it follows that the orbits of σ partition the set $\{1, \ldots, n\}$ into disjoint subsets.   ∎

Note that the σ-orbit of a number i is trivial, that is, it contains only the number i, if and only if σ fixes i; that is, $O_\sigma(i) = \{i\}$ if and only if $\sigma(i) = i$. The number of trivial σ-orbits is therefore the same as the number of fixed points of the permutation σ. A permutation that has exactly one nontrivial orbit is called a *cycle*, in which case the number of elements in the nontrivial orbit is called the *length of the cycle*. If σ is a cycle of length s, σ is called an *s-cycle*; in this case, we write $\sigma = (i_1 \, i_2 \cdots i_s)$ to mean that the set $\{i_1, i_2, \ldots, i_s\}$ is the nontrivial orbit of σ and that $\sigma(i_1) = i_2, \sigma(i_2) = i_3, \ldots,$ and $\sigma(i_s) = i_1$. As a permutation of the set $\{1, \ldots, n\}$, the cycle σ cyclically permutes the numbers $i_1, i_2, \ldots, i_s$, as shown in the marginal diagram, while the remaining numbers in the set $\{1, \ldots, n\}$, if any, are fixed. For example, the permutation $\sigma = \left(\begin{smallmatrix} 1 & 2 & 3 & 4 \\ 4 & 2 & 1 & 3 \end{smallmatrix}\right)$ has one nontrivial orbit, namely, $O_\sigma(1) = \{1, 4, 3\}$. Hence, σ is a 3-cycle and we write $\sigma = (1 \, 4 \, 3)$. On the other hand, if $\tau = (2 \, 5 \, 3 \, 4)$, then τ is a 4-cycle in Sym(5) and may be written as $\tau = \left(\begin{smallmatrix} 1 & 2 & 3 & 4 & 5 \\ 1 & 5 & 4 & 2 & 3 \end{smallmatrix}\right)$. Notice that τ may also be regarded as a 4-cycle in Sym(6) or in any of the groups Sym(n) for $n \geq 6$; as such, it fixes the numbers $1, 6, \ldots, n$. Finally, two cycles are said to be *disjoint* if their nontrivial orbits have no elements in common; for example, (1 2) and (3 4) are disjoint cycles of length 2 in Sym(4).

cyclically permuted

$\sigma(j) = j \quad j \notin \{i_1, \ldots, i_s\}$

Now, multiplication of permutations is, in general, a noncommutative operation; for example, in the symmetric group Sym(3) discussed in Example 1, Section 1, we found that

$$\sigma\tau = \begin{pmatrix} 1 & 2 & 3 \\ 2 & 1 & 3 \end{pmatrix} \quad \text{while} \quad \tau\sigma = \begin{pmatrix} 1 & 2 & 3 \\ 3 & 2 & 1 \end{pmatrix}.$$

For disjoint cycles, however, the situation is different.

PROPOSITION 2. Let σ and τ be disjoint cycles in Sym(n). Then $\sigma\tau = \tau\sigma$. Thus, disjoint cycles commute.

Proof. Let I and J stand for the nontrivial orbits of σ and τ, respectively. Then I and J are disjoint and hence, for any number $i \in \{1, \ldots, n\}$,

$$(\sigma\tau)(i) = \begin{cases} \sigma(i) & \text{if } i \in I \text{ since } i \notin J \\ \tau(i) & \text{if } i \in J \text{ since } \tau(i) \notin I \\ i & \text{if } i \notin I \cup J \text{ since } i \text{ is fixed by both } \sigma \text{ and } \tau. \end{cases}$$

Similarly,

$$(\tau\sigma)(i) = \begin{cases} \sigma(i) & \text{if } i \in I \\ \tau(i) & \text{if } i \in J \\ i & \text{if } i \notin I \cup J. \end{cases}$$

Therefore, $\sigma\tau = \tau\sigma$. ∎

We now claim that every permutation may be written as a product of disjoint cycles. For example, the permutation $\sigma = \left(\begin{smallmatrix} 1 & 2 & 3 & 4 & 5 \\ 3 & 5 & 4 & 1 & 2 \end{smallmatrix}\right)$ has two nontrivial orbits, $\{1, 3, 4\}$ and $\{2, 5\}$, whose corresponding cycles are (1 3 4) and (2 5), and we find that σ is the product of these two cycles: $(1\ 3\ 4)(2\ 5) = \left(\begin{smallmatrix} 1 & 2 & 3 & 4 & 5 \\ 3 & 5 & 4 & 1 & 2 \end{smallmatrix}\right)$. Moreover, since (1 3 4) and (2 5) are disjoint, they commute; hence, $\sigma = (1\ 3\ 4)(2\ 5) = (2\ 5)(1\ 3\ 4)$. Thus, σ is factored into a product of two disjoint cycles. Let us now carry out this procedure in general.

PROPOSITION 3. Every permutation in Sym(n) except the identity permutation may be written as a product of disjoint cycles, and this decomposition is unique except for the order in which the cycles appear.

Proof.   Let $\sigma \in \text{Sym}(n)$, $\sigma \neq 1$. Then σ has at least one nontrivial orbit. Let $I_1, \ldots, I_t$ stand for the nontrivial orbits of σ. For $i = 1, \ldots, t$, let $\sigma_i = \sigma_{I_i}$ be the cycle in Sym(n) corresponding to the orbit I_i. We claim that $\sigma = \sigma_1 \cdots \sigma_t$. To show this, we must show that the two functions σ and $\sigma_1 \cdots \sigma_t$ have the same value on each number in the set $\{1, \ldots, n\}$. To this end, let k be any number in $\{1, \ldots, n\}$. If k does not belong to any of the orbits $I_1, \ldots, I_t$, then k is fixed by σ and therefore $\sigma(k) = k = (\sigma_1 \cdots \sigma_t)(k)$. Otherwise, k belongs to exactly one on these orbits, say I_j, and hence $(\sigma_1 \cdots \sigma_t)(k) = \sigma_j(k) = \sigma(k)$. It now follows that $\sigma = \sigma_1 \cdots \sigma_t$. To prove the uniqueness of this decomposition, suppose that $\sigma = \tau_1 \cdots \tau_s$ is any representation of σ as a product of disjoint cycles $\tau_1, \ldots, \tau_s$. We must show that $\{\tau_1, \ldots, \tau_s\} = \{\sigma_1, \ldots, \sigma_t\}$. Let T_1 stand for the nontrivial orbit of τ_1 and let $i \in T_1$. Then $\sigma(i) = (\tau_1 \cdots \tau_s)(i) = \tau_1(i)$ since the orbits of $\tau_1, \ldots, \tau_s$ are disjoint and each τ_j is the identity off of its orbit. It follows that $T_1 = \{i, \tau_1(i), \ldots\} = \{i, \sigma(i), \ldots\} = O_\sigma(i)$. Hence, $T_1 = I$, where I is the σ-orbit of i, and therefore $\tau_1 = \sigma_I$. Similarly, each of $\tau_2, \ldots, \tau_s$ is equal to some σ_j. Hence, $\{\tau_1, \ldots, \tau_s\} \subseteq \{\sigma_1, \ldots, \sigma_t\}$. Applying this same argument to each of the orbits $I_1, \ldots, I_t$, it follows that $\{\sigma_1, \ldots, \sigma_t\} \subseteq \{\tau_1, \ldots, \tau_s\}$. Therefore, the two sets of cycles are the same, and hence we conclude that the cycle decomposition of σ is unique except for the order in which the cycles appear. ∎

The factorization of a permutation into a product of cycles is called the *cycle decomposition* of the permutation. For example, if $\sigma = \left(\begin{smallmatrix} 1 & 2 & 3 & 4 & 5 & 6 \\ 4 & 3 & 2 & 5 & 6 & 1 \end{smallmatrix}\right)$, then the cycles of σ are (1 4 5 6) and (2 3); hence, $\sigma = (1\ 4\ 5\ 6)(2\ 3) = (2\ 3)(1\ 4\ 5\ 6)$ is the cycle decomposition of σ. Using the cycle decomposition, we can now find the order of σ. For the cycles (1 4 5 6) and (2 3) are disjoint and hence commute. Since $|(1\ 4\ 5\ 6)| = 4$ and $|(2\ 3)| = 2$, it follows that σ^4 $= (1\ 4\ 5\ 6)^4(2\ 3)^4 = 1$ and hence $|\sigma| = 4$. Similarly, we find that

$$\sigma = \begin{pmatrix} 1 & 2 & 3 & 4 & 5 & 6 \\ 4 & 5 & 6 & 1 & 2 & 3 \end{pmatrix} = (1\ 4)(2\ 5)(3\ 6), \quad |\sigma| = 2;$$

$$\sigma = \begin{pmatrix} 1 & 2 & 3 & 4 & 5 & 6 & 7 \\ 3 & 6 & 2 & 7 & 4 & 1 & 5 \end{pmatrix} = (1\ 3\ 2\ 6)(4\ 7\ 5), \quad |\sigma| = 12;$$

$$\sigma = \begin{pmatrix} 1 & 2 & 3 & 4 & 5 & 6 & 7 & 8 \\ 3 & 7 & 6 & 5 & 2 & 8 & 4 & 1 \end{pmatrix} = (1\ 3\ 6\ 8)(2\ 7\ 4\ 5), \quad |\sigma| = 4.$$

In the exercises, we ask the reader to show, in general, that the order of any permutation is equal to the least common multiple of the lengths of its cycles.

Let $\sigma \in \text{Sym}(n)$, $\sigma \neq 1$, and let $\sigma = \sigma_1 \cdots \sigma_s$ be the cycle decomposition of σ. Since the cycles that occur in this decomposition are unique, except for order, it is meaningful to speak of the number of 2-cycles, 3-cycles, . . . , n-cycles that occur among $\sigma_1, \ldots, \sigma_s$. It is also convenient, when counting cycles, to refer to any fixed point of σ as a 1-cycle; 1-cycles correspond to the trivial orbits of σ. The *cycle structure* of σ is then defined to be the n-tuple $(c_1, \ldots, c_n)$, where c_s is the number of s-cycles in the cycle decomposition of σ for $s = 1, \ldots, n$. For example, the cycle structure of the permutation $\begin{pmatrix} 1 & 2 & 3 & 4 & 5 & 6 \\ 4 & 5 & 6 & 1 & 2 & 3 \end{pmatrix} = (1\ 4)(2\ 5)(3\ 6)$ is $(0, 3, 0, 0, 0, 0)$, while the cycle structure of $\begin{pmatrix} 1 & 2 & 3 & 4 & 5 & 6 & 7 \\ 2 & 1 & 3 & 6 & 4 & 5 & 7 \end{pmatrix}$ is $(2, 1, 1, 0, 0, 0, 0)$. The cycle structure of the identity permutation in $\text{Sym}(n)$ is $(n, 0, 0, \ldots, 0)$.

Exercises

1. Calculate the following products.
   (a) $(1\ 2\ 3\ 4\ 5)(1\ 4)(3\ 5)$
   (b) $[(1\ 5)(2\ 3)]^2$
   (c) $(1\ 5\ 6\ 4\ 2)^{-1}$
   (d) $(2\ 1\ 5\ 3)(4\ 3\ 1)$
   (e) $(1\ 3\ 6)(2\ 1\ 4)(4\ 2\ 3\ 5)$
   (f) $(1\ 2)(1\ 3)(1\ 4)(1\ 5)$
   (g) $(1\ 5)(1\ 4)(1\ 3)(1\ 2)$
   (h) $(2\ 3\ 4)^{-1}(3\ 5\ 2)^{-1}$

2. Find the cycle decomposition, cycle structure, and order of the following permutations.

   (a) $\begin{pmatrix} 1 & 2 & 3 & 4 & 5 \\ 2 & 3 & 4 & 5 & 1 \end{pmatrix}$ (c) $\begin{pmatrix} 1 & 2 & 3 & 4 & 5 & 6 \\ 5 & 1 & 3 & 2 & 6 & 4 \end{pmatrix}$

   (b) $\begin{pmatrix} 1 & 2 & 3 & 4 & 5 & 6 \\ 5 & 3 & 2 & 4 & 1 & 6 \end{pmatrix}$ (d) $\begin{pmatrix} 1 & 2 & 3 & 4 & 5 & 6 \\ 5 & 6 & 1 & 2 & 3 & 4 \end{pmatrix}$

3. Write out the cycle decomposition, cycle structure, and order of each element in the group $\text{Sym}(4)$.

4. Find two different permutations in Sym(3) that have the same cycle structure.

5. If σ is a cycle in Sym(n), is σ^m a cycle for every integer m?

6. Let σ be a permutation in Sym(n).
   (a) If σ is an s-cycle, show that the order of σ is s.
   (b) In general, let $(c_1, \ldots, c_n)$ be the cycle structure of σ and let $c_{i_1}, \ldots, c_{i_t}$ be those c_i that are not zero. Show that $|\sigma| = \text{lcm}(i_1, \ldots, i_t)$. Thus, the order of a permutation is the least common multiple of the lengths of its cycles.

7. Let $1 \le s \le n$. How many s-cycles are in the group Sym(n)?

8. Let $(c_1, \ldots, c_n)$ be the cycle structure of a permutation in Sym(n). How many permutations in Sym(n) have this same cycle structure?

9. Determine the subgroup of Sym(4) generated by the following permutations.
   (a) (1 3) and (1 2)   (c) (1 2)(3 4) and (1 3)
   (b) (1 3) and (2 4)   (d) (2 3) and (1 3 4)

10. Let σ be a permutation in Sym(n).
    (a) If the cycle structure of σ is $(c_1, \ldots, c_n)$, show that $c_1 + 2c_2 + \cdots + nc_n = n$.
    (b) A *partition* of n is any representation of n as a sum of positive integers that are less than or equal to n. By part (a), every permutation in Sym(n) defines a partition of n by means of its cycle structure:

$$n = \underbrace{(1 + \cdots + 1)}_{c_1} + \underbrace{(2 + \cdots + 2)}_{c_2} + \cdots + \underbrace{(n + \cdots + n)}_{c_n}.$$

For example, if $\sigma = (1\ 4\ 3)(2\ 5) \in$ Sym(6), then the cycle structure of σ is $(1, 1, 1, 0, 0, 0)$ and the corresponding partition of the number 6 is $1 + 2 + 3 = 6$. Find the partitions of the number 4 that correspond to each permutation in Sym(4).

    (c) Let $n = n_1 + \cdots + n_s$ be a partition of n. Construct a permutation σ in Sym(n) such that the partition of n corresponding to σ is $n_1 + \cdots + n_s$.

    (d) Show that two permutations in Sym(n) determine the same partition of n if and only if they have the same cycle structure.

11. Let X be a set and let $P = \{X_\alpha | \alpha \in A\}$ be a partition of X; that is, every element in X belongs to exactly one of the subsets X_α. If $\sigma \in$ Sym(X), then it is not true, in general, that $\sigma(X_\alpha) = X_\alpha$. Let $S_P = \{\sigma \in \text{Sym}(X) | \sigma(X_\alpha) = X_\alpha \text{ for every } X_\alpha \in P\}$.
    (a) Show that $S_P \le$ Sym(X). S_P is called the *partition subgroup* of Sym(X) associated with the partition P.
    (b) Let $X = \{1, 2, 3, 4, 5\}$. Find the partition subgroup associated with the partition $\{\{1, 2\}, \{3, 4, 5\}\}$.
    (c) If $\sigma \in S_P$, let $\sigma|X_\alpha$ stand for the restriction of σ to the subset X_α. Show that $\sigma|X_\alpha \in$ Sym(X_α) for every $X_\alpha \in P$.

(d) Show that $S_P = (1)$ if and only if every subset X_α contains exactly one element.

(e) Show that $S_P = \text{Sym}(X)$ if and only if $P = \{X\}$.

12. Let X be a set and let $P = \{X_\alpha | \alpha \in A\}$ and $Q = \{Y_\beta | \beta \in B\}$ be partitions of X. We say that these two partitions are *equal* if they are equal as sets; that is, every X_α is equal to some Y_β, and vice versa.

(a) Let $\sigma \in \text{Sym}(X)$ and let $\sigma P = \{\sigma X_\alpha | \alpha \in A\}$. Show that σP is a partition of X.

(b) Q is said to be *equivalent* to P if $Q = \sigma P$ for some $\sigma \in \text{Sym}(X)$. Show that this is an equivalence relation on the collection of all partitions of X.

(c) Show that the two partitions $\{\{1, 2\}, \{3, 4, 5\}\}$ and $\{\{2, 3\}, \{1, 4, 5\}\}$ of the set $\{1, 2, 3, 4, 5\}$ are equivalent.

(d) Show that the two partitions $\{\{1, 3\}, \{2, 4, 5\}\}$ and $\{\{2\}, \{3\}, \{1, 4, 5\}\}$ are not equivalent.

(e) Assume that X is a finite set, and let $P = \{X_1, \ldots, X_s\}$ and $Q = \{Y_1, \ldots, Y_t\}$ be partitions of X. Show that P and Q are equivalent if and only if $s = t$ and $|X_1| = |Y_{i_1}|, \ldots, |X_s| = |Y_{i_s}|$ for some permutation $\begin{pmatrix} 1 & 2 & \cdots & s \\ i_1 & i_2 & \cdots & i_s \end{pmatrix}$.

(f) If σ is a permutation in $\text{Sym}(n)$, then the orbits of σ partition the set $\{1, \ldots, n\}$. Show that two permutations have the same cycle structure if and only if the corresponding partitions of $\{1, \ldots, n\}$ are equivalent. Conclude that a given cycle structure completely determines an equivalence class of partitions of X.

13. Let $P = \{X_\alpha | \alpha \in A\}$ and $Q = \{Y_\beta | \beta \in B\}$ be equivalent partitions of a set X. Let $Q = \sigma P$, $\sigma \in \text{Sym}(X)$.

(a) If $Y_\beta = \sigma X_\alpha$ and $\tau X_\alpha = X_\alpha$, show that $\sigma \tau \sigma^{-1} Y_\beta = Y_\beta$.

(b) Let S_P and S_Q stand for the partition subgroups of P and Q, respectively. Define $f : S_P \to S_Q$ by setting $f(\tau) = \sigma \tau \sigma^{-1}$ for every element $\tau \in S_P$. Show that $f(\tau) \in S_Q$ for every $\tau \in S_P$ and that f is a group-isomorphism. Thus, equivalent partitions of X have isomorphic partition subgroups.

14. Let σ be a permutation in $\text{Sym}(n)$ and let S_σ stand for the partition subgroup of $\text{Sym}(n)$ corresponding to the partition of $\{1, \ldots, n\}$ defined by the orbits of σ.

(a) Let I be any orbit of σ. Show that $\tau \in S_\sigma$ if and only if $\tau | I \in \text{Sym}(I)$.

(b) Let $I_1, \ldots, I_t$ stand for the orbits of σ. By part (a), if $\tau \in S_\sigma$, then $\tau | I_j \in \text{Sym}(I_j)$ for $j = 1, \ldots, t$. Define the function $f : S_\sigma \to \text{Sym}(I_1) \times \cdots \times \text{Sym}(I_t)$ by setting $f(\tau) = (\tau | I_1, \ldots, \tau | I_t)$ for every $\tau \in S_\sigma$. Show that f is a group-isomorphism. Thus, $S_\sigma \cong \text{Sym}(I_1) \times \cdots \times \text{Sym}(I_t)$.

(c) Let σ' be a permutation in $\text{Sym}(n)$ that has the same cycle structure as σ. Show that there is some permutation $\tau \in \text{Sym}(n)$ such that $\sigma' = \tau \sigma \tau^{-1}$.

(d) Let $\sigma = \begin{pmatrix} 1 & 2 & 3 & 4 & 5 \\ 2 & 1 & 4 & 5 & 3 \end{pmatrix}$ and $\sigma' = \begin{pmatrix} 1 & 2 & 3 & 4 & 5 \\ 3 & 4 & 5 & 2 & 1 \end{pmatrix}$. Show that σ and σ' have the same cycle structure and find a permutation τ such that $\sigma' = \tau \sigma \tau^{-1}$.

3. TRANSPOSITIONS AND THE ALTERNATING SUBGROUP

In the previous section, we showed that every permutation of a finite set may be written as the product of disjoint cycles in an essentially unique way. In this section, our goal is to show that every permutation may in fact be written as a product of 2-cycles. While the factorization of a permutation as a product of 2-cycles is no longer unique, the number of such factors is always even or always odd and hence we may distinguish between permutations according to whether their factorization involves an even or odd number of 2-cycles. We then discuss some consequences of this result.

Definition

A 2-cycle is called a *transposition*.

PROPOSITION 1. If $n \geq 2$, every permutation in $\text{Sym}(n)$ may be written as a product of transpositions.

Proof. Let $\sigma \in \text{Sym}(n)$. If $\sigma = 1$, write $\sigma = (1\ 2)(1\ 2)$. If $\sigma \neq 1$ but is a cycle, let $\sigma = (i_1\ i_2 \cdots i_s)$. Then we find by direct calculation that $\sigma = (i_1\ i_s)(i_1\ i_{s-1}) \cdots (i_1\ i_2)$. Hence, every cycle is a product of transpositions. Finally, if σ is any permutation in $\text{Sym}(n)$, first factor σ into a product of cycles and then factor each cycle as above ∎

We refer to the factorization of a permutation into a product of transpositions as the *transposition decomposition* of the permutation. For example, following the procedure outlined in the proof of Proposition 1, we find that

$$(1\ 4\ 3\ 2) = (1\ 2)(1\ 3)(1\ 4)$$

and

$$\begin{pmatrix} 1 & 2 & 3 & 4 & 5 & 6 & 7 \\ 5 & 1 & 2 & 7 & 3 & 4 & 6 \end{pmatrix} = (1\ 5\ 3\ 2)(4\ 7\ 6)$$

$$= (1\ 2)(1\ 3)(1\ 5)(4\ 6)(4\ 7).$$

Notice that we may also write $(1\ 4\ 3\ 2) = (2\ 1\ 4\ 3) = (2\ 3)(2\ 4)(2\ 1)$ or $(1\ 4\ 3\ 2) = (1\ 2)(1\ 3)(1\ 4)(1\ 2)(1\ 2)$, which are different factorizations from the previous one. Thus, unlike the cycle decomposition, the transposition decomposition of a permutation is not unique. Note also that the transpositions occurring in such a factorization are not necessarily disjoint.

EXAMPLE 1

Let us use the transposition decomposition of permutations to show that every rotation in the group of the triangle is a product of symmetries. Recall that a symmetry of the real plane is a reflection of the plane about a line passing through the origin. Referring to the figure in Section 1, we see that the group T of the triangle contains three symmetries:

$$S_1 = \begin{pmatrix} 1 & 2 & 3 \\ 1 & 3 & 2 \end{pmatrix} = (2\ 3), \qquad S_2 = \begin{pmatrix} 1 & 2 & 3 \\ 3 & 2 & 1 \end{pmatrix} = (1\ 3), \qquad S_8 = \begin{pmatrix} 1 & 2 & 3 \\ 2 & 1 & 3 \end{pmatrix} = (1\ 2),$$

each of which is a transposition of the vertices of the triangle. There are three remaining elements in the group T, namely, the rotations 1, R, and R^2. Taking a transposition decomposition for each of these rotations, we find that

$$1 = \begin{pmatrix} 1 & 2 & 3 \\ 1 & 2 & 3 \end{pmatrix} = (2\ 3)(2\ 3) = S_1 S_1$$

$$R = \begin{pmatrix} 1 & 2 & 3 \\ 2 & 3 & 1 \end{pmatrix} = (1\ 2\ 3) = (1\ 3)(1\ 2) = S_2 S_3$$

$$R^2 = \begin{pmatrix} 1 & 2 & 3 \\ 3 & 1 & 2 \end{pmatrix} = (1\ 3\ 2) = (1\ 2)(1\ 3) = S_3 S_2.$$

Thus, the rotations in the group of the triangle may be written as a product of symmetries.

We now have two methods for factoring a permutation: the cycle decomposition, which factors the permutation into a product of disjoint cycles that are uniquely determined, and the transposition decomposition, which factors the permutation into a product of transpositions that are not unique. The number of transpositions occurring in a transposition decomposition is not uniquely determined, as the previous examples show. Nevertheless, the parity of this number—that is, its evenness or oddness—is uniquely determined. To show this we will use some basic properties of determinants. Recall that the determinant of a square matrix may be defined without making use of permutations. Let M stand for any $n \times n$ matrix with real number entries. If $\sigma \in \text{Sym}(n)$, let σM stand for the matrix obtained from M by permuting the rows of M according to σ: row 1 in M becomes row $\sigma(1)$ in $\sigma M, \ldots$, and row n in M becomes row $\sigma(n)$ in σM. For example, if $\sigma = (1\ 2\ 3)$ and

$$M = \begin{pmatrix} 1 & 2 & 3 \\ 4 & 5 & 6 \\ 7 & 8 & 9 \end{pmatrix},$$

then

$$\sigma M = \begin{pmatrix} 7 & 8 & 9 \\ 1 & 2 & 3 \\ 4 & 5 & 6 \end{pmatrix}.$$

Let us now show that the parity of any transposition decomposition of a permutation is always the same.

PROPOSITION 2. Let σ be a permutation in Sym(n). If σ may be written as a product of an even(odd) number of transpositions, then every decomposition of σ as a product of transpositions will involve an even(odd) number of transpositions.

Proof.   First observe that if τ is any transposition in Sym(n) and M is any $n \times n$ real matrix, then $\det(\tau M) = -\det(M)$ since τ simply interchanges two rows of M. Now, let $\sigma = \tau_1 \cdots \tau_s = \tau'_1 \cdots \tau'_t$ be transposition decompositions of σ. Then $\tau_1 \cdots \tau_s \tau'_t \cdots \tau'_1 = 1$ since each τ'_i is its own inverse. It follows, therefore, that if I_n is the $n \times n$ identity matrix, then $1 = \det(I_n) = \det[\tau_1(\cdots \tau_s(\tau'_t(\cdots (\tau'_1 I_n) \cdots)) \cdots)] = (-1)^{s+t}$. Hence, $s + t$ is even, and therefore s and t are both even or both odd, as required.   ∎

In view of this result, we may separate permutations into two classes according to the parity of their transposition decompositions.

For example, $\sigma = (\begin{smallmatrix} 1 & 2 & 3 & 4 & 5 & 6 & 7 \\ 5 & 1 & 2 & 7 & 3 & 4 & 6 \end{smallmatrix}) = (1\ 5\ 3\ 2)(4\ 7\ 6) = (1\ 2)(1\ 3)(1\ 5)(4\ 6)(4\ 7)$ is a transposition decomposition of σ involving an odd number of transpositions. Hence, σ is an odd permutation. The permutation $(1\ 5\ 3)(2\ 6\ 4)$, on the other hand, is an even permutation since $(1\ 5\ 3)(2\ 6\ 4) = (1\ 3)(1\ 5)(2\ 4)(2\ 6)$, which is a transposition decomposition involving an even number of transpositions. Observe that the identity permutation is always an even permutation since $1 = (1\ 2)(1\ 2)$.

PROPOSITION 3. Let Alt(n) stand for the set of even permutations in Sym(n). Then Alt(n) is a subgroup of Sym(n) and has order $\frac{1}{2}n!$.

Proof.   As we noted above, the identity permutation is even. Hence, $1 \in$ Alt(n). Now, let σ and σ' be even permutations and let $\sigma = \tau_1 \cdots \tau_s$ and $\sigma' = \tau'_1 \cdots \tau'_t$ be transposition decompositions of σ and σ', where s and t are even. Then $\sigma\sigma' = \tau_1 \cdots \tau_s \tau'_1 \cdots \tau'_t$ is a transposition decomposition of $\sigma\sigma'$ having $s + t$ factors. Since $s + t$ is even, $\sigma\sigma' \in$ Alt(n). It now follows that Alt(n) $\leq$ Sym(n) since Sym(n) is a finite group. To show that the order of Alt(n) is $\frac{1}{2}n!$, let τ be any odd permutation in Sym(n), say $\tau = (1\ n)$, and let τ Alt(n) $= \{\tau\sigma \in$ Sym(n)$\,|\,\sigma \in$ Alt(n)$\}$. We claim that τ Alt(n) is precisely the set of odd permutations in Sym(n). For clearly, every permutation in τ Alt(n) is odd. Moreover, if τ' is any odd permutation, then $\tau^{-1}\tau'$ is even and hence $\tau' = \tau(\tau^{-1}\tau') \in \tau$ Alt(n). Therefore, τ Alt(n) is the set of odd permutations in Sym(n). It follows that Alt(n) and τ Alt(n) are disjoint sets of Sym(n) and that Sym(n) = Alt(n) $\cup$ τ Alt(n). Finally, there is a 1–1 correspondence $\sigma \leftrightarrow \tau\sigma$

between the sets Alt(n) and τ Alt(n), and hence these two sets contain the same number of elements. Therefore, $|\text{Alt}(n)| = \frac{1}{2}n!$.   ∎

Definition

The subgroup Alt(n) is called the *alternating subgroup* of Sym(n).

To find the alternating subgroup of Sym(3), for example, we first determine the parity of each permutation in Sym(3):

$$1 = \begin{pmatrix} 1 & 2 & 3 \\ 1 & 2 & 3 \end{pmatrix} = (1\ 2)(1\ 2) \qquad\qquad \text{even}$$

$$\sigma = \begin{pmatrix} 1 & 2 & 3 \\ 2 & 3 & 1 \end{pmatrix} = (1\ 2\ 3) = (1\ 3)(1\ 2) \quad \text{even}$$

$$\sigma^2 = \begin{pmatrix} 1 & 2 & 3 \\ 3 & 1 & 2 \end{pmatrix} = (1\ 3\ 2) = (1\ 2)(1\ 3) \quad \text{even}$$

$$\tau = \begin{pmatrix} 1 & 2 & 3 \\ 1 & 3 & 2 \end{pmatrix} = (2\ 3) \qquad\qquad \text{odd}$$

$$\sigma\tau = \begin{pmatrix} 1 & 2 & 3 \\ 2 & 1 & 3 \end{pmatrix} = (1\ 2) \qquad\qquad \text{odd}$$

$$\tau\sigma = \begin{pmatrix} 1 & 2 & 3 \\ 3 & 2 & 1 \end{pmatrix} = (1\ 3) \qquad\qquad \text{odd}$$

Hence, Alt(3) = $\{1, \sigma, \sigma^2\} = (\sigma) \cong \mathbb{Z}_3$. From a geometrical point of view, this shows that the three rotations of an equilateral triangle define even permutations of the vertices, while the three symmetries define odd permutations of the vertices.

Exercises

1. Find a transposition decomposition for each of the following permutations. In each case determine the parity of the permutation and whether it is even or odd.

   (a) $\begin{pmatrix} 1 & 2 & 3 & 4 & 5 \\ 3 & 4 & 1 & 2 & 5 \end{pmatrix}$ (c) $(2\ 3\ 5\ 1)(6\ 3\ 5\ 7\ 1)$

   (b) $\begin{pmatrix} 1 & 2 & 3 & 4 & 5 & 6 \\ 4 & 1 & 2 & 3 & 5 & 6 \end{pmatrix}^{-1}$ (d) $(2\ 3\ 5\ 1\ 7)^{-1}$

2. Find a transposition decomposition for each permutation in the group Sym(4) and determine the alternating subgroup Alt(4). Is Alt(4) a cyclic group? Is Alt(4) abelian?

3. Does every rotation of a square define an even permutation of the vertices? Is every symmetry of a square an odd permutation of the vertices?

4. Show that the inverse of an even permutation is even and that the inverse of an odd permutation is odd.

5. Let $n \geq 2$.
   (a) Show that every transposition in $\text{Sym}(n)$ may be written as a product of the transpositions $(1\ 2), (1\ 3), \ldots, (1\ n)$.
   (b) Show that the transpositions $(1\ 2), (1\ 3), \ldots, (1\ n)$ generate the group $\text{Sym}(n)$.

6. Let $n \geq 3$.
   (a) Show that every 3-cycle in $\text{Sym}(n)$ lies in $\text{Alt}(n)$.
   (b) Show that every permutation in $\text{Alt}(n)$ is a product of 3-cycles by verifying that $(a\ b)(b\ c) = (a\ b\ c)$ and $(a\ b)(c\ d) = (a\ b)(b\ c)(b\ c)(c\ d)$ $= (a\ b\ c)(b\ c\ d)$ for any distinct numbers a, b, c, and d in the set $\{1, \ldots, n\}$.
   (c) Show that $\text{Alt}(n)$ is generated by the 3-cycles in $\text{Sym}(n)$.

7. Let $H \leq \text{Sym}(n)$.
   (a) Show that either $H \leq \text{Alt}(n)$ or $|H \cap \text{Alt}(n)| = \frac{1}{2}|H|$.
   (b) Show that every subgroup of $\text{Sym}(n)$ that has odd order is a subgroup of $\text{Alt}(n)$.

8. Let σ be a permutation in $\text{Sym}(n)$ and let $(c_1, \ldots, c_n)$ be the cycle structure of σ. Show that σ is an even permutation if and only if $n - (c_1 + \cdots + c_n)$ is an even integer.

9. Let $V = \mathbb{R}^n$ and let $(e_1, \ldots, e_n)$ be the standard coordinate system for V, where e_i is the vector all of whose entries are zero except in position i where the entry is 1. Let $\sigma \in \text{Sym}(n)$. Define the mapping $L_\sigma : V \to V$ by setting $L_\sigma(a_1, \ldots, a_n) = (a_{\sigma(1)}, \ldots, a_{\sigma(n)})$ for all $(a_1, \ldots, a_n) \in \mathbb{R}^n$; that is, if $x = (a_1, \ldots, a_n)$, $L_\sigma(x)$ is just x with its coordinate entries permuted according to σ.
   (a) Show that $L_\sigma(e_i) = e_{\sigma(i)}$ for $i = 1, \ldots, n$.
   (b) Show that the matrix of L_σ relative to the coordinate system $(e_1, \ldots, e_n)$ is σI_n.
   (c) Show that $(\sigma\tau)I_n = (\sigma I_n)(\tau I_n)$ for all $\sigma, \tau \in \text{Sym}(n)$.

10. Matrices of the form σI_n, for $\sigma \in \text{Sym}(n)$, are called *permutation matrices*. Let $\text{Perm}(n)$ stand for the set of all $n \times n$ permutation matrices.
    (a) Show that $\text{Perm}(n) \leq \text{Mat}_n^*(\mathbb{R})$.
    (b) Show that $\text{Perm}(n) \cong \text{Sym}(n)$.
    (c) Show that the group $\text{Sym}(n)$ may be embedded in $\text{Mat}_n^*(\mathbb{R})$.

11. Let $L : \mathbb{R}^3 \to \mathbb{R}^3$ be the linear transformation defined by setting

$$L\begin{pmatrix} x_1 \\ x_2 \\ x_3 \end{pmatrix} = \begin{pmatrix} 3 & 2 & -3 \\ -2 & -1 & 2 \\ 3 & 3 & -2 \end{pmatrix} \begin{pmatrix} x_1 \\ x_2 \\ x_3 \end{pmatrix}$$

for all $(x_1, x_2, x_3) \in \mathbb{R}^3$, and let M stand for the matrix of L in the standard coordinate system (e_1, e_2, e_3). Define a new coordinate system (B_1, B_2, B_3), where $B_1 = -e_1 + e_2 - e_3$, $B_2 = -2e_1 + e_2 - e_3$, and $B_3 = 2e_1 - e_2 + 2e_3$.

(a) Find the matrix representation of L in the new coordinate system and show that it is a permutation matrix. Find the permutation $\sigma \in \mathrm{Sym}(3)$ for this permutation matrix.

(b) Show that there is a 3×3 nonsingular matrix S such that

$$S^{-1}\begin{pmatrix} 3 & 2 & -3 \\ -2 & -1 & 2 \\ 3 & 3 & -2 \end{pmatrix} S = \sigma I_3.$$

12. Let $L : V \to V$ be a nonsingular linear transformation on a finite-dimensional vector space V such that the matrix representation of L relative to some coordinate system is a permutation matrix. Show that $\det(L) = \pm 1$. Thus, the determinant of every permutation matrix is ± 1.

4. THE CAYLEY REPRESENTATION

We showed in the previous section that the transformations in the group of the triangle could be regarded as permutations of the vertices of a triangle. But to what extent is this possible for an arbitrary group? The Cayley representation answers this question by showing that it is always possible to represent the elements of a group as permutations. As such, it provides a way to use the language and properties of permutations in dealing with arbitrary groups. Although we are mainly interested in this representation for finite groups, our discussion applies to finite as well as infinite groups.

Let G be a group and let x be an element in G. Define the map $\pi_x : G \to G$ by setting $\pi_x(g) = xg$ for all elements $g \in G$.

PROPOSITION 1. π_x is a permutation of G.

Proof. To show that π_x is a 1–1 function, suppose that $\pi_x(g) = \pi_x(h)$ for some elements $g, h \in G$. Then $xg = xh$. Hence, $g = h$ and therefore π_x is 1–1. To show that π_x is an onto mapping, let y be any element in G. Then $x^{-1}y \in G$ and $\pi_x(x^{-1}y) = x(x^{-1}y) = y$. Hence, π_x is onto. It follows that π_x is a 1–1 correspondence of G with itself and thus is a permutation of G. ∎

Observe that the function π_x is left multiplication of the elements in G by x and hence is just the row of the multiplication table for G beginning with x. If $G = \{1, g_2, \ldots, g_n\}$ is finite, we display this row as indicated in the table.

	1	g_2	$\cdots$	g_n
$\vdots$				
x	$x = \pi_x(1)$	$xg_2 = \pi_x(g_2)$		$xg_n = \pi_x(g_n)$
$\vdots$				

PROPOSITION 2. Let G be a group and let $G_L = \{\pi_x \in \mathrm{Sym}(G) \mid x \in G\}$. Then G_L is a subgroup of $\mathrm{Sym}(G)$ and $G_L \cong G$.

Proof. G_L is clearly nonempty. To show that G_L is a subgroup of $\mathrm{Sym}(G)$, it remains to verify that $\pi_x \pi_y$ and π_x^{-1} belong to G_L for all elements, $x, y \in G$. We claim, in fact, that $\pi_x \pi_y = \pi_{xy}$ and $\pi_x^{-1} = \pi_{x^{-1}}$. For let $g \in G$. Then $\pi_x \pi_y(g) = \pi_x(yg) = x(yg) = (xy)g = \pi_{xy}(g)$ and hence $\pi_x \pi_y = \pi_{xy}$. Moreover, since $\pi_1 = 1$, the identity permutation, it follows that $\pi_x \pi_{x^{-1}} = \pi_{x^{-1}} \pi_x = 1$ and therefore $\pi_x^{-1} = \pi_{x^{-1}}$. Hence, $\pi_x \pi_y$ and π_x^{-1} belong to G_L for all $x, y \in G$ and therefore G_L is a subgroup of $\mathrm{Sym}(G)$. Now, define the function $f : G \to G_L$ by setting $f(x) = \pi_x$ for all elements $x \in G$. Then f is a group-isomorphism. For f is 1–1 since, if $\pi_x = \pi_y$ for $x, y \in G$, then $x = \pi_x(1) = \pi_y(1) = y$, and f is obviously onto. Finally, $f(xy) = \pi_{xy} = \pi_x \pi_y = f(x) f(y)$ for all $x, y \in G$. Thus, f is a group-isomorphism and hence $G_L \cong G$. ∎

The group G_L is called the *Cayley representation* of G; each element $x \in G$ is represented by the permutation $\pi_x \in G_L$. The representation is important because it shows that the elements of any group may be represented as permutations of the group itself and that the correspondence $x \leftrightarrow \pi_x$ between the elements in G and their representation in terms of left multiplication preserves the arithmetic on each group; that is, the arithmetic of permutations in G_L is exactly the same as the arithmetic of the corresponding elements in G. Let us mention that there are other ways to represent the elements of a group as permutations in addition to the Cayley representation. We discuss these methods in more detail in Chapter 5.

EXAMPLE 1

Let us find the Cayley representation of a cyclic group of order 4. Let $G = \{1, x, x^2, x^3\}$ stand for the cyclic group of order 4 generated by an element x. Then

$$\pi_1 = \begin{pmatrix} 1 & x & x^2 & x^3 \\ 1 & x & x^2 & x^3 \end{pmatrix} = 1 \qquad \pi_x^2 = \begin{pmatrix} 1 & x & x^2 & x^3 \\ x^2 & x^3 & 1 & x \end{pmatrix}$$

$$\pi_x = \begin{pmatrix} 1 & x & x^2 & x^3 \\ x & x^2 & x^3 & 1 \end{pmatrix} \qquad \pi_x^3 = \begin{pmatrix} 1 & x & x^2 & x^3 \\ x^3 & 1 & x & x^2 \end{pmatrix}.$$

Hence, $G_L = \{\pi_1, \pi_x, \pi_{x^2}, \pi_{x^3}\}$ is the Cayley representation of G as a group of permutations; $\pi_1 = 1$ is the identity, $\pi_x = (1\ x\ x^2\ x^3)$ is a 4-cycle, $\pi_{x^2} = (1\ x^2)(x\ x^3)$ is a product of two disjoint and hence commuting transpositions, and $\pi_{x^3} = (1\ x^3\ x^2\ x)$ is a 4-cycle. If we label the four elements in G as $1 \leftrightarrow 1$, $x \leftrightarrow 2$, $x^2 \leftrightarrow 3$, $x^3 \leftrightarrow 4$, then $\pi_1 = 1$, $\pi_x = (1\ 2\ 3\ 4)$, $\pi_{x^2} = (1\ 3)(2\ 4)$, and $\pi_{x^3} = (1\ 4\ 3\ 2)$, which shows that G is embedded in $\mathrm{Sym}(4)$ as the subgroup $H = \{1, (1\ 2\ 3\ 4), (1\ 3)(2\ 4), (1\ 4\ 3\ 2)\}$. On the other hand, if we relabel the elements in G, we obtain a different embedding; for example, the labeling $1 \leftrightarrow 1$, $x \leftrightarrow 3$, $x^2 \leftrightarrow 2$, $x^3 \leftrightarrow 4$ gives $\pi_1 = 1$, $\pi_x = (1\ 3\ 2\ 4)$, $\pi_{x^2} = (1\ 2)(3\ 4)$, and $\pi_{x^3} = (1\ 4\ 3\ 2)$, and hence G is embedded in Sym(4) as the subgroup $K = \{1, (1\ 3\ 2\ 4), (1\ 2)(3\ 4), (1\ 4\ 2\ 3)\}$.

EXAMPLE 2. The Cayley representation of the real plane

The real plane $\mathbb{R}^2$ is an abelian group under vector addition. To determine its Cayley representation, let $A \in \mathbb{R}^2$ be an arbitrary vector in the plane. Then the permutation $\pi_A : \mathbb{R}^2 \to \mathbb{R}^2$ maps a typical vector $X \in \mathbb{R}^2$ to the vector $\pi_A(X) = X + A$. But $X + A$ is just the translation of X by A, as illustrated in the marginal figure. Thus, the Cayley representation of the plane is the group of translations of the plane under function composition, each vector $A \in \mathbb{R}^2$ corresponding to translation of the plane by A.

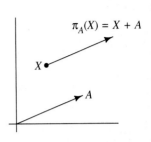

Exercises

1. Let $G = \{1, x, x^2\}$ be a cyclic group of order 3.
   (a) Find the Cayley representation of G.
   (b) Below are two labelings of the group G. Use the Cayley representation in part (a) to find the embedding of G into Sym(3) corresponding to each of these labelings.

$$1 \leftrightarrow 1 \qquad 1 \leftrightarrow 1$$
$$x \leftrightarrow 2 \qquad x \leftrightarrow 3$$
$$x^2 \leftrightarrow 3 \qquad x^2 \leftrightarrow 2$$

   (c) Show that every labeling of G leads to the same embedded subgroup of Sym(3).
2. Find the Cayley representation of the group D_4 of the square.
3. Find the Cayley representation of the additive group $\mathbb{Z}$ of integers.
4. Let G be a finite group and let x be an element in G.
   (a) Show that every orbit of the permutation π_x has the same length and that this common length is equal to $|x|$.
   (b) Show that $|x|$ divides $|G|$.
   (c) If x has odd order, show that π_x is an even permutation of G.
5. **The right-regular representation of a group.** Let G be a group. If $x \in G$, define the function $\mu_x : G \to G$ by setting $\mu_x(g) = gx$ for every element $g \in G$.
   (a) Show that $\mu_x \in \text{Sym}(G)$ for every $x \in G$.
   (b) Let $G_R = \{\mu_x \in \text{Sym}(G) \,|\, x \in G\}$. Show that $G_R \leq \text{Sym}(G)$.
   (c) Show that $G_R \cong G$. The subgroup G_R is called the *right-regular representation* of G; each element in G is represented as a permutation of G by means of right multiplication. This is in contrast to the Cayley representation G_L, which is also called the *left-regular representation* of G since each element is represented as a permutation of G by means of left multiplication.

(d) Show that $G_L \cap G_R = \{\pi_x \in \text{Sym}(G) \,|\, x \in Z(G)\}$, where $Z(G)$ stands for the center of G.

(e) Show that $G_L = G_R$ if and only if G is abelian.

(f) Let $\pi_x \in G_L$ and $\mu_y \in G_R$. Show that $\pi_x \mu_y = \mu_y \pi_x$.

6. We have used the idea of labeling, together with the Cayley representation, to embed a finite group of order n in the symmetric group $\text{Sym}(n)$. The purpose of this exercise is to characterize when two such labelings give rise to the same embedded subgroup of $\text{Sym}(n)$. Let G be a finite group of order n and let $f:G \to \{1, \ldots, n\}$ be a labeling of G. The map $f^*:\text{Sym}(G) \to \text{Sym}(n)$ defined in Exercise 12, Section 1 of this chapter then embeds G_L in $\text{Sym}(n)$ as the subgroup $f^*(G_L) = \{f^*(\pi_x) \in \text{Sym}(n) \,|\, x \in G\}$.

(a) For each element $\sigma \in \text{Sym}(n)$, let $\sigma f^*(G_L)\sigma^{-1} = \{\sigma f^*(\pi_x)\sigma^{-1} \in \text{Sym}(n) \,|\, x \in G\}$, and let $N_f = \{\sigma \in \text{Sym}(n) \,|\, \sigma f^*(G_L)\sigma^{-1} = f^*(G_L)\}$. Show that $f^*(G_L) \leq N_f \leq \text{Sym}(n)$.

(b) Let $f, g:G \to \{1, \ldots, n\}$ be labelings of G and let $f^*(G_L)$ and $g^*(G_L)$ be the corresponding embedded subgroups of $\text{Sym}(n)$. Show that $f^*(G_L) = g^*(G_L)$ if the only if there is some permutation $\sigma \in N_f$ such that $g = \sigma f$. Conclude that the labelings of G that give rise to the same embedding of G in $\text{Sym}(n)$ as f does are precisely those labelings of the form σf, where σ is any element of the subgroup N_f.

(c) If $N_f = \text{Sym}(n)$, show that every labeling of G defines the same embedding of G_L in $\text{Sym}(n)$.

(d) Show that $N_f = \text{Sym}(3)$ for every labeling f of a cyclic group of order 3.

(e) Let $G = \{1, x, x^2, x^3\}$ be a cyclic group of order 4 generated by x and let $f:G \to \{1, 2, 3, 4\}$ be the labeling of G defined by setting $f(1) = 1$, $f(x) = 2$, $f(x^2) = 3$, and $f(x^3) - 4$. Show that $N_f \neq \text{Sym}(4)$.

7. Let G and G' be isomorphic finite groups of order n and let $f:G \to \{1, \ldots, n\}$ and $g:G' \to \{1, \ldots, n\}$ be labelings of G and G'. Let $\alpha:G \to G'$ be a group-isomorphism.

(a) Set $\sigma = g\alpha f^{-1}$. Show that $\sigma \in \text{Sym}(n)$ and that $\sigma f^*(G_L)\sigma^{-1} = g^*(G'_L)$.

(b) If τ is any permutation in $\text{Sym}(n)$, show that the map $\beta_\tau:\text{Sym}(n) \to \text{Sym}(n)$ defined by setting $\beta_\tau(\pi) = \tau\pi\tau^{-1}$ for all $\pi \in \text{Sym}(n)$ is an isomorphism of the group $\text{Sym}(n)$ with itself.

(c) Using the results in parts (a) and (b), show that $\beta_\sigma | f^*(G_L):f^*(G_L) \to f^*(G'_L)$ is a group-isomorphism between the two embeddings of G in $\text{Sym}(n)$. Thus, the isomorphism $\alpha:G \to G'$ is transferred, via the Cayley representation, to the isomorphism $\beta_\sigma:\text{Sym}(n) \to \text{Sym}(n)$ that maps the embedded subgroup $f^*(G_L)$ isomorphically onto the embedded subgroup $f^*(G'_L)$.

4

Quotient Groups

Let us return for a moment to the construction of the group $\mathbb{Z}_n$ of integers modulo n. When we constructed this group in Chapter 2, our main purpose was to obtain an arithmetic model for finite cyclic groups. Recall that we first defined the congruence relation modulo n on the group of integers, showed that it is an equivalence relation, and then showed that the congruence classes forms a cyclic group of order n under addition of classes. In this chapter our goal is to extend this construction to arbitrary groups.

We begin with a subgroup H of an arbitrary group G. Instead of defining congruence modulo an integer n, where $s \equiv t \pmod{n}$ means that $t = s + kn$ for some integer k, we now define congruence modulo H by setting $x \equiv y \pmod{H}$ whenever $y = xh$ for some element $h \in H$. Congruence modulo H is then an equivalence relation on G. The equivalence class of an element $x \in G$ has the form $xH = \{xh \in G \mid h \in H\}$ and is called the left coset of H determined by x. Left cosets are multiplied elementwise by taking the product of all elements in one coset with those in another; thus, if xH and yH are typical left cosets, their product is the set $xH\,yH = \{xhyh' \in G \mid h, h' \in H\}$. But $xH\,yH$ is not necessarily a left coset of H in G, and consequently, multiplication of cosets is not, in general, a binary operation on the set G/H of all left cosets of H in G. To ensure that the product of left cosets is in fact a left coset—and hence that coset multiplication is a binary operation on the collection of left cosets—the subgroup H must satisfy a special condition called normality. When this is the case, the set G/H of left cosets is a group under coset multiplication called the quotient group of G modulo H. The construction of quotient groups from normal subgroups represents a new and important method for obtaining groups. But unlike the groups $\mathbb{Z}_n$, the general quotient groups G/H need not be cyclic; on the contrary, their properties depend upon those of G as well as those of H.

One of the important properties of cosets is that any two distinct left cosets are in fact disjoint. Moreover, all left cosets have the same size in the sense that there is a 1–1 correspondence between any two of them. Thus, the left cosets of a subgroup H partition the group G into disjoint subsets all of which have the same size. It follows that if G is finite, then the number of elements in G must be a multiple of the number of elements in H; in other words, the order of H must divide the order of G. This result, known as Lagrange's theorem, is among the most important results in elementary group theory.

In the second part of the chapter, we turn our attention to mappings from one group to another that preserve the arithmetic on the groups. By this we mean functions f that map the elements of a group G to those in a group G' in such a way that if $z = xy$ in G, then $f(z) = f(x)f(y)$ in G'. These mappings are called group-homomorphisms and are important because they provide a way of transferring information from one group to another. But they are also important to us for another reason, namely, that they define group-isomorphisms. To understand the relationship between a group-homomorphism and its corresponding isomorphism, suppose that $f : G \rightarrow G'$ is a group-homomorphism. Then G is partitioned into disjoint subsets according to the image of elements under f; two elements lie in the same subset of G if and only if they have the same image under f. This is where cosets come into the picture, for it turns out that the subset containing the identity element is a normal subgroup of G called the kernel of f and denoted by Ker f, while the other subsets are cosets of the kernel. Thus, the cosets of the kernel are in 1–1 correspondence with the images of f. But more importantly, this correspondence between cosets and images of f is in fact a group-isomorphism. Thus, the quotient group $G/\text{Ker} f$ is isomorphic to the image of f. This basic fact is called the fundamental theorem of group-homomorphism and is one of the most useful results in elementary group theory. It not only shows that every group-homomorphism determines a group-isomorphism, it also gives a better understanding of quotient groups by interpreting them as homomorphic images. In the end, it is the fundamental theorem of group-homomorphism that this chapter is really all about since it brings together in one simple statement the three basic ideas of the chapter: cosets, quotient groups, and group-homomorphism.

1. COSETS AND THE LAGRANGE THEOREM

In this section we discuss left cosets defined by a subgroup of a group, illustrate the partitioning of the group which they define, and then use this partitioning to prove Lagrange's theorem. We assume, initially, that all groups are written multiplicatively.

Let G be a group, H a subgroup of G, and let x and y be elements in G. We say that x is *congruent to y modulo H* if $y = xh$ for some element $h \in H$ and indicate this by writing $x \equiv y \pmod{H}$.

PROPOSITION 1. Congruence modulo H is an equivalence relation on G. The equivalence class of an element x is the set $\{xh \in G \mid h \in H\}$.

Proof.   Let $x \in G$. Then $x \equiv x \pmod{H}$ since $x = x1$ and $1 \in H$. Therefore, congruence mod H is a reflexive relation. Now, if x and y are elements in G such that $x \equiv y \pmod{H}$, then $y = xh$ for some element $h \in H$. Therefore, $x = yh^{-1}$ and hence $y \equiv x \pmod{H}$ since $h^{-1} \in H$. Hence, congruence mod H is symmetric. Finally, suppose that $x, y, z \in G$ are such that $x \equiv y \pmod{H}$ and $y \equiv z \pmod{H}$. Then $y = xh$ and $z = yh'$ for some elements $h, h' \in H$. Therefore, $z = yh' = x(hh')$ and hence $x \equiv z \pmod{H}$ since $hh' \in H$. Thus, congruence mod H is transitive and is therefore an equivalence relation on G. By definition, the equivalence class of an element x is the set of elements y that are congruent to x modulo H. This set is clearly $\{xh \in G \mid h \in H\}$, as required.   ■

Definition

Let x be an element in G and let $xH = \{xh \in G \mid h \in H\}$. The set xH is called the *left coset* of H in G determined by the element x. The collection of all left cosets of H in G is denoted by G/H. The number of left cosets of H in G is called the *index* of H in G and is denoted by the symbol $[G:H]$; thus, $[G:H] = |G/H|$.

Let us make a few observations. According to Proposition 1, the left cosets of H in G are the equivalence classes of the congruence relation modulo H. Consequently, the left cosets of H partition the group G into disjoint subsets. We call this partitioning the *left coset decomposition* of G by H: every element $x \in G$ lies in exactly one left coset of H, namely, xH, and two elements x and y determine the same left coset if and only if $x \equiv y \pmod{H}$. But $x \equiv y \pmod{H}$ if and only if $y = xh$ for some element $h \in H$, which is the same as saying that $x^{-1}y \in H$. Hence, $xH = yH$ if and only if $x^{-1}y \in H$. Let us emphasize that distinct elements in G do not necessarily determine distinct left cosets of H; the fact that $xH = yH$ if and only if $x^{-1}y \in H$ is the basic criterion for when two elements $x, y \in G$ determine the same left coset.

For example, consider the symmetric group Sym(3) and the cyclic subgroup (σ) generated by the 3-cycle $\sigma = (1\ 2\ 3)$. The left coset determined by the identity element 1 is $1(\sigma) = (\sigma) = \{1, \sigma, \sigma^2\}$, while the left coset determined by $\tau = (1\ 2)$ is $\tau(\sigma) = \tau\{1, \sigma, \sigma^2\} = \{\tau, \tau\sigma, \tau\sigma^2\} = \{(1\ 2), (2\ 3), (1\ 3)\}$. Since every element of Sym(3) belongs to one of these two cosets, it follows that (σ) and

| · 1 | · σ | · σ^2 |
| · τ | · $\tau\sigma$ | · $\tau\sigma^2$ |

Sym(3) = $(\sigma) \cup \tau(\sigma)$

$\tau(\sigma)$ are the only left cosets of H in Sym(3). Therefore, Sym(3)/(σ) = $\{(\sigma), \tau(\sigma)\}$ and hence $[\text{Sym}(3):(\sigma)] = 2$. In this case, the left coset decomposition of Sym(3) by (σ) is

$$\text{Sym}(3) = (\sigma) \cup \tau(\sigma) = \{1, \sigma, \sigma^2\} \cup \{\tau, \tau\sigma, \tau\sigma^2\}.$$

We find, for example, that $\sigma \equiv \sigma^2 \pmod{(\sigma)}$ since $\sigma^{-1}\sigma^2 = \sigma \in (\sigma)$, and similarly $\tau\sigma \equiv \tau\sigma^2 \pmod{(\sigma)}$, but $\sigma \not\equiv \tau\sigma \pmod{(\sigma)}$ since $\sigma^{-1}(\tau\sigma) = \tau\sigma^2 \notin (\sigma)$.

EXAMPLE 1

Let $H = \{\sigma \in \text{Sym}(5) \mid \sigma(2) = 2\}$ stand for the set of permutations in Sym(5) that fix the number 2. Then H is clearly a subgroup of Sym(5). Let us discuss the left cosets of H in Sym(5).

(A) We claim first that if σ and τ are typical permutations in Sym(5), then $\sigma \equiv \tau \pmod{H}$ if and only if $\sigma(2) = \tau(2)$; that is, two permutations in Sym(5) determine the same left coset of H if and only if they have the same value on the number 2. For if $\sigma \equiv \tau \pmod{H}$, then $\sigma = \tau\pi$ for some $\pi \in H$ and hence $\sigma(2) = (\tau\pi)(2) = \tau(\pi(2)) = \tau(2)$, while conversely if $\sigma(2) = \tau(2)$, then $(\sigma^{-1}\tau)(2) = \sigma^{-1}(\tau(2)) = \sigma^{-1}(\sigma(2)) = 2$, in which case $\sigma^{-1}\tau \in H$ and hence $\sigma \equiv \tau \pmod{H}$. Thus, all permutations in a left coset of H have the same value on the number 2, but permutations in different left cosets have different values.

(B) It follows from part (A) that a typical left coset σH consists of those permutations that map the number 2 to $\sigma(2)$. Since there are only five possible values for $\sigma(2)$, we conclude that there are five left cosets of H in Sym(5). Hence, $[\text{Sym}(5):H] = 5$. To describe the five cosets explicitly, we need only observe that the five permutations 1, (2 1), (2 3), (2 4), and (2 5) map the number 2 to five distinct values and hence determine five cosets of H in Sym(5), namely, H, (2 1)H, (2 3)H, (2 4)H, and (2 5)H, as shown here.

Sym(5)

H	} permutations mapping $2 \mapsto 2$
(2 1)H	} permutations mapping $2 \mapsto 1$
(2 3)H	} permutations mapping $2 \mapsto 3$
(2 4)H	} permutations mapping $2 \mapsto 4$
(2 5)H	} permutations mapping $2 \mapsto 5$

It follows, for example, that if $\sigma = (1\ 3\ 2\ 5\ 4)$, then $\sigma(2) = 5$ and hence $\sigma H = (2\ 5)H$; indeed, we find by direct calculation that $\sigma H = (2\ 5)H$ since $\sigma^{-1}(2\ 5) = (1\ 4\ 5\ 3)$, which lies in H since it fixes the number 2.

Up to this point we have been using the multiplicative notation xH for left cosets. In the case of additive groups, however, left cosets are written additively as $x + H$. Thus, $x + H = \{x + h \in G \mid h \in H\}$, and $x \equiv y \pmod{H}$ means that $y = x + h$ for some element $h \in H$, or equivalently, that $y - x \in H$. Here are some examples of left cosets in additive groups.

EXAMPLE 2

Let n be an integer and let $n\mathbb{Z}$ stand for the set of integer multiples of n. Then $n\mathbb{Z}$ is a subgroup of the additive group $\mathbb{Z}$ of integers. If $s, t \in \mathbb{Z}$, then $s + n\mathbb{Z} = t + n\mathbb{Z}$ if and only if $s - t \in n\mathbb{Z}$, or equivalently, $s \equiv t \pmod{n}$. Hence, the left cosets of $n\mathbb{Z}$ in $\mathbb{Z}$ are just the congruence classes of integers modulo n; that is, $s + n\mathbb{Z} = [s] = \{s + nk \in \mathbb{Z} \mid k \in \mathbb{Z}\}$ for all integers $s \in \mathbb{Z}$. Since there are n such classes, namely, $[0], [1], \ldots$, and $[n - 1]$, it follows that $\mathbb{Z}/n\mathbb{Z} = \{[0], [1], \ldots, [n - 1]\}$ and hence $[\mathbb{Z}:n\mathbb{Z}] = |\mathbb{Z}/n\mathbb{Z}| = n$. Hence, in this case the left coset decomposition of $\mathbb{Z}$ by $n\mathbb{Z}$ is just the partitioning of $\mathbb{Z}$ into congruence classes modulo n. For example, the subgroup $3\mathbb{Z}$ partitions $\mathbb{Z}$ into three cosets $[0]$, $[1]$, and $[2]$, as shown here.

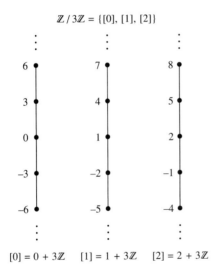

$$\mathbb{Z}/3\mathbb{Z} = \{[0], [1], [2]\}$$

$$[0] = 0 + 3\mathbb{Z} \qquad [1] = 1 + 3\mathbb{Z} \qquad [2] = 2 + 3\mathbb{Z}$$

EXAMPLE 3

Let A be a nonzero vector in real 2-space $\mathbb{R}^2$ and let $U = \langle A \rangle$ stand for the line spanned by A. Then U is a subgroup of the additive group of $\mathbb{R}^2$. Let us use the parallelogram law to describe geometrically the left cosets of U in $\mathbb{R}^2$. Let $B \in \mathbb{R}^2$. Then the left coset of U determined by B is the set $B + U = \{B + cA \in \mathbb{R}^2 \mid c \in \mathbb{R}\}$. But for each nonzero scalar c, the vector $B + cA$ is the diagonal of the parallelogram having B and cA as sides. Thus, as c assumes all real values the vectors $B + cA$ form the line through B parallel to U. If B and B' are vectors in $\mathbb{R}^2$, then $B + U = B' + U$ if and only if

$B' - B \in U$, or equivalently, if and only if the line through B and B' is parallel to U. The set $\mathbb{R}^2/U$ of left cosets therefore consists of all lines in $\mathbb{R}^2$ that are parallel to U and hence, in particular, the index $[\mathbb{R}^2 : U]$ is infinite.

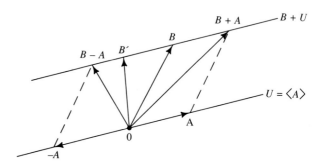

EXAMPLE 4

Let U stand for the set of solutions of the following system of homogeneous linear equations:

$$7x + 2y - z = 0$$

$$x - y - z = 0.$$

Then U is a subspace of $\mathbb{R}^3$ and we find that $U = \{(t, -2t, 3t) \in \mathbb{R}^3 \,|\, t \in \mathbb{R}\}$ $= \langle (1, -2, 3) \rangle$, the subspace spanned by the vector $(1, -2, 3)$. U is also a subgroup of the additive group of $\mathbb{R}^3$. To describe the left cosets of U in $\mathbb{R}^3$, let $B, B' \in \mathbb{R}^3$ and set $B = (b_1, b_2, b_3)$ and $B' = (b'_1, b'_2, b'_3)$. Then $B + U = B' + U$ if and only if $B' - B \in U$. But this is true if and only if

$$\begin{pmatrix} 7 & 2 & -1 \\ 1 & -1 & -1 \end{pmatrix} \begin{pmatrix} b'_1 - b_1 \\ b'_2 - b_2 \\ b'_3 - b_3 \end{pmatrix} = \begin{pmatrix} 0 \\ 0 \end{pmatrix},$$

or equivalently,

$$\begin{pmatrix} 7 & 2 & -1 \\ 1 & -1 & -1 \end{pmatrix} \begin{pmatrix} b'_1 \\ b'_2 \\ b'_3 \end{pmatrix} = \begin{pmatrix} 7 & 2 & -1 \\ 1 & -1 & -1 \end{pmatrix} \begin{pmatrix} b_1 \\ b_2 \\ b_3 \end{pmatrix} = \begin{pmatrix} c_1 \\ c_2 \end{pmatrix},$$

where $c_1 = 7b_1 + 2b_2 - b_3$ and $c_2 = b_1 - b_2 - b_3$ are the values obtained by substituting the coordinates of B into the system. It follows that the left coset $B + U$ consists of all solutions B' of the system

$$7x + 2y - z = c_1$$

$$x - y - z = c_2.$$

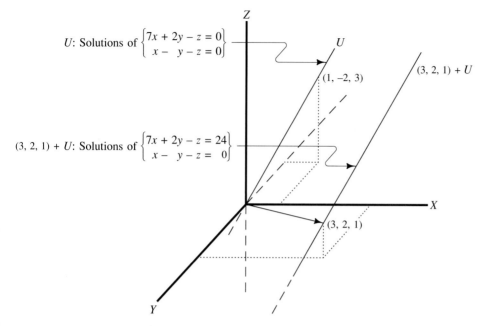

U: Solutions of $\begin{cases} 7x + 2y - z = 0 \\ x - y - z = 0 \end{cases}$

$(3, 2, 1) + U$: Solutions of $\begin{cases} 7x + 2y - z = 24 \\ x - y - z = 0 \end{cases}$

FIGURE 1.   Cosets in $\mathbb{R}^3/U$. Cosets are lines parallel to U and represent solutions to the corresponding system of linear equations.

Geometrically, $B + U$ is the line in $\mathbb{R}^3$ obtained by translating U to the vector B. In this case the left coset decomposition of $\mathbb{R}^3$ by U consists of all lines in $\mathbb{R}^3$ parallel to U, each line representing the solutions to the corresponding system of linear equations. The diagram in Figure 1 illustrates two of these lines: the coset U, which represents the solutions of the original homogeneous system, and the coset $(3, 2, 1) + U$, which represents the solutions of the system $7x + 2y - z = 24$, $x - y - z = 0$.

The previous two examples give a geometric interpretation of left cosets using translation of vectors in $\mathbb{R}^2$ and $\mathbb{R}^3$. In general, if x is an element of an arbitrary group G and H is a subgroup of G, then the permutation $g \mapsto xg$ maps H to the left coset xH and hence may be thought of as "translating" H to xH. In this manner the left cosets of H are distributed throughout G. But the mapping $g \mapsto xg$, being a permutation, also shows that the left cosets of H are in 1–1 correspondence with each other and hence must have the same cardinality. Let us now use this fact to prove Lagrange's theorem, one of the most important results in elementary group theory.

PROPOSITION 2.   (LAGRANGE) Let G be a finite group and let H be a subgroup of G. Then $|G| = |H|[G:H]$. In particular, the order of H divides the order of G and $[G:H] = |G|/|H|$.

Proof. Let $m = [G:H]$ and let $H, x_2H, \ldots, x_mH$ stand for the distinct left cosets of H in G. Since every element of G lies in some left coset and distinct left cosets are disjoint, it follows that $|G| = |H| + |x_2H| + \cdots + |x_mH|$. But all of these cosets contain the same number of elements as H. Hence, $|G| = m|H| = |H|[G:H]$. ■

COROLLARY 1. If G is a finite group, then the order of every element in G must divide the order of G.

Proof. Let $x \in G$. Then the order of the cyclic subgroup (x) divides $|G|$ by Lagrange's theorem. But $|(x)| = |x|$. Hence, $|x|$ divides $|G|$. ■

COROLLARY 2. A group of prime order must be cyclic and generated by any nonidentity element of the group.

Proof. Let G be a group of prime order p and let x be any nonidentity element in G. Then $|x|$ divides p and hence $|x| = p$. Therefore, $G = (x)$. ■

COROLLARY 3. Let G be a finite group of order n and let x be an element in G. Then $x^n = 1$.

Proof. Since $|x|$ divides n, $n = |x|k$ for some integer k. Hence, $x^n = (x^{|x|})^k = 1$. ■

The importance of Lagrange's theorem in the study of finite groups cannot be overestimated; indeed, the above corollaries give three elementary but exceptionally important results that come directly from this theorem. They show, for example, that the symmetric group Sym(5) cannot contain an element of order 7 since $|\text{Sym}(5)| = 5! = 2^3 \cdot 3 \cdot 5$ is not divisible by 7, but it might contain an element of order 5. In fact, it does: the 5-cycle (1 2 3 4 5). Similarly, it might contain a subgroup of order 15, but it cannot contain one of order 18.

EXAMPLE 5

Consider the alternating group Alt(4) of even permutations of the set $\{1, 2, 3, 4\}$. Since $|\text{Alt}(4)| = \frac{1}{2}(4!) = 12$, it follows from Lagrange's theorem that

$$[\text{Sym}(4):\text{Alt}(4)] = \frac{|\text{Sym}(4)|}{|\text{Alt}(4)|} = \frac{24}{12} = 2.$$

Hence, the subgroup Alt(4) has two left cosets in Sym(4). One of them, of course, is Alt(4). To find the other coset, we need only observe that the transposition (1 2) is an odd permutation and therefore the left coset (1 2) Alt(4) consists of all 12 odd permutations. Hence, Sym(4)/Alt(4) = {Alt(4), (1 2) Alt(4)}. Figure 2 shows the left coset decomposition of Sym(4) by Alt(4),

$$\text{Sym}(4) = \text{Alt}(4) \cup (1\ 2)\,\text{Alt}(4)$$

Alt(4)	(1 2) Alt(4)

1

$\left.\begin{array}{l}(1\ 2)(3\ 4)\\(1\ 3)(2\ 4)\\(1\ 4)(2\ 3)\end{array}\right\}$ order 2

$\left.\begin{array}{l}(1\ 2\ 3) = (1\ 3)(1\ 2)\\(1\ 3\ 2) = (1\ 2)(1\ 3)\\(1\ 2\ 4) = (1\ 4)(1\ 2)\\(1\ 4\ 2) = (1\ 2)(1\ 4)\\(1\ 3\ 4) = (1\ 4)(1\ 3)\\(1\ 4\ 3) = (1\ 3)(1\ 4)\\(2\ 3\ 4) = (2\ 4)(2\ 3)\\(2\ 4\ 3) = (2\ 3)(2\ 4)\end{array}\right\}$ order 3

$\left.\begin{array}{l}(1\ 2)\\(1\ 3)\\(1\ 4)\\(2\ 3)\\(2\ 4)\\(3\ 4)\end{array}\right\}$ order 2

$\left.\begin{array}{l}(1\ 2\ 3\ 4) = (1\ 4)(1\ 3)(1\ 2)\\(1\ 2\ 4\ 3) = (1\ 3)(1\ 4)(1\ 2)\\(1\ 3\ 2\ 4) = (1\ 4)(1\ 2)(1\ 3)\\(1\ 3\ 4\ 2) = (1\ 2)(1\ 4)(1\ 3)\\(1\ 4\ 3\ 2) = (1\ 2)(1\ 3)(1\ 4)\\(1\ 4\ 2\ 3) = (1\ 3)(1\ 2)(1\ 4)\end{array}\right\}$ order 4

FIGURE 2.   Left coset decomposition of Sym(4) by the alternating subgroup Alt(4).

with elements in the cosets arranged by order. Note that the order of each permutation in Alt(4) is a divisor of 12, which agrees with Corollary 1, although there is no element of order 4 even though 4 divides 12. Thus, the converse of Corollary 1 is not true; that is, if the order of a group is divisible by an integer d, the group does not necessarily contain an element of order d.

EXAMPLE 6

Consider the cyclic group $\mathbb{Z}_{12}$ of congruence classes of integers modulo 12 under addition. Let us find the left cosets of the cyclic subgroup $([8])$ generated by the element $[8]$. Since $|([8])| = |[8]| = 12/\gcd(12, 8) = 3$, it follows from Lagrange's theorem that $[\mathbb{Z}_{12}:([8])] = |\mathbb{Z}_{12}|/|([8])| = 12/3 = 4$. Therefore, $([8])$ has 4 left cosets in $\mathbb{Z}_{12}$:

$$\mathbb{Z}_{12}/([8]) = \{([8]), [1] + ([8]), [2] + ([8]), [3] + ([8])\},$$

where $([8]) = \{[0], [8], [4]\}$, $[1] + ([8]) = \{[1], [9], [5]\}$, $[2] + ([8]) = \{[2], [10], [6]\}$, and $[3] + ([8]) = \{[3], [11], [7]\}$.

EXAMPLE 7. Fermat's Little Theorem

Let us use our results on finite groups to prove an important result in number theory known as Fermat's Little Theorem: if p is any prime, then $n^p \equiv n$ (mod p) for every positive integer n. We do this by first using the group $\mathbb{Z}_n$ of integers modulo n to construct a new group, the multiplicative group $U(\mathbb{Z}_n)$ of generators of $\mathbb{Z}_n$, and then apply our results on finite groups to $U(\mathbb{Z}_n)$.

(A) Recall that for any positive integer n, the group $\mathbb{Z}_n$ is a finite cyclic group of order n generated by congruence cases of the form $[s]$, where s and n are relatively prime. Let $U(\mathbb{Z}_n) = \{[s] \in \mathbb{Z}_n | \gcd(s, n) = 1\}$ stand for the set of generators of $\mathbb{Z}_n$. Define multiplication on the set $U(\mathbb{Z}_n)$ as follows: if $[s], [t] \in U(\mathbb{Z}_n)$, set $[s][t] = [st]$. Then $[s][t]$ is independent of the numbers s and t used to represent the classes $[s]$ and $[t]$. For if $[s'] = [s]$ and $[t'] = [t]$ for some integers s, s', t, and t', then $s' \equiv s \pmod{n}$ and $t' \equiv t \pmod{n}$ and hence $s' = s + kn$ and $t' = t + qn$ for some integers k and q. Therefore, $s't' = st + n(kt + qs + kqn)$ and hence $s't' \equiv st \pmod{n}$. Thus, $[s't'] = [st]$, which shows that the product $[s][t]$ does not depend on the numbers s and t but only on their classes. Now, $[s][t] \in U(\mathbb{Z}_n)$; for if s and t are relatively prime to n, there are integers a, b, c, and d such that $as + bn = ct + dn = 1$. Hence, $(as + bn)(ct + dn) = (ac)st + (bct + asd + bdn)n = 1$, which shows that st and n are relatively prime and hence that $[st]$ is an element of $U(\mathbb{Z}_n)$. Therefore, multiplication is a binary operation on $U(\mathbb{Z}_n)$.

(B) We now claim that $U(\mathbb{Z}_n)$ is a group under the multiplication defined in part (A). The proof that multiplication is associative and has the class $[1]$ as identity element is left as an exercise for the reader. To show that every element $[s]$ in $U(\mathbb{Z}_n)$ has an inverse, note that since s and n are relatively prime, $as + bn = 1$ for some integers a and b. Therefore, $[1] = [as + bn] = [as] + [bn] = [a][s] + [b][n] = [a][s]$ and hence $[a] = [s]^{-1}$. Thus, every element in $U(\mathbb{Z}_n)$ has an inverse, and hence $U(\mathbb{Z}_n)$ is a group under multiplication of classes.

(C) Since $\mathbb{Z}_n$ has $\varphi(n)$ generators, where φ is the Euler phi-function discussed in Chapter 2, it follows that $|U(\mathbb{Z}_n)| = \varphi(n)$. Hence, by Corollary 3, $[s]^{\varphi(n)} = [1]$ for every element $[s] \in U(\mathbb{Z}_n)$. In the language of number theory, this means that if s and n are relatively prime integers, then $s^{\varphi(n)} \equiv 1 \pmod{n}$. In particular, if $n = p$ is prime, then $s^{p-1} \equiv 1 \pmod{p}$ since $\varphi(p) = p - 1$. Multiplying both sides by s, we obtain $s^p \equiv s \pmod{p}$. But this last congruence is also true when p divides s. Therefore, $s^p \equiv s \pmod{p}$ for all positive integers s, as required. For example, $4^7 \equiv 4 \pmod{7}$, which means that the remainder of 4^7 upon division by 7 is 4, while the remainder of 12^5 upon division by 5 is 2 since $12^5 \equiv 12 \pmod{5} \equiv 2 \pmod{5}$.

We mentioned earlier that Lagrange's theorem is an exceptionally important result since it shows that the order of every element and every subgroup of a finite group must divide the order of the group. But what can be said about the converse of this theorem? Is it true, for example, that if a number d divides the order of a group, then the group contains an element and a subgroup of order d? We showed in Example 5 that the alternating group Alt(4) has order 12, but does not contain an element of order 4. But it does contain a subgroup of order 4 since the set $\{1, (1\ 2)(3\ 4), (1\ 3)(2\ 4), (1\ 4)(2\ 3)\}$, for example, is easily verified to be a subgroup of Alt(4). The group Sym(4), on the other hand, is a group of order 24 that does not contain a subgroup of order 6. Thus, the converse of Lagrange's theorem is simply not true in general, although there are certain cases when the converse, or at least a partial converse, is true. For example, if G is a cyclic group of order n and d divides n, then G contains an element as well as a subgroup of order d. More generally, we will show in Chapter 7 that if p^s is a prime power dividing the order of an arbitrary finite group G, then G contains a subgroup of order p^s as well as an element of order p.

Exercises

1. Let $\tau = (1\ 2) \in \text{Sym}(3)$. Determine the left cosets of (τ) in $\text{Sym}(3)$ and find the index $[\text{Sym}(3):(\tau)]$.
2. Determine the left cosets of the cyclic subgroup $([12])$ in the group $\mathbb{Z}_{20}$.
3. Show that $[\text{Sym}(n):\text{Alt}(n)] = 2$ for every integer $n \geq 2$.
4. Calculate the following indices:
   (a) $[\mathbb{Z}_{60}:([6])]$
   (b) $[\mathbb{Z}_{20}:([3])]$
   (c) $[([18]):([54])]$ in the group $\mathbb{Z}_{60}$
   (d) $[\text{Sym}(6):H]$, where $H = \{\sigma \in \text{Sym}(6) \mid \sigma(2) = 2 \text{ and } \sigma(3) = 3\}$
5. Below are listed various groups and subgroups. In each case determine if the indicated elements lie in the same left coset of the subgroup.
   (a) $G = \text{Sym}(5)$, $H = \text{Alt}(5)$; $\sigma = (1\ 2\ 3)$, $\tau = (2\ 3)$
   (b) $G = \text{Sym}(5)$, $H = \text{Alt}(5)$; $\sigma = (2\ 4\ 5\ 1)$, $\tau = (2\ 4)(2\ 3)(5\ 2)$
   (c) $G = \mathbb{Z}_{20}$, $H = ([15])$; $x = [1]$, $y = [6]$
   (d) $G = \mathbb{Z}_{20}$, $H = ([15])$; $x = [8]$, $y = [12]$
   (e) $G = \mathbb{R}^*$, $H = (2)$; $x = 14/5$, $y = 20/7$
   (f) $G = \mathbb{R}^3$, $H = \langle(1, -2, 1)\rangle$; $A = (5, -2, 1)$, $B = (3, 2, -1)$
   (g) $G = \mathbb{R}^3$, $H = \langle(1, -2, 1), (2, 1, 1)\rangle$; $A = (0, 1, 1)$, $B = (8, -1, 5)$
6. Let H be the subgroup of $\text{Sym}(5)$ generated by the permutations $(1\ 2\ 3)$ and $(1\ 3\ 5)$. Do the permutations $(1\ 2\ 3\ 4\ 5)$ and $(2\ 4\ 3\ 1)$ determine the same left coset of H?
7. Let $U = \langle(0, 1, 1), (1, 0, 1)\rangle$ stand for the subspace of $\mathbb{R}^3$ spanned by the vectors $(0, 1, 1)$ and $(1, 0, 1)$. Which of the following left cosets are the same: $(1, 2, 2) + U$, $(3, 2, 5) + U$, $(-1, 0, 1) + U$, $(2, 2, 3) + U$, U?
8. Let $SL_2(\mathbb{R})$ stand for the special linear subgroup of $\text{Mat}_2^*(\mathbb{R})$ discussed in Exercise 17, Section 3, Chapter 2.
   (a) If $M, N \in \text{Mat}_2^*(\mathbb{R})$, show that $M\,SL_2(\mathbb{R}) = N\,SL_2(\mathbb{R})$ if and only if $\det(M) = \det(N)$.
   (b) Which of the following left cosets are the same: $\left(\begin{smallmatrix}2&1\\1&3\end{smallmatrix}\right)SL_2(\mathbb{R})$, $\left(\begin{smallmatrix}1&-1\\0&-1\end{smallmatrix}\right)SL_2(\mathbb{R})$, $\left(\begin{smallmatrix}1&2\\2&3\end{smallmatrix}\right)SL_2(\mathbb{R})$, $\left(\begin{smallmatrix}5&-1\\0&-1\end{smallmatrix}\right)SL_2(\mathbb{R})$, $\left(\begin{smallmatrix}3&1\\4&3\end{smallmatrix}\right)SL_2(\mathbb{R})$, and $\left(\begin{smallmatrix}4&1\\9&2\end{smallmatrix}\right)SL_2(\mathbb{R})$.
   (c) Show that every left coset of $SL_2(\mathbb{R})$ in $\text{Mat}_2^*(\mathbb{R})$ contains a unique matrix of the form $\left(\begin{smallmatrix}c&0\\0&1\end{smallmatrix}\right)$, where $c \neq 0$.
   (d) Show that every invertible 2×2 real matrix may be written in the form $\left(\begin{smallmatrix}c&0\\0&1\end{smallmatrix}\right)S$, where $c \neq 0$ and $\det(S) = 1$.
   (e) Factor the matrix $\left(\begin{smallmatrix}2&3\\1&4\end{smallmatrix}\right)$ into the form described in part (d).

9. Let $\mathbb{Q}$ stand for the additive group of rational numbers and $\mathbb{Z}$ the subgroup of integers.
   (a) Show that two rational numbers x and y determine the same left coset of $\mathbb{Z}$ if and only if $x - y$ is an integer.
   (b) Show that every left coset in $\mathbb{Q}/\mathbb{Z}$ contains a unique rational number r such that $0 \le r < 1$.

10. Let $F(\mathbb{R})$ stand for the additive group of functions from $\mathbb{R}$ to $\mathbb{R}$. Let $x_0 \in \mathbb{R}$ and set $F_{x_0}(\mathbb{R}) = \{f \in F(\mathbb{R}) | f(x_0) = 0\}$.
    (a) Show that $F_{x_0}(\mathbb{R}) \le F(\mathbb{R})$.
    (b) Let $f, g \in F(\mathbb{R})$. Show that $f \equiv g \pmod{F_{x_0}(\mathbb{R})}$ if and only if $f(x_0) = g(x_0)$.
    (c) Show that every left coset in $F(\mathbb{R})/F_{x_0}(\mathbb{R})$ contains a unique constant function.
    (d) Interpret the cosets in $F(\mathbb{R})/F_{x_0}(\mathbb{R})$ geometrically.

11. Let $\mathbb{C}^*$ stand for the multiplicative group of nonzero complex numbers and let $U = \{z \in \mathbb{C}^* | \|z\| = 1\}$ stand for the subgroup of complex numbers of unit norm.
    (a) Let $z, z' \in \mathbb{C}^*$. Show that $z \equiv z' \pmod{U}$ if and only if $\|z\| = \|z'\|$.
    (b) Show that the left coset zU is the circle in the complex plane centered at the origin and passing through z.
    (c) Show that every left coset in $\mathbb{C}^*/U$ contains a unique positive real number.

12. Let $AX = O$ be a system of n homogeneous linear equations in m variables expressed in matrix form, where O stands for the $n \times 1$ zero matrix and A the $n \times m$ coefficient matrix. Let U stand for the set of solutions of this system.
    (a) Show that U is a subgroup of the additive group of the vector space $\mathbb{R}^m$.
    (b) Let $B, B' \in \mathbb{R}^m$. Show that $B \equiv B' \pmod{U}$ if and only if $AB = AB'$.
    (c) Let $B \in \mathbb{R}^m$. Show that the left coset $B + U = \{X \in \mathbb{R}^m | AX = AB\}$.

13. Let G be a finite group and let x be an element in G. Show that $[G:(x)]$ is equal to the number of orbits of the permutation $\pi_x : G \to G$ defined by $\pi_x(g) = xg$, $g \in G$.

14. Let G be a group and let H be a subgroup of G.
    (a) If $g \in G$, define the map $\sigma_g : G/H \to G/H$ by setting $\sigma_g(xH) = gxH$ for all left cosets $xH \in G/H$. Show that σ_g is a permutation of G/H.
    (b) Let $xH, yH \in G/H$. Construct a 1–1 correspondence from xH to yH.

15. Let H, K, and G be finite groups with $H \le K \le G$. Show that $[G:H] = [G:K][K:H]$.

16. Let G be a finite group having a subgroup H such that $[G:H] = p$, a prime. Show that the only subgroups of G that contain H are H and G.

17. Let n be a positive integer.
    (a) Show that the group $U(\mathbb{Z}_n)$ discussed in Example 7 is abelian.
    (b) Find $U(\mathbb{Z}_2)$, $U(\mathbb{Z}_3)$, $U(\mathbb{Z}_4)$, $U(\mathbb{Z}_5)$, and $U(\mathbb{Z}_8)$.
    (c) Is $U(\mathbb{Z}_n)$ necessarily cyclic?

18. Let G be a finite abelian group and let n be a positive integer.
    (a) Let $G_n = \{x \in G \mid x^n = 1\}$. Show that $G_n \leq G$.
    (b) Show that the number of solutions of the equation $X^n = 1$ in G must divide the order of G.
    (c) Find all solutions of the equations $X^2 = 1$ in the group $U(\mathbb{Z}_{60})$.
    (d) If $x, y \in G$, show that $x \equiv y \pmod{G_n}$ if and only if $x^n = y^n$.
    (e) Let $x \in G$. An element $g \in G$ is called an *nth root of* x if $g^n = x$. If x has an nth root in G, show that it has exactly $|G_n|$ nth roots.
    (f) Show that $[G:G_n]$ is equal to the number of elements in G that have an nth root.
    (g) Determine G_2 for the group $U(\mathbb{Z}_{60})$. Which elements in $U(\mathbb{Z}_{60})$ have a square root?
19. Let G be a finite group. For each divisor d of $|G|$, let $G_d = \{x \in G \mid x^d = 1\}$ and let N_d stand for the number of elements of order d in G.
    (a) If G is cyclic, show that $|G_d| \leq d$.
    (b) Suppose, conversely, that G has the property that $|G_d| \leq d$ for each divisor d of $|G|$.
        (1) Show that N_d is either zero or $\varphi(d)$.
        (2) Show that $\Sigma N_d = |G|$ and $\Sigma (\varphi(d) - N_d) = 0$, where the summations are taken over all divisors d of $|G|$.
        (3) Show that G is cyclic.
    (c) Conclude that a finite group G is cyclic if and only if $|G_d| \leq d$ for each divisor d of $|G|$.
20. Let G be a group of order 6.
    (a) Show that G must contain at least one element of order 2 and at least one element of order 3.
    (b) Let x and y be elements in G such that $|x| = 2$ and $|y| = 3$. If x and y commute, show that $G \cong \mathbb{Z}_6$.
    (c) If x and y do not commute, show that $G \cong \mathrm{Sym}(3)$.
    (d) Show that there are two classes of nonisomorphic groups of order 6.
21. Let $G = U(\mathbb{Z}_{18})$.
    (a) Find $|G|$.
    (b) Determine the structure of G.
22. **Right cosets.** Let G be a group and let H be a subgroup of G. If x and y are elements of G, we say that x is *right congruent to* y *modulo* H if $y = hx$ for some element $h \in H$ and indicate this by writing $x \equiv_r y \pmod{H}$.
    (a) Show that right congruence modulo H is an equivalence relation on G.
    (b) Show that the equivalence class of an element $x \in G$ is the set $Hx = \{hx \in G \mid h \in H\}$. Hx is called the *right coset* of H in G determined by the element x.
    (c) Find all right cosets of (τ) in $\mathrm{Sym}(3)$, where $\tau = (1\ 2)$.
    (d) Find all right cosets of (σ) in $\mathrm{Sym}(3)$, where $\sigma = (1\ 2\ 3)$.
    (e) Show that there is a 1–1 correspondence between any two right cosets of H in G.

(f) Show that the number of right cosets of H in G is the same as the number of left cosets of H in G.

(g) If G is abelian, show that $xH = Hx$ for every element $x \in G$.

(h) If $[G:H] = 2$, show that $xH = Hx$ for every element $x \in G$.

(i) Is it true that for every $x \in G$, there is some $y \in G$ such that $xH = Hy$?

2.   QUOTIENT GROUPS AND NORMAL SUBGROUPS

Let us now turn our attention to the construction of quotient groups. As we mentioned earlier, this construction is simply an extension of the method we used to obtain the group of integers modulo n. Instead of using congruence classes of integers mod n, however, we now begin with an arbitrary group G and use the left cosets of a subgroup H.

To begin the construction, let G be a group and let H be a subgroup of G. Then the congruence classes of G modulo H are the left cosets of H and have the form $xH = \{xh \in G \mid h \in H\}$, where $x \in G$. We multiply left cosets elementwise by taking the product of all elements in one coset with those in another; thus, if xH and yH are typical left cosets, their product is the set $xH\,yH = \{xhyh' \in G \mid h, h' \in H\}$. But this product set is not necessarily a left coset of H in G, and consequently, coset multiplication is not, in general, a binary operation on the set G/H of left cosets. Observe, however, that if $xhyh'$ is a typical element in $xH\,yH$, then we may write $xhyh' = xy(y^{-1}hy)h'$. It follows that if the subgroup H has the property that $y^{-1}hy$ is an element in H for every $y \in G$ and every $h \in H$, then each of the products $xhyh'$ lies in the coset xyH and therefore $xH\,yH = xyH$. With this observation in mind, we make the following definition.

> **Definition**
>
> Let G be a group and let H be a subgroup of G. H is called a *normal subgroup* of G if $g^{-1}hg \in H$ for all elements $g \in G$ and $h \in H$. We indicate that H is a normal subgroup of G by writing $H \lhd G$.

Thus, if H is a normal subgroup of G, then $xH\,yH = xyH$ for all left cosets xH and yH, and hence coset multiplication is a binary operation on the set G/H of all left cosets. Let us now show that in this case, G/H is a group under coset multiplication.

PROPOSITION 1. Let H be a subgroup of a group G. Then the set G/H of left cosets is a group under coset multiplication if and only if H is a normal subgroup of G. In this case, $xH\,yH = xyH$ for all cosets xH and yH, the identity element is the subgroup H, and $(xH)^{-1} = x^{-1}H$ for every left coset xH.

Proof.   If $H \lhd G$, then, as we noted above, coset multiplication is a binary operation on G/H. To show that G/H is a group, it remains to verify that it is an associative operation having an identity element and that every element in G/H has an inverse. Let xH, yH, zH be left cosets of H in G. Then
$$xH[(yH)(zH)] = xH(yzH) = x(yz)H = (xy)zH = (xy)H\,zH = [(xH)(yH)]zH.$$
Hence, coset multiplication is associative. Now, if 1 stands for the identity element in G, then $(1H)(xH) = xH = (xH)(1H)$ and hence $1H$, or simply H, is the identity element for coset multiplication. Finally, $(xH)^{-1} = x^{-1}H$ since $(xH)(x^{-1}H) = (xx^{-1})H = H = (x^{-1}H)(xH)$, and hence every element in G/H has an inverse with respect to coset multiplication. Thus, G/H is a group under coset multiplication. Conversely, suppose that G/H is a group under coset multiplication and let $h \in H$ and $y \in G$. Then $yhy^{-1}H = (yH)(hH)(y^{-1}H)$ $= (yH)(y^{-1}H) = H$ and therefore $yhy^{-1} \in H$. Hence, H is a normal subgroup of G.   ■

The group G/H is called the *quotient group of G modulo H*. Let us emphasize that the construction of the quotient group G/H is possible only when H is a normal subgroup of G since otherwise coset multiplication is not a binary operation on G/H. The elements in G/H are the left cosets of H in G and are multiplied according to the formula $xH\,yH = xyH$. The identity element is the subgroup H, and the inverse of a typical left coset xH is given by $(xH)^{-1} = x^{-1}H$. Moreover, $|G/H| = [G:H]$ since the number of left cosets of H in G is, by definition, the index of H in G. Note that when the binary operation on G is written additively, which is usually the case when G is abelian, the cosets in G/H have the form $x + H$ and coset addition is expressed by the formula $(x + H) + (y + H) = (x + y) + H$. In this case, inverses are given by the formula $-(x + H) = -x + H$. Before we construct some quotient groups, let us first show that in an abelian group every subgroup is normal and has an abelian quotient group and that every quotient group of a cyclic group is cyclic.

COROLLARY.   Let G be an abelian group and let H be a subgroup of G. Then H is a normal subgroup of G and the quotient group G/H is abelian. If in addition G is cyclic, then the quotient group G/H is also cyclic.

Proof.   Let $h \in H$ and $g \in G$. Then $g^{-1}hg = hg^{-1}g = h \in H$ since G is abelian. Hence, H is a normal subgroup of G. To show that G/H is abelian, let xH, $yH \in G/H$. Then $xy = yx$ and hence $(xH)(yH) = xyH = yxH = (yH)(xH)$. Therefore, G/H is abelian. Finally, suppose that G is a cyclic group with generator x. Then G/H is cyclic with generator xH. For let $yH \in G/H$. Then $y \in G$ and hence $y = x^s$ for some integer s. Therefore, $yH = x^sH = (xH)^s$ and hence G/H is cyclic with generator xH.   ■

For example, consider the group Z_n of integers modulo n. We claim that $\mathbb{Z}_n$ is just the quotient group $\mathbb{Z}/n\mathbb{Z}$. For clearly, the subgroup $n\mathbb{Z}$ is a normal subgroup of $\mathbb{Z}$ since $\mathbb{Z}$ is abelian. Therefore, $\mathbb{Z}/n\mathbb{Z}$ is a group under coset addition defined by $(s + n\mathbb{Z}) + (t + n\mathbb{Z}) = (s + t) + n\mathbb{Z}$. But $s + n\mathbb{Z} = [s]$, the congruence class of s mod n, and $[s] + [t] = [s + t]$ for all classes $[s]$, $[t]$. Thus, the cosets in $\mathbb{Z}/n\mathbb{Z}$ are the same as congruence classes in $\mathbb{Z}_n$ and are added in the same way that congruence classes are added. Therefore, $\mathbb{Z}/n\mathbb{Z} = \mathbb{Z}_n$.

If $H \lhd G$, a convenient way to interpret the group G/H is by means of the multiplication table for G. In the following figure, we have partitioned G into left cosets of H. The block in the table that corresponds to the product of the elements in the row coset xH and the column coset yH is the set $\{xhyh' \in G \mid h, h' \in H\}$. Since H is a normal subgroup of G, this product set is the left coset xyH. The multiplication table of G is therefore partitioned into left cosets of H and these blocks, or cosets, then form the individual elements of the quotient group G/H.

EXAMPLE 1

Let $H = \{1, (1\ 2\ 3), (1\ 3\ 2)\}$ stand for the cyclic subgroup of Sym(3) generated by the 3-cycle $(1\ 2\ 3)$. Let us show that H is a normal subgroup of Sym(3) and then discuss the quotient group Sym(3)/H.

(A) To show that $H \lhd \text{Sym}(3)$, we must verify that the products $\sigma^{-1}1\sigma$, $\sigma^{-1}(1\ 2\ 3)\sigma$, and $\sigma^{-1}(1\ 3\ 2)\sigma$ lie in H for every element $\sigma \in \text{Sym}(3)$. Now $\sigma^{-1}1\sigma = 1$, and $\sigma^{-1}(1\ 3\ 2)\sigma$ is the inverse of $\sigma^{-1}(1\ 2\ 3)\sigma$. Hence, normality will follow as soon as we show that $\sigma^{-1}(1\ 2\ 3)\sigma \in H$ for every $\sigma \in \text{Sym}(3)$. Here are the calculations:

$$1^{-1}(1\ 2\ 3)1 = (1\ 2\ 3) \in H$$

$$(1\ 2\ 3)^{-1}(1\ 2\ 3)(1\ 2\ 3) = (1\ 2\ 3) \in H$$

$$(1\ 3\ 2)^{-1}(1\ 2\ 3)(1\ 3\ 2) = (1\ 2\ 3) \in H$$

$$(1\ 2)^{-1}(1\ 2\ 3)(1\ 2) = (1\ 3\ 2) \in H$$

$$(1\ 3)^{-1}(1\ 2\ 3)(1\ 3) = (1\ 3\ 2) \in H$$

$$(2\ 3)^{-1}(1\ 2\ 3)(2\ 3) = (1\ 3\ 2) \in H$$

Therefore, $H \lhd \text{Sym}(3)$.

(B) Since $[\text{Sym}(3):H] = 2$, it follows that Sym(3)/H is a group of order 2 under coset multiplication and hence is cyclic. Therefore, $\text{Sym}(3)/H \cong \mathbb{Z}_2$.

Using H and $(12)H$ as the two distinct left cosets of H in Sym(3), we find that $\text{Sym}(3)/H = \{H, (12)H\}$. The coset $(12)H$ has order 2 as an element of $\text{Sym}(3)/H$ since $(12)H(12)H = [(12)(12)]H = H$; hence, $(12)H$ generates $\text{Sym}(3)/H$. The figure below shows the multiplication table of Sym(3) partitioned by the cosets of H and the corresponding multiplication table for the quotient group $\text{Sym}(3)/H$. Using coset multiplication we find, for example, that $[(13)H][(123)H] = (13)(123)H = (12)H$ and $[(23)H][(12)H] = (23)(12)H = (132)H = H$. Note that the last product shows that $[(23)H]^{-1} = (12)H$, which also follows from the fact that $[(23)H]^{-1} = (23)^{-1}H = (23)H = (12)H$.

$$\text{Sym}(3)/H \cong \mathbb{Z}_2$$

Sym(3)	1	(1 2 3)	(1 3 2)	(1 2)	(1 3)	(2 3)
1	1	(1 2 3)	(1 3 2)	(1 2)	(1 3)	(2 3)
(1 2 3)	(1 2 3)	(1 3 2)	1	(1 3)	(2 3)	(1 2)
(1 3 2)	(1 3 2)	1	(1 2 3)	(2 3)	(1 2)	(1 3)
(1 2)	(1 2)	(2 3)	(1 3)	1	(1 3 2)	(1 2 3)
(1 3)	(1 3)	(1 2)	(2 3)	(1 2 3)	1	(1 3 2)
(2 3)	(2 3)	(1 3)	(1 2)	(1 3 2)	(1 2 3)	1

Sym(3)/H	H	(12)H
H	H	(12)H
(12)H	(12)H	H

EXAMPLE 2

Let D_4 stand for the group of the square. Recall that D_4 is generated by the counterclockwise rotation r of the square through $\pi/2$ radians about its center and by the reflection s about a line through the midpoints of two opposite

$$D_4/\langle r^2\rangle \cong \{\langle r^2\rangle, r\langle r^2\rangle\} \times \{\langle r^2\rangle, s\langle r^2\rangle\} \cong \mathbb{Z}_2 \times \mathbb{Z}_2$$

D_4	1	r^2	r	r^3	s	sr^2	sr	sr^3
1	1	r^2	r	r^3	s	sr^2	sr	sr^3
r^2	r^2	1	r^3	r	sr^2	s	sr^3	sr
r	r	r^3	r^2	1	sr^3	sr	s	sr^2
r^3	r^3	r	1	r^2	sr	sr^3	sr^2	s
s	s	sr^2	sr	sr^3	1	r^2	r	r^3
sr^2	sr^2	s	sr^3	sr	r^2	1	r^3	r
sr	sr	sr^3	sr^2	s	r^3	r	1	r^2
sr^3	sr^3	sr	s	sr^2	r	r^3	r^2	1

$D_4/\langle r^2\rangle$	$\langle r^2\rangle$	$r\langle r^2\rangle$	$s\langle r^2\rangle$	$sr\langle r^2\rangle$
$\langle r^2\rangle$	$\langle r^2\rangle$	$r\langle r^2\rangle$	$s\langle r^2\rangle$	$sr\langle r^2\rangle$
$r\langle r^2\rangle$	$r\langle r^2\rangle$	$\langle r^2\rangle$	$sr\langle r^2\rangle$	$s\langle r^2\rangle$
$s\langle r^2\rangle$	$s\langle r^2\rangle$	$sr\langle r^2\rangle$	$\langle r^2\rangle$	$r\langle r^2\rangle$
$sr\langle r^2\rangle$	$sr\langle r^2\rangle$	$s\langle r^2\rangle$	$r\langle r^2\rangle$	$\langle r^2\rangle$

sides. The multiplication table for D_4 shows that r^2, the $180°$ rotation, commutes with every element in D_4. Hence, if $y \in D_4$, $y^{-1}r^2y = r^2$ and therefore the subgroup $(r^2) = \{1, r^2\}$ is a normal subgroup of D_4. In this case the quotient group $D_4/(r^2)$ is a group of order 4 since $[D_4:(r^2)] = 4$. It is clear from the multiplication table for D_4 partitioned into the left cosets of (r^2) that $D_4/(r^2) \cong \mathbb{Z}_2 \times \mathbb{Z}_2$.

EXAMPLE 3

Consider the alternating group Alt(4). Alt(4), we recall, consists of the 12 even permutations in Sym(4). In Example 5, Section 1, we listed these permutations and showed that the nonidentity elements fall into two classes, those of order 2 and those of order 3:

order 3: (1 2 3), (1 3 2), (1 2 4), (1 4 2), (2 3 4), (2 4 3),

(1 3 4), (1 4 3)

order 2: (1 2)(3 4), (1 3)(2 4), (1 4)(2 3).

Let $H = \{1, (1\ 2)(3\ 4), (1\ 3)(2\ 4), (1\ 4)(2\ 3)\}$. Let us show that H is a normal subgroup of Alt(4) and then construct the quotient group Alt(4)/H.

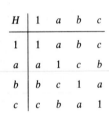

(A) To show that H is in fact a subgroup of Alt(4), we first construct the multiplication table for H as indicated in the margin, where $a = (1\ 2)(3\ 4)$, $b = (1\ 3)(2\ 4)$, and $c = (1\ 4)(2\ 3)$. It follows from the table that H is closed under multiplication and hence is a subgroup of Alt(4).

(B) We now claim that $H \lhd$ Alt(4). To show this, we must verify that $\sigma^{-1}\tau\sigma \in H$ for all elements $\tau \in H$ and $\sigma \in$ Alt(4). Rather than verify this for every possible combination of elements, we need only observe that H consists of those elements in Alt(4) that have order 1 or 2. But if $\tau \in H$, every element of the form $\sigma^{-1}\tau\sigma$ also has order 1 or 2 since $(\sigma^{-1}\tau\sigma)^2 = \sigma^{-1}\tau\sigma\sigma^{-1}\tau\sigma = \sigma^{-1}\tau^2\sigma = \sigma^{-1}1\sigma = 1$. Therefore, $\sigma^{-1}\tau\sigma \in H$ and hence $H \lhd$ Alt(4).

(C) Since $[\text{Alt}(4):H] = 3$, the quotient group Alt(4)/H has order 3 and hence is cyclic. Therefore, Alt(4)/$H \cong \mathbb{Z}_3$. Moreover, H, (1 2 3)H, and (1 3 2)H are three distinct left cosets of H in Alt(4). Therefore, Alt(4)/$H = \{H,$ (1 2 3)H, (1 3 2)$H\}$. Observe that both (1 2 3)H and (1 3 2)H have order 3 and hence generate Alt(4)/H. This agrees with the fact that a cyclic group of order 3 is generated by any one of its nonidentity elements.

Alt(4)/H	H	(1 2 3)H	(1 3 2)H
H	H	(1 2 3)H	(1 3 2)H
(1 2 3)H	(1 2 3)H	(1 3 2)H	H
(1 3 2)H	(1 3 2)H	H	(1 2 3)H

EXAMPLE 4

Consider the group $\mathbb{Z}_{12}$ of congruence classes of integers modulo 12. Since $\mathbb{Z}_{12}$ is cyclic and hence abelian, every element generates a normal subgroup whose quotient group is also cyclic. For example, the element $[4]$ generates the normal cyclic subgroup $([4])$ of order 3. Therefore, the quotient group $\mathbb{Z}_{12}/([4])$ is a cyclic group of order 4 and hence $\mathbb{Z}_{12}/([4]) \cong \mathbb{Z}_4$. In the group $\mathbb{Z}_{12}/([4])$ we find, for example, that

$$([2] + ([4])) + ([3] + ([4])) = [2] + [3] + ([4]) = [5] + ([4]) = [1] + ([4]),$$

$$-([1] + ([4])) = -[1] + ([4]) = [11] + ([4]) = [3] + ([4]),$$

$$-([2] + ([4])) = -[2] + ([4]) = [10] + ([4]) = [2] + ([4]),$$

$$-([3] + ([4])) = -[3] + ([4]) = [9] + ([4]) = [1] + ([4]).$$

Furthermore, $|[1] + ([4])| = |[3] + ([4])| = 4$ and hence both $[1] + ([4])$ and $[3] + ([4])$ are generators of the group $\mathbb{Z}_{12}/([4])$, which agrees with the fact that a cyclic group of order 4 has $\varphi(4)$, or 2, generators. On the other hand, $|[2] + ([4])| = 2$. Finally, observe that the generators of $\mathbb{Z}_{12}$ are the classes $[1]$, $[5]$, $[7]$, and $[11]$ and that $[5] + ([4]) = [1] + ([4])$ and $[7] + ([4]) = [11] + ([4]) = [3] + ([4])$; that is, the four generators of $\mathbb{Z}_{12}$ determine two cosets in $\mathbb{Z}_{12}/([4])$ both of which generate the quotient group. This agrees with the result proved in Corollary 2 that the coset xH generates the quotient group G/H whenever x generates G.

$\mathbb{Z}_{12}/([4])$	$([4])$	$[1] + ([4])$	$[2] + ([4])$	$[3] + (4])$
$([4])$	$([4])$	$[1] + ([4])$	$[2] + ([4])$	$[3] + ([4])$
$[1] + ([4])$	$[1] + ([4])$	$[2] + ([4])$	$[3] + ([4])$	$([4])$
$[2] + ([4])$	$[2] + ([4])$	$[3] + ([4])$	$([4])$	$[1] + ([4])$
$[3] + ([4])$	$[3] + ([4])$	$([4])$	$[1] + ([4])$	$[2] + ([4])$

EXAMPLE 5

Let $U = \{(a, 2a) \in \mathbb{R}^2 \mid a \in \mathbb{R}\}$ stand for the subspace of real 2-space $\mathbb{R}^2$ spanned by the vector $(1, 2)$. Then U is a normal subgroup of the additive group of $\mathbb{R}^2$ since $\mathbb{R}^2$ is abelian. In this case, the quotient group $\mathbb{R}^2/U$ consists of all cosets $A + U$, where $A \in \mathbb{R}^2$. If $A = (a, b)$ and $B = (a', b')$ are typical vectors in $\mathbb{R}^2$, coset addition in the group $\mathbb{R}^2/U$ is defined by $(A + U) + (B + U) = A + B + U = (a + a', b + b') + U$. Finally, since the coset $A + U$ is the line through A parallel to U, the quotient group $\mathbb{R}^2/U$ may be regarded as the set of lines in the plane that are parallel to U.

If a group is abelian, then, as we have seen, every subgroup is normal and hence we may form the corresponding quotient group. But finding normal subgroups of nonabelian groups is not as simple since we must first verify that certain products lie in the subgroup. Let us look at this requirement more closely. If H is a subgroup of group G, then, in order that H be a normal subgroup of G, all products of the form $g^{-1}hg$ must lie in H for all elements $h \in H$ and $g \in G$. This suggests that we form the set $g^{-1}Hg = \{g^{-1}hg \in G \mid h \in H\}$ consisting of all such products, where g is a fixed element in G. We claim that $g^{-1}Hg$ is, in fact, a subgroup of G. For $g^{-1}Hg$ is clearly a nonempty subset of G, and if $g^{-1}hg$ and $g^{-1}h'g$ are typical elements in $g^{-1}Hg$, then their product $(g^{-1}hg)(g^{-1}h'g) = g^{-1}hh'g \in g^{-1}Hg$ and the inverse $(g^{-1}hg)^{-1} = g^{-1}h^{-1}g \in g^{-1}Hg$. Therefore, $g^{-1}Hg$ is closed under products and inverses and hence is a subgroup of G. The following result shows that the subgroups $g^{-1}Hg$ for $g \in G$ provide a convenient way to characterize normality of H in G.

PROPOSITION 2. Let H be a subgroup of a group G. Then the following statements are equivalent:

(1) $H \lhd G$
(2) $g^{-1}Hg = H$ for every element $g \in G$
(3) $gH = Hg$ for every element $g \in G$.

Proof. We show that statement (1) implies statement (2) and leave the proof that (2) implies (3) and (3) implies (1) as an exercise for the reader. Let $H \lhd G$ and let $g \in G$. Then $g^{-1}hg \in H$ for every element $h \in H$ and hence $g^{-1}Hg \subseteq H$. Conversely, if $h \in H$, then $h = g^{-1}(ghg^{-1})g = g^{-1}[(g^{-1})^{-1}h(g^{-1})]g$. But $(g^{-1})^{-1}h(g^{-1}) \in H$ since $H \lhd G$. Hence, $h \in g^{-1}Hg$ and therefore $H \subseteq g^{-1}Hg$. It now follows that $g^{-1}Hg = H$. ∎

In view of statement (3), normality of a subgroup H is the same as saying that the left coset determined by an arbitrary element g in the group is the same as the right coset determined by g. Thus, there is no need to distinguish between right and left cosets when dealing with normal subgroups, and we will therefore delete the adjective "right" or "left" in this case. Let us also emphasize, in connection with statement (2), that the equation $g^{-1}Hg = H$ does not mean that $g^{-1}hg = h$ for all elements $g \in G$ and $h \in H$. It means simply that if $g \in G$ and $h \in H$, then $g^{-1}hg$ is some element in H, not necessarily h. For example, $H = \{1, (1\ 2\ 3), (1\ 3\ 2)\}$ is a normal subgroup of Sym(3), but $(1\ 2)^{-1}(1\ 2\ 3)(1\ 2) = (1\ 3\ 2)$, not $(1\ 2\ 3)$. Of course, if the group G is abelian, then $g^{-1}hg = hg^{-1}g = h$ for all $g \in G$ and $h \in H$, and therefore every subgroup of an abelian group is a normal subgroup, which agrees with Corollary 2.

As we have seen, some of the subgroups of a given group are normal while others are not. The trivial subgroups (1) and G, for example, are normal subgroups of any group G; in this case, the corresponding quotient groups are $G/(1)$, which is isomorphic to G itself, and G/G, which is the trivial group.

In the following example we discuss an important class of groups, called simple groups, for which the trivial subgroups are the only normal subgroups.

EXAMPLE 6. Simple groups

A group G is called a *simple group* if the only normal subgroups of G are the trivial subgroups. For example, the cyclic groups $\mathbb{Z}_2$ and $\mathbb{Z}_3$ are simple groups. Let us look at some other examples of simple groups.

(A) Every finite cyclic group of prime order is simple since, by Lagrange's theorem, a cyclic group of prime order has only the trivial subgroups. In fact, the finite cyclic groups of prime order are the only abelian simple groups; for if G is an abelian simple group, then G must be cyclic since every nonidentity element generates a nontrivial normal subgroup, which, since the group is simple, must be the entire group. Since a cyclic group has a subgroup of order d for each divisor d of its order, it follows that G must have prime order. Thus, the groups $\mathbb{Z}_p$, for p a prime, are the only abelian simple groups.

(B) The alternating group Alt(3) is simple since it is a cyclic group of order 3. The alternating group Alt(4) is not simple, however, since it contains the proper normal subgroup $H = \{1, (1\ 2)(3\ 4), (1\ 3)(2\ 4), (1\ 4)(2\ 3)\}$ discussed in Example 3.

(C) Let us show that the alternating group Alt(5) is also a simple group. Recall first that Alt(5) consists of all even permutations in Sym(5), that it contains all 3-cycles, and that the 3-cycles in fact generate Alt(5) (see Exercise 6, Section 3, Chapter 3). Now, suppose that N is a normal subgroup of Alt(5), $N \neq (1)$. Then N must contain some 3-cycle. For let $\sigma \in N$, $\sigma \neq 1$. Then the cycle decomposition of σ is either a product of two disjoint 2-cycles, a single 3-cycle, or a single 5-cycle. Suppose that $\sigma = (a\ b)(c\ d)$ is a product of two disjoint 2-cycles. Let e be a number other than a, b, c, or d, and let $\tau = (a\ e\ b)$. Then $\tau \in$ Alt(5) since τ is an even permutation. Consider the element $\sigma\tau\sigma^{-1}\tau^{-1}$. Since $N \lhd$ Alt(5) and $\tau \in$ Alt(5), $\tau\sigma^{-1}\tau^{-1} \in N$ and hence $\sigma\tau\sigma^{-1}\tau^{-1} \in N$. But we find by direct calculation that $\sigma\tau\sigma^{-1}\tau^{-1} = (a\ c\ b)$, a 3-cycle. Thus, N contains a 3-cycle if $\sigma = (a\ b)(c\ d)$. On the other hand, if $\sigma = (a\ b\ c\ d\ e)$ is a 5-cycle, let $\tau = (a\ b\ c)$. Then $\sigma\tau\sigma^{-1}\tau^{-1} = (a\ b\ d)$ and, as before, this 3-cycle must lie in N. Thus, N must contain some 3-cycle. To complete the argument that $N =$ Alt(5), we must show that N in fact contains all 3-cycles. Let $\sigma = (a\ b\ c)$ be a 3-cycle in N and let σ' be any 3-cycle. If $\sigma' = \sigma$, then $\sigma' \in N$. Otherwise the orbits of σ and σ' have either one or two numbers in common. Thus, σ' is either $(a\ d\ e)$ or $(a\ b\ e)$. Now,

$$[(b\ d)(c\ e)]\sigma[(b\ d)(c\ e)]^{-1} = (a\ d\ e) \quad \text{and}$$

$$[(c\ d)(c\ e)]\sigma[(c\ d)(c\ e)]^{-1} = (a\ b\ e).$$

Sym(5)

|

Alt(5) = (3-cycles)

$\triangledown$

$\sigma \in N: (a\ b)(c\ d), (a\ b\ c),$
$\qquad (a\ b\ c\ d\ e)$

Since $(b\ d)(c\ e)$, $(c\ d)(c\ e) \in \text{Alt}(5)$ and $N \lhd \text{Alt}(5)$, it follows that $\sigma' \in N$. Therefore, N contains all 3-cycles and hence $N = \text{Alt}(5)$ since the 3-cycles generate $\text{Alt}(5)$. It follows that the group $\text{Alt}(5)$ contains no nontrivial normal subgroups and is, therefore, a simple group. In the exercises we ask the reader to modify the above argument to show that the alternating group $\text{Alt}(n)$ is a simple group for all $n \geq 5$.

(D) Finally, let us make an observation about simple groups. Suppose that G is a simple group and let x be an element in G, $x \neq 1$. Let N stand for the subgroup generated by all elements of the form $g^{-1}xg$, where $g \in G$. A typical element in N may then be written in the form $(g_1^{-1}xg_1)^{\pm 1} \cdots (g_n^{-1}xg_n)^{\pm 1}$, where $g_1, \ldots, g_n \in G$. We claim that $N \lhd G$. For, if $g \in G$, then

$$g^{-1}[(g_1^{-1}xg_1) \cdots (g_n^{-1}xg_n)]g = [g^{-1}(g_1^{-1}xg_1)g] \cdots [g^{-1}(g_n^{-1}xg_n)g]$$
$$= [(g_1g)^{-1}x(g_1g)] \cdots [(g_ng)^{-1}x(g_ng)] \in N.$$

Thus, $N \lhd G$. But G is simple and hence has no nontrivial normal subgroups. Therefore, $N = G$. Thus, if G is a simple group and x is any nonidentity element in G, then every element in G may be written in the form $(g_1^{-1}xg_1)^{\pm 1} \cdots (g_n^{-1}xg_n)^{\pm 1}$ for some $g_1, \ldots, g_n \in G$. For example, if x is a 3-cycle in the group $\text{Alt}(5)$, then elements of the form $(g^{-1}xg)^{\pm 1}$ are 3-cycles and hence every element in $\text{Alt}(5)$ may be written as a product of 3-cycles.

As we have seen, normal subgroups are an important part of group theory since they allow us to construct the corresponding quotient group. They are also important from the point of view of group structure since, in many cases, they tell us how the group is constructed from its normal subgroups. For example, if G is a finite group having normal subgroups H and K such that $H \cap K = (1)$ and $|G| = |H||K|$, then $G \cong H \times K$, the direct product of H and K. There is a more general form of the direct product called the semi-direct product, denoted by $H \times K$, in which H is a normal subgroup of G and in which ordered pairs are multiplied using an automorphism of H. If H is a normal subgroup of G, then under certain conditions, there is a subgroup K such that $G \cong H \times K$. We discuss some of these results in more detail in the exercises in Section 4.

Exercises

1. Let D_4 stand for the group of the square and H the subgroup of rotations in D_4. Show that $H \lhd D_4$. Construct the quotient group D_4/H and show that $D_4/H \cong \mathbb{Z}_2$.

2. Let $\mathbb{C}^*$ stand for the multiplicative group of nonzero complex numbers and let $U = \{z \in \mathbb{C}^* \,||z|| = 1\}$.
   (a) Show that $U \lhd \mathbb{C}^*$.

(b) Show that each element of the quotient group $\mathbb{C}*/U$ corresponds to a circle in the complex plane.

(c) Show that the only element in $\mathbb{C}*/U$ that has finite order is the identity element U.

3. Construct the addition table for the group $\mathbb{Z}_{20}/([5])$ and verify that $\mathbb{Z}_{20}/([5]) \cong \mathbb{Z}_5$.

4. In (a) through (e) that follow, we have listed a group G, subgroup H, and element x in G. In each case show that H is a normal subgroup of G and find the order of the element xH in the quotient group G/H.

   (a) $G = \text{Sym}(3)$, $H = \{1, (1\ 2\ 3), (1\ 3\ 2)\}$, $x = (1\ 2)$
   (b) $G = \text{Sym}(3)$, $H = \{1, (1\ 2\ 3), (1\ 3\ 2)\}$, $x = (1\ 2)(1\ 3)$
   (c) $G = \text{Alt}(4)$, $H = \{1, (1\ 2)(3\ 4), (1\ 3)(2\ 4), (1\ 4)(2\ 3)\}$, $x = (1\ 2\ 3)$
   (d) $G = \mathbb{Z}_{20}$, $H = ([6])$, $x = [5]$
   (e) $G = \mathbb{Z}_{20}$, $H = ([5])$, $x = [2]$

6. Let $H = \{1, (1\ 2)\}$ be the cyclic subgroup of $\text{Sym}(3)$ generated by the transposition $(1\ 2)$.

   (a) Show that H is not a normal subgroup of $\text{Sym}(3)$.
   (b) The quotient set $\text{Sym}(3)/H$ contains three left cosets: H, $(1\ 3)H$, and $(2\ 3)H$. Multiply the cosets $(1\ 3)H$ and $(2\ 3)H$ elementwise and verify that this product set is not a left coset of H.
   (c) Partition the multiplication table of $\text{Sym}(3)$ into left cosets of H and verify that of the nine product sets, only three are left cosets.

7. Let G be a group and let $N \triangleleft G$.

   (a) If $H \leq G$, show that $H \cap N \triangleleft H$.
   (b) If $H \leq G$ and H is a simple group, show that either $H \leq N$ or $H \cap N = (1)$.

8. Let G be a group. If N and M are normal subgroups of G, show that $N \cap M \triangleleft G$.

9. Let G be a group and let H be a subgroup of G such that $[G:H] = 2$. Show that $H \triangleleft G$.

10. Let n be a positive integer.

    (a) Show that $\text{Alt}(n) \triangleleft \text{Sym}(n)$.
    (b) If $n \geq 2$, show that $\text{Sym}(n)/\text{Alt}(n) \cong \mathbb{Z}_2$.

11. Let $SL_2(\mathbb{R})$ stand for the special linear subgroup of $\text{Mat}_2^*(\mathbb{R})$ discussed in Exercise 17, Section 3, Chapter 2.

    (a) Show that $SL_2(\mathbb{R}) \triangleleft \text{Mat}_2^*(\mathbb{R})$.
    (b) Let $H = \{\left(\begin{smallmatrix} a & 0 \\ 0 & a \end{smallmatrix}\right) \in \text{Mat}_2^*(\mathbb{R}) \mid a \in \mathbb{R}^*\}$. Show that $H \triangleleft \text{Mat}_2^*(\mathbb{R})$.
    (c) Let $H = \{\left(\begin{smallmatrix} a & 0 \\ 0 & b \end{smallmatrix}\right) \in \text{Mat}_2^*(\mathbb{R}) \mid a, b \in \mathbb{R}^*\}$. Show that H is a subgroup of $\text{Mat}_2^*(\mathbb{R})$, but not a normal subgroup.

12. Let $H = \{1, (1\ 2)(3\ 4), (1\ 3)(2\ 4), (1\ 4)(2\ 3)\}$. Show that $H \triangleleft \text{Alt}(4) \triangleleft \text{Sym}(4)$ but that H is not a normal subgroup of $\text{Sym}(4)$. Thus, the normality relation on subgroups is not a transitive relation.

13. Let G be a group and let x be an element of G.

    (a) Show that $(x) \triangleleft G$ if and only if for every element $g \in G$, there is some integer n such that $gxg^{-1} = x^n$.

(b) Show that $|gxg^{-1}| = |x|$ for every element $g \in G$.

(c) If x has finite order and $gxg^{-1} = x^n$ for some integer n, show that n and $|x|$ are relatively prime.

14. Let $H = \{\sigma \in \text{Sym}(5) \,|\, \sigma(2) = 2\}$.

(a) Show that $\tau H \tau^{-1} = \{\sigma \in \text{Sym}(5) \,|\, \sigma(\tau(2)) = \tau(2)\}$ for every element $\tau \in \text{Sym}(5)$. Thus, $\tau H \tau^{-1}$ consists of those permutations in $\text{Sym}(5)$ that fix the number $\tau(2)$.

(b) Although τ may be any one of 120 different permutations, show that there are only 5 different subgroups of the form $\tau H \tau^{-1}$.

(c) Determine $\bigcap \tau H \tau^{-1}$, where the intersection is taken over all elements $\tau \in \text{Sym}(5)$.

15. Let G be a group and let $Z(G)$ stand for the center of G discussed in Exercise 15, Section 3, Chapter 2.

(a) Show that $Z(G) \triangleleft G$.

(b) If the group $G/Z(G)$ is cyclic, show that G must be abelian.

(c) Show that the index $[G:Z(G)]$ cannot be a prime number.

16. Let G be an abelian group and let $T = \{x \in G \,|\, x \text{ has finite order}\}$.

(a) Show that $T \triangleleft G$. T is called the *torsion subgroup* of G.

(b) Show that no element in the quotient group G/T has finite order except the identity element. G/T is said to be *torsion free*.

(c) Find the torsion subgroup of the multiplicative group $\mathbb{C}^*$. Describe this subgroup geometrically using the complex plane.

17. Let G be a group. If H and K are subgroups of G, let $HK = \{hk \in G \,|\, h \in H, k \in K\}$.

(a) Find an example to show that HK need not be a subgroup of G.

(b) If either H or K is a normal subgroup of G, show that HK is a subgroup of G and that $HK = KH$.

18. Let G be a group. A *normal factorization* of G is a pair of subgroups H, K of G such that $G = HK$, and at least one of H or K is normal in G.

(a) Find a normal factorization of the group $\text{Sym}(n)$.

(b) Find a normal factorization of the group $\text{Mat}_2^*(\mathbb{R})$.

(c) Find a normal factorization of the group $\mathbb{Z}$.

(d) Let $G = NH$ be a normal factorization of G such that $N \triangleleft G$ and $N \cap H = (1)$. Show that $G/N \cong H$.

19. Let G be a group and let X be a subset of G having the property that $y^{-1}xy \in X$ for every $x \in X$ and every $y \in G$. Show that the subgroup (X) generated by X is a normal subgroup of G.

20. Let G be a group and let G' stand for the subgroup generated by all elements of the form $xyx^{-1}y^{-1}$, where $x, y \in G$.

(a) Show that $G' \triangleleft G$. G' is called the *commutator subgroup* of G.

(b) Show that G is abelian if and only if $G' = (1)$.

(c) Show that G/G' is an abelian group for every group G.

(d) Suppose that N is a normal subgroup of G such that G/N is abelian. Show that $G' \leq N$.

(e) Suppose that N is a subgroup of G such that $G' \leq N$. Show that $N \triangleleft G$ and that G/N is abelian. It follows from this result and the result in part (d) that the normal subgroups of G for which the quotient group is abelian are precisely those subgroups that contain the commutator subgroup.

(f) Show that $\text{Sym}(n)' = \text{Alt}(n)$ for all positive integers n.

21. Let G be a finite group and let $N \triangleleft G$.
    (a) If $x \in G$, show that the order of the element xN in G/N is the smallest positive integer n such that $x^n \in N$.
    (b) Show that $|xN|$ divides $|x|$.

22. Show that $\text{Alt}(n)$ is a simple group for all integers $n \geq 5$ by modifying the proof in Example 6 that $\text{Alt}(5)$ is simple.

23. Let V be a real vector space and let U be a subspace of V. Then U is a normal subgroup of the additive group of V. Define scalar multiplication on the quotient group V/U by setting $c(A + U) = \{cx \in V \mid x \in A + U\}$ for all scalars $c \in \mathbb{R}$ and every coset $A + U \in V/U$.
    (a) Show that $c(A + U) = cA + U$.
    (b) Show that V/U is a real vector space under coset addition and scalar multiplication.
    (c) If V is finite-dimensional, show that V/U is also finite-dimensional and that $\dim V/U = \dim V - \dim U$.

3. GROUP-HOMOMORPHISM

Mappings between mathematical structures play an important role in all areas of mathematics. In calculus, for example, we deal with continuous functions, while in linear algebra it is linear transformations that are important. These functions are important because they preserve basic properties of the structure and, as such, may be used to transfer information from the one structure to the other in such a way that the information is preserved. In this section, our purpose is to discuss functions from one group to another that preserve the arithmetic on each group. These functions, called group-homomorphisms, therefore provide a method for transferring algebraic information from one group to another. We first discuss the basic properties of group-homomorphisms and show that every such mapping $f : G \to H$ has associated with it two subgroups: the image of f, denoted by Im f, which is a subgroup of H; and the kernel of f, denoted by Ker f, which is a normal subgroup of G. Following this, our main goal is to prove the fundamental theorem of group-homomorphism, which states that the quotient group $G/\text{Ker } f$ is isomorphic to Im f. It follows, then, that every group-homomorphism f determines a normal subgroup, Ker f, and the fundamental theorem interprets, or represents, the corresponding quotient group $G/\text{Ker } f$ as the image of f. Thus, from the point of view of mappings, quotient groups may be interpreted as homomorphic images of groups.

Definition

Let G and H be groups. A *group-homomorphism* is a function $f : G \to H$ such that $f(xy) = f(x)f(y)$ for all elements $x, y \in G$.

We think of a group-homomorphism informally as a mapping from one group to another that preserves the arithmetic on the groups; that is, the image of a product is the product of the images. If G is any group, the identity map $1 : G \to G$ is clearly a group-homomorphism since $1(xy) = xy = 1(x)1(y)$ for all elements $x, y \in G$. More generally, it is clear that any group-isomorphism is a group-homomorphism. Let us emphasize that, as with isomorphisms, the equation $f(xy) = f(x)f(y)$ must be interpreted within the context of the group operations. If the operations on G and H are written additively, for example, then the equation becomes $f(x + y) = f(x) + f(y)$. Let us look in more detail at some examples of group-homomorphisms.

EXAMPLE 1

Let G and H be groups. Define the function $f : G \to H$ by setting $f(x) = 1$, the identity element of H, for every element $x \in G$. Then f is a group-homomorphism since $f(xy) = 1 = f(x)f(y)$ for all $x, y \in G$. The mapping f is called the *trivial homomorphism* from G to H; it maps every element in G to the identity element of H.

EXAMPLE 2

Let n be an integer. Define the mapping $f_n : \mathbb{Z} \to \mathbb{Z}$ by setting $f_n(s) = ns$ for every integer $s \in \mathbb{Z}$. Then f_n is a group-homomorphism since $f_n(s + t) = n(s + t) = ns + nt = f_n(s) + f_n(t)$ for all integers s and t. For example, f_1 is the identity map since $f_1(s) = s$ for all $s \in \mathbb{Z}$, while f_0 is the trivial homomorphism since $f_0(s) = 0$ for all $s \in \mathbb{Z}$.

EXAMPLE 3. The determinant homomorphism

Let $\mathrm{Mat}_2^*(\mathbb{R})$ stand for the multiplicative group of invertible 2×2 real matrices and $\mathbb{R}^*$ the multiplicative group of nonzero real numbers. Define the mapping $f : \mathrm{Mat}_2^*(\mathbb{R}) \to \mathbb{R}^*$ by setting $f(M) = \det(M)$, the determinant of M, for every matrix $M \in \mathrm{Mat}_2^*(\mathbb{R})$; note that $f(M)$ is in fact an element of $\mathbb{R}^*$ since every invertible matrix has a nonzero determinant. Then f is a group-homomorphism since $f(MN) = \det(MN) = \det(M)\det(N) = f(M)f(N)$ for all matrices $M, N \in \mathrm{Mat}_2^*(\mathbb{R})$.

EXAMPLE 4

Let V and W be real vector spaces and let $L:V \to W$ be a linear transformation. Then L is a group-homomorphism from the additive group of V into the additive group of W since $L(x + y) = L(x) + L(y)$ for all vectors $x, y \in V$. For example, the mapping $L:\mathbb{R}^2 \to \mathbb{R}^2$ defined by setting $L(a, b) = (a, 0)$ for every $(a, b) \in \mathbb{R}^2$ is a linear transformation and hence a homomorphism; it projects the plane $\mathbb{R}^2$ onto the X-axis, as illustrated in the marginal figure.

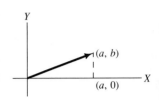

EXAMPLE 5

Define the function $f:\mathrm{Sym}(3) \to \mathrm{Sym}(3)$ by setting $f(1) = 1$, $f((1\ 2)) = (1\ 2)$, $f((1\ 2\ 3)) = 1$, $f((1\ 3)) = (1\ 2)$, $f((1\ 3\ 2)) = 1$, $f((2\ 3)) = (1\ 2)$. Then f is a group-homomorphism. To show that $f(xy) = f(x)f(y)$ for all $x, y \in \mathrm{Sym}(3)$, we calculate the image of the multiplication table for $\mathrm{Sym}(3)$ under f and verify that the entry in row $f(x)$ and column $f(y)$ is, in fact, $f(xy)$.

Sym(3)	1	(1 2 3)	(1 3 2)	(1 2)	(1 3)	(2 3)
1	1	(1 2 3)	(1 3 2)	(1 2)	(1 3)	(2 3)
(1 2 3)	(1 2 3)	(1 3 2)	1	(1 3)	(2 3)	(1 2)
(1 3 2)	(1 3 2)	1	(1 2 3)	(2 3)	(1 2)	(1 3)
(1 2)	(1 2)	(2 3)	(1 3)	1	(1 3 2)	(1 2 3)
(1 3)	(1 3)	(1 2)	(2 3)	(1 2 3)	1	(1 3 2)
(2 3)	(2 3)	(1 3)	(1 2)	(1 3 2)	(1 2 3)	1

	1	1	1	(1 2)	(1 2)	(1 2)
1	1	1	1	(1 2)	(1 2)	(1 2)
1	1	1	1	(1 2)	(1 2)	(1 2)
1	1	1	1	(1 2)	(1 2)	(1 2)
(1 2)	(1 2)	(1 2)	(1 2)	1	1	1
(1 2)	(1 2)	(1 2)	(1 2)	1	1	1
(1 2)	(1 2)	(1 2)	(1 2)	1	1	1

The table on the left is the multiplication table for $\mathrm{Sym}(3)$, while that on the right is its image under f. Since the image table is the multiplication table for 1 and (1 2), it follows that $f(xy) = f(x)(f(y)$ for all elements $x, y \in \mathrm{Sym}(3)$. Therefore, f is a group-homomorphism. On the other hand, if we define the function $g:\mathrm{Sym}(3) \to \mathrm{Sym}(3)$ by setting $g(1) = 1$, $g((1\ 2)) = 1$, $g((1\ 2\ 3)) = (1\ 2\ 3)$, $g((1\ 3)) = 1$, $g((1\ 3\ 2)) = (1\ 3\ 2)$, and $g((2\ 3)) = 1$, then g is not a group-homomorphism since, for example, $g((1\ 3))g((1\ 2)) = 1$ but $g((1\ 3)(1\ 2)) = g((1\ 2\ 3)) = (1\ 2\ 3)$.

EXAMPLE 6. Canonical homomorphisms

Let G be a group, N a normal subgroup of G, and let G/N be the corresponding quotient group. Define the function $f:G \to G/N$ by setting $f(x) = xN$ for every $x \in G$. Then $f(xy) = xyN = xNyN = f(x)f(y)$ for all elements $x, y \in G$ and hence f is a group-homomorphism. The homomorphism f is called the *canonical homomorphism* from G onto the quotient group G/N; it maps every element in G onto its left coset modulo N. For example, if n is a positive

$f:\mathbb{Z} \to \mathbb{Z}/4\mathbb{Z}$
$-1 \mapsto [3]$
$0 \mapsto [0]$
$1 \mapsto [1]$
$2 \mapsto [2]$
$3 \mapsto [3]$
$4 \mapsto [0]$

integer, then $n\mathbb{Z}$ is a normal subgroup of the additive group $\mathbb{Z}$ of integers and the canonical homomorphism $f:\mathbb{Z} \to \mathbb{Z}/n\mathbb{Z}$ maps a typical integer s to the coset $s + n\mathbb{Z}$. But $s + n\mathbb{Z} = [s]$, the congruence class of s mod n. Therefore, $f(s) = [s]$ and hence the canonical homomorphism $f:\mathbb{Z} \to \mathbb{Z}/n\mathbb{Z}$ maps every integer to its congruence class mod n. The table in the margin illustrates the images of some integers under the canonical homomorphism $f:\mathbb{Z} \to \mathbb{Z}/4\mathbb{Z}$.

These examples illustrate different types of group-homomorphisms, some between multiplicative groups and some between additive groups. But whatever the case, all group-homomorphisms have certain basic properties in common. The following result summarizes these properties.

PROPOSITION 1. Let $f:G \to H$ be a group-homomorphism. Then:

(1) $f(1) = 1$; that is, f maps the identity of G to the identity of H;

(2) $f(x^{-1}) = f(x)^{-1}$ for every element $x \in G$; that is, the image of an inverse is the inverse of the image;

(3) $f(x_1 \cdots x_n) = f(x_1) \cdots f(x_n)$ for any finite number of elements $x_1, \ldots, x_n \in G$;

(4) If x is an element of finite order in G, then its image $f(x)$ has finite order and $|f(x)|$ divides $|x|$.

Proof. Since $f(1) = f(1 \cdot 1) = f(1)f(1)$, it follows that $f(1) = 1$. Now, let $x \in G$. Then $xx^{-1} = x^{-1}x = 1$ and therefore $f(x)f(x^{-1}) = f(x^{-1})f(x) = f(1) = 1$. Hence, $f(x^{-1}) = f(x)^{-1}$, which proves statement (2). To prove statement (3), we use induction on n. If $n = 1$, the statement is trivial, while if $n = 2$, it is true since f is a group-homomorphism. Now assume the statement is true for any collection of n elements in G, $n > 2$, and let $x_1, \ldots, x_{n+1}$ be $n + 1$ elements in G. Then

$$f(x_1 \cdots x_{n+1}) = f(x_1(x_2 \cdots x_{n+1})) = f(x_1)f(x_2 \cdots x_{n+1}).$$

But this last factor is equal to $f(x_2) \cdots f(x_{n+1})$ by the induction hypothesis. Hence, $f(x_1 \cdots x_{n+1}) = f(x_1) \cdots f(x_{n+1})$. Therefore, statement (3) is true for all integers $n \geq 1$. Finally, to prove statement (4), let $n = |x|$, where x is an element of finite order in G. Then $x^n = 1$ and hence $f(x)^n = f(x^n) = f(1) = 1$. Therefore, $f(x)$ has finite order and its order divides n. ∎

Let us make a few comments about these results. The equation $f(1) = 1$ means that a group-homomorphism $f:G \to H$ maps the identity element of G to the identity element of H. If the identity elements happen to be denoted by the symbol 0, as is the case for additive groups, then the equation becomes $f(0) = 0$. This is the case, for example, in vector spaces: if $L:V \to W$ is a linear transformation of vector spaces, then $L(0) = 0$, where 0 stands for the zero vector of the appropriate space. Similar remarks also apply to statements (2) and (3); using additive notation, these results are written as $f(-x) = -f(x)$ and $f(x_1 + \cdots + x_n) = f(x_1) + \cdots + f(x_n)$. Finally, statement (4) shows that under a homomorphism, the order of an element cannot increase; it either remains the same or decreases and, in fact, divides the order of the original element. For example, the homomorphism $f:\text{Sym}(3) \to \text{Sym}(3)$ discussed in Example 5 maps the 3-cycle (1 2 3), whose order is 3, to the identity element, whose order is 1. This result is also useful for showing that certain homomorphisms cannot exist; for example, there is no homomorphism of Sym(3) mapping (1 2 3) to the 2-cycle (1 2) since $|(1\ 2)| = 2$, which does not divide 3.

We now turn our attention to two important subgroups associated with every group-homomorphism.

Definition

Let $f:G \to H$ be a group-homomorphism. The *image* of f is the subset $\text{Im } f = \{ f(x) \in H \mid x \in G \}$, consisting of the images under f of all elements in G. The *kernel* of f is the subset $\text{Ker } f = \{x \in G \mid f(x) = 1\}$, consisting of those elements in G that f maps to the identity element.

PROPOSITION 2. Let $f:G \to H$ be a group-homomorphism. Then:

(1) $\text{Ker } f \lhd G$; f is 1–1 if and only if $\text{Ker } f = (1)$.
(2) $\text{Im } f \le H$; f is onto if and only if $\text{Im } f = H$.

Proof. Let us prove statement (1) and leave the proof of (2) as an exercise for the reader. To show that $\text{Ker } f$ is a subgroup of G, first note that $1 \in \text{Ker } f$ since $f(1) = 1$. Now, let $x, y \in \text{Ker } f$. Then $f(x) = f(y) = 1$ and therefore $f(xy) = f(x)f(y) = 1$. Hence, $xy \in \text{Ker } f$. Similarly, $f(x^{-1}) = f(x)^{-1} = 1$ and hence $x^{-1} \in \text{Ker } f$. It now follows that $\text{Ker } f \le G$. To show that $\text{Ker } f$ is, in fact, a normal subgroup of G, let $y \in G$ and let $x \in \text{Ker } f$. Then $f(y^{-1}xy) = f(y^{-1})f(x)f(y) = f(y)^{-1}\, 1\, f(y) = 1$ and hence $y^{-1}xy \in \text{Ker } f$. Therefore, $\text{Ker } f \lhd G$. Finally, we claim that f is 1–1 if and only if $\text{Ker } f = (1)$. For if f is 1–1 and $x \in \text{Ker } f$, then $f(x) = 1 = f(1)$ and hence $x = 1$; thus $\text{Ker } f = (1)$. Conversely, if $\text{Ker } f = (1)$ and $x, y \in G$ are such that $f(x) = f(y)$, then $f(x^{-1}y) = f(x)^{-1}f(y) = 1$ and hence $x^{-1}y \in \text{Ker } f$, which shows that $x^{-1}y = 1$ or, equivalently, that $x = y$. Hence, f is 1–1. $\blacksquare$

For example, let $f:\mathbb{Z} \to \mathbb{Z}_n$ stand for the canonical homomorphism that maps every integer s to its congruence class mod n; that is, $f(s) = [s]$ for every $s \in \mathbb{Z}$. Clearly, f maps $\mathbb{Z}$ onto $\mathbb{Z}_n$ and hence $\text{Im } f = \mathbb{Z}_n$. On the other hand, $\text{Ker } f = \{s \in \mathbb{Z} \mid [s] = [0]\}$. Since $[s] = [0]$ if and only if $s \in n\mathbb{Z}$, it follows that $\text{Ker } f = n\mathbb{Z}$. In general, if N is a normal subgroup of a group G, then the canonical homomorphism $f:G \to G/N$ maps G onto the quotient group G/N and hence $\text{Im } f = G/N$, while $\text{Ker } f = \{x \in G \mid xN = N\} = N$. Thus, every normal subgroup of a group is the kernel of some homomorphism, namely, the canonical homomorphism mapping the group onto the corresponding quotient group. Since the kernel of a homomorphism is always a normal subgroup by Proposition 2, it follows that normal subgroups are the same as kernels of group-homomorphisms.

EXAMPLE 7

Let $L:V \to W$ be a linear transformation from a vector space V to a vector space W. Then Ker $L = \{x \in V \mid L(x) = 0\}$ and Im $L = \{L(x) \in W \mid x \in V\}$. In this case, the kernel and image of L are the same kernel and image that occur in linear algebra; not only is Ker L a normal subgroup of the additive group of V, it is also a subspace of V. Similarly, Im L is both a subgroup of the additive group of W as well as a subspace of W. For example, let $L:\mathbb{R}^2 \to \mathbb{R}^2$ be the linear transformation defined by setting $L(a, b) = (a, 0)$ for all vectors $(a, b) \in \mathbb{R}^2$. Then

$$\text{Ker } L = \{(a, b) \in \mathbb{R}^2 \mid (a, 0) = (0, 0)\} = \{(0, b) \in \mathbb{R}^2 \mid b \in \mathbb{R}\},$$

which is represented geometrically as the Y-axis. Similarly, Im $L = \{(a, 0) \in \mathbb{R}^2 \mid a \in \mathbb{R}\}$, which is the X-axis. In this case the homomorphism L is projecting the real plane $\mathbb{R}^2$ orthogonally onto the X-axis; under this projection, it is precisely the vectors on the Y-axis that collapse to the zero vector.

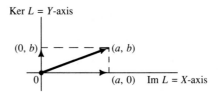

EXAMPLE 8

Let $f:\text{Sym}(3) \to \text{Sym}(3)$ stand for the group-homomorphism discussed in Example 5. Then Ker $f = \{1, (1\ 2\ 3), (1\ 3\ 2)\}$ and Im $f = \{1, (1\ 2)\}$. Thus, Ker f is the cyclic subgroup generated by $(1\ 2\ 3)$, which we showed is a normal subgroup of Sym(3) in Example 1, Section 2, and Im f is the cyclic subgroup generated by $(1\ 2)$.

If $f:G \to H$ is a group-homomorphism, there is an important relationship between the group G and the subgroups Ker f and Im f known as the fundamental theorem of group-homomorphism. Before we prove this result, let us first study the homomorphism $f:\text{Sym}(3) \to \text{Sym}(3)$ previously discussed in Examples 5 and 8. Let $K = \text{Ker } f = \{1, (1\ 2\ 3), (1\ 3\ 2)\}$. Then K is a normal subgroup of Sym(3) and hence partitions the multiplication table of Sym(3) into left cosets of K, as shown in Figure 1. Observe that every element in a given left coset maps to the same element in Sym(3); in other

Sym(3)	1	(1 2 3)	(1 3 2)	(1 2)	(1 3)	(2 3)
1	1	(1 2 3)	(1 3 2)	(1 2)	(1 3)	(2 3)
(1 2 3)	(1 2 3)	(1 3 2)	1	(1 3)	(2 3)	(1 2)
(1 3 2)	(1 3 2)	1	(1 2 3)	(2 3)	(1 2)	(1 3)
(1 2)	(1 2)	(2 3)	(1 3)	1	(1 3 2)	(1 2 3)
(1 3)	(1 3)	(1 2)	(2 3)	(1 2 3)	1	(1 3 2)
(2 3)	(2 3)	(1 3)	(1 2)	(1 3 2)	(1 2 3)	1

$f \longrightarrow$

f	1	1	1	(1 2)	(1 2)	(1 2)
1	1	1	1	(1 2)	(1 2)	(1 2)
1	1	1	1	(1 2)	(1 2)	(1 2)
1	1	1	1	(1 2)	(1 2)	(1 2)
(1 2)	(1 2)	(1 2)	(1 2)	1	1	1
(1 2)	(1 2)	(1 2)	(1 2)	1	1	1
(1 2)	(1 2)	(1 2)	(1 2)	1	1	1

Sym(3)/K	K	(12)K
K	K	(12)K
(12)K	(12)K	K

$\cong$

Im f	1	(1 2)
1	1	(1 2)
(1 2)	(1 2)	1

FIGURE 1. Multiplication table for Sym(3) partitioned by cosets of the kernel of the homomorphism f and the corresponding isomorphism between Sym(3)/K and Im f.

words, f is constant on any particular left coset of K, while distinct left cosets have distinct images in Sym(3). Thus, there is a 1–1 correspondence between the two left cosets of K and the two images in Sym(3). Also observe that the partitioned multiplication table for Sym(3) is the same as the multiplication table for Im f. Consequently, the correspondence between left cosets of K and their images is not only a 1–1 correspondence, but in fact preserves the arithmetic on each group and is therefore an isomorphism. Hence, Sym(3)/$K \cong$ Im f. In this way, the group-homomorphism f defines, or induces, an isomorphism between its image, Im f, and the quotient group Sym(3)/K. Let us now imitate this procedure and show that it is true for any group-homomorphism.

PROPOSITION 3. (FUNDAMENTAL THEOREM OF GROUP-HOMOMORPHISM) Let $f:G \to H$ be a group-homomorphism. Then $G/\mathrm{Ker}\ f \cong$ Im f.

Proof.   Let $K = \mathrm{Ker}\ f$. We first show that f maps every element of an arbitrary left coset of K onto the same element in Im f, use this fact to define a mapping $f^*:G/K \to$ Im f, and conclude by showing that f^* is a group-isomorphism.

(A) Let xK be an arbitrary left coset of K and let $u \in xK$. Then $u = xk$ for some element $k \in K$. Since f is a group-homomorphism and $f(k) = 1$, we have that $f(u) = f(xk) = f(x)f(k) = f(x)$. Thus, f maps every element in xK to the same element $f(x)$ in Im f.

(B) Now, define the mapping $f^*:G/K \to$ Im f by setting $f^*(xK) = f(x)$ for every left coset $xK \in G/K$. Then f^* is a function since, by part (A), it does not matter which element in xK we use to define $f^*(xK)$. We claim that f^* is a group-isomorphism. For let $xK, yK \in G/K$. Then

$$f^*(xKyK) = f^*(xyK) = f(xy) = f(x)f(y) = f^*(xK)f^*(yK).$$

Hence, f^* is a group-homomorphism. To show that f^* is 1–1, we need only verify that its kernel is trivial, that is, it contains only the identity element K of the group G/K. Let $xK \in \mathrm{Ker}\ f^*$. Then $f^*(xK) = f(x) = 1$. Hence, $x \in K$ and therefore $xK = K$. Thus, $\mathrm{Ker}\ f^* = \{K\}$ is trivial and hence f^* is a 1–1 mapping. Finally, the map f^* is onto since, for any element $f(x) \in$ Im f, we have that $f^*(xK) = f(x)$. It now follows that f^* is a group-isomorphism and the proof is complete. ■

For example, consider the determinant homomorphism $f:\mathrm{Mat}_2^*(\mathbb{R}) \to \mathbb{R}^*$, where $f(M) = \det(M)$ for every invertible 2×2 real matrix M. In this case, the kernel of f consists of all 2×2 matrices whose determinant is 1. Thus, $\mathrm{Ker}\ f = \mathrm{SL}_2(\mathbb{R})$, the 2-dimensional special linear subgroup (see Chapter 2, Section 3, Exercise 17). Moreover, the mapping f is onto; for if $c \in \mathbb{R}^*$ is any nonzero scalar, then $\begin{pmatrix} c & 0 \\ 0 & 1 \end{pmatrix}$ is an invertible matrix whose determinant is c. Therefore, Im $f = \mathbb{R}^*$ and hence, by the fundamental theorem,

$\text{Mat}_2^*(\mathbb{R})/SL_2(\mathbb{R}) \cong \mathbb{R}^*$. In this case, if $M \in \text{Mat}_2^*(\mathbb{R})$, then every matrix in the left coset $M\ SL_2(\mathbb{R})$ has the same determinant as M, and the isomorphism $f^*: \text{Mat}_2^*(\mathbb{R})/SL_2(\mathbb{R}) \to \mathbb{R}^*$ maps the coset $M\ SL_2(\mathbb{R})$ to $\det(M)$.

EXAMPLE 8. Evaluation homomorphisms

Let $F(\mathbb{R})$ stand for the set of functions from $\mathbb{R}$ into $\mathbb{R}$. We recall that $F(\mathbb{R})$ is a group under addition of functions. If $c \in \mathbb{R}$, define the function $\varphi_c: F(\mathbb{R}) \to \mathbb{R}$ by setting $\varphi_c(f) = f(c)$ for every function $f \in F(\mathbb{R})$. Then φ_c is a group-homomorphism since $\varphi_c(f + g) = (f + g)(c) = f(c) + g(c) = \varphi_c(f) + \varphi_c(g)$ for every $f, g \in F(\mathbb{R})$. The mapping φ_c is called the *evaluation homomorphism at c* since it evaluates a given function at the number c. Now, Ker $\varphi_c = \{f \in F(\mathbb{R}) \mid f(c) = 0\}$. This is the subgroup of all functions that are zero at c; geometrically, they are the functions whose graph passes through the point $(c, 0)$, as illustrated in the figure. On the other hand, the image of φ_c is the entire set of real numbers; for if $a \in \mathbb{R}$ is any real number and $f_a: \mathbb{R} \to \mathbb{R}$ is the constant function defined by $f_a(x) = a$ for all $x \in \mathbb{R}$, then $\varphi_c(f_a) = f_a(c) = a$. Thus, Im $\varphi_c = \mathbb{R}$. Hence, by the fundamental theorem, $F(\mathbb{R})/\text{Ker } \varphi_c \cong \mathbb{R}$. Observe that the coset $f_a + \text{Ker } \varphi_c$ consists of those functions whose value at c is a and hence may be thought of as the translation of Ker φ_c to the point $(a, 0)$. Under the isomorphism $F(\mathbb{R})/\text{Ker } \varphi_c \cong \mathbb{R}$, the coset $f_a + \text{Ker } \varphi_c$ corresponds to the real number a.

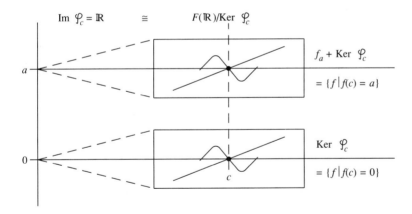

EXAMPLE 9. Homomorphisms of cyclic groups

Let us use our results about group-homomorphisms and our knowledge of cyclic groups to determine all homomorphisms from one cyclic group to another. Let n and m be positive integers, and let $\mathbb{Z}_n = (x)$ and $\mathbb{Z}_m = (y)$ be cyclic groups of orders n and m, respectively. If $f: \mathbb{Z}_n \to \mathbb{Z}_m$ is a homomorphism, then the image of f is the cyclic subgroup of $\mathbb{Z}_m$ generated by $f(x)$.

Let $\text{Im } f = (f(x)) = \mathbb{Z}_d$ and $\text{Ker } f = K_d$. Then, by the fundamental theorem, $\mathbb{Z}_n/K_d \cong \mathbb{Z}_d$.

(A) We claim first that the number d divides both n and m. The fact that d divides m is clear from Lagrange's theorem since $\mathbb{Z}_d$ has order d and is a subgroup of $\mathbb{Z}_m$. To show that d divides n, note that, by the fundamental theorem, $\mathbb{Z}_n/K_d \cong \mathbb{Z}_d$. Therefore, $|\mathbb{Z}_n| = |\mathbb{Z}_d|\,|K_d|$. Hence, $n = d|K_d|$, which shows that d divides n.

$$\mathbb{Z}_n \xrightarrow{\quad f \quad} \mathbb{Z}_m$$
$$\text{Ker } f = K_d \qquad \text{Im } f = \mathbb{Z}_d \cong \mathbb{Z}_n/K_d$$

It is now clear from the result in part (A) that if there is a homomorphism f mapping $\mathbb{Z}_n$ into $\mathbb{Z}_m$, then the order of $f(x)$, which is equal to d, must be a divisor of both n and m. Let us now show that every common divisor d of n and m in fact determines exactly $\varphi(d)$ such homomorphisms, where φ stands for the Euler phi-function.

(B) Let d be a divisor of both n and m. Then $\mathbb{Z}_m$ has a unique subgroup $\mathbb{Z}_d$ of order d, and we claim there are exactly $\varphi(d)$ homomorphisms from $\mathbb{Z}_n$ into $\mathbb{Z}_m$ whose image is $\mathbb{Z}_d$. To construct these homomorphisms, we map x to each of the possible generators of $\mathbb{Z}_d$. Recall that $\mathbb{Z}_d$ has $\varphi(d)$ generators, say $y_1, \ldots, y_{\varphi(d)}$. For $i = 1, \ldots, \varphi(d)$, define the mapping $f_i : \mathbb{Z}_n \to \mathbb{Z}_m$ by setting $f_i(x) = y_i$, and extend f_i to all of $\mathbb{Z}_n$ by setting $f_i(x^s) = y_i^s$ for every element $x^s \in \mathbb{Z}_n$. We must be careful to verify that f_i is, in fact, a function since x^s and x^t may represent the same element in $\mathbb{Z}_n$, even though s and t are different. If $x^s = x^t$, then $s \equiv t \pmod{n}$ and hence $s \equiv t \pmod{d}$ since d divides n. Therefore, $y_i^s = y_i^t$. Thus, f_i is indeed a function. Moreover, it is clear that each f_i is a group-homomorphism since $f_i(x^s x^t) = f_i(x^{s+t}) = y_i^{s+t} = y_i^s y_i^t = f_i(x^s) f_i(x^t)$ for all elements $x^s, x^t \in \mathbb{Z}_n$, and that $\text{Im } f_i = (y_i) = \mathbb{Z}_d$. Thus, we have constructed $\varphi(d)$ homomorphisms $f_1, \ldots, f_{\varphi(d)}$ from $\mathbb{Z}_n$ into $\mathbb{Z}_m$ having $\mathbb{Z}_d$ as image, and we claim, finally, that there are no other such homomorphisms. For suppose that $f : \mathbb{Z}_n \to \mathbb{Z}_m$ is any homomorphism whose image is $\mathbb{Z}_d$. Since the subgroup $\mathbb{Z}_d$ is unique and generated by $f(x)$, we must have that $f(x) = y_i$ for some i. It follows that $f(x^s) = y_i^s$ for all $x^s \in \mathbb{Z}_n$ and therefore $f = f_i$.

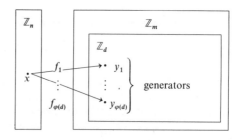

(C) The number of homomorphisms from $\mathbb{Z}_n$ into $\mathbb{Z}_m$ is equal to $\gcd(n, m)$. For, by parts (A) and (B), the number of such mappings is $\Sigma\ \varphi(d)$, where the summation is over all divisors d of $\gcd(n, m)$. But we showed in Chapter 1, Section 4, that $\Sigma\ \varphi(d) = \gcd(n, m)$. Hence, there are $\gcd(n, m)$ such mappings.

(D) Let us illustrate the preceding results by determining all homomorphisms $f: \mathbb{Z}_8 \to \mathbb{Z}_4$. Since $\gcd(8, 4) = 4$, there are four such homomorphisms. In this case, $m = 4$ and hence the possible values of d are 1, 2, and 4. Suppose that $d = 1$. Then $\varphi(d) = 1$ and hence there is only one possible homomorphism whose image is $\mathbb{Z}_1$, namely, the map obtained by setting $f(x) = 1$. Then $f(x^s) = 1$ for all $x^s \in \mathbb{Z}_8$ and hence f is the trivial homomorphism. Next, suppose that $d = 2$. Then $\varphi(d) = 1$ and hence there is one homomorphism whose image is $\mathbb{Z}_2$. The single element in $\mathbb{Z}_4$ having order 2 is y^2 and hence we set $f(x) = y^2$. Then $f(x^s) = y^{2s}$ for all $x^s \in \mathbb{Z}_8$. The table in the margin shows the image of each element in $\mathbb{Z}_8$ under the homomorphism f. In this case, $\text{Ker } f = \{1, x^2, x^4, x^6\}$ $= (x^2) \cong \mathbb{Z}_4$, $\text{Im } f = \{1, y^2\} \cong \mathbb{Z}_2$, and therefore the isomorphism corresponding to f is $\mathbb{Z}_8/\mathbb{Z}_4 \cong \mathbb{Z}_2$. Finally, suppose that $d = 4$. Then $\varphi(d) = 2$ and hence there are two homomorphisms whose image is $\mathbb{Z}_4$. The two generators of $\mathbb{Z}_4$ are y and y^3, and thus the two homomorphisms are defined by mapping $x \mapsto y$ and $x \mapsto y^3$. The figure here illustrates these two homomorphisms, their kernels, and the corresponding isomorphism.

$f: \mathbb{Z}_8 \mapsto \mathbb{Z}_4$
$f(x) = y^2$
$1 \mapsto 1$
$x \mapsto y^2$
$x^2 \mapsto 1$
$x^3 \mapsto y^2$
$x^4 \mapsto 1$
$x^5 \mapsto y^2$
$x^6 \mapsto 1$
$x^7 \mapsto y^2$

$\mathbb{Z}_8/\mathbb{Z}_2 \cong \mathbb{Z}_4$	
$f_1: \mathbb{Z}_8 \to \mathbb{Z}_4$	$f_2: \mathbb{Z}_8 \to \mathbb{Z}_4$
$f_1(x) = y$	$f_2(x) = y^3$
$1 \mapsto 1$	$1 \mapsto 1$
$x \mapsto y$	$x \mapsto y^3$
$x^2 \mapsto y^2$	$x^2 \mapsto y^2$
$x^3 \mapsto y^3$	$x^3 \mapsto y$
$x^4 \mapsto 1$	$x^4 \mapsto 1$
$x^5 \mapsto y$	$x^5 \mapsto y^3$
$x^6 \mapsto y^2$	$x^6 \mapsto y^2$
$x^7 \mapsto y^3$	$x^7 \mapsto y$
$\text{Ker } f_1 = \text{Ker } f_2 = \{1, x^4\} = \mathbb{Z}_2$	

EXAMPLE 10

Let $L: \mathbb{R}^2 \to \mathbb{R}^2$ stand for the linear transformation defined by setting $L(x, y) = (4x + 2y, 2x + y)$ for all vectors $(x, y) \in \mathbb{R}^2$. Then L is a group-homomorphism from the additive group of $\mathbb{R}^2$ into itself. Let $W = \text{Ker } L$

and $U = \text{Im } L$. Let us determine the kernel and image of L, and then discuss the corresponding isomorphism geometrically.

(A) We find that

$$W = \text{Ker } L = \{(x, y) \in \mathbb{R}^2 \,|\, 4x + 2y = 2x + y = 0\}$$
$$= \{(x, -2x) \in \mathbb{R}^2 \,|\, x \in \mathbb{R}\}$$
$$= \langle(1, -2)\rangle,$$

which is the line spanned by the vector $(1, -2)$, and, similarly,

$$U = \text{Im } L = \{(4x + 2y, 2x + y) \in \mathbb{R}^2 \,|\, x, y \in \mathbb{R}\}$$
$$= \{(2x + y)(2, 1) \in \mathbb{R}^2 \,|\, x, y \in \mathbb{R}\}$$
$$= \langle(2, 1)\rangle.$$

It now follows from the fundamental theorem that $\mathbb{R}^2/W \cong U$.

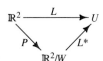

(B) We may interpret this isomorphism geometrically as follows. Let $P: \mathbb{R}^2 \to \mathbb{R}^2/W$ stand for the canonical homomorphism defined by $P(A) = A + W$ for all $A \in \mathbb{R}^2$, and let $L^*: \mathbb{R}^2/W \to U$ be the isomorphism associated with L, where $L^*(A + W) = L(A)$ for all $A \in \mathbb{R}^2$. Then $L(A) = L^*(A + W) = L^*(P(A))$ for all $A \in \mathbb{R}^2$. Thus, $L = L^*P$. We refer to the equation $L = L^*P$ as a factorization of the homomorphism L through the quotient group $\mathbb{R}^2/W$, as illustrated in the marginal diagram, and say that the diagram is commutative since either of the two paths from $\mathbb{R}^2$ to U result in the same image. The map P assigns to a typical vector $A \in \mathbb{R}^2$ the coset $A + W$, which is the line through A parallel to W, and every vector in $A + W$ is mapped by L onto $L(A)$.

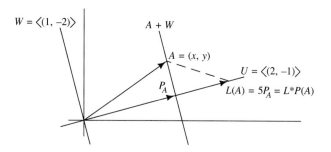

Now, in each coset $A + W$, we single out the unique vector P_A that lies on the line U. Since the dot product of $(1, -2)$ and $(2, 1)$ is zero, the lines $A + W$ and U are perpendicular. Therefore, P_A is the perpendicular projection of A onto U. Thus, we may think of the canonical homomorphism P as the perpendicular projection of the plane $\mathbb{R}^2$ onto U. L^* is then an isomorphism that maps each vector P_A to $L(A)$. Since U is a line, L^* must be a scalar magnification. To determine the scalar, let A

$= (x, y)$. We first calculate P_A by setting $P_A = (x, y) + \lambda(1, -2)$ and requiring λ to be such that $P_A \in U = \langle(2, 1)\rangle$. It now follows easily that $\lambda = \frac{1}{5}(2y - x)$ and hence that $P_A = \frac{1}{5}(4x + 2y, 2x + y)$. Therefore,

$$L^*(P_A) = L(A) = (4x + 2y, 2x + y) = 5P_A.$$

The unknown scalar is therefore equal to 5, and hence the isomorphism L^*, when regarded as an isomorphism from U to U, is scalar multiplication by a factor of 5. Thus, from the geometric point of view, the factorization $L = L^*P$ first projects the plane $\mathbb{R}^2$ orthogonally onto the line U and then magnifies by a factor of 5. It is the orthogonal projection that gives the nontrivial kernel W, and the scalar magnification that gives the entire line U as image.

EXAMPLE 11

Let $\mathbb{R}$ stand for the additive group of real numbers and $\mathbb{C}^*$ the multiplicative group of nonzero complex numbers. Define the function $W:\mathbb{R} \to \mathbb{C}^*$ by setting $W(x) = \cos x + i \sin x$ for all $x \in \mathbb{R}$. Then W is a group-homomorphism since $W(x + y) = \cos(x + y) + i \sin(x + y) = \cos x \cos y - \sin x \sin y + i \sin x \cos y + i \cos x \sin y = (\cos x + i \sin x)(\cos y + i \sin y) = W(x)W(y)$ for all $x, y \in \mathbb{R}$.

(A) The image of W is the unit circle in the complex plane. For let $U = \{z \in \mathbb{C}^* \,\|z\| = 1\}$ stand for the unit complex circle. Then $W(x) \in U$ since $\|W(x)\| = \sqrt{\cos^2 x + \sin^2 x} = 1$ for all $x \in \mathbb{R}$, and hence $\operatorname{Im} W \subseteq U$. On the other hand, U is a circle and therefore every point on U has the form $(\cos x, \sin x)$ for some $x \in \mathbb{R}$. Hence, $\operatorname{Im} W = U$.

(B) The kernel of W is the subgroup $2\pi\mathbb{Z}$ consisting of all integer multiples of 2π. For clearly, if $n \in \mathbb{Z}$, $W(2\pi n) = \cos 2\pi n + i \sin 2\pi n = 1$ and hence $2\pi\mathbb{Z} \subseteq \operatorname{Ker} W$. Conversely, if $x \in \operatorname{Ker} W$, then $W(x) = 1$ and hence $\cos x + i \sin x = 0$. Therefore, $x = 2\pi n$ for some integer n. Hence, $\operatorname{Ker} W = 2\pi\mathbb{Z}$.

(C) It follows from the fundamental theorem that $\mathbb{R}/2\pi\mathbb{Z} \cong U$. Let us interpret the homomorphism W geometrically. Attach a line tangent to the unit complex circle at the point $(1, 0) = W(0)$. When the line is "wrapped" around the circle, the number x on the line corresponds to the point $(\cos x, \sin x)$ on the circle. Thus, W is the "wrapping" function—that is, it wraps the line around the circle, as illustrated in the marginal figure. The fact that $\operatorname{Ker} W = 2\pi\mathbb{Z}$ corresponds to the statement that every interval of length 2π on the line is wrapped around the circle exactly once or, equivalently, that W is periodic with period 2π.

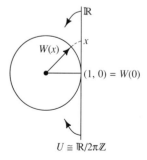

As we have seen throughout this section, a group-homomorphism $f:G \to H$ not only provides an important way of transferring information from G to H, it also determines the isomorphism $G/\operatorname{Ker} f \cong \operatorname{Im} f$. This, then, gives us a new and exceptionally important interpretation of quotient groups, namely, as homomorphic images.

Exercises

1. Determine which of the following functions are group-homomorphisms.
   (a) $f:\mathbb{Z} \to \mathbb{R}^*, f(n) = 2^n$

   (b) $f:\mathbb{R}^* \to \mathbb{R}^*, f(x) = \dfrac{1}{x}$

   (c) $f:\mathbb{Z} \to \mathbb{Z}, f(n) = 2n + 1$
   (d) $f:\mathbb{R} \to \mathbb{R}, f(x) = [x]$, the greatest integer in x
   (e) $f:\mathbb{Z} \to \mathbb{Z}, f(n) = n^2$
   (f) $\exp:\mathbb{R} \to \mathbb{R}^*, \exp(x) = e^x$

2. Let $\{1, -1\}$ be the multiplicative group consisting of the integers ± 1. Define the function $f:\text{Sym}(n) \to \{1, -1\}$ by setting $f(\sigma) = 1$ if σ is an even permutation, -1 if σ is odd.
   (a) Show that f is a group-homomorphism.
   (b) Find the kernel and image of f.
   (c) What isomorphism corresponds to f?

3. Let $r = (1\ 2\ 3\ 4)$ and $s = (1\ 4)(2\ 3)$ be permutations in $\text{Sym}(4)$ and let (r, s) stand for the subgroup generated by r and s.
   (a) Show that $(r, s) \cong D_4$, the group of the square.
   (b) Define the function $f:(r, s) \to (r, s)$ by setting

$$f(1) = 1 \qquad f(s) = s$$
$$f(r) = r^2 \qquad f(sr) = sr^2$$
$$f(r^2) = 1 \qquad f(sr^2) = s$$
$$f(r^3) = r^2 \qquad f(sr^3) = sr^2.$$

   Show that f is a group-homomorphism.
   (c) Find the kernel and image of f.
   (d) Illustrate the isomorphism $(r, s)/\text{Ker } f \cong \text{Im } f$ by partitioning the multiplication table for the group (r, s).

4. Define $f:\mathbb{C}^* \to \mathbb{C}^*$ by setting $f(z) = z^2$ for all complex numbers $z \in \mathbb{C}^*$.
   (a) Show that f is a group-homomorphism.
   (b) Show that $\mathbb{C}^*/\{1, -1\} \cong \mathbb{C}^*$.
   (c) It follows from part (b) that the group $\mathbb{C}^*$ has a proper quotient group that is isomorphic to $\mathbb{C}^*$. Is this possible for finite groups? That is, is it possible for a finite group G to have a proper normal subgroup N such that $G/N \cong G$?

5. Let $L:\mathbb{R}^2 \to \mathbb{R}^2$ stand for the linear transformation defined by setting $L(x, y) = (0, 3x - 2y)$ for all vectors $(x, y) \in \mathbb{R}^2$.

(a) Find the kernel and image of L and sketch them on the standard coordinate system for $\mathbb{R}^2$.

(b) Both $\mathbb{R}^2/\mathrm{Ker}\ L$ and $\mathrm{Im}\ L$ are one-dimensional real vector spaces. Let $L^*:\mathbb{R}^2/\mathrm{Ker}\ L \to \mathrm{Im}\ L$ stand for the isomorphism associated with L, where $L^*(A + \mathrm{Ker}\ L) = L(A)$ for all cosets $A + \mathrm{Ker}\ L \in \mathbb{R}^2/\mathrm{Ker}\ L$. Then L^* is a scalar transformation; that is, $L^*(P_A) = cP_A$ for some scalar c, where P_A is the unique vector in $A + \mathrm{Ker}\ L$ that lies on $\mathrm{Im}\ L$. Find the scalar c.

6. Find all group-homomorphisms $f:\mathbb{Z}_{12} \to \mathbb{Z}_8$.

7. Find all group-homomorphisms $f:\mathbb{Z}_4 \to \mathbb{Z}_4$.

8. Let n and d be positive integers such that d divides n. Show that $\mathbb{Z}_n/\mathbb{Z}_d \cong \mathbb{Z}_{n/d}$.

9. Let n be a positive integer. If two elements $[s]$ and $[t]$ in the group $\mathbb{Z}_n$ have the same order, show that there is a homomorphism $f:\mathbb{Z}_n \to \mathbb{Z}_n$ such that $f([s]) = [t]$.

10. Let G be an abelian group and let n be an integer. Define the function $f_n:G \to G$ by setting $f_n(x) = x^n$ for every element $x \in G$.

    (a) Show that f_n is a group-homomorphism.

    (b) If G is a finite group and n is relatively prime to the order of G, show that f_n is a group-isomorphism. In this case, show that every element in G has a unique nth root; that is, given $x \in G$, there is a unique element $y \in G$ such that $y^n = x$.

11. Let x be a generator of $\mathbb{Z}_{12}$. Define the function $f:\mathbb{Z}_{12} \to \mathbb{Z}_3 \times \mathbb{Z}_4$ by setting $f(x) = (x^4, x^3)$.

    (a) Show that f is a group-homomorphism.

    (b) Find the kernel and image of f.

    (c) Show that $\mathbb{Z}_{12} \cong \mathbb{Z}_3 \times \mathbb{Z}_4$.

12. Let x be a generator of $\mathbb{Z}_{12}$. Define the function $f:\mathbb{Z}_{12} \to \mathbb{Z}_2 \times \mathbb{Z}_6$ by setting $f(x) = (x^6, x^2)$.

    (a) Show that f is a group-homomorphism.

    (b) Find the kernel and image of f.

    (c) Is f a group-isomorphism?

13. Let $\mathbb{Z}_6 = (x)$ and $\mathbb{Z}_9 = (y)$, and let K stand for the cyclic subgroup of $\mathbb{Z}_6 \times \mathbb{Z}_9$ generated by the element (x^3, y^3). Show that $K \triangleleft \mathbb{Z}_6 \times \mathbb{Z}_9$ and that $\mathbb{Z}_6 \times \mathbb{Z}_9/K \cong \mathbb{Z}_3 \times \mathbb{Z}_3$.

14. If n and m are positive integers that are relatively prime, show that $\mathbb{Z}_n \times \mathbb{Z}_m \cong \mathbb{Z}_{nm}$.

15. Let G and H be finite groups having the same order and let $f:G \to H$ be a group-homomorphism. Show that f is 1–1 if and only if f is onto.

16. Let G and H be groups and let $G^* = \{(g, 1) \in G \times H | g \in G\}$ and $H^* = \{(1, h) \in G \times H | h \in H\}$.

    (a) Define the function $P_G:G \times H \to G$ by setting $P_G(g, h) = g$ for all pairs $(g, h) \in G \times H$. Show that P_G is a group-homomorphism whose kernel is H^* and that $G \times H/H^* \cong G$.

(b) Show that $G \times H/G^* \cong H$.

17. Define the function $f : \mathbb{Z}_4 \times \mathbb{Z}_4 \to \mathbb{Z}_4$ by setting $f(x, y) = xy^{-1}$ for all pairs $(x, y) \in \mathbb{Z}_4 \times \mathbb{Z}_4$. Show that f is a group-homomorphism and find its kernel and image.

18. Let G be a finite group and let H be a subgroup of G. For each element $x \in G$, define the mapping $\pi_x : G/H \to G/H$ by setting $\pi_x(gH) = xgH$ for every coset $gH \in G/H$.
    (a) Show that $\pi_x \in \text{Sym}(G/H)$ for every $x \in G$.
    (b) Define the function $f : G \to \text{Sym}(G/H)$ by setting $f(x) = \pi_x$ for every element $x \in G$. Show that f is a group-homomorphism.
    (c) Show that $[G : \text{Ker } f]$ divides $[G : H]!$.
    (d) If $|G|$ does not divide $[G : H]!$, show that G must contain a proper normal subgroup.
    (e) Using part (d) and the fact that the alternating group Alt(5) is simple, show that Alt(5) cannot have a subgroup of order greater than 12.

19. Let a and b be real numbers with $a \neq 0$. Define the mapping $T_{(a,b)} : \mathbb{R} \to \mathbb{R}$ by setting $T_{(a,b)}(x) = ax + b$ for all numbers $x \in \mathbb{R}$.
    (a) Show that $T_{(a,b)} \circ T_{(c,d)} = T_{(ac, ad+b)}$.
    (b) Show that $T_{(a,b)}$ is a permutation of $\mathbb{R}$. $T_{(a,b)}$ is called an *affine transformation* of $\mathbb{R}$.
    (c) Let $\text{Aff}(\mathbb{R}) = \{T_{(a,b)} \mid a, b \in \mathbb{R}, a \neq 0\}$. Show that $\text{Aff}(\mathbb{R})$ is a group under function composition. $\text{Aff}(\mathbb{R})$ is called the *affine group of the real line*.
    (d) Let $M = \{T_{(a,0)} \in \text{Aff}(\mathbb{R}) \mid a \in \mathbb{R}, a \neq 0\}$ and $T = \{T_{(1,b)} \in \text{Aff}(\mathbb{R}) \mid b \in \mathbb{R}\}$. Show that $M \leq \text{Aff}(\mathbb{R})$, $T \lhd \text{Aff}(\mathbb{R})$, and that $T \cap M = (1)$. The transformations in M are called *magnifications* since $T_{(a,0)}$ maps a number x to ax, while those in T are called *translations* since $T_{(1,b)}$ maps x to $x + b$.
    (e) Show that $\text{Aff}(\mathbb{R}) = TM$. That is, every affine transformation of $\mathbb{R}$ may be written as a product of a translation and a magnification.
    (f) Show that $\text{Aff}(\mathbb{R})/T \cong M$ by using the function $f : \text{Aff}(\mathbb{R}) \to M$ defined by setting $f(T_{(a,b)}) = a$ for all $T_{(a,b)} \in \text{Aff}(\mathbb{R})$.
    (g) Show that $M \cong \mathbb{R}^*$, the multiplicative group of nonzero real numbers, and that $T \cong \mathbb{R}$, the additive group of real numbers.

20. Let G be a group and suppose that $G = NH$ is a normal factorization of G, where $N \lhd G$. Show that $NH/N \cong H/(H \cap N)$.

21. Let G be a group with normal subgroups N and K such that $N \leq K \leq G$.
    (a) Show that $K/N \lhd G/N$.
    (b) Show that $(G/N)/(K/N) \cong G/K$.

22. Let $f : G \to H$ be a group-homomorphism and let H_1 be a subgroup of H.
    (a) Let $f^{-1}(H_1) = \{x \in G \mid f(x) \in H_1\}$ stand for the preimage of H_1 under f. Show that $f^{-1}(H_1) \leq G$ and that $\text{Ker } f \lhd f^{-1}(H_1)$.
    (b) If $H_1 \lhd H$, show that $f^{-1}(H_1) \lhd G$.
    (c) If f maps G onto H, show that $f^{-1}(H_1)/\text{Ker } f \cong H_1$.

(d) Let G_1 be any subgroup of G such that Ker $f \leq G_1$. Show that Ker $f \lhd G_1$ and that $G_1/\text{Ker } f \cong f(G_1)$.

23. Let G be a group and let $N \lhd G$.
   (a) Let $\bar{K} \leq G/N$. Show that there is a subgroup K of G such that $N \leq K$ and $\bar{K} = K/N$.
   (b) Show that there is a 1–1 correspondence between the set of all subgroups of G/N and the set of all subgroups of G that contain N.
   (c) Find all subgroups of $\mathbb{Z}_{12}$ that contain $\mathbb{Z}_3$, and all subgroups of $\mathbb{Z}_{12}/\mathbb{Z}_3$. Illustrate the 1–1 correspondence between these two collections of subgroups.
24. Let $f:G_1 \to G_2$ and $g:G_2 \to G_3$ be group-homomorphisms. Show that the composition $g \circ f:G_1 \to G_3$ is a group-homomorphism.

4. AUTOMORPHISMS AND AUTOMORPHISM GROUPS

In this section we turn our attention to a special type of group-homomorphism called an automorphism. Automorphisms are simply isomorphisms of a group with itself. Our goal is to first show that the automorphisms of a given group G themselves form a group, called the automorphism group of G and denoted by Aut(G), and then discuss the relationship between G and its automorphism group Aut(G). Automorphisms are important not only because they give rise to a new group—the automorphism group—but also because they reveal symmetries within the given group that are otherwise not apparent. Such symmetries are exceptionally important in understanding the structure and properties of the original group.

> **Definition**
>
> Let G be a group. An *automorphism* of G is any group-isomorphism $f:G \to G$. We let Aut(G) stand for the set of all automorphisms of G.

PROPOSITION 1. Let G be a group. Then Aut(G) is a subgroup of the symmetric group Sym(G).

Proof.   Let us first note that if f is an automorphism of G, then f is a 1–1 correspondence of the set G with itself and is therefore a permutation of G. Hence, Aut(G) is, in fact, a subset of Sym(G). Moreover, Aut(G) is nonempty since the identity map is clearly an automorphism of G. Now, let $f, g:G \to G$ be automorphisms of G. Then the composition $f \circ g$ is 1–1 and onto, and for every $x, y \in G$, $(f \circ g)(xy) = f(g(xy)) = f(g(x)g(y)) = f(g(x))f(g(y)) = (f \circ g)(x)(f \circ g)(y)$. Hence, $f \circ g$ is an automorphism of G. To show that f^{-1} is an automorphism, recall that f^{-1} is both 1–1 and onto.

Moreover, since $f(f^{-1}(xy)) = xy = f(f^{-1}(x))f(f^{-1}(y)) = f(f^{-1}(x)f^{-1}(y))$ for all x, $y \in G$, it follows that $f^{-1}(xy) = f^{-1}(x)f^{-1}(y)$ since f is 1–1. Thus, f^{-1} is an automorphism of G and we conclude, therefore, that $\text{Aut}(G)$ is a subgroup of $\text{Sym}(G)$. ∎

The group $\text{Aut}(G)$ is called the *automorphism group* of G. Since $\text{Aut}(G)$ is a subgroup of the symmetric group $\text{Sym}(G)$, we may think of the automorphisms of G as those permutations that preserve the arithmetic on the group. The identity map $1:G \to G$, for example, is an automorphism of any group G. As another example of an automorphism, suppose that G is an abelian group and define the map $f:G \to G$ by setting $f(x) = x^{-1}$ for all elements $x \in G$. Then f is an automorphism of G. For f is 1–1 since, if $x^{-1} = y^{-1}$, then $x = y$, and is onto since $f(x^{-1}) = (x^{-1})^{-1} = x$ for all $x \in G$. Finally, f preserves the arithmetic on G since $f(xy) = (xy)^{-1} = y^{-1}x^{-1} = x^{-1}y^{-1} = f(x)f(y)$ for all x, $y \in G$. Therefore, f is an automorphism of G. It follows, therefore, that every abelian group has at least one automorphism, namely, the function that maps every element to its inverse. Note, however, that if every element in the group has order 2, then this inversion automorphism is the identity map on the group; this is the case, for example, in the cyclic group $\mathbb{Z}_2$.

EXAMPLE 1. Automorphisms of cyclic groups

Let us determine the automorphism group of a finite cyclic group. Let $G = (x)$ stand for a cyclic group of order n generated by an element x.

(A) We recall from our discussion of cyclic groups in Chapter 1 that the isomorphisms mapping G to G are those functions of the form $f_s:G \to G$, where s is an integer relatively prime to n and where $f_s(x) = x^s$. Thus,

$$\text{Aut}(G) = \{f_s \mid 1 \le s \le n, \gcd(s, n) = 1\}.$$

Since the number of integers between 1 and n that are relatively prime to n is $\varphi(n)$, it follows that $|\text{Aut}(G)| = \varphi(n)$. Moreover, if s and t are relatively prime to n, then $(f_s \circ f_t)(x) = x^{st} = f_{st}(x)$ and hence multiplication in $\text{Aut}(G)$ satisfies the formula $f_s \circ f_t = f_{st}$. Thus, the automorphism group of a finite cyclic group of order n has order $\varphi(n)$ and consists of those mappings of the form f_s, where s is relatively prime to n, and products may be calculated using the formula $f_s f_t = f_{st}$.

(B) For example, let $G = \{1, x, x^2, x^3, x^4, x^5, x^6, x^7\}$ be a cyclic group of order 8. Then $|\text{Aut}(G)| = \varphi(8) = 4$ and hence G has four automorphisms, namely, f_1, f_3, f_5, and f_7, on page 151. To calculate the products $f_s f_t$, we use the rule that $f_s f_t = f_{st}$. For example, $f_3 f_5 = f_{15}$. But $f_{15} = f_7$ since $15 \equiv 7 \pmod 8$. Hence, $f_3 f_5 = f_7$. The complete multiplication table for $\text{Aut}(G)$ shows, for example, that all elements in $\text{Aut}(G)$ have order 2.

$$G = \{1, x, x^2, x^3, x^4, x^5, x^6, x^7\} \cong \mathbb{Z}_8$$

$$\mathrm{Aut}(G) = \{f_1, f_3, f_5, f_7\}$$

$$f_1 = \begin{pmatrix} 1 & x & x^2 & x^3 & x^4 & x^5 & x^6 & x^7 \\ 1 & x & x^2 & x^3 & x^4 & x^5 & x^6 & x^7 \end{pmatrix}$$

$$f_3 = \begin{pmatrix} 1 & x & x^2 & x^3 & x^4 & x^5 & x^6 & x^7 \\ 1 & x^3 & x^6 & x & x^4 & x^7 & x^2 & x^5 \end{pmatrix}$$

$$f_5 = \begin{pmatrix} 1 & x & x^2 & x^3 & x^4 & x^5 & x^6 & x^7 \\ 1 & x^5 & x^2 & x^7 & x^4 & x & x^6 & x^3 \end{pmatrix}$$

$$f_7 = \begin{pmatrix} 1 & x & x^2 & x^3 & x^4 & x^5 & x^6 & x^7 \\ 1 & x^7 & x^6 & x^5 & x^4 & x^3 & x^2 & x \end{pmatrix}$$

$\mathrm{Aut}(G)$	1	f_3	f_5	f_7
1	1	f_3	f_5	f_7
f_3	f_3	1	f_7	f_5
f_5	f_5	f_7	1	f_3
f_7	f_7	f_5	f_3	1

(C) The fact that $f_s f_t = f_{st}$ for all integers s and t suggests that there is a relationship between the group $\mathrm{Aut}(G)$ and the multiplicative group $U(\mathbb{Z}_8)$ discussed in Example 7 of Section 1. Recall that $U(\mathbb{Z}_8) = \{[1], [3], [5], [7]\}$ consists of those congruence classes $[s]$ for which s is relatively prime to 8 and that these classes multiply according to the rule $[s][t] = [st]$. Let us define a function $\alpha : U(\mathbb{Z}_8) \to \mathrm{Aut}(G)$ by setting $\alpha([s]) = f_s$ for all classes $[s] \in U(\mathbb{Z}_8)$. Then it is easily verified that α is a group-isomorphism and hence $\mathrm{Aut}(G) \cong U(\mathbb{Z}_8)$. Since the group $U(\mathbb{Z}_8)$ is the direct product of the subgroups $\{[1], [3]\}$ and $\{[1], [5]\}$, each of which is isomorphic to the cyclic group $\mathbb{Z}_2$, it follows that $\mathrm{Aut}(G) \cong U(\mathbb{Z}_8) \cong \mathbb{Z}_2 \times \mathbb{Z}_2$. The table in Figure 1 displays these isomorphisms. In the exercises we ask

$\mathrm{Aut}(G)$	1	f_3	f_5	f_7	$U(\mathbb{Z}_8)$	[1]	[3]	[5]	[7]	$\mathbb{Z}_2 \times \mathbb{Z}_2$	([1],[1])	([3],[1])	([1],[5])	([3],[5])
1	1	f_3	f_5	f_7	[1]	[1]	[3]	[5]	[7]	([1],[1])	([1],[1])	([3],[1])	([1],[5])	([3],[5])
f_3	f_3	1	f_7	f_5	[3]	[3]	[1]	[7]	[5]	([3],[1])	([3],[1])	([1],[1])	([3],[5])	([1],[5])
f_5	f_5	f_7	1	f_3	[5]	[5]	[7]	[1]	[3]	([1],[5])	([1],[5])	([3],[5])	([1],[1])	([3],[1])
f_7	f_7	f_5	f_3	1	[7]	[7]	[5]	[3]	[1]	([3],[5])	([3],[5])	([1],[5])	([3],[1])	([1],[1])
	$\mathrm{Aut}(G)$		$\cong$		$U(\mathbb{Z}_8)$		$\cong$			$\mathbb{Z}_2 \times \mathbb{Z}_2$				

FIGURE 1. Isomorphisms between $\mathrm{Aut}(G)$, $U(\mathbb{Z}_8)$, and $\mathbb{Z}_2 \times \mathbb{Z}_2$.

the reader to generalize this example and show that $\text{Aut}(\mathbb{Z}_n) \cong U(\mathbb{Z}_n)$ for all positive integers n.

The preceding examples illustrate several methods for obtaining automorphisms of abelian groups and cyclic groups. But if the group is nonabelian, the power mapping $x \mapsto x^n$ cannot be an automorphism since $(xy)^n$ is not, in general, equal to the product $x^n y^n$ for all elements in the group. For nonabelian groups, we must look elsewhere for automorphisms. It is here that inner automorphisms are useful.

Let G be a group and let g be an arbitrary element in G. Define the function $f_g: G \to G$ by setting $f_g(x) = gxg^{-1}$ for all elements $x \in G$. We claim that f_g is an automorphism of G. For, if x, $y \in G$, then $f_g(xy) = g(xy)g^{-1}$ $= (gxg^{-1})(gyg^{-1}) = f_g(x)f_g(y)$ and hence f_g is a group-homomorphism. Moreover, $f_g(x) = 1$ if and only if $x = 1$, which shows that $\text{Ker} f_g = (1)$ and hence that f_g is 1–1, and $f_g(g^{-1}xg) = x$, which shows that f_g is onto. Therefore, f_g is an automorphism of G.

> **Definition**
>
> Let G be a group. An *inner automorphism* of G is any automorphism of the form f_g for some element $g \in G$. We let $\text{Inn}(G)$ stand for the set of inner automorphisms of G.

PROPOSITION 2. Let G be a group. Then:

(1) $\text{Inn}(G) \lhd \text{Aut}(G)$
(2) $\text{Inn}(G) \cong G/Z(G)$

Proof. Define the function $\alpha: G \to \text{Aut}(G)$ by setting $\alpha(g) = f_g$, the inner automorphism defined by g, for every element $g \in G$. Then α is a group-homomorphism. For if g, $h \in G$ and $x \in G$, we have that $f_{gh}(x) = (gh)x(gh)^{-1}$ $= g(hxh^{-1})g^{-1} = f_g(f_h(x)) = (f_g f_h)(x)$ and hence $f_{gh} = f_g f_h$. Therefore, $\alpha(gh)$ $= f_{gh} = f_g f_h = \alpha(g)\alpha(h)$ and hence α is a homomorphism. Since the image of α is clearly equal to the set $\text{Inn}(G)$ of inner automorphisms of G, it follows that $\text{Inn}(G)$ is a subgroup of $\text{Aut}(G)$. For the kernel of α, we find that $\text{Ker}\,\alpha$ $= \{g \in G \mid f_g = 1\} = \{g \in G \mid gxg^{-1} = x \text{ for all } x \in G\} = Z(G)$, the center of G. Thus, since α is a homomorphism and $\text{Im}\,\alpha = \text{Inn}(G)$, it follows from the fundamental theorem of group-homomorphism that $G/Z(G) \cong \text{Inn}(G)$. Finally, to show that $\text{Inn}(G) \lhd \text{Aut}(G)$, let $f_g \in \text{Inn}(G)$ and $f \in \text{Aut}(G)$. We must show that $ff_g f^{-1} \in \text{Inn}(G)$. We claim, in fact, that $ff_g f^{-1} = f_{f(g)}$. For let $x \in G$. Then $(ff_g f^{-1})(x) = f(gf^{-1}(x)g^{-1}) = f(g)ff^{-1}(x)f(g^{-1}) = f(g)xf(g)^{-1} = f_{f(g)}(x)$. Hence, $ff_g f^{-1} = f_{f(g)}$ and therefore $\text{Inn}(G) \lhd \text{Aut}(G)$. ∎

EXAMPLE 2

Let us determine the automorphism group of Sym(3). Observe first that since the center of Sym(3) is trivial, it follows from Proposition 2 that Inn(Sym(3)) $\cong$ Sym(3). Hence, there are six inner automorphisms of Sym(3), one for each of the six elements of the group. For example, if $\sigma = (1\ 2\ 3)$, we find that

$$f_\sigma(1) = \sigma 1 \sigma^{-1} = 1 \qquad\qquad f_\sigma((1\ 2)) = \sigma(1\ 2)\sigma^{-1} = (2\ 3)$$

$$f_\sigma((1\ 2\ 3)) = \sigma(1\ 2\ 3)\sigma^{-1} = (1\ 2\ 3) \qquad f_\sigma((1\ 3)) = \sigma(1\ 3)\sigma^{-1} = (1\ 2)$$

$$f_\sigma((1\ 3\ 2)) = \sigma(1\ 3\ 2)\sigma^{-1} = (1\ 3\ 2) \qquad f_\sigma((2\ 3)) = \sigma(2\ 3)\sigma^{-1} = (1\ 3).$$

We display f_σ as follows:

$$f_\sigma = \begin{pmatrix} 1 & (1\ 2\ 3) & (1\ 3\ 2) & (1\ 2) & (1\ 3) & (2\ 3) \\ 1 & (1\ 2\ 3) & (1\ 3\ 2) & (2\ 3) & (1\ 2) & (1\ 3) \end{pmatrix}.$$

Notice that the automorphism f_σ expresses an internal symmetry of the group Sym(3) in the sense that it fixes the three elements 1, (1 2 3), and (1 3 2), while cyclically permuting the three 2-cycles. We display this symmetry by means of the following diagram.

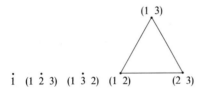

Each of the remaining inner automorphisms of Sym(3) are found similarly and are shown in the following table. We now claim that these six inner automorphisms account for all the automorphisms of Sym(3). To show this, suppose that f is any automorphism of Sym(3). Then $f((1\ 2\ 3))$ must have order 3 and consequently there are at most two choices for $f((1\ 2\ 3))$: either (1 2 3) or (1 3 2). Similarly, $f((1\ 2))$ must have order 2 and hence is either (1 2), (1 3), or (2 3). Since (1 2 3) and (1 2) generate Sym(3), f is completely determined by where it maps (1 2 3) and (1 2). Thus, there are at most six possible automorphisms of Sym(3). Since we have already found six automorphisms, we conclude that Aut(Sym(3)) $= \{1, f_{(123)}, f_{(132)}, f_{(12)}, f_{(13)}, f_{(23)}\}$. In particular, Aut(Sym(3)) $\cong$ Sym(3).

$$\mathrm{Inn}(\mathrm{Sym}(3))$$

$$f_1 = \begin{pmatrix} 1 & (1\ 2\ 3) & (1\ 3\ 2) & (1\ 2) & (1\ 3) & (2\ 3) \\ 1 & (1\ 2\ 3) & (1\ 3\ 2) & (1\ 2) & (1\ 3) & (2\ 3) \end{pmatrix} \quad \cdot \cdot \cdot \cdot \cdot \cdot$$

$$f_{(123)} = \begin{pmatrix} 1 & (1\ 2\ 3) & (1\ 3\ 2) & (1\ 2) & (1\ 3) & (2\ 3) \\ 1 & (1\ 2\ 3) & (1\ 3\ 2) & (2\ 3) & (1\ 2) & (1\ 3) \end{pmatrix} \quad \cdot \cdot \cdot \triangle$$

$$f_{(132)} = \begin{pmatrix} 1 & (1\ 2\ 3) & (1\ 3\ 2) & (1\ 2) & (1\ 3) & (2\ 3) \\ 1 & (1\ 2\ 3) & (1\ 3\ 2) & (1\ 3) & (2\ 3) & (1\ 2) \end{pmatrix} \quad \cdot \cdot \cdot \triangle$$

$$f_{(12)} = \begin{pmatrix} 1 & (1\ 2\ 3) & (1\ 3\ 2) & (1\ 2) & (1\ 3) & (2\ 3) \\ 1 & (1\ 3\ 2) & (1\ 2\ 3) & (1\ 2) & (2\ 3) & (1\ 3) \end{pmatrix} \quad \cdot \cdot \,\rule{0.5cm}{0.4pt}\,\rule{0.5cm}{0.4pt}$$

$$f_{(13)} = \begin{pmatrix} 1 & (1\ 2\ 3) & (1\ 3\ 2) & (1\ 2) & (1\ 3) & (2\ 3) \\ 1 & (1\ 3\ 2) & (1\ 2\ 3) & (2\ 3) & (1\ 3) & (1\ 2) \end{pmatrix} \quad \cdot \cdot \,\rule{0.5cm}{0.4pt}\,\rule{0.5cm}{0.4pt}$$

$$f_{(23)} = \begin{pmatrix} 1 & (1\ 2\ 3) & (1\ 3\ 2) & (1\ 2) & (1\ 3) & (2\ 3) \\ 1 & (1\ 3\ 2) & (1\ 2\ 3) & (1\ 3) & (1\ 2) & (2\ 3) \end{pmatrix} \quad \cdot \cdot \,\rule{0.5cm}{0.4pt}\,\rule{0.5cm}{0.4pt}$$

An automorphism of a group that is not an inner automorphism is called an *outer automorphism* of the group. Although the inner automorphisms of a group G form a normal subgroup $\mathrm{Inn}(G)$ of $\mathrm{Aut}(G)$, which is related structurally to G by means of the isomorphism $\mathrm{Inn}(G) \cong G/Z(G)$, the outer automorphisms do not form a group. They correspond, instead, to the cosets in the quotient group $\mathrm{Aut}(G)/\mathrm{Inn}(G)$, although this correspondence is not usually 1–1. The quotient group $\mathrm{Aut}(G)/\mathrm{Inn}(G)$ nevertheless provides one way to deal with the outer automorphisms of G and for this reason is usually referred to as the *group of outer automorphisms of G*. In the case of the group $\mathrm{Sym}(3)$, for example, the group of outer automorphisms is trivial—a statement that we interpret as meaning that there are no outer automorphisms of $\mathrm{Sym}(3)$. On the other hand, if G is abelian, then every automorphism other than the identity automorphism is an outer automorphism, and therefore $\mathrm{Aut}(G)/\mathrm{Inn}(G) = \mathrm{Aut}(G)/(1) \cong \mathrm{Aut}(G)$.

Exercises

1. Let f be an automorphism of a group G and let $x \in G$. Show that $|f(x)| = |x|$.
2. Show that $\mathrm{Aut}(\mathbb{Z}) = \{1, -1\}$, where 1 stands for the identity automorphism and -1 the inversion automorphism defined by $-1(n) = -n$ for every integer n. Show that $\mathrm{Aut}(\mathbb{Z}) \cong \mathbb{Z}_2$.
3. Let V_4 stand for the Klein 4-group (see Chapter 2, Section 2, Exercise 3). Show that $\mathrm{Aut}(V_4) \cong \mathrm{Sym}(3)$ and $\mathrm{Inn}(V_4) \cong (1)$.
4. Let D_4 stand for the group of the square.
   (a) Show that $\mathrm{Inn}(D_4) \cong \mathbb{Z}_2 \times \mathbb{Z}_2$.

(b) Show that $\text{Aut}(D_4) \cong D_4$.

(c) Show that $\text{Aut}(D_4)/\text{Inn}(D_4) \cong \mathbb{Z}_2$

(d) Find the outer automorphisms of D_4.

5. Let n be a positive integer. Show that $\text{Aut}(\mathbb{Z}_n) \cong U(\mathbb{Z}_n)$.

6. Let n be a positive integer.

   (a) If f is any automorphism of the symmetric group $\text{Sym}(n)$, show that $f(\sigma)$ and σ have the same cycle structure for every element $\sigma \in \text{Sym}(n)$.

   (b) Let σ and σ' be permutations in $\text{Sym}(n)$ that have the same cycle structure. Show that there is an inner automorphism f of $\text{Sym}(n)$ such that $f(\sigma) = \sigma'$. Thus, in the symmetric groups, there is an automorphism mapping one element to another if and only if the two elements have the same cycle structure.

7. Let $\mathbb{R}^2$ stand for the additive group of the two-dimensional real plane.

   (a) Show that $GL(\mathbb{R}^2) \le \text{Aut}(\mathbb{R}^2)$. Thus, every nonsingular linear transformation of $\mathbb{R}^2$ is an automorphism of the additive group of $\mathbb{R}^2$.

   (b) Let $\left(\begin{smallmatrix} a & b \\ c & d \end{smallmatrix}\right)$ be an invertible 2×2 real matrix. Define the function $L: \mathbb{R}^2 \to \mathbb{R}^2$ by setting $L(x, y) = (ax + by, cx + dy)$ for all vectors $(x, y) \in \mathbb{R}^2$.

   Show that L is an automorphism of $\mathbb{R}^2$.

   (c) Referring to part (b), illustrate geometrically the automorphisms of $\mathbb{R}^2$ defined by matrices $\left(\begin{smallmatrix} 0 & 1 \\ 1 & 0 \end{smallmatrix}\right)$ and $\left(\begin{smallmatrix} -1 & 0 \\ 0 & -1 \end{smallmatrix}\right)$.

8. Let $\mathbb{C}$ stand for the additive group of complex numbers.

   (a) Complex conjugation is the function $f: \mathbb{C} \to \mathbb{C}$ defined by $f(a + bi) = a - bi$ for all $a + bi \in \mathbb{C}$. Show that complex conjugation is an automorphism of $\mathbb{C}$ that has order 2. Illustrate this automorphism geometrically using the complex plane.

   (b) Is complex conjugation an automorphism of the multiplicative group $\mathbb{C}^*$ of nonzero complex numbers?

9. Let N be a normal subgroup of a group G.

   (a) Show that $f(N) = N$ for every inner automorphism f of G.

   (b) Let f be an inner automorphism of G and let $f|N: N \to N$ stand for the restriction of f to N. Show that $f|N$ is an automorphism of N.

   (c) Define the map $\text{Res}: \text{Inn}(G) \to \text{Aut}(N)$ by setting $\text{Res}(f) = f|N$ for all $f \in \text{Inn}(G)$. Show that Res is a group-homomorphism.

   (d) Give an example of a group G having an inner automorphism f and a normal subgroup N such that $f|N$ is an outer automorphism of N.

10. Let G be a group. A subgroup H of G is called a *characteristic subgroup* if $f(H) = H$ for every automorphism f of G.

    (a) Show that a characteristic subgroup of G is a normal subgroup of G.

    (b) Show that the center $Z(G)$ is a characteristic subgroup of G.

    (c) Show that the commutator subgroup G' discussed in Exercise 20, Section 2, is a characteristic subgroup of G.

    (d) Let G be a cyclic group. Show that every subgroup of G is a characteristic subgroup.

(e) Show that the group $\mathbb{Z}_2 \times \mathbb{Z}_2$ has a normal subgroup that is not characteristic.

(f) Let H be a characteristic subgroup of G. Define the map $\text{Res}: \text{Aut}(G) \rightarrow \text{Aut}(H)$ by setting $\text{Res}(f) = f|H$, the restriction of f to H, for all automorphisms $f \in \text{Aut}(G)$. Show that $\text{Res}(f)$ is, in fact, an automorphism of H for all $f \in \text{Aut}(G)$ and that the map Res is a group-homomorphism.

(g) Referring to part (f), show that $\text{Ker}(\text{Res}) = \{f \in \text{Aut}(G) | f(x) = x$ for every $x \in H\}$; that is, the kernel of Res consists of all automorphisms of G that leave every element in H fixed.

11. Let G be a group. The purpose of this exercise is to show that G may be embedded in a larger group in such a way that every automorphism of G is the restriction of some inner automorphism of the larger group. Let f be an automorphism of G.

(a) If $\pi \in \text{Sym}(G)$, show that $f\pi f^{-1} \in \text{Sym}(G)$.

(b) Define the mapping $\alpha_f : \text{Sym}(G) \rightarrow \text{Sym}(G)$ by setting $\alpha_f(\pi) = f\pi f^{-1}$ for all $\pi \in \text{Sym}(G)$. Show that α_f is an inner automorphism of $\text{Sym}(G)$.

(c) Let $G_L = \{\pi_x \in \text{Sym}(G) | x \in G\}$ stand for the Cayley embedding of G into $\text{Sym}(G)$. Show that $\alpha_f(\pi_x) = \pi_{f(x)}$ for every element $x \in G$. Thus, if the automorphism f of G is regarded as the automorphism α_f of G_L, then f is the restriction to G_L of the inner automorphism α_f of $\text{Sym}(G)$.

(d) Define the function $\gamma : \text{Aut}(G) \rightarrow \text{Aut}(\text{Sym}(G))$ by setting $\gamma(f) = \alpha_f$ for every automorphism f of G. Show that γ is a 1–1 group-homomorphism. It follows that the automorphism group of any group G may be embedded in the automorphism group of $\text{Sym}(G)$ in such a way that every automorphism of G is the restriction of some inner automorphism of $\text{Sym}(G)$.

12. Let f be an automorphism of a group G. An element $x \in G$ is called a *fixed-point* of f if $f(x) = x$. If G is a finite group, show that the number of fixed-points of an automorphism must divide the order of G.

13. An automorphism f of a group G is said to be *fixed-point free* if the identity element of G is the only fixed-point of f. Let G be a finite group and suppose that G has a fixed-point free automorphism f.

(a) Show that every element in G may be written in the form $x^{-1}f(x)$ for some element $x \in G$ by setting up a 1–1 correspondence between G and the set $\{x^{-1}f(x) \in G | x \in G\}$.

(b) If f has order 2, show that $f(x) = x^{-1}$ for every element $x \in G$.

(c) If f has order 2, show that G is abelian.

14. Let G be a finite abelian group of even order. Show that G contains an element of order 2.

15. **Semi-direct products.** Let G and H be groups and let $\alpha : H \rightarrow \text{Aut}(G)$ be a group-homomorphism; that is, $\alpha(h)$ is an automorphism of G for each element $h \in H$, and $\alpha(hh') = \alpha(h)\alpha(h')$ for all $h, h' \in H$. Define a binary operation on the Cartesian product $G \times H$ as follows: if (g, h), (g', h')

$\in G \times H$, set $(g, h)(g', h') = (g[\alpha(h)](g'), hh')$, where $g[\alpha(h)](g')$ means g multiplied by the image of g' under the automorphism $\alpha(h)$.

(a) Show that $G \times H$ is a group under this binary operation. It is called the *semi-direct product* of G and H with respect to the automorphism α and is denoted by the symbol $G \rtimes H$.

(b) Let $G^* = \{(g, 1) \in G \rtimes H \mid g \in G\}$. Show that $G^* \lhd G \rtimes H$ and $G^* \cong G$.

(c) Choose the trivial homomorphism for α; that is, $\alpha(h) = 1$, the identity automorphism of G, for all $h \in H$. Show that $G \rtimes H = G \times H$, the direct product of G and H.

(d) Let r stand for the rotation of order 4 in the group D_4 of the square and let s be any reflection in D_4. Then $(r) \lhd D_4$. Define the map $\alpha:(s) \to \mathrm{Aut}((r))$ by setting $\alpha(s) = f_s|(r)$; that is, $[\alpha(s)](r) = srs^{-1}$. Show that $(r) \rtimes (s) \cong D_4$. Thus, the group D_4 of the square is a semi-direct product of two of its subgroups.

(e) Show that the group D_4 of the square is not the direct product of any two of its subgroups.

5

Permutation Representations
of Groups

Recall that in Chapter 3 we showed that an element x of an arbitrary group G can always be represented by the permutation $\pi_x : g \mapsto xg$ which multiplies every element of G on the left by x. The mapping $x \leftrightarrow \pi_x$ is then an isomorphism between the group G and its representation as a group of permutations. This representation, called the Cayley representation, provides one way to represent the elements of G as permutations. But there are other ways to represent the elements of a group as permutations. The transformations in the group of the triangle, for example, may be represented as permutations of the vertices of an equilateral triangle. Our goal in this chapter is to discuss these and other such methods for representing the elements of a group as permutations. In the first section we discuss the basic concepts and terminology of permutation representations, while the second section deals with an application of these ideas to the Burnside Counting Theorem, a theorem that is especially useful for counting the number of ways of choosing objects from a set when several choices are considered to be the same.

1. REPRESENTING GROUPS AS PERMUTATION GROUPS

Let us begin with an observation. As previously noted, when we represent a group by means of permutations, our basic goal is to associate a permutation with each element of the group. Moreover, we require that the permutation corresponding to the product of two group elements should be the product of the corresponding permutations. Thus, the association between group elements and permutations is to be a group-homomorphism. With this observation in mind, we give the following definition.

Definition

Let G be a group and let X be a nonempty set. A *permutation representation* of G on X is a group-homomorphism $f : G \to \mathrm{Sym}(X)$. The representation is said to be *faithful* if its kernel contains only the identity element. If X is finite, the number of elements in X is called the *degree* of the representation.

If $f : G \to \mathrm{Sym}(X)$ is a permutation representation of G on a set X and $\sigma \in G$, we write f_σ for the permutation of X corresponding to σ instead of $f(\sigma)$, and say that f *represents* the element σ by means of the permutation f_σ on X. The set $\{ f_\sigma \in \mathrm{Sym}(X) \mid \sigma \in G \}$ of all such permutations is the image of the homomorphism f and hence is a subgroup of $\mathrm{Sym}(X)$. When G is regarded in this way as a group of permutations, we refer to it as a *permutation group acting on the set* X: each element $\sigma \in G$ acts on, or permutes, each element $x \in X$ according to the permutation $x \mapsto f_\sigma(x)$. If the representation f is faithful, then the correspondence $\sigma \leftrightarrow f_\sigma$ between elements in G and the permutations that represent them is a $1-1$ correspondence and hence a group-isomorphism.

The Cayley representation, for example, is the permutation representation $\pi : G \to \mathrm{Sym}(G)$ defined by setting $\pi_\sigma(x) = \sigma x$ for all $\sigma, x \in G$. In this case, an element $\sigma \in G$ is represented by the permutation π_σ that multiplies every element in the group on the left by σ. The Cayley representation is faithful since $\pi_\sigma = 1$, the identity permutation, if and only if $\sigma = 1$, and hence distinct elements of G correspond to distinct permutations.

In general, the kernel of a permutation representation f is the subgroup $\mathrm{Ker}\, f = \{ \sigma \in G \mid f_\sigma = 1 \}$ consisting of all elements $\sigma \in G$ that fix every element in X, and it is a normal subgroup of G since f is a group-homomorphism. If f is not faithful, then $\mathrm{Ker}\, f \neq (1)$, and hence there are at least two distinct elements $\sigma, \tau \in G$ that are represented by the same permutation; that is, $f_\sigma = f_\tau$. For example, if X is any nonempty set, the *trivial representation of G on X* is the homomorphism $f : G \to \mathrm{Sym}(X)$ that maps every element in G to the identity permutation on X; that is, $f_\sigma = 1$, the identity permutation on X, for all $\sigma \in G$. In this case, every element in G is represented by the same permutation, and hence the representation is not faithful unless, of course, the group G is trivial.

EXAMPLE 1

Let us show that the group of the triangle acts as a faithful permutation group of degree 3 on the vertices of the triangle. Let V stand for the set of vertices of an equilateral triangle and T for the group of the triangle. Then the elements in T are rigid motions of the plane that bring the triangle into coincidence with itself, and consequently, they permute the vertices of the triangle. Thus, we define the mapping $f: T \rightarrow \mathrm{Sym}(V)$ by setting $f_\sigma(v) = \sigma(v)$ for all vertices $v \in V$ and all motions $\sigma \in T$; f_σ is the permutation of the vertices defined by the motion σ. In the marginal figure we have labeled the three vertices of the triangle with the numbers 1, 2, and 3. The six permutations corresponding to the six motions in T are

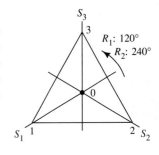

$$f_1 = \begin{pmatrix} 1 & 2 & 3 \\ 1 & 2 & 3 \end{pmatrix} \qquad f_{S_1} = \begin{pmatrix} 1 & 2 & 3 \\ 1 & 3 & 2 \end{pmatrix}$$

$$f_{R_1} = \begin{pmatrix} 1 & 2 & 3 \\ 2 & 3 & 1 \end{pmatrix} \qquad f_{S_2} = \begin{pmatrix} 1 & 2 & 3 \\ 3 & 2 & 1 \end{pmatrix}$$

$$f_{R_2} = \begin{pmatrix} 1 & 2 & 3 \\ 3 & 1 & 2 \end{pmatrix} \qquad f_{S_3} = \begin{pmatrix} 1 & 2 & 3 \\ 2 & 1 & 3 \end{pmatrix}.$$

We now claim that f is a faithful permutation representation of T of degree 3. To show this, let σ and τ be elements in T and let v be any vertex in V. Then $f_{\sigma\tau}(v) = (\sigma\tau)(v) = \sigma(\tau(v)) = f_\sigma(f_\tau(v)) = (f_\sigma f_\tau)(v)$. Hence, $f_{\sigma\tau} = f_\sigma f_\tau$. Therefore, f is a group-homomorphism and hence a permutation representation of T. Since $|V| = 3$, f has degree 3. Finally, f is faithful since the only rigid motion of the triangle that fixes every vertex is the identity transformation.

EXAMPLE 2

Let n be a positive integer. Every element in the symmetric group $\mathrm{Sym}(n)$ is, by definition, a permutation of the set $\{1, \ldots, n\}$, and hence the group $\mathrm{Sym}(n)$ is represented as a permutation group on this set; in this case each element of $\mathrm{Sym}(n)$ represents itself. This is called the *canonical representation* of $\mathrm{Sym}(n)$ and corresponds to the identity homomorphism $1: \mathrm{Sym}(n) \rightarrow \mathrm{Sym}(n)$; $1_\sigma = \sigma$ for all $\sigma \in \mathrm{Sym}(n)$. Clearly, the canonical representation is faithful and has degree n.

The previous examples described permutation representations of specific groups. In the following two examples, we illustrate several methods for obtaining permutation representations of any group.

EXAMPLE 3. The inner automorphism representation

Let G be an arbitrary group. Define the function $f: G \rightarrow \mathrm{Sym}(G)$ by letting $f_\sigma: G \rightarrow G$ stand for the inner automorphism of G defined by $f_\sigma(x) = \sigma x \sigma^{-1}$ for all $\sigma, x \in G$.

(A) We claim that f is a permutation representation of G. To show this, first observe that if $\sigma \in G$, then f_σ is in fact a permutation of G since it is the inner automorphism of G defined by σ. Now, let σ and σ' be elements in G. Then $f_{\sigma\sigma'}(x) = (\sigma\sigma')x(\sigma\sigma')^{-1} = \sigma[\sigma'x\sigma'^{-1}]\sigma^{-1} = f_{\sigma}f_{\sigma'}(x)$ for all $x \in G$ and hence $f_{\sigma\sigma'} = f_{\sigma}f_{\sigma'}$. Therefore, f is a homomorphism and hence a permutation representation of G. The representation f is called the *inner automorphism representation* of G since it represents each element of the group by the inner automorphism determined by that element.

(B) Ker $f = Z(G)$, the center of G. To show this, recall that the kernel of the representation f consists of those elements $\sigma \in G$ for which $f_{\sigma} = 1$. Since $f_{\sigma}(x) = \sigma x \sigma^{-1} = x$ for every $x \in G$ if and only if σ lies in the center of G, it follows that Ker $f = Z(G)$. Thus, in general, the inner automorphism representation of a group is not a faithful representation; indeed, it is a faithful representation if and only if $Z(G) = (1)$. On the other hand, if G is abelian, then Ker $f = G$ and therefore $f_{\sigma} = 1$ for every $\sigma \in G$; in this case f is the trivial representation of G.

(C) Let us determine the inner automorphism representation of the symmetric group Sym(3) (Figure 1). This representation has degree 6 since $|\text{Sym}(3)| = 6$, and is faithful since $Z(\text{Sym}(3)) = (1)$. For convenience, let $a = 1, b = (1\ 2\ 3), c = (1\ 3\ 2), d = (1\ 2), e = (1\ 3),$ and $f = (2\ 3)$. To determine the permutation $f_{(123)}$ corresponding to the element $(1\ 2\ 3)$, for example, we find that

$$f_{(123)}(a) = (1\ 2\ 3)\ 1\ (1\ 2\ 3)^{-1} = 1 = a$$

$$f_{(123)}(b) = (1\ 2\ 3)(1\ 2\ 3)(1\ 2\ 3)^{-1} = (1\ 2\ 3) = b$$

$$f_{(123)}(c) = (1\ 2\ 3)(1\ 3\ 2)(1\ 2\ 3)^{-1} = (1\ 3\ 2) = c$$

$$f_{(123)}(d) = (1\ 2\ 3)(1\ 2)(1\ 2\ 3)^{-1} = f$$

$$f_{(123)}(e) = (1\ 2\ 3)(1\ 3)(1\ 2\ 3)^{-1} = d$$

$$f_{(123)}(f) = (1\ 2\ 3)(2\ 3)(1\ 2\ 3)^{-1} = e.$$

Therefore,

$$f_{(123)} = \begin{pmatrix} a & b & c & d & e & f \\ a & b & c & f & d & e \end{pmatrix} = (d\ f\ e).$$

The remaining permutations are found similarly. Observe that each of the 3-cycles $(1\ 2\ 3)$ and $(1\ 3\ 2)$ in Sym(3) are represented by f as a 3-cycle on the set $\{a, b, c, d, e, f\}$, and hence as even permutations, and the 2-cycles $(1\ 2), (1\ 3),$ and $(2\ 3)$, which are odd permutations, are represented as a product of two 2-cycles and hence also are represented as even permutations. Thus, the inner automorphism f faithfully represents all elements in Sym(3) as even permutations of six letters and hence embeds the group Sym(3) into the alternating group Alt(6) as the subgroup $\{1, (d\ f\ e), (d\ e\ f), (b\ c)(e\ f), (b\ c)(d\ f), (b\ c)(d\ e)\}$.

$$\text{Sym}(3) \xrightarrow{f} \text{Sym}(\{a, b, c, d, e, f\})$$
$$\sigma \mapsto f_\sigma : x \mapsto \sigma x \sigma^{-1}$$

$$1 \mapsto f_1 = \begin{pmatrix} a & b & c & d & e & f \\ a & b & c & d & e & f \end{pmatrix} = 1$$

$$(1\ 2\ 3) \mapsto f_{(123)} = \begin{pmatrix} a & b & c & d & e & f \\ a & b & c & f & d & e \end{pmatrix} = (d\ f\ e)$$

$$(1\ 3\ 2) \mapsto f_{(132)} = \begin{pmatrix} a & b & c & d & e & f \\ a & b & c & e & f & d \end{pmatrix} = (d\ e\ f)$$

$$(12) \mapsto f_{(12)} = \begin{pmatrix} a & b & c & d & e & f \\ a & c & b & d & f & e \end{pmatrix} = (b\ c)(e\ f)$$

$$(13) \mapsto f_{(13)} = \begin{pmatrix} a & b & c & d & e & f \\ a & c & b & f & e & d \end{pmatrix} = (b\ c)(d\ f)$$

$$(2\ 3) \mapsto f_{(23)} = \begin{pmatrix} a & b & c & d & e & f \\ a & c & b & e & d & f \end{pmatrix} = (b\ c)(d\ e)$$

FIGURE 1.   Inner automorphism representation of Sym(3).

EXAMPLE 4. The left regular representation

Let G be an arbitrary group and let H be a subgroup of G. For each element $\sigma \in G$, define the function $r_\sigma : G/H \to G/H$ by setting $r_\sigma(xH) = \sigma xH$ for every left coset $xH \in G/H$.

(A) We claim that each of the mappings r_σ is a permutation of G/H. To show this, let $\sigma \in G$. If $r_\sigma(xH) = r_\sigma(yH)$ for some cosets $xH, yH \in G/H$, then $\sigma xH = \sigma yH$ and hence $xH = yH$. Therefore, r_σ is 1–1. To show that r_σ is onto, we need only observe that if xH is a typical left coset in G/H, then $r_\sigma(\sigma^{-1} xH) = xH$. Therefore, r_σ is a 1–1 and onto mapping and hence is a permutation of G/H.

(B) Now, define the function $\text{reg}_H : G \to \text{Sym}(G/H)$ by setting $\text{reg}_H(\sigma) = r_\sigma$ for every $\sigma \in G$. Then reg_H is a permutation representation of G having degree equal to the index $[G:H]$. For if σ and σ' are elements in G and xH is a typical left coset in G/H, $r_{\sigma\sigma'}(xH) = \sigma\sigma'xH = r_\sigma r_{\sigma'}(x)$. Hence, $r_{\sigma\sigma'} = r_\sigma r_{\sigma'}$. Therefore, reg_H is a group-homomorphism and hence a permutation representation of G on the set G/H of left cosets. Since the degree of reg_H is the number of left cosets in G/H, it follows that $\deg \text{reg}_H = [G:H]$. The representation reg_H is called the *left regular representation of G on G/H*; it represents each element of G by left multiplication on the left cosets of H.

(C) $\text{Ker reg}_H = \bigcap xHx^{-1}$, where the intersection is taken over all elements $x \in G$. For, by definition, $\text{Ker reg}_H = \{\sigma \in G \mid r_\sigma = 1\}$. But $r_\sigma = 1$ if and only if $r_\sigma(xH) = \sigma xH = xH$ for all cosets $xH \in G/H$. Since $\sigma xH = xH$ if and only if $\sigma \in xHx^{-1}$, it follows that Ker $\text{reg}_H = \{\sigma \in G \mid \sigma \in xHx^{-1}$ for all $x \in G\} = \bigcap xHx^{-1}$.

(D) Let us illustrate the preceding discussion by determining the left regular representation of the group T of the triangle on the left cosets of the subgroup $\{1, S_1\}$ generated by the symmetry S_1. Let $H = \{1, S_1\}$. Then $T/H = \{H, R_1 H, R_2 H\}$, and hence the representation reg_H has degree 3. Now, $r_{R_1}(H) = R_1 H$, $r_{R_1}(R_1 H) = R_1^2 H = R_2 H$, and $r_{R_1}(R_2 H) = R_1 R_2 H = H$. Setting $a = H$, $b = R_1 H$, and $c = R_2 H$, it follows that $r_{R_1} = \begin{pmatrix} a & b & c \\ b & c & a \end{pmatrix}$ $= (a\ b\ c)$. Similarly, we find that

$$r_1 = \begin{pmatrix} a & b & c \\ a & b & c \end{pmatrix} = 1 \qquad r_{S_1} = \begin{pmatrix} a & b & c \\ a & c & b \end{pmatrix} = (b\ c)$$

$$r_{R_1} = \begin{pmatrix} a & b & c \\ b & c & a \end{pmatrix} = (a\ b\ c) \qquad r_{S_2} = \begin{pmatrix} a & b & c \\ c & b & a \end{pmatrix} = (a\ c)$$

$$r_{R_2} = \begin{pmatrix} a & b & c \\ c & a & b \end{pmatrix} = (a\ c\ b) \qquad r_{S_3} = \begin{pmatrix} a & b & c \\ b & a & c \end{pmatrix} = (a\ b).$$

It follows easily from these calculations that reg_H is a faithful representation of T, which agrees with the result in part (C) since

$$\text{Ker reg}_H = \bigcap xHx^{-1} = \{1, S_1\} \cap \{1, S_2\} \cap \{1, S_3\} = (1).$$

(E) Let us now determine the left regular representation of the group T of the triangle on the left cosets of the subgroup (R_1). Let $K = (R_1)$ $= \{1, R_1, R_1{}^2\}$. Then $T/K = \{K, S_1 K\}$ and hence reg_K has degree 2. Letting $a = K$ and $b = S_1 K$, we find that

$$r_1 = \begin{pmatrix} a & b \\ a & b \end{pmatrix} = 1 \qquad r_{S_1} = \begin{pmatrix} a & b \\ b & a \end{pmatrix} = (a\ b)$$

$$r_{R_1} = \begin{pmatrix} a & b \\ a & b \end{pmatrix} = 1 \qquad r_{S_2} = \begin{pmatrix} a & b \\ b & a \end{pmatrix} = (a\ b)$$

$$r_{R_2} = \begin{pmatrix} a & b \\ a & b \end{pmatrix} = 1 \qquad r_{S_3} = \begin{pmatrix} a & b \\ b & a \end{pmatrix} = (a\ b).$$

Clearly reg_K is not faithful since Ker $\text{reg}_K = \{1, R_1, R_2\} = K$, which agrees with the result in part (C) since $K \triangleleft T$ and hence $\bigcap xKx^{-1} = K$.

(F) Finally, we mention that if H is any subgroup of a group G, the subgroup $\bigcap xHx^{-1}$ is called the *core* of H in G. The core of H is clearly a normal subgroup of G and is the kernel of the left regular representation reg_H. For example, if G is a simple group, then the core of every subgroup is trivial, and hence the left regular representation of the group on the cosets of any subgroup is always a faithful representation of the group.

The above examples have illustrated several types of permutation representations. In the first two examples, we discussed representations of specific groups, while the last two examples illustrate general methods for obtaining representations. Some of these representations are faithful, in which case the group is isomorphic to the corresponding permutation group, while in other cases they are not. We claim, however, that every permutation representation always determines a faithful representation. For if $f: G \to \text{Sym}(X)$ is a representation of a group G on a set X with kernel K, then by the fundamental theorem of group-homomorphism, f may be factored through the quotient group G/K to give a group-isomorphism $f^*: G/K \to \text{Sym}(X)$ in which $f^*_{\sigma K}(x) = f_\sigma(x)$ for all elements $x \in X$ and all cosets $\sigma K \in G/K$. The mapping f^* is therefore a faithful permutation representation of G/K on X. Thus, if the original representation f is faithful, then the corresponding permutation group $f(G)$ is isomorphic to G itself, while if f is not faithful, the permutation group is isomorphic to the quotient group of G modulo the kernel of the representation.

Let us now turn our attention to the basic properties of permutation representations. Our goal is to show that every permutation representation partitions the underlying set into disjoint subsets called the orbits of the representation and then to describe the number of elements in an orbit. If there is only one orbit, then it is possible to permute any given element in the set to any other element; in this case we say that the representation is transitive. Transitive representations are important because they form the basic building blocks from which all other representations are constructed in the sense that every representation may be regarded as the sum of its transitive constituents.

Definition

Let $f: G \to \text{Sym}(X)$ be a permutation representation of a group G and let $x \in X$. The *orbit of* x is the set $O(x) = \{f_\sigma(x) \in X \mid \sigma \in G\}$ of images of x under all permutations in G. The sets $O(x)$, for $x \in X$, are called the *orbits of* f. If $O(x) = X$ for some element $x \in X$, f is called a *transitive representation*.

PROPOSITION 1. Let $f: G \to \text{Sym}(X)$ be a permutation representation of a group G. Then the orbits of f partition the set X into disjoint subsets.

Proof.   Define a relation $\sim$ on the set X as follows: if $x, y \in X$, then $x \sim y$ if there is some element $\sigma \in G$ such that $y = f_\sigma(x)$. Then $\sim$ is an equivalence relation on X. For if $x \in X$, then $f_1(x) = x$ since $f_1 = 1$, the identity map, and therefore $x \sim x$. Hence, $\sim$ is reflexive. Now suppose that $x \sim y$ for some elements $x, y \in X$, and let $y = f_\sigma(x)$, $\sigma \in G$. Then $x = f_{\sigma^{-1}}(y)$ since $f_\sigma^{-1} = f_{\sigma^{-1}}$, and therefore $y \sim x$. Hence, $\sim$ is symmetric. Finally, to show that the relation $\sim$ is transitive, suppose that $x \sim y$ and $y \sim z$ for some elements $x, y, z \in X$, and let $y = f_\sigma(x)$ and $z = f_\tau(y)$, $\sigma, \tau \in G$. Then $z = f_\tau(f_\sigma(x)) = f_{\tau\sigma}(x)$ and therefore $x \sim z$. Hence, $\sim$ is transitive. It now follows that $\sim$ is an equivalence relation on the set X. If $x \in X$, then those elements in X that are equivalent to x have the form $f_\sigma(x)$, where $\sigma \in G$, and hence the equivalence class of x is the orbit $O(x)$. Hence, the orbits of f partition the set X into disjoint subsets, as required. ∎

COROLLARY. A permutation representation $f: G \rightarrow \text{Sym}(X)$ is transitive if and only if it has the property that for any pair of elements $x, y \in X$, there is some element $\sigma \in G$ such that $y = f_\sigma(x)$.

Proof. If f is transitive and $x, y \in X$, then $X = O(x)$ and hence $y = f_\sigma(x)$ for some element $\sigma \in G$. On the other hand, if f has the property that for any pair of elements $x, y \in X$, there is some element $\sigma \in G$ such that $y = f_\sigma(x)$, then $O(x) = X$ for any element $x \in X$ and hence f is transitive. ∎

For example, the group T of the triangle acts transitively on the vertices of the triangle since there is some rotation of the triangle that maps any given vertex to any other vertex. The left regular representations of a group are also transitive representations; for if $\text{reg}_H: G \rightarrow \text{Sym}(G/H)$ is the left regular representation of a group G on the left cosets of a subgroup H, and σH and τH are typical left cosets of H, then $r_{\tau \sigma^{-1}}(\sigma H) = \tau H$. Hence, there is some element of G that permutes any given left coset to any other coset and therefore reg_H is transitive.

Observe that whenever a group G acts as a permutation group on a set X, it acts transitively on all of the orbits. For let $O(x)$ be an orbit of f and let $y, y' \in O(x)$. Then $y = f_\sigma(x)$ and $y' = f_\tau(x)$ for some $\sigma, \tau \in G$. Therefore, $f_{\tau \sigma^{-1}}(y) = f_\tau f_{\sigma^{-1}}(f_\sigma(x)) = y'$ and hence G acts transitively on the orbit $O(x)$. The representation of G on the orbit $O(x)$ is called the *transitive constituent* of f on $O(x)$. We think of f as the sum of its transitive constituents since X is the disjoint union of the orbits of f. For example, the inner automorphism representation $f: \text{Sym}(3) \rightarrow \{a, b, c, d, e, f\}$ discussed in Example 3 is not a transitive representation since, for example, there is no element in $\text{Sym}(3)$ that permutes a to b. In fact, this representation has three orbits: $O(a) = \{a\}$, $O(b) = \{b, c\}$, and $O(d) = \{d, e, f\}$. Thus, the transitive constituent on $O(a)$ has degree 1, the transitive constituent on $O(b)$ has degree 2, and the transitive constitutent on $O(d)$ has degree 3.

Definition

Let $f: G \rightarrow \text{Sym}(X)$ be a permutation representation of a group G on a set X and let $x \in X$. The *stabilizer of* x is the subset $G_x = \{\sigma \in G \mid f_\sigma(x) = x\}$ consisting of all elements in G that fix x.

PROPOSITION 2. Let $f: G \rightarrow \text{Sym}(X)$ be a permutation representation of a group G and let $x \in X$. Then the stabilizer G_x is a subgroup of G. Furthermore, for every element $\sigma \in G$, we have that $G_{f_\sigma(x)} = \sigma G_x \sigma^{-1}$.

Proof. Clearly $1 \in G_x$ since $f_1 = 1$, the identity permutation, which fixes all elements in X. Now, let $\sigma, \tau \in G_x$. Then $f_\sigma(x) = f_\tau(x) = x$ and hence $f_{\sigma\tau}(x) = f_\sigma(f_\tau(x)) = x$, which shows that $\sigma\tau \in G_x$, and $f_{\sigma^{-1}}(x) = x$, which shows that

$\sigma^{-1} \in G_x$. Therefore, G_x is a subgroup of G. To prove the second statement, let $\sigma \in G$. Then

$$
\begin{aligned}
G_{f_\sigma(x)} &= \{\tau \in G \mid f_\tau(f_\sigma(x)) = f_\sigma(x)\} \\
&= \{\tau \in G \mid f_{\sigma^{-1}\tau\sigma}(x) = x\} \\
&= \{\tau \in G \mid \sigma^{-1}\tau\sigma \in G_x\} \\
&= \{\tau \in G \mid \tau \in \sigma G_x \sigma^{-1}\} \\
&= \sigma G_x \sigma^{-1}. \quad \blacksquare
\end{aligned}
$$

PROPOSITION 3. Let $f : G \to \text{Sym}(X)$ be a transitive permutation representation of a group G and let $x \in X$. Then there is a 1–1 correspondence $G/G_x \leftrightarrow X$ between the left cosets of the stabilizer G_x and the elements in X; under this correspondence, a typical left coset σG_x corresponds to the element $f_\sigma(x)$. In particular, $|X| = [G : G_x]$.

Proof. Define the function $\alpha : G \to X$ by setting $\alpha(\sigma) = f_\sigma(x)$ for every element $\sigma \in G$. We show that the mapping α is constant on the left cosets of G_x in G and then use it to define a 1–1 correspondence between the left cosets and the elements in X.

(A) α is constant on the left cosets of G_x in G; that is, if σG_x is an arbitrary left coset in G/G_x, then α maps every element in σG_x to the same element in X. To show this, let $\sigma\tau$ be a typical element in σG_x, where $\tau \in G_x$. Then $\alpha(\sigma\tau) = f_{\sigma\tau}(x) = f_\sigma(f_\tau(x))$. But $f_\tau(x) = x$ since $\tau \in G_x$. Hence, $\alpha(\sigma\tau) = f_\sigma(x)$. Therefore, α maps every element in the coset σG_x to the element $f_\sigma(x)$ and hence is constant on σG_x.

(B) Since α is constant on left cosets of G_x by part (A), we may define a mapping $\alpha^* : G/G_x \to X$ by setting $\alpha^*(\sigma G_x) = \alpha(\sigma) = f_\sigma(x)$ for every coset $\sigma G_x \in G/G_x$. We claim that α^* is a 1–1 correspondence. To show this, suppose that σG_x and τG_x are left cosets such that $\alpha^*(\sigma G_x) = \alpha^*(\tau G_x)$. Then $f_\sigma(x) = f_\tau(x)$. Therefore, $f_{\sigma^{-1}\tau}(x) = x$ and hence $\sigma^{-1}\tau \in G_x$. It follows that $\sigma G_x = \tau G_x$ and hence α^* is 1–1. Finally, if y is an arbitrary element in X, then $y = f_\sigma(x)$ for some $\sigma \in G$ since f is transitive. Therefore, $\alpha^*(\sigma G_x) = f_\sigma(x) = y$ and hence α^* is an onto mapping. It now follows that α^* is a 1–1 correspondence, and the proof is complete. $\blacksquare$

COROLLARY. If G acts as a group of permutations on a set X and $x \in X$, then $|O(x)| = [G : G_x]$.

Proof. Let $x \in X$. Since G acts transitively on the orbit $O(x)$, it follows from Proposition 3 that $|O(x)| = [G : G_x]$. $\blacksquare$

Before we illustrate these results, let us first take a moment to put them into some perspective. When a group G acts as a group of permutations on a set X, it decomposes X into disjoint orbits and acts transitively on each of these orbits. If $O(x)$ is a typical orbit, there is a 1–1 correspondence between $O(x)$ and the left cosets in G/G_x, where G_x is the stabilizer of x; under this correspondence, an element $f_\sigma(x) \in O(x)$ corresponds to the coset σG_x. Thus, we may interpret the left coset decomposition of G by G_x as a partitioning of G by the possible images of x. It follows, in particular, that $|O(x)| = [G:G_x]$. Finally, if two elements $x, y \in X$ belong to the same orbit, then their stabilizers are related by the equation $G_y = \sigma G_x \sigma^{-1}$, where $y = f_\sigma(x)$; in other words, if σ permutes x to y, then the inner automorphism determined by σ maps G_x to G_y.

EXAMPLE 5

Consider the inner automorphism representation $f: \text{Sym}(3) \to \text{Sym}(\{a, b, c, d, e, f\})$ of the symmetric group $\text{Sym}(3)$ discussed in Example 3. As we noted earlier, the representation f is not transitive but has three orbits: $O(a) = \{a\}$, $O(b) = \{b, c\}$, and $O(d) = \{d, e, f\}$. Let us discuss the stabilizers associated with these elements.

(A) We begin by first finding the stabilizer of each element in the set $\{a, b, c, d, e, f\}$. Since a has a trivial orbit, it is fixed by all of the permutations f_σ, $\sigma \in \text{Sym}(3)$, and hence its stabilizer $\text{Sym}(3)_a = \text{Sym}(3)$. Notice that this agrees with the fact that $[\text{Sym}(3):\text{Sym}(3)_a] = |O(a)| = 1$. The element b, however, is fixed only by the permutations f_1, $f_{(123)}$, and $f_{(132)}$. Therefore, $\text{Sym}(3)_b = \{1, (1\ 2\ 3), (1\ 3\ 2)\}$. In this case, $[\text{Sym}(3):\text{Sym}(3)_b] = |O(b)| = 2$. The remaining stabilizers are found the same way and are listed here.

$$\text{Sym}(3)_a = \text{Sym}(3) \qquad [\text{Sym}(3):\text{Sym}(3)_a] = |O(a)| = 1$$

$$\left.\begin{aligned}\text{Sym}(3)_b &= \{1, (1\ 2\ 3), (1\ 3\ 2)\} \\ \text{Sym}(3)_c &= \{1, (1\ 2\ 3), (1\ 3\ 2)\}\end{aligned}\right\} \quad [\text{Sym}(3):\text{Sym}(3)_b] = |O(b)| = 2$$

$$\left.\begin{aligned}\text{Sym}(3)_d &= \{1, (1\ 2)\} \\ \text{Sym}(3)_e &= \{1, (1\ 3)\} \\ \text{Sym}(3)_f &= \{1, (2\ 3)\}\end{aligned}\right\} \quad [\text{Sym}(3):\text{Sym}(3)_d] = |O(d)| = 3$$

(B) Let us now compare the stabilizers of the three points in $O(d)$. Consider d and e. Since $f_{(132)}(d) = e$, it follows that

$$\text{Sym}(3)_e = \text{Sym}(3)_{f_{(132)}(d)} = (1\ 3\ 2)\text{Sym}(3)_d(1\ 3\ 2)^{-1}.$$

Indeed, we find by direct calculation that

$$(1\ 3\ 2)\text{Sym}(3)_d(1\ 3\ 2)^{-1} = (1\ 3\ 2)\{1, (1\ 2)\}(1\ 3\ 2)^{-1}$$
$$= \{1, (1\ 3)\} = \text{Sym}(3)_e.$$

Similarly, for the points d and f we have that $f_{(13)}(d) = f$ and therefore $\text{Sym}(3)_f = (1\ 3)\text{Sym}(3)_d(1\ 3)^{-1}$.

(C) Finally, let us discuss the 1–1 correspondence between the elements in $O(d)$ and the left cosets of the stabilizer $\text{Sym}(3)_d$. The subgroup $\text{Sym}(3)_d$

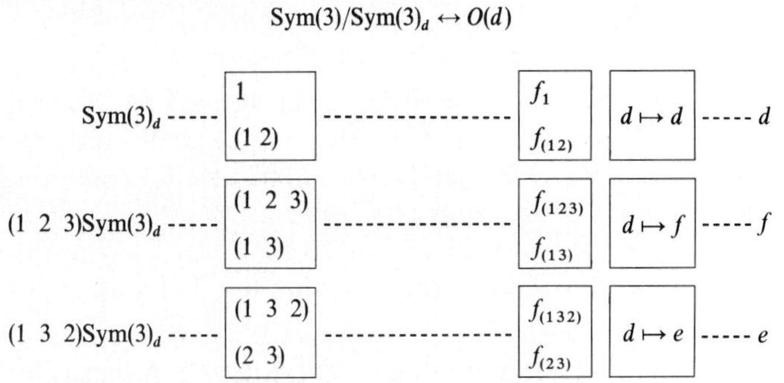

$$\text{Sym}(3)/\text{Sym}(3)_d \leftrightarrow O(d)$$

FIGURE 2.   The 1–1 correspondence between cosets of the stabilizer $\text{Sym}(3)_d$ and points in the orbit $O(d)$ for the inner automorphism representation of $\text{Sym}(3)$.

has three left cosets in $\text{Sym}(3)$:

$$\text{Sym}(3)_d = \{1, (1\ 2)\}$$

$$(1\ 2\ 3)\text{Sym}(3)_d = \{(1\ 2\ 3), (1\ 3)\}$$

$$(1\ 3\ 2)\text{Sym}(3)_d = \{(1\ 3\ 2), (2\ 3)\}.$$

Observe that all elements in a given left coset map the point d to the same element. For example, in the coset $(1\ 2\ 3)\text{Sym}(3)_d$, both $f_{(123)}$ and $f_{(13)}$ map d to f; the coset $(1\ 2\ 3)\text{Sym}(3)_d$ corresponds to the element f. In $(1\ 3\ 2)\text{Sym}(3)_d$, both $f_{(132)}$ and $f_{(23)}$ map d to e, and $(1\ 3\ 2)\text{Sym}(3)_d$ corresponds to e. The diagram in Figure 2 summarizes the 1–1 correspondence between the left cosets of $\text{Sym}(3)_d$ and the elements in $O(d)$. Note that the column of cosets on the left is the left coset decomposition of $\text{Sym}(3)$ by the stabilizer $\text{Sym}(3)_d$.

In Example 2 we noted that the symmetric group $\text{Sym}(n)$ acts naturally as a permutation group on the set $\{1, \ldots, n\}$; if σ is a permutation in $\text{Sym}(n)$, σ maps a number i to the number $\sigma(i)$. It follows, therefore, that every permutation in $\text{Sym}(n)$ maps any subset $\{i_1, i_2, \ldots, i_s\}$ of such numbers to the subset $\{\sigma(i_1), \ldots, \sigma(i_s)\}$, and hence $\text{Sym}(n)$ acts as a permutation group on the collection of all subsets of $\{1, 2, \ldots, n\}$ containing s elements. We conclude this section with an example that illustrates this idea in more detail for 3-element subsets of $\{1, 2, 3, 4, 5\}$.

EXAMPLE 6

Let $P_3(5)$ stand for the collection of all subsets of the set $\{1, 2, 3, 4, 5\}$ that contain three elements. If $A = \{i, j, k\}$ is a typical 3-element subset and $\sigma \in \text{Sym}(5)$, let $\sigma(A) = \{\sigma(i), \sigma(j), \sigma(k)\}$. Define the function $f : \text{Sym}(5) \rightarrow \text{Sym}(P_3(5))$ by setting $f_\sigma(A) = \sigma(A)$ for every subset $A \in P_3(5)$. Clearly, f_σ is a permutation of the set $P_3(5)$ for every $\sigma \in \text{Sym}(5)$, and hence f is a permutation representation of the group $\text{Sym}(5)$ on the set $P_3(5)$. Let us show that f is in fact a faithful and transitive permutation representation of Sym(5) and then discuss the orbits and stabilizers of subsets in $P_3(5)$.

(A) f is faithful. For suppose that σ lies in the kernel of f. Then $f_\sigma(A) = A$ for every 3-element subset A. Now, if $\sigma \neq 1$, there is some number i in the set $\{1, 2, 3, 4, 5\}$ such that $\sigma(i) \neq i$. Let j and k be any numbers in $\{1, 2, 3, 4, 5\}$ other than i and $\sigma(i)$. Then $f_\sigma(\{i, j, k\}) = \{\sigma(i), \sigma(j), \sigma(k)\}$. But this set is not equal to $\{i, j, k\}$ since $\sigma(i) \notin \{i, j, k\}$. Thus, we have a contradiction and conclude, therefore, that $\sigma = 1$. Hence, f is faithful.

(B) f is transitive. For let $A = \{1, 2, 3\}$, and let $B = \{i, j, k\}$ be any 3-element subset in $P_3(5)$. Let a and b stand for the two numbers in the set $\{1, 2, 3, 4, 5\}$ other than i, j, or k, and set $\sigma = \begin{pmatrix} 1 & 2 & 3 & 4 & 5 \\ i & j & k & a & b \end{pmatrix}$. Then $f_\sigma(A) = \{\sigma(1), \sigma(2), \sigma(3)\} = \{i, j, k\}$. Thus, the set A may be permuted to any 3-element set B in $P_3(5)$, and hence f is a transitive representation.

(C) Let us discuss the stabilizer Sym(5)$_A$ of the 3-element subset $A = \{1, 2, 3\}$. To find Sym(5)$_A$, we first observe that $[\mathrm{Sym}(5):\mathrm{Sym}(5)_A] = |P_3(5)|$ since f is transitive. Now, the number of 3-element subsets that may be chosen from the set $\{1, 2, 3, 4, 5\}$ is equal to the binomial coefficient $\binom{5}{3}$; that is, $|P_3(5)| = \binom{5}{3} = 10$. Therefore,

$$|\mathrm{Sym}(5)_A| = \frac{|\mathrm{Sym}(5)|}{|P_3(5)|} = \frac{5!}{10} = 12.$$

To determine these 12 permutations, observe that any permutation $\sigma \in \mathrm{Sym}(5)_A$ permutes the numbers 1, 2, 3 among themselves and the numbers 4, 5 among themselves. Since there are exactly $3!2! = 12$ such permutations, it follows that Sym(5)$_A$ consists of precisely those permutations in Sym(5) that permute 1, 2, 3 among themselves and 4, 5 among themselves. Thus,

$$\begin{aligned} \mathrm{Sym}(5)_A \ = \ & \{1, (1\ 2\ 3), (1\ 3\ 2), (1\ 2), (1\ 3), (2\ 3), (4\ 5), \\ & (1\ 2\ 3)(4\ 5), (1\ 3\ 2)(4\ 5), (1\ 2)(4\ 5), \\ & (1\ 3)(4\ 5), (2\ 3)(4\ 5)\}. \end{aligned}$$

To determine the structure of this group, let $H = \{1, (1\ 2\ 3), (1\ 3\ 2), (1\ 2), (1\ 3), (2\ 3)\}$ and $K = \{1, (4\ 5)\}$. Then H and K are subgroups of Sym(5), and it follows that Sym(5)$_A = HK$. In fact, since $H \cap K = (1)$ and since every element in H commutes with every element in K, we have that

$$\mathrm{Sym}(5)_A \cong H \times K \cong \mathrm{Sym}(3) \times \mathrm{Sym}(2).$$

Finally, note that the stabilizer of any 3-element subset B is related to Sym(5)$_A$ by means of the equation Sym(5)$_B = \sigma\, \mathrm{Sym}(5)_A \sigma^{-1}$, for some $\sigma \in \mathrm{Sym}(5)$, since Sym(5) acts transitively on $P_3(5)$. For example, if $B = \{2, 3, 5\}$, then $f_{(1235)}(A) = B$ and hence

$$\mathrm{Sym}(5)_B \ = \ (1\ 2\ 3\ 5)\mathrm{Sym}(5)_A(1\ 2\ 3\ 5)^{-1}.$$

(D) Let us interpret the left cosets of the stabilizer Sym(5)$_A$, where $A = \{1, 2, 3\}$, in terms of mappings (Figure 3). We begin by recalling that since Sym(5) acts transitively on $P_3(5)$, the orbit of the subset A is

$$\text{Sym}(5)/\text{Sym}(5)_A \leftrightarrow P_3(5)$$

$\text{Sym}(5)_A$	$\{1, 2, 3\} \mapsto \{1, 2, 3\}$
$(1\ 4)\text{Sym}(5)_A$	$\{1, 2, 3\} \mapsto \{2, 3, 4\}$
$(1\ 5)\text{Sym}(5)_A$	$\{1, 2, 3\} \mapsto \{2, 3, 5\}$
$(2\ 4)\text{Sym}(5)_A$	$\{1, 2, 3\} \mapsto \{1, 3, 4\}$
$(2\ 5)\text{Sym}(5)_A$	$\{1, 2, 3\} \mapsto \{1, 3, 5\}$
$(3\ 4)\text{Sym}(5)_A$	$\{1, 2, 3\} \mapsto \{1, 2, 4\}$
$(3\ 5)\text{Sym}(5)_A$	$\{1, 2, 3\} \mapsto \{1, 2, 5\}$
$(1\ 4)(2\ 5)\text{Sym}(5)_A$	$\{1, 2, 3\} \mapsto \{3, 4, 5\}$
$(2\ 4)(3\ 5)\text{Sym}(5)_A$	$\{1, 2, 3\} \mapsto \{2, 4, 5\}$
$(1\ 4)(3\ 5)\text{Sym}(5)_A$	$\{1, 2, 3\} \mapsto \{1, 4, 5\}$

FIGURE 3.

the entire set $P_3(5)$. Thus, we have a $1-1$ correspondence $\text{Sym}(5)/\text{Sym}(5)_A$ $\leftrightarrow P_3(5)$ between the left cosets of $\text{Sym}(5)_A$ and the collection of all 3-element subsets. Under this correspondence, a typical left coset $\sigma\,\text{Sym}(5)_A$ corresponds to the 3-element subset $f_\sigma(A) = \sigma(A) = \{\sigma(1),$ $\sigma(2), \sigma(3)\}$; $\sigma\,\text{Sym}(5)_A$ contains precisely those permutations that map A to $\sigma(A)$. For example, the coset $(14)\text{Sym}(5)_A$ contains those permutations that map $\{1, 2, 3\}$ to $\{4, 2, 3\}$. On the other hand, to find those permutations that map $\{1, 2, 3\}$ to $\{2, 4, 5\}$, for example, we set σ $= \left(\begin{smallmatrix} 1 & 2 & 3 & 4 & 5 \\ 2 & 4 & 5 & 1 & 3 \end{smallmatrix}\right) = (1\ 2\ 4)(3\ 5)$. Then $\sigma(A) = \{2, 4, 5\}$ and therefore the coset $(1\ 2\ 4)(3\ 5)\text{Sym}(5)_A$ consists of those permutations that map A to $\{2, 4, 5\}$. The table gives a complete list of the 10 cosets of $\text{Sym}(5)_A$ in $\text{Sym}(5)$ and the mappings associated with each coset. Note that if $\sigma, \tau \in \text{Sym}(5)$, then $\sigma\,\text{Sym}(5)_A = \tau\,\text{Sym}(5)_A$ if and only if $\sigma(A) = \tau(A)$; this provides a useful way of determining whether or not σ and τ lie in the same left coset of $\text{Sym}(5)_A$. For example, the permutations $(1\ 5\ 4\ 3\ 2)$ and $(1\ 3\ 2\ 5\ 4)$ do not lie in the same left coset since they map the set $\{1, 2, 3\}$ to different sets: $f_{(15432)}(\{1, 2, 3\}) = \{1, 2, 5\}$, while $f_{(13254)}(\{1, 2, 3\}) = \{2, 3, 5\}$. Hence, $(1\ 5\ 4\ 3\ 2)\text{Sym}(5)_A \neq (1\ 3\ 2\ 5\ 4)\text{Sym}(5)_A$. On the other hand, $(1\ 3\ 4\ 5\ 2)\text{Sym}(5)_A = (2\ 4)\text{Sym}(5)_A$ since $f_{(13452)}(A) = f_{(24)}(A) = \{1, 3, 4\}$. This completes the example.

Exercises

1. Let D_4 stand for the group of the square and V the set of vertices of the square. Then every element in D_4 is a rigid motion of the plane that maps vertices to vertices. Define the function $f : D_4 \to \text{Sym}(V)$ by setting $f_\sigma(v) = \sigma(v)$ for all $\sigma \in D_4$ and $v \in V$; $\sigma(v)$ is the image of vertex v under the motion σ.

    (a) Show that f is a faithful and transitive permutation representation of D_4; f represents D_4 as a permutation group on the four vertices of the square.

(b) Label the vertices of the square as shown in the marginal figure. Write out the permutations in Sym(4) corresponding to each element in D_4.

(c) Let H stand for the stabilizer of vertex 1. Find H, and write out the 1–1 correspondence α between the left cosets in D_4/H and the vertices in V. Verify that the mapping α is constant on the left cosets of H in D_4.

2. Let $X = \{v, h, d_1, d_2\}$ stand for the set of vertical, horizontal, and diagonal lines of a square, as indicated in the marginal figure. Define the function $f : D_4 \rightarrow \text{Sym}(X)$ by setting $f_\sigma(x) = \sigma(x)$ for all $x \in X$ and $\sigma \in D_4$.

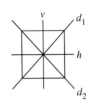

(a) Show that f is a permutation representation of D_4; f represents D_4 as a permutation group on the four lines.

(b) Determine the kernel of f. Is f faithful?

(c) Show that f has two orbits and determine the transitive constituents of f on each orbit.

(d) Let $H = (D_4)_v$ and $K = (D_4)_{d_1}$ stand for the stabilizers of the lines v and d_1, respectively. Show that $H \cong K \cong \mathbb{Z}_2 \times \mathbb{Z}_2$.

(e) Is there an element $\sigma \in D_4$ such that $\sigma H \sigma^{-1} = K$, where H and K stand for the stabilizers in part (d)?

3. Partition an equilateral triangle into six congruent regions by means of the perpendicular bisectors of the sides, as indicated in the figure, and let $X = \{c_1, c_2, c_3, c_4, c_5, c_6\}$ stand for the set of six regions. Show that the group T of the triangle acts as a faithful and transitive permutation group on the set X.

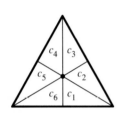

4. Let $\text{reg}_H : \mathbb{Z}_8 \rightarrow \text{Sym}(\mathbb{Z}_8/H)$ be the left regular representation of the cyclic group $\mathbb{Z}_8$ on the cosets of the subgroup H, where $\mathbb{Z}_8 = \{[0], [1], \ldots, [7]\}$ and $H = \{[0], [4]\}$.

(a) Write out the permutations $r_{[s]}$ for each element $[s] \in \mathbb{Z}_8$.

(b) Find the stabilizer $(\mathbb{Z}_8)_H$ of the coset H.

(c) If $[s] + H$ is an arbitrary coset in $\mathbb{Z}_8/H$, show that $(\mathbb{Z}_8)_{[s]+H} = (\mathbb{Z}_8)_H$. Thus, in the left regular representation of $\mathbb{Z}_8$ on $\mathbb{Z}_8/H$, all cosets have the same stabilizer.

(d) Is reg_H a faithful representation?

5. Let G be a group and let $\text{Aut}(G)$ stand for the automorphism group of G. Define the function $f : \text{Aut}(G) \rightarrow \text{Sym}(G)$ by setting $f_\sigma(x) = \sigma(x)$ for all $x \in G$ and $\sigma \in \text{Aut}(G)$.

(a) Show that f is a faithful permutation representation of G. The representation f is called the *canonical representation of* $\text{Aut}(G)$ since each automorphism permutes the elements of G and is represented by itself.

(b) Show that f is transitive if and only if $G = (1)$.

(c) Let $f : \text{Aut}(\mathbb{Z}_8) \rightarrow \text{Sym}(\mathbb{Z}_8)$ stand for the canonical representation of $\text{Aut}(\mathbb{Z}_8)$.

   (1) Write out the permutations of $\mathbb{Z}_8$ corresponding to each automorphism in $\text{Aut}(\mathbb{Z}_8)$.

(2) Find the orbits of f and the stabilizer of each element in $\mathbb{Z}_8$.

(3) Show that f has four transitive constituents.

6. Let $f:G \to \text{Sym}(X)$ be a permutation representation of a group G on a finite set X and let $P(X)$ stand for the collection of all subsets of X. Define the function $P(f):G \to \text{Sym}(P(X))$ by letting $P(f)_\sigma(A) = f_\sigma(A) = \{f_\sigma(a) \mid a \in A\}$ for every $A \in P(X)$ and $\sigma \in G$.

(a) Show that $P(f)$ is a permutation representation of G having degree $2^{|X|}$. Thus, any representation of G as a permutation group on a set X extends to a representation of G as a permutation group on the subsets of X.

(b) If f is faithful, show that $P(f)$ is faithful.

(c) Let n be a positive integer and let f stand for the canonical representation of the group $\text{Sym}(n)$ on the set $\{1, \ldots, n\}$. Then f is a faithful representation of $\text{Sym}(n)$ having degree n, and hence, by parts (a) and (b), $P(f)$ is a faithful representation of $\text{Sym}(n)$ having degree 2^n. If A is a k-element subset of $\{1, \ldots, n\}$, show that the orbit of A under $P(f)$ consists of all k-element subsets of $\{1, \ldots, n\}$ and that

$$|O(A)| = \binom{n}{k} = \frac{na}{k!(n-k)!}.$$

How many orbits does the representation $P(f)$ have?

(d) Let $P(f):\text{Sym}(3) \to \text{Sym}(P(3))$ stand for the representation of the group $\text{Sym}(3)$ acting on all subsets of $\{1, 2, 3\}$.

(1) Write out the permutations $P(f)_\sigma$ for the representation $P(f)$.

(2) Find the transitive constituents of $P(f)$.

(e) Let $A = \{2, 3, 4, 6\}$. Do the permutations $(1\ 3\ 5\ 2)$ and $(2\ 4\ 5)(1\ 3\ 6)$ in $\text{Sym}(6)$ lie in the same left coset of the stabilizer $\text{Sym}(6)_A$ of the representation $P(f):\text{Sym}(6) \to \text{Sym}(P(6))$?

(f) Let $A = \{1, 3, 4, 6\}$ and $B = \{2, 4, 5, 6\}$, and let $H = \text{Sym}(6)_A$ and $K = \text{Sym}(6)_B$ stand for the stabilizers of A and B, respectively. Find an element $\sigma \in \text{Sym}(6)$ such that $K = \sigma H \sigma^{-1}$.

7. Let $f:G \to \text{Sym}(X)$ be a permutation representation of a group G.

(a) Show that $\text{Ker } f = \bigcap G_x$, where the intersection is taken over all elements $x \in X$.

(b) If X is a finite set, show that $[G:\text{Ker } f]$ divides $|X|!$.

8. Let G be a finite group that acts nontrivially as a group of permutations on a finite set X. If $|G|$ does not divide $|X|!$, show that G contains a proper normal subgroup.

9. Let G be a finite simple group. Show that every permutation representation of G is either trivial or faithful.

10. Let G be a group, H a subgroup of G, and reg_H the left regular representation of G on the left cosets of H in G.

(a) If σH is an arbitrary left coset of H in G, show that the stabilizer $G_{\sigma H}$ is the subgroup $\sigma H \sigma^{-1}$.

(b) If $H \lhd G$, show that $\text{Ker}(\text{reg}_H) = H$.

(c) Let $H \lhd G$ and let $\text{reg}_H^*:G/H \to \text{Sym}(G/H)$ stand for the faithful representation associated with reg_H. Show that $\text{reg}_H^* = \pi$, the Cayley representation of the group G/H.

11. Let $f:G \to \text{Sym}(X)$ be a permutation representation of a group G.
    (a) If $H \leq G$, show that the restriction $f|H:H \to \text{Sym}(X)$ of f to H is a permutation representation of H.
    (b) Let $f:\text{Alt}(n) \to \text{Sym}(n)$ be the restriction of the canonical representation of $\text{Sym}(n)$ to the alternating subgroup $\text{Alt}(n)$. Show that f is faithful and transitive for all positive integers $n \geq 3$. Is f transitive when $n = 2$?
    (c) Let H stand for the subgroup of $\text{Sym}(5)$ generated by the permutations (2 3 5) and (1 3 4). Show that H acts faithfully and transitively as a group of permutations on the set $\{1, 2, 3, 4, 5\}$.
    (d) Let H stand for the subgroup of $\text{Sym}(5)$ generated by the permutations (1 2) and (3 5). Show that H acts as a group of permutations on the set $\{1, 2, 3, 4, 5\}$, but that the action is neither faithful nor transitive.
    (e) Let σ be a permutation in $\text{Sym}(n)$ and consider the cyclic subgroup (σ) acting as a group of permutations on the set $\{1, \ldots, n\}$.
        (1) Show that the orbits of this representation are just the usual orbits of the permutation σ as discussed in Chapter 3.
        (2) Let i be any one of the numbers $1, \ldots, n$ and let $(\sigma)_i$ stand for the stabilizer of i. Show that $(\sigma)_i = (\sigma^s)$, where s is the least positive integer such that $\sigma^s(i) = i$.

12. Let V be a real vector space and $GL(V)$ be the general linear group of V. Show that $GL(V)$ acts as a faithful permutation group on the set V. Is the action of $GL(V)$ on V transitive?

13. Define the function $f:GL(\mathbb{R}^2) \to \text{Sym}(\mathbb{R}^2)$ by setting $f_\sigma(A) = \sigma(A)$ for all vectors $A \in \mathbb{R}^2$ and all $\sigma \in GL(\mathbb{R}^2)$; f is the representation of $GL(\mathbb{R}^2)$ as a permutation group on the vector space $\mathbb{R}^2$ and hence, by Exercise 12, is a faithful representation of $GL(\mathbb{R}^2)$.
    (a) Let $A = (a, b)$ be an arbitrary nonzero vector in $\mathbb{R}^2$. Define $\sigma_A:\mathbb{R}^2 \to \mathbb{R}^2$ by setting $\sigma_A(x, y) = (ax - by, bx + ay)$ for all vectors $(x, y) \in \mathbb{R}^2$. Show that $\sigma_A \in GL(\mathbb{R}^2)$ and that $\sigma_A(e_1) = A$, where $e_1 = (1, 0)$.
    (b) Show that the representation f has two orbits, one consisting of the zero vector, the other of all nonzero vectors.
    (c) Show that the group $GL(\mathbb{R}^2)$ acts as a faithful transitive permutation group on the set of all nonzero vectors in $\mathbb{R}^2$.
    (d) Let A and B be arbitrary nonzero vectors in $\mathbb{R}^2$. Show that there is some element $\sigma \in GL(\mathbb{R}^2)$ such that $GL(\mathbb{R}^2)_B = \sigma GL(\mathbb{R}^2)_A \sigma^{-1}$.

14. Let $G = \{(\begin{smallmatrix}1&0\\0&1\end{smallmatrix}), (\begin{smallmatrix}0&-1\\1&-1\end{smallmatrix}), (\begin{smallmatrix}1&1\\-1&0\end{smallmatrix}), (\begin{smallmatrix}0&1\\1&0\end{smallmatrix}), (\begin{smallmatrix}1&-1\\0&-1\end{smallmatrix}), (\begin{smallmatrix}-1&0\\-1&1\end{smallmatrix})\}$ be the matrix representation of the group of the triangle discussed in Example 2, Section 2, Chapter 2, and let $X = \{(\begin{smallmatrix}0\\0\end{smallmatrix}), (\begin{smallmatrix}1\\0\end{smallmatrix}), (\begin{smallmatrix}0\\1\end{smallmatrix}), (\begin{smallmatrix}1\\1\end{smallmatrix}), (\begin{smallmatrix}-1\\0\end{smallmatrix}), (\begin{smallmatrix}0\\-1\end{smallmatrix}), (\begin{smallmatrix}-1\\-1\end{smallmatrix})\}$. Define the function $f:G \to \text{Sym}(X)$ by setting $f_M(\begin{smallmatrix}a\\b\end{smallmatrix}) = M(\begin{smallmatrix}a\\b\end{smallmatrix})$ for all vectors $(\begin{smallmatrix}a\\b\end{smallmatrix}) \in X$ and all matrices $M \in G$.

(a) Show that f is a faithful permutation representation of G of degree 7.

(b) Show that f has three transitive constituents.

15. Let $\mathbb{Z}_n$ be the cyclic group of integers modulo n, where n is a positive integer, and let d be any positive divisor of n. Set $q = n/d$. Then $\mathbb{Z}_n$ has a unique subgroup $\mathbb{Z}_d$ of order d. Define the function $\Delta(d):\mathbb{Z}_n \to \mathrm{Sym}(\mathbb{Z}_d)$ by letting $\Delta(d)_{[s]}([x]) = [x] + q[s]$, where $[x] \in \mathbb{Z}_d$ and $[s]$ $\mathbb{Z}_n$.

(a) Show that $\Delta(d)_{[s]}$ is a permutation of $\mathbb{Z}_d$ for every element $[s] \in \mathbb{Z}_n$.

(b) Show that $\Delta(d)$ is a permutation representation of $\mathbb{Z}_n$ having degree equal to d.

(c) Show that $\Delta(d)$ is transitive.

(d) Write out the representations $\Delta(1)$, $\Delta(2)$, $\Delta(3)$, and $\Delta(6)$ for the cyclic group $\mathbb{Z}_6$.

2. THE BURNSIDE COUNTING THEOREM

There are many problems in combinatorics that ask for the number of ways of choosing objects from a set when certain choices are considered to be the same. The dinner-seating problem is typical: in how many ways can 10 guests be seated around a circular table if two seatings are considered the same whenever the one can be rotated into the other? Another example is the cube-painting problem: in how many ways can the faces of a cube be painted with three colors if two such paintings are considered to be the same whenever the one can be obtained from the other by some rotation of the cube? In each of these problems we are asked to find the number of ways of choosing objects from a set when certain choices are considered to be the same. In this section we describe a method for finding this number using basic properties of permutation representations.

We begin by first showing that the similar objects in the underlying set—the objects that we are being asked to consider as a single choice—may in fact be considered as orbits of a permutation group acting on the set of all possible choices. It follows that the number of distinct choices is equal to the number of orbits of the representation. We then prove the Burnside counting theorem, a theorem that describes the number of such orbits explicitly in terms of the fixed-points of the permutations. Finally, we apply the Burnside theorem to counting coloring patterns, that is, counting the number of distinct ways of coloring an object when certain colorings are considered to be the same.

Definition

Let $f:G \to \mathrm{Sym}(X)$ be a permutation representation of a group G and let σ be an element in G. The *fixed-point set of* σ is the set $F(\sigma) = \{x \in X \mid f_\sigma(x) = x\}$ consisting of all elements in X fixed by σ.

For example, consider the group T of the triangle acting as a permutation group on the set of vertices of the triangle as labeled in Example 1, Section 1. Clearly, every vertex is fixed by the identity element and therefore $F(1) = \{1, 2, 3\}$. On the other hand, the only vertex fixed by the symmetry S_1 is vertex 1; hence $F(S_1) = \{1\}$. Similarly, we find that $F(R_1) = F(R_2) = \varnothing$, $F(S_1) = \{1\}$, $F(S_2) = \{2\}$, and $F(S_3) = \{3\}$.

Let us now prove our main result, the Burnside counting theorem, which shows that the number of orbits of a permutation representation is the average size of the fixed-point sets.

PROPOSITION 1. (BURNSIDE COUNTING THEOREM) Let $f : G \to \text{Sym}(X)$ be a permutation representation of a finite group G. Then the number of orbits of f is equal to $(1/|G|) \Sigma |F(\sigma)|$, where the summation is over all elements $\sigma \in G$.

Proof. Let $S = \{(\sigma, x) \in G \times X \mid \sigma \in G, x \in F(\sigma)\}$. We will prove the theorem by counting the number of pairs in S in two different ways. Let $\sigma \in G$. Then the number of pairs $(\sigma, x) \in S$ that have σ in the first component is equal to $|F(\sigma)|$. Therefore, $|S| = \Sigma |F(\sigma)|$, where the summation is over all $\sigma \in G$. On the other hand, if $x \in X$, the number of pairs $(\sigma, x) \in S$ having x in the second component is equal to $|G_x|$ since $(\sigma, x) \in S$ if and only if $\sigma(x) = x$. But if x and y belong to the same orbit in X, then $|G_y| = |G_x|$ since $G_y = \tau G_x \tau^{-1}$ for some $\tau \in G$. Thus, the number of pairs in S all of whose second components belong to the orbit $O(x)$ is equal to $|G_x||O(x)|$. Since $|G_x||O(x)| = |G_x|[G : G_x] = |G|$ is constant, it follows that the total number of pairs in S is the sum of $|G|$ with itself t times, where t is the number of orbits. Therefore, $t|G| = |S| = \Sigma |F(\sigma)|$, and the formula for t follows immediately. ∎

EXAMPLE 1

Consider the inner automorphism representation of the group $\text{Sym}(3)$. If $\sigma \in \text{Sym}(3)$, then $F(\sigma) = \{\tau \in \text{Sym}(3) \mid \tau \sigma \tau^{-1} = \sigma\}$ and hence $F(\sigma)$ consists of those elements in $\text{Sym}(3)$ that commute with σ. We find that

$$F(1) = \text{Sym}(3)$$

$$F((1\ 2\ 3)) = F((1\ 3\ 2)) = \{1, (1\ 2\ 3), (1\ 3\ 2)\}$$

$$F((1\ 2)) = \{1, (1\ 2)\}$$

$$F((1\ 3)) = \{1, (1\ 3)\}$$

$$F((2\ 3)) = \{1, (2\ 3)\}.$$

Therefore the number of orbits of the inner automorphism representation of $\text{Sym}(3)$ is

$$\frac{1}{|\text{Sym}(3)|} \Sigma |F(\sigma)| = \frac{1}{6}(6 + 3 + 3 + 2 + 2 + 2) = 3.$$

The three orbits are $\{1\}$, $\{(1\ 2\ 3), (1\ 3\ 2)\}$, and $\{(1\ 2), (1\ 3), (2\ 3)\}$.

EXAMPLE 2

Let us use the Burnside counting theorem to find the number of ways of seating 10 people around a circular table if two seating arrangements are considered the same whenever one can be rotated into the other. In this case, the underlying set X consists of the 10! possible ways to seat 10 people around the table. Let G be the group of rotations of the table that maps seats to seats. Then G is a cyclic group of order 10 and acts as a permutation group on the set X. If x is a particular seating arrangement, then the orbit of x consists of the 10 seating arrangements that are to be considered the same as x. Hence, the number of distinct seating arrangements is equal to the number of orbits in X. Now, the identity rotation fixes all 10! seating arrangements, while any nonidentity rotation fixes none. Thus, $|F(1)| = 10!$, while $|F(\sigma)| = 0$ for every nonidentity element $\sigma \in G$. Hence, by the Burnside formula, the number of orbits in X is equal to $(1/10)(10!) = 9!$, and thus there are 9! ways of seating 10 people around the table if we assume two such seatings are the same whenever one can be rotated into the other.

An especially useful application of the Burnside counting theorem is in counting coloring patterns. Intuitively, we color a set by assigning any one of several colors to each of its elements. More formally, let X be a finite set and let C be a finite set of colors. A *coloring of X by C* is any function $\alpha: X \to C$; if $x \in X$, $\alpha(x)$ is the color assigned to x. The set of all colorings of X by C is denoted by Color$(X;C)$. Our goal is to first show that if a group G acts as a permutation group on X, then G also permutes the colorings of X and hence acts as a permutation group on the set Color$(X;C)$. We then consider two colorings to be the same, or equivalent, if one coloring can be permuted to the other. Thus, the number of distinct colorings of X by C is equal to the number of orbits of G acting on the set Color$(X;C)$. We then use the Burnside counting theorem to obtain an explicit formula for this number.

To begin, let $f: G \to \text{Sym}(X)$ be a permutation representation of a group G and let α be a coloring of X by C. For each element $\sigma \in G$, define the coloring $\alpha_\sigma: X \to C$ by setting $\alpha_\sigma(x) = \alpha(f_{\sigma^{-1}}(x))$ for every $x \in X$. We interpret this by saying that the new color of x, $\alpha_\sigma(x)$, is the same as the original color of the point that is permuted to x; it is more convenient, however, to say that σ permutes the color of $f_{\sigma^{-1}}(x)$ to x. Now, define the function $C(f): G \to \text{Sym}(\text{Color}(X;C))$ by setting $C(f)_\sigma(\alpha) = \alpha_\sigma$ for every coloring $\alpha \in \text{Color}(X;C)$ and every $\sigma \in G$.

PROPOSITION 2. Let f be a permutation representation of G on a set X. Then $C(f)$ is a permutation representation of G on the set Color$(X;C)$ of all colorings of X by C.

Proof.   We must verify that each of the mappings $C(f)_\sigma$, $\sigma \in G$, is a permutation of the set Color$(X;C)$ and that $C(f)$ is a group-homomorphism.

(A) Let $\sigma \in G$. To show that $C(f)_\sigma$ is a permutation of Color($X;C$), suppose first that α and β are colorings such that $C(f)_\sigma(\alpha) = C(f)_\sigma(\beta)$. Then $\alpha(f_{\sigma^{-1}}(x)) = \beta(f_{\sigma^{-1}}(x))$ for every $x \in X$. But $f_{\sigma^{-1}}$ is a permutation of X, and hence every element in X may be written in the form $f_{\sigma^{-1}}(x)$ for some $x \in X$. Therefore, $\alpha = \beta$ and hence $C(f)_\sigma$ is 1–1. To show that $C(f)_\sigma$ is an onto mapping, let $\alpha \in$ Color($X;C$). We claim that $C(f)_\sigma(\alpha_{\sigma^{-1}}) = \alpha$. For $\alpha_{\sigma^{-1}}$ is a coloring of X by C, and, for any $x \in X$, $C(f)_\sigma(\alpha_{\sigma^{-1}}(x) = \alpha_{\sigma^{-1}}(f_{\sigma^{-1}}(x)) = \alpha(f_\sigma f_{\sigma^{-1}}(x)) = \alpha(x)$. Hence, $C(f)_\sigma(\alpha_{\sigma^{-1}}) = \alpha$. It follows that $C(f)_\sigma$ is an onto mapping and hence a permutation of Color($X;C$).

(B) We now show that the function $C(f)$ is a group-homomorphism. Let $\sigma, \tau \in G$ and let $\alpha \in$ Color($X;C$). Then $C(f)_{\sigma\tau}(\alpha) = \alpha_{\sigma\tau}$. Now, if $x \in X$, $\alpha_{\sigma\tau}(x) = \alpha(f_{(\sigma\tau)^{-1}}(x)) = \alpha(f_{\tau^{-1}\sigma^{-1}}(x)) = \alpha(f_{\tau^{-1}}f_{\sigma^{-1}}(x)) = \alpha_\tau(f_{\sigma^{-1}}(x)) = (\alpha_\tau)_\sigma(x)$. Therefore, $C(f)_{\sigma\tau}(\alpha) = \alpha_{\sigma\tau} = (\alpha_\tau)_\sigma = C(f)_\sigma C(f)_\tau(\alpha)$. Hence, $C(f)$ is a group-homomorphism and therefore a permutation representation of G. ∎

We say that two colorings of X by C are *equivalent* with respect to a permutation representation f if they lie in the same orbit of the representation $C(f)$; that is, two colorings α and β are equivalent if there is some element $\sigma \in G$ such that $\beta = \alpha_\sigma$. The orbits in Color($X;C$) under the action of the group G are frequently called *G-coloring patterns*; all colorings within a given pattern are considered to be the same since they are just permutations of each other. Determining the number of inequivalent colorings of X by C with respect to the action of G is therefore the same as counting the number of orbits of the permutation group G acting on the set Color($X;C$). Let us now use the Burnside counting theorem to obtain an explicit formula for this number.

COROLLARY. Let $f: G \to$ Sym(X) be a permutation representation of a finite group G acting on a finite set X and let C be a finite set of colors. Then the number of inequivalent colorings of X by C with respect to f is equal to $(1/|G|) \Sigma |C|^{N(f_\sigma)}$, where the summation is over all elements $\sigma \in G$ and $N(f_\sigma)$ is the number of cycles in the permutation f_σ.

Proof. Let $\sigma \in G$. Recall that a cycle of f_σ is a subset of X having the form $\{x, f_\sigma(x), f_{\sigma^2}(x), \ldots\}$, where $x \in X$. Now, let $F_C(\sigma) = \{\alpha \in$ Color($X;C$) $| C(f)_\sigma(\alpha) = \alpha\}$ stand for the fixed-point set of σ in the representation $C(f)$. We calculate $|F_C(\sigma)|$ as follows. Since $C(f)_\sigma(\alpha) = \alpha$ if and only if $\alpha(f_{\sigma^{-1}}(x)) = \alpha(x)$ for all $x \in X$, and since this last equation is true if and only if $\alpha(x) = \alpha(f_\sigma(x))$ for all $x \in X$, it follows that $\alpha \in F_C(\sigma)$ if and only if α is constant on every cycle in f_σ. But the number of functions $\alpha: X \to C$ that are constant on each cycle in f_σ is equal to the product of $|C|$ with itself $N(f_\sigma)$ times since there are $|C|$ choices for the image of α on each cycle and there are $N(f_\sigma)$ cycles. Hence, $|F_C(\sigma)| = |C|^{N(f_\sigma)}$. It now follows from the Burnside counting theorem that the number of inequivalent colorings of X by C with respect to the representation f is given by the preceding formula.   ∎

EXAMPLE 3

Six beads are equally spaced around a circular necklace and each bead is colored either white or black. Two colored necklaces are considered to be the same if it is possible to rotate the one into the other. Let us determine the total number of distinct colored necklaces that can be designed. In this case, the underlying set X consists of the six beads, say $B_1, \ldots, B_6$, equally spaced around a circle, and a coloring of X is any function $\alpha: \{B_1, \ldots, B_6\} \to \{W, B\}$, where W stands for white and B for black. Let G be the group of all rotations of the circle that maps beads to beads. Then G is a cyclic group of order 6 and acts as a permutation group on the set X. Consequently, G acts as a permutation group on the set of colorings of X by $\{W, B\}$. To illustrate this action, let σ stand for the counterclockwise rotation of the circle through an angle of $2\pi/6$ radians about its center, and let α stand for the coloring of the beads defined by $\alpha(B_1) = \alpha(B_2) = \alpha(B_4) = B$ and $\alpha(B_3) = \alpha(B_5) = \alpha(B_6) = W$. Then

$$\alpha_\sigma(B_1) = \alpha(f_{\sigma^{-1}}(B_1)) = \alpha(B_6) = W$$

$$\alpha_\sigma(B_2) = B$$

$$\alpha_\sigma(B_3) = B$$

$$\alpha_\sigma(B_4) = W$$

$$\alpha_\sigma(B_5) = B$$

$$\alpha_\sigma(B_6) = W.$$

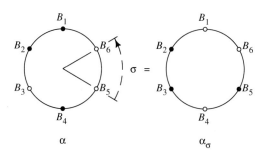

$$\alpha \qquad\qquad \alpha_\sigma$$

Let us emphasize that it is the coloring that is being rotated in this figure and not the beads. Now, to determine the number of inequivalent colored necklaces that can be designed, we first determine the cycle structure of each element in G as a permutation on the set of beads. The following table summarizes the cycle structure for each element in G.

$\tau \in G$	Cycle decomposition of f_τ	$N(f_\tau)$
1	1	6
σ	$(B_1\ B_2\ B_3\ B_4\ B_5\ B_6)$	1
σ^2	$(B_1\ B_3\ B_5)(B_2\ B_4\ B_6)$	2
σ^3	$(B_1\ B_4)(B_2\ B_5)(B_3\ B_6)$	3
σ^4	$(B_1\ B_5\ B_3)(B_2\ B_6\ B_4)$	2
σ^5	$(B_1\ B_6\ B_5\ B_4\ B_3\ B_2)$	1

Since there are two colors, it now follows from the corollary to Proposition 2 that the number of distinct colored necklaces is equal to $(1/6)(2^6 + 2^1 + 2^2 + 2^3 + 2^2 + 2^1) = 14$. Thus, the 2^6, or 64, possible colored necklaces fall into 14 equivalence classes under the action of the rotation group G. Figure 1 lists the 14 inequivalent necklaces, one from each of the classes, according to the number of black beads in the necklace. This concludes the example.

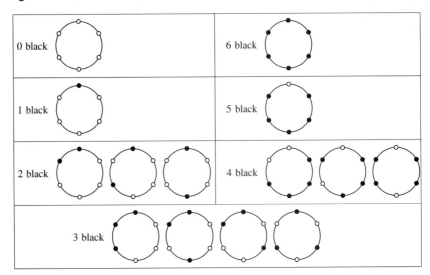

FIGURE 1. The 14 inequivalent colorings of a 6-bead necklace using two colors.

EXAMPLE 4

The faces of a tetrahedron are colored red, white, or blue. Let us determine the number of distinct colored tetrahedrons if two such figures are considered the same whenever the one can be rotated into the other. In this case the set X consists of the four faces of a tetrahedron, each of which is colored with one of three colors. Let G stand for the group of rotations of the tetrahedron, that is, the group of rotations of 3-space that bring the tetrahedron into coincidence with itself. In Chapter 8 we will show that the group G contains twelve rotations, classified as follows:

(a) one identity rotation;

(b) eight face–vertex rotations, each of which rotates the tetrahedron 120° or 240° about an axis passing through any vertex and the midpoint of the opposite face;

(c) three edge–edge rotations, each of which rotates the tetrahedron 180° about an axis passing through the midpoint of any edge and the midpoint of the opposite edge.

Since every rotation of the tetrahedron permutes the faces of the tetrahedron, it follows that G acts as a permutation group on the set of faces. If 1 stands for the identity rotation and f_1, f_2, f_3, and f_4 for the four faces of the tetrahedron, then 1 corresponds to the permutation $(f_1)(f_2)(f_3)(f_4)$. On the other hand, if σ is a typical face–vertex rotation, then σ fixes one of the faces, say f_1, and cyclically permutes f_2, f_3, and f_4. Hence, $\sigma = (f_1)(f_2\, f_3\, f_4)$, which shows that each of the eight vertex–face rotations consists of two cycles. Similarly, each edge–edge rotation interchanges pairs of faces and hence has cycle structure $(f_1\, f_2)(f_3\, f_4)$, which contains two cycles. It now follows that the number of inequivalent colored tetrahedrons is equal to

$$\frac{1}{12}(3^4 + 8 \cdot 3^2 + 3 \cdot 3^2) = 15.$$

At the beginning of this section we mentioned the cube-painting problem: in how many ways can the faces of a cube be painted with three colors if two such paintings are considered the same whenever the one can be rotated into the other? This problem is the same as the tetrahedron coloring problem in Example 4, except that we are now dealing with the six faces of a cube and the group of rotations of the cube instead of those of a tetrahedron. It turns out that the rotation group of the cube has order 24. We leave the reader to identify these 24 rotations of a cube and analyze them as permutations on the six faces of the cube, and then apply the Burnside theorem to find the number of inequivalent painted cubes using three colors.

In addition to using the Burnside counting theorem for counting the distinct number of color patterns in the usual sense of the word *color*, it is also possible to think of coloring in a more general sense. For example, we can think of a color as any one of a finite number of positive integers, say 0 and 1. In this case, coloring a set simply assigns either 0 or 1 to each object in the set. A graph with edges may then be thought of as a coloring of the set of all pairs of vertices: a pair of vertices is assigned the color 1 if the vertices are connected by an edge, 0 otherwise. Two simple graphs are considered to be the same if there is some permutation of the vertices that maps edges to edges. Using the Burnside counting theorem, it is then possible to describe the number of distinct simple graphs having a given number of vertices. We discuss these ideas in more detail in the exercises.

Exercises

1. Determine the number of inequivalent necklaces that can be made using five beads, each colored red, white, or blue.

2. Determine the number of inequivalent necklaces that can be made using eight beads, each colored black or white.

3. Let p be a prime.
   (a) Show that the number of inequivalent necklaces that can be made using p beads, each colored black or white, is equal to $(1/p)(2^p - 2) + 2$.
   (b) Let k be an integer such that $1 \le k \le p - 1$. Show that the number of inequivalent necklaces in part (a) that contain exactly k white beads is equal to $(1/p)\binom{p}{k}$.

4. Let p be a prime and n be a positive integer. Show that the number of inequivalent necklaces that may be made using p beads, each colored with one of n colors, is equal to $(1/p)[n^p + (p - 1)n]$. Is this formula valid if p is not prime?

5. The edges of an equilateral triangle are colored red, white, or blue. Two such colored triangles are considered to be the same if it is possible to rotate the one triangle into the other triangle. Determine the number of inequivalent colored triangles.

6. An equilateral triangle is divided into six congruent regions by the side bisectors, as indicated in the marginal figure. Each region is colored with one of four colors, two colorings being considered the same if there is a rotation of the triangle that maps the one colored triangle to the other. Determine the number of inequivalent colored triangles.

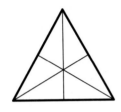

7. Let n be a positive integer.
   (a) Show that the number of inequivalent colorings of a tetrahedron using n colors is equal to $(1/12)n^2(n^2 + 11)$.
   (b) It follows from part (a) that 12 divides the product $n^2(n^2 + 11)$ for every positive integer n. Prove this statement directly without using the Burnside counting theorem.

8. Let $f: G \to \text{Sym}(X)$ be a permutation representation of a finite group G, and let σ be an element in G. Show that the number of orbits of the permutation f_σ on X is equal to

$$(1/|G|)[|F(1)| + |F(\sigma)| + \cdots + |F(\sigma^{n-1})|],$$

where $n = |G|$.

9. Let G be a finite group of order n and let s stand for the number of orbits of the inner automorphism representation of G. Show that $s = (1/n) \sum N(\sigma)$, where the summation is over all elements $\sigma \in G$ and where $N(\sigma)$ is the number of elements in G that commute with σ.

10. The six faces of a cube are painted with three colors, two colorings being considered the same if there is some rotation of the cube mapping the one into the other. How many distinct colorings of the cube are possible?

11. **Simple graphs.** A *simple graph* is a pair $\Gamma = (V, E)$ consisting of a finite set V and a set E of unordered pairs $\{i, j\}$ of elements in V; the elements in V are called *vertices*, while those in E are called *edges*. The purpose of this exercise is to use the theory of coloring patterns developed in this

section to determine the number of nonisomorphic simple graphs that have a given number of vertices.

Let $\Gamma = (V, E)$ be a simple graph. If σ is a permutation of V and $e = \{i, j\}$ is an edge in E, let $\sigma(e) = \{\sigma(i), \sigma(j)\}$. Let $\sigma(\Gamma) = (V, \sigma(E))$ stand for the simple graph whose vertices are the elements in V and whose edges are the elements of the set $\sigma(E) = \{\sigma(e) | e \in E\}$. We say that a simple graph Γ' is *isomorphic* to Γ if $\Gamma' = \sigma(\Gamma)$ for some permutation σ of V; that is, if there is some permutation of the vertices that maps the edges of the one graph to the edges of the other graph. For example, the following two graphs are isomorphic:

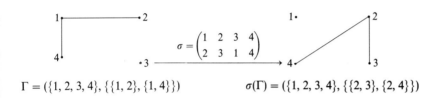

$$\sigma = \begin{pmatrix} 1 & 2 & 3 & 4 \\ 2 & 3 & 1 & 4 \end{pmatrix}$$

$\Gamma = (\{1, 2, 3, 4\}, \{\{1, 2\}, \{1, 4\}\})$ $\qquad$ $\sigma(\Gamma) = (\{1, 2, 3, 4\}, \{\{2, 3\}, \{2, 4\}\})$

(a) Show that isomorphism is an equivalence relation on the set of all simple graphs that have the same set of vertices.

Let $V = \{1, \ldots, n\}$ and let $P_2(n)$ stand for the set of all 2-element subsets of V. By definition, a coloring of $P_2(n)$ by the set $\{0, 1\}$ is a function $\alpha : P_2(n) \to \{0, 1\}$. We may interpret such a function α as the simple graph $\Gamma = (V, E)$, where $e = \{i, j\}$ is an edge in E if and only if $\alpha(\{i, j\}) = 1$. Conversely, every simple graph $\Gamma = (V, E)$ corresponds uniquely to the coloring $\alpha : P_2(n) \to \{0, 1\}$ defined by setting $\alpha(e) = 1$ when $e \in E$ and $\alpha(e) = 0$ otherwise. It follows, therefore, that the set $\mathrm{Color}(P_2(n); \{0, 1\})$ of all colorings of $P_2(n)$ by $\{0, 1\}$ is just the set of simple graphs on the n vertices $1, \ldots, n$. If $\sigma \in \mathrm{Sym}(n)$, let σ^* stand for the permutation of $P_2(n)$ defined by setting $\sigma^*(\{i, j\}) = \{\sigma(i), \sigma(j)\}$ for all sets $\{i, j\} \in P_2(n)$.

(b) Let Γ and Γ' be simple graphs on the vertices in V and let α and α' be the corresponding colorings. Show that Γ and Γ' are isomorphic if and only if $\alpha' = \alpha_{\sigma^*}$ for some $\sigma \in \mathrm{Sym}(n)$.

(c) Show that the number of nonisomorphic simple graphs on n vertices is equal to $(1/n!) \sum 2^{N(\sigma^*)}$, where the summation is over all $\sigma \in \mathrm{Sym}(n)$ and where $N(\sigma^*)$ is the number of cycles in the permutation σ^*.

(d) Determine the number of nonisomorphic simple graphs on three vertices and draw one graph for each isomorphism type.

(e) Determine the number of nonisomorphic simple graphs on four vertices and draw one graph for each isomorphism type.

(f) Show that the total number of simple graphs on n vertices is equal to $2^{\binom{n}{2}}$.

6

Conjugation

There are many instances in group theory when two elements x, y of a group are related to each other by means of an equation of the form $y = gxg^{-1}$ for some element g in the group. For example, if M and M' are matrix representations of a linear transformation relative to different coordinate systems and P is the transition matrix, then $M' = PMP^{-1}$. This relationship is called conjugation and arises naturally, as previously noted, when we use matrices to represent linear transformations relative to different coordinate systems. In this chapter our purpose is to discuss the conjugation relation in more detail, to establish its basic properties, and to give both combinatorial and geometric interpretations of this important concept. The first section deals with conjugation of group elements, and the second with conjugation of subgroups.

1. CONJUGATION OF ELEMENTS

Let us begin by showing that conjugation is an equivalence relation on the elements of a group and then determine the number of distinct conjugates of a given element.

Definition

Let G be a group and let x and y be elements in G. We say that x is *conjugate* to y, and write $x \sim y$, if $y = gxg^{-1}$ for some element $g \in G$. The *conjugacy class* of x is the set $K(x) = \{gxg^{-1} \in G \mid g \in G\}$ consisting of all conjugates of x.

PROPOSITION 1. Conjugation is an equivalence relation on a group. Under this relation, the equivalence class of an element x is its conjugacy class $K(x)$.

Proof. Let $x \in G$. Then $x = xxx^{-1}$ and hence $x \sim x$. Therefore, the conjugation relation is reflexive. Moreover, if $x \sim y$ for some $x, y \in G$, with $y = gxg^{-1}$, $g \in G$, then $x = g^{-1}yg$ and hence $y \sim x$. Thus, conjugation is a symmetric relation. Finally, if $x, y, z \in G$ are such that $x \sim y$ and $y \sim z$, with $y = gxg^{-1}$ and $z = hyh^{-1}$, $g, h \in G$, then $z = h(gxg^{-1})h^{-1} = (hg)x(hg)^{-1}$ and hence $x \sim z$. Thus, the conjugation relation is transitive and is therefore an equivalence relation on G. Finally, the equivalence class of an element $x \in G$ is, by definition, the set of all conjugates of x and hence is the conjugacy class $K(x)$. ∎

Since conjugation is an equivalence relation on G, it follows that the conjugacy classes $K(x)$, for $x \in G$, partition the group into disjoint subsets. Observe that in any group G, $K(1) = \{1\}$ since the only element in G conjugate to the identity element is the identity element itself. Moreover, since a group G is abelian if and only if $gxg^{-1} = x$ for all elements $g, x \in G$, it follows that G is abelian if and only $K(x) = \{x\}$ for all $x \in G$. Thus, for abelian groups, conjugation is not giving us any new information about the group. But for nonabelian groups, the situation is different. For example, in the group T of the triangle we find that

$$1R1^{-1} = R \qquad\qquad SRS^{-1} = R^2$$

$$RRR^{-1} = R \qquad (SR)R(SR)^{-1} = R^2$$

$$R^2RR^{-2} = R \qquad (SR^2)R(SR^2)^{-1} = R^2.$$

The element R therefore has two distinct conjugates in T, namely, R and R^2, and hence $K(R) = \{R, R^2\}$. Similarly, we find that $K(1) = \{1\}$ and $K(S) = \{S, SR, SR^2\}$. The group T therefore has three distinct conjugacy classes: $K(1)$, $K(R)$, and $K(S)$.

Let us also emphasize that conjugates of an element are relative to the group in which they are calculated. In particular, if H is a subgroup of G and $x \in H$, then the conjugates of x in H are part of but not necessarily all of the conjugates of x in G.

Now, let G be an arbitrary group and let x be an element of G. In general, there is duplication among the conjugates gxg^{-1}; that is, distinct elements g may determine the same conjugate gxg^{-1}. To determine the number of conjugates of x, observe that if gxg^{-1} and hxh^{-1} are typical conjugates of x, where $g, h \in G$, then $gxg^{-1} = hxh^{-1}$ if and only if $(h^{-1}g)x(h^{-1}g)^{-1} = x$; that is, the element $h^{-1}g$ commutes with x. With this observation in mind, we let $C(x) = \{u \in G \mid uxu^{-1} = x\}$ stand for the set of all elements in G that commute with x; $C(x)$ is called the *centralizer* of x. We claim that $C(x)$ is a subgroup of G and that the left cosets of $C(x)$ measure the number of distinct conjugates of x.

PROPOSITION 2. Let G be a group and let x be an element in G. Then the centralizer $C(x)$ is a subgroup of G. Moreover, the mapping $K(x) \to G/C(x)$

defined by $gxg^{-1} \mapsto gC(x)$ is a 1–1 correspondence between the conjugates of x and the left cosets of $C(x)$ in G. Thus, in particular, $|K(x)| = [G:C(x)]$.

Proof.   The fact that $C(x)$ is a subgroup of G is easily verified and is left as an exercise for the reader. To show that the mapping $gxg^{-1} \mapsto gC(x)$ is a 1–1 correspondence between $K(x)$ and the cosets in $G/C(x)$, we need only observe that $gxg^{-1} = hxh^{-1}$ for some $g, h \in G$ if and only if $h^{-1}g \in C(x)$, which is equivalent to saying that $gC(x) = hC(x)$. Thus, two elements in G determine the same conjugate of x if and only if they determine the same left coset of the centralizer $C(x)$. The mapping is therefore well defined and 1–1. Since every left coset corresponds to some conjugate of x, we conclude that the mapping is onto and hence a 1–1 correspondence. It follows, in particular, that the number of distinct conjugates of x is the same as the number of left cosets of $C(x)$ in G, that is, $|K(x)| = [G:C(x)]$, and the proof is complete.   ∎

Before we illustrate these results, let us make a few observations. As we have shown, two elements g and h determine the same conjugate of an element $x \in G$ if and only if they determine the same left coset of its centralizer; that is, $gxg^{-1} = hxh^{-1}$ if and only if $gC(x) = hC(x)$. Consequently, the elements in $gC(x)$ are precisely those elements that conjugate x to gxg^{-1}. For example, we showed earlier by direct calculation that the rotation R has two distinct conjugates in the group T of the triangle, namely, R and R^2. Here we find that $C(R) = \{1, R, R^2\}$ and hence $|K(R)| = [T:C(R)] = 2$, which agrees with this result. In this case, $T/C(R) = \{C(R), SC(R)\}$; the elements in the coset $C(R)$ conjugate R to R, while those in $SC(R)$ conjugate R to R^2, as shown in Figure 1. Similarly, to find the conjugates of the reflection S, let us reason as follows. The only elements in T that commute with S are S and 1. Therefore, $C(S) = \{1, S\}$ and hence $|K(S)| = [T:C(S)] = 6/2 = 3$. Thus, S has three conjugates. Since $RSR^{-1} = SR$ and $R^2SR^{-2} = SR^2$, we conclude that

$T/C(R) \leftrightarrow K(R)$		
$C(R)$ ---- $\begin{array}{c}1\\R\\R^2\end{array}$ ---- $\begin{array}{c}1R1^{-1}=R\\RRR^{-1}=R\\R^2RR^{-1}=R\end{array}$ ---- R		
$SC(R)$ ---- $\begin{array}{c}S\\SR\\SR^2\end{array}$ ---- $\begin{array}{c}SRS^{-1}=R^2\\(SR)R(SR)^{-1}=R^2\\(SR^2)R(SR^2)^{-1}=R^2\end{array}$ ---- R^2		

FIGURE 1.   The 1–1 correspondence between cosets of $C(R)$ and conjugates of R.

$K(x)$	$C(x)$	$|K(x)|$
1	T	$|K(1)| = [T:T] = 1$
R	$\{1, R, R^2\}$	$|K(R)| = [T:C(R)] = 2$
R^2	$\{1, R, R^2\}$	$|K(R^2)| = [T:C(R^2)] = 2$
S	$\{1, S\}$	$|K(S)| = [T:C(S)] = 3$
SR	$\{1, SR\}$	$|K(SR)| = [T:C(SR)] = 3$
SR^2	$\{1, SR^2\}$	$|K(SR^2)| = [T:C(SR^2)] = 3$

FIGURE 2. Conjugacy classes and centralizers of elements in the group T of the triangle.

$K(S) = \{S, SR, SR^2\}$. The conjugates and centralizers of the remaining elements in T are found similarly and are summarized in Figure 2.

EXAMPLE 1. Conjugacy classes in the symmetric groups

Let us use the previous results to describe the conjugacy classes in the symmetric groups. Our goal is to show that two permutations are conjugate if
and only if they have the same cycle structure. To begin, let n be a positive
integer and let $\sigma \in \text{Sym}(n)$. Recall that the cycle structure of σ is the n-tuple
$(c_1, \ldots, c_n)$, where c_s is the number of s-cycles in the cycle decomposition
of σ.

(A) We first show that if two permutations in $\text{Sym}(n)$ are conjugate, then
they have the same cycle structure. To show this, first observe that every
conjugate of a cycle is a cycle; for if $(i_1 \cdots i_s)$ is any s-cycle in $\text{Sym}(n)$ and
$\tau \in \text{Sym}(n)$, then $\tau(i_1 \cdots i_s)\tau^{-1} = (\tau(i_1) \cdots \tau(i_s))$ since $\tau(i_1 \cdots i_s)\tau^{-1}$ maps
$\tau(i_1)$ to $\tau(i_2)$, $\tau(i_2)$ to $\tau(i_3), \ldots, \tau(i_s)$ to $\tau(i_1)$, and fixes all other numbers
in the set $\{1, \ldots, n\}$. Now, suppose that σ and σ' are conjugate permu-
tations in $\text{Sym}(n)$, where $\sigma' = \tau\sigma\tau^{-1}$, $\tau \in \text{Sym}(n)$, and let $\sigma = \sigma_1 \cdots \sigma_t$ be
the cycle decomposition of σ. Then

$$\sigma' = \tau\sigma\tau^{-1} = \tau(\sigma_1 \cdots \sigma_t)\tau^{-1} = (\tau\sigma_1\tau^{-1}) \cdots (\tau\sigma_t\tau^{-1}).$$

Since $\sigma_1, \ldots, \sigma_t$ are disjoint cycles, it follows from the previous observation that $\tau\sigma_1\tau^{-1}, \ldots, \tau\sigma_t\tau^{-1}$ are also disjoint cycles. Therefore, the
product $(\tau\sigma_1\tau^{-1}) \cdots (\tau\sigma_t\tau^{-1})$ is the cycle decomposition of σ', and hence
σ and σ' have the same cycle structure. Thus, conjugate permutations
have the same cycle structure.

(B) Now suppose, conversely, that σ and σ' are permutations in $\text{Sym}(n)$
having the same cycle structure. Then we may write

$$\sigma = (i_{11} \cdots i_{1N_1})(i_{21} \cdots i_{2N_2}) \cdots (i_{t1} \cdots i_{tN_t})$$

and

$$\sigma' = (j_{11} \cdots j_{1N_1})(j_{21} \cdots j_{2N_2}) \cdots (j_{t1} \cdots j_{tN_t}),$$

where t is the number of cycles in σ and σ', including any trivial cycles, and $N_1, \ldots, N_t$ are the lengths of the cycles. Now, define $\tau \in \mathrm{Sym}(n)$ by setting $\tau(i_{uv}) = j_{uv}$ for all indices u, v. Then

$$\tau(i_{u1} \cdots i_{uN_u})\tau^{-1} = (\tau(i_{u1}) \cdots \tau(i_{uN_u})) = (j_{u1} \cdots j_{uN_u})$$

and therefore

$$\tau\sigma\tau^{-1} = [\tau(i_{11} \cdots i_{1N_1})\tau^{-1}] \cdots [\tau(i_{t1} \cdots i_{tN_t})\tau^{-1}]$$
$$= (j_{11} \cdots j_{1N_1}) \cdots (j_{t1} \cdots i_{tN_t})$$
$$= \sigma'.$$

Therefore, σ and σ' are conjugate. It now follows from parts (A) and (B) that two permutations are conjugate if and only if they have the same cycle structure. Observe that if σ and σ' have the same cycle structure, the above proof also indicates how to construct a permutation τ such that $\sigma' = \tau\sigma\tau^{-1}$.

(C) To illustrate the preceding results, let $\sigma = (1\ 2\ 3\ 4)$ and $\tau = (1\ 4)(2\ 3)$ be typical permutations in the group $\mathrm{Sym}(4)$. Conjugating σ by τ, we find that $\tau\sigma\tau^{-1} = (\tau(1)\ \tau(2)\ \tau(3)\ \tau(4)) = (4\ 3\ 2\ 1) = (1\ 4\ 3\ 2)$. Similarly,

$$(2\ 5\ 3)(1\ 4\ 3\ 5\ 2)(2\ 5\ 3)^{-1} = (1\ 4\ 2\ 3\ 5)$$

$$[(1\ 3)(2\ 4\ 5)](1\ 4\ 2)(3\ 5)[(1\ 3)(2\ 4\ 5)]^{-1} = (3\ 5\ 4)(1\ 2)$$

$$(2\ 6\ 3\ 4)[(2\ 3\ 5)(1\ 2\ 3)(4\ 5)](2\ 6\ 3\ 4)^{-1} = (6\ 4\ 5)(1\ 6\ 4)(2\ 5)$$
$$= (1\ 4)(2\ 6\ 5).$$

Now, let $\sigma = (1\ 2\ 3\ 4)$ and $\sigma' = (1\ 4\ 3\ 2)$. Then σ and σ' have the same cycle structure and hence are conjugate in $\mathrm{Sym}(4)$. To find a permutation τ such that $\tau\sigma\tau^{-1} = \sigma'$, we use the method described in part (B) and set $\tau(1) = 1$, $\tau(2) = 4$, $\tau(3) = 3$, and $\tau(4) = 2$. Then $\tau = (2\ 4)$ and we have that $\tau\sigma\tau^{-1} = (2\ 4)(1\ 2\ 3\ 4)(2\ 4)^{-1} = (1\ 4\ 3\ 2) = \sigma'$. Here are some other examples that illustrate this method.

$$\sigma = (1 2 3 4)$$
$$\downarrow\downarrow\downarrow\downarrow$$
$$\sigma' = (1 4 3 2)$$

$$\left.\begin{array}{l}\sigma = (1\ 2) = (1\ 2)(3)(4)\\ \downarrow\ \downarrow\ \downarrow\ \downarrow\\ \sigma' = (2\ 4) = (2\ 4)(1)(3)\end{array}\right\}\ \tau = \begin{pmatrix}1\ 2\ 3\ 4\\ 2\ 4\ 1\ 3\end{pmatrix} = (1\ 2\ 4\ 3),\quad \tau\sigma\tau^{-1} = \sigma'$$

$$\left.\begin{array}{l}\sigma = (1\ 3\ 2\ 5)(4\ 6)\\ \downarrow\ \downarrow\ \downarrow\ \downarrow\ \downarrow\ \downarrow\\ \sigma' = (2\ 3\ 1\ 6)(4\ 5)\end{array}\right\}\ \tau = \begin{pmatrix}1\ 2\ 3\ 4\ 5\ 6\\ 2\ 1\ 3\ 4\ 6\ 5\end{pmatrix} = (1\ 2)(5\ 6),\quad \tau\sigma\tau^{-1} = \sigma'$$

$$\left.\begin{array}{l}\sigma = (1\ 6\ 4)(8\ 9\ 3)(2\ 7)\\ \downarrow\ \downarrow\ \downarrow\ \downarrow\ \downarrow\ \downarrow\ \downarrow\ \downarrow\\ \sigma' = (2\ 7\ 4)(5\ 8\ 9)(3\ 6)\end{array}\right\}\ \tau = \begin{pmatrix}1\ 2\ 3\ 4\ 5\ 6\ 7\ 8\ 9\\ 2\ 3\ 9\ 4\ 1\ 7\ 6\ 5\ 8\end{pmatrix}$$
$$= (1\ 2\ 3\ 9\ 8\ 5)(6\ 7),\quad \tau\sigma\tau^{-1} = \sigma'.$$

(D) If σ is a permutation in Sym(n) having cycle structure $(c_1, \ldots, c_n)$, then by parts (A) and (B), the conjugacy class of σ consists of precisely those permutations in Sym(n) that have the same cycle structure $(c_1, \ldots, c_n)$. We denote this conjugacy class by the symbol $1^{c_1} 2^{c_2} \cdots n^{c_n}$ and call it the *type* of σ. Let us emphasize that this symbol is not a product of integers but merely a symbol for the conjugacy class of permutations in Sym(n) having cycle structure $(c_1, \ldots, c_n)$. It is customary not to include s^{c_s} in the symbol if $c_s = 0$, that is, if the permutation contains no s-cycles. For example, the type of the 3-cycle (1 2 3) in Sym(3) is 3^1, while its type as a permutation in Sym(5) is $1^2 3^1$. The type of the permutation (1 3 8)(2 7)(4 6) in Sym(8) is $1^1 2^2 3^1$.

EXAMPLE 2

Let us use the results in Example 1 to determine the conjugacy classes and centralizers of the permutations in the group Sym(4).

(A) There are five possible cycle structures for the permutations in Sym(4): $(4, 0, 0, 0)$, $(1, 0, 1, 0)$, $(2, 1, 0, 0)$, $(0, 2, 0, 0)$, and $(0, 0, 0, 1)$. The corresponding class symbols are 1^4, $1^1 3^1$, $1^2 2^1$, 2^2, and 4^1. The class $1^1 3^1$, for example, consists of all 3-cycles in Sym(4), while 2^2 consists of those permutations that are a product of two disjoint transpositions. Here is a list of the permutations in each class:

$$1^4 = \{1\};$$
$$1^1 3^1 = \{(1\ 2\ 3), (1\ 3\ 2), (1\ 2\ 4), (1\ 4\ 2), (1\ 3\ 4), (1\ 4\ 3), (2\ 3\ 4), (2\ 4\ 3)\};$$
$$1^2 2^1 = \{(1\ 2), (1\ 3), (1\ 4), (2\ 3), (2\ 4), (3\ 4)\};$$
$$2^2 = \{(1\ 2)(3\ 4), (1\ 3)(2\ 4), (1\ 4)(2\ 3)\};$$
$$4^1 = \{(1\ 2\ 3\ 4), (1\ 2\ 4\ 3), (1\ 3\ 2\ 4), (1\ 3\ 4\ 2), (1\ 4\ 2\ 3), (1\ 4\ 3\ 2)\}.$$

The group Sym(4) therefore has five conjugacy classes.

(B) Let us now find the centralizer of one permutation in each of the conjugacy classes of Sym(4). For the class 1^4, it is clear that $C(1) = $ Sym(4). Now consider the 3-cycle (1 2 3). Since

$$|K((1\ 2\ 3))| = |1^1 3^1| = 8 = [\text{Sym}(4) : C((1\ 2\ 3))],$$

it follows that $|C((1\ 2\ 3))| = 24/8 = 3$. That is, there are only three elements in the group that commute with (1 2 3). Since 1, (1 2 3) and (1 3 2) are three elements that commute with (1 2 3), it follows that $C((1\ 2\ 3)) = ((1\ 2\ 3)) = \{1, (1\ 2\ 3), (1\ 3\ 2)\}$. Thus, the only permutations in Sym(4) that commute with (1 2 3) are its powers. Similarly, we find that

$$|C((1\ 2))| = \frac{24}{6} = 4,$$

$$C((1\ 2)) = \{1, (1\ 2), (3\ 4), (1\ 2)(3\ 4)\};$$

$$|C((1\ 2)(3\ 4))| = \frac{24}{3} = 8,$$

$$C((1\ 2)(3\ 4)) = \{1, (1\ 2)(3\ 4), (1\ 2), (3\ 4), (1\ 3)(2\ 4), (1\ 4)(2\ 3),$$
$$(1\ 3\ 2\ 4), (1\ 4\ 2\ 3)\};$$

$$|C((1\ 2\ 3\ 4))| = \frac{24}{6} = 4,$$

$$C((1\ 2\ 3\ 4)) = \{1, (1\ 2\ 3\ 4), (1\ 3)(2\ 4), (1\ 4\ 3\ 2)\}.$$

The marginal table summarizes these results.

$K(\sigma)$	$|K(\sigma)|$	$|C(\sigma)|$
1^4	1	24
$1^1\,3^1$	8	3
$1^2\,2^1$	6	4
2^2	3	8
4^1	6	4

EXAMPLE 3

Let $\sigma = (1\ 2\ 3)(4\ 5\ 6)(7\ 8)$. Let us find the number of permutations in Sym(8) that commute with σ. We begin by first finding the size of the conjugacy class $K(\sigma)$, and then use the fact that $|K(\sigma)| = [\mathrm{Sym}(8):C(\sigma)]$ to determine $|C(\sigma)|$.

(A) Since σ has type $2^1\,3^2$, the class $K(\sigma)$ consists of all permutations in Sym(8) that have type $2^1\,3^2$. A typical permutation in $K(\sigma)$ therefore has the form $(a\ b\ c)(d\ e\ f)(g\ h)$, where $a, b, \ldots, h$ are distinct choices from the set $\{1, \ldots, 8\}$. Now, the number of choices for $(a\ b\ c)$ is $8 \cdot 7 \cdot 6/3$, where we divide by 3 since $(a\ b\ c) = (b\ c\ a) = (c\ a\ b)$. Similarly, the number of remaining choices for $(d\ e\ f)$ is $5 \cdot 4 \cdot 3/3$. Hence, the number of choices for the product $(a\ b\ c)(d\ e\ f)$ is equal to

$$\frac{\dfrac{8 \cdot 7 \cdot 6}{3} \cdot \dfrac{5 \cdot 4 \cdot 3}{3}}{2}$$

since $(a\ b\ c)(d\ e\ f) = (d\ e\ f)(a\ b\ c)$. Finally, the number of choices remaining for $(g\ h)$ is $2 \cdot 1/2$. The total number of choices for the product $(a\ b\ c)(d\ e\ f)(g\ h)$ is therefore equal to

$$\frac{\dfrac{8 \cdot 7 \cdot 6}{3} \cdot \dfrac{5 \cdot 4 \cdot 3}{3}}{2} \cdot \frac{2 \cdot 1}{2} = \frac{8!}{36} = 1120.$$

Hence, $|K(\sigma)| = 1120$; that is, there are 1120 permutations in the group Sym(8) that are conjugate to (1 2 3)(4 5 6)(7 8).

(B) Since $[\text{Sym}(8):C(\sigma)] = |K(\sigma)| = 1120 = 8!/36$, it follows that $|C(\sigma)| = 36$. Thus, there are 36 permutations in Sym(8) that commute with (1 2 3)(4 5 6)(7 8). Six of these, of course, are 1, σ, $\sigma^2 = (1\ 3\ 2)(4\ 6\ 5)$, $\sigma^3 = (7\ 8)$, $\sigma^4 = (1\ 2\ 3)(4\ 5\ 6)$, and $\sigma^5 = (1\ 3\ 2)(4\ 6\ 5)(7\ 8)$. To find a permutation in $C(\sigma)$ other than a power of σ, we first rewrite σ in the form $\sigma = (4\ 5\ 6)(1\ 2\ 3)(7\ 8)$ and then use the method of Example 1 to find a permutation τ such that $\tau\sigma\tau^{-1} = \sigma$:

$$\left.\begin{array}{c}\sigma = (1\ \ 2\ \ 3)(4\ \ 5\ \ 6)(7\ \ 8) \\ \downarrow\ \downarrow\ \ \downarrow\downarrow\ \ \downarrow\ \ \downarrow\downarrow\ \ \downarrow \\ \sigma = (4\ \ 5\ \ 6)(1\ \ 2\ \ 3)(7\ \ 8)\end{array}\right\} \tau = \begin{pmatrix}1 & 2 & 3 & 4 & 5 & 6 & 7 & 8 \\ 4 & 5 & 6 & 1 & 2 & 3 & 7 & 8\end{pmatrix}$$
$$= (1\ 4)(2\ 5)(3\ 6), \quad \tau\sigma\tau^{-1} = \sigma$$

Thus, the permutation (1 4)(2 5)(3 6) commutes with σ but is not a power of σ.

EXAMPLE 4. An algebraic and geometric analysis of the group of the square

Recall that the group D_4 of the square consists of four rotations 1, R, R^2, and R^3, and four reflections S, SR, SR^2, and SR^3, where R is the 90° counterclockwise rotation of the square about its center and S is a reflection of the square about a line bisecting opposite sides, as shown in the figure. Let us

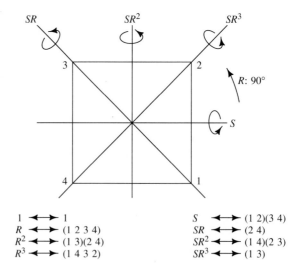

1	⟷	1	S	⟷	(1 2)(3 4)
R	⟷	(1 2 3 4)	SR	⟷	(2 4)
R^2	⟷	(1 3)(2 4)	SR^2	⟷	(1 4)(2 3)
R^3	⟷	(1 4 3 2)	SR^3	⟷	(1 3)

find the conjugacy class and centralizer of each element in D_4 and then describe these classes from an algebraic and geometric point of view.

(A) Let us first find the conjugacy classes in D_4, describe them geometrically, and compare them with the corresponding classes in Sym(4). Recall that the basic equation relating R and S is $SRS^{-1} = R^3$.

The class of R: Since $SRS^{-1} = R^3$, we have that $R \sim R^3$. Now, the only elements in D_4 that commute with R are $1, R, R^2,$ and R^3. Hence, $C(R) = \{1, R, R^2, R^3\}$, and therefore R has only two conjugates since $[D_4:C(R)] = 2$. Thus, $K(R) = \{R, R^3\}$. Geometrically, R and R^3 are the rotations of the square about its center through angles of $\pm 90°$. As permutations of the vertices, R and R^3 have type 4^1 and are odd permutations.

The class of R^2:   Since $SRS^{-1} = R^3$, it follows that $SR^2S^{-1} = R^6 = R^2$ and hence R^2 commutes with S. Therefore, R^2 commutes with every element in D_4 and hence is in the center of D_4. Thus, $C(R^2) = D_4$. It follows that $[D_4:C(R^2)] = 1$ and hence that $K(R^2) = \{R^2\}$. Geometrically, R^2 is the $180°$ rotation of the square about its center. As a vertex permutation, R^2 has type 2^2 and is an even permutation.

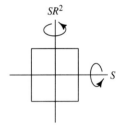

SR^2

The class of S: Since $RSR^{-1} = SR^{-1}R^{-1} = SR^2$, S has at least two conjugates, namely, S and SR^2. But $1, S, R^2,$ and SR^2 are the only elements in D_4 that commute with S. Therefore, $C(S) = \{1, S, R^2, SR^2\}$ and hence $K(S) = \{S, SR^2\}$. Geometrically, S and SR^2 are reflections of the square about lines that bisect opposite sides. As vertex permutations, S and SR^2 have type 2^2 and hence are even permutations.

S

$K(S) = \{S, SR^2\}$

The class of SR: Since $R(SR)R^{-1} = RS = SR^3$, we have that $SR \sim SR^3$. The only elements in D_4 that commute with SR are $1, SR, R^2,$ and SR^3. Hence, $C(SR) = \{1, SR, R^2, SR^3\}$ and therefore $K(SR) = \{SR, SR^3\}$. Geometrically, SR and SR^3 are reflections of the square about lines connecting diagonally opposite vertices. As vertex permutations, SR and SR^3 have type $1^2 \, 2^1$ and are odd permutations.

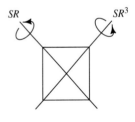

SR $\qquad$ SR^3

$K(SR) = \{SR, SR^3\}$

(B) These classes account for all elements in D_4 and are thus all of the conjugacy classes of D_4. Figure 3 (page 194) summarizes these results. Observe that R^2 and S have the same cycle type and hence are conjugate as permutations of the vertices, but they are not conjugate as geometric transformations in the group D_4. In other words, there are permutations in Sym(4) that conjugate (1 3)(2 4) to (1 4)(2 3), but none of them correspond to a congruence motion of the square. The four reflections in D_4, on the other hand, fall into two conjugacy classes, $\{S, SR^2\}$ and $\{SR, SR^3\}$, according to whether or not their reflecting lines are perpendicular; reflections in the same class have perpendicular reflecting lines. This concludes the example.

$K(\sigma)$	Vertex permutation	$\lvert K(\sigma)\rvert$	$\lvert C(\sigma)\rvert$	Geometric description	Type as a permutation
1	1	1	8	Identity transformation	1^4
R^2	(1 3)(2 4)	1	8	180° rotation	2^2
R R^3	(1 2 3 4) (1 4 3 2)	2	4	±90° rotation	4^1
S SR^2	(1 4)(2 3) (1 2)(3 4)	2	4	Reflections about lines bisecting opposite sides	2^2
SR SR^3	(1 3) (2 4)	2	4	Reflections about lines through opposite vertices	$1^2\,2^1$

FIGURE 3.   Conjugacy class table for the group D_4 of the square.

The above examples illustrate conjugation in a variety of algebraic and geometric settings. In the symmetric groups, for example, we showed that conjugation is characterized by cycle structure, two permutations being conjugate if and only if they have the cycle structure. This is an important combinatorial interpretation of conjugation that allows us to calculate rather easily the number of conjugates of any given permutation. Knowing the number of such conjugates, we can then use the equation $[\mathrm{Sym}(n):C(\sigma)]$ $= \lvert K(\sigma)\rvert$ to find the number of permutations that commute with the given permutation. Our analysis of the group of the square, on the other hand, illustrates some of the geometric aspects of conjugation; in this case, conjugacy classes correspond to the geometric classification of the transformations as summarized in the class table for D_4. Interpreting conjugation geometrically is important since it provides one more way to understand the algebraic properties of a group.

Let us now use the conjugacy classes of a finite group to derive an important formula that relates the order of the group to the order of its center.

COROLLARY.   Let G be a finite group. Then $\lvert G\rvert = \lvert Z(G)\rvert + \Sigma^*\,[G:C(x)]$, where Σ^* stands for the summation over one element x from each nontrivial conjugacy class in G.

Proof.   Let $K(x_1), \ldots, K(x_s)$ stand for the distinct conjugacy classes in G. Since these classes partition G into disjoint subsets, it follows that $\lvert G\rvert = \Sigma\,\lvert K(x_i)\rvert$. But $\lvert K(x_i)\rvert = [G:C(x_i)]$ for $i = 1, \ldots, s$. Since a conjugacy class $K(x_i)$ is trivial, that is, $\lvert K(x_i)\rvert = 1$, if and only if $x_i \in Z(G)$, it follows that $\lvert G\rvert = \lvert Z(G)\rvert + \Sigma^*\,[G:C(x_i)]$, where the summation is over one element x_i from each nontrivial conjugacy class in G.   ∎

The equation

$$|G| = |Z(G)| + \sum{}^* [G:C(x)]$$

is called the *class equation* of G. It is an especially important equation since it gives a numerical relationship between the orders of certain subgroups— the centralizers—and the total order of the group. In this sense it complements Lagrange's theorem. Lagrange's theorem, we recall, states that $|C(x)|$ must divide $|G|$. The class equation, on the other hand, shows that these divisors must add up to give the total order of the group. Thus, Lagrange's theorem and the class equation, when taken together, provide a great deal of information about possible orders of centralizers.

The group T of the triangle, for example, has two nontrivial classes, $K(R)$ and $K(S)$, and hence its class equation is $6 = |Z(T)| + [T:C(R)] + [T:C(S)] = 1 + 2 + 3$. For the group Sym(4), the class equation is $24 = 1 + 8 + 6 + 3 + 6$, while for the group of the square it is $8 = 2 + 2 + 2 + 2$. In each case, the class equation may be obtained directly from the class table for the group.

The class equation of a group is also useful for showing that certain groups must have a nontrivial center. For example, suppose that G is a nonabelian group of order 8. Then G contains at least one element that has a nontrivial conjugacy class. If x is any such element, then $[G:C(x)] = |K(x)| \neq 1$, and therefore the index $[G:C(x)]$ must be divisible by 2. Since, by the class equation, $8 = |G| = |Z(G)| + \sum{}^* [G:C(x)]$, and since 2 divides each term in the sum $\sum{}^*$, it follows that 2 must divide $|Z(G)|$. Hence, $Z(G)$ is nontrivial, thus showing that every nonabelian group of order 8 contains at least one element, other than the identity, that commutes with every element in the group. In the group of the square, for example, $Z(D_4) = \{1, R^2\}$, where R^2 is the $180°$ rotation of the square.

Exercises

1. Determine whether or not the following permutations are conjugate. If they are conjugate, find a permutation τ such that $\sigma' = \tau\sigma\tau^{-1}$.
   (a) $\sigma = (1\ 3\ 4)(2\ 5)$, $\sigma' = (2\ 5\ 4)(1\ 3)$
   (b) $\sigma = (2\ 4\ 3\ 5)(1\ 7\ 8)$, $\sigma' = (7\ 3\ 2\ 1)(4\ 6\ 8)$
   (c) $\sigma = (1\ 6\ 2)(2\ 3\ 4)(1\ 3)$, $\sigma' = (2\ 1\ 3)(1\ 5\ 4)(3\ 4)$
   (d) $\sigma = (1\ 3\ 5\ 2)$, $\sigma' = \sigma^{-1}$

   (e) $\sigma = \begin{pmatrix} 1 & 2 & 3 & 4 & 5 & 6 \\ 4 & 3 & 6 & 1 & 5 & 2 \end{pmatrix}$, $\sigma' = \begin{pmatrix} 1 & 2 & 3 & 4 & 5 & 6 \\ 1 & 5 & 6 & 3 & 2 & 4 \end{pmatrix}$.

2. Let $\sigma = (1\ 3\ 5\ 2)$ and $\sigma' = (2\ 5\ 4\ 3)$ in Sym(5).
   (a) Show that $\sigma \sim \sigma'$.
   (b) Find $C(\sigma)$ and $C(\sigma')$.
   (c) Find all permutations τ such that $\sigma' = \tau\sigma\tau^{-1}$.
   (d) Find all permutations τ such that $\sigma = \tau\sigma'\tau^{-1}$.

3. Let $\sigma = (1\ 2)(3\ 4)(5\ 6\ 7)$.
   (a) How many permutations in Sym(7) are conjugate to σ?
   (b) How many permutations in Sym(7) commute with σ?
   (c) Find a permutation that commutes with σ but is not a power of σ.

4. Determine the number of permutations in Sym(8) that commute with the permutation $(1\ 2)(3\ 4)(5\ 6)$.

5. Let W_X and W_Y stand for the reflections of the real plane $\mathbb{R}^2$ about the X- and Y-axes, respectively. Show that $W_X \sim W_Y$ in the group $GL(\mathbb{R}^2)$.

6. Let $\sigma \in$ Sym(n). Show that σ is an even permutation if and only if every conjugate of σ is an even permutation. Is it true that all even permutations are conjugate?

7. Let G be a group.
   (a) Show that all elements in a given conjugacy class of G have the same order.
   (b) Give an example of a group containing two elements of the same order that are not conjugate.
   (c) If x and y are elements in G, show that $|xy| = |yx|$.

8. Let G be an abelian group. Show that $K(x) = \{x\}$ and $C(x) = G$ for every element $x \in G$.

9. Let G be a group. Show that $(x) \lhd C(x)$ and $Z(G) \lhd C(x)$ for every element $x \in G$.

10. Let G be a group and let x be an element in G.
    (a) Show that $K(x^n) = \{u^n \in G \mid u \in K(x)\}$ for every integer n.
    (b) Show that $C(x) \leq C(x^n)$ for every integer n.
    (c) If x has finite order and n is an integer relatively prime to $|x|$, show that $C(x^n) = C(x)$ and $|K(x^n)| = |K(x)|$.
    (d) Is it true that $|K(x^n)| = |K(x)|$ for every integer n?

11. Let G be a group, K a conjugacy class of elements in G, and (K) the subgroup of G generated by K.
    (a) Show that $(K) \lhd G$.
    (b) If G is a simple group and x is any nonidentity element in G, show that the conjugates of x generate G.

12. Let G be a group and let N be a normal subgroup of G. If $x \in N$, let $K_G(x)$ and $K_N(x)$ stand for the conjugacy classes of x in G and N, respectively.
    (a) Show that $K_N(x) \subseteq K_G(x) \subseteq N$.
    (b) Show that $K_G(x)$ is the disjoint union of one or more conjugacy classes in N.
    (c) Is it true that $K_N(x) = N \cap K_G(x)$?
    (d) Show that $K_{Sym(4)}(\sigma) = K_{Alt(4)}(\sigma)$ for all permutations $\sigma \in$ Alt(4) except the 3-cycles. Show that there is one conjugacy class of 3-cycles in Sym(4) but two conjugacy classes of 3-cycles in Alt(4).

13. Suppose that G is a group of order 27.
    (a) Show that $Z(G) \neq (1)$.
    (b) If N is a nontrivial normal subgroup of G, show that $N \cap Z(G) \neq (1)$.

Thus, every nontrivial normal subgroup of G contains some element other than the identity that commutes with every element in G.

(c) Let $\sigma_4 : \mathbb{Z}_9 \to \mathbb{Z}_9$ stand for the automorphism of $\mathbb{Z}_9$ defined by setting $\sigma_4(x) = x^4$, where x is a generator of $\mathbb{Z}_9$, and let $G = \mathbb{Z}_9 \rtimes (\sigma_4)$ stand for the semidirect product of $\mathbb{Z}_9$ and the cyclic subgroup generated by σ_4 (see Exercise 15, Section 4, Chapter 4).
  (1) Show that $|\sigma_4| = 3$.
  (2) Show that $|G| = 27$ and $\mathbb{Z}_9 \lhd G$.
  (3) Find a nonidentity element in $\mathbb{Z}_9$ that commutes with every element in G.

14. Let $f : G \to G'$ be a group-homomorphism and let x and y be elements in G.
  (a) If $x \sim y$ in G, show that $f(x) \sim f(y)$ in G'; that is, group-homomorphism preserves conjugation.
  (b) Show by means of an example that $f(x)$ and $f(y)$ may be conjugate in G' even though x and y are not conjugate in G.

15. Let G be a group. Show that $Z(G) = \bigcap C(x)$, where the intersection is taken over all elements $x \in G$.

16. **Center of the symmetric groups.** This exercise outlines another proof that the center of the symmetric group $\mathrm{Sym}(n)$ is trivial when $n \geq 3$ (see Exercise 5, Section 1, Chapter 3). Let $n \geq 3$ and let $\sigma = (1\ 2\ \cdots\ n)$ and $\tau = (1\ 2\ \cdots\ n-1)$.
  (a) Show that $C(\sigma) = (\sigma)$ and $C(\tau) = (\tau)$.
  (b) Show that $Z(\mathrm{Sym}(n)) \leq C(\sigma) \cap C(\tau)$.
  (c) Show that $\gcd(|C(\sigma)|, |C(\tau)|) = 1$.
  (d) Show that $Z(\mathrm{Sym}(n)) = (1)$.

17. **Class functions.** Let G be a group. A *class function* on G is any real-valued function $f : G \to \mathbb{R}$ such that $f(x) = f(y)$ whenever x and y are conjugate in G.
  (a) Show that a function $f : G \to \mathbb{R}$ is a class function on G if and only if $f(gxg^{-1}) = f(x)$ for all elements $g, x \in G$. Thus, the class functions on G are those real-valued functions on G that are constant on the conjugacy classes of G.
  (b) Show that the determinant function $\det : \mathrm{Mat}_2^*(\mathbb{R}) \to \mathbb{R}$ and the trace function $\mathrm{Tr} : \mathrm{Mat}_2^*(\mathbb{R}) \to \mathbb{R}$ are class functions on the group $\mathrm{Mat}_2^*(\mathbb{R})$.
  (c) Let G be a group and let $f : G \to \mathrm{Sym}(X)$ be a permutation representation of G. The *character of* f is the function $\chi : G \to \mathbb{R}$ defined by setting $\chi(\sigma) = |F(\sigma)|$ for every $\sigma \in G$, where $F(\sigma)$ is the fixed-point set of σ. Show that χ is a class function on G.
  (d) Calculate the character of the representation of the group of the triangle acting on the vertices of the triangle.
  (e) Calculate the character of the representation of the group of the square acting on the vertices of the square.
  (f) Calculate the character of the inner automorphism representation of $\mathrm{Sym}(3)$.

18. Let G be a finite group and let $\chi: G \to \mathbb{R}$ stand for the character of the inner automorphism representation of G.
    (a) Show that $\chi(\sigma) = |C(\sigma)|$ for all $\sigma \in G$.
    (b) Show that the number of conjugacy classes in G is equal to $(1/|G|) \Sigma \chi(\sigma)$, where the summation is over all $\sigma \in G$.

2. CONJUGATION OF SUBGROUPS

Let us now turn our attention to conjugate subgroups. If H is a subgroup of a group G, then, like the elements of a group, we may also form the conjugates gHg^{-1} of H for each element $g \in G$. The reader may recall that we already introduced this idea in Chapter 4 to characterize normal subgroups. At that time, we showed that the set $gHg^{-1} = \{ghg^{-1} \in G \,|\, h \in H\}$ is a subgroup of G for every element $g \in G$ and that $H \lhd G$ if and only if $gHg^{-1} = H$ for all $g \in G$. In this section, we discuss the properties of conjugate subgroups in more detail. Our goal is first to show that subgroup conjugation is an equivalence relation on the set of subgroups of a group and then discuss the number of conjugates of a given subgroup.

> **Definition**
>
> Let G be a group and let H and K be subgroups of G. We say that H is *conjugate* to K, and write $H \sim K$, if $K = gHg^{-1}$ for some element $g \in G$. The *conjugacy class* of H is the set $K(H) = \{gHg^{-1} \,|\, g \in G\}$ of all conjugates of H.

PROPOSITION 1. Conjugation of subgroups is an equivalence relation on the set of all subgroups of a group. Under this relation, the equivalence class of a subgroup H is its conjugacy class $K(H)$.

Proof.   The proof is similar to that of Proposition 1, Section 1, and is left as an exercise for the reader. ∎

For example, the trivial subgroup (1) of any group is conjugate only to itself and therefore $K((1)) = \{(1)\}$; similarly, $K(G) = \{G\}$. On the other hand, in the group of the triangle we find that the conjugates of the cyclic subgroup (S) generated by the reflection S are

$$1(S)1^{-1} = \{1, S\} = (S);$$

$$S(S)S^{-1} = \{1, S\} = (S);$$

$$R(S)R^{-1} = \{1, RSR^{-1}\} = \{1, SR\} = (SR);$$

$$R^2(S)R^{-2} = \{1, R^2SR^{-2}\} = \{1, SR^2\} = (SR^2);$$

$$SR(S)(SR)^{-1} = \{1, SR(S)(SR)^{-1}\} = \{1, SR^2\} = (SR^2);$$

$$SR^2(S)(SR^2)^{-1} = \{1, SR^2S(SR^2)^{-1}\} = \{1, SR\} = (SR).$$

Thus, the cyclic subgroup (S) has three distinct conjugates, namely, (S), (SR), and (SR^2), and therefore $K((S)) = \{(S), (SR), (SR^2)\}$.

Counting the conjugates of a subgroup is similar to counting those of an element. Let G be an arbitrary group and let $g_1 H g_1^{-1}$ and $g_2 H g_2^{-1}$ be conjugates of a subgroup H, where $g_1, g_2 \in G$. Then $g_1 H g_1^{-1} = g_2 H g_2^{-1}$ if and only if $(g_2^{-1}g_1)H(g_2^{-1}g_1)^{-1} = H$. Thus, to count the number of distinct conjugates of H we first let $N(H) = \{g \in G \mid gHg^{-1} = H\}$ stand for the set of elements in G that conjugate H to itself. The set $N(H)$ is called the *normalizer* of the subgroup H. Let us now show that $N(H)$ is a subgroup of G and that it plays the same role for subgroup conjugation that the centralizer plays for element conjugation, namely, that it measures the number of conjugates of H.

PROPOSITION 2. Let G be a group and let H be a subgroup of G. Then the normalizer $N(H)$ is a subgroup of G and the mapping $K(H) \rightarrow G/N(H)$ defined by $gHg^{-1} \mapsto gN(H)$ is a 1–1 correspondence between the conjugates of H and the left cosets of $N(H)$ in G. Thus, in particular, $|K(H)| = [G:N(H)]$.

Proof. The proof is similar to that of Proposition 2, Section 1, and is left as an exercise for the reader. ■

In the group T of the triangle, for example, there are six subgroups, as illustrated in the figure here. Clearly $N(T) = T$ and $N((1)) = T$. Furthermore, since $g(S)g^{-1} = (S)$ if and only if g commutes with S, it follows that $N((S)) = C(S)$. Therefore, $N((S)) = \{1, S\}$ and hence $|K((S))| = [T:N((S))] = 3$. Thus, (S) has three conjugates in T, namely, (S), $R(S)R^{-1} = (SR)$, and $R^2(S)R^{-1} = (SR^2)$, which agrees with our previous result in which we found these conjugates by direct calculation.

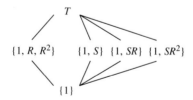

EXAMPLE 1

Let $\sigma = (1\ 2\ 3\ 4)$. Let us find the normalizer and conjugacy class of the cyclic subgroup (σ) generated by σ in $\mathrm{Sym}(4)$.

(A) To determine $N((\sigma))$, first note that $\tau(\sigma)\tau^{-1} = (\tau\sigma\tau^{-1})$ for every $\tau \in \mathrm{Sym}(4)$. Therefore, $\tau \in N((\sigma))$ if and only if $\tau\sigma\tau^{-1}$ is a generator of (σ) and hence if and only if $\tau\sigma\tau^{-1} = \sigma$ or $\tau\sigma\tau^{-1} = \sigma^3$. Those elements τ for which $\tau\sigma\tau^{-1} = \sigma$ form the centralizer $C(\sigma)$, while those τ for which

$\tau\sigma\tau^{-1} = \sigma^3$ form the left coset $(2\ 4)C(\sigma)$ since $(2\ 4)\sigma(2\ 4)^{-1} = \sigma^3$. Hence,

$$N((\sigma)) = C(\sigma) \cup (2\ 4)C(\sigma)$$

$$= \{1, (1\ 2\ 3\ 4), (1\ 3)(2\ 4), (1\ 4\ 3\ 2), (2\ 4), (1\ 4)(2\ 3), (1\ 3), (1\ 2)(3\ 4)\}.$$

In particular, $N((\sigma))$ is a group of order 8.

(B) It follows from part (A) that $|K((\sigma))| = [\mathrm{Sym}(4):N((\sigma))] = 24/8 = 3$. Therefore, the subgroup (σ) has three conjugates in $\mathrm{Sym}(4)$. We can find these conjugates by using the 1–1 correspondence between conjugates of (σ) and left cosets of $N((\sigma))$. First, choose permutations in $\mathrm{Sym}(4)$ that represent the left cosets of $N((\sigma))$, for example, 1, (1 2), and (1 4), which represent $N((\sigma))$, $(1\ 2)N((\sigma))$, and $(1\ 4)N((\sigma))$, respectively. The conjugates associated with these cosets are then (σ), $(1\ 2)(\sigma)(1\ 2)^{-1}$, and $(1\ 4)(\sigma)(1\ 4)^{-1}$. Hence, $K((\sigma)) = \{((1\ 2\ 3\ 4)), ((2\ 1\ 3\ 4)), ((4\ 2\ 3\ 1))\}$.

$\mathrm{Sym}(4)/N((\sigma)) \leftrightarrow K((\sigma))$
$N((\sigma)) \leftrightarrow (\sigma) = ((1\ 2\ 3\ 4))$
$(1\ 2)N((\sigma)) \leftrightarrow (1\ 2)(\sigma)(1\ 2)^{-1} = ((2\ 1\ 3\ 4))$
$(1\ 4)N((\sigma)) \leftrightarrow (1\ 4)(\sigma)(1\ 4)^{-1} = ((4\ 2\ 3\ 1))$

Thus, the cyclic subgroup $((1\ 2\ 3\ 4))$ has three distinct conjugates in $\mathrm{Sym}(4)$, while the generators of these subgroups form six distinct conjugates of the element (1 2 3 4):

$$((1\ 2\ 3\ 4)) \qquad\qquad ((2\ 1\ 3\ 4)) \qquad\qquad ((4\ 2\ 3\ 1))$$

$$\diagup\diagdown \qquad\qquad\qquad \diagup\diagdown \qquad\qquad\qquad \diagup\diagdown$$

$$(1\ 2\ 3\ 4)\ (1\ 4\ 3\ 2) \qquad (2\ 1\ 3\ 4)\ (2\ 4\ 3\ 1) \qquad (4\ 2\ 3\ 1)\ (4\ 1\ 3\ 2)$$

Said another way, the element (1 2 3 4) has six distinct conjugates that pair off to generate three distinct but conjugate subgroups.

An especially useful way to obtain conjugate subgroups is by means of permutation representations. Recall that if $f: G \to \mathrm{Sym}(X)$ is a permutation representation of a group G and if x and y are elements in X that belong to the same orbit of f, say $y = f_\sigma(x)$, then the stabilizers G_x and G_y are subgroups of G and $G_y = \sigma G_x \sigma - 1$. Thus, G_x and G_y are conjugate subgroups of G. It follows, in particular, that if G acts transitively on X, then the stabilizers of any two points in X are conjugate subgroups of X.

EXAMPLE 2

Let n be a positive integer. For $i = 1, \ldots, n$, let $H_i = \{\sigma \in \mathrm{Sym}(n) \mid \sigma(i) = i\}$ stand for the set of permutations in $\mathrm{Sym}(n)$ that fix the number i.

(A) We claim that $H_1, \ldots, H_n$ are subgroups of $\mathrm{Sym}(n)$ and form a single conjugacy class of subgroups; that is, $K(H_1) = \{H_1, \ldots, H_n\}$. For H_i is the stabilizer of the number i in the canonical representation of the group $\text{Sym}(n)$ on the set $\{1, \ldots, n\}$. Therefore, $H_1, \ldots, H_n$ are subgroups of $\text{Sym}(n)$ and, since this action is transitive, they are conjugate subgroups of $\text{Sym}(n)$. Now, suppose that H is any conjugate of H_1. Then $H = \sigma H_1 \sigma^{-1}$ for some $\sigma \in \text{Sym}(n)$. But clearly $\sigma H_1 \sigma^{-1} = H_{\sigma(1)}$ and therefore H is one of the subgroups $H_1, \ldots, H_n$. Hence, $K(H_1) = \{H_1, \ldots, H_n\}$.

(B) Let us now show that $N(H_i) = H_i$ for $i = 1, \ldots, n$. It follows from part (A) that $K(H_i) = \{H_1, \ldots, H_n\}$. Therefore, $[\text{Sym}(n):N(H_i)] = |K(H_i)| = n$ and hence $|N(H_i)| = (n - 1)!$. On the other hand, the group $\text{Sym}(n)$ permutes the numbers $1, \ldots, n$ transitively and H_i is the stabilizer of i. Hence, $[\text{Sym}(n):H_i] = n$, from which it follows that $|H_i| = (n - 1)!$. Since $H_i \leq N(H_i)$ and since both groups have the same order, we conclude that $N(H_i) = H_i$.

(C) For example, in $\text{Sym}(4)$, H_1 consists of all permutations that fix the number 1. H_1 has order 3! and $N(H_1) = H_1$. Thus, the only elements $\sigma \in \text{Sym}(4)$ such that $\sigma H_1 \sigma^{-1} = H_1$ are precisely those elements in H_1.

We now conclude our discussion of conjugation with two examples that illustrate geometrically how conjugation is related to coordinate transformations.

EXAMPLE 3

Let $\mathbb{R}^2$ stand for real 2-space. Recall that every coordinate system X for $\mathbb{R}^2$ defines a group-isomorphism $f_X : GL(\mathbb{R}^2) \rightarrow \text{Mat}_2^*(\mathbb{R})$ by letting $f_X(\sigma)$ stand for the matrix representation of the transformation σ relative to X.

(A) If X and Y are coordinate systems for $\mathbb{R}^2$ and $\sigma \in GL(\mathbb{R}^2)$, we claim that the matrices $f_X(\sigma)$ and $f_Y(\sigma)$ that represent σ are conjugate matrices in the group $\text{Mat}_2^*(\mathbb{R})$. To show this, let P stand for the transition matrix from X to Y. Then P is an invertible matrix and $f_Y(\sigma) = Pf_X(\sigma)P^{-1}$. Hence, $f_X(\sigma)$ and $f_Y(\sigma)$ are conjugate matrices in the group $\text{Mat}_2^*(\mathbb{R})$. It follows from this result that all matrix representations of a nonsingular linear transformation on the real plane $\mathbb{R}^2$ are conjugate in $\text{Mat}_2^*(\mathbb{R})$. Consequently, a conjugacy class in $\text{Mat}_2^*(\mathbb{R})$ consists of the various matrix representations of exactly one nonsingular linear transformation on $\mathbb{R}^2$ relative to different coordinate systems.

(B) It follows from part (A) that if G is any subgroup of $GL(\mathbb{R}^2)$, then the groups $f_X(G)$ and $f_Y(G)$ of matrix representations of G relative to different coordinate systems are conjugate subgroups of $\text{Mat}_2^*(\mathbb{R})$. For, by part (A), $f_Y(\sigma) = Pf_X(\sigma)P^{-1}$ for every element $\sigma \in G$ and therefore $f_Y(G) = Pf_X(G)P^{-1}$. Hence, $f_X(G)$ and $f_Y(G)$ are conjugate subgroups of $\text{Mat}_2^*(\mathbb{R})$. Figure 1 of Section 5, Chapter 2 illustrates these subgroups for the group T of the triangle.

EXAMPLE 4

Let D_4 stand for the group of the square. If we label the vertices of the square with the numbers 1, 2, 3, 4, then every transformation in D_4 determines a permutation of 1, 2, 3, 4 and hence a permutation representation $f: D_4 \to \text{Sym}(4)$. In general, the permutation f_σ that represents a given transformation σ is not always the same and depends upon the particular labeling of the vertices.

(A) We claim that all permutations that represent a given transformation $\sigma \in D_4$ relative to different labelings of the vertices are in fact conjugate in the group Sym(4). To show this, let the vertices of the square be labeled as indicated in the figure here and let f and g stand for the corresponding permutation representations.

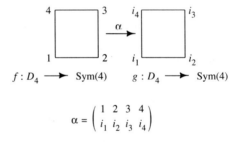

$$f: D_4 \longrightarrow \text{Sym}(4) \qquad g: D_4 \longrightarrow \text{Sym}(4)$$

$$\alpha = \begin{pmatrix} 1 & 2 & 3 & 4 \\ i_1 & i_2 & i_3 & i_4 \end{pmatrix}$$

Let $\alpha: j \to i_j$, $j = 1, 2, 3, 4$ stand for the permutation that changes the labeling on the left square to the labeling on the right. Then $g_\sigma = \alpha f_\sigma \alpha^{-1}$ for all $\sigma \in D_4$. To show this, let $\sigma \in D_4$ and let i_j be a typical vertex on the right square. Then i_j was originally labeled $\alpha^{-1}(i_j)$ on the left and hence is permuted by σ to $f_\sigma(\alpha^{-1}(i_j))$. This vertex is then labeled $\alpha(f_\sigma(\alpha^{-1}(i_j)))$ on the right. Therefore, $g_\sigma(i_j) = \alpha(f_\sigma(\alpha^{-1}(i_j)))$ for all vertices i_j. Hence, $g_\sigma = \alpha f_\sigma \alpha^{-1}$ and therefore g_σ and f_σ are conjugate permutations in Sym(4). Thus, if we think of f_σ as the standard representation of σ, then every representation of σ is conjugate to f_σ. It follows, therefore, that permutations representing a given congruence motion of the square relative to different labelings are conjugate in the group Sym(4), and the permutation that changes labelings conjugates the one representation to the other.

(B) It follows from part (A) that the permutation groups $f(D_4)$ and $g(D_4)$ which represent D_4 relative to different labelings of the vertices of the square are conjugate subgroups of Sym(4) since $g(D_4) = \{g_\sigma \in \text{Sym}(4) \mid \sigma \in D_4\} = \{\alpha f_\sigma \alpha^{-1} \mid \sigma \in D_4\} = \alpha f(D_4)\alpha^{-1}$. For example, consider the two labelings of the squares indicated in the following figure. In this case,

$$f: D_4 \to \text{Sym}(4) \qquad\qquad g: D_4 \to \text{Sym}(4)$$
$$\sigma \mapsto f_\sigma \qquad\qquad\qquad \sigma \mapsto g_\sigma = \alpha f_\sigma \alpha^{-1}$$

$1 \mapsto$	1		$1 \mapsto$	1			1	
$R \mapsto$	$(1\ 2\ 3\ 4)$		$R \mapsto$	$(1\ 3\ 4\ 2)$			$(1\ 2\ 3\ 4)$	
$R^2 \mapsto$	$(1\ 3)(2\ 4)$		$R^2 \mapsto$	$(1\ 4)(2\ 3)$			$(1\ 3)(2\ 4)$	
$R^3 \mapsto$	$(1\ 4\ 3\ 2)$		$R^3 \mapsto$	$(1\ 2\ 4\ 3)$	$= (1\ 2)$		$(1\ 4\ 3\ 2)$	$(1\ 2)^{-1}$
$S \mapsto$	$(1\ 4)(2\ 3)$		$S \mapsto$	$(1\ 3)(2\ 4)$			$(1\ 4)(2\ 3)$	
$SR \mapsto$	$(1\ 3)$		$SR \mapsto$	$(2\ 3)$			$(1\ 3)$	
$SR^2 \mapsto$	$(1\ 2)(3\ 4)$		$SR^2 \mapsto$	$(1\ 2)(3\ 4)$			$(1\ 2)(3\ 4)$	
$SR^3 \mapsto$	$(2\ 4)$		$SR^3 \mapsto$	$(1\ 4)$			$(2\ 4)$	

$$f(D_4) \qquad\qquad g(D_4) \quad = \quad (1\ 2)f(D_4)(1\ 2)^{-1}$$

FIGURE 1. The permutation representations f and g corresponding to the two labelings of the square and the conjugate subgroups $f(D_4)$ and $g(D_4)$ of Sym(4) associated with these labelings.

the permutation that changes labelings is $\alpha = (1\ 2)$. Figure 1 lists the permutations f_σ and g_σ for the eight elements $\sigma \in D_4$. In this case, $g(D_4) = (1\ 2)f(D_4)(1\ 2)^{-1} \sim f(D_4)$.

$$f: D_4 \longrightarrow \text{Sym}(4) \qquad g: D_4 \longrightarrow \text{Sym}(4)$$

$$\alpha = \begin{pmatrix} 1 & 2 & 3 & 4 \\ 2 & 1 & 3 & 4 \end{pmatrix} = (1\ 2)$$

Exercises

1. Let $\sigma = (1\ 2)(3\ 4) \in \text{Sym}(4)$.
   (a) Find the conjugacy class of the cyclic subgroup (σ) in Sym(4).
   (b) Find the normalizer $N((\sigma))$ of the cyclic subgroup generated by σ and show that $N((\sigma)) = C(\sigma)$, the centralizer of the element σ.
2. Let $\sigma = (1\ 2)(3\ 4)(5\ 6\ 7) \in \text{Sym}(7)$.
   (a) Find the conjugacy class $K((\sigma))$ of the cyclic subgroup generated by σ.
   (b) Find the normalizer $N((\sigma))$ of the cyclic subgroup generated by σ.
   (c) Find a permutation τ such that $N((\sigma)) = (\tau)C(\sigma)$.

3. Determine the conjugacy class and normalizer of each subgroup of the group D_4 of the square.

4. Let $H = \{1, (1\ 2)(3\ 4), (1\ 3)(2\ 4), (1\ 4)(2\ 3)\}$.
   (a) Show that H is a subgroup of Sym(5).
   (b) Find the conjugacy class $K(H)$ in Sym(5).
   (c) Find the normalizer $N(H)$ in Sym(5).
   (d) Illustrate the 1–1 correspondence between left cosets of $N(H)$ in Sym(5) and the conjugates of H in Sym(5).

5. Let $H = ((1\ 2), (3\ 4))$ stand for the subgroup of Sym(4) generated by the transpositions $(1\ 2)$ and $(3\ 4)$, and let $K = ((2\ 3), (1\ 4))$. Show that H and K are conjugate subgroups of Sym(4).

6. Let $H = ((1\ 2\ 3\ 4\ 5))$ stand for the cyclic subgroup of Sym(5) generated by $(1\ 2\ 3\ 4\ 5)$. Find the conjugacy class of H.

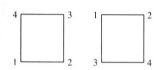

7. Label the vertices of a square as indicated in the marginal figures, and let $f:D_4 \to$ Sym(4) and $g:D_4 \to$ Sym(4) stand for the corresponding representation of D_4 as a permutation group on the four vertices. Find a permutation $\alpha \in$ Sym(4) such that $g(D_4) = \alpha f(D_4)\alpha^{-1}$.

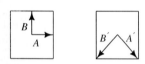

8. Let (A, B) and (A', B') stand for the two coordinate systems of the real plane as illustrated in the marginal figures, and let H and H' stand for the groups of matrix representations of the group D_4 of the square relative to (A, B) and (A', B'), respectively. Show that $H \sim H'$ and find a 2×2 matrix P such that $H' = PHP^{-1}$.

9. Let x and y be elements of a group G. If $x \sim y$, show that $(x) \sim (y)$. Is it true that if $(x) \sim (y)$, then $x \sim y$?

10. Let G be an abelian group. Show that $K(H) = \{H\}$ and $N(H) = G$ for every subgroup H of G.

11. Let G be a group.
    (a) Show that $H \lhd N(H)$ for every subgroup H of G.
    (b) If H and L are subgroups of G such that $H \lhd L$, show that $L \leq N(H)$. Thus, $N(H)$ is the largest subgroup of G containing H as a normal subgroup.
    (c) Show that $N(H) = \bigcap K$, where the intersection is taken over all subgroups K of G such that $H \lhd K$.

12. Let G be a finite group and let x be an element in G.
    (a) Show that $N((x)) = \{y \in G \mid yxy^{-1} = x^n \text{ for some integer } n \text{ relatively prime to } |x|\}$.
    (b) Show that $C(x) \lhd N((x))$.
    (c) Let $y \in N((x))$. Define the function $\sigma_y:(x) \to (x)$ by setting $\sigma_y(x^s) = yx^sy^{-1}$ for every element $x^s \in (x)$. Show that σ_y is an automorphism of (x).
    (d) Define the function $f:N((x)) \to$ Aut$((x))$ by setting $f(y) = \sigma_y$ for every element $y \in N((x))$.
        (1) Show that f is a group-homomorphism.
        (2) Show that Ker $f = C(x)$.
        (3) Show that Im $f = \{\alpha \in$ Aut$((x)) \mid \alpha(x) \sim x\}$.

(e) Show that $N((x))/C(x)$ is isomorphic to a subgroup of $\text{Aut}((x))$.

(f) Show that the index $[N((x)):C(x)]$ divides $\varphi(|x|)$, where φ is the Euler phi-function.

(g) Suppose that G has the property that for every $x \in G$, $x^n \sim x$ whenever n is relatively prime to $|x|$. Show that $N((x))/C(x) \cong \text{Aut}((x))$ for every element $x \in G$.

(h) Show that $N((x))' \lhd C(x)$, where $N((x))'$ is the commutator subgroup of $N((x))$ (Exercise 20, Section 2, Chapter 4).

13. Let σ be a permutation in $\text{Sym}(n)$ and let s be a positive integer relatively prime to n.

(a) Show that $\sigma^s \sim \sigma$.

(b) Show that $N((\sigma))/C(\sigma) \cong \text{Aut}((\sigma))$.

(c) Show that $[N((\sigma)):C(\sigma)] = \varphi(|\sigma|)$.

(d) Let $\sigma = (1\ 2\ 3\ 4\ 5) \in \text{Sym}(5)$. Find $|N((\sigma))|$.

(e) Let $\sigma = (1\ 2\ 4)(3\ 5\ 6\ 7) \in \text{Sym}(7)$. Find $|N((\sigma))|$.

14. Let G be a group and let H be a subgroup of G.

(a) Let $C(H) = \{x \in G \mid xh = hx \text{ for every } h \in H\}$. Show that $Z(G) \lhd C(H) \leq G$ and that $C(H) \lhd N(H)$. $C(H)$ is called the *centralizer* of H in G.

(b) If $H = (x)$ is a cyclic subgroup of G, show that $C(H) = C(x)$.

(c) Let $y \in N(H)$. Define the function $\sigma_y : H \to H$ by setting $\sigma_y(h) = yhy^{-1}$ for every $h \in H$. Show that σ_y is an automorphism of H.

(d) Show that $N(H)/C(H)$ is isomorphic to a subgroup of $\text{Aut}(H)$.

15. Let n be a positive integer and let M be a nonempty subset of $\{1, \ldots, n\}$. Set $H_M = \{\sigma \in \text{Sym}(n) \mid \sigma(i) = i \text{ for all } i \in M\}$.

(a) Show that $H_M \leq \text{Sym}(n)$.

(b) If M and M' are nonempty subsets of $\{1, \ldots, n\}$ such that $|M| = |M'|$, show that $H_M \sim H_{M'}$.

(c) If $|M| = n - 1$, show that $H_M = (1)$ and $N(H_M) = \text{Sym}(n)$.

(d) Let $|M| = m < n - 1$. Show that:

(1) $N(H_M) = \{\sigma \in \text{Sym}(n) \mid \sigma(M) = M\}$

(2) $[\text{Sym}(n):N(H_M)] = \binom{n}{m}$

(3) $N(H_M)/H_M \cong \text{Sym}(m)$.

16. Let G be a group of permutations that acts transitively on a set X and let $x, y \in X$. Show that the stabilizers G_x and G_y are conjugate subgroups of G.

7

p-Groups and the Sylow Theory

Recall that in a finite group, Lagrange's theorem shows that the order of every subgroup must divide the order of the group. But the theorem says nothing about the existence of subgroups whose order is a given divisor of the group order. Indeed, we have seen several examples of groups that do not have a subgroup whose order is a given divisor of the group order; the alternating group Alt(4) has order 12, for example, but does not have a subgroup of order 6. In this chapter our goal is to prove a number of remarkable theorems dealing with the existence of subgroups in a finite group and which represent, to a certain extent, a partial converse of Lagrange's theorem. These theorems were first proved in 1872 by the Norwegian mathematician Ludvig Sylow (1832–1918) and are known collectively as the Sylow theorems. They show that if the order of a finite group G is divisible by a prime power p^n, then G must contain a subgroup of order p^n. In addition, if p^n is the largest power of p that divides the order of G, then all subgroups of order p^n are in fact conjugate and the number of such subgroups is congruent to 1 modulo p. Taken as a whole, the Sylow theorems are among the most useful and powerful results in finite group theory.

1. *p*-GROUPS

As we just noted, groups of prime power order are central to our discussion throughout this chapter. Let us begin by discussing the basic properties of these groups.

Definition

Let p be a prime. A *p-group* is a finite group whose order is a power of p.

The cyclic group $\mathbb{Z}_{p^n}$, for example, is an abelian p-group of order p^n for any prime p; $\mathbb{Z}_8$, for example, is an abelian 2-group of order 8 and $\mathbb{Z}_{27}$ is an abelian 3-group of order 27. The group D_4 of the square, on the other hand, is a nonabelian 2-group of order 8.

PROPOSITION 1. Let G be a p-group. Then:

(1) every element in G has order a power of p;
(2) every subgroup of G is a p-group;
(3) if H is a subgroup of G, the index $[G:H]$ is a power of p;
(4) every quotient group of G is a p-group.

Proof. All of these statements follow immediately from Lagrange's theorem.
∎

Recall that in Chapter 6 we used the class equation to show that every group of order 8 must contain some element other than the identity that commutes with every element in the group. Let us now show that this is true for any p-group.

PROPOSITION 2. The center of a nontrivial p-group is nontrivial.

Proof. Let G be a nontrivial p-group and consider the class equation $|G| = |Z(G)| + \Sigma^* [G:C(x)]$ of G. Every index $[G:C(x)]$ in the summation Σ^* is not equal to 1 and hence is divisible by p. Since p also divides $|G|$, it follows that p divides $|Z(G)|$. Thus, the center $Z(G)$ is nontrivial. ∎

In view of Proposition 2, a nontrivial p-group must contain some element other than the identity that commutes with every element in the group. For example, we previously noted that the group D_4 of the square is a nonabelian 2-group of order 8; in this case $Z(D_4) = \{1, R^2\}$, where R^2 is the 180° rotation of the square. The following example illustrates another important group of order 8.

EXAMPLE 1. The group of quaternions

Let Q_8 stand for the subgroup of Sym(8) generated by the permutations $i = (1\ 3\ 2\ 6)(4\ 5\ 7\ 8)$ and $j = (1\ 4\ 2\ 7)(3\ 8\ 6\ 5)$. We claim that Q_8 is a non-abelian group of order 8 whose generators i and j satisfy the relations $jij^{-1} = i^{-1}$ and $i^2 = j^2 \neq 1$, that these relations characterize the group Q_8 up to isomorphism, and finally, that every nonabelian group of order 8 is isomorphic to either Q_8 or the group D_4 of the square.

(A) Let us begin by constructing the multiplication table for Q_8. Let $k = ij = (1\ 5\ 2\ 8)(3\ 4\ 6\ 7)$. Then we find that $i^2 = j^2 = k^2 = (1\ 2)(3\ 6)(4\ 7)(5\ 8)$,

$i = (1326)(4578)$
$j = (1427)(3865)$
$k = (1528)(3467)$
$z = (12)(36)(47)(58)$
$zi = (1623)(4875)$
$zj = (1724)(3568)$
$zk = (1825)(3764)$

Q_8	1	z	i	j	k	zi	zj	zk
1	1	z	i	j	k	zi	zj	zk
z	z	1	zi	zj	zk	i	j	k
i	i	zi	z	k	zj	1	zk	j
j	j	zj	zk	z	i	k	1	zi
k	k	zk	j	zi	z	zj	i	1
zi	zi	i	1	zk	j	z	k	zj
zj	zj	j	k	1	zi	zk	z	i
zk	zk	k	zj	i	1	j	zi	z

FIGURE 1. The group Q_8 of quaternions and its multiplication table.

and we denote this common element by z. Now, the eight permutations 1, z, i, j, k, zi, zj, zk are clearly elements in Q_8. The multiplication table in Figure 1 shows that these eight elements are in fact closed under multiplication and are therefore all of the elements in Q_8. From the multiplication table we find that $i^{-1} = zi, j^{-1} = zj, k^{-1} = zk, jij^{-1} = i^{-1}$, and $Z(Q_8) = \{1, z\}$. Thus, Q_8 is a nonabelian group of order 8 generated by the two elements i and j which satisfy the relations $jij^{-1} = i^{-1}$ and $i^2 = j^2$. The group Q_8 is called the *group of quaternions*. Observe that Q_8 contains exactly one element of order 2, namely, z, and is therefore not isomorphic to the group D_4 of the square since D_4 contains five such elements.

(B) Here is a more convenient way to write the elements in Q_8. Let -1 stand for z and let $-i = zi$, $-j = zj$, and $-k = zk$. Then $Q_8 = \{\pm 1, \pm i, \pm j, \pm k\}$, where $i^{-1} = -i$, $j^{-1} = -j$, $k^{-1} = -k$, $jij^{-1} = i^{-1}$, $i^2 = j^2 = k^2 = -1$, and $Z(Q_8) = \{1, -1\}$. The multiplication table for Q_8 then appears as shown here.

Q_8	1	-1	i	j	k	$-i$	$-j$	$-k$
1	1	-1	i	j	k	$-i$	$-j$	$-k$
-1	-1	1	$-i$	$-j$	$-k$	i	j	k
i	i	$-i$	-1	k	$-j$	1	$-k$	j
j	j	$-j$	$-k$	-1	i	k	1	$-i$
k	k	$-k$	j	$-i$	-1	$-j$	i	1
$-i$	$-i$	i	1	$-k$	j	-1	k	$-j$
$-j$	$-j$	j	k	1	$-i$	$-k$	-1	i
$-k$	$-k$	k	$-j$	i	1	j	$-i$	-1

Using this notation, we find that if x, $y \in \{i, j, k\}$, then $yx = -xy$. This fact is summarized by placing the three symbols i, j, and k on a circle

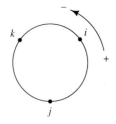

and noting that the product of any two is the third one, with a plus sign if the rotation is clockwise and a minus sign if counterclockwise. For example, $ij = k$, while $ji = -k$.

(C) We now claim that if G is any group generated by two elements x and y such that $yxy^{-1} = x^{-1}$ and $x^2 = y^2 \neq 1$, then G is in fact isomorphic to the group Q_8. For let $w = xy$. Then $w^2 = xyxy = x(x^{-1}y)y = y^2$ and hence $x^2 = y^2 = w^2$. Now, let $s = x^2 \neq 1$. Then 1, s, x, y, w, sx, sy, and sw are eight elements in G. To show that they are distinct, suppose, for example, that $sx = y$. Then $yxy^{-1} = x^{-1} = y^{-1}s = y^{-1}y^2 = y$ and hence $xy^{-1} = 1$, or $x = y$, a contradiction. Thus, $sx \neq y$. Similarly, no other two elements are equal. It is now straightforward to construct the multiplication table for these eight elements, from which it follows that $G = \{1, s, x, y, w, sx, sy, sw\}$. Finally, define the function $f : Q_8 \to G$ by setting $f(1) = 1$, $f(-1) = s$, $f(i) = x$, $f(j) = y$, $f(k) = w$, $f(-i) = sx$, $f(-j) = sy$, and $f(-k) = sw$. Then the multiplication tables for Q_8 and G correspond under f, and f is therefore a group-isomorphism. Hence, $G \cong Q_8$. Thus, the generator relations $yxy^{-1} = x^{-1}$ and $x^2 = y^2 \neq 1$ characterize the group of quaternions up to isomorphism.

(D) Let us use the isomorphism in part (C) to describe a matrix representation for the group of quaternions. Let $\text{Mat}_2^*(\mathbb{C})$ stand for the group of 2×2 invertible matrices with complex number entries, and let $x = \begin{pmatrix} i & 0 \\ 0 & -i \end{pmatrix}$ and $y = \begin{pmatrix} 0 & i \\ i & 0 \end{pmatrix}$, where $i = \sqrt{-1}$. Then $\det(x) = \det(y) = 1$ and therefore x and y are elements in the group $\text{Mat}_2^*(\mathbb{C})$. Let G be the subgroup of $\text{Mat}_2^*(\mathbb{C})$ generated by x and y. We claim that $G \cong Q_8$. For, by direct calculation, we find that

$$yxy^{-1} = \begin{pmatrix} 0 & i \\ i & 0 \end{pmatrix}\begin{pmatrix} i & 0 \\ 0 & -i \end{pmatrix}\begin{pmatrix} 0 & -i \\ -i & 0 \end{pmatrix} = \begin{pmatrix} -i & 0 \\ 0 & i \end{pmatrix} = x^{-1}$$

and

$$x^2 = \begin{pmatrix} -1 & 0 \\ 0 & -1 \end{pmatrix} = y^2.$$

Since, by part (C), there is only one group up to isomorphism whose generators satisfy these relations, we conclude that $G \cong Q_8$. The following table shows the isomorphism between Q_8 and G.

$Q_8 \leftrightarrow G$	
$1 \leftrightarrow 1 = \begin{pmatrix} 1 & 0 \\ 0 & 1 \end{pmatrix}$	$z \leftrightarrow s = \begin{pmatrix} -1 & 0 \\ 0 & -1 \end{pmatrix}$
$i \leftrightarrow x = \begin{pmatrix} i & 0 \\ 0 & -i \end{pmatrix}$	$zi \leftrightarrow sx = \begin{pmatrix} -i & 0 \\ 0 & i \end{pmatrix}$
$j \leftrightarrow y = \begin{pmatrix} 0 & i \\ i & 0 \end{pmatrix}$	$zj \leftrightarrow sy = \begin{pmatrix} 0 & -i \\ -i & 0 \end{pmatrix}$
$k \leftrightarrow w = \begin{pmatrix} 0 & -1 \\ 1 & 0 \end{pmatrix}$	$zk \leftrightarrow sw = \begin{pmatrix} 0 & 1 \\ -1 & 0 \end{pmatrix}$

Note that the matrix of z is $-I_2$, which commutes with all 2×2 matrices. The group G is important because it gives us a representation for the group of quaternions in terms of 2×2 complex matrices. In contrast to this, we ask the reader to show in the exercises that Q_8 does not have a 2-dimensional real representation; that is, there is no group of 2×2 matrices with real number entries that is isomorphic to Q_8.

(E) We noted earlier that the group of quaternions is different from the group of the square since these two groups contain a different number of elements of order 2. These groups therefore represent two nonisomorphic, nonabelian groups of order 8. Let us conclude this example by showing that, up to isomorphism, these are the only two nonabelian groups of order eight; that is, that every nonabelian group of order eight is isomorphic to either D_4 or Q_8. To this end, let G be a nonabelian group of order 8. Then G contains no element of order 8, and not every element has order 2 since a group in which every element has order 2 is necessarily abelian. It follows that G contains some element x of order 4, and therefore $(x) \triangleleft G$ since $[G{:}(x)] = 2$. Let $G/(x) = \{(x), y(x)\}$. Now, $yxy^{-1} \in (x)$ and therefore $yxy^{-1} = x$ or $yxy^{-1} = x^{-1}$. If $yxy^{-1} = x$, then x and y commute and therefore G is abelian, a contradiction. Thus, $yxy^{-1} = x^{-1}$. Now, $y^2 \in (x)$ since $[y(x)]^2 = (x)$. Therefore, $|y|$ is either 2 or 4. If $|y| = 2$, then $G = \{1, x, x^2, x^3, y, yx, yx^2, yx^3\}$ and it follows that $G \cong D_4$. On the other hand, if $|y| = 4$, then $y^2 = x^2$ since x^2 is the only element of order 2 in (x); in this case, G is generated by two elements, x and y, satisfying the relations $yxy^{-1} = x^{-1}$ and $y^2 = x^2$, and hence $G \cong Q_8$ by part (C). Thus, we conclude that up to isomorphism there are only two nonabelian groups of order 8: the group D_4 of the square and the group Q_8 of quaternions. This concludes the example.

EXAMPLE 2

We claim that if p is a prime, then every group of order p or p^2 is abelian. For suppose that G is a group of order p. Then G is cyclic and therefore abelian; in this case $G \cong \mathbb{Z}_p$. Now suppose that G is a group of order p^2. Then the center $Z(G)$ is nontrivial and hence must have order p or p^2. If $|Z(G)| = p$, then G contains an element $x \notin Z(G)$ and therefore $Z(G)$ is a proper subgroup of the centralizer $C(x)$. But $[G{:}Z(G)] = p$. Therefore, $C(x) = G$ and hence $x \in Z(G)$, a contradiction. It follows that $|Z(G)| = p^2$ and hence $Z(G) = G$, or equivalently, G is abelian. We ask the reader to show in the exercises that G is in fact isomorphic to either the cyclic group $\mathbb{Z}_{p^2}$ or the direct product $\mathbb{Z}_p \times \mathbb{Z}_p$.

These examples have illustrated some of the elementary properties of *p*-groups. Before we discuss some of their deeper properties, let us first take a moment to generalize the class equation of a finite group. The class equation, we recall, is the equation $|G| = |Z(G)| + \Sigma^* [G{:}C(x)]$. It gives a numerical relationship between the order of a group, the order of its center, and the indices of its nontrivial centralizers, and it comes directly from the fact that the conjugacy classes decompose the group into disjoint subsets. Now, whenever a group G acts as a permutation group on a set X, it decomposes X into disjoint orbits and hence has associated with it a decomposition equation similar to the class equation.

To obtain this equation, let $f:G \to \mathrm{Sym}(X)$ be a permutation representation of G acting on a finite set X and let $X^G = \{x \in X \mid f_\sigma(x) = x$ for all $\sigma \in G\}$ stand for the set of elements in X fixed by every element $\sigma \in G$. The set X^G is called the *fixed point set of f*. Since the orbits of G acting on X decompose X into disjoint subsets, it follows that $|X| = \Sigma\, |O(x)|$, where the summation is over one element from each orbit. Since $|O(x)| = [G:G_x]$ for every element $x \in X$, and $x \in X^G$ if and only if $[G:G_x] = 1$, it follows that $|X| = |X^G| + \Sigma^*\, [G:G_x]$, where Σ^* stands for the summation over one element from each nontrivial orbit. This equation is called the *decomposition equation of f*. The class equation, for example, is just the decomposition equation of the inner automorphism representation $f:G \to \mathrm{Sym}(G)$; in this case $f_\sigma(x) = \sigma x \sigma^{-1}$, and hence the fixed-point set $G^G = Z(G)$ and $G_x = C(x)$.

The decomposition equation of a permutation representation is an important combinatorial result that is especially useful in the study of *p*-groups. For if G is a *p*-group, then p divides each of the indices $[G:G_x]$ in the sum Σ^* and therefore $|X| \equiv |X^G| \pmod p$. It follows, in particular, that if G is a *p*-group acting as a permutation group on a set X and p does not divide $|X|$, then the fixed-point set X^G cannot be empty; that is, there must be some element $x \in X$ that is fixed by every permutation in G. The following proposition summarizes this result, which we refer to as the *fixed-point property* of *p*-groups.

PROPOSITION 3. (FIXED-POINT PROPERTY OF *p*-GROUPS) Let G be a *p*-group. If G acts as a permutation group on a finite set X and p does not divide $|X|$, then X contains some element that is fixed by every permutation in G.   ■

Let us illustrate the fixed-point property of *p*-groups by using it to give another proof of the fact that a nontrivial *p*-group must have a nontrivial center. Let G be a nontrivial *p*-group and let $G - \{1\}$ stand for the set of nonidentity elements in G. Then every conjugate of an element in $G - \{1\}$ is an element in $G - \{1\}$, and hence the group G acts on the set $G - \{1\}$ by means of the inner automorphism representation. Since $|G - \{1\}| = |G| - 1$, which is not divisible by p, it follows from the fixed-point property that there is some element $x \in G - \{1\}$ that is fixed by every inner automorphism of G. The element x is therefore a nonidentity element in the center $Z(G)$ and hence $Z(G)$ is nontrivial.

Using the fixed-point property, let us now develop some of the deeper properties of *p*-groups.

PROPOSITION 4. Let G be a *p*-group and let N be a nontrivial normal subgroup of G. Then $N \cap Z(G) \neq (1)$. Thus, in a *p*-group, every nontrivial normal subgroup contains some element other than the identity that commutes with every element in the *p*-group.

Proof. Let $N - \{1\}$ stand for the set of nonidentity elements in N. Since $N \lhd G$, $gng^{-1} \in N$ for all elements $g \in G$ and $n \in N$. Therefore, G acts as a permutation group on the set $N - \{1\}$ by means of the inner automorphism representation. But $|N - \{1\}| = |N| - 1$ and p does not divide $|N - \{1\}|$. Hence, by the fixed-point property, there is some element $n \in N - \{1\}$ such that $gng^{-1} = n$ for all $g \in G$. Therefore, $n \in N \cap Z(G)$ and hence $N \cap Z(G) \neq (1)$. ∎

Now recall that if H is a subgroup of an arbitrary group G, an element $g \in G$ is said to normalize H if $gHg^{-1} = H$; that is, g conjugates every element in H to an element in H. The normalizer of H is the subgroup $N(H) = \{g \in G \mid gHg^{-1} = H\}$ consisting of all such elements. Clearly, $H \leq N(H)$. In general, $N(H)$ need not be larger than H; that is, the elements in H may be the only elements in G that normalize H. For example, we showed that in the group T of the triangle $N((S)) = (S)$. The situation for *p*-groups is different, however, for in this case we claim that every proper subgroup must in fact be normalized by some element not in the subgroup.

PROPOSITION 5. Let G be a *p*-group and let H be a proper subgroup of G. Then $H \subsetneq N(H)$; that is, H is normalized by at least one element not in H.

Proof. If $H \lhd G$, then $N(H) = G$ and we are done. Assume therefore that H is not a normal subgroup of G. Then $N(H) \neq G$ and hence H has at least one conjugate other than itself. Let $X = K(H) - \{H\}$ stand for the set of conjugates of H other than H itself. Then H acts by conjugation on the set X; for if $xHx^{-1} \in K(H) - \{H\}$ and $h \in H$, then $h(xHx^{-1})h^{-1} \neq H$. Now, $|X| = |K(H)| - 1 = [G:N(H)] - 1$, and this number is not divisible by p since $N(H) \neq G$. Hence, by the fixed-point property of *p*-groups, there is some element $xHx^{-1} \in X$ such that $h(xHx^{-1})h^{-1} = xHx^{-1}$ for all $h \in H$. Consequently, $x^{-1}hx$ normalizes H for every $h \in H$. But there is some element of the form $x^{-1}hx$ that is not in H since $xHx^{-1} \neq H$. Therefore, H is normalized by some element not in H and hence $H \subsetneq N(H)$, as required. ∎

PROPOSITION 6. Let G be a *p*-group of order p^n. Then G contains a subgroup of order p^{n-1} and every such subgroup is a normal subgroup of G.

Proof. First note that if H is any subgroup of order p^{n-1}, then $H \lhd G$; for, by Proposition 5, $H \subsetneq N(H)$ and therefore $N(H) = G$ since $[G:H] = p$ is prime. To show that G in fact has a subgroup of order p^{n-1}, first observe that since G is finite, it has only a finite number of subgroups and hence has a maximal subgroup; that is, a subgroup M not properly contained in any subgroup of G other than G itself. Since M is a proper subgroup of its normalizer $N(M)$ by Proposition 5, it follows that $N(M) = G$. Hence, $M \lhd G$.

Now consider the quotient group G/M. Recall that the subgroups of G/M are in 1–1 correspondence with the subgroups of G that contain M. Since M is maximal, G/M has no proper subgroups and hence is a cyclic group of order p. Therefore, $[G:M] = p$. It follows that $|M| = p^{n-1}$ and the proof is complete. ∎

COROLLARY. Let G be a *p*-group of order p^n. Then there are subgroups $H_1, \ldots, H_n$ such that (1) $\lhd H_1 \lhd H_2 \lhd \cdots \lhd H_{n-1} \lhd H_n = G$, where $|H_i| = p^i$ for $i = 1, \ldots, n$.

Proof. This follows immediately from Proposition 6 by induction. We leave the details as an exercise for the reader. ∎

It follows from the corollary that every *p*-group contains an ascending chain of subgroups, each a normal subgroup of index p in its successor. In particular, a *p*-group contains a subgroup of every possible order. In this sense, the structure of a *p*-group is similar to that of a cyclic group since cyclic groups also contain a subgroup of every possible order. In the case of cyclic groups, however, there is only one subgroup of each order, while for *p*-groups, there is usually more than one. This result also shows that the converse of Lagrange's theorem is true for *p*-groups: if G is a *p*-group and d divides $|G|$, then G contains a subgroup of order d.

Consider the group D_4 of the square, for example. D_4 contains three subgroups of order 4: $\{1, R, R^2, R^3\}$, $\{1, R^2, S, SR^2\}$, and $\{1, R^2, SR, SR^3\}$. Both $\{1, R^2, S, SR^2\}$ and $\{1, R^2, SR, SR^3\}$ contain three subgroups of order 2, namely, $\{1, R^2\}$, $\{1, S\}$, and $\{1, SR^2\}$ and $\{1, R^2\}$, $\{1, SR\}$, and $\{1, SR^3\}$, respectively, while the cyclic subgroup $\{1, R, R^2, R^3\}$ contains only one such subgroup, $\{1, R^2\}$. The diagram in Figure 2 displays these subgroups and

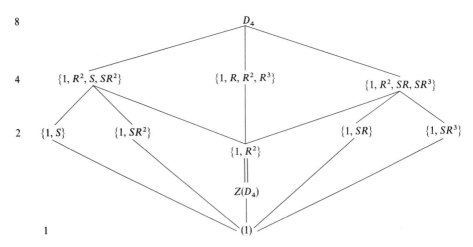

FIGURE 2. Lattice of subgroups of the group of the square.

their relationship to each other. A typical chain of subgroups having orders 1, 2, 4, and 8, respectively, is $(1) \lhd \{1, SR\} \lhd \{1, R^2, SR, SR^3\} \lhd D_4$. Observe that the proper normal subgroups of D_4 are the three subgroups of order 4 and the center $Z(D_4)$. According to Proposition 4, each of these subgroups must intersect the center, and it is clear that they do in fact intersect $Z(D_4)$ since each of them contain the element R^2. Finally, each of the eight proper subgroups of D_4 must be normalized by some element not in the subgroup; for example, we find that $N(\{1, S\}) = \{1, R^2, S, SR^2\}$ and hence both R^2 and SR^2 normalize the subgroup $\{1, S\}$ but lie outside of it.

EXAMPLE 3

Let us discuss the subgroup structure of the group Q_8 of quaternions. The group Q_8 has three subgroups of order 4: $\{1, -1, i, -i\}$, $\{1, -1, j, -j\}$, and $\{1, -1, k, -k\}$. Since each of these subgroups is cyclic, each has only one subgroup of order 2, namely, the center $\{1, -1\}$. In this case, a typical chain of subgroups having orders 1, 2, 4, and 8 is $(1) \lhd \{1, -1\} \lhd \{1, -1, i, -i\} \lhd Q_8$. Finally, observe that every subgroup of Q_8 is a normal subgroup.

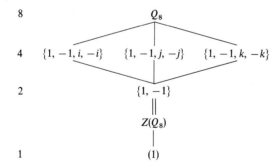

Exercises

1. Determine the conjugacy classes and centralizers of all elements in the group Q_8 of quaternions.
2. Determine the conjugacy classes and normalizers of all subgroups of the group Q_8 of quaternions.
3. Show that $D_4/Z(D_4) \cong Q_8/Z(Q_8) \cong \mathbb{Z}_2 \times \mathbb{Z}_2$.
4. Let G be a group of order p^2, where p is a prime, and suppose that G contains no element of order p^2.
   (a) Show that G contains elements x and y, $y \notin (x)$, such that $|x| = |y| = p$.
   (b) Define the function $f:(x) \times (y) \to G$ by setting $f(x^s, y^t) = x^s y^t$ for all integers s and t. Show that f is a group-isomorphism and conclude therefore that $G \cong \mathbb{Z}_p \times \mathbb{Z}_p$.
5. Let G be a group of order 8. Show that G is isomorphic to exactly one of the following groups: $\mathbb{Z}_8$, $\mathbb{Z}_4 \times \mathbb{Z}_2$, $\mathbb{Z}_2 \times \mathbb{Z}_2 \times \mathbb{Z}_2$, D_4, or Q_8.

6. Let G be a nontrivial p-group. Show that the commutator subgroup G' is a proper normal subgroup of G (see Exercise 20, Section 2, Chapter 4).

7. Let G be an abelian group of order p^n, $n \geq 1$, and suppose that every nonidentity element in G has order p. Show that $G \cong \mathbb{Z}_p \times \cdots \times \mathbb{Z}_p$, the direct product of n copies of $\mathbb{Z}_p$.

8. The purpose of this exercise is to show that the group of quaternions does not have a 2-dimensional real representation. Suppose, to the contrary, that the group $\text{Mat}_2^*(\mathbb{R})$ contains matrices x and y such that $yxy^{-1} = x^{-1}$ and $x^2 = y^2 \neq 1$. Let $x = \begin{pmatrix} a & b \\ c & d \end{pmatrix}$ and $y = \begin{pmatrix} e & f \\ g & h \end{pmatrix}$. Then the relations $yxy^{-1} = x^{-1}$ and $x^2 = y^2 \neq 1$ determine algebraic equations among the numbers $a, \ldots, h$. Show that these algebraic equations do not have a solution in the set $\mathbb{R}$ of real numbers and conclude, therefore, that the group Q_8 of quaternions does not have a 2-dimensional real representation.

2. THE SYLOW THEORY

As we noted in the introduction, the Sylow theory is a collection of three fundamental theorems dealing with the existence and properties of subgroups of prime power order in any finite group. Our goal in this section is to prove and illustrate these theorems. We begin with the following basic result due to Cauchy.

PROPOSITION 1. (CAUCHY) Every finite abelian group whose order is divisible by a prime p contains an element of order p.

Proof. Let G be a finite abelian group of order n. We prove the statement by induction on n. If $n = 1$, there is nothing to prove. Assume therefore that $n > 1$ and that every abelian group of order $< n$ whose order is divisible by p contains an element of order p. Let x be a nonidentity element in G. If p divides $|x|$, say $|x| = pt$, then $|x^t| = p$ and therefore x^t is the required element of order p. On the other hand, if p does not divide $|x|$, then p divides $[G:(x)]$. But $(x) \triangleleft G$ since G is abelian. Therefore, $G/(x)$ is an abelian group whose order is $< n$ and divisible by p. Thus, by assumption, $G/(x)$ contains an element $y(x)$ of order p. Since $|y(x)|$ divides $|y|$, it now follows that p divides $|y|$ and hence some power of y has order p. Hence, in both cases, G contains an element of order p and the proof by induction is therefore complete. ∎

THE FIRST SYLOW THEOREM. Let G be a finite group and let p^n be the largest power of a prime p that divides the order of G. Then G contains a subgroup of order p^n.

Proof. Let $m = |G|$. We prove the statement by induction on m. If $m = 1$, the trivial subgroup satisfies the statement. Assume therefore that $m > 1$ and that every finite group of order $< m$ contains a subgroup of order p^n whenever p^n is the largest power of p dividing its order. Consider the center $Z(G)$.

(A) If p divides $|Z(G)|$, then $Z(G)$ is an abelian group whose order is divisible by p and hence, by Proposition 1, $Z(G)$ contains an element x of order p. Then $(x) \lhd G$ and the quotient group $G/(x)$ has order $< m$. Hence, by the induction assumption, $G/(x)$ contains a subgroup H^* of order p^{n-1} since p^{n-1} is the largest power of p dividing $|G/(x)|$. Let $H = \{y \in G \mid y(x) \in H^*\}$ stand for the preimage of H^* under the homomorphism $G \to G/(x)$. Then H is a subgroup of G and, by the fundamental theorem of group-homomorphism, $H/(x) \cong H^*$. Therefore, $|H| = |H^*||x| = p^n$ and hence the subgroup H satisfies the statement.

(B) If, on the other hand, p does not divide $|Z(G)|$, then it follows from the class equation $|G| = |Z(G)| + \Sigma^* [G:C(x)]$ that p does not divide some index $[G:C(x)]$. In this case, p^n is the largest power of p dividing $|C(x)|$. But $|C(x)| < m$ since $[G:C(x)] \neq 1$ in the sum Σ^*. Hence, by the induction assumption, $C(x)$ contains a subgroup H of order p^n. The subgroup H then satisfies the requirements of the theorem and the proof by induction is complete. ∎

COROLLARY. Let G be a finite group and let p^s be any power of a prime p that divides the order of G. Then G contains a subgroup of order p^s. In particular, G contains an element of order p whenever p divides $|G|$.

Proof. Let p^n be the largest power of p dividing $|G|$. Then, by the First Sylow Theorem, G contains a subgroup P of order p^n. Since P is a p-group, it contains a subgroup of order p^s which clearly is a subgroup of G of order p^s, as required. In particular, if p divides $|G|$, then G contains a subgroup of order p and any one of its generators is therefore an element of order p. ∎

Definition

Let G be a finite group and let p be a prime. A *p-subgroup* of G is any subgroup of G whose order is a power of p. A *Sylow p-subgroup* of G is any subgroup of G whose order is the largest power of p that divides the order of G. We let $\mathrm{Syl}_p(G)$ stand for the set of all Sylow p-subgroups of G.

Thus, if p^n is the largest power of a prime p that divides the order of a group G, then G has a Sylow p-subgroup whose order is p^n, as well as p-subgroups of orders $p, p^2, \ldots, p^{n-1}$. In this sense, the converse of Lagrange's theorem is true for any finite group G: if p^s is any prime power divisor of $|G|$, then G contains a subgroup of order p^s. For example, in the group T of the triangle, the Sylow 2-subgroups have order 2 since $|T| = 6 = 2 \cdot 3$, and there are three such subgroups, namely, $\{1, S\}$, $\{1, SR\}$, and $\{1, SR^2\}$. The Sylow 3-subgroups of T have order 3, and there is only one of them, namely, $\{1, R, R^2\}$. Thus,

$$\mathrm{Syl}_2(T) = \{\{1, S\}, \{1, SR\}, \{1, SR^2\}\}$$

$$\mathrm{Syl}_3(T) = \{\{1, R, R^2\}\}.$$

Moreover, observe that the three Sylow 2-subgroups of T are all conjugate and that the Sylow 3-subgroup is a normal subgroup of T.

EXAMPLE 1

Let us find a Sylow 2-subgroup and a Sylow 3-subgroup of the symmetric group Sym(4). Since $|\text{Sym}(4)| = 24 = 2^3 \cdot 3$, the Sylow 2-subgroups have order 8, while the Sylow 3-subgroups have order 3.

(A) To find a Sylow 2-subgroup of Sym(4), recall that the group D_4 of the square acts as a permutation group on the four vertices of the square and hence defines a permutation representation $f : D_4 \to \text{Sym}(4)$. Since the representation is faithful, f embeds the group D_4 as a subgroup of Sym(4) of order 8. The image $f(D_4)$ is therefore a Sylow 2-subgroup of Sym(4). For example, if we label the vertices of the square as indicated in Example 4, Section 2, Chapter 6, then

$$f(D_4) = \{1, (1\ 2\ 3\ 4), (1\ 3)(2\ 4), (1\ 4\ 3\ 2), (1\ 4)(2\ 3), (1\ 3), (1\ 2)(3\ 4), (2\ 4)\}.$$

(B) A Sylow 3-subgroup of Sym(4) has order 3 and is therefore a cyclic subgroup generated by any 3-cycle in Sym(4). For example, $((1\ 2\ 3))$ $= \{1, (1\ 2\ 3), (1\ 3\ 2)\}$ is a typical Sylow 3-subgroup of Sym(4).

The remaining Sylow theorems deal with the nature and number of Sylow subgroups of a group. Recall that the group T of the triangle has three Sylow 2-subgroups which, we observed, are conjugate subgroups of T. In general, if P is a Sylow p-subgroup of a group G, then every conjugate of P is also a Sylow p-subgroup of G. For if xPx^{-1} is a typical conjugate of P, then xPx^{-1} has the same order as P and hence is a Sylow p-subgroup of G. Let us now show that the conjugates of P are in fact all of the Sylow p-subgroups of G and that the number of such subgroups is congruent to 1 modulo p.

THE SECOND SYLOW THEOREM. Let G be a finite group and let p be a prime. Then all Sylow p-subgroups of G are conjugate. Moreover, every p-subgroup of G is contained in some Sylow p-subgroup.

Proof. We first show that every p-subgroup of G is contained in some Sylow p-subgroup. Let H be a p-subgroup of G and let P be any Sylow p-subgroup of G. Then H acts as a group of permutations on the left cosets in G/P by left multiplication; if $h \in H$ and xP is a typical left coset of P, then h permutes xP to hxP. Since H is a p-group and $|G/P|$ is not divisible by p, it follows from the fixed-point property of p-groups that there is a left coset $xP \in G/P$ such that $hxP = xP$ for every element $h \in H$. Therefore, $H \leq xPx^{-1}$. Since xPx^{-1} is a Sylow p-subgroup of G, it follows that H is contained in a Sylow p-subgroup of G. In particular, if P^* is any Sylow p-subgroup of G, then $P^* \leq xPx^{-1}$ for some element $x \in G$. But both P^* and xPx^{-1} have the same order. Therefore, $P^* = xPx^{-1}$ and hence P and P^* are conjugate. Thus, all Sylow p-subgroups of G are conjugate and the proof is complete.   ∎

COROLLARY.   Let P be a Sylow p-subgroup of a finite group G and let $N(P)$ stand for the normalizer of P in G. Then P is the unique Sylow p-subgroup of its normalizer $N(P)$. In particular, if P is a normal subgroup of G, then P is the only Sylow p-subgroup of G.

Proof.   Clearly, P is a Sylow p-subgroup of its normalizer $N(P)$. Moreover, by the Second Sylow Theorem, every Sylow p-subgroup of $N(P)$ is conjugate to P and hence has the form xPx^{-1} for some element $x \in N(P)$. But $xPx^{-1} = P$ for every $x \in N(P)$. Thus, P is the only Sylow p-subgroup of $N(P)$.   ∎

Now recall that if H is any subgroup of a group G, there is a 1–1 correspondence $K(H) \leftrightarrow G/N(H)$ between the conjugates of H and the left cosets of the normalizer $N(H)$ in G. Under this correspondence, a typical conjugate xHx^{-1} corresponds to the left coset xH. The number of conjugates of H in G is therefore equal to the index $[G:N(H)]$. It follows, in particular, that since all of the Sylow p-subgroups of G are conjugate, the number of Sylow p-subgroups is equal to $[G:N(P)]$, where P is any Sylow p-subgroup of G; that is, $|\text{Syl}_p(G)| = [G:N(P)]$. Let us now prove the third and final Sylow theorem which shows that this number is congruent to 1 modulo p.

THE THIRD SYLOW THEOREM.   The number of Sylow p-subgroups of a finite group is congruent to 1 modulo p.

Proof.   Let G be a finite group and let P be a Sylow p-subgroup of G. Since every conjugate of a Sylow p-subgroup of G is a Sylow p-subgroup, it follows that P acts as a permutation group on the set $\text{Syl}_p(G)$ by conjugation. Therefore $|\text{Syl}_p(G)| \equiv |\text{Syl}_p(G)^P| \pmod{p}$, where $\text{Syl}_p(G)^P$ stands for the fixed-point set of the action. Now, $P^* \in \text{Syl}_p(G)^P$ if and only if $xP^*x^{-1} = P^*$ for all $x \in P$. In this case, $P \le N(P^*)$ and therefore $P = P^*$ since P^* is the unique Sylow p-subgroup of its normalizer. It follows that $\text{Syl}_p(G)^P = \{P\}$ and hence that $|\text{Syl}_p(G)| \equiv 1 \pmod{p}$, as required.   ∎

Thus, if p^n is the largest power of a prime p dividing the order of a finite group G, then G contains a subgroup of order p^n and all such subgroups are conjugate and contain all of the p-subgroups of G. Moreover, the number of such subgroups is congruent to 1 modulo p. For example, the group T of the triangle has three Sylow 2-subgroups, namely, $\{1, S\}$, $\{1, SR\}$, and $\{1, RS\}$, and these subgroups are indeed conjugate subgroups of T. In this case, $|\text{Syl}_2(T)| = 3$ and $3 \equiv 1 \pmod{2}$, which agrees with the third Sylow theorem. On the other hand, there is only one Sylow 3-subgroup, namely, $(R) = \{1, R, R^2\}$, which is a normal subgroup of T; in this case, $|\text{Syl}_3(T)| = 1$.

EXAMPLE 2

Let us find the Sylow subgroups of the symmetric group Sym(4). Since $|\text{Sym}(4)|$ $= 24 = 2^3 \cdot 3$, the Sylow 2-subgroups have order 8, the Sylow 3-subgroups have order 3, and all other Sylow subgroups are trivial.

(A) The Sylow 2-subgroups.   Let

$$P = \{1, (1\ 2\ 3\ 4), (1\ 3)(2\ 4), (1\ 4\ 3\ 2), (1\ 4)(2\ 3), (1\ 3), (1\ 2)(3\ 4), (2\ 4)\}$$

stand for the Sylow 2-subgroup of Sym(4) found in Example 1. Then $[\text{Sym}(4):P] = 3$. Since $P \leq N(P) \leq \text{Sym}(4)$, it follows that $N(P)$ is either P or Sym(4). But $N(P) \neq \text{Sym}(4)$ since P is not a normal subgroup of Sym(4). Therefore, $N(P) = P$. Hence, $[\text{Sym}(4):N(P)] = 3$ and therefore Sym(4) contains three Sylow 2-subgroups, all of which are conjugate. Since P, $(1\ 2)P(1\ 2)^{-1}$, and $(1\ 4)P(1\ 4)^{-1}$ are three distinct conjugates of P in Sym(4), it follows that

$$\text{Syl}_2(\text{Sym}(4)) = \{P, (1\ 2)P(1\ 2)^{-1}, (1\ 4)P(1\ 4)^{-1}\}.$$

The diagram in Figure 1 illustrates these subgroups.

(B) The Sylow 3-subgroups.   Let $Q = ((1\ 2\ 3)) = \{1, (1\ 2\ 3), (1\ 3\ 2)\}$ stand for the Sylow 3-subgroup of Sym(4) found in Example 1. Then $[\text{Sym}(4):Q]$ $= 8$ and hence $[\text{Sym}(4):N(Q)]$ is either 8, 4, 2, or 1. Since $[\text{Sym}(4):N(Q)]$ $\equiv 1 \pmod 3$, $[\text{Sym}(4):N(Q)]$ is either 4 or 1. But $[\text{Sym}(4):N(Q)] = 1$ means that $N(Q) = \text{Sym}(4)$ and hence that $Q \lhd \text{Sym}(4)$, which is clearly not the case. Therefore, $[\text{Sym}(4):N(Q)] = 4$ and hence Sym(4) has four Sylow 3-subgroups. As before, we find that Q, $(3\ 4)Q(3\ 4)^{-1}$, $(1\ 4)Q(1\ 4)^{-1}$,

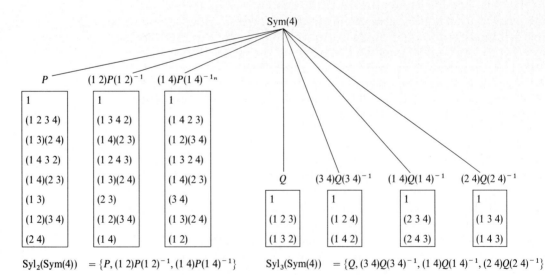

$$\text{Syl}_2(\text{Sym}(4)) = \{P, (1\ 2)P(1\ 2)^{-1}, (1\ 4)P(1\ 4)^{-1}\} \qquad \text{Syl}_3(\text{Sym}(4)) = \{Q, (3\ 4)Q(3\ 4)^{-1}, (1\ 4)Q(1\ 4)^{-1}, (2\ 4)Q(2\ 4)^{-1}\}$$

FIGURE 1.   The Sylow 2-subgroups and Sylow 3-subgroups of the group Sym(4).

and $(2\ 4)Q(2\ 4)^{-1}$ are distinct conjugates of Q and therefore

$$\mathrm{Syl}_3(\mathrm{Sym}(4)) = \{Q, (3\ 4)Q(3\ 4)^{-1}, (1\ 4)Q(1\ 4)^{-1}, (2\ 4)Q(2\ 4)^{-1}\},$$

as shown in Figure 1.

EXAMPLE 3. The Sylow subgroups of a finite cyclic group

Let $G = (x)$ be a finite cyclic group of order m and let p^n be the largest power of a prime p that divides m. Set $m = p^n t$. Then $|x^t| = p^n$ and therefore (x^t) is a Sylow p-subgroup of G. Moreover, since $(x^t) \lhd G$, (x^t) is the unique Sylow p-subgroup of G. For example, in the additive cyclic group $\mathbb{Z}_{60}$, $60 = 2^2 \cdot 15$ and hence $[15]$ generates the Sylow 2-subgroup of $\mathbb{Z}_{60}$. Similarly, $[20]$ generates the Sylow 3-subgroup since $60 = 3 \cdot 20$, and $[12]$ generates the Sylow 5-subgroup since $60 = 5 \cdot 12$. Here is the complete list of Sylow subgroups of $\mathbb{Z}_{60}$:

$$\mathrm{Syl}_2(\mathbb{Z}_{60}) = \{([15])\} = \{\{[0], [15], [30], [45]\}\}$$

$$\mathrm{Syl}_3(\mathbb{Z}_{60}) = \{([20])\} = \{\{[0], [20], [40]\}\}$$

$$\mathrm{Syl}_5(\mathbb{Z}_{60}) = \{([12])\} = \{\{[0], [12], [24], [36], [48]\}\}.$$

EXAMPLE 4

As our final example, let us use the Sylow theory to show that every group of order 15 is cyclic. Let G be a group of order 15 and let P be a Sylow 3-subgroup of G and Q be a Sylow 5-subgroup of G. Then $|P| = 3$. Therefore, $[G:P] = 5$ and hence $N(P)$ is either P or G. If $N(P) = P$, then $[G:N(P)] = 5$, which is impossible since $5 \not\equiv 1 \pmod 3$. Hence, $N(P) = G$ and therefore $P \lhd G$. Similarly, $Q \lhd G$. Now, choose elements $x \in P$ and $y \in Q$ such that $|x| = 3$ and $|y| = 5$ and consider the element $xyx^{-1}y^{-1}$. On the one hand, $xyx^{-1}y^{-1} = (xyx^{-1})y^{-1} \in Q$ since $y \in Q$ and $Q \lhd G$; on the other hand, $xyx^{-1}y^{-1} = x(yx^{-1}y^{-1}) \in P$ since $x \in P$ and $P \lhd G$. Therefore, $xyx^{-1}y^{-1} \in P \cap Q$. But $P \cap Q = (1)$ since the orders of P and Q are relatively prime. Hence, $xyx^{-1}y^{-1} = 1$, or equivalently, $xy = yx$. It follows that $|xy| = 15$ and hence $G = (xy) \cong \mathbb{Z}_{15}$. Thus, up to isomorphism there is only one group of order 15, namely, the cyclic group $\mathbb{Z}_{15}$.

Exercises

1. Find all Sylow subgroups of the alternating group Alt(4) and show that the Sylow 2-subgroup is a normal subgroup.
2. Find all Sylow subgroups of the alternating group Alt(5).
3. Find all Sylow subgroups of $\mathbb{Z}_{63}$. Draw the lattice of subgroups and indicate the Sylow subgroups.
4. Let G be a group of order 20. Show that G has a unique Sylow 5-subgroup.

5. Let $U(\mathbb{Z}_{20})$ stand for the multiplicative group of generators of $\mathbb{Z}_{20}$ discussed in Example 7, Section 1, Chapter 4. Find all Sylow subgroups of $U(\mathbb{Z}_{20})$.

6. Find all Sylow subgroups of the multiplicative group $U(\mathbb{Z}_{216})$.

7. Let p be a prime. Show that a finite group G is a p-group if and only if p divides the order of every nonidentity element in G.

8. Let G and H be finite groups and let p be a prime. If G_p and H_p are Sylow p-subgroups of G and H, respectively, show that $G_p \times H_p$ is a Sylow p-subgroup of the direct product $G \times H$.

9. Let G be a group. An *involution* in G is any element that has order 2.
   (a) If G is finite, show that G contains an involution if and only if $|G|$ is even.
   (b) If G contains exactly one involution x, show that $x \in Z(G)$.
   (c) Suppose that $|G| = 2m$, where m is odd. Show that all involutions in G are conjugate.
   (d) Illustrate the results in part (c) using the group T of the triangle.

10. Let G be a finite group and suppose that $|G| = p^n q^m$ for distinct primes p and q. Let P and Q stand for a Sylow p-subgroup and Sylow q-subgroup of G, respectively, and assume that $P \lhd G$ and $Q \lhd G$.
    (a) If x, $x' \in P$ and y, $y' \in Q$, show that $xy = x'y'$ if and only if $x = x'$ and $y = y'$.
    (b) Show that $G = PQ$.
    (c) If $x \in P$ and $y \in Q$, show that $xy = yx$.
    (d) Show that $G \cong P \times Q$.
    (e) Show that $\text{Aut}(G) \cong \text{Aut}(P) \times \text{Aut}(Q)$.
    (f) Show that $\mathbb{Z}_{20} \cong \mathbb{Z}_4 \times \mathbb{Z}_5$.
    (g) Show that $\text{Aut}(\mathbb{Z}_{20}) \cong \mathbb{Z}_2 \times \mathbb{Z}_4$.

11. Show that a finite group is isomorphic to the direct product of its Sylow subgroups if and only if every Sylow subgroup is a normal subgroup.

12. Let G be a finite abelian group and let p be a prime. Let $G_p = \{x \in G \,|\, |x|$ is a power of $p\}$ stand for the set of elements in G having p-power order.
    (a) Show that G_p is a p-subgroup of G.
    (b) Show that G_p is the unique Sylow p-subgroup of G.
    (c) Let $p_1, \ldots, p_s$ be the prime divisors of $|G|$. Show that $G \cong G_{p_1} \times \cdots \times G_{p_s}$.
    (d) Show that $\text{Aut}(G) \cong \text{Aut}(G_{p_1}) \times \cdots \times \text{Aut}(G_{p_s})$.

13. A finite group G is called an *elementary abelian p-group* if $G \cong \mathbb{Z}_p \times \cdots \times \mathbb{Z}_p$, the direct product of a finite number of copies of $\mathbb{Z}_p$, for some prime p.
    (a) Show that a finite abelian group is an elementary abelian p-group if and only if every nonidentity element has order p.
    (b) Show that every subgroup of an elementary abelian p-group is an elementary abelian p-group.
    (c) Show that every quotient group of an elementary abelian p-group is an elementary abelian p-group.

14. Let G be a group. A subgroup H of G is said to be *self-normalizing* if $N(H) = H$.
    (a) Let P be a Sylow p-subgroup of G. Show that the normalizer $N(P)$ is a self-normalizing subgroup of G.
    (b) Let H be a self-normalizing subgroup of G and let $G/H = \{H, x_2 H, \ldots, x_m H\}$. Show that $H, x_2 H x_2^{-1}, \ldots, x_m H x_m^{-1}$ are the distinct conjugates of H.

15. Let H be a subgroup of a finite group G and let p be a prime such that p does not divide the index $[G:H]$.
    (a) Show that every Sylow p-subgroup of H is a Sylow p-subgroup of G.
    (b) Is it true that every Sylow p-subgroup of G is a Sylow p-subgroup of H?

16. Let G be a finite group, p a prime, and N a normal subgroup of G.
    (a) Show that there are Sylow p-subgroups N_p and G_p of N and G, respectively, such that $N_p \lhd G_p$.
    (b) Show that G_p/N_p is a Sylow p-subgroup of G/N.

17. Let G be a finite group and N a normal subgroup of G. Suppose that for some prime p, G has a Sylow p-subgroup P such that $P \leq N$.
    (a) Show that every Sylow p-subgroup of G is a Sylow p-subgroup of N.
    (b) If $x \in G$, show that $xPx^{-1} = nPn^{-1}$ for some element $n \in N$.
    (c) Show that $G = NN(P)$.

18. Let P be a Sylow p-subgroup of a finite group G. Show that the only elements in the normalizer $N(P)$ that have p-power order are the elements in P.

19. Let P be a Sylow p-subgroup of a finite group G. If H is any p-subgroup of G, show that $H \cap N(P) = H \cap P$.

8

The Geometric Groups

Throughout the first part of this book, we have frequently used the group of the triangle and other groups of geometric transformations to illustrate basic concepts of group theory. These groups are useful because they provide a way to visualize and interpret algebraic concepts from a geometric point of view. In this chapter our goal is to extend these ideas to groups of geometric transformations in the more general setting of an arbitrary Euclidean space.

Recall that Euclidean spaces combine the structure of a vector space together with the Euclidean distance function. As such, they provide a natural setting in which to discuss distance-preserving transformations such as those in the group of the triangle. There are two types of distance-preserving transformations that are of primary concern to us in this chapter: orthogonal transformations and Euclidean transformations. Orthogonal transformations are linear isomorphisms of the space that preserve the distance between points; they generalize the notions of rotation and reflection from ordinary plane geometry to higher dimensional Euclidean spaces and they form a group called the orthogonal group of the space. Euclidean transformations, on the other hand, are all distance-preserving mappings of the space onto itself. Euclidean motions are also called rigid motions, and like orthogonal transformations, they form a group called the Euclidean group of the space. Our goal in this chapter is to discuss and illustrate the basic properties of these geometric transformations and their groups. By doing so, we not only obtain a new and important collection of groups—the geometric groups— but also gain a better understanding of the groups of ordinary plane geometry and how these groups extend to higher dimensions, as well as a better understanding of the interaction between group theory and geometry.

1. EUCLIDEAN SPACES

Since we will be dealing with Euclidean spaces throughout this chapter, let us take a moment to review, without proof, the basic terminology and properties of these spaces. Proofs can be found in any standard linear algebra book.

Definition

A *Euclidean space* is a finite-dimensional real vector space V together with a function $\langle \, , \, \rangle : V \times V \to \mathbb{R}$, called the *inner product* on V, that satisfies the following conditions:

(1) $\langle A, B \rangle = \langle B, A \rangle$ for all vectors $A, B \in V$;

(2) $\langle A + B, C \rangle = \langle A, C \rangle + \langle B, C \rangle$ for all vectors $A, B, C \in V$;

(3) $\langle cA, B \rangle = \langle A, cB \rangle = c \langle A, B \rangle$ for all vectors $A, B \in V$ and all scalars $c \in \mathbb{R}$;

(4) $\langle A, A \rangle \geq 0$ for all vectors $A \in V$, and $\langle A, A \rangle = 0$ if and only if $A - 0$, the zero vector of V.

If V is a Euclidean space with inner product $\langle \, , \, \rangle$ and $A \in V$, the *length* of the vector A is the number $\|A\| = \sqrt{\langle A, A \rangle}$. The *distance function* on V is the function $d : V \times V \to \mathbb{R}$ defined by setting $d(A, B) = \|A - B\|$ for all vectors $A, B \in V$. For example, the Euclidean plane is the real 2-dimensional space $\mathbb{R}^2$ together with the inner product $\langle \, , \, \rangle : \mathbb{R}^2 \times \mathbb{R}^2 \to \mathbb{R}$ defined by setting $\langle (x_1, y_1), (x_2, y_2) \rangle = x_1 x_2 + y_1 y_2$ for all vectors $(x_1, y_1), (x_2, y_2) \in \mathbb{R}^2$. In this space, the length of the vector $(1, 3)$, for example, is $\|(1, 3)\| = \sqrt{10}$, while the inner product of the vectors $(1, 3)$ and $(-2, 1)$ is $\langle (1, 3), (-2, 1) \rangle = 1$. The inner product of two vectors A, B in the Euclidean plane is frequently called their dot product and written $A \cdot B$.

Let us also mention an important relationship between the inner product on a Euclidean space V and the distance function: if $A, B \in V$, then

$$\langle A, B \rangle = \frac{1}{2}[d(A, 0)^2 + d(B, 0)^2 - d(A, B)^2].$$

This formula expresses the inner product in terms of the distance function and is easily verified by direct calculation.

Let V be a Euclidean space. Two vectors $A, B \in V$ are said to be *orthogonal* if $\langle A, B \rangle = 0$, in which case we write $A \perp B$. If U is a subspace of V, the *orthogonal complement* of U is the set $U^\perp = \{X \in V \mid X \perp A \text{ for every vector } A \in U\}$ consisting of all vectors in V orthogonal to every vector in U. It is easily verified that $U^\perp$ is a subspace of V. Moreover, $V = U \oplus U^\perp$; that is, every vector in V may be written uniquely in the form $A + B$, where $A \in U$ and $B \in U^\perp$. An *orthonormal basis* for V is any basis $\{A_1, \ldots, A_n\}$ such that $\langle A_i, A_j \rangle = \delta_{ij}$ for all indices i, j, where δ_{ij} is the Kronecker symbol; δ_{ij} equals

1 when $i = j$ and is 0 otherwise. Thus, an orthonormal basis is any basis consisting of unit vectors that are pairwise orthogonal. Every nontrivial subspace of V has an orthonormal basis.

Now recall from Chapter 1 that a *linear isomorphism* of a vector space V is any nonsingular linear transformation mapping V onto V. The set $GL(V)$ of all linear isomorphisms on V is a group under function composition called the *general linear group* of V. But linear isomorphisms do not, in general, preserve the inner product on Euclidean spaces; that is, the inner product of two vectors is not necessarily the same as the inner product of their images under a linear isomorphism. To distinguish those linear isomorphisms that preserve inner products from those which do not, we introduce the concept of isometry. An *isometry* from a Euclidean space V with inner product $\langle \, , \, \rangle$ to a Euclidean space V' with inner product $\langle \, , \, \rangle'$ is any linear isomorphism $\sigma: V \to V'$ such that $\langle \sigma(A), \sigma(B) \rangle' = \langle A, B \rangle$ for all vectors $A, B \in V$; that is, a nonsingular linear transformation mapping V onto V' that preserves the inner product on the two spaces. Isometry is an equivalence relation on the collection of all Euclidean spaces, and two Euclidean spaces are isometric if and only if they have the same dimension. Thus, up to isometry, there is only one Euclidean space of any given dimension.

The basic model of an n-dimensional Euclidean space is ordinary Euclidean n-space $\mathbb{E}^n$. $\mathbb{E}^n$, we recall, is the n-dimensional real vector space $\mathbb{R}^n$ together with the Euclidean inner product $\langle \, , \, \rangle: \mathbb{R}^n \times \mathbb{R}^n \to \mathbb{R}^n$ defined by setting

$$\langle (a_1, \dots, a_n), (b_1, \dots, b_n) \rangle = a_1 b_1 + \cdots + a_n b_n$$

for all vectors $(a_1, \dots, a_n), (b_1, \dots, b_n) \in \mathbb{R}^n$. If $A = (a_1, \dots, a_n)$ and $B = (b_1, \dots, b_n)$ are typical vectors in $\mathbb{E}^n$, then

$$\|A\| = \sqrt{a_1^2 + \cdots + a_n^2},$$

$$d(A, B) = \sqrt{(a_1 - b_1)^2 + \cdots + (a_n - b_n)^2},$$

and

$$\langle A, B \rangle = \|A\| \, \|B\| \cos \theta,$$

where θ is the angle between A and B. The *standard orthonormal basis* for $\mathbb{E}^n$ is $\{e_1, \dots, e_n\}$, where $e_i = (0, \dots, 1, \dots, 0)$ is the vector all of whose entries are 0 except the ith entry which is 1. Finally, if U is any k-dimensional subspace of $\mathbb{E}^n$, then U is a k-dimensional Euclidean space and hence is isometric to $\mathbb{E}^k$.

Let us illustrate these ideas with an example. Let U stand for the subspace of Euclidean 3-space $\mathbb{E}^3$ spanned by the vectors $A = (-3, 1, 2)$ and $B = (2, -3, 1)$, as illustrated in Figure 1 (page 228). Then A and B are linearly independent and hence the set $\{A, B\}$ is a basis for U. In particular, dim $U = 2$. To find the orthogonal complement $U^\perp$, observe that a typical vector (a, b, c) lies in $U^\perp$ if and only if it is orthogonal to both A and B; that is, if and only if

$$\langle (a, b, c), A \rangle = -3a + b + 2c = 0$$

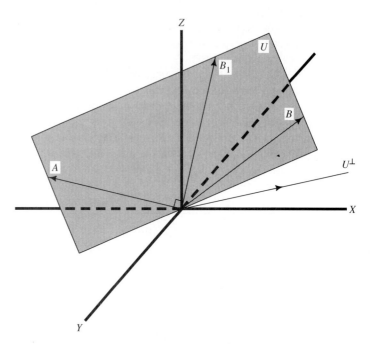

FIGURE 1. The subspace U of $\mathbb{E}^3$ spanned by vectors $A = (-3, 1, 2)$ and $B = (2, -3, 1)$.

and

$$\langle(a, b, c), B\rangle = 2a - 3b + c = 0.$$

Solving these equations, we find that $a = b = c$. Therefore, $U^{\perp} = \{(a, a, a) \in \mathbb{E}^3 \,|\, a \in \mathbb{R}\} = \{a(1, 1, 1) \in \mathbb{E}^3 \,|\, a \in \mathbb{R}\} = \langle(1, 1, 1)\rangle$; that is, $U^{\perp}$ is the line spanned by the vector $(1, 1, 1)$. Moreover, every vector in $\mathbb{E}^3$ may be written uniquely in the form $C + D$ for some vectors $C \in U$ and $D \in U^{\perp}$ since $\{A, B, (1, 1, 1)\}$ is a basis for $\mathbb{E}^3$. Hence, $\mathbb{E}^3 = U \oplus U^{\perp}$. Note, however, that the vectors A and B are not orthogonal since $\langle A, B\rangle = -7 \neq 0$, and hence do not form an orthonormal basis for U.

Now, we noted earlier that U is a 2-dimensional subspace of $\mathbb{E}^3$. U is therefore isometric to $\mathbb{E}^2$. Let us conclude this example by constructing a specific isometry $\sigma : \mathbb{E}^2 \to U$. To define such a mapping, we cannot simply map a basis for $\mathbb{E}^2$ to a basis for U, but must first find an orthonormal basis for U. To this end, let $A_1 = A/\|A\| = 1/\sqrt{14}(-3, 1, 2)$. Then A_1 is a unit vector in U. Now, a typical vector $aA + bB \in U$ is orthogonal to A_1 if and only if

$$\langle aA + bB, (-3, 1, 2)\rangle = \langle(-3a + 2b, a - 3b, 2a + b), (-3, 1, 2)\rangle$$
$$= 14a - 7b = 0.$$

Thus, if we let $a = 1$ and $b = 2$, then $aA + bB = A + 2B = (1, -5, 4)$ and hence the vector

$$B_1 = \frac{A + 2B}{\|A + 2B\|} = \frac{1}{\sqrt{42}}(1, -5, 4)$$

is a unit vector in U orthogonal to A_1. Therefore, $\{A_1, B_1\}$ is an orthonormal basis for U. Finally, define $\sigma: \mathbb{E}^2 \to U$ by setting $\sigma(e_1) = A_1$ and $\sigma(e_2) = B_1$. Then, extending σ linearly to $\mathbb{E}^2$, we find that

$$\sigma(x, y) = xA_1 + yB_1 = \left(\frac{-3x}{\sqrt{14}} + \frac{y}{\sqrt{42}}, \frac{x}{\sqrt{14}} - \frac{5y}{\sqrt{42}}, \frac{2x}{\sqrt{14}} + \frac{4y}{\sqrt{42}} \right)$$

for all vectors $(x, y) \in \mathbb{E}^2$. The mapping σ is clearly a nonsingular linear transformation mapping $\mathbb{E}^2$ onto U. Moreover, if (x, y) and (x', y') are typical vectors in $\mathbb{E}^2$, then

$$\begin{aligned}
\langle \sigma(x, y), \sigma(x', y') \rangle &= \langle xA_1 + yA_2, x'A_1 + y'A_2 \rangle \\
&= xx'\langle A_1, A_1 \rangle + (xy' + yx')\langle A_1, A_2 \rangle + yy'\langle A_2, A_2 \rangle \\
&= xx' + yy' \\
&= \langle (x, y), (x', y') \rangle.
\end{aligned}$$

Therefore, σ preserves the inner product and hence is an isometry mapping $\mathbb{E}^2$ onto U. Let us emphasize that the basis $\{A_1, B_1\}$ used in this example to define the function σ must be an orthonormal basis for U; for although any basis can be used to define a linear isomorphism from $\mathbb{E}^2$ onto U, such a mapping will not preserve the inner product and hence will not be an isometry unless the basis is orthonormal.

Exercises

1. Find the lengths of the following vectors and determine which vectors are orthogonal to each other: $(1, 2, 1), (-1, 0, 1), (2, 1, 0), (1, 1, 1), (0, 3, 1)$.
2. Let $A = 1/\sqrt{13}\,(2, 3) \in \mathbb{E}^2$. Find a vector B such that the set $\{A, B\}$ is an orthonormal basis for $\mathbb{E}^2$.
3. Let $U = \langle (2, 0, 2), (-1, 1, 1) \rangle$ stand for the subspace of $\mathbb{E}^3$ spanned by the vectors $(2, 0, 2)$ and $(-1, 1, 1)$.
   (a) Find an orthonormal basis for U.
   (b) Find an isometry $\sigma: \mathbb{E}^2 \to U$.
4. Let $A = (1, 2, 1)$ and let $U = \langle A \rangle^\perp$ stand for the orthogonal complement of the line spanned by A.
   (a) Find an orthonormal basis for U.
   (b) Find an isometry $\sigma: \mathbb{E}^2 \to U$.
5. Let $U = \langle (2, 1, -1), (1, 0, 1) \rangle$ stand for the subspace of $\mathbb{E}^3$ spanned by the vectors $(2, 1, -1)$ and $(1, 0, 1)$.
   (a) Find an isometry $\sigma: \mathbb{E}^2 \to U$.
   (b) Find an orthonormal basis for $U^\perp$.
   (c) Find an isometry $\tau: \mathbb{E}^1 \to U^\perp$.

(d) Show by direct calculation that $\mathbb{E}^3 = U \oplus U^\perp$; that is, if (x, y, z) is a typical vector in $\mathbb{E}^3$, show that there are unique vectors $A \in U$ and $B \in U^\perp$ such that $(x, y, z) = A + B$.

2.   ORTHOGONAL TRANSFORMATIONS AND THE ORTHOGONAL GROUP

We begin our study of the geometric transformations of a Euclidean space by discussing the orthogonal transformations of the space, which, as we noted at the beginning of the chapter, extend the idea of rotations and reflections of ordinary plane geometry to higher dimensional Euclidean spaces. We first characterize these transformations geometrically and show that they form a group called the orthogonal group of the space. Our main goal is to then identify a special type of reflection called a symmetry and show that every orthogonal transformation may be written as the product of a finite number of symmetries. This result, called the Cartan theorem, shows that the symmetries of a Euclidean space generate the orthogonal group and hence are the basic building blocks from which all orthogonal transformations are constructed.

Let V be a Euclidean space with inner product $\langle\ ,\ \rangle$ and let $d : V \times V \to \mathbb{R}$ stand for the distance function on V.

Definition

An *orthogonal transformation* of V is any isometry mapping V onto V; that is, a linear isomorphism $\sigma : V \to V$ such that $\langle \sigma(A), \sigma(B) \rangle = \langle A, B \rangle$ for all vectors $A, B \in V$

Thus, an orthogonal transformation of V is any linear isomorphism that preserves the inner product on V. The following result gives three geometric descriptions of orthogonal transformations.

PROPOSITION 1. Let σ be a linear isomorphism on a Euclidean space V. Then the following statements are equivalent:

(1) σ is an orthogonal transformation;

(2) $d(\sigma(A), \sigma(B)) = d(A, B)$ for all vectors $A, B \in V$; that is, σ preserves the distance between all points in V;

(3) $\|\sigma(A)\| = \|A\|$ for every vector $A \in V$; that is, σ preserves the length of all vectors in V;

(4) If $A, B \in V$, then $A \perp B$ if and only if $\sigma(A) \perp \sigma(B)$; that is, σ preserves orthogonality of vectors in V.

Proof.   Let us show that statements (1) and (2) are equivalent and leave the remaining proofs as an exercise for the reader. Suppose that σ is an orthogonal transformation of V and let A, B be arbitrary vectors in V. Then

$$
\begin{aligned}
d(\sigma(A), \sigma(B))^2 &= \|\sigma(A) - \sigma(B)\|^2 \\
&= \langle \sigma(A) - \sigma(B), \sigma(A) - \sigma(B) \rangle \\
&= \langle \sigma(A - B), \sigma(A - B) \rangle \\
&= \langle A - B, A - B \rangle \\
&= \|A - B\|^2 = d(A, B)^2.
\end{aligned}
$$

It follows that $d(\sigma(A), \sigma(B)) = d(A, B)$, which proves statement (2). Now suppose, conversely, that statement (2) is true; that is, that σ preserves the distance between any two vectors in V. To show that σ is an isometry of V, recall that $\langle A, B \rangle = \frac{1}{2}[d(A, 0)^2 + d(B, 0)^2 - d(A, B)^2]$ for all vectors $A, B \in V$. Using this equation, the fact that $\sigma(0) = 0$, and the fact that σ preserves the distance between any two vectors, it follows that for any two vectors $A, B \in V$,

$$
\langle \sigma(A), \sigma(B) \rangle = \frac{1}{2}[d(\sigma(A), \sigma(0))^2 + d(\sigma(B), \sigma(0))^2 - d(\sigma(A), \sigma(B))^2]
$$

$$
= \frac{1}{2}[d(A, 0)^2 + d(B, 0)^2 - d(A, B)^2]
$$

$$
= \langle A, B \rangle.
$$

Therefore, σ preserves the inner product on V and hence is an orthogonal transformation of V, as required.   ∎

   In view of Proposition 1, we may think of orthogonal transformations as those linear isomorphisms of the space that preserve either the inner product, the distance function, the length of vectors, or finally, the orthogonality relation. For example, the identity map $1_V : V \to V$ is an orthogonal transformation of any Euclidean space V. Or consider the function $\sigma : \mathbb{E}^2 \to \mathbb{E}^2$ defined by setting $\sigma(x, y) = (y, -x)$ for all vectors $(x, y) \in \mathbb{E}^2$. The mapping σ is clearly a linear isomorphism of $\mathbb{E}^2$, and since $\|\sigma(x, y)\| = \sqrt{\langle (y, -x), (y, -x) \rangle} = \sqrt{x^2 + y^2} = \|(x, y)\|$ for all vectors $(x, y) \in \mathbb{E}^2$, it follows that σ preserves the length of vectors and hence is an orthogonal transformation of $\mathbb{E}^2$. Let us now look more closely at some other examples of orthogonal transformations.

EXAMPLE 1.   The inversion transformation

Let V be a Euclidean space. The *inversion transformation* on V is the function $-1_V : V \to V$ defined by setting $-1_V(A) = -A$ for all vectors $A \in V$. The mapping -1_V is clearly a linear isomorphism on V and is an orthogonal transformation of V since $\langle -1_V(A), -1_V(B) \rangle = \langle -A, -B \rangle = \langle A, B \rangle$ for all vectors $A, B \in V$. Geometrically, the inversion transformation maps every

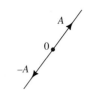

$-1_{\mathbb{E}^1} : \mathbb{E}^1 \rightarrow \mathbb{E}^1$

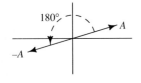

$-1_{\mathbb{E}^2} : \mathbb{E}^2 \rightarrow \mathbb{E}^2$

vector to its inverse. The figures in the margin illustrate the inversion transformation on the Euclidean line $\mathbb{E}^1$, where each vector is inverted through the origin to its inverse $-A$, and on the Euclidean plane $\mathbb{E}^2$, where each vector A is rotated around the origin to its inverse $-A$. The inversion transformation of the Euclidean plane is therefore the 180° rotation of the plane about the origin.

Before we continue with the examples, let us mention that a useful way to show that a linear isomorphism is or is not an orthogonal transformation is by means of matrices. For suppose that $\sigma : V \rightarrow V$ is a linear isomorphism of an n-dimensional Euclidean space V and let $(A_1, \ldots, A_n)$ be an orthonormal coordinate system for V. Let M stand for the matrix of σ relative to $(A_1, \ldots, A_n)$ and M^T its transpose. Then the ith row of M^T is the ith column of M, which is just the coordinates of $\sigma(A_i)$, and the jth column of M is $\sigma(A_j)$. Therefore, the (i, j)th entry of the product $M^T M$ is the inner product $\langle \sigma(A_i), \sigma(A_j) \rangle$. Since σ is orthogonal if and only if $\langle \sigma(A_i), \sigma(A_j) \rangle = \langle A_i, A_j \rangle = \delta_{ij}$ for all indices i, j, it follows that σ is orthogonal if and only if its matrix M has the property that $M^T M = (\delta_{ij}) = I_n$, the $n \times n$ identity matrix, or equivalently, that $M^{-1} = M^T$. An invertible matrix M for which $M^{-1} = M^T$ is called an *orthogonal matrix*. Thus, a linear isomorphism is an orthogonal transformation if and only if its matrix representation relative to an orthonormal basis is an orthogonal matrix. Let us emphasize that when the matrix of a linear transformation is used to decide the question of orthogonality, the matrix representation must be relative to an orthonormal basis of the underlying Euclidean space. The matrix of an orthogonal transformation relative to a nonorthonormal basis is not, in general, an orthogonal matrix.

EXAMPLE 2

Let us use the matrix approach to show that every rotation of the Euclidean plane about the origin is an orthogonal transformation of the plane. Recall that the rotation of the plane through an angle of θ radians about the origin is the mapping $\rho_\theta : \mathbb{E}^2 \rightarrow \mathbb{E}^2$, where $\rho_\theta(x, y) = (x \cos \theta - y \sin \theta, x \sin \theta + y \cos \theta)$ for all vectors $(x, y) \in \mathbb{E}^2$. It follows easily from this formula that ρ_θ is a linear transformation of $\mathbb{E}^2$ whose matrix representation relative to the standard orthonormal coordinate system $\{e_1, e_2\}$ is

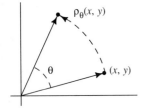

$$M = \begin{pmatrix} \cos \theta & -\sin \theta \\ \sin \theta & \cos \theta \end{pmatrix}.$$

Since the inverse of a rotation is the rotation in the opposite direction, $\rho_\theta^{-1} = \rho_{-\theta}$. Therefore,

$$M^{-1} = \begin{pmatrix} \cos(-\theta) & -\sin(-\theta) \\ \sin(-\theta) & \cos(-\theta) \end{pmatrix} = \begin{pmatrix} \cos \theta & \sin \theta \\ -\sin \theta & \cos \theta \end{pmatrix} = M^T.$$

It follows that M is an orthogonal matrix and hence that the rotation ρ_θ is an orthogonal transformation of the Euclidean plane.

EXAMPLE 3. Symmetries

Let V be a Euclidean space having dimension $n \geq 1$. Recall that a hyperplane of V is an $(n-1)$-dimensional subspace. A *symmetry* of V is a reflection of V through any hyperplane. In this example we show that every symmetry of V is an orthogonal transformation, find a matrix representation for symmetries, and derive a formula for calculating the image of any vector under a symmetry.

(A) Let us begin by first recalling what is meant by the expression "reflection through a hyperplane." If H is a hyperplane in V, then its orthogonal complement $H^{\perp}$ is a 1-dimensional subspace and hence is a line. Let $L = H^{\perp}$. Then $V = H \oplus L$ and therefore every vector $X \in V$ may be written uniquely in the form $X = X_H + X_L$, where $X_H \in H$ is the component of X lying in H and $X_L \in L$ is the component lying on the line L. The reflection of V through the hyperplane H is the function $S: V \to V$ defined by setting $S(X) = -X_L + X_H$ for every vector $X \in V$. S is the symmetry that reflects V through the hyperplane H. Figure 1 illustrates a typical symmetry of Euclidean 3-space.

(B) Let us now show that every symmetry of a Euclidean space is an orthogonal transformation of the space. Let V be a Euclidean space and let $S: V \to V$ be the symmetry that reflects V through a hyperplane H. As before, let $L = H^{\perp}$ stand for the line orthogonal to H. Then, for arbitrary

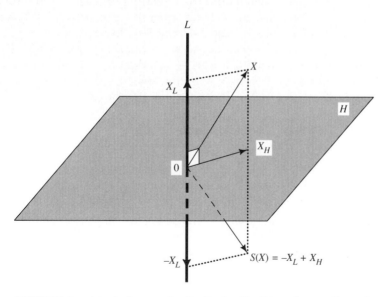

FIGURE 1.   A typical symmetry S reflecting Euclidean 3-space through the plane H.

vectors $X, Y \in V$, we find that

$$S(X + Y) = -(X + Y)_L + (X + Y)_H$$
$$= -X_L - Y_L + X_H + Y_H$$
$$= (-X_L + X_H) + (-Y_L + Y_H)$$
$$= S(X) + S(Y).$$

Similarly, $S(cX) = cS(X)$ for all scalars $c \in \mathbb{R}$ and vectors $X \in V$. Therefore, S is a linear transformation on V. Now, to find a matrix representation of S let $\{B_1, \ldots, B_{n-1}\}$ be an orthonormal basis for H and let A be a unit vector on L. Then $\{A, B_1, \ldots, B_{n-1}\}$ is an orthonormal basis for V. Since $S(A) = -A$ and $S(B_i) = B_i$ for $i = 1, \ldots, n-1$, the matrix of S relative to the orthonormal basis $\{A, B_1, \ldots, B_{n-1}\}$ is

$$\operatorname{diag}(-1, 1, \ldots, 1) = \begin{pmatrix} -1 & & & & \\ & 1 & & \mathbf{O} & \\ & & 1 & & \\ & \mathbf{O} & & \ddots & \\ & & & & 1 \end{pmatrix}.$$

Since this matrix is invertible and clearly orthogonal, it follows that S is an orthogonal transformation of V.

(C) The image of a vector under a symmetry may be calculated as follows. Let V be a Euclidean space, let A be a nonzero vector in V, and let $S_A : V \to V$ be the symmetry of V that reflects V through the hyperplane $H = \langle A \rangle^\perp$ orthogonal to A. Note that every symmetry of V has the form S_A for some nonzero vector A since every hyperplane is the orthogonal complement of some line. Now, to calculate $S_A(X)$ for a typical vector $X \in V$, we first write $X = cA + X_H$, where c is a scalar and $X_H \in H$. Then $S_A(X) = -cA + X_H = X - 2cA$. But $\langle X, A \rangle = \langle cA + X_H, A \rangle = c\langle A, A \rangle$ since $A \perp X_H$. Therefore, $c = \langle X, A \rangle / \langle A, A \rangle$ and hence

$$S_A(X) = X - 2\frac{\langle X, A \rangle}{\langle A, A \rangle} A.$$

Using this formula, it is now easy to calculate the image of any vector under a symmetry. Observe that if we choose A to be a unit vector, which is always possible, the formula becomes simply $S_A(X) = X - 2\langle X, A \rangle A$.

(D) Let us illustrate these results by discussing symmetries of the Euclidean plane. In this case hyperplanes are lines passing through the origin, and hence symmetries are just reflections of the plane about such lines. Thus, symmetries of the plane are the familiar reflections of ordinary plane geometry. For example, the symmetry $S_{(1,2)} : \mathbb{E}^2 \to \mathbb{E}^2$ reflects the plane through the line $\langle (1, 2) \rangle^\perp$. If (x, y) is a typical point in the plane, then,

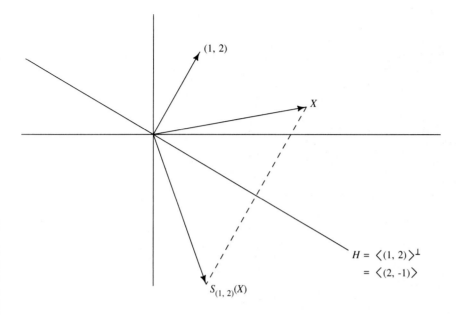

using the formula in part (C), we find that

$$S_{(1,2)}(x, y) = (x, y) - 2 \frac{\langle (1, 2), (x, y) \rangle}{\langle (1, 2), (1, 2) \rangle} (1, 2)$$

$$= (x, y) - \frac{2}{5}(x + 2y)(1, 2)$$

$$= \left(\frac{3}{5}x - \frac{4}{5}y, -\frac{4}{5}x - \frac{3}{5}y \right).$$

In particular, $S_{(1,2)}(e_1) = (3/5, -4/5)$ and $S_{(1,2)}(e_2) = (-4/5, -3/5)$, and hence the matrix of $S_{(1,2)}$ relative to the standard orthonormal coordinate system $\{e_1, e_2\}$ is $M = \left(\begin{smallmatrix} 3/5 & -4/5 \\ -4/5 & -3/5 \end{smallmatrix} \right)$. Let us verify by direct calculation that M is an orthogonal matrix:

$$M^T M = \begin{pmatrix} \dfrac{3}{5} & -\dfrac{4}{5} \\ \dfrac{4}{5} & -\dfrac{3}{5} \end{pmatrix} \begin{pmatrix} \dfrac{3}{5} & -\dfrac{4}{5} \\ -\dfrac{4}{5} & -\dfrac{3}{5} \end{pmatrix} = \begin{pmatrix} 1 & 0 \\ 0 & 1 \end{pmatrix}.$$

Therefore, $M^{-1} = M^T$ and hence M is orthogonal. Finally, if we let $A = 1/\sqrt{5}(1, 2)$ and $B = 1/\sqrt{5}(2, -1)$, then (A, B) is an orthonormal coordinate system for $\mathbb{E}^2$ and the matrix of $S_{(1,2)}$ relative to this coordinate system is $\operatorname{diag}(-1, 1) = \left(\begin{smallmatrix} -1 & 0 \\ 0 & 1 \end{smallmatrix} \right)$. This concludes the example.

The preceding examples illustrate different types of orthogonal transformations. They show, in particular, that the ordinary rotations and reflections of the Euclidean plane are orthogonal transformations and that the symmetries of the plane are just reflections about a line passing through the origin. The symmetries of a Euclidean space are especially important because, as we will show later in this section, every orthogonal transformation of the space may be written as a product of symmetries. This result, known as Cartan's theorem, shows that the symmetries of a Euclidean space generate all of its orthogonal transformations. For the moment, however, let us first extend the notion of rotation and reflection from the Euclidean plane to arbitrary Euclidean spaces.

We begin by observing that in each of the preceding examples, the determinant of the orthogonal transformation is ± 1. For example, if ρ_θ is a typical rotation of the plane, then ρ_θ has a matrix representation of the form

$$\begin{pmatrix} \cos\theta & -\sin\theta \\ \sin\theta & \cos\theta \end{pmatrix}$$

and therefore $\det(\rho_\theta) = \cos^2\theta + \sin^2\theta = 1$. On the other hand, if S is a symmetry of the plane, then S has a matrix representation of the form $\begin{pmatrix} -1 & 0 \\ 0 & 1 \end{pmatrix}$ and therefore $\det(S) = -1$. Let us now show that the determinant of an orthogonal transformation is always ± 1 and then use this fact to extend the notion of rotation and reflection.

PROPOSITION 2. Let σ be an orthogonal transformation of a Euclidean space V. Then $\det(\sigma) = \pm 1$.

Proof. Recall that the determinant of a linear transformation is defined to be the determinant of any matrix representation of the transformation and is independent of the matrix used to represent the transformation. Thus, if M stands for the matrix of σ relative to an orthonormal coordinate system, then $\det(\sigma) = \det(M)$. But $M^{-1} = M^T$ since M is an orthogonal matrix. Therefore, $(\det(M))^{-1} = \det(M)$ and hence $(\det(M))^2 = 1$, or equivalently, $\det(M) = \pm 1$, as required. ∎

Since every orthogonal transformation has determinant ± 1 and since, as we previously noted, rotations in the plane have determinant $+1$ while reflections have determinant -1, we now make the following definition.

Definition

Let V be a Euclidean space. An orthogonal transformation σ of V is called a *rotation* if $\det(\sigma) = 1$ and a *reflection* if $\det(\sigma) = -1$.

The terms *rotation* and *reflection*, as just defined for arbitrary Euclidean spaces, extend the usual notion of rotation and reflection from plane geometry to Euclidean spaces of any dimension. It is important to realize, however, that when we use these terms in the context of an arbitrary Euclidean space, they simply express a formal property of an orthogonal transformation and should not mislead us into thinking, for example, that a rotation "rotates" the space about an "axis" or that a reflection "reflects" the space through some hyperplane. The terms simply provide a convenient way to classify the orthogonal transformations of an arbitrary Euclidean space. Let us illustrate these remarks more carefully.

EXAMPLE 4

Consider the inversion transformation $\sigma = -1_{\mathbb{E}^4}$ on Euclidean 4-space. Since $\det(\sigma) = (-1)^4 = 1$, σ is a rotation of $\mathbb{E}^4$. But σ does not "rotate" the space about a fixed axis since it clearly fixes no vector except the origin. Thus, the inversion transformation on $\mathbb{E}^4$ is an example of a rotation, although it does not rotate the space about an axis.

EXAMPLE 5

Consider the inversion transformation $\sigma = -1_{\mathbb{E}^3}$ on Euclidean 3-space. Since $\det(\sigma) = (-1)^3 = -1$, σ is a reflection of $\mathbb{E}^3$. But σ does not reflect the space through a hyperplane since every such reflection fixes the vectors on the hyperplane while σ fixes no vectors except the origin. Thus, the inversion transformation on $\mathbb{E}^3$ is an example of a reflection in the general sense of the word, although it does not reflect the space through a hyperplane.

Example 5 shows that the inversion transformation on $\mathbb{E}^3$ is a reflection of the space but is not a symmetry. In this regard, let us note that every symmetry of a Euclidean space is a reflection since every symmetry has a matrix representation of the form $\text{diag}(-1, 1, \ldots, 1)$ and hence its determinant is equal to -1. The symmetries of a Euclidean space are therefore always reflections, but in general, there are reflections that are not symmetries. To summarize, let us say that the classification of orthogonal transformations into rotations and reflections is a convenient geometric language for describing these geometric transformations, provided we do not infer more about them than is stated in the definition.

Let us now turn our attention to the orthogonal group of a Euclidean space.

PROPOSITION 3. Let V be a Euclidean space and let $O(V)$ stand for the set of orthogonal transformations of V and $O^+(V)$ for the subset of rotations of V. Then $O(V)$ is a group under function composition and $O^+(V)$ is a normal subgroup of $O(V)$. Moreover, $O(V)/O^+(V) \cong \mathbb{Z}_2$ and therefore $[O(V):O^+(V)] = 2$.

Proof.   Clearly, the identity map $1_V \in O(V)$ since it is an orthogonal transformation of V. Furthermore, if $\sigma \in O(V)$, then σ preserves the inner product on V and hence $\langle A, B \rangle = \langle \sigma\sigma^{-1}(A), \sigma\sigma^{-1}(B) \rangle = \langle \sigma^{-1}(A), \sigma^{-1}(B) \rangle$ for all vectors $A, B \in V$. Therefore, σ^{-1} is an orthogonal transformation on V and hence is an element of $O(V)$. Finally, if $\sigma, \tau \in O(V)$, then $\langle \sigma\tau(A), \sigma\tau(B) \rangle = \langle \tau(A), \tau(B) \rangle = \langle A, B \rangle$ for all $A, B \in V$ and hence $\sigma\tau \in O(V)$. It follows, therefore, that $O(V)$ is a group under function composition. Now, let $\det: O(V) \to \mathbb{R}^*$ stand for the function that maps an orthogonal transformation σ to its determinant $\det(\sigma)$. Since $\det(\sigma\tau) = \det(\sigma)\det(\tau)$ for all $\sigma, \tau \in O(V)$, it follows that det is a group-homomorphism from $O(V)$ to the multiplicative group $\mathbb{R}^*$ of nonzero real numbers. By definition, the kernel of det is the subset $O^+(V)$ of rotations of V and, by Proposition 2, the image of det is the subgroup $\{1, -1\}$. Therefore, $O^+(V) \lhd O(V)$ and $O(V)/O^+(V) \cong \mathbb{Z}_2$. ∎

The group $O(V)$ of orthogonal transformations of the Euclidean space V is called the *orthogonal group* of V, while the subgroup $O^+(V)$ of rotations is called the *rotation subgroup*. $O(V)$ is clearly a subgroup of the general linear group $GL(V)$ and consists of those linear isomorphisms of V that preserve the inner product on V. Since $[O(V):O^+(V)] = 2$, the rotation subgroup $O^+(V)$ partitions $O(V)$ into two cosets. One of these, of course, is the rotation subgroup $O^+(V)$ consisting of those transformations having determinant $+1$; the other coset is the set of reflections of V and consists of those transformations having determinant -1. And finally, observe that in the group $O(V)$ the product of any two reflections is always a rotation, while the product of a rotation and a reflection is always a reflection.

Consider the orthogonal group $O(\mathbb{E}^1)$ of the Euclidean line. We recall that the general linear group $GL(\mathbb{E}^1)$ consists of all scalar transformations $\sigma_c: \mathbb{E}^1 \to \mathbb{E}^1$, where c is any nonzero scalar and $\sigma_c(x) = cx$ for all $x \in \mathbb{F}^1$. To determine which scalar transformations are orthogonal, observe that $\langle \sigma_c(x), \sigma_c(y) \rangle = c^2 \langle x, y \rangle$ for all $x, y \in \mathbb{E}^1$, and hence σ_c is orthogonal if and only if $c^2 = 1$, or equivalently, $c = \pm 1$. Therefore, $O(\mathbb{E}^1) = \{1, -1\}$ and $O^+(\mathbb{E}^1) = \{1\}$. That is, the orthogonal group of the Euclidean line consists only of the identity map and the inversion transformation, the rotation subgroup being trivial.

EXAMPLE 6. The orthogonal group of the Euclidean plane

We showed earlier that the familiar rotations and reflections of the Euclidean plane are orthogonal transformations and hence are elements of the orthogonal group $O(\mathbb{E}^2)$. Let us now show that $O(\mathbb{E}^2)$ consists of precisely these transformations and that the rotation subgroup $O^+(\mathbb{E}^2)$ is isomorphic to the additive quotient group $\mathbb{R}/2\pi\mathbb{Z}$.

(A) Let σ be an orthogonal transformation of $\mathbb{E}^2$ and let $\{e_1, e_2\}$ stand for the standard orthonormal coordinate system of $\mathbb{E}^2$. Then σ preserves the length of the vector e_1 and hence $\sigma(e_1)$ lies on the unit circle in $\mathbb{E}^2$. Therefore, $\sigma(e_1) = (\cos\theta, \sin\theta)$ for some real number θ. Now, $\sigma(e_2) \perp \sigma(e_1)$ since $e_2 \perp e_1$. But all vectors on the line orthogonal to $(\cos\theta, \sin\theta)$ lie on the line spanned by $(-\sin\theta, \cos\theta)$. Thus, $\sigma(e_2) = c(-\sin\theta, \cos\theta)$ for

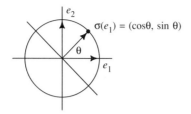

some scalar c, and since $\sigma(e_2)$ is also a unit vector, $c = \pm 1$. Therefore, $\sigma(e_1) = (\cos \theta, \sin \theta)$ and $\sigma(e_2) = \pm(-\sin \theta, \cos \theta)$. It follows that the matrix of σ relative to the coordinate system $\{e_1, e_2\}$ is either

$$M = \begin{pmatrix} \cos \theta & -\sin \theta \\ \sin \theta & \cos \theta \end{pmatrix} \quad \text{or} \quad M' = \begin{pmatrix} \cos \theta & \sin \theta \\ \sin \theta & -\cos \theta \end{pmatrix}.$$

Now, $\det(M) = 1$ and $\det(M') = -1$. Hence, if σ is a rotation, then its matrix must be M and therefore $\sigma = p_\theta$, the rotation of the plane through an angle of θ radians about the origin. If, on the other hand, σ is a reflection, then its matrix must be M', in which case we claim that $\sigma = S_A$, the symmetry determined by the vector A, where $A = (-\sin \frac{1}{2}\theta, \cos \frac{1}{2}\theta)$. To show this, we need only verify that σ and S_A agree on the basis $\{e_1, e_2\}$ since both mappings are linear transformations. Using the formula obtained in Example 3 and noting that A is a unit vector, we find that

$$S_A(e_1) = e_1 - 2\langle e_1, A\rangle A = \left(1 - 2 \sin^2 \frac{1}{2}\theta, 2 \sin \frac{1}{2}\theta \cos \frac{1}{2}\theta\right)$$
$$= (\cos \theta, \sin \theta)$$
$$= \sigma(e_1)$$

and, similarly, $S_A(e_2) = \sigma(e_2)$. Hence, $\sigma = S_A$. It follows, therefore, that the orthogonal group $O(\mathbb{E}^2)$ consists of just the usual rotations and reflections of the plane. In particular, the preceding discussion shows that every reflection of $\mathbb{E}^2$ is a symmetry; that is, a reflection of the plane through some line passing through the origin.

(B) It follows from part (A) that the rotation subgroup $O^+(\mathbb{E}^2)$ consists of the rotations p_θ; that is, $O^+(\mathbb{E}^2) = \{p_\theta \in GL(\mathbb{E}^2) | \theta \in \mathbb{R}\}$. We now claim that $O^+(\mathbb{E}^2)$ is an abelian group and that $O^+(\mathbb{E}^2) \cong \mathbb{R}/2\pi\mathbb{Z}$. For clearly, if p_θ and $p_{\theta'}$ are typical rotations of $\mathbb{E}^2$, then $p_\theta p_{\theta'} = p_{\theta+\theta'} = p_{\theta'} p_\theta$ and therefore $O^+(\mathbb{E}^2)$ is abelian. Now, define the function $f: \mathbb{R} \to O^+(\mathbb{E}^2)$ by setting $f(\theta) = p_\theta$ for every $\theta \in \mathbb{R}$. Then f is a group-homomorphism from the additive group of real numbers into the rotation subgroup $O^+(\mathbb{E}^2)$, and f is onto since, as we showed in part (A), every rotation of $\mathbb{E}^2$ has the form p_θ for some $\theta \in \mathbb{R}$. To determine the kernel of f, observe that $p_\theta = 1$ if and only if $\theta = 2\pi k$ for some integer k. Therefore, $\text{Ker } f = 2\pi\mathbb{Z}$

and hence, by the fundamental theorem of group-homomorphism, $O^+(\mathbb{E}^2) \cong \mathbb{R}/2\pi\mathbb{Z}$.

(C) Let us conclude this example by interpreting the isomorphism $O^+(\mathbb{E}^2) \cong \mathbb{R}/2\pi\mathbb{Z}$ from another point of view. On the one hand, the isomorphism associates with each rotation ρ_θ the coset $\theta + 2\pi\mathbb{Z}$. But recall from Example 11, Section 3, Chapter 4, that $\mathbb{R}/2\pi\mathbb{Z} \cong U$, the multiplicative group of unit complex numbers; under this isomorphism, a typical coset $\theta + 2\pi\mathbb{Z}$ corresponds to the complex number $\cos\theta + i\sin\theta$. Thus, the isomorphisms $O^+(\mathbb{E}^2) \cong \mathbb{R}/2\pi\mathbb{Z} \cong U$ show that every rotation of the plane has associated with it a uniquely determined unit complex number, as indicated in the diagram.

$$
\begin{array}{ccc}
GL(\mathbb{E}^2) & & \mathbb{C}^* \\
| & & | \\
O^+(\mathbb{E}^2) \xrightarrow{\cong} & \mathbb{R}/2\pi\mathbb{Z} \xrightarrow{\cong} & U \\
\rho_\theta \quad \leftrightarrow & \theta + 2\pi\mathbb{Z} \leftrightarrow & \cos\theta + i\sin\theta
\end{array}
$$

Consequently, we may think of the rotations of the Euclidean plane as corresponding to unit complex numbers in $\mathbb{C}^*$. In fact, we leave the reader to verify that if z is any complex number, then $\rho_\theta(z) = z(\cos\theta + i\sin\theta)$. Thus, the rotation ρ_θ corresponds to the unit complex number $\cos\theta + i\sin\theta$, and multiplication of complex numbers by $\cos\theta + i\sin\theta$ corresponds to the rotation of the complex plane through an angle of θ radians. This concludes the example.

Up to this point in our discussion, we have shown that the orthogonal transformations of a Euclidean space form a group, the orthogonal group, and are classified as either rotations or reflections depending upon the value of their determinant. Reflections are further classified as either symmetries or nonsymmetries. Moreover, in the case of the Euclidean plane, every reflection is a symmetry, although we emphasize that this is not usually true in higher dimensions. Let us now discuss the symmetries of a Euclidean space in more detail. Our goal is to first show that the symmetries form a single conjugacy class in the orthogonal group. We then conclude the section by proving an important theorem, due to Cartan, showing that the symmetries in fact generate the orthogonal group. To begin, recall from Example 3 that every symmetry of a Euclidean space has the form S_A, where A is a nonzero vector and S_A is the reflection of the space through the hyperplane orthogonal to A.

PROPOSITION 4. Let V be a Euclidean space and let S_A be a symmetry of V, where A is any nonzero vector in V. Then $\sigma S_A \sigma^{-1} = S_{\sigma(A)}$ for every element $\sigma \in O(V)$. Thus, in particular, every conjugate of a symmetry is a symmetry.

Proof.   We may assume that A is a unit vector since the symmetry determined by any vector on the line spanned by A is the same as S_A. Then, for any vector $X \in V$,

$$(\sigma S_A \sigma^{-1})(X) = \sigma[\sigma^{-1}(X) - 2\langle \sigma^{-1}(X), A \rangle A]$$
$$= X - 2\langle X, \sigma(A) \rangle \sigma(A)$$
$$= S_{\sigma(A)}(X).$$

Therefore, $\sigma S_A \sigma^{-1} = S_{\sigma(A)}$, as required.   ■

PROPOSITION 5. Let V be a Euclidean space. Then all symmetries of V are conjugate in the orthogonal group $O(V)$ and form a single conjugacy class.

Proof.   It follows from Proposition 4 that the conjugacy class of a symmetry in $O(V)$ contains only symmetries. To complete the proof, we must show that any two symmetries of V are in fact conjugate. Thus, let S_A and S_B be typical symmetries in $O(V)$, where A and B are unit vectors in V, and let $\rho = S_{A+B} S_A$. Then $\rho \in O(V)$ and we find that

$$\rho(A) = S_{A+B}(-A) = -A - 2 \frac{\langle -A, A + B \rangle}{\langle A + B, A + B \rangle}(A + B)$$
$$= -A + \frac{2(1 + \langle A, B \rangle)}{2(1 + \langle A, B \rangle)}(A + B)$$
$$= B.$$

It now follows from Proposition 4 that $\rho S_A \rho^{-1} = S_{\rho(A)} = S_B$. Therefore, S_A and S_B are conjugate in $O(V)$ and the proof is complete.   ■

Geometrically, the equation $\sigma S_A \sigma^{-1} = S_{\sigma(A)}$ shows that if S is the symmetry that reflects a Euclidean space V through a hyperplane H orthogonal to A and σ is an orthogonal transformation of V, then $\sigma S \sigma^{-1}$ is the symmetry that reflects V through the hyperplane $\sigma(H)$ orthogonal to $\sigma(A)$. The fact that all such symmetries are conjugate is simply saying that if H and H' are any hyperplanes in V, then there is some orthogonal transformation σ of the space such that $\sigma(H) = H'$. Thus, in particular, the orthogonal group $O(V)$ acts transitively on the hyperplanes in V. This geometric statement is expressed algebraically by the fact that all symmetries of V are conjugate in $O(V)$.

We have already discovered these results when we discussed the conjugacy classes of the group D_4 of the square in Example 4, Section 1, Chapter 6. At that time we observed that D_4 contains four symmetries, S_A, S_B, S_C, and S_D, as indicated in the marginal figure and that these four symmetries are conjugate as elements in $O(\mathbb{E}^2)$ since there is some orthogonal transformation that maps any one of the lines $\langle A \rangle$, $\langle B \rangle$, $\langle C \rangle$, or $\langle D \rangle$ to any other line.

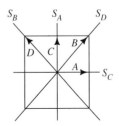

For example, the $45°$ rotation $\rho_{\pi/4}$ maps C to D and hence $\rho_{\pi/4} S_C \rho_{\pi/4}^{-1} = S_D$. Note, however, that the symmetries in D_4 are not all conjugate as elements of the group D_4 since, for example, there is no element in D_4 that maps $\langle C \rangle$ to $\langle D \rangle$. In fact, the symmetries fall into two conjugacy classes in D_4 according to whether their reflecting lines are perpendicular or not, namely, $\{S_A, S_C\}$ and $\{S_B, S_D\}$.

Let us now conclude our discussion of orthogonal transformations by showing that the symmetries of a Euclidean space are in fact the basic building blocks from which all orthogonal transformations are constructed.

PROPOSITION 6 (CARTAN). Let V be a Euclidean space of dimension $n \geq 1$. Then every orthogonal transformation of V may be written as a product of at most n symmetries of V. Hence, in particular, the symmetries of V generate the orthogonal group $O(V)$.

Proof. We prove this statement by induction on the dimension n of the space. Suppose that $n = 1$. Then, as we showed earlier, $O(V) = \{1_V, -1_V\}$. In this case the only symmetry is -1_V, and since $1_V = (-1_V)^0$ and $-1_V = (-1_V)^1$, it follows that every orthogonal transformation of V is a product of at most one symmetry, as required. Now suppose that $n > 1$ and assume that every orthogonal transformation of a Euclidean space of dimension $k < n$ may be written as a product of at most k symmetries of the space. Let $\sigma \in O(V)$. There are two cases to consider.

(A) Suppose first that σ fixes some nonzero vector $A \in V$. Let $H = \langle A \rangle^{\perp}$ stand for the hyperplane orthogonal to A. Then H is a Euclidean space of dimension $n - 1$ and the restriction $\sigma | H$ of σ to H is an orthogonal transformation of H. Hence, by assumption, there are symmetries $S_{A_1}^*, \ldots, S_{A_s}^*$ of H such that $\sigma | H = S_{A_1}^* \cdots S_{A_s}^*$, where $s \leq n - 1$ and $A_1, \ldots, A_s \in H$. Let $S_{A_1}, \ldots, S_{A_s}$ stand for the symmetries of V corresponding to the vectors $A_1, \ldots, A_s$. We claim that $\sigma = S_{A_1} \cdots S_{A_s}$. For let X be a typical vector in V and write $X = X_A + X_H$, where $X_A \in \langle A \rangle$ and $X_H \in H$. Then $\sigma(X_A) = X_A$ since $\sigma(A) = A$, and $\sigma(X_H) = (\sigma | H)(X_H) = S_{A_1}^* \cdots S_{A_s}^*(X_H)$. But $S_{A_i} | H = S_{A_i}^*$, and $S_{A_i}(X_A) = X_A$ since $X_A \perp A_i$. Therefore,

$$\sigma(X) = \sigma(X_A) + \sigma(X_H) = X_A + S_{A_1}^* \cdots S_{A_s}^*(X_H)$$
$$= X_A + S_{A_1} \cdots S_{A_s}(X_H)$$
$$= S_{A_1} \cdots S_{A_s}(X_A + X_H)$$
$$= S_{A_1} \cdots S_{A_s}(X).$$

Thus, $\sigma = S_{A_1} \cdots S_{A_s}$ and hence σ is a product of at most $n - 1$ symmetries of V.

(B) Finally, suppose that σ fixes no vector in V other than the zero vector. Let A be any nonzero vector in V and let $B = \sigma(A) - A$. Then $B \neq 0$

and we find that

$$S_B\sigma(A) = \sigma(A) - 2\,\frac{\langle \sigma(A), B \rangle}{\langle B, B \rangle}\,B$$

$$= \sigma(A) - 2\,\frac{\langle \sigma(A), \sigma(A) - A \rangle}{\langle \sigma(A) - A, \sigma(A) - A \rangle}\,(\sigma(A) - A)$$

$$= \sigma(A) - \frac{2(\langle A, A \rangle - \langle \sigma(A), A \rangle)}{2(\langle A, A \rangle - \langle \sigma(A), A \rangle)}\,(\sigma(A) - A)$$

$$= A.$$

Therefore, $S_B\sigma$ is an orthogonal transformation of V that fixes the nonzero vector A and hence, by part (A), there are symmetries $S_{A_1}, \ldots, S_{A_s}$ such that $S_B\sigma = S_{A_1} \cdots S_{A_s}$, where $s \le n - 1$. It follows that $\sigma = S_B S_{A_1} \cdots S_{A_s}$ and therefore σ is the product of at most n symmetries, as required.   ∎

When we apply the Cartan theorem to the Euclidean plane, where the dimension is 2, it shows that every orthogonal transformation of the plane is a product of either one or two symmetries. In particular, every reflection is a symmetry, a result that we verified by direct calculation in Example 6, and every rotation is a product of two symmetries. In the case of Euclidean 3-space, however, the theorem shows that every orthogonal transformation is a product of one, two, or three symmetries. Consequently, there are two types of reflections of $\mathbb{E}^3$: those that are symmetries and those that are a product of three symmetries. The inversion transformation, for example, is of the latter type since it fixes no vector except the origin and hence cannot be a symmetry. We claim, in fact, that $-1_{\mathbb{E}^3} = S_{e_1}S_{e_2}S_{e_3}$. To show this, observe that $S_{e_i}(e_j) = e_j$ whenever $i \ne j$ since $e_i \perp e_j$, while $S_{e_i}(e_i) = -e_i$. Therefore,

$$S_{e_1}S_{e_2}S_{e_3}(e_1) = S_{e_1}(e_1) = -e_1$$
$$S_{e_1}S_{e_2}S_{e_3}(e_2) = S_{e_1}(-e_2) = -e_2$$
$$S_{e_1}S_{e_2}S_{e_3}(e_3) = S_{e_1}S_{e_2}(e_3) = -e_3,$$

and hence $-1_{\mathbb{E}^3} = S_{e_1}S_{e_2}S_{e_3}$. The rotations of $\mathbb{E}^3$, on the other hand, are those orthogonal transformations that may be written as a product of two symmetries. It turns out—and we ask the reader to show this in the exercises—that every rotation of $\mathbb{E}^3$ has a nonzero fixed vector and therefore an "axis" of rotation in the usual sense of the word.

In conclusion, let us make two remarks about the Cartan theorem. First, the theorem says nothing about the uniqueness of the symmetries that occur in the factorization of an orthogonal transformation. In general, there are many choices for such factors. And second, the theorem says nothing about the least number of symmetries into which an orthogonal transformation may be factored; it only guarantees that not more than n symmetries are required, where n is the dimension of the space. It is in fact possible to determine the least number of symmetries needed to factor an orthogonal transformation: if σ is an orthogonal transformation of a Euclidean space and $F(\sigma)$ is the subspace of vectors fixed by σ, then the least number of symmetries into which σ may be factored is equal to the dimension of the orthogonal complement $F(\sigma)^{\perp}$. This result is known as Scherk's theorem and we have outlined a proof of it in the exercises.

Exercises

1. Determine which of the following mappings are orthogonal transformations. If the mapping is orthogonal, determine if it is a rotation or reflection.
   (a) $\sigma: \mathbb{E}^2 \to \mathbb{E}^2$, $\sigma(x, y) = (x, -y)$
   (b) $\sigma: \mathbb{E}^2 \to \mathbb{E}^2$, $\sigma(x, y) = (y, x)$
   (c) $\sigma: \mathbb{E}^2 \to \mathbb{E}^2$, $\sigma(x, y) = (x - y, x + y)$
   (d) $\sigma: \mathbb{E}^2 \to \mathbb{E}^2$, $\sigma(x, y) = (x + 1, y - 2)$
   (e) $\sigma: \mathbb{E}^3 \to \mathbb{E}^3$, $\sigma(x, y, z) = (-x, -y, z)$
   (f) $\sigma: \mathbb{E}^3 \to \mathbb{E}^3$, $\sigma(x, y, z) = (\frac{1}{2}\sqrt{3}x - \frac{1}{2}z, y, \frac{1}{2}x + \frac{1}{2}\sqrt{3}z)$
   (g) $\sigma: \mathbb{E}^3 \to \mathbb{E}^3$, $\sigma(x, y, z) = (z, -x, y)$
2. Let $S_A: \mathbb{E}^2 \to \mathbb{E}^2$ be the symmetry of the Euclidean plane determined by the vector $A - (3, 4)$.
   (a) Find $S_A(x, y)$ for any vector $(x, y) \in \mathbb{E}^2$. Illustrate the mapping S_A geometrically.
   (b) Find $S_A(4, 1)$.
   (c) Find a vector $X \in \mathbb{E}^2$ such that $S_A(X) = (-1, 1)$.
3. Let $A = (-1, 2)$ and $B = (-1 - 2\sqrt{3}, 2 - \sqrt{3})$.
   (a) Describe the rotation $S_A S_B$ of the plane.
   (b) Describe the rotation $S_B S_A$ of the plane.
   (c) Find a rotation ρ of the plane such that $\rho S_A \rho^{-1} = S_B$.
4. Let $A = (\cos \theta(A), \sin \theta(A))$ and $B = (\cos \theta(B), \sin \theta(B))$ be unit vectors in the plane and let S_A and S_B be the symmetries determined by these vectors. Show that $S_A S_B = \rho_{2[\theta(A) - \theta(B)]}$. Thus, $S_A S_B$ is the rotation of the plane through twice the angle between A and B.
5. Let V be a Euclidean space and let S_A and S_B be symmetries of V.
   (a) Show that $S_A = S_B$ if and only if A and B are collinear.
   (b) Show that $S_A S_B = S_B S_A$ if and only if $A \perp B$ or A and B are collinear.
   (c) Show that $S_A S_B$ and $S_B S_A$ are conjugate in the group $O(V)$.
   (d) Show that every rotation of V that is a product of two symmetries is conjugate to its inverse.
   (e) Show that every rotation of the Euclidean plane is conjugate to its inverse.
6. Let V be a Euclidean space and let U be a subspace of V. If A is a non-zero vector in U, let S_A^* and S_A stand for the symmetries of U and V, respectively, determined by the vector A.
   (a) Show that $S_A|U = S_A^*$, where $S_A|U$ stands for the restriction of S_A to U.
   (b) Let U stand for the xy-plane in $\mathbb{E}^3$. Illustrate, geometrically, the fact that $S_{e_1}|U = S_{e_1}^*$.

7. Let V be a Euclidean space of dimension 2 and let ρ be a rotation of V.
   (a) If S_A is any symmetry of V, show that there is a symmetry S_B such that $\rho = S_A S_B$.
   (b) Let $\rho_{\pi/6}$ stand for the $30°$ counterclockwise rotation of the plane. Find a symmetry S_B of the plane such that $\rho_{\pi/6} = S_{e_1} S_B$.

8. Let U be a subspace of a Euclidean space V and suppose that σ is an orthogonal transformation of V such that $\sigma(U) = U$. Show that $\sigma(U^\perp) = U^\perp$.

9. Let U be a subspace of a Euclidean space V, and let $\sigma: U \to U$ and $\tau: U^\perp \to U^\perp$ be orthogonal transformations of U and $U^\perp$, respectively. Define the function $\sigma \oplus \tau: V \to V$ by setting $(\sigma \oplus \tau)(A + B) = \sigma(A) + \tau(B)$ for all vectors $A \in U$ and $B \in U^\perp$.
   (a) Show that $\sigma \oplus \tau$ is an orthogonal transformation of V.
   (b) Let M_σ and M_τ be matrix representations of σ and τ, respectively. Show that $\sigma \oplus \tau$ has a matrix representation of the form $M = \begin{pmatrix} M_\sigma & O \\ O & M_\tau \end{pmatrix}$, and that $\det(M) = \det(M_\sigma)\det(M_\tau)$.
   (c) Show that $\sigma \oplus \tau$ is a rotation of V if and only if either σ is a rotation of U and τ is a rotation of $U^\perp$ or σ is a reflection of U and τ a reflection of $U^\perp$.
   (d) Show that $\sigma \oplus \tau$ is a reflection of V if and only if either σ is a reflection of U and τ a rotation of $U^\perp$ or σ is a rotation of U and τ a reflection of $U^\perp$.
   (e) Define the function $f: O(U) \times O(U^\perp) \to O(V)$ by setting $f(\sigma, \tau) = \sigma \oplus \tau$ for all $\sigma \in O(U)$ and all $\tau \in O(U^\perp)$. Show that f is a 1–1 group-homomorphism. Thus, the direct product $O(U) \times O(U^\perp)$ is embedded as a subgroup of $O(V)$.
   (f) Let $O(V)_U = \{\sigma \in O(V) \mid \sigma(U) = U\}$. Show that $O(V)_U \le O(V)$.
   (g) Show that $O(U) \times O(U^\perp) \cong O(V)_U$.

10. Let U be a subspace of a Euclidean space V. Define the function $f: O(U) \to O(V)$ by setting $f(\sigma) = \sigma \oplus 1$, where 1 stands for the identity map on $U^\perp$.
    (a) Show that f is a 1–1 group-homomorphism.
    (b) Show that $\operatorname{Im} f = \{\sigma \in O(V) \mid \sigma(X) = X \text{ for every } X \in U^\perp\}$.

11. Let S_A be a symmetry of a Euclidean space V. Show that $S_A = -1_{\langle A \rangle} \oplus 1_H$, where $-1_{\langle A \rangle}$ is the inversion transformation of the line $\langle A \rangle$ and $H = \langle A \rangle^\perp$.

12. **Rotations of Euclidean 3-space.** In this exercise we describe the rotations of Euclidean 3-space $\mathbb{E}^3$.
    (a) Let H be a hyperplane in $\mathbb{E}^3$ and let $l = H^\perp$ stand for the line orthogonal to H. If ρ_H is a rotation of H, show that $1_l \oplus \rho_H$ is a rotation of $\mathbb{E}^3$. $1_l \oplus \rho_H$ is the familiar rotation of Euclidean 3-space about the axis l; it rotates points in $\mathbb{E}^3$ around the axis l according to ρ_H.
    (b) Let σ be a rotation of $\mathbb{E}^3$, $\sigma \ne 1$. It follows from the Cartan theorem that there are symmetries S_A and S_B of $\mathbb{E}^3$ such that $\sigma = S_A S_B$. Let $l = \langle A \rangle^\perp \cap \langle B \rangle^\perp$ stand for the intersection of the planes orthogonal

to $\langle A \rangle$ and $\langle B \rangle$. Show that l is a 1-dimensional subspace of $\mathbb{E}^3$ and that σ leaves every vector on l fixed. If $\rho = \sigma | l^{\perp}$, show that ρ is a rotation of the hyperplane $l^{\perp}$ and that $\sigma = 1_l \oplus \rho$.

(c) Using parts (a) and (b), show that the rotations of $\mathbb{E}^3$ are those orthogonal transformations of the form $1_l \oplus \rho_H$, where ρ_H is a rotation of some hyperplane H and 1_l is the identity map on the line l orthogonal to H. The line l is called the *axis of rotation*. Show that every rotation of $\mathbb{E}^3$ has a matrix representation of the form

$$\begin{pmatrix} 1 & 0 & 0 \\ 0 & \cos\theta & -\sin\theta \\ 0 & \sin\theta & \cos\theta \end{pmatrix}.$$

(d) Let $\sigma = S_{e_1} S_{e_2}$. Find the axis of rotation for σ and the angle of rotation. Illustrate σ geometrically and find an explicit formula for $\sigma(x, y, z)$.

(e) Define the function $\sigma: \mathbb{E}^3 \to \mathbb{E}^3$ by setting

$$\sigma(x, y, z) = \left(\frac{2}{3}x - \frac{1}{3}y + \frac{2}{3}z, \frac{2}{3}x + \frac{2}{3}y - \frac{1}{3}z, -\frac{1}{3}x + \frac{2}{3}y + \frac{2}{3}z \right)$$

for all $(x, y, z) \in \mathbb{E}^3$. Show that σ is a rotation of $\mathbb{E}^3$. Find the axis of rotation and the angle of rotation. Find a matrix representation for σ of the form described in part (c), and find symmetries S_A and S_B such that $\sigma = S_A S_B$.

(f) Show that the rotation subgroup $O^+(\mathbb{E}^3)$ is nonabelian.

13. Let n be a positive integer and let $\sigma \in \text{Sym}(n)$. Define the function σ^*: $\mathbb{E}^n \to \mathbb{E}^n$ by setting $\sigma^*(x_1, \ldots, x_n) = (x_{\sigma(1)}, \ldots, x_{\sigma(n)})$ for all points $(x_1, \ldots, x_n) \in \mathbb{E}^n$.

(a) Show that σ^* is an orthogonal transformation of $\mathbb{E}^n$.

(b) Show that $\sigma^*(e_i) = e_{\sigma(i)}$ for $i = 1, \ldots, n$.

(c) Define the function $f: \text{Sym}(n) \to O(\mathbb{E}^n)$ by setting $f(\sigma) = \sigma^*$ for every $\sigma \in \text{Sym}(n)$. Show that f is a 1–1 group-homomorphism. Thus, the symmetric group $\text{Sym}(n)$ may be embedded as a subgroup of the orthogonal group $O(\mathbb{E}^n)$.

(d) Let $\tau = (i\,j)$ stand for the transposition that interchanges i and j. Show that $\tau^* = S_{e_i - e_j}$.

(e) Show that an even permutation $\sigma \in \text{Sym}(n)$ corresponds to a rotation $\sigma^* \in O(\mathbb{E}^n)$, and that odd permutations corresponds to reflections.

(f) Describe, geometrically, the orthogonal transformations of $\mathbb{E}^3$ that correspond to each of the permutations in $\text{Sym}(3)$.

(g) Let G be a finite group of order n. Show that G is isomorphic to a subgroup of $O(\mathbb{E}^n)$. Thus, every finite group may be represented as a group of orthogonal transformations of some Euclidean space.

14. Let V be a Euclidean space and let G be a subgroup of $O(V)$.

(a) Let G^+ stand for the set of rotations in G. Show that $G^+ \lhd G$ and that either $G^+ = G$ or $[G:G^+] = 2$. Thus, every group of orthogonal transformations of a Euclidean space is either a group of rotations or its rotation subgroup is a normal subgroup of index 2.

(b) If the only rotation in G is the identity map 1, show that either $G = (1)$ or $G = \{1, \tau\}$ for some reflection τ of V.

15. Show that isometric Euclidean spaces have isomorphic orthogonal groups.

16. Let $H_1, \ldots, H_t$ be a finite collection of hyperplanes in a Euclidean space V and let $A_1, \ldots, A_t$ be vectors such that $H_i = \langle A_i \rangle^\perp$ for $i = 1, \ldots, t$.
    (a) Show that $H_1 \cap \cdots \cap H_t = \langle A_1, \ldots, A_t \rangle^\perp$.
    (b) Show that $\dim (H_1 \cap \cdots \cap H_t) = n - s$, where s is the dimension of the subspace spanned by $A_1, \ldots, A_t$.

17. Let σ be an orthogonal transformation of a Euclidean space V and let $\mathrm{Im}(\sigma - 1)$ and $\mathrm{Ker}(\sigma - 1)$ stand for the image and kernel of the mapping $\sigma - 1$.
    (a) Show that $\mathrm{Im}(\sigma - 1)$ and $\mathrm{Ker}(\sigma - 1)$ are subspaces of V.
    (b) Show that $\mathrm{Im}(\sigma - 1)^\perp = \mathrm{Ker}(\sigma - 1)$.
    (c) Let $F(\sigma) = \{X \in V \,|\, \sigma(X) = X\}$ stand for the set of vectors in V fixed by σ. Show that $F(\sigma)$ is a subspace of V and that $F(\sigma) = \mathrm{Ker}(\sigma - 1)$.
    (d) Show that $F(\sigma)^\perp = \mathrm{Im}(\sigma - 1)$.

18. Let $S_{A_1}, \ldots, S_{A_t}$ be a finite number of symmetries of a Euclidean space V.
    (a) If $X \in V$, show that there are scalars $c_1, \ldots, c_t$ such that $S_{A_1} \cdots S_{A_t}(X) = X + c_1 A_1 + \cdots + c_t A_t$.
    (b) Show that $\mathrm{Im}(S_{A_1} \cdots S_{A_t} - 1) \subseteq \langle A_1, \ldots, A_t \rangle$
    (c) Show that $\langle A_1, \ldots, A_t \rangle^\perp \subseteq \mathrm{Ker}(S_{A_1} \cdots S_{A_t} - 1)$

19. **Scherk's Theorem.** In this exercise we outline a proof of Scherk's theorem: if σ is an orthogonal transformation of a Euclidean space and $F(\sigma)$ is the subspace of vectors fixed by σ, then the least number of symmetries into which σ may be factored is equal to $\dim F(\sigma)^\perp$.
    (a) Let σ be an orthogonal transformation of a Euclidean space V and let s be the least number of symmetries into which σ may be factored. Show that $s \leq \dim F(\sigma)^\perp$ by applying Cartan's theorem to the transformation $\sigma \,|\, F(\sigma)^\perp$.
    (b) Let $\sigma = S_{A_1} \cdots S_{A_s}$ be a factorization of σ into s symmetries of V. Using Exercise 18, show that $F(\sigma)^\perp$ is a subspace of $\langle A_1, \ldots, A_s \rangle$ and conclude therefore that $\dim F(\sigma)^\perp \leq s$.
    (c) It follows from parts (a) and (b) that $s = \dim F(\sigma)^\perp$, which proves Scherk's theorem. Any factorization of σ into a product of s symmetries, where $s = \dim F(\sigma)^\perp$, is called a *minimal factorization* of σ. Show that $\sigma = S_{A_1} \cdots S_{A_s}$ is a minimal factorization of σ if and only if the vectors $A_1, \ldots, A_s$ are linearly independent.
    (d) Let V be a Euclidean space of odd dimension. Using Scherk's theorem, show that every rotation of V has an axis of rotation, that is, a 1-dimensional subspace that is fixed by the rotation.

3. EUCLIDEAN MOTIONS AND THE EUCLIDEAN GROUP

In the previous section we discussed orthogonal transformations of a Euclidean space, which, as we showed, are the distance-preserving linear isomorphisms of the space. Let us now turn our attention to the class of all distance-preserving mappings on a Euclidean space. These mappings are called Euclidean motions of the space and include, for example, all translations of the space. Our goal is to first show that the Euclidean motions of a space form a group under function composition called the Euclidean group of the space and that every such motion may be written as the composition of an orthogonal transformation followed by a translation of the space. We then conclude the section by describing in detail the Euclidean motions of the plane.

Let V be a Euclidean space and let $d: V \times V \to \mathbb{R}$ stand for the distance function on V.

> **Definition**
>
> A *Euclidean motion* of V is any function $f: V \to V$ that maps V onto V and is such that $d(f(A), f(B)) = d(A, B)$ for all vectors $A, B \in V$.

Euclidean motions are the distance-preserving mappings of a Euclidean space onto itself. Observe that every Euclidean motion is necessarily a 1–1 correspondence of the space with itself; for if f is a Euclidean motion of V and $f(A) = f(B)$ for some vectors $A, B \in V$, then $d(f(A), f(B)) = 0$ and hence $d(A, B) = 0$. Therefore, $A = B$ and hence f is a 1–1 mapping. Thus, the Euclidean motions of a space are those 1–1 correspondences of the space with itself that preserve the distance between points. For this reason, they are frequently called rigid motions of the space. The group of the triangle, for example, consists of the six rigid motions of the plane that map the triangle to itself, while the group of the square consists of the eight rigid motions of the plane that map the square to itself. Clearly, every orthogonal transformation of V is a Euclidean motion of V. In general, however, there are Euclidean motions that are not linear and hence not orthogonal transformations. The following example illustrates a Euclidean motion of this form.

EXAMPLE 1. Translations

Let V be a Euclidean space and let A be any vector in V. The *translation* of V by A is the function $T_A: V \to V$ defined by setting $T_A(X) = X + A$ for every vector $X \in V$. We claim that every translation of V is a Euclidean motion. To show this, let T_A be the translation of V by the vector A and let X and Y be typical vectors in V. Then

$$
\begin{aligned}
d(T_A(X), T_A(Y))^2 &= \|T_A(X) - T_A(Y)\|^2 \\
&= \|(X + A) - (Y + A)\|^2 \\
&= \|X - Y\|^2 \\
&= d(X, Y)^2.
\end{aligned}
$$

Therefore, $d(T_A(X), T_A(Y)) = d(X, Y)$ and hence T_A is a Euclidean motion of V. T_A "shifts" the space in the direction of the vector A by an amount equal to the magnitude of A. For example, the translation $T_{(1,2)} \colon \mathbb{E}^2 \to \mathbb{E}^2$ of the Euclidean plane maps an arbitrary point (x, y) to the point $T_{(1,2)}(x, y) = (x, y) + (1, 2) = (x + 1, y + 2)$, as shown in the figure. Observe that the identity map 1_V of any Euclidean space V is a translation of V since $1_V = T_0$, the translation of V by the zero vector. And finally, note that translations of V are not, in general, linear transformations since $T_A(X + Y) \neq T_A(X) + T_A(Y)$ for every $X, Y \in V$.

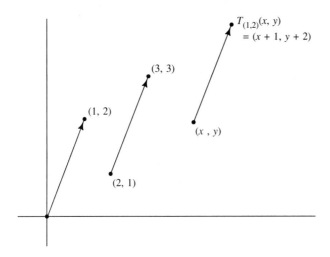

Now, let $E(V)$ stand for the set of Euclidean motions of V, $T(V)$ the subset of translations of V, and $O(V)$ the orthogonal group of V. Since every Euclidean motion of V is a permutation of V, it follows that $E(V) \subseteq \mathrm{Sym}(V)$, the symmetric group of V. Let us now show that $E(V)$ is in fact a subgroup of $\mathrm{Sym}(V)$, and hence is a group under function composition, and that both $T(V)$ and $O(V)$ are subgroups of $E(V)$.

PROPOSITION 1. Let V be a Euclidean space. Then the set $E(V)$ of Euclidean motions of V is a group under function composition and is a subgroup of the symmetric group $\mathrm{Sym}(V)$. Moreover, both $O(V)$ and $T(V)$ are subgroups of $E(V)$.

Proof. Clearly, the identity map $1_V \in E(V)$. Moreover, if f and g are Euclidean motions of V, it follows easily that the inverse f^{-1} and composition $f \circ g$ are also Euclidean motions of V and hence are elements of $E(V)$. Therefore, $E(V) \leq \text{Sym}(V)$. Since every orthogonal transformation of V is a Euclidean motion, $O(V) \leq E(V)$. Finally, to show that the translations $T(V)$ form a subgroup of $E(V)$, first recall that $1_V \in T(V)$ since $1_V = T_0$. Now, if T_A and T_B are typical translations of V, we claim that $T_A \circ T_B = T_{A+B}$ and that $T_A^{-1} = T_{-A}$. For let $X \in V$. Then $T_{A+B}(X) = A + B + X = T_A(X + B) = T_A(T_B(X)) = (T_A \circ T_B)(X)$. Therefore, $T_A \circ T_B = T_{A+B}$. In particular, $T_A \circ T_{-A} = T_0 = 1_V$ and hence $T_A^{-1} = T_{-A}$. It follows that $T(V)$ is closed under composition and inverses and is, therefore, a subgroup of $E(V)$. ∎

The group $E(V)$ of Euclidean motions of V is called the *Euclidean group* of V and the subgroup $T(V)$ of translations the *translation subgroup*. For convenience, we denote the Euclidean group of $\mathbb{E}^n$ by $\mathbb{E}(n)$ and call it the *n-dimensional Euclidean group*. Before we look at other examples of Euclidean motions, let us first show that every such motion may be written as the composition of an orthogonal transformation followed by a translation, and then use this fact to show that the translation subgroup is a normal subgroup of $E(V)$. We begin with the following preliminary result.

PROPOSITION 2. Let V be a Euclidean space and let f be a Euclidean motion of V that fixes the zero vector. Then f is an orthogonal transformation of V.

Proof. Since f is a 1–1 correspondence, it remains to show that f is a linear transformation that preserves the inner product on V.

(A) We first show that f preserves the inner product on V. Let $X, Y \in V$. Now recall from Section 2 that

$$\langle f(X), f(Y) \rangle = \frac{1}{2}[d(f(X), 0)^2 + d(f(Y), 0)^2 - d(f(X), f(Y))^2].$$

Since f preserves the distance function d and since $f(0) = 0$ by assumption, it follows that

$$\langle f(X), f(Y) \rangle = \frac{1}{2}[d(f(X), f(0))^2 + d(f(Y), f(0))^2 - d(F(X), f(Y))^2]$$

$$= \frac{1}{2}[d(X, 0)^2 + d(Y, 0)^2 - d(X, Y)^2]$$

$$= \langle X, Y \rangle.$$

Therefore, f preserves the inner product on V.

(B) To show that f is a linear transformation, let $X, Y \in V$ and consider the inner product $\langle f(X + Y) - f(X) - f(Y), Z \rangle$, where Z is any vector in V. Since f is a 1–1 correspondence on V, there is some vector $Z' \in V$

such that $Z = f(Z')$. Then, using the fact that f preserves the inner product, we find that

$$\langle f(X+Y) - f(X) - f(Y), Z\rangle = \langle f(X+Y) - f(X) - f(Y), f(Z')\rangle$$
$$= \langle f(X+Y), f(Z')\rangle - \langle f(X), f(Z')\rangle - \langle f(Y), f(Z')\rangle$$
$$= \langle X+Y, Z'\rangle - \langle X, Z'\rangle - \langle Y, Z'\rangle$$
$$= 0.$$

Therefore, $f(X+Y) - f(X) - f(Y)$ is orthogonal to every vector in V. But the only vector in V orthogonal to every vector in V is the zero vector. Hence, $f(X+Y) - f(X) - f(Y) = 0$, or equivalently, $f(X+Y) = f(X) + f(Y)$. Similarly, we find that $f(cX) = cf(X)$ for all scalars $c \in \mathbb{R}$ and all vectors $X \in V$. Therefore, f is a linear transformation and the proof is complete. ∎

COROLLARY.   Let f be a Euclidean motion of V. Then $f = T_A \sigma$ for some orthogonal transformation σ and some translation T_A of V. Moreover, the factors σ and T_A are uniquely determined by f.

Proof.   Let $A = f(0)$ and $\sigma = T_{-A} f$. Then σ is a Euclidean motion since it is the product of two such motions. Since $\sigma(0) = T_{-A} f(0) = T_{-A}(A) = 0$, σ fixes the origin and hence is an orthogonal transformation by Proposition 1. Therefore, $f = T_{-A}^{-1} \sigma = T_A \sigma$, where σ is an orthogonal transformation of V. Now, to show that the factors σ and T_A are uniquely determined, suppose that $T_A \sigma = T_B \sigma'$ for some orthogonal transformation σ' and some vector $B \in V$. Then $\sigma' \sigma^{-1} = T_A T_B^{-1} = T_{A-B}$, from which it follows that T_{A-B} is a linear transformation. But the only translation that is a linear transformation is the identity map. Hence, $\sigma' \sigma^{-1} = 1_V = T_{A-B}$ and therefore $\sigma' = \sigma$ and $A = B$. Thus, the factors σ and T_A are uniquely determined by f. ∎

COROLLARY. $T(V) \lhd E(V)$. Moreover, $E(V) = T(V)O(V)$ and $E(V)/T(V) \cong O(V)$.

Proof.   We have already shown that $T(V)$ is a subgroup of $E(V)$. To prove normality, let $T_A \in T(V)$ and $f \in E(V)$. Then $f = T_B \sigma$ for some element $\sigma \in O(V)$ and therefore $fT_A f^{-1} = T_B \sigma T_A \sigma^{-1} T_{-B}$. Now, $\sigma T_A \sigma^{-1} = T_{\sigma(A)}$; for if $X \in V$, then $(\sigma T_A \sigma^{-1})(X) = \sigma(\sigma^{-1}(X) + A) = X + \sigma(A) = T_{\sigma(A)}(X)$ and hence $\sigma T_A \sigma^{-1} = T_{\sigma(A)}$. It follows, then, that $fT_A f^{-1} = T_B(\sigma T_A \sigma^{-1})T_{-B} = T_B T_{\sigma(A)} T_{-B} = T_{\sigma(A)}$ and hence that $fT_A f^{-1} \in T(V)$. Therefore, $T(V) \lhd E(V)$. The fact that $E(V) = T(V)O(V)$ follows immediately from Proposition 1. Finally, to show that $E(V)/T(V) \cong O(V)$, define the function $\alpha : E(V) \to O(V)$ by setting $\alpha(T_A \sigma) = \sigma$ for every $T_A \in T(V)$ and every $\sigma \in O(V)$. We leave it as an exercise for the reader to verify that α is a group-homomorphism mapping $E(V)$ onto $O(V)$ whose kernel is $T(V)$. Therefore, $E(V)/T(V) \cong O(V)$ and the proof is complete. ∎

The preceding results now give us a complete description of the Euclidean motions of a Euclidean space V: they are precisely the orthogonal transformations of the space followed by any translation of the space. The orthogonal transformations and translations of the space therefore generate the Euclidean group $E(V)$. Moreover, the translation subgroup $T(V)$ is a normal subgroup of $E(V)$, and the quotient group $E(V)/T(V)$ is isomorphic to the orthogonal group $O(V)$.

Consider the Euclidean line $\mathbb{E}^1$, for example. We showed in the previous section that $O(\mathbb{E}^1) = \{1, -1\}$. The translations of $\mathbb{E}^1$, on the other hand, are the mappings $T_a: \mathbb{E}^1 \to \mathbb{E}^1$, where $a \in \mathbb{R}$ and $T_a(x) = x + a$ for all $x \in \mathbb{R}$. Now observe that the composition $T_a \circ -1 = -T_{-a}$ since $T_a(-1(x)) = -x + a = -T_{-a}(x)$ for all $x \in \mathbb{R}$. It follows, therefore, that the Euclidean line $\mathbb{E}^1$ has two types of Euclidean motions: those of the form T_a, which are translations and include the identity map 1 since $1 = T_0$, and those of the form $-T_a$, which are their negatives and include the inversion map -1 since $-1 = -T_0$. Thus, $\mathbb{E}(1) = \{\pm T_a | a \in \mathbb{R}\} = \{T_a | a \in \mathbb{R}\}\{1_{\mathbb{E}^1}, -1_{\mathbb{E}^1}\} = T(\mathbb{E}^1)O(\mathbb{E}^1)$ and $\mathbb{E}(1)/T(\mathbb{E}^1) = \{\pm 1\} \cong O(\mathbb{E}^1)$.

All of these results follow from the fact, proved in Proposition 1, that the only Euclidean motions of a space that fix the zero vector are the orthogonal transformations. Let us now extend this result by showing that the only Euclidean motions that fix an arbitrary vector are in fact conjugates of orthogonal transformations.

COROLLARY.   Let f be a Euclidean motion of V that fixes some vector $A \in V$. Then $f = T_A \sigma T_A^{-1}$ for some orthogonal transformation σ of V.

Proof.   Let $\sigma = T_A^{-1} f T_A$. Then σ is a Euclidean motion of V since it is the product of three elements in $E(V)$ and fixes the zero vector since $\sigma(0) = T_A^{-1} f T_A(0) = T_A^{-1}(A) = 0$. Therefore, σ is an orthogonal transformation of V. Solving the equation for f, we find that $f = T_A \sigma T_A^{-1}$, as required.   ∎

COROLLARY.   Let $A \in V$ and let $E(V)_A = \{f \in E(V) | f(A) = A\}$ stand for the set of Euclidean motions that fix the vector A. Then $E(V)_A = T_A O(V) T_A^{-1}$.

Proof.   It follows from the previous corollary that $E(V)_A \subseteq T_A O(V) T_A^{-1}$. On the other hand, it is easily verified that every motion in $T_A O(V) T_A^{-1}$ fixes the vector A and hence lies in $E(V)_A$. Thus, $E(V)_A = T_A O(V) T_A^{-1}$.   ∎

Thus, the Euclidean motions of V that fix an arbitrary vector A have the form $T_A \sigma T_A^{-1}$ for some orthogonal transformation σ and are therefore conjugate to the orthogonal transformations of V. Consequently, there is a 1–1 correspondence $\sigma \leftrightarrow T_A \sigma T_A^{-1}$ between Euclidean motions that fix the origin and those that fix an arbitrary vector A. We think of $T_A \sigma T_A^{-1}$ as representing the orthogonal transformation σ, but with the origin of the coordinate system translated to the point A. For example, if σ is a rotation of the plane through an angle of θ radians about the origin, then $T_A\sigma T_A^{-1}$ is a rotation of the plane through an angle of θ radians about the point A. The following example illustrates these ideas in more detail.

EXAMPLE 2

Define the function $f:\mathbb{E}^2 \to \mathbb{E}^2$ by setting $f(x, y) = (y + 2, -x + 4)$ for all points $(x, y) \in \mathbb{E}^2$. Let us show that f is a Euclidean motion of the plane and then interpret it geometrically.

(A) Let $A = (x, y)$ and $B = (x', y')$ be arbitrary points in the plane. Then

$$
\begin{aligned}
d(f(A), f(B))^2 &= d((y + 2, -x + 4), (y' + 2, -x' + 4))^2 \\
&= [(y + 2) - (y' + 2)]^2 + [(-x + 4) - (-x' + 4)]^2 \\
&= (y - y')^2 + (-x + x')^2 \\
&= d(A, B)^2.
\end{aligned}
$$

Hence, $d(f(A), f(B)) = d(A, B)$ and therefore f is a Euclidean motion of $\mathbb{E}^2$.

(B) To factor f into the product of an orthogonal transformation and a translation, we let $A = f(0) = (2, 4)$ and set $\sigma = T_A^{-1}f$, as in the corollary following Proposition 2. Then σ is an orthogonal transformation and, for any point $(x, y) \in \mathbb{E}^2$,

$$
\sigma(x, y) = T_{(2,4)}^{-1}f(x, y) = T_{(-2,-4)}(y + 2, -x + 4) = (y, -x).
$$

The matrix of σ relative to the standard orthonormal coordinate system is therefore $\begin{pmatrix} 0 & 1 \\ -1 & 0 \end{pmatrix}$. Since $\det(\sigma) = 1$, σ is a rotation of the plane. In fact, $\sigma = \rho_{-\pi/2}$, the clockwise rotation of the plane through an angle of $90°$ about the origin. It now follows that $f = T_{(2,4)}\sigma$. In terms of matrices,

$$
f\begin{pmatrix} x \\ y \end{pmatrix} = \begin{pmatrix} 0 & 1 \\ -1 & 0 \end{pmatrix}\begin{pmatrix} x \\ y \end{pmatrix} + \begin{pmatrix} 2 \\ 4 \end{pmatrix}.
$$

(C) To determine if f fixes any vectors in $\mathbb{E}^2$, set $f(x, y) = (x, y)$. Then $(y + 2, -x + 4) = (x, y)$ and hence $y + 2 = x$ and $-x + 4 = y$. These equations have the solution $x = 3$, $y = 1$, and therefore the vector $(3, 1)$, and only this vector, is fixed by f. Hence, $f = T_{(3,1)}\sigma'T_{(3,1)}^{-1}$ for some orthogonal transformation σ'. To find σ', let (x, y) be an arbitrary point in the plane. Then

$$
\begin{aligned}
\sigma'(x, y) &= T_{(3,1)}^{-1}fT_{(3,1)}(x, y) \\
&= T_{(3,1)}^{-1}f(x + 3, y + 1) \\
&= T_{(-3,-1)}(y + 3, -x + 1) \\
&= (y, -x).
\end{aligned}
$$

Hence, $\sigma' = \rho_{-\pi/2}$. Therefore, $f = T_{(3,1)}\rho_{-\pi/2}T_{(3,1)}$ and hence f is the $90°$ clockwise rotation of the plane about the point $(3, 1)$. Figure 1 (page 254) illustrates these two rotations.

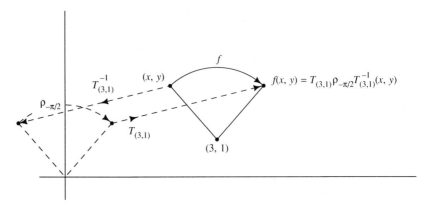

FIGURE 1. Structure of the Euclidean motion $f:f$ is the 90° clockwise rotation of the plane about the point (3, 1).

In the previous section we described the orthogonal transformations of the Euclidean plane. Let us now conclude this section by using those results together with the preceding results to give a complete description of the Euclidean motions of the plane.

EXAMPLE 3. Euclidean motions of the plane

We begin by first identifying four specific types of Euclidean motions of the plane—translations, rotations, reflections, and glide reflections—and then conclude by giving an algebraic analysis of the Euclidean group $\mathbb{E}(2)$, showing that these four types of motions account for all Euclidean motions of the plane.

(A) *Translations.* These are the mappings of the form $T_A:\mathbb{E}^2 \to \mathbb{E}^2$, where A is any vector in $\mathbb{E}^2$. If $A = (a, b)$, $T_A(x, y) = (x + a, y + b)$ for all $(x, y) \in \mathbb{E}^2$. Geometrically, T_A translates the plane along the vector A. If $A \neq 0$, the translation T_A is an example of a Euclidean motion of the plane that fixes no points.

(B) *Rotations about an arbitrary point.* Let $A \in \mathbb{E}^2$ and let $\rho_{\theta;A}:\mathbb{E}^2 \to \mathbb{E}^2$ stand for the counterclockwise rotation of the plane through an angle of θ radians about the point A, where $\theta \in \mathbb{R}$. The function $\rho_{\theta;A}$ corresponds to the rotation ρ_θ of the plane through an angle of θ radians about the origin and hence $\rho_{\theta;A} = T_A\rho_\theta T_A^{-1}$. To factor $\rho_{\theta;A}$ into the product of an orthogonal transformation followed by a translation, we write

$$\rho_{\theta;A} = T_A\rho_\theta T_A^{-1} = T_A(\rho_\theta T_A^{-1}\rho_\theta^{-1})\rho_\theta$$
$$= T_A T_{-\rho_\theta(A)}\rho_\theta$$
$$= T_{A-\rho_\theta(A)}\rho_\theta.$$

The rotation $\rho_{\theta;A}$ may therefore be obtained by first rotating the plane through an angle of θ radians about the origin and then translating to the point $A - \rho_\theta(A)$. The motion f discussed in Example 2 is a Euclidean motion of this type: $f = \rho_{-\pi/2;(3,1)}$. Finally, observe that the rotations $\rho_{\theta;A}$, for $\theta \neq 0$, are examples of Euclidean motions of the plane that have exactly one fixed point.

(C) *Reflections about an arbitrary line.* Let l be any line in the plane and let $R_l: \mathbb{E}^2 \to \mathbb{E}^2$ stand for the reflection of $\mathbb{E}^2$ through the line l. R_l corresponds to reflection of the plane about the line passing through the origin parallel to l, as shown in Figure 2. Note that if l passes through the origin, then $R_l = S_A$, the symmetry of the plane determined by any nonzero vector A perpendicular to l. If l does not pass through the origin, there is a unique point A on l such that A is perpendicular to l and therefore $R_l = T_A S_A T_A^{-1}$; in this case $R_l = T_A(S_A T_A^{-1} S_A)S_A = T_A T_{S_A(-A)}S_A = T_{2A}S_A$ is the factorization of R_l as the product of the symmetry S_A followed by translation to the point $2A$. Reflections of the plane about arbitrary lines are examples of Euclidean motions that leave exactly one line fixed. Let us illustrate this type of motion by describing

Translation to an arbitrary point A

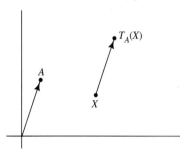

$$T_A(X) = X + A$$

Rotation about an arbitrary point A

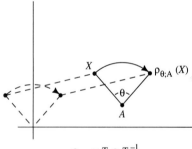

$$\rho_{\theta;A} = T_A \rho_\theta T_A^{-1}$$

Reflection about an arbitrary line

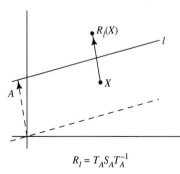

$$R_l = T_A S_A T_A^{-1}$$

Glide reflection in direction of A about line l

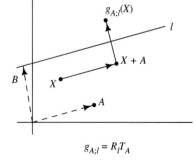

$$g_{A;l} = R_l T_A$$

FIGURE 2. The four types of Euclidean motions of the plane.

the reflection of the plane about the line l having equation $y = x + 2$. In this case we find that $A = (-1, 1)$ is the unique vector perpendicular to l and hence, for all points $(x, y) \in \mathbb{E}^2$,

$$R_l(x, y) = T_{(-1,1)}S_{(-1,1)}T_{(-1,1)}^{-1}(x, y)$$

$$= T_{(-1,1)}S_{(-1,1)}(x + 1, y - 1)$$

$$= T_{(-1,1)}\left((x + 1, y - 1) - 2\,\frac{\langle(x + 1, y - 1), (-1, 1)\rangle}{\langle(-1, 1), (-1, 1)\rangle}(-1, 1)\right)$$

$$= T_{(-1,1)}(y - 1, x + 1)$$

$$= (y - 2, x + 2).$$

The origin, for example, is reflected to the point $(-2, 2)$.

(D) *Glide reflections.* A *glide reflection* of the plane is any translation of the plane by a nonzero vector followed by a reflection about any line parallel to the translating vector. If A stands for the nonzero translation vector and l stands for any line parallel to A, then the glide reflection determined by A and l is the function $g_{A;l}:\mathbb{E}^2 \to \mathbb{E}^2$ defined by setting $g_{A;l} = R_lT_A$, as shown in Figure 2. If the reflecting line l passes through the origin, then $R_l = S_B$ for any vector B perpendicular to A and therefore

$$g_{A;l} = R_lT_A = S_BT_A = (S_BT_AS_B)S_B = T_{S_B(A)}S_B.$$

If l does not pass through the origin, however, then we first use part (C) to write $R_l = T_{2B}S_B$, where B is the unique vector on l perpendicular to A, and find that

$$g_{A;l} = T_{2B}S_BT_A = T_{2B}(S_BT_AS_B)S_B = T_{2B + S_B(A)}S_B.$$

Glide reflections of the plane are examples of Euclidean motions of the plane that leave no points fixed, although they leave the reflecting line

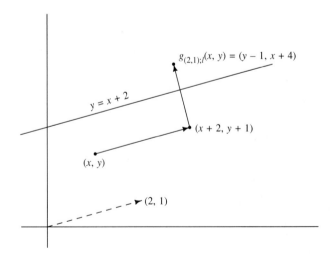

l invariant, that is, points on l are mapped to points on l. Let us illustrate this type of motion by describing the glide reflection that translates the plane in the direction of the vector $(2, 1)$ and then reflects it through the line l having equation $y = x + 2$. Here $A = (2, 1)$ and R_l is the reflection described in part (C). Using the formula for R_l obtained in part (C), we find that

$$g_{(2,1);l}(x, y) = R_l T_{(2,1)}(x, y)$$
$$= R_l(x + 2, y + 1)$$
$$= (y + 1 - 2, x + 2 + 2)$$
$$= (y - 1, x + 4)$$

for all points $(x, y) \in \mathbb{E}^2$. In this case the origin is mapped to the point $(-1, 4)$.

(E) This completes our description of the four basic types of Euclidean motions of the plane: translations, rotations, reflections, and glide reflections. We now claim that every Euclidean motion of the plane is one of these four types. To show this, recall that every Euclidean motion of the plane may be written in the form $T_A \sigma$, where σ is an orthogonal transformation. Since the orthogonal transformations of the plane are either rotations ρ_θ or symmetries S_B, it follows that the Euclidean motions have the form T_A, ρ_θ, S_A, $T_A \rho_\theta$, or $T_A S_B$. Motions of the form T_A, ρ_θ, and S_A are translations, rotations, and symmetries, respectively. Now, consider a Euclidean motion of the form $T_A \rho_\theta$. We claim that $T_A \rho_\theta = \rho_{\theta;B}$, the rotation of the plane about some point B. To show this, first observe that ρ_θ fixes only the zero vector and therefore $1 - \rho_\theta$ is a linear transformation that fixes no vector. Thus, $\text{Ker}(1 - \rho_\theta) = \{0\}$. Consequently, $1 - \rho_\theta$ is a linear isomorphism of the plane and hence $(1 - \rho_\theta)(B) = A$ for some vector $B \in \mathbb{E}^2$. Therefore, $A = B - \rho_\theta(B)$ and hence

$$T_A \rho_\theta = T_{B - \rho_0(B)} \rho_\theta = T_B T_{\rho_\theta(B)}^{-1} \rho_\theta$$
$$= T_B(\rho_\theta T_B^{-1} \rho_\theta^{-1}) \rho_\theta = T_B \rho_\theta T_B^{-1} = \rho_{\theta;B}.$$

Thus, $T_A \rho_\theta$ is the rotation of the plane about the point B through an angle of θ radians. Finally, consider a Euclidean motion of the form $T_A S_B$, where $A, B \in \mathbb{E}^2$ with $B \neq 0$. If $A = 0$, then $T_A S_B = S_B$ and hence is a symmetry. Assume, therefore, that $A \neq 0$ and let

$$X = A - \frac{\langle A, B \rangle}{\langle B, B \rangle} B$$

$$Y = \frac{\langle A, B \rangle}{2 \langle B, B \rangle} B.$$

If $X = 0$, then B is a scalar multiple of A; in this case $S_B = S_A$ and therefore $T_A S_A = T_{2(1/2A)} S_{1/2A}$, which, by part (C), is the reflection of the plane about the line $l = 1/2A + \langle A \rangle^\perp$. If, on the other hand, $X \neq 0$, then we find by direct calculation that $X \perp Y$. In this case let $l = Y + \langle X \rangle$ stand for the line through Y parallel to X. Then $T_A S_B = g_{X;l}$, the glide reflection along X reflecting about l. To show this, recall from part (D) that $g_{X;l} = T_{2Y + S_Y(X)} S_Y$. But $2Y + S_Y(X) = A - X + X = A$ since $X \perp Y$, while $S_Y = S_B$ since Y and B are collinear. Therefore, $g_{X;l} = T_A S_B$, thus showing that $T_A S_B$ is a glide reflection. We conclude, therefore, that every Euclidean motion of the plane is either a translation, rotation about an arbitrary point, reflection about an arbitrary line, or a glide reflection. This concludes the example.

Exercises

1. Define the function $f: \mathbb{E}^2 \to \mathbb{E}^2$ by setting $f(x, y) = (3 - y, 1 + x)$ for all points $(x, y) \in \mathbb{E}^2$.
   (a) Show that f is a Euclidean motion of the plane.
   (b) Factor f into the product of an orthogonal transformation followed by a translation.
   (c) Is the orthogonal transformation obtained in part (b) a rotation or reflection?
   (d) Does f have any fixed points?
   (e) Determine if f is a translation, rotation, reflection, or glide reflection of the plane.

2. Define the function $f: \mathbb{E}^2 \to \mathbb{E}^2$ by setting $f(x, y) = (-\frac{3}{5}x - \frac{4}{5}y + \frac{8}{5}, -\frac{4}{5}x + \frac{3}{5}y + \frac{4}{5})$ for all points $(x, y) \in \mathbb{E}^2$.
   (a) Show that f is a Euclidean motion of the plane.
   (b) Factor f into the product of an orthogonal transformation and a translation.
   (c) Is the orthogonal transformation obtained in part (b) a rotation or a reflection?
   (d) Does f have any fixed points?
   (e) Determine if f is a translation, rotation, reflection, or glide reflection of the plane.

3. Let $f: \mathbb{E}^2 \to \mathbb{E}^2$ be the reflection of the plane about the line having equation $2x + 3y = 6$. Find $f(x, y)$ for any point (x, y) in the plane.

4. Let $f: \mathbb{E}^2 \to \mathbb{E}^2$ be the glide reflection that translates the plane along the vector $(-1, 3)$ and then reflects it through the line having equation $-3x + y = 1$. Find $f(x, y)$ for any point (x, y) in the plane.

5. Let V be a Euclidean space. An *affine transformation* of V is any function of the form $f_{\sigma;A}: V \to V$, where σ is any linear isomorphism of V, $A \in V$, and $f_{\sigma;A}(X) = \sigma(X) + A$ for all vectors $X \in V$.
   (a) Show that every Euclidean motion of V is an affine transformation of V.
   (b) Show that an affine transformation $f_{\sigma;A}$ is a Euclidean motion if and only if σ is an orthogonal transformation.
   (c) Show that every affine transformation of V is a permutation of V.

(d) Let $f_{\sigma;A}$ and $f_{\tau;B}$ be affine transformations of V. Show that

$$f_{\sigma;A}f_{\tau;B} = f_{\sigma\tau;\sigma(B)+A}$$

$$f_{\sigma;A}^{-1} = f_{\sigma^{-1};-\sigma^{-1}(A)}$$

(e) Let Aff(V) stand for the set of affine transformations of V. Show that Aff(V) is a group under function composition and is a subgroup of the symmetric group Sym(V). Aff(V) is called the *affine group* of V.

(f) Show that $E(V) \leq$ Aff(V).

(g) Show that $T(V) = \{f_{1;A} \in$ Aff(V)$\,|\,A \in V\}$ and that $T(V) \lhd$ Aff(V).

(h) Show that $GL(V) \leq$ Aff(V).

(i) Show that Aff(V) $= T(V)GL(V)$ and that Aff(V)/$T(V) \cong GL(V)$.

6. Show that Aff($\mathbb{E}^1$) $= \{f \in$ Sym($\mathbb{E}^1$)$\,|\,f(x) = mx + b$ for all $x \in \mathbb{E}^1$, where $m, b \in \mathbb{R}, m \neq 0\}$. Thus, the affine group of the Euclidean line consists of those linear functions on $\mathbb{E}^1$ whose graph has a nonzero slope.

7. Let R and R' be reflections of the plane about any two lines. Show that R and R' are conjugate as elements of the group $\mathbb{E}(2)$.

4. SYMMETRY GROUPS

As we noted in the introduction to this chapter, we have frequently used the group of the triangle and the group of the square to illustrate basic concepts in group theory. These groups are examples of symmetry groups; that is, groups of rigid motions of the plane that bring a given figure into coincidence with itself. In this section our goal is to extend the idea of a symmetry group to arbitrary Euclidean spaces. Symmetry groups may be associated with any geometric configuration and are especially important because they reflect geometric properties of the configuration. After developing the general properties of symmetry groups, we analyze in detail the general dihedral groups, which are the symmetry groups of regular n-gons in the plane, and the group of the tetrahedron, which is the symmetry group of a regular tetrahedron in Euclidean 3-space.

We begin with a nonempty subset S of Euclidean n-space $\mathbb{E}^n$. A Euclidean motion f of $\mathbb{E}^n$ is said to leave S *invariant* if $f(S) = S$; that is, if f maps every point in S to a point in S. The identity map, for example, leaves S invariant. More generally, it is easy to see that if f and g are Euclidean motions leaving S invariant, then both the composition $f \circ g$ and inverse f^{-1} leave S invariant. Thus, the Euclidean motions of $\mathbb{E}^n$ that leave S invariant form a group.

Definition

Let S be a nonempty subset of $\mathbb{E}^n$. The group of Euclidean motions that leave S invariant is called the *symmetry group of S* and is denoted by G_S.

For example, suppose that $S = \{A\}$ consists of a single point $A \in \mathbb{E}^n$. Then the symmetry group G_S consists of all Euclidean motions that fix the point

A, and hence, using the notation from the previous section, $G_S = \mathbb{E}(n)_A$. Since $\mathbb{E}(n)_A = T_A O(\mathbb{E}^n) T_A^{-1}$, it follows that $G_S \cong O(\mathbb{E}^n)$; indeed, G_S consists of the rotations and reflections of the space, with the point A considered as origin. On the other hand, if S is an equilateral triangle in the plane, then G_S consists of all Euclidean motions of the plane that leave the triangle invariant and hence is the group T of the triangle. Similarly, the group of the square is the symmetry group of a square in the plane.

To what extent does the location of a geometric figure determine its symmetry group? For example, it is clear that if two equilateral triangles in the plane are congruent but located in different positions, their symmetry groups should be essentially the same. The groups will consist of different motions, of course, but they should be isomorphic. Let us now extend the notion of congruence to arbitrary Euclidean spaces and then show that congruent figures indeed have isomorphic symmetry groups.

Let S and S' be nonempty subsets of $\mathbb{E}^n$. We say that S and S' are *congruent* if there is a Euclidean motion f of $\mathbb{E}^n$ such that $S' = f(S)$; that is, if there is some rigid motion of the space that brings the one subset into coincidence with the other. A Euclidean motion that leaves S invariant, that is, a motion f such that $f(S) = S$, is then called a *congruence motion of S*. The symmetry group G_S may therefore be described as the group of congruence motions of S. We now claim that congruent subsets of $\mathbb{E}^n$ have isomorphic symmetry groups.

PROPOSITION 1. Let S and S' be congruent subsets of $\mathbb{E}^n$. Then $G_S \cong G_{S'}$.

Proof.   Let f be a Euclidean motion of $\mathbb{E}^n$ such that $S' = f(S)$. Then, for any Euclidean motion g, $g(S) = S$ if and only if $fgf^{-1}(S') = S'$. Therefore, $G_{S'} = fG_S f^{-1}$ and hence $G_{S'} \cong G_S$.   ∎

It follows, therefore, that a symmetry group G_S does not depend upon the location of the set S in the space in the sense that congruent figures have isomorphic, in fact conjugate, symmetry groups. In the exercises we define the notion of a similarity transformation, which involves scaling the figure, and ask the reader to show that similar figures also have conjugate symmetry groups.

EXAMPLE 1. The dihedral groups

Let n be a positive integer ≥ 3. The symmetry group of a regular n-gon in the plane is called the *dihedral group of degree n* and is denoted by D_n. D_n consists of all rigid motions of the plane that map a given n-gon to itself. The dihedral group D_3, for example, is the group of the triangle, while D_4 is the group of the square. In this example we show that D_n is a finite group of order $2n$ that acts transitively as a permutation group on the set of vertices of the n-gon and then find a matrix representation for the transformations in D_n.

(A) We begin by observing that since the transformations in D_n are congruence motions of an n-gon, they permute the vertices of the n-gon and hence D_n acts as a group of permutations on the set of n vertices of the n-gon. Moreover, this representation is transitive since, given any two vertices, there is some rotation that maps the one to the other. It follows, therefore, that if v is any vertex of the n-gon and $(D_n)_v$ is the stabilizer of v, then $[D_n:(D_n)_v] = n$. Now, the only congruence motions of the n-gon that leave the vertex v fixed are the identity map and the reflection of the n-gon about the line connecting v with the center of the n-gon. Therefore, $|(D_n)_v| = 2$. Since $[D_n:(D_n)_v] = n$, it follows that $|D_n| = 2n$. Thus, D_n is a finite group of order $2n$ that acts transitively as a permutation group on the vertices of the n-gon.

(B) To find the $2n$ elements in D_n, let R stand for the counterclockwise rotation of the plane about the center of the n-gon through an angle of $2\pi/n$ radians, and let S stand for the reflection of the plane about a line connecting the center of the n-gon to any vertex. We claim that R and S generate the group D_n and satisfy the relations $R^n = S^2 = 1$ and $SRS^{-1} = R^{-1}$. For it is clear that R and S are, in fact, congruence motions of the n-gon and $R^n = S^2 = 1$. Hence, $1, R, \ldots, R^{n-1}, S, SR, \ldots, SR^{n-1}$ are $2n$ distinct elements in D_n. Since D_n contains only $2n$ elements, it follows that $D_n = \{1, R, \ldots, R^{n-1}, S, SR, \ldots, SR^{n-1}\}$. Therefore, R and S generate D_n. In particular, D_n contains n rotations, namely $1, R, \ldots, R^{n-1}$, and n reflections, $S, SR, \ldots, SR^{n-1}$. Now, to show that $SRS^{-1} = R^{-1}$ we need only verify that $SRS^{-1}(v) = R^{-1}(v)$ for a particular vertex v of the n-gon since SRS^{-1} is a rotation and hence must be a power of R. Let v be the vertex of the n-gon lying on the line of reflection of S, and let v' and v'' be the vertices adjacent to v, as illustrated in the figure. Then $SRS^{-1}(v) = SR(v) = S(v') = v'' = R^{-1}(v)$ and hence $SRS^{-1} = R^{-1}$. We conclude, therefore, that D_n is a group of order $2n$ generated by two elements R and S that satisfy the relations $R^n = S^2 = 1$ and $SRS^{-1} = R^{-1}$. In the exercises we ask the reader to show that any group generated by two elements satisfying these relations is in fact isomorphic to D_n. Thus, the dihedral group D_n is characterized, up to isomorphism, by these relations.

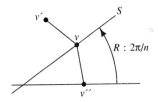

(C) Geometrically, the rotations $R, R^2, \ldots, R^{n-1}$ rotate the n-gon about its center through angles of $2\pi/n, 2(2\pi/n), \ldots, (n-1)(2\pi/n)$ radians, respectively. To describe the reflections $S, SR, \ldots, SR^{n-1}$ geometrically, let k be a positive integer, $1 \leq k \leq n-1$, and let θ be the angle between the reflecting lines of SR^k and S. Since $(SR^k)S = (SRS^{-1})^k = R^{-k}$, it follows that twice the angle between SR^k and S is equal to the angle of the rotation R^{-k} (see Exercise 4, Section 2). Therefore, $2\theta = -(2\pi/n)k$ and hence $\theta = -(\pi/n)k$. Thus, the reflecting lines for $SR, \ldots, SR^{n-1}$ are obtained by rotating the reflecting line of S clockwise through angles of $\pi/n, 2(\pi/n), \ldots, (n-1)(\pi/n)$ radians, respectively, as illustrated in the margin. Note that when n is odd, the reflecting line of each reflection

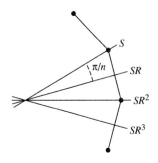

SR^k passes through a vertex and the center of the n-gon, but if n is even, half of the reflecting lines pass through a vertex and center while the other half pass through the midpoint of a side and the center.

(D) Let us use the geometric description of the reflections in D_n to show that the n reflections $S, SR, \ldots, SR^{n-1}$ form a single conjugacy class in D_n when n is odd, but form two conjugacy classes when n is even. To do this, we first observe that when n is even, the reflecting line of any reflection either passes through two diagonally opposite vertices or through the midpoints of opposite sides, while when n is odd, the reflecting line passes through a vertex and the midpoint of the diagonally opposite side, as illustrated in the margin for $n = 3$ and $n = 4$. Now, if W is any symmetry of the plane about a line l and $\sigma \in D_n$, recall that $\sigma W \sigma^{-1}$ is the symmetry of the plane about the line $\sigma(l)$. Thus, if l connects opposite vertices or opposite midpoints of the n-gon, $\sigma(l)$ does likewise. Hence, there are two conjugacy classes of reflections in D_n when n is even: those whose reflecting line connects opposite vertices and those whose reflecting line connects opposite midpoints. If n is odd, however, the reflecting lines all pass through a vertex and hence the rotations in D_n permute the reflecting lines transitively; in this case, the reflections form a single conjugacy class in D_n. We conclude, therefore, that the reflections in D_n fall into either one or two conjugacy classes, depending on whether n is odd or even, respectively.

(E) Let us conclude our discussion of the dihedral groups by describing a matrix representation for the transformations in D_n. For this purpose it is convenient to choose the n-gon whose vertices are the points $(\cos(2\pi/n)k, \sin(2\pi/n)k)$, where $k = 1, \ldots, n$. Let S be the reflection of the plane about the X-axis. Then the matrix of S relative to the standard orthonormal coordinate system $\{e_1, e_2\}$ is $\left(\begin{smallmatrix} 1 & 0 \\ 0 & -1 \end{smallmatrix}\right)$. On the other hand, a typical rotation R^k maps e_1 to $(\cos(2\pi/n)k, \sin(2\pi/n)k)$ and e_2 to $(\cos(\pi/2 + (2\pi/n)k), \sin(\pi/2 + (2\pi/n)k)) = (-\sin(2\pi/n)k, \cos(2\pi/n)k)$. We then obtain the following matrix representation for the elements in D_n:

$n = 3$

$n = 4$

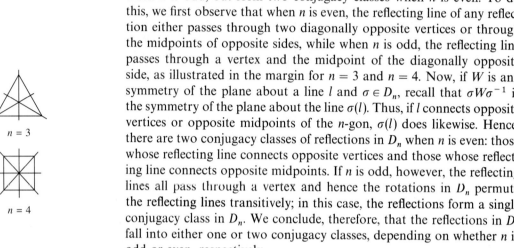

$$R^k \leftrightarrow \begin{pmatrix} \cos\dfrac{2\pi}{n}k & -\sin\dfrac{2\pi}{n}k \\ \sin\dfrac{2\pi}{n}k & \cos\dfrac{2\pi}{n}k \end{pmatrix}, \quad SR^k \leftrightarrow \begin{pmatrix} \cos\dfrac{2\pi}{n}k & -\sin\dfrac{2\pi}{n}k \\ -\sin\dfrac{2\pi}{n}k & -\cos\dfrac{2\pi}{n}k \end{pmatrix}.$$

For example, if $n = 3$ we obtain the following matrix representations for the elements in D_3:

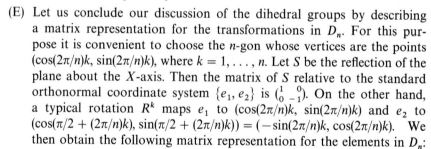

$$1 \leftrightarrow \begin{pmatrix} 1 & 0 \\ 0 & 1 \end{pmatrix}, \quad R \leftrightarrow \begin{pmatrix} \frac{1}{2} & -\frac{1}{2}\sqrt{3} \\ \frac{1}{2}\sqrt{3} & \frac{1}{2} \end{pmatrix}, \quad R^2 \leftrightarrow \begin{pmatrix} -\frac{1}{2} & -\frac{1}{2}\sqrt{3} \\ \frac{1}{2}\sqrt{3} & -\frac{1}{2} \end{pmatrix}$$

$$S \leftrightarrow \begin{pmatrix} 1 & 0 \\ 0 & -1 \end{pmatrix}, \quad SR \leftrightarrow \begin{pmatrix} \frac{1}{2} & -\frac{1}{2}\sqrt{3} \\ -\frac{1}{2}\sqrt{3} & -\frac{1}{2} \end{pmatrix}, \quad SR^2 \leftrightarrow \begin{pmatrix} -\frac{1}{2} & -\frac{1}{2}\sqrt{3} \\ -\frac{1}{2}\sqrt{3} & \frac{1}{2} \end{pmatrix}.$$

Note that these matrices are different from those obtained in Example 2, Section 2, Chapter 2 for the group of the triangle since the coordinate systems used to represent the transformations are different. This concludes our discussion of the dihedral groups.

EXAMPLE 2. The group of the tetrahedron

The symmetry group of a regular tetrahedron in Euclidean 3-space is called the *group of the tetrahedron* and is denoted by Tet; Tet consists of all rigid motions of $\mathbb{E}^3$ that map a regular tetrahedron to itself. The purpose of this example is to show that Tet consists of 24 rigid motions, to describe these motions geometrically, and to then show that Tet $\cong$ Sym(4) while the rotation subgroup Tet$^+$ $\cong$ Alt(4). Throughout this discussion we refer the reader to Figure 1.

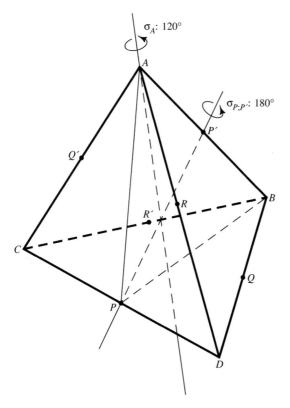

FIGURE 1. The regular tetrahedron and its congruence motions.

(A) Let us begin by first identifying four specific types of Euclidean motions that leave the tetrahedron invariant.

The identity transformation: $1 = 1_{\mathbb{E}^3}$

Vertex–face rotations: These are rotations of $\mathbb{E}^3$ whose axis passes through a vertex and the center of the opposite face, through an angle of either 120° or 240°. If σ_v stands for a 120° rotation of the tetrahedron about an axis passing through vertex v and the center of the face opposite v, then there are eight such vertex–face rotations: σ_A, σ_A^2, σ_B, σ_B^2, σ_C, σ_C^2, σ_D, σ_D^2.

Edge–edge rotations: These are 180° rotations about an axis passing through the midpoints of opposite edges. There are three such pairs of opposite edge midpoints: (P, P'), (Q, Q'), and (R, R'). We denote the corresponding edge–edge rotations by $\sigma_{P;P'}$, $\sigma_{Q;Q'}$, and $\sigma_{R;R'}$.

Symmetries: The symmetries of $\mathbb{E}^3$ that leave the tetrahedron invariant are those whose reflecting plane bisects the tetrahedron. Such a plane is determined by any edge and the midpoint of the opposite edge. For example, midpoint P is opposite edge AB and determines the plane $(P;AB)$. Let τ_P stand for the symmetry that reflects the tetrahedron through the plane $(P;AB)$. Using similar notation, there are six such symmetries leaving the tetrahedron invariant: τ_P, τ_Q, τ_R, $\tau_{P'}$, $\tau_{Q'}$, $\tau_{R'}$.

The rigid motions just listed account for 18 elements in the group Tet: 1 identity transformation, 8 vertex–face rotations, 3 edge–edge rotations, and 6 symmetries.

(B) Let us now show that the group Tet has order 24 and then describe the remaining 6 congruence motions of the tetrahedron. Clearly, Tet acts as a transitive permutation group on the set of four vertices of the tetrahedron. Moreover, the stabilizer Tet_A of vertex A is just the group of the triangular face BCD opposite A and hence $|\mathrm{Tet}_A| = 6$. It follows that $[\mathrm{Tet}:\mathrm{Tet}_A] = 4$ and therefore $|\mathrm{Tet}| = 24$. In particular, Tet contains 12 rotations; these are the 12 rotations listed above in part (A). The remaining six motions in Tet are reflections of $\mathbb{E}^3$ that are not symmetries; we will refer to such a motion as a *twist*. For example, the product $\sigma_A \tau_{Q'}$ is a typical twist of the tetrahedron. As a permutation of the vertices, $\sigma_A \tau_{Q'}$ is equal to $(BCD)(AC) = (ADBC)$. The six twists of the tetrahedron correspond to the six 4-cycles in $\mathrm{Sym}(\{A, B, C, D\})$; geometrically, they are products of three symmetries. For example, $(ADBC) = (AC)(AB)(AD)$, and hence the twist $\sigma_A \tau_{Q'} = \tau_{Q'} \tau_P \tau_R$.

(C) We now claim that $\mathrm{Tet} \cong \mathrm{Sym}(4)$. To show this, observe that Tet acts as a transitive permutation group on the set of four vertices of the tetrahedron. Moreover, this representation is in fact faithful: if $\sigma \in \mathrm{Tet}$ leaves every vertex fixed, then $\sigma = 1$. It follows, therefore, that Tet is isomorphic to a subgroup of $\mathrm{Sym}(4)$. But $|\mathrm{Tet}| = 24$. Hence, $\mathrm{Tet} \cong \mathrm{Sym}(4)$. This

Tet	Geometric Description	Vertex Permutation	Type
1	Identity transformation	1	1^4
σ_A, σ_A^2	Vertex–face	$(B\ C\ D),\quad (B\ D\ C)$	$1^1 3^1$
σ_B, σ_B^2	Rotations: 120°, 240°	$(A\ D\ C),\quad (A\ C\ D)$	
σ_C, σ_C^2		$(A\ B\ D),\quad (A\ D\ B)$	
σ_D, σ_D^2		$(A\ C\ B),\quad (A\ B\ C)$	
$\sigma_{P,P'}$	Edge–edge	$(A\ B)(C\ D)$	2^2
$\sigma_{Q,Q'}$	Rotations: 180°	$(A\ C)(B\ D)$	
$\sigma_{R,R'}$		$(B\ C)(A\ D)$	
$\tau_P, \tau_{P'}$	Symmetries	$(C\ D),\quad (A\ B)$	$1^2 2^1$
$\tau_Q, \tau_{Q'}$		$(B\ D),\quad (A\ C)$	
$\tau_R, \tau_{R'}$		$(A\ D),\quad (B\ C)$	
$\tau_{Q'}\tau_P\tau_R$	Twists	$(A\ D\ B\ C)$	4^1
$\tau_R\tau_{P'}\tau_{Q'}$		$(A\ C\ B\ D)$	
$\tau_{P'}\tau_{Q'}\tau_R$		$(A\ D\ C\ B)$	
$\tau_R\tau_{Q'}\tau_{P'}$		$(A\ B\ C\ D)$	
$\tau_{Q'}\tau_R\tau_{P'}$		$(A\ B\ D\ C)$	
$\tau_{P'}\tau_R\tau_{Q'}$		$(A\ C\ D\ B)$	

FIGURE 2.   Class table for the group of the tetrahedron.

shows, in particular, that every permutation of the vertices of the tetrahedron is represented by some congruence motion of the tetrahedron.
Figure 2 summarizes the 24 elements in the group Tet, describes them
both geometrically and as permutations of the vertices, and gives their
type as vertex permutations. It is clear from this figure that each distinct
type of motion, such as the vertex–face rotations, all have the same type
as permutations and hence are conjugate in Sym(4). Since every permutation in Sym(4) corresponds to some congruence motion of the tetrahedron, all motions of the same type are in fact conjugate in the group
Tet itself. Thus, conjugacy classes in the group Tet may be described
geometrically as the classes of vertex–face rotations, edge–edge rotations, symmetries, twists, and, of course, the identity transformation.

(D) Let us conclude our discussion of the group of the tetrahedron by showing that the rotation subgroup Tet^+ is isomorphic to the alternating
subgroup Alt(4). The group Tet^+ has order 12 and consists of the identity
transformation, the eight vertex–face rotations, and the three edge–edge

rotations. From Figure 2, each of these motions corresponds to an even permutation of the vertices. Therefore, $\text{Tet}^+ \cong \text{Alt}(4)$. This concludes the example.

Exercises

1. Let S stand for a rectangle in the plane that is not a square. Find the symmetry group G_S and show that $G_S \cong \mathbb{Z}_2 \times \mathbb{Z}_2$.

2. (a) Find the symmetry group of a line in the plane.
   (b) Show that the symmetry group of any two lines in the plane are isomorphic.

3. Find the symmetry group of a circle in the plane.

4. Find the symmetry group of a sphere in Euclidean 3-space.

5. Write out the elements in the dihedral group D_5 and construct the class table for the group. Indicate each of the conjugacy classes, their size, and the centralizer of each element.

6. Let n be an integer, $n \geq 3$, D_n the dihedral group of degree n, and $Z(D_n)$ the center of D_n.
   (a) If n is even, show that $R^{n/2} = -1$, the $180°$ rotation of the plane.
   (b) If n is even, show that $Z(D_n) = \{1, -1\}$.
   (c) If n is odd, show that $Z(D_n) = \{1\}$.
   (d) Determine the Sylow p-subgroups of D_n for all primes p.
   (e) Show that every odd order subgroup of D_n is cyclic.

7. Let n be an integer, $n \geq 3$, R the counterclockwise rotation of the plane about the origin through an angle of $2\pi/n$ radians, and S the reflection of the plane about the X-axis. For $k = 1, \ldots, n$, let $A_k = (-\sin(\pi/n)k, \cos(\pi/n)k)$. Show that $SR^{-k} = S_{A_k}$, the reflection of the plane about the line perpendicular to A_k, for $k = 1, \ldots, n$.

8. Let us set $D_1 = \mathbb{Z}_2$ and $D_2 = \mathbb{Z}_2 \times \mathbb{Z}_2$. With this convention, show that if G is a group generated by two elements x and y satisfying the relations $x^n = y^2 = 1$ and $yxy^{-1} = x^{-1}$, then $G \cong D_n$.

9. Let G be a finite group. An element in G having order 2 is called an *involution*.
   (a) Show that G contains an involution if and only if $|G|$ is even.
   (b) Let $|G|$ be even and let τ and τ' be involutions in G. Set $x = \tau\tau'$ and $y = \tau$. Show that $yxy^{-1} = x^{-1}$.
   (c) Let $|G|$ be even. Show that any two involutions in G generate a subgroup isomorphic to the dihedral group D_n for some integer $n \geq 1$. We express this fact by saying that two involutions generate a subgroup of *dihedral type*.
   (d) Determine the subgroup of dihedral type generated by the following pairs of involutions:
   (1) $\text{Sym}(3)$; $\tau = (1\ 2)$, $\tau' = (1\ 3)$
   (2) $\text{Sym}(4)$; $\tau = (1\ 2)$, $\tau' = (2\ 3)$
   (3) $\text{Sym}(4)$; $\tau = (1\ 2)$, $\tau' = (3\ 4)$

(4) Sym(5); $\tau = (2\ 5)(3\ 4)$, $\tau' = (1\ 5)(2\ 4)$

(5) D_6; $\tau = SR$, $\tau' = SR^3$.

(e) Let SR^s and SR^t be symmetries in the dihedral group D_n, where $0 \le t < s$. Show that the subgroup of D_n generated by SR^s and SR^t is isomorphic to D_m, where $m = \dfrac{n}{(n,\ t - s)}$.

(f) Let G be a finite group of even order and let n be the number of involutions in G. Suppose that n is odd and that there are two involutions in G whose product has order n. Show that the involutions in G form a single conjugacy class.

10. Let G be a group of dihedral type. Show that every subgroup of G is either cyclic or of dihedral type.

11. Let $M = \left(\begin{smallmatrix} 0 & -1 \\ -1 & 0 \end{smallmatrix}\right)$ and $N = \left(\begin{smallmatrix} -1 & -1 \\ 0 & 1 \end{smallmatrix}\right)$. Show that the subgroup of $\mathrm{Mat}_2^*(\mathbb{R})$ generated by M and N is of dihedral type.

12. This exercise refers to the labeling of the regular tetrahedron in Example 2. Let $[Q, R, Q', R']$ stand for the quadrilateral whose vertices are the points Q, R, Q', and R'.

(a) Show that $[Q, R, Q', R']$ is a square.

(b) Let $\mathrm{Tet}_{[Q,R,Q',R']}$ stand for the subgroup of motions in Tet that leave the square $[Q, R, Q', R']$ invariant. Show that $\mathrm{Tet}_{[Q,R,Q',R']}$ is isomorphic to the group of the square $[Q, R, Q', R']$.

13. Let F stand for the set of faces of a regular tetrahedron and V the set of vertices.

(a) Show that the group Tet acts as a faithful and transitive permutation group on F. Write out the permutations of F corresponding to each motion in Tet.

(b) Show that the permutation representations Tet $\to$ Sym(V) and Tet $\to$ Sym(F) of Tet acting on the faces and vertices, respectively, are isomorphic representations.

14. Let E stand for the set of edges of a regular tetrahedron.

(a) Show that Tet acts as a faithful and transitive permutation group on E. Write out the permutations of E corresponding to each motion in Tet.

(b) Find a permutation of the edges that does not correspond to a congruence motion of the tetrahedron.

15. Let N stand for the subset of the group Tet consisting of the identity transformation and the three edge–edge rotations.

(a) Show that $N \lhd \mathrm{Tet}$.

(b) Show that $\mathrm{Tet}/N \cong \mathrm{Tet}_A$, the stabilizer of vertex A.

16. **The group of the cube.** Let C stand for the symmetry group of a cube.

(a) Show that C acts a faithful and transitive permutation group on the set of faces of the cube.

(b) If F is any face of the cube and C_F the stabilizer of F, show that $C_F \cong D_4$.

(c) Show that $|C| = 48$.

17. The diagram here illustrates a regular tetrahedron inscribed in a cube. Let Tet stand for the group of the tetrahedron and C the group of the cube.

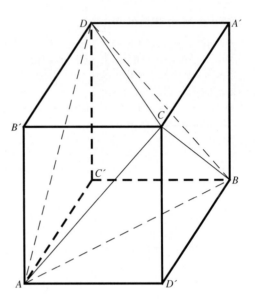

(a) Show that Tet $\leq C$ and $[C:\text{Tet}] = 2$.

(b) Show that $C \cong \text{Tet} \times (-1)$.

(c) Is it true that Tet $= C^+$, the rotation subgroup of C?

(d) Let τ_{AB}, τ_{BD}, and τ_{AD} be the symmetries of $\mathbb{E}^3$ through the planes $ABA'B'$, $BDB'D'$, and $ADA'D'$, respectively. Show that Tet is generated by τ_{AB}, τ_{BD}, and τ_{AD}.

(e) Show that C is generated by τ_{AB}, τ_{BD}, and τ_{AD} and the inversion transformation -1.

(f) Choosing the center of the cube as origin of coordinates, assign the following coordinates to the vertices of the cube:

$$A = \frac{1}{\sqrt{3}}(-1, 1, -1), \quad C = \frac{1}{\sqrt{3}}(1, 1, 1),$$

$$B = \frac{1}{\sqrt{3}}(1, -1, -1), \quad D = \frac{1}{\sqrt{3}}(-1, -1, -1),$$

$$A' = \frac{-1}{\sqrt{3}}(-1, 1, -1), \quad C' = \frac{-1}{\sqrt{3}}(1, 1, 1),$$

$$B' = \frac{-1}{\sqrt{3}}(1, -1, -1), \quad D' = \frac{-1}{\sqrt{3}}(-1, -1, -1).$$

Show that

$$\tau_{AB} = S_{C+D} = S_{(1/\sqrt{2})(1,1,0)},$$

$$\tau_{BD} = S_{A'+C} = S_{(1/\sqrt{2})(1,0,1)},$$

$$\tau_{AD} = S_{B'+C} = S_{(1/\sqrt{2})(0,1,1)}.$$

(g) Using part (f), find the matrix representation for τ_{AB}, τ_{BD}, τ_{AD}, and -1 relative to the coordinate system used in part (f).

(h) Find a matrix representation for the rotation of the cube that corresponds to the vertex permutation (BCD).

18. A *similarity transformation* of Euclidean n-space is any function $f:\mathbb{E}^n \to \mathbb{E}^n$ that maps $\mathbb{E}^n$ onto itself and for which there is some scalar $\lambda \neq 0$ such that $d(f(A), f(B)) = \lambda d(A, B)$ for all vectors $A, B \in \mathbb{E}^n$; that is, the distance between the image of any two points is a fixed scalar multiple of the distance between the points. The scalar λ is called the *ratio* of the transformation f. Let $\mathrm{Sim}(\mathbb{E}^n)$ stand for the set of similarity transformations of $\mathbb{E}^n$.

   (a) Show that $\mathrm{Sim}(\mathbb{E}^n)$ is a group under function composition.
   (b) Show that every similarity transformation of $\mathbb{E}^n$ is an affine transformation and that $\mathrm{Sim}(\mathbb{E}^n) \leq \mathrm{Aff}(\mathbb{E}^n)$.
   (c) Let f be a similarity transformation of $\mathbb{E}^n$ with ratio λ and let $\sigma_\lambda:\mathbb{E}^n \to \mathbb{E}^n$ be the scalar transformation defined by setting $\sigma_\lambda(A) = \lambda A$ for all vectors $A \in \mathbb{E}^n$. Show that $f = \sigma_\lambda g$ for some Euclidean motion g of $\mathbb{E}^n$.
   (d) Let S be a nonempty subset of $\mathbb{E}^n$ and let f be a similarity transformation of $\mathbb{E}^n$. Show that $G_{f(S)} \cong G_S$. Thus, similar figures in $\mathbb{E}^n$ have isomorphic symmetry groups.

19. Let $S = \{(x, y) \in \mathbb{E}^2 \mid y = x^2\}$ stand for the graph of the parabola $y = x^2$. Find the symmetry group G_S.

20. Let $S = \{(x, y) \in \mathbb{E}^2 \mid xy = 1\}$ stand for the graph of the hyperbola $xy = 1$. Find the symmetry group G_S.

TWO

BASIC RING THEORY

9

Rings

In the first part of this book we studied groups, an algebraic structure having a single binary operation. There are many algebraic structures, however, in which elements may be combined using two binary operations rather than one; numbers, matrices and functions, for example, may be added as well as multiplied. While group theory can deal with each of these operations separately, it is not designed to handle both operations in the same structure. What is needed is a new structure having two binary operations, each with its own individual properties yet at the same time interacting with each other to give additional properties. This new structure is called a ring, a concept that evolved out of late nineteenth-century attempts to extend properties of the integers to larger and more complex types of number systems. Today, ring theory not only provides the basic algebraic model for various types of number systems, including matrices, but has also become an essential part of the algebraic study of curves and surfaces. In the second part of this book our goal is to introduce the reader to the basic concepts of ring theory.

In this chapter we define and illustrate the basic concepts of ring theory: rings and subrings, ideals and quotient rings, ring-homomorphisms and ring-isomorphisms. We will introduce several important types of rings, including the ring of ordinary integers, the ring of Gaussian integers, and the ring of intergers modulo n, as well as matrix rings, rings of functions, and the ring of quaternions, discuss the properties of these rings, and then establish a few properties that are common to all rings. We then turn our attention to a special type of subring called an ideal. Ideals are analogous to normal subgroups in group theory in the sense that they are used to construct quotient rings. As with groups we first discuss cosets of a ring modulo an ideal, and then use the cosets of an ideal to define and illustrate quotient rings.

Finally, we conclude the chapter by discussing mappings from one ring to another that preserve the binary operations on the rings. These mappings, called ring-homomorphisms, are analogous to group-homomorphisms in that they allow us to transfer information from one ring to another in such a way that the algebraic properties of the ring are preserved. As with group-homomorphisms, we will show that every ring-homomorphism determines a ring-isomorphism between the image of the homomorphism and the quotient ring modulo its kernel. This result, which is called the fundamental theorem of ring-homomorphism, shows that quotient rings may therefore be interpreted as images of ring-homomorphisms.

As these remarks indicate, many of the fundamental concepts of ring theory are already familiar to us from group theory. Indeed, part of our goal in this chapter is to reformulate the basic algebraic concepts of group theory within the framework of ring theory.

1. RINGS AND THEIR ARITHMETIC PROPERTIES

Up to this point in our study of algebra we have discussed four basic number systems: the integers, rational numbers, real numbers, and complex numbers. In each of these number systems there are two binary operations, addition and multiplication, and the number systems, together with these operations, have three basic properties in common: first, each of the number systems is an abelian group under addition; second, multiplication is an associative binary operation; and third, multiplication is related to addition by means of the distributive law. The set of $n \times n$ matrices with real number entries also forms an abelian group under addition for a given positive integer n, and has a multiplication operation that is associative and which distributes over addition. With these observations in mind, we now make the following definition.

> **Definition**
>
> A *ring* is a set R with two binary operations, addition and multiplication—the sum and product of two elements x and y being written $x + y$ and xy, respectively—such that the following three conditions are satisfied:
>
> **(1)** R is an abelian group under addition;
>
> **(2)** multiplication is associative; that is, $x(yz) = (xy)z$ for all elements x, y, $z \in R$;
>
> **(3)** multiplication distributes over addition; that is, $x(y + z) = xy + xz$ and $(x + y)z = xz + yz$ for all elements $x, y, z \in R$.

If R is a ring, an *identity* for R is any element $e \in R$ such that $ex = xe = x$ for all elements $x \in R$. If a ring R has an identity, then it is easy to see that it has only one such element; for if both e and e' are identities for R, then $e = ee' = e'$. In this case we denote the unique identity element by the symbol

1_R, or simply 1 if there is no confusion about the underlying ring R. A *commutative ring* is a ring in which multiplication is commutative. Thus, R is a commutative ring if, in addition to the ring axioms listed above, $xy = yx$ for all elements $x, y \in R$. The reader should note that the terms "identity" and "commutative," when applied to a ring, refer to multiplication, not addition. And finally, a *subring* of a ring R is any nonempty subset that inherits both addition and multiplication and is a ring under these operations. Clearly, if S is a nonempty subset of a ring R, then S is a subring of R if and only if the following three conditions are satisfied:

(1) closure under addition: if $x, y \in S$, then $x + y \in S$;
(2) closure under products: if $x, y \in S$, then $xy \in S$;
(3) closure under additive inverses: if $x \in S$, then $-x \in S$.

For example, consider the set $\mathbb{Z}$ of integers under addition and multiplication. Since $\mathbb{Z}$ is an abelian group under addition and since multiplication is associative and distributes over addition, it follows that $\mathbb{Z}$ is a ring under addition and multiplication. Moreover, since multiplication of integers is commutative and the number 1 is an identity for multiplication, $\mathbb{Z}$ is in fact a commutative ring with identity. Similarly, the set $\mathbb{Q}$ of rational numbers, the set $\mathbb{R}$ of real numbers, and the set $\mathbb{C}$ of complex numbers are all commutative rings with identity.

Now, let $2\mathbb{Z} = \{2n \in \mathbb{Z} \mid n \in \mathbb{Z}\}$ stand for the subset of even integers. Then $2\mathbb{Z}$ is a subring of $\mathbb{Z}$; for if x and y are typical even integers, then the sum $x + y$, the product xy, and the inverse $-x$ are also even integers and hence elements of $2\mathbb{Z}$. Thus, by the three conditions listed above, $2\mathbb{Z}$ is a subring of $\mathbb{Z}$. Similarly, $\mathbb{Z}$ is a subring of $\mathbb{Q}$, $\mathbb{Q}$ is a subring of $\mathbb{R}$, and $\mathbb{R}$ is a subring of $\mathbb{C}$. Note, however, that although $2\mathbb{Z}$ is a commutative subring of $\mathbb{Z}$, it does not have an identity element since $1 \notin 2\mathbb{Z}$. Thus it is possible for a ring with identity to have a subring that does not have an identity; in fact, we will give an example below of a ring with identity having a subring with identity, but the identity of the subring is different from that of the ring itself.

EXAMPLE 1. The ring of Gaussian integers

A *Gaussian integer* is a complex number of the form $n + mi$, where $n, m \in \mathbb{Z}$. Let $\mathbb{Z}[i] = \{n + mi \in \mathbb{C} \mid n, m \in \mathbb{Z}\}$ stand for the set of Gaussian integers. We claim that $\mathbb{Z}[i]$ is a commutative ring with identity under addition and multiplication of complex numbers. To show this, let $x = n + mi$ and $y = s + ti$ be typical Gaussian integers. Then

$$x + y = (n + mi) + (s + ti) = (n + s) + (m + t)i \in \mathbb{Z}[i],$$

$$xy = (n + mi)(s + ti) = (ns - mt) + (nt + ms)i \in \mathbb{Z}[i],$$

$$-x = -(n + mi) = (-n) + (-m)i \in \mathbb{Z}[i].$$

Therefore $\mathbb{Z}[i]$ is a subring of $\mathbb{C}$ and hence is a ring under addition and multiplication of complex numbers. Since multiplication of complex numbers is commutative, $\mathbb{Z}[i]$ inherits commutativity from $\mathbb{C}$, and since the integer $1 = 1 + 0i$ is a Gaussian integer, it is clearly the identity for $\mathbb{Z}[i]$. Thus $\mathbb{Z}[i]$ is a commutative ring with identity. The ring $\mathbb{Z}[i]$ is called the *ring of Gaussian integers*. Note, finally, that $\mathbb{Z}$ is a subring of $\mathbb{Z}[i]$.

EXAMPLE 2. The ring of integers modulo *n*

Let n be a positive integer and let $\mathbb{Z}_n = \{[s] \mid s \in \mathbb{Z}\}$ stand for the set of congruence classes of integers modulo n. Recall that $\mathbb{Z}_n$ has two binary operations defined on it: if $[s], [t] \in \mathbb{Z}_n$, addition of classes is defined by setting $[s] + [t] = [s + t]$, while multiplication of classes is defined by $[s][t] = [st]$. We claim that $\mathbb{Z}_n$ is a commutative ring with identity under addition and multiplication of congruence classes. To show this, recall that $\mathbb{Z}_n$ is an abelian group under addition. It remains to show, therefore, that multiplication is associative, commutative, has an identity, and distributes over addition. To this end, let $[s], [t], [r] \in \mathbb{Z}_n$. Then

$$[s]([t][r]) = [s][tr] = [s(tr)] = [(st)r] = [st][r] = ([s][t])[r],$$

$$[s][t] = [st] = [ts] = [t][s],$$

$$[s]([t] + [r]) = [s][t + r] = [s(t + r)] = [st + sr] = [s][t] + [s][r].$$

Thus, multiplication of classes is associative, commutative, and distributes over addition, and therefore $\mathbb{Z}_n$ is a commutative ring. Note that since multiplication is commutative, we only needed to verify one of the two requirements for distributivity. Finally, since $[1][s] = [s][1] = [s]$ for all $[s] \in \mathbb{Z}_n$, it follows that $[1]$ is the identity for $\mathbb{Z}_n$ and therefore $\mathbb{Z}_n$ is a commutative ring with identity. We call $\mathbb{Z}_n$ the *ring of integers modulo n*. Observe that $\mathbb{Z}_n$ is an example of a finite ring since it contains only n elements, namely $[0]$, $[1], \ldots, [n-1]$. For example, $\mathbb{Z}_3 = \{[0], [1], [2]\}$, where all calculations are performed modulo 3, and $\mathbb{Z}_4 = \{[0], [1], [2], [3]\}$, where all calculations are modulo 4. Figures 1 and 2 illustrate the addition and multiplication tables for the rings $\mathbb{Z}_3$ and $\mathbb{Z}_4$.

$\mathbb{Z}_3 = \{[0], [1], [2]\}$							
+	[0]	[1]	[2]	$\cdot$	[0]	[1]	[2]
[0]	[0]	[1]	[2]	[0]	[0]	[0]	[0]
[1]	[1]	[2]	[0]	[1]	[0]	[1]	[2]
[2]	[2]	[0]	[1]	[2]	[0]	[2]	[1]

FIGURE 1. Addition and multiplication tables for the ring $\mathbb{Z}_3$ of integers modulo 3.

$\mathbb{Z}_4 = \{[0], [1], [2], [3]\}$									
+	[0]	[1]	[2]	[3]	$\cdot$	[0]	[1]	[2]	[3]
[0]	[0]	[1]	[2]	[3]	[0]	[0]	[0]	[0]	[0]
[1]	[1]	[2]	[3]	[0]	[1]	[0]	[1]	[2]	[3]
[2]	[2]	[3]	[0]	[1]	[2]	[0]	[2]	[0]	[2]
[3]	[3]	[0]	[1]	[2]	[3]	[0]	[3]	[2]	[1]

FIGURE 2. Addition and multiplication tables for the ring $\mathbb{Z}_4$ of integers modulo 4.

EXAMPLE 3. The matrix rings

Let n be a positive integer and let $\mathrm{Mat}_n(\mathbb{R})$ stand for the set of $n \times n$ matrices with real number entries. Then matrix addition and matrix multiplication are binary operations on $\mathrm{Mat}_n(\mathbb{R})$. Since $\mathrm{Mat}_n(\mathbb{R})$ is an abelian group under matrix addition and since matrix multiplication is associative and distributes over addition, it follows that $\mathrm{Mat}_n(\mathbb{R})$ is a ring under these two operations. We refer to these rings as *matrix rings*. Let us discuss some of the properties of matrix rings in more detail.

(A) Every matrix ring has an identity. Clearly, the $n \times n$ identity matrix I_n is the identity for $\mathrm{Mat}_n(\mathbb{R})$.

(B) If $n = 1$, $\mathrm{Mat}_1(\mathbb{R}) = \{(a) \,|\, a \in \mathbb{R}\}$ consists of all 1×1 real matrices and hence is a commutative ring since $(a)(b) = (ab) = (ba) = (b)(a)$ for all $a, b \in \mathbb{R}$. If $n > 1$, however, $\mathrm{Mat}_n(\mathbb{R})$ is a noncommutative ring since there is always a pair of $n \times n$ matrices that do not commute; for example, in $\mathrm{Mat}_2(\mathbb{R})$ we find that $\left(\begin{smallmatrix} 1 & 0 \\ 0 & 0 \end{smallmatrix}\right)\left(\begin{smallmatrix} 0 & 1 \\ 0 & 0 \end{smallmatrix}\right) = \left(\begin{smallmatrix} 0 & 1 \\ 0 & 0 \end{smallmatrix}\right)$, while $\left(\begin{smallmatrix} 0 & 1 \\ 0 & 0 \end{smallmatrix}\right)\left(\begin{smallmatrix} 1 & 0 \\ 0 & 0 \end{smallmatrix}\right) = \left(\begin{smallmatrix} 0 & 0 \\ 0 & 0 \end{smallmatrix}\right)$.

(C) Let $S = \{\left(\begin{smallmatrix} a & 0 \\ 0 & a \end{smallmatrix}\right) \in \mathrm{Mat}_2(\mathbb{R}) \,|\, a \in \mathbb{R}\}$ stand for the subset of 2×2 scalar matrices in $\mathrm{Mat}_2(\mathbb{R})$. We claim that S is a commutative subring of $\mathrm{Mat}_2(\mathbb{R})$ with identity. To show this, let $A = \left(\begin{smallmatrix} a & 0 \\ 0 & a \end{smallmatrix}\right)$ and $B = \left(\begin{smallmatrix} b & 0 \\ 0 & b \end{smallmatrix}\right)$ be typical matrices in S. Then

$$A + B = \begin{pmatrix} a + b & 0 \\ 0 & a + b \end{pmatrix} \in S,$$

$$AB = \begin{pmatrix} ab & 0 \\ 0 & ab \end{pmatrix} \in S,$$

$$-A = \begin{pmatrix} -a & 0 \\ 0 & -a \end{pmatrix} \in S.$$

Therefore, S is a subring of $\mathrm{Mat}_2(\mathbb{R})$. Moreover, since

$$BA = \begin{pmatrix} ba & 0 \\ 0 & ba \end{pmatrix} = \begin{pmatrix} ab & 0 \\ 0 & ab \end{pmatrix} = AB,$$

it follows that S is commutative. Finally, the 2×2 identity matrix $I_2 = \left(\begin{smallmatrix} 1 & 0 \\ 0 & 1 \end{smallmatrix}\right)$ is an element of S and hence is the identity of S. Thus S is a commutative subring of $\mathrm{Mat}_2(\mathbb{R})$ with identity.

(D) Let $T = \{\left(\begin{smallmatrix} a & 0 \\ 0 & 0 \end{smallmatrix}\right) \in \mathrm{Mat}_2(\mathbb{R}) \,|\, a \in \mathbb{R}\}$. We claim that T is a commutative subring of $\mathrm{Mat}_2(\mathbb{R})$ with identity, but the identity of T is not the 2×2 identity matrix. For if we let $A = \left(\begin{smallmatrix} a & 0 \\ 0 & 0 \end{smallmatrix}\right)$ and $B = \left(\begin{smallmatrix} b & 0 \\ 0 & 0 \end{smallmatrix}\right)$ stand for typical matrices in T, it follows, as in part (C), that $A + B$, AB, and $-A$ are also in T, and that $AB = BA$. Hence T is a commutative subring of $\mathrm{Mat}_2(\mathbb{R})$. But T does not contain the identity matrix I_2 of $\mathrm{Mat}_2(\mathbb{R})$. Nevertheless, the ring T has an identity, namely the matrix $\left(\begin{smallmatrix} 1 & 0 \\ 0 & 0 \end{smallmatrix}\right)$, since $\left(\begin{smallmatrix} 1 & 0 \\ 0 & 0 \end{smallmatrix}\right)\left(\begin{smallmatrix} a & 0 \\ 0 & 0 \end{smallmatrix}\right) = \left(\begin{smallmatrix} a & 0 \\ 0 & 0 \end{smallmatrix}\right) = \left(\begin{smallmatrix} a & 0 \\ 0 & 0 \end{smallmatrix}\right)\left(\begin{smallmatrix} 1 & 0 \\ 0 & 0 \end{smallmatrix}\right)$ for all $a \in \mathbb{R}$. Therefore T is a commutative subring of $\mathrm{Mat}_2(\mathbb{R})$ with identity, but the identity of T is not the same as that of the ring $\mathrm{Mat}_2(\mathbb{R})$.

EXAMPLE 4. The ring of quaternions

Let $\mathrm{Mat}_2(\mathbb{C})$ stand for the set of 2×2 matrices with complex number entries. Then it follows as in Example 3 that $\mathrm{Mat}_2(\mathbb{C})$ is a noncommutative ring with identity I_2 under matrix addition and multiplication. Now, let

$$x = \begin{pmatrix} i & 0 \\ 0 & -i \end{pmatrix}, \quad y = \begin{pmatrix} 0 & i \\ i & 0 \end{pmatrix}, \quad z = xy = \begin{pmatrix} 0 & -1 \\ 1 & 0 \end{pmatrix},$$

and let $\mathbb{H} = \{aI_2 + bx + cy + dz \in \mathrm{Mat}_2(\mathbb{C}) \,|\, a, b, c, d \in \mathbb{R}\}$ stand for the subset of all linear combinations of the matrices I_2, x, y, z, with coefficients in the ring $\mathbb{R}$ of real numbers. We claim that $\mathbb{H}$ is a noncommutative ring with identity under matrix addition and multiplication.

(A) To show that $\mathbb{H}$ is a ring, let $\alpha = aI_2 + bx + cy + dz$ and $\beta = a'I_2 + b'x + c'y + d'z$ stand for arbitrary matrices in $\mathbb{H}$, where $a, b, c, d, a', b', c', d' \in \mathbb{R}$. Then, using standard matrix properties, we find that

$$\alpha + \beta = (a + a')I_2 + (b + b')x + (c + c')y + (d + d')z$$

$$-\alpha = (-a)I_2 + (-b)x + (-c)y + (-d)z.$$

Therefore, $\alpha + \beta \in \mathbb{H}$ and $-\alpha \in \mathbb{H}$. Now, to show that the product $\alpha\beta$ is also in $\mathbb{H}$, we need only verify that the nine products x^2, xy, xz, yx, y^2, yz, zx, zy, and z^2 lie in $\mathbb{H}$ since $\alpha\beta$ is a linear combination of these nine matrices and $\mathbb{H}$ is clearly closed under linear combinations. We find, for example, that

$$x^2 = \begin{pmatrix} i & 0 \\ 0 & -i \end{pmatrix}\begin{pmatrix} i & 0 \\ 0 & -i \end{pmatrix} = \begin{pmatrix} -1 & 0 \\ 0 & -1 \end{pmatrix} = (-1)I_2 \in \mathbb{H}$$

$$yx = \begin{pmatrix} 0 & i \\ i & 0 \end{pmatrix}\begin{pmatrix} i & 0 \\ 0 & -i \end{pmatrix} = \begin{pmatrix} 0 & 1 \\ -1 & 0 \end{pmatrix} = (-1)z \in \mathbb{H}.$$

Similarly, $y^2 = z^2 = (-1)I_2$, $zx = y$, $xz = -y$, $yz = x$, and $zy = -x$. Thus the nine products lie in $\mathbb{H}$ and therefore $\alpha\beta \in \mathbb{H}$. It now follows that $\mathbb{H}$ is a subring of $\mathrm{Mat}_2(\mathbb{C})$ and hence is a ring under matrix addition and multiplication. The ring $\mathbb{H}$ is called the *ring of quaternions*. Observe, finally, that the ring $\mathbb{H}$ of quaternions has the identity matrix I_2 as its identity, and is a noncommutative ring since $xy = z$, for example, while $yx = -z$.

(B) Here is an easy way to calculate products in the ring of quaternions. Place the symbols x, y, and z on a circle as indicated in the marginal figure. Then the calculations in part (A) show that the product of any two symbols on the circle is the third one, with a $+$ sign if the rotation is clockwise and a $-$ sign if it is counterclockwise. Then, to calculate the product of two linear combinations of I_2, x, y, and z, simply use distributivity. For example, let $\alpha = 2I_2 - x + y$ and $\beta = x + z$. Then

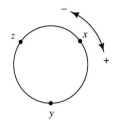

$$\alpha\beta = (2I_2 - x + y)(x + z) = 2x + 2z - x^2 - xz + yx + yz$$
$$= 2x + 2z - (-I_2) - (-y) + (-z) + x = I_2 + 3x + y + z.$$

EXAMPLE 5. Rings of Functions

Let R be a ring and let $F(R)$ stand for the set of functions mapping R to R. If $f, g \in F(R)$, define the sum $f + g : R \to R$ by setting $(f + g)(x) = f(x) + g(x)$ for all $x \in R$ and the product $fg : R \to R$ by setting $(fg)(x) = f(x)g(x)$ for all $x \in R$. Then the sum and product of functions are binary operations on $F(R)$ and it is easily verified that $F(R)$ is a ring under these operations. $F(R)$ is called the *ring of functions on R*. Let us make a few general observations about the ring $F(R)$ and then discuss in detail the ring $F(\mathbb{Z}_2)$, where $\mathbb{Z}_2$ is the ring of integers modulo 2.

(A) The zero element in $F(R)$ is the function $f_0 : R \to R$ that maps every element in R to the zero element: $f_0(x) = 0$ for all $x \in R$. To show this, we need only note that if f is an arbitrary function in $F(R)$, then $(f + f_0)(x) = f(x) + f_0(x) = f(x) = (f_0 + f)(x)$ for all $x \in R$ and hence $f + f_0 = f_0 + f = f$. Therefore f_0 is the zero element of $F(R)$.

(B) If $f \in F(R)$, the additive inverse of f is the function $-f : R \to R$ defined by setting $(-f)(x) = -f(x)$ for all $x \in R$. For, as in part (A), it is easily verified that $f + (-f) = (-f) + f = f_0$.

(C) We emphasize that the multiplication operation on $F(R)$ is not function composition. It is, instead, the pointwise product of function values in R. Consequently the identity function on R is not the identity element for multiplication. If R is a ring with identity 1, then the constant function $f_1 : R \to R$ defined by setting $f_1(x) = 1$ for all $x \in R$ is clearly the identity for $F(R)$. If R does not have an identity, we leave the reader to show that $F(R)$ does not have an identity. Thus, the ring of functions $F(R)$ has an identity if and only if R has an identity.

(D) Finally, if the ring R is commutative, then so is $F(R)$. For if R is commutative and $f, g \in F(R)$, then $(fg)(x) = f(x)g(x) = g(x)f(x) = (gf)(x)$ for all $x \in R$ and hence $fg = gf$. Therefore $F(R)$ is commutative if R is commutative.

(E) Let us illustrate the preceding discussion by describing in detail the ring of functions $F(\mathbb{Z}_2)$, where $\mathbb{Z}_2 = \{[0], [1]\}$ is the ring of integers modulo 2. In this case there are four functions mapping $\mathbb{Z}_2$ to $\mathbb{Z}_2$, as shown here.

$F(\mathbb{Z}_2) = (f_0, f_1, g, h)$									
$+$	f_0	f_1	g	h	$\cdot$	f_0	f_1	g	h
f_0	f_0	f_1	g	h	f_0	f_0	f_0	f_0	f_0
f_1	f_1	f_0	h	g	f_1	f_0	f_1	g	h
g	g	h	f_0	f_1	g	f_0	g	g	f_0
h	h	g	f_1	f_0	h	f_0	h	f_0	h

$f_0 : \mathbb{Z}_2 \to \mathbb{Z}_2$	$f_1 : \mathbb{Z}_2 \to \mathbb{Z}_2$	$g : \mathbb{Z}_2 \to \mathbb{Z}_2$	$h : \mathbb{Z}_2 \to \mathbb{Z}_2$
$f_0([0]) = [0]$	$f_1([0]) = [1]$	$g([0]) = [0]$	$h([0]) = [1]$
$f_0([1]) = [0]$	$f_1([1]) = [1]$	$g([1]) = [1]$	$h([1]) = [0]$

To calculate the product gh, for example, we find that $(gh)([0]) = g([0])h([0]) = [0][1] = [0]$ and $(gh)([1]) = g([1])h([1]) = [1][0] = [0]$. Hence $gh = f_0$. The remaining sums and products are found similarly. Observe that

$$-f_1 = f_1, \quad -g = g, \quad -h = h,$$
$$f_1^2 = f_1, \quad g^2 = g, \quad h^2 = h.$$

It follows, therefore, that $-x = x$ and $x^2 = x$ for all $x \in F(\mathbb{Z}_2)$. A ring having this property is called a *Boolean ring*. In the exercises we indicate how the power set $\mathscr{P}(X)$ of all subsets of a given set X may be made into a ring using the union and intersection operations, and ask the reader to show that $\mathscr{P}(X)$ is also a Boolean ring. There is, in fact, an important relationship between the ring $\mathscr{P}(X)$ and the ring of functions discussed in this example which we discuss in more detail in the next section. This concludes the example.

Let us now turn our attention to the basic arithmetic properties common to all rings. Let R be a ring. Then R is an abelian group under addition and, as such, contains a unique zero element which we denote by 0. Moreover, every element $x \in R$ has a unique additive inverse, denoted by $-x$.

PROPOSITION 1. Let R be a ring. Then $0x = x0 = 0$ for every element $x \in R$; that is, the product of the zero element with any element of a ring is always the zero element.

Proof.   Let $x \in R$. Then $0x = (0 + 0)x = 0x + 0x$ and therefore $0x = 0$ since R is a group under addition. Similarly, $x0 = 0$.   ∎

Now, since addition and multiplication are associative binary operations on R, it follows from our discussion in Chapter 2, Section 1, that addition and multiplication satisfy general associativity; that is, the sum or product of any finite number of elements in R may be calculated by inserting parentheses in any meaningful way. Similarly, if multiplication in R is commutative, then multiplication satisfies general commutativity; in this case the product of any finite number of elements in R is the same regardless of how the factors are rearranged. Let us summarize these results and show, additionally, that there is general distributivity in a ring.

PROPOSITION 2.   Let R be a ring. Then:

(1)   addition and multiplication satisfy general associativity;

(2)   addition and multiplication satisfy general distributivity; that is, if $x, y_1, \ldots, y_n$ are any finite number of elements in R, then

$$x(y_1 + \cdots + y_n) = xy_1 + \cdots + xy_n$$

$$(y_1 + \cdots + y_n)x = y_1 x + \cdots + y_n x;$$

(3)   if R is commutative, then multiplication satisfies general commutativity.

Proof.   It remains to prove general distributivity. Let us show that $x(y_1 + \cdots + y_n) = xy_1 + \cdots + xy_n$ for any finite number of elements $x, y_1, \ldots, y_n \in R$ by using induction on n, and leave the proof of the second equation in (2) as an exercise for the reader. If $n = 1$ the equation is trivial, while if $n = 2$ it is true since multiplication distributes over addition in R. Now suppose that $n > 2$ and that $x(y_1 + \cdots + y_n) = xy_1 + \cdots + xy_n$ for every $x \in R$ and any n elements $y_1, \ldots, y_n \in R$. Then if $x \in R$ and $y_1, \ldots, y_{n+1}$ are any $n + 1$ elements in R, we have that

$$x(y_1 + \cdots + y_{n+1}) = x[(y_1 + \cdots + y_n) + y_{n+1}] = x(y_1 + \cdots + y_n) + xy_{n+1}.$$

Since this last sum is equal to $xy_1 + \cdots + xy_n + xy_{n+1}$ by the induction hypothesis, it follows that $x(y_1 + \cdots + y_{n+1}) = xy_1 + \cdots + xy_{n+1}$. We conclude, therefore, that addition and multiplication in a ring satisfy general distributivity.   ∎

Let R be a ring and let x be an element in R. If n is a positive integer, we let $nx = x + \cdots + x$ and $x^n = x \cdots x$ stand for the sum and product of x with itself n times; note that these are well-defined elements in R since addition and multiplication satisfy general associativity. If $n = 0$, set $0x = 0$, the zero element in R. And finally, if n is negative, we set $nx = (-n)(-x)$ $= -x + \cdots + -x$, where $-x$ stands for the unique additive inverse of x. Note, in particular, that $(-1)x = -x$. Let us emphasize that the symbol nx is simply a convenient way to represent the repeated sum of an element in R and is not, in general, the product of two elements in R. For example, in the ring $\mathbb{Z}_6$, we find that $2[4] = [4] + [4] = [8] = [2]$ and $-3[4] = 3[-4]$ $= 3[2] = [2] + [2] + [2] = [0]$. Let us also mention that the symbol x^n cannot always be defined in a ring when $n \leq 0$ since the ring may not have an identity and the element x may not have a multiplicative inverse. We will discuss this point in more detail shortly. For the moment, however, let us summarize the basic arithmetic properties of the symbols nx and x^n.

PROPOSITION 3. Let R be a ring and let x and y be elements in R. Then, for any integers n and m, we have that

(1) $n(x + y) = nx + ny$;
(2) $(n + m)x = nx + mx$;
(3) $n(mx) = (nm)x$;
(4) $(nx)y = x(ny) = n(xy)$.

If in addition R is commutative and $n > 0$, then $(xy)^n = x^n y^n$ and

$$(x + y)^n = x^n + \binom{n}{1}x^{n-1}y + \cdots + \binom{n}{k}x^{n-k}y^k + \cdots + y^n,$$

where $\binom{n}{k} = \dfrac{n!}{k!(n-k)!}$ stands for the binomial coefficient for $k = 1, \ldots, n$.

Proof. Statements (1)–(3) follow from the fact that R is an abelian group under addition, while statement (4) and the last statement are easily proved by induction. We leave the details as an exercise for the reader. ∎

These results summarize the basic arithmetic properties common to all rings. Let us make two observations about these results. First, it follows from statements (3) and (4) of Proposition 3 that $(nx)(my) = (nm)xy$ for all elements x, y in a ring and all integers n, m. In particular, since $(-1)x = -x$, $(-x)(-y) = xy$. And second, observe that the last statement in Proposition 3 is just the usual binomial theorem for expanding $(x + y)^n$, where n is a positive integer. The binomial theorem is therefore a valid method for obtaining powers of a sum in any commutative ring.

We mentioned earlier that the symbol x^n cannot always be defined in a ring when $n \leq 0$. The problem, of course, is that the ring may not have an identity—which is what we want x^0 to equal—and the element x may not have a multiplicative inverse—which is what we want x^{-1} to equal. If R is a ring with identity 1 we set $x^0 = 1$ for all elements $x \in R$. For inverses, we make the following definition.

Definition

Let R be a ring with identity 1 and let x be an element in R. A *multiplicative inverse* for x is any element $y \in R$ such that $xy = yx = 1$. An element in R that has a multiplicative inverse is called a *unit* of R. We let $U(R)$ stand for the set of units of R.

It is clear that every ring with identity has at least one unit, namely the identity element, and that if x has a multiplicative inverse, then it has only one such inverse. In this case we denote the unique multiplicative inverse of x by x^{-1}. If x has a multiplicative inverse x^{-1} and n is any negative integer, we set $x^n = (x^{-1})^{-n}$. If x is not a unit of R, however, the symbol x^n is undefined whenever n is negative. With these conventions, it follows easily by induction that if x is a unit of R, then $x^n x^m = x^{n+m}$ and $(x^n)^m = x^{nm}$ for all integers n and m. If, in addition, R is commutative, then $(xy)^n = x^n y^n$ for all units x, y and all integers n. Let us emphasize that the concept of a unit is defined relative to a particular ring. For example, in the ring $\mathbb{Z}$ of integers, the only integers that have a multiplicative inverse are ± 1 and hence $U(\mathbb{Z}) = \{1, -1\}$; we find that $1^{-1} = 1$ and $(-1)^{-1} = -1$. In the ring $\mathbb{Q}$ of rational numbers, however, every nonzero element is a unit and hence $U(\mathbb{Q}) = \mathbb{Q}^*$, the set of nonzero rational numbers. Similarly, $U(\mathbb{R}) = \mathbb{R}^*$ and $U(\mathbb{C}) = \mathbb{C}^*$. Thus it is meaningless to say that the number 2, for example, is a unit, without any reference to the underlying ring; for although it is a unit of the rings $\mathbb{Q}$, $\mathbb{R}$, and $\mathbb{C}$, it is not a unit of the ring $\mathbb{Z}$.

We conclude our discussion of the basic properties of rings by showing that in a ring with identity, the units form a group under multiplication.

PROPOSITION 4. Let R be a ring with identity. Then $U(R)$ is a group under multiplication in R.

Proof. Let 1 stand for the identity of R and let x, $y \in U(R)$. Then x^{-1} and y^{-1} exist in R and $(xy)(y^{-1}x^{-1}) = (y^{-1}x^{-1})(xy) = 1$. Therefore xy has a multiplicative inverse and hence is a unit of R. It follows, therefore, that multiplication is a binary operation on $U(R)$. Moreover, multiplication is associative since it inherits associativity from multiplication on R. Since $1 \in U(R)$, 1 is the identity for multiplication on $U(R)$. And finally, to show that every element in $U(R)$ has a multiplicative inverse, let $x \in U(R)$. Then $xx^{-1} = x^{-1}x = 1$ and hence $(x^{-1})^{-1} = x$. Therefore the multiplicative inverse x^{-1} is itself a unit of R and hence an element of $U(R)$. It now follows that $U(R)$ is a group under multiplication. ∎

The group $U(R)$ is called the *group of units* of R. Clearly, if R is a commutative ring with identity, $U(R)$ is an abelian group; for example, $U(\mathbb{Z}) = \{1, -1\}$ is a cyclic group of order 2, while $U(\mathbb{Q}) = \mathbb{Q}^*$, the multiplicative group of non-zero rational numbers, $U(\mathbb{R}) = \mathbb{R}^*$, the multiplicative group of nonzero real numbers, and $U(\mathbb{C}) = \mathbb{C}^*$, the multiplicative group of nonzero complex numbers. If the ring R is noncommutative, however, its group of units may or may not be commutative. In the ring $\mathbb{H}$ of quaternions, for example, the elements x and y are both units since $(-x)(x) = 1$ and $(-y)(y) = 1$, but $xy \neq yx$.

$U(\mathbb{Z}_8)$	[1]	[3]	[5]	[7]
[1]	[1]	[3]	[5]	[7]
[3]	[3]	[1]	[7]	[5]
[5]	[5]	[7]	[1]	[3]
[7]	[7]	[5]	[3]	[1]

EXAMPLE 6.   The group of units of the ring $\mathbb{Z}_n$

Let n be a positive integer. Recall that a congruence class $[s] \in \mathbb{Z}_n$ has a multiplicative inverse if and only if s and n are relatively prime (Example 7, Section 1, Chapter 4). It follows, therefore, that $U(\mathbb{Z}_n) = \{[s] \in \mathbb{Z}_n \,|\, gcd(s, n) = 1\}$. In this case $U(\mathbb{Z}_n)$ is a finite abelian group of order $\varphi(n)$, where φ stands for the Euler phi-function. For example, $U(\mathbb{Z}_3) = \{[1], [2]\}$ is a cyclic group of order 2 generated by $[2]$ and hence $U(\mathbb{Z}_3) \cong \mathbb{Z}_2$. Similarly, $U(\mathbb{Z}_4) = \{[1], [3]\}$ $= ([3]) \cong \mathbb{Z}_2$ and $U(\mathbb{Z}_5) = \{[1], [2], [3], [4]\} = ([2]) \cong \mathbb{Z}_4$. And finally, let us consider the group $U(\mathbb{Z}_8)$ of units of the ring $\mathbb{Z}_8$. Here we find that $U(\mathbb{Z}_8) = \{[1], [3], [5], [7]\}$. This group is not cyclic, however, since each of its non-identity elements has order 2: $[3]^2 = [9] = [1]$, $[5]^2 = [25] = [1]$, and $[7]^2 = [49] = [1]$. In fact, $U(\mathbb{Z}_8) \cong \{[1], [3]\} \times \{[1], [5]\} \cong \mathbb{Z}_2 \times \mathbb{Z}_2$ (Example 1, Section 4 Chapter 4).

Exercises

1. Below are several subsets of the ring $\mathbb{Q}$ of rational numbers. In each case determine if the subset is a subring, and, if so, whether or not it has an identity.

   (a) $\left\{ \dfrac{n}{2^m} \in \mathbb{Q} \,\middle|\, n, m \in \mathbb{Z}, m \geq 0 \right\}$

   (b) $\left\{ \dfrac{2^n}{m} \in \mathbb{Q} \,\middle|\, n, m \in \mathbb{Z}, n \geq 0, m \neq 0 \right\}$

   (c) $\left\{ \dfrac{n}{m} \in \mathbb{Q} \,\middle|\, n, m \in \mathbb{Z}, m \text{ odd} \right\}$

2. Below are several subsets of the ring $\mathbb{R}$ of real numbers. In each case determine if the subset is a subring, and, if so, whether or not it has an identity.
   (a) $\{n\sqrt{2} \in \mathbb{R} \,|\, n \in \mathbb{Z}\}$
   (b) $\{x \in \mathbb{R} \,|\, x \text{ is irrational}\}$
   (c) $\{x \in \mathbb{R} \,|\, |x| \geq 1 \text{ or } x = 0\}$
3. Below are several subsets of the ring $\mathbb{C}$ of complex numbers. In each case determine if the subset is a subring, and, if so, whether or not it has an identity.
   (a) $\{a + bi \in \mathbb{C} \,|\, a, b \in \mathbb{Q}\}$

(b) $\{a + bi \in \mathbb{C} \mid a \in \mathbb{Z}, b \in \mathbb{R}\}$

(c) $\{bi \in \mathbb{C} \mid b \in \mathbb{R}\}$

4. Below are several subsets of the matrix ring $\mathrm{Mat}_2(\mathbb{R})$. In each case determine if the subset is a subring; if it is a subring, determine if it is commutative or not, and whether or not it has an identity.

   (a) $\{\begin{pmatrix} a & 0 \\ 0 & b \end{pmatrix} \in \mathrm{Mat}_2(\mathbb{R}) \mid a, b \in \mathbb{R}\}$

   (b) $\{\begin{pmatrix} a & b \\ 0 & c \end{pmatrix} \in \mathrm{Mat}_2(\mathbb{R}) \mid a, b, c \in \mathbb{R}\}$

   (c) $\{A \in \mathrm{Mat}_2(\mathbb{R}) \mid \det(A) = 0\}$

5. Below are several subsets of the ring $\mathbb{H}$ of quaternions. In each case determine if the subset is a subring; if it is a subring, determine if it is commutative or not, and whether or not it has an identity.

   (a) $\{aI_2 \in \mathbb{H} \mid a \in \mathbb{R}\}$

   (b) $\{aI_2 + bx + cy + dz \in \mathbb{H} \mid a, b, c, d \in \mathbb{Z}\}$

   (c) $\{bx + cy + dz \in \mathbb{H} \mid b, c, d \in \mathbb{R}\}$

6. Let n be a positive integer.

   (a) Show that the subrings of the ring $\mathbb{Z}_n$ of integers modulo n are just the subgroups of the additive group of $\mathbb{Z}_n$.

   (b) Determine all subrings of $\mathbb{Z}_6$.

7. Let $\mathbb{Z}[\sqrt{2}] = \{n + m\sqrt{2} \in \mathbb{R} \mid n, m \in \mathbb{Z}\}$.

   (a) Show that $\mathbb{Z}[\sqrt{2}]$ is a commutative ring with identity under addition and multiplication of real numbers.

   (b) Show that $U(\mathbb{Z}[\sqrt{2}]) = \{n + m\sqrt{2} \in \mathbb{Z}[\sqrt{2}] \mid n^2 - 2m^2 = \pm 1\}$.

8. Let $\mathbb{Z}[i]$ stand for the ring of Gaussian integers. Define the mapping $N : \mathbb{Z}[i] \to \mathbb{Z}$ by setting $N(n + mi) = n^2 + m^2$ for all Gaussian integers $n + mi$.

   (a) Show that $N(\alpha\beta) = N(\alpha)N(\beta)$ for all $\alpha, \beta \in \mathbb{Z}[i]$.

   (b) If α is a unit of $\mathbb{Z}[i]$, show that $N(\alpha) = 1$.

   (c) Show that the units of $\mathbb{Z}[i]$ are $1, -1, i$, and $-i$, and that the group $U(\mathbb{Z}[i])$ is a cyclic group of order 4.

9. Let $\mathrm{Mat}_n(\mathbb{R})$ stand for the ring of $n \times n$ matrices with real number entries.

   (a) Let $A \in \mathrm{Mat}_n(\mathbb{R})$. Show that A is a unit of $\mathrm{Mat}_n(\mathbb{R})$ if and only if $\det(A) \neq 0$.

   (b) Show that $U(\mathrm{Mat}_n(\mathbb{R})) = \mathrm{Mat}_n^*(\mathbb{R})$, the multiplicative group of $n \times n$ invertible matrices over $\mathbb{R}$.

10. Let $\mathbb{H}$ stand for the ring of quaternions.

    (a) Calculate the following sums and products in $\mathbb{H}$:

        (1) $(-I_2 + y + z) + (3I_2 - x + 4y - z)$

        (2) $(x + y)(x - y)$

        (3) $(x^3 - xy)(I_2 + x + y)$

        (4) $(I_2 - x)(I_2 - y)(I_2 - z)$

        (5) $(x + y)^2$

    (b) Show that $\mathbb{H}$ consists of precisely those matrices in $\mathrm{Mat}_2(\mathbb{C})$ that have the form $\begin{pmatrix} a+bi & -c+di \\ c+di & a-bi \end{pmatrix}$, where $a, b, c, d \in \mathbb{R}$.

    (c) If α stands for the matrix in part (b), show that $\det(\alpha) = a^2 + b^2 c^2 + d^2$.

(d) Show that every non-zero element in $\mathbb{H}$ is a unit. If α stands for the matrix in part (b) and $\alpha \neq 0$, find an explicit matrix for α^{-1}.

(e) Let $\alpha = aI_2 + bx + cy + dz$ be a typical non-zero element in $\mathbb{H}$. Show that

$$\alpha^{-1} = \frac{1}{\det(\alpha)}(aI_2 - bx - cy - dz).$$

11. Let $F(\mathbb{R})$ stand for the ring of functions on $\mathbb{R}$, the ring of real numbers.
    (a) Let $S = \{f \in F(\mathbb{R}) \mid f(0) = 0\}$. Is S a subring of $F(\mathbb{R})$?
    (b) Let $T = \{f \in F(\mathbb{R}) \mid f(0) = 1\}$. Is T a subring of $F(\mathbb{R})$?
    (c) Let $f \in F(\mathbb{R})$. Show that f is a unit of $F(\mathbb{R})$ if and only if $f(x) \neq 0$ for all $x \in \mathbb{R}$.

12. Let R be a ring.
    (a) Show that the subset $\{0\}$ is a commutative subring with identity of R.
    (b) If $x, y \in R$, show that $(nx)(my) = (nm)xy$ for all integers n and m.

13. Let R be a ring with identity 1.
    (a) Show that the product $(-1)x = -x$ for every element $x \in R$, where -1 stands for the additive inverse of 1.
    (b) Show that $nx = (n1)x$, the product of the elements $n1$ and x, for every integer n and every element $x \in R$.
    (c) If R contains more than one element, show that $1 \neq 0$.

14. Let G be an additive abelian group. Define multiplication on G by setting $xy = 0$ of all elements $x, y \in G$, where 0 stands for the zero element of G. Show that G is a commutative ring under addition and multiplication.

15. Let R be a ring and $F(R)$ the ring of functions on R.
    (a) Show that $F(R)$ has an identity if and only if R has an identity.
    (b) Show that $F(R)$ is commutative if and only if R is commutative.

16. Let R be a ring. The *characteristic* of R is the smallest positive integer n such that $nx = 0$ for every element $x \in R$. If no such integer exists, the characteristic is defined to be zero.
    (a) Find the characteristic of the following rings: $\mathbb{Z}_2, \mathbb{Z}_3, \mathbb{Z}_6, \mathbb{Z}$, and $\mathbb{R}$.
    (b) If R is a ring with identity 1, show that the characteristic of R is either 0 or the smallest positive integer n such that $n1 = 0$.
    (c) Let n be a positive integer. Find the characteristic of the ring $\mathbb{Z}_n$.
    (d) Let n and s be positive integers and let $S = \{k[s] \in \mathbb{Z}_n \mid k \in \mathbb{Z}\}$ stand for the set of integral multiples of the element $[s] \in \mathbb{Z}_n$. Show that S is a subring of $\mathbb{Z}_n$ and find the characteristic of S.
    (e) If S is a subring of R, is it true that the characteristic of S is the same as the characteristic of R?

17. Let R be a ring and let $Z(R) = \{x \in R \mid xy = yx \text{ for all } y \in R\}$ stand for the set of elements in R that commute with every element in R.
    (a) Show that $Z(R)$ is a commutative subring of R. $Z(R)$ is called the *center* of R.
    (b) Show that $Z(\text{Mat}_n(\mathbb{R})) = \{aI_n \in \text{Mat}_n(\mathbb{R}) \mid a \in \mathbb{R}\}$ consists of all $n \times n$ scalar matrices.

(c) Show that $Z(\mathbb{H}) = \{aI_2 \in \text{Mat}_2(\mathbb{C}) \mid a \in \mathbb{C}\}$, where $\mathbb{H}$ is the ring of quaternions.

18. **Algebra of subsets.** Let X be a nonempty set and let $\mathscr{P}(X)$ stand for the set of subsets of X. Define addition and multiplication on $\mathscr{P}(X)$ as follows: if $A, B \in \mathscr{P}(X)$, let $A + B = A \cup B - A \cap B$ and $AB = A \cap B$, where $\cup$ and $\cap$ stand for set union and set intersection, respectively.

    (a) Show that addition and multiplication are associative binary operations on $\mathscr{P}(X)$ and that $\mathscr{P}(X)$ is a commutative ring with identity under these operations. What is the zero element and identity element of the ring $\mathscr{P}(X)$? The ring $\mathscr{P}(X)$ is called the *algebra of subsets of X*.

    (b) Write out the addition and multiplication tables for $\mathscr{P}(X)$ if $X = \{1, 2\}$.

    (c) Determine the units of the ring $\mathscr{P}(X)$.

    (d) Show that $A^2 = A$ and $2A = 0$ for every element $A \in \mathscr{P}(X)$.

    (e) Is the set $\mathscr{P}(X)$ a ring if addition is defined to be set union and multiplication is defined as set intersection?

19. **Boolean rings.** A *Boolean ring* is any ring R in which $x^2 = x$ for every element $x \in R$.

    (a) Show that the ring of functions $F(\mathbb{Z}_2)$ is a Boolean ring.

    (b) If X is a nonempty set, show that the ring $\mathscr{P}(X)$ discussed in Exercise 18 is a Boolean ring.

    (c) Let R be a Boolean ring. Show that:

    (1) $2x = 0$ for every element $x \in R$.

    (2) R is commutative.

    (3) Every subring of R is a Boolean ring.

20. **Direct product of rings.** Let R and S be rings and let $R \times S = \{(x, y) \mid x \in R, y \in S\}$ stand for the Cartesian product of the sets R and S. Define addition and multiplication on $R \times S$ as follows: if $(x, y), (x', y') \in R \times S$, let

    $$(x, y) + (x', y') = (x + x', y + y')$$

    $$(x, y)(x', y') = (xx', yy').$$

    (a) Show that $R \times S$ is a ring under these operations. $R \times S$ is called the *direct product* of R and S.

    (b) Show that $R \times S$ is commutative if and only if both R and S are commutative.

    (c) If both R and S have an identity, does $R \times S$ have an identity?

    (d) Write out the addition and multiplication tables for the ring $\mathbb{Z}_2 \times \mathbb{Z}_3$. What is the characteristic of the ring $\mathbb{Z}_2 \times \mathbb{Z}_3$?

21. **Nilpotent elements and the nilradical of a ring.** Let R be a commutative ring with identity. An element $x \in R$ is said to be *nilpotent* if $x^n = 0$ for some positive integer n.

    (a) Let x be a nilpotent element of R and let y be any element in R. Show that xy is nilpotent.

    (b) If x and y are nilpotent elements of R, show that $x + y$ is also nilpotent.

(c) Let $N(R) = \{x \in R \,|\, x \text{ is nilpotent}\}$ stand for the set of nilpotent elements of R. Show that $N(R)$ is a subring of R. $N(R)$ is called the *nilradical* of R.

(d) Find $N(\mathbb{Z}_4)$, $N(\mathbb{Z}_5)$, $N(\mathbb{Z}_{27})$, and $N(\mathbb{Z}_{72})$.

(e) Let x be a nilpotent element of R and u any unit of R. Show that $u + x$ is a unit of R.

(f) Let x be a nilpotent element of R with $x^n = 0$, n a positive integer, and let 1 stand for the identity of R. Then by part (e) $1 + x$ is a unit of R. Show that $(1 + x)^{-1} = 1 + x + \cdots + x^{n-1}$.

2.   IDEALS, QUOTIENT RINGS, AND RING-HOMOMORPHISMS

We continue our discussion of the basic concepts of ring theory by introducing a special type of subring called an ideal. Our goal is to first define and illustrate ideals, and then use them to construct quotient rings.

Let us begin with an observation. As we showed in the previous section, the set $2\mathbb{Z}$ of even integers is a subring of the ring $\mathbb{Z}$ of integers; that is, it is closed under sums, products, and additive inverses. But it is closed under products in a more general sense: if we multiply an even integer by any integer, not just an even integer, the product is always an even integer. In other words, the product of any element in $2\mathbb{Z}$ with any element in the ring $\mathbb{Z}$ is again an element in $2\mathbb{Z}$. This property, when extended to arbitrary rings, leads us to the concept of an ideal.

> **Definition**
>
> Let R be a ring. An *ideal* of R is any nonempty subset I of R that satisfies the following three conditions:
>
> **(1)** if $a, b \in I$, then $a + b \in I$;
> **(2)** if $a \in I$, then $-a \in I$;
> **(3)** if $a \in I$ and $x \in R$, then $ax \in I$ and $xa \in I$.

Note that condition (3) requires that the product of any element in an ideal I with any element of the ring must always belong to I. If the ring is commutative, then we need only verify that one of these products, say ax, lies in I since $ax = xa$. Finally, observe that every ideal is a subring since condition (1) takes care of closure under addition, (2) takes care of additive inverses, and (3) takes care of closure under products. It follows, in particular, that every ideal is a subgroup of the additive group of the ring. The subset $2\mathbb{Z}$ of even integers, for example, is an ideal of the ring $\mathbb{Z}$ of integers since the sum of two even integers is even, and the product of an even integer with any integer is always an even integer. However, the ring $\mathbb{Z}$ is not an ideal of the ring $\mathbb{Q}$ of rational numbers since the product of an integer and an arbitrary rational number is not, in general, an integer, although $\mathbb{Z}$ is, of course, a subring of $\mathbb{Q}$. Thus, while ideals are subrings, not all subrings are ideals. Finally, observe that if R is any ring, then both R and $\{0\}$ are ideals of R; we refer to these ideals as the *trivial ideals* of R.

Let us now look more closely at some other examples of ideals. We begin by first imitating the construction of the ideal $2\mathbb{Z}$: instead of forming all integral multiples of 2 to form the ideal $2\mathbb{Z}$, we show that the set of all integral multiples of an arbitrary integer is also an ideal of the ring $\mathbb{Z}$, and that every ideal of $\mathbb{Z}$ in fact has this form. In Example 2 we use this same idea to construct ideals in the rings $\mathbb{Z}_n$, while Example 3 extends this idea to arbitrary commutative rings by showing that the set of all multiples of any given element is always an ideal of the ring.

EXAMPLE 1. Ideals of $\mathbb{Z}$

Let n be an integer and let $n\mathbb{Z} = \{nk \in \mathbb{Z} \mid k \in \mathbb{Z}\}$ stand for the set of integral multiples of n. We claim that $n\mathbb{Z}$ is an ideal of the ring $\mathbb{Z}$ and that every ideal of $\mathbb{Z}$ has this form for some integer n.

(A) Let us first show that $n\mathbb{Z}$ is an ideal of $\mathbb{Z}$. To this end, let ns and nt be typical elements in $n\mathbb{Z}$ and let k be an arbitrary integer. Then

$$ns + nt = n(s + t) \in n\mathbb{Z},$$

$$-ns = n(-s) \in n\mathbb{Z},$$

$$k(ns) = n(ks) \in n\mathbb{Z}.$$

Hence, $n\mathbb{Z}$ is an ideal of $\mathbb{Z}$.

(B) To show that every ideal of $\mathbb{Z}$ has the form $n\mathbb{Z}$ for some integer n, let I be an arbitrary ideal of $\mathbb{Z}$. Then I is a subgroup of the additive group of integers. But the additive group of integers is cyclic and hence all of its subgroups are cyclic. Therefore, $I = n\mathbb{Z}$ for some integer n, as required. The reader will recall from our discussion of cyclic groups that if $I \neq \{0\}$, the integer n may be chosen as the smallest positive integer in I. Note that $\mathbb{Z} = 1\mathbb{Z}$ and $\{0\} = 0\mathbb{Z}$.

EXAMPLE 2. Ideals of $\mathbb{Z}_n$

Let n be a positive integer. For each congruence class $[m] \in \mathbb{Z}_n$, let $[m]\mathbb{Z}_n = \{[m][s] \in \mathbb{Z}_n \mid [s] \in \mathbb{Z}_n\}$ stand for the set of all multiples of $[m]$ by elements in $\mathbb{Z}_n$. We claim that $[m]\mathbb{Z}_n$ is an ideal of $\mathbb{Z}_n$ and that every ideal of $\mathbb{Z}_n$ has this form.

(A) To show that $[m]\mathbb{Z}_n$ is an ideal of $\mathbb{Z}_n$, let $[m][s]$ and $[m][t]$ be typical elements in $[m]\mathbb{Z}_n$ and let $[k]$ be an arbitrary element in $\mathbb{Z}_n$. Then

$$[m][s] + [m][t] = [m]([s + t]) \in [m]\mathbb{Z}_n$$

$$-[m][s] = [m][-s] \in [m]\mathbb{Z}_n$$

$$[k]([m][s]) = [m][ks] \in [m]\mathbb{Z}_n.$$

Therefore, $[m]\mathbb{Z}_n$ is an ideal of $\mathbb{Z}_n$.

(B) Now if I is an arbitrary ideal of $\mathbb{Z}_n$, then I is a subgroup of the additive group of $\mathbb{Z}_n$. But $\mathbb{Z}_n$ is a cyclic group under addition and hence its subgroups are cyclic. Thus, $I = [m]\mathbb{Z}_n$ for some congruence class $[m]$, as required. Note, as above, that $\mathbb{Z}_n = [1]\mathbb{Z}_n$ and $\{[0]\} = [0]\mathbb{Z}_n$.

(C) Let us determine the ideals $[2]\mathbb{Z}_6$ and $[5]\mathbb{Z}_6$ of the ring $\mathbb{Z}_6$. We find by direct calculation that

$$[2]\mathbb{Z}_6 = \{[2][0], [2][1], [2][2], [2][3], [2][4], [2][5]\}$$
$$= \{[0], [2], [4]\}$$
$$\text{and } [5]\mathbb{Z}_6 - \{[5][0], [5][1], [5][2], [5][3], [5][4], [5][5]\}$$
$$= \{[0], [5], [4], [3], [2], [1]\} = \mathbb{Z}_6.$$

EXAMPLE 3. Principal ideals

Let R be a commutative ring with identity, $a \in R$, and let $aR = \{ax \in R \mid x \in R\}$ stand for the set of all multiples of a by elements in R.

(A) We claim that aR is an ideal of R. To show this, let ax and ay be typical elements in aR and let w be any element in R. Then

$$ax + ay = a(x + y) \in aR,$$

$$-ax = a(-x) \in aR,$$

$$w(ax) = a(wx) \in aR,$$

$$(ax)w = a(xw) \in aR.$$

Therefore aR is an ideal of R. The ideal aR is called the *principal ideal of R generated by a*. Note that the trivial ideals of R are principal ideals since $R = 1R$, where 1 stands for the identity of R, $\{0\} = 0R$, and that $a \in aR$ since $a = a1$.

(B) For example, if n is an integer, the ideal $n\mathbb{Z}$ discussed in Example 1 is just the principal ideal of $\mathbb{Z}$ generated by n. Since every ideal of $\mathbb{Z}$ has this form for some integer n, every ideal of $\mathbb{Z}$ is a principal ideal. Moreover, since $n\mathbb{Z} = -n\mathbb{Z}$, different elements of a ring may in fact generate the same principal ideal.

(C) If $[m] \in \mathbb{Z}_n$, then the ideal $[m]\mathbb{Z}_n$ discussed in Example 2 is the principal ideal of $\mathbb{Z}_n$ generated by $[m]$. And, since every ideal of $\mathbb{Z}_n$ has this form for some element $[m] \in \mathbb{Z}_n$, every ideal of $\mathbb{Z}_n$ is a principal ideal. For example, in the ring $\mathbb{Z}_6$ we find that:

$$[0]\mathbb{Z}_6 = \{[0]\}$$

$$[1]\mathbb{Z}_6 = \{[0], [1], [2], [3], [4], [5]\} = \mathbb{Z}_6$$

$$[2]\mathbb{Z}_6 = \{[0], [2], [4]\}$$

$$[3]\mathbb{Z}_6 = \{[0], [3]\}$$

$$[4]\mathbb{Z}_6 = \{[0], [4], [2]\} = [2]\mathbb{Z}_6$$

$$[5]\mathbb{Z}_6 = \{[0], [5], [4], [3], [2], [1]\} = \mathbb{Z}_6.$$

Since every ideal of $\mathbb{Z}_6$ is principal, we conclude that $\mathbb{Z}_6$ has four ideals, namely, $\{[0]\}$, $[2]\mathbb{Z}_6$, $[3]\mathbb{Z}_6$, and $\mathbb{Z}_6$.

EXAMPLE 4

Let R be a ring and let $F(R)$ stand for the ring of functions on R. For each element $c \in R$, let $I_c = \{f \in F(R) | f(c) = 0\}$ stand for the subset of functions whose value at c is zero. Let us show that the sets I_c are ideals of $F(R)$ and then give a geometric interpretation of these ideals.

(A) To show that I_c is an ideal of $F(R)$, let $f, g \in I_c$ and let h be an arbitrary function in $F(R)$. Then $f(c) = g(c) = 0$ and hence

$$(f + g)(c) = f(c) + g(c) = 0,$$

$$(-f)(c) = -f(c) = 0,$$

$$(fh)(c) = f(c)h(c) = 0,$$

$$(hf)(c) = h(c)f(c) = 0.$$

Therefore, $f + g$, $-f$, fh, and hf are elements in I_c and hence I_c is an ideal of $F(R)$.

(B) For example, in the function ring $F(\mathbb{Z}_2)$ discussed in Example 5, Section 1, the only functions that vanish at $[0]$ are f_0 and g. Hence $I_{[0]} = \{f_0, g\}$. Similarly, we find that $I_{[1]} = \{f_0, h\}$.

(C) Let us interpret the ideal I_c geometrically in the case where $R = \mathbb{R}$, the ring of real numbers. In this case the functions in $F(\mathbb{R})$ may be represented geometrically by means of their graphs, and I_c consists of those functions $f : \mathbb{R} \to \mathbb{R}$ whose graphs intersect the x-axis at the point $x = c$. The figure to the right illustrates several functions in I_c.

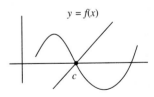

EXAMPLE 5

Let $\mathrm{Mat}_2(\mathbb{R})$ stand for the ring of 2×2 real matrices and let $R = \{\left(\begin{smallmatrix} a & 0 \\ 0 & b \end{smallmatrix}\right) \in \mathrm{Mat}_2(\mathbb{R}) | a, b \in \mathbb{R}\}$ stand for the subset of diagonal matrices. Then R is a commutative subring of $\text{Mat}_2(\mathbb{R})$ with identity I_2; for if A and B are typical matrices in R, it is easily verified that $A + B$, AB, and $-A$ are elements in R. Now, let $I = \{ \left(\begin{smallmatrix} a & 0 \\ 0 & 0 \end{smallmatrix}\right) \in R \,|\, a \in R \}$. Then I is the principal ideal of R generated by the matrix $\left(\begin{smallmatrix} 1 & 0 \\ 0 & 0 \end{smallmatrix}\right)$ since

$$\begin{pmatrix} 1 & 0 \\ 0 & 0 \end{pmatrix} R = \left\{ \begin{pmatrix} 1 & 0 \\ 0 & 0 \end{pmatrix} \begin{pmatrix} a & 0 \\ 0 & b \end{pmatrix} \in R \,\middle|\, a, b \in \mathbb{R} \right\} = \left\{ \begin{pmatrix} a & 0 \\ 0 & 0 \end{pmatrix} \in R \,\middle|\, a \in \mathbb{R} \right\} = I.$$

But I is not an ideal of the ring $\text{Mat}_2(\mathbb{R})$ since the product of a matrix in I with an arbitrary matrix is not necessarily in I; for example, $\left(\begin{smallmatrix} 1 & 0 \\ 0 & 0 \end{smallmatrix}\right) \in I$ and $\left(\begin{smallmatrix} 1 & 2 \\ 3 & 4 \end{smallmatrix}\right) \in \text{Mat}_2(\mathbb{R})$, but $\left(\begin{smallmatrix} 1 & 0 \\ 0 & 0 \end{smallmatrix}\right)\left(\begin{smallmatrix} 1 & 2 \\ 3 & 4 \end{smallmatrix}\right) = \left(\begin{smallmatrix} 1 & 2 \\ 0 & 0 \end{smallmatrix}\right) \notin I$.

Observe that if R is a ring with identity and I is an ideal of R that contains a unit of R, then $I = R$. For if $u \in I$ is a unit of R and $x \in R$, then $x = u(u^{-1}x) \in I$. Therefore $I = R$. Thus, if an ideal contains a unit of the ring, then the ideal is the entire ring. For example, $[5]$ is a unit of the ring $\mathbb{Z}_6$ since $\gcd(5, 6) = 1$; hence $[5]\mathbb{Z}_6 = \mathbb{Z}_6$. Indeed, we verified this result by direct calculation in Example 2. In the ring $\mathbb{Q}$ of rational numbers, every nonzero element is a unit and hence the only ideals of $\mathbb{Q}$ are the trivial ideals $\{0\}$ and $\mathbb{Q}$. Similarly, the only ideals of the rings $\mathbb{R}$ and $\mathbb{C}$ are the trivial ideals.

Let us now turn our attention to the construction of quotient rings. Let R be a ring and let I be an ideal of R. Then R is an abelian group under addition and I is a normal subgroup of R. Thus, we may form the additive quotient group R/I. Recall that the elements of R/I are cosets $x + I = \{x + a \in R \,|\, a \in I\}$, where $x \in R$, and that addition of cosets is defined by setting $(x + I) + (y + I) = (x + y) + I$ for all cosets $x + I, y + I \in R/I$. Recall also that $x' + I = x + I$ if and only if $x' - x \in I$, or equivalently, $x' = x + a$ for some element $a \in I$. We now define multiplication of cosets as follows:

$$\text{if } x + I, y + I \in R/I, \quad \text{let } (x + I)(y + I) = xy + I.$$

Since the definition of the product $(x + I)(y + I)$ depends upon the representatives x and y, we must first verify that this definition makes sense—that is, that coset multiplication is well defined. To this end, suppose that $x' + I = x + I$ and $y' + I = y + I$ are different ways of representing the cosets $x + I$ and $y + I$, where $x, x', y, y' \in R$. Then, $x' = x + a$ and $y' = y + b$ for some elements $a, b \in I$ and therefore $x'y' = (x + a)(y + b) = xy + ay + xb + ab$. But ay, xb, and ab are elements of I since I is an ideal containing a and b and ideals are closed under products by arbitrary elements of the ring. Therefore, $x'y' - xy \in I$ and hence $x'y' + I = xy + I$. Thus, the formula for coset multiplication is a well-defined binary operation on the quotient group R/I. Let use now show that R/I is a ring under coset addition and multiplication.

PROPOSITION 1. Let R be a ring and let I be an ideal of R. Then R/I is a ring under coset addition and multiplication. Furthermore, if R is commutative, then R/I is commutative, and if R has an identity 1, R/I has identity $1 + I$.

Proof.   As we noted above, R/I is an abelian group under coset addition. It remains to show, therefore, that coset multiplication is associative and distributes over coset addition. To prove associativity, let $x + I$, $y + I$, and $z + I$ be typical cosets in R/I. Then

$$(x + I)[(y + I)(z + I)] = (x + I)(yz + I) = x(yz) + I.$$

But $x(yz) = (xy)z$ since multiplication on R is associative. Therefore

$$(x + I)[(y + I)(z + I)] = x(yz) + I = (xy)z + I = [(x + I)(y + I)](z + I).$$

Thus coset multiplication is associative. The proof that multiplication distributes over addition is similar and is left as an exercise for the reader. We conclude, therefore, that R/I is a ring under coset addition and multiplication. Finally, if R is commutative and $x, y \in R$, then $xy = yx$ and hence $(x + I)(y + I) = xy + I = yx + I = (y + I)(x + I)$, which shows that R/I is commutative, while if R has an identity 1, then $(x + I)(1 + I) = x1 + I = x + I = (1 + I)(x + I)$, which shows that $1 + I$ is the identity for R/I.   ∎

The ring R/I is called the *quotient ring of R modulo I*. For example, consider the quotient ring $\mathbb{Z}/n\mathbb{Z}$, where n is a positive integer. In this case a typical coset $s + n\mathbb{Z}$ is the same as the congruence class $[s]$ of s modulo n. Moreover, addition and multiplication of cosets is the same as addition and multiplication of congruence classes:

$$(s + n\mathbb{Z}) + (t + n\mathbb{Z}) = (s + t) + n\mathbb{Z} \qquad (s + n\mathbb{Z})(t + n\mathbb{Z}) = st + n\mathbb{Z}$$

$$[s] \quad + \quad [t] \quad = \quad [s + t] \qquad\qquad [s] \quad [t] \quad = \quad [st].$$

It follows, therefore, that $\mathbb{Z}/n\mathbb{Z} = \mathbb{Z}_n$, the ring of integers modulo n.

EXAMPLE 6

Let us construct the quotient ring $\mathbb{Z}_6/[3]\mathbb{Z}_6$, illustrate addition and multiplication of cosets in this ring, and then determine its group of units.

(A) Since $[3]\mathbb{Z}_6 = \{[0], [3]\}$, the cosets in $\mathbb{Z}_6/[3]\mathbb{Z}_6$ are

$$[0] + [3]\mathbb{Z}_6 = \{[0], [3]\} \qquad [3] + [3]\mathbb{Z}_6 = \{[3], [0]\} = [0] + [3]\mathbb{Z}_6$$

$$[1] + [3]\mathbb{Z}_6 = \{[1], [4]\} \qquad [4] + [3]\mathbb{Z}_6 = \{[4], [1]\} = [1] + [3]\mathbb{Z}_6$$

$$[2] + [3]\mathbb{Z}_6 = \{[2], [5]\} \qquad [5] + [3]\mathbb{Z}_6 = \{[5], [2]\} = [2] + [3]\mathbb{Z}_6.$$

Hence, $\mathbb{Z}_6/[3]\mathbb{Z}_6 = \{[0] + [3]\mathbb{Z}_6, [1] + [3]\mathbb{Z}_6, [2] + [3]\mathbb{Z}_6\}$. In particular, $\mathbb{Z}_6/[3]\mathbb{Z}_6$ is a finite ring containing three elements. Moreover, since $\mathbb{Z}_6$ is a commutative ring with identity $[1]$, $\mathbb{Z}_6/[3]\mathbb{Z}_6$ is a commutative ring with identity $[1] + [3]\mathbb{Z}_6$. Note that we may also use Lagrange's theorem to find the number of elements in $\mathbb{Z}_6/[3]\mathbb{Z}_6$: since $[3]\mathbb{Z}_6$ is a subgroup of $\mathbb{Z}_6$ having order 2, $|\mathbb{Z}_6/[3]\mathbb{Z}_6| = [\mathbb{Z}_6:[3]\mathbb{Z}_6] = 6/2 = 3$.

(B) Let us illustrate coset operations in $\mathbb{Z}_6/[3]\mathbb{Z}_6$ by calculating the sum $([1] + [3]\mathbb{Z}_6) + ([2] + [3]\mathbb{Z}_6)$ and the product $([2] + [3]\mathbb{Z}_6)([2] + [3]\mathbb{Z}_6)$. We find that

$$([1] + [3]\mathbb{Z}_6) + ([2] + [3]\mathbb{Z}_6) = ([1] + [2]) + [3]\mathbb{Z}_6$$
$$= [3] + [3]\mathbb{Z}_6 = [0] + [3]\mathbb{Z}_6,$$

the zero element of $\mathbb{Z}_6/[3]\mathbb{Z}_6$, while

$$([2] + [3]\mathbb{Z}_6)([2] + [3]\mathbb{Z}_6) = [2][2] + [3]\mathbb{Z}_6 = [4] + [3]\mathbb{Z}_6 = [1] + [3]\mathbb{Z}_6,$$

the identity of $\mathbb{Z}_6/[3]\mathbb{Z}_6$. The first equation shows that the additive inverse of $[2] + [3]\mathbb{Z}_6$ is $[1] + [3]\mathbb{Z}_6$, that is, $-([2] + [3]\mathbb{Z}_6) = [1] + [3]\mathbb{Z}_6$, while the second equation shows that $[2] + [3]\mathbb{Z}_6$ is a unit of $\mathbb{Z}_6/[3]\mathbb{Z}_6$ that is its own multiplicative inverse, that is, $([2] + [3]\mathbb{Z}_6)^{-1} = [2] + [3]\mathbb{Z}_6$. The remaining sums and products in $\mathbb{Z}_6/[3]\mathbb{Z}_6$ are calculated similarly and are shown in Figure 1.

(C) Referring to the multiplication table in Figure 1, we find that the ring $\mathbb{Z}_6/[3]\mathbb{Z}_6$ has two units, namely $[1] + [3]\mathbb{Z}_6$ and $[2] + [3]\mathbb{Z}_6$. Therefore $U(\mathbb{Z}_6/[3]\mathbb{Z}_6) = \{[1] + [3]\mathbb{Z}_6, [2] + [3]\mathbb{Z}_6\}$ and hence $U(\mathbb{Z}_6/[3]\mathbb{Z}_6)$ is a cyclic group of order 2.

We conclude this section by discussing mappings from one ring to another that preserve addition and multiplication on the rings. These mappings are called ring-homomorphisms and are similar to group-homomorphisms in that they allow us to transfer information from one ring to another in such a way that the algebraic properties of the rings are preserved. Our goal is to

$$\mathbb{Z}_6/[3]\mathbb{Z}_6 = \{[0] + [3]\mathbb{Z}_6, [1] + [3]\mathbb{Z}_6, [2] + [3]\mathbb{Z}_6\}$$

+	$[0] + [3]\mathbb{Z}_6$	$[1] + [3]\mathbb{Z}_6$	$[2] + [3]\mathbb{Z}_6$
$[0] + [3]\mathbb{Z}_6$	$[0] + [3]\mathbb{Z}_6$	$[1] + [3]\mathbb{Z}_6$	$[2] + [3]\mathbb{Z}_6$
$[1] + [3]\mathbb{Z}_6$	$[1] + [3]\mathbb{Z}_6$	$[2] + [3]\mathbb{Z}_6$	$[0] + [3]\mathbb{Z}_6$
$[2] + [3]\mathbb{Z}_6$	$[2] + [3]\mathbb{Z}_6$	$[0] + [3]\mathbb{Z}_6$	$[1] + [3]\mathbb{Z}_6$

$\cdot$	$[0] + [3]\mathbb{Z}_6$	$[1] + [3]\mathbb{Z}_6$	$[2] + [3]\mathbb{Z}_6$
$[0] + [3]\mathbb{Z}_6$	$[0] + [3]\mathbb{Z}_6$	$[0] + [3]\mathbb{Z}_6$	$[0] + [3]\mathbb{Z}_6$
$[1] + [3]\mathbb{Z}_6$	$[0] + [3]\mathbb{Z}_6$	$[1] + [3]\mathbb{Z}_6$	$[2] + [3]\mathbb{Z}_6$
$[2] + [3]\mathbb{Z}_6$	$[0] + [3]\mathbb{Z}_6$	$[2] + [3]\mathbb{Z}_6$	$[1] + [3]\mathbb{Z}_6$

FIGURE 1. Addition and multiplication tables for the quotient ring $\mathbb{Z}_6/[3]\mathbb{Z}_6$.

first show that if R and S are rings, then every ring-homomorphism $f:R \to S$ has associated with it its image, Im f, which is a subring of S, and its kernel, Ker f, which is an ideal of R, and then prove the fundamental theorem of ring-homomorphism, which shows that the quotient ring $R/\text{Ker } f \cong \text{Im } f$.

Definition

Let R and S be rings. A *ring-homomorphism* is a function $f:R \to S$ such that $f(x + y) = f(x) + f(y)$ and $f(xy) = f(x)f(y)$ for all elements $x, y \in R$. If $f:R \to S$ is a ring-homomorphism, the *kernel* of f is the subset Ker f $= \{x \in R \,|\, f(x) = 0\}$ consisting of all elements in R that map to the zero element in S, and the *image* of f is the subset Im $f = \{f(x) \in S \,|\, x \in R\}$ consisting of the images of all elements in R under f. A ring-homomorphism that is both 1–1 and onto is called a *ring-isomorphism*.

Said informally, a ring-homomorphism is a function from one ring to another that preserves both addition and multiplication on the rings. If $f:R \to S$ is a ring-homomorphism, then f is clearly a group-homomorphism from the additive group of R to the additive group of S since $f(x + y) = f(x) + f(y)$ for all $x, y \in R$. It follows, therefore, that $f(0) = 0$, $f(-x) = -f(x)$ for all $x \in R$, $f(x - y) = f(x) - f(y)$ for all $x, y \in R$, and $f(x_1 + \cdots + x_n) = f(x_1)$ $+ \cdots + f(x_n)$ for any finite collection of elements $x_1, \ldots, x_n \in R$. In addition, it follows easily by induction that $f(x_1 \cdots x_n) = f(x_1) \cdots f(x_n)$ for any finite collection of elements $x_1, \ldots, x_n \in R$. Note, however, that if R and S are rings with identity, then f need not map the identity of R to the identity of S; for example, the function $f:R \to S$ defined by setting $f(x) = 0$ for all $x \in R$ is clearly a ring-homomorphism called the *zero-homomorphism*, but it does not map the identity of R to that of S. Finally, ring-isomorphism is an equivalence relation on the collection of rings. We write $R \cong S$ to indicate that R and S are isomorphic rings. The identity map $1_R:R \to R$, for example, is a ring-isomorphism from any ring R to itself.

For example, let n be a positive integer and let $f:\mathbb{Z} \to \mathbb{Z}_n$ stand for the function defined by setting $f(s) = [s]$ for every integer $s \in \mathbb{Z}$; that is, f maps every integer to its congruence class modulo n. Then f is a ring-homomorphism since $f(s + t) = [s + t] = [s] + [t] = f(s) + f(t)$ and $f(st) = [st]$ $= [s][t] = f(s)f(t)$ for all integers $s, t \in \mathbb{Z}$. The mapping f is clearly onto, but it is not 1–1 since $f(n) = [n] = [0] = f(0)$, for example, but $n \neq 0$. In this case we find that Ker $f = \{s \in \mathbb{Z} \,|\, [s] = [0]\} = n\mathbb{Z}$ and Im $f = \{[s] \in \mathbb{Z}_n \,|\, s \in \mathbb{Z}\} = \mathbb{Z}_n$.

PROPOSITION 2. Let $f:R \to S$ be a ring-homomorphism. Then:

(1) Ker f is an ideal of R; f is 1–1 if and only if Ker $f = \{0\}$;

(2) Im f is a subring of S; f is onto if and only if Im $f = S$.

Proof. To prove statement (1), let $x, y \in \mathrm{Ker}\, f$ and let r be an arbitrary element in R. Then $f(x) = f(y) = 0$ and hence

$$f(x + y) = f(x) + f(y) = 0,$$

$$f(-x) = -f(x) = 0,$$

$$f(rx) = f(r)f(x) = 0,$$

$$f(xr) = f(x)f(r) = 0.$$

Therefore, $x + y, -x, rx, xr \in \mathrm{Ker}\, f$ and hence $\mathrm{Ker}\, f$ is an ideal of R. Clearly, f is 1–1 if and only if $\mathrm{Ker}\, f = \{0\}$. The proof of statment (2) is similar and is left as an exercise for the reader. ∎

Now, if $f : R \rightarrow S$ is a ring-homomorphism, then f is also a group-homomorphism from the additive group of R to that of S and hence, as we showed in our study of group-homomorphisms, maps all elements of an arbitrary coset $x + \mathrm{Ker}\, f$ to the same element in S, namely $f(x)$. The mapping f therefore defines a mapping $f^* : R/\mathrm{Ker}\, f \rightarrow \mathrm{Im}\, f$ by setting $f^*(x + \mathrm{Ker}\, f) = f(x)$, and f^* is a group-isomorphism. To show that f^* is in fact a ring-isomorphism, we need only verify that f^* preserves multiplication.

PROPOSITION 3. (FUNDAMENTAL THEOREM OF RING-HOMOMORPHISM) Let $f : R \rightarrow S$ be a ring-homomorphism. Then $R/\mathrm{Ker}\, f \cong \mathrm{Im}\, f$.

Proof. Let $K = \mathrm{Ker}\, f$. As we noted above, f defines a group-isomorphism $f^* : R/\mathrm{Ker}\, f \rightarrow \mathrm{Im}\, f$ by setting $f^*(x + K) = f(x)$ for every coset $x + K \in R/K$. To show that f^* is a ring-isomorphism, it remains only to verify that f^* preserves multiplication. To this end, let $x + K$ and $y + K$ be typical cosets in R/K. Then $f^*((x + K)(y + K)) = f^*(xy + K) = f(xy)$. But $f(xy) = f(x)f(y)$ since f is a ring-homomorphism. Therefore $f^*((x + K)(y + K)) = f(xy) = f(x)f(y) = f^*(x + K)f^*(y + K)$ and hence f^* preserves multiplication. Thus f^* is a ring-isomorphism and hence $R/K \cong \mathrm{Im}\, f$, as required. ∎

EXAMPLE 7

Define the function $f : \mathbb{Z}_6 \rightarrow \mathbb{Z}_6$ by setting $f([s]) = [4s]$ for every element $[s] \in \mathbb{Z}_6$. Let us show that f is a ring-homomorphism, find its kernel and image, and then discuss the corresponding ring-isomorphism.

(A) To show that f is a ring-homomorphism, let $[s], [t] \in \mathbb{Z}_6$. Then

$$f([s] + [t]) = f([s + t]) = [4(s + t)] = [4s] + [4t] = f([s]) + f([t]),$$

$$f([s][t]) = f([st]) = [4st] = [16st] = [4s][4t] = f([s])f([t]).$$

Therefore, f is a ring-homomorphism. Note that f preserves multiplication on $\mathbb{Z}_6$ because $[4]^2 = [16] = [4]$ in the ring $\mathbb{Z}_6$. The marginal table

$\mathbb{Z}_6 \rightarrow \mathbb{Z}_6$ $[s] \mapsto [4s]$
$[0] \mapsto [0]$
$[1] \mapsto [4]$
$[2] \mapsto [8] = [2]$
$[3] \mapsto [12] = [0]$
$[4] \mapsto [16] = [4]$
$[5] \mapsto [20] = [2]$

lists the image of each element under f; note, in particular, that $f([1]) = [4]$ and hence f does not map the identity of $\mathbb{Z}_6$ to itself.

(B) It is clear from the table of values in part (A) that Ker $f = \{[0], [3]\}$ $= [3]\mathbb{Z}_6$ and Im $f = \{[0], [2], [4]\} = [2]\mathbb{Z}_6$. Hence, by the fundamental theorem, $\mathbb{Z}_6/[3]\mathbb{Z}_6 \cong [2]\mathbb{Z}_6$. In this case the isomorphism $f^* : \mathbb{Z}_6/[3]\mathbb{Z}_6$ $\to [2]\mathbb{Z}_6$ is defined by setting $f^*([s] + [3]\mathbb{Z}_6) = f([s]) = [4s]$ for all cosets $[s] + [3]\mathbb{Z}_6 \in \mathbb{Z}_6/[3]\mathbb{Z}_6$. The marginal table shows the image of each coset in $\mathbb{Z}_6/[3]\mathbb{Z}_6$ under f^*.

$\mathbb{Z}_6/[3]\mathbb{Z}_6 \to [2]\mathbb{Z}_6$
$[s] + [3]\mathbb{Z}_6 \mapsto [4s]$
$[0] + [3]\mathbb{Z}_6 \mapsto [0]$
$[1] + [3]\mathbb{Z}_6 \mapsto [4]$
$[2] + [3]\mathbb{Z}_6 \mapsto [2]$

(C) Let us conclude this example with an observation. Since $[4s] = [4][s]$ for all $[s] \in \mathbb{Z}_6$, the homomorphism f may be thought of as multiplication by the element $[4]$; f preserves multiplication since, as we noted above, $[4]^2 = [4]$. Similarly, if we define the function $g : \mathbb{Z}_6 \to \mathbb{Z}_6$ by setting $g([s]) = [3s] = [3][s]$ for all $[s] \in \mathbb{Z}_6$, then it is easily verified that g is also a ring-homomorphism since $[3]^2 = [3]$. In this case Ker g $= [2]\mathbb{Z}_6$, Im $g = [3]\mathbb{Z}_6$, and the corresponding ring-isomorphism is $\mathbb{Z}_6/[2]\mathbb{Z}_6 \cong [3]\mathbb{Z}_6$. In the exercises we ask the reader to show, more generally, that for any positive integer n, the ring-homomorphisms mapping $\mathbb{Z}_n$ to itself are precisely those functions $f : \mathbb{Z}_n \to \mathbb{Z}_n$ defined by $f([s]) = [cs]$ for all $[s] \in \mathbb{Z}_n$, where $[c]$ is a fixed element in $\mathbb{Z}_n$ such that $[c]^2 = [c]$. Any such element $[c]$ is called an idempotent of the ring.

EXAMPLE 8. Complex conjugation

Let $\mathbb{C}$ stand for the ring of complex numbers. Define the function $f : \mathbb{C} \to \mathbb{C}$ by setting $f(a + bi) = a - bi$ for every complex number $a + bi$; that is, f maps every complex number to its complex conjugate. We claim that f is a ring-isomorphism. To show this, let $z = a + bi$ and $z' = a' + b'i$ be typical complex numbers. Then

$$f(z + z') = f((a + a') + (b + b')i)$$
$$= (a + a') - (b + b')i$$
$$= (a - bi) + (a' - b'i)$$
$$= f(a + bi) + f(a' + b'i)$$
$$= f(z) + f(z'),$$

and, similarly, $f(zz') = f(z)f(z')$. Thus f is a ring-homomorphism. Moreover, f is 1–1 since $f(a + bi) = a - bi = 0$ if and only if $a = b = 0$. Since f is clearly an onto mapping, we conclude, therefore, that f is a ring-isomorphism. We refer to f as the *complex conjugation isomorphism*. Geometrically, f reflects each point in the complex plane about the real axis.

The following example deals with a particularly useful type of ring-homomorphism called an evaluation homomorphism. Whenever the elements of a ring are functions on some underlying ring, we may evaluate the functions at a given point in the ring. The example shows that the process of evaluating functions at a given element of the ring preserves addition and multiplication of the functions and hence is a ring-homomorphism.

EXAMPLE 9. Evaluation homomorphisms

Let R be a ring and let $F(R)$ stand for the ring of functions on R. Let $c \in R$. Define the function $\varphi_c : F(R) \to R$ by setting $\varphi_c(f) = f(c)$ for all functions $f \in F(R)$. Let us show that φ_c is a ring-homomorphism mapping $F(R)$ onto R, discuss the corresponding ring-isomorphism, and then illustrate these homomorphisms with some examples.

(A) To show that φ_c is a ring-homomorphism, let $f, g \in F(R)$. Then

$$\varphi_c(f + g) = (f + g)(c) = f(c) + g(c) = \varphi_c(f) + \varphi_c(g)$$

$$\varphi_c(fg) = (fg)(c) = f(c)g(c) = \varphi_c(f)\varphi_c(g).$$

Therefore, φ_c is a ring-homomorphism. We call φ_c the *evaluation homomorphism at c* since it evaluates every function in $F(R)$ at the point c. Clearly, the evaluation homomorphism φ_c is an onto mapping; for if r is a typical element in R and $f_r : R \to R$ stands for the constant function defined by setting $f_r(x) = r$ for all $x \in R$, then $\varphi_c(f_r) = f_r(c) = r$. But φ_c is not, in general, a 1–1 mapping. In fact, Ker $\varphi_c = \{f \in F(R) \,|\, f(c) = 0\}$; that is, the kernel of the evaluation homomorphism φ_c consists of all functions that vanish at the point c. Recall that this is just the ideal I_c discussed in Example 4 of this section. It follows from the fundamental theorem, therefore, that $F(R)/I_c \cong R$. Under this isomorphism, a typical coset $f + I_c$ corresponds to the value $f(c)$. Thus, there is a 1–1 correspondence between the cosets in $F(R)/I_c$ and the elements in the ring R, and this correspondence preserves both addition and multiplication on the two rings.

(B) Consider the ring of functions $F(\mathbb{R})$, where $\mathbb{R}$ stands for the ring of real numbers. Let $\varphi_0 : F(\mathbb{R}) \to \mathbb{R}$ be the evaluation homomorphism on $F(\mathbb{R})$ at the point 0. Then the kernel I_0 of φ_0 consists of all real-valued functions that vanish at 0 and $F(\mathbb{R})/I_0 \cong \mathbb{R}$.

(C) As a second example, consider the ring of functions $F(\mathbb{Z}_2)$ discussed in Example 5, Section 1. Referring to Figure 3, Section 1, we found that $F(\mathbb{Z}_2) = \{f_0, f_1, g, h\}$. Now, let $\varphi_0 : F(\mathbb{Z}_2) \to \mathbb{Z}_2$ stand for the evaluation homomorphism at 0. Then $I_0 = $ Ker $\varphi_0 = \{f_0, g\}$ and hence $F(\mathbb{Z}_2)/I_0 \cong \mathbb{Z}_2$, as illustrated below:

$$f_0 + I_0 = g + I_0 = \{f_0, g\} \leftrightarrow [0]$$

$$f_1 + I_0 = h + I_0 = \{f_1, h\} \leftrightarrow [1].$$

Similarly, for the evaluation homomorphism $\varphi_1 : F(\mathbb{Z}_2) \to \mathbb{Z}_2$ we find that $I_1 = $ Ker $\varphi_1 = \{f_0, h\}$ and $F(\mathbb{Z}_2)/I_1 \cong \mathbb{Z}_2$, where

$$f_0 + I_1 = h + I_1 = \{f_0, h\} \leftrightarrow [0]$$

$$f_1 + I_1 = g + I_1 = \{f_1, g\} \leftrightarrow [1].$$

As our final example, we discuss the matrix representation of linear transformations. The reader will recall that the set $\mathrm{Mat}_n^*(\mathbb{R})$ of invertible $n \times n$ real matrices is a group under matrix multiplication and is isomorphic to the general linear group $GL(\mathbb{R}^n)$: every coordinate system X of $\mathbb{R}^n$ defines a group-isomorphism $f_X: GL(\mathbb{R}^n) \to \text{Mat}_n^*(\mathbb{R})$ that maps each nonsingular linear transformation to its matrix representation relative to X (Example 2, Chapter 2, Section 5). In the following example we show that the collection of all linear transformations on $\mathbb{R}^n$ form a ring under addition and multiplication of transformations, and that this ring is in fact isomorphic to matrix ring $\text{Mat}_n(\mathbb{R})$.

EXAMPLE 10

Let V be an n-dimensional real vector space and let $L(V)$ stand for the set of linear transformations mapping V to V. Let us show that the linear transformations in $L(V)$ may be added and multiplied, and that $L(V)$ is a ring under these operations that is isomorphic to the matrix ring $\text{Mat}_n(\mathbb{R})$.

(A) Let $\sigma, \tau \in L(V)$. Define the sum $\sigma + \tau : V \to V$ by setting $(\sigma + \tau)(A) = \sigma(A) + \tau(A)$ for all vectors $A \in V$, and the product $\sigma\tau : V \to V$ by setting $\sigma\tau = \sigma \circ \tau$, the composition of σ and τ. Then it is easily verified that $\sigma + \tau$ and $\sigma\tau$ are also linear transformations on V and therefore addition and multiplication of linear transformations are binary operations on $L(V)$. We leave the reader to verify that $L(V)$ is a ring with identity under these two operations. Note that the zero element is the zero transformation $0 : V \to V$ that maps every vector in V to the zero vector, while the identity is the identity transformation $1_V : V \to V$ that maps every vector to itself. $L(V)$ is a noncommutative ring since function composition on V is, in general, a noncommutative operation. Observe also that the units of the ring $L(V)$ are the nonsingular linear transformations on V, that is, $U(L(V)) = GL(V)$.

(B) Now, to show that $L(V) \cong \text{Mat}_n(\mathbb{R})$, let X be a coordinate system for V and let $f_X : L(V) \to \text{Mat}_n(\mathbb{R})$ stand for the function that maps each linear transformation σ to its matrix representation $f_X(\sigma)$ relative to X. Then f_X is a ring-isomorphism. To show this, recall that the matrix representation of the sum of two linear transformations is equal to the sum of the matrices of each transformation, and the representation of the composition is the product of the two matrices. Therefore, $f_X(\sigma + \tau) = f_X(\sigma) + f_X(\tau)$ and $f_X(\sigma\tau) = f_X(\sigma)f_X(\tau)$ for all $\sigma, \tau \in L(V)$. Moreover, since every $n \times n$ matrix M determines a linear transformation on V whose matrix is M relative to X, the mapping f_X is onto. And finally, to show that f_X is 1–1, we need only recall that if the matrix representation $f_X(\sigma)$ of σ is the zero matrix, then σ must be the zero transformation on V. Therefore f_X has a trivial kernel and hence is 1–1. It now follows that f_X is a ring-isomorphism and hence $L(V) \cong \text{Mat}_n(\mathbb{R})$.

(C) Let us illustrate the isomorphism in part (B) by discussing the case where $n = 1$. In this case V is 1-dimensional, and hence $V = \langle A \rangle_{\mathbb{R}}$ for any non-zero vector $A \in V$. The linear transformations on V are then the scalar mappings $\sigma_c : V \to V$, where $c \in \mathbb{R}$ and $\sigma_c(B) = cB$ for every vector $B \in V$.

Thus $L(V) = \{\sigma_c \,|\, c \in \mathbb{R}\}$. In this case we find that addition and multiplication on $L(V)$ are given by the formulas $\sigma_c + \sigma_d = \sigma_{c+d}$ and $\sigma_c\sigma_d = \sigma_{cd}$ for all scalars $c, d \in \mathbb{R}$. It follows, in particular, that $L(V)$ is a commutative ring since $\sigma_c\sigma_d = \sigma_{cd} = \sigma_{dc} = \sigma_d\sigma_c$ for all $c, d \in \mathbb{R}$. If we choose $X = \{A\}$, then the matrix representation of σ_c relative to X is just the 1×1 matrix (c) and hence $f_X(\sigma_c) = (c)$ for all $\sigma_c \in L(V)$. Thus, the three rings $L(V)$, $\mathrm{Mat}_1(\mathbb{R})$ and $\mathbb{R}$ are all isomorphic:

$$L(V) \cong \mathrm{Mat}_1(\mathbb{R}) \cong \mathbb{R}$$

$$\sigma_c \leftrightarrow (c) \leftrightarrow c$$

(D) Let us conclude this example by choosing a coordinate system for the 2-dimensional real plane $\mathbb{R}^2$ and describing, explicitly, the ring-isomorphism $L(\mathbb{R}^2) \cong \mathrm{Mat}_2(\mathbb{R})$. Let $X = \{e_1, e_2\}$ stand for the standard coordinate system for $\mathbb{R}^2$, where $e_1 = (1, 0)$ and $e_2 = (0, 1)$, and let $\sigma : \mathbb{R}^2 \to \mathbb{R}^2$ be a linear transformation. Then $\sigma(e_1) = (a, b)$ and $\sigma(e_2) = (c, d)$ for some scalars a, b, c, d, and hence the matrix representation of σ relative to X is $f_X(\sigma) = \begin{pmatrix} a & c \\ b & d \end{pmatrix}$.

$$L(\mathbb{R}^2) \to \mathrm{Mat}_2(\mathbb{R})$$

$$\sigma \mapsto \begin{pmatrix} a & c \\ b & d \end{pmatrix}$$

Thus, a typical matrix $\begin{pmatrix} a & c \\ b & d \end{pmatrix}$ in $\mathrm{Mat}_2(\mathbb{R})$ corresponds to the linear transformation σ, where $\sigma(x, y) = (ax + cy, bx + dy)$ for all vectors $(x, y) \in \mathbb{R}^2$. For example, if σ is the $90°$ counterclockwise rotation of the plane about the origin, then $\sigma(e_1) = (0, 1)$ and $\sigma(e_2) = (-1, 0)$, and hence σ corresponds to the matrix $\begin{pmatrix} 0 & -1 \\ 1 & 0 \end{pmatrix}$.

Exercises

1. Let $I = \{(2n - m) + (n + 2m)i \in \mathbb{Z}[i] \,|\, n, m \in \mathbb{Z}\}$. Show that I is an ideal of the ring $\mathbb{Z}[i]$ of Gaussian integers.
2. Let $I = \{(3n - 4m) + (-2n + 3m)\sqrt{2} \in \mathbb{Z}[\sqrt{2}] \,|\, n, m \in \mathbb{Z}\}$, where $\mathbb{Z}[\sqrt{2}]$ stands for the ring discussed in Exercise 7, Section 1. Show that I is an ideal of $\mathbb{Z}[\sqrt{2}]$.
3. Find all ideals of the ring $\mathbb{Z}_{12}$ of integers modulo 12.
4. Show that the only ideals of the ring $\mathbb{Z}_p$ of integers modulo a prime p are the trivial ideals $\{[0]\}$ and $\mathbb{Z}_p$.
5. Let $\mathbb{H}$ stand for the ring of quaternions. Show that the only ideals of $\mathbb{H}$ are the trivial ideals $\{0\}$ and $\mathbb{H}$.
6. Let $R = \left\{ \dfrac{n}{m} \in \mathbb{Q} \,|\, m \text{ is odd} \right\}$ stand for the set of rational numbers that may be written in the form $\dfrac{n}{m}$, where m is an odd integer.

   (a) Show that R is a commutative ring with identity under addition and multiplication of rational numbers.

(b) Let $I = \left\{\dfrac{n}{m} \in R \,\middle|\, n \text{ is even}\right\}$. Show that I is an ideal of R.

(c) If I stands for the ideal in part (b) and x is any element in R such that $x \notin I$, show that x is a unit of R.

(d) Let $U(R)$ stand for the group of units of R. Show that $U(R) = R - I = \{x \in R \,|\, x \notin I\}$; that is, the units of R are all elements not in the ideal I.

7. Let $\alpha = n + mi$ be a Gaussian integer and let $\alpha\mathbb{Z}[i]$ stand for the principal ideal of $\mathbb{Z}[i]$ generated by α.

   (a) Show that a typical element in $\alpha\mathbb{Z}[i]$ may be written in the form $(na - mb) + (ma + nb)i$ for some integers $a, b \in \mathbb{Z}$.

   (b) Let $N(\alpha) = n^2 + m^2$. Show that $N(\alpha) \in \alpha\mathbb{Z}[i]$.

8. Let R be a ring and let $I_1, \ldots, I_n$ be a finite collection of ideals of R.

   (a) Show that $I_1 \cap \cdots \cap I_n$ is an ideal of R.

   (b) Find a generator for each of the following ideals:

   (1) $2\mathbb{Z} \cap 3\mathbb{Z}$ (3) $12\mathbb{Z} \cap 8\mathbb{Z} \cap 3\mathbb{Z}$

   (2) $4\mathbb{Z} \cap 6\mathbb{Z}$ (4) $(2 + i)\mathbb{Z}[i] \cap (1 + i)\mathbb{Z}[i]$

9. Let S be a subring of a ring R.

   (a) If I is an ideal of R, show that $I \cap S$ is an ideal of S.

   (b) Show that $(2 + i)\mathbb{Z}[i] \cap \mathbb{Z} = 5\mathbb{Z}$.

   (c) Show that $i\mathbb{Z}[i] \cap \mathbb{Z} = \mathbb{Z}$.

   (d) Find a generator of the ideal $(2 + 3i)\mathbb{Z}[i] \cap \mathbb{Z}$.

10. Write out the elements in the quotient ring $\mathbb{Z}_{12}/[4]\mathbb{Z}_{12}$ and construct the addition and multiplication tables for this ring. Find the group of units $U(\mathbb{Z}_{12}/[4]\mathbb{Z}_{12})$.

11. Let $F(\mathbb{R})$ stand for the ring of functions on $\mathbb{R}$, the ring of real numbers, and let I_c be the ideal of functions in $F(\mathbb{R})$ that are zero at c, where $c \in \mathbb{R}$.

   (a) If $f + I_c$ is a typical coset in the quotient ring $F(\mathbb{R})/I_c$, show that $f + I_c = f_{f(c)} + I_c$, where $f_{f(c)}$ is the constant function whose value at every real number is $f(c)$.

   (b) Show that every nonzero element in $F(\mathbb{R})/I_c$ is a unit.

12. Let R be a ring with identity, I an ideal of R, and $x + I$ a coset in R/I.

   (a) Show that $x + I$ is a unit of R/I if and only if there is some element $y \in R$ such that $xy - 1 \in I$.

   (b) If every nonzero coset in R/I contains some unit of R, show that the only ideals of R/I are the trivial ideals I and R/I.

13. Define the function $f : \mathbb{Z}_{10} \to \mathbb{Z}_{10}$ by setting $f([s]) = [5s]$ for all $[s] \in \mathbb{Z}_{10}$.

   (a) Show that f is a ring-homomorphism.

   (b) Show that $\operatorname{Ker} f = [2]\mathbb{Z}_{10}$.

   (c) Show that $\operatorname{Im} f = [5]\mathbb{Z}_{10}$.

   (d) Show that $\mathbb{Z}_{10}/[2]\mathbb{Z}_{10} \cong [5]\mathbb{Z}_{10}$.

14. Define the function $f : \mathbb{Z}_{10} \to \mathbb{Z}_2$ by setting $f([s]) = [s]_2$ for all $[s] \in \mathbb{Z}_{10}$, where $[s]_2$ stands for the congruence class of s modulo 2.

   (a) Show that f is well defined; that is, if $[s] = [t]$ in $\mathbb{Z}_{10}$, then $f([s]) = f([t])$.

(b) Show that f is a ring-homomorphism.

(c) Show that Ker $f = [2]\mathbb{Z}_{10}$ and Im $f = \mathbb{Z}_2$.

(d) Show that $\mathbb{Z}_{10}/[2]\mathbb{Z}_{10} \cong \mathbb{Z}_2$.

15. Let n be a positive integer and let $[m]$ be a congruence class in $\mathbb{Z}_n$, where m is positive. Show that $\mathbb{Z}_n/[m]\mathbb{Z}_n \cong \mathbb{Z}_d$, where $d = \gcd(n, m)$.

16. Let n be a positive integer and let $[c] \in \mathbb{Z}_n$. Define the function $f_{[c]}$: $\mathbb{Z}_n \to \mathbb{Z}_n$ by setting $f_{[c]}([s]) = [cs]$ for every element $[s] \in \mathbb{Z}_n$.

(a) Show that $f_{[c]}$ is a ring-homomorphism if and only if $[c]^2 = [c]$.

(b) If $f : \mathbb{Z}_n \to \mathbb{Z}_n$ is a ring-homomorphism, show that there is a uniquely determined element $[c] \in \mathbb{Z}_n$ such that $f = f_{[c]}$ and $[c]^2 = [c]$.

(c) Show that the ring-homomorphisms from $\mathbb{Z}_n$ to $\mathbb{Z}_n$ are precisely those functions of the form $f_{[c]}$, where $[c]^2 = [c]$.

(d) Show that $f_{[c]}$ is a ring-isomorphism if and only if $[c] = [1]$. Thus, the only ring-isomorphism mapping $\mathbb{Z}_n$ onto $\mathbb{Z}_n$ is the identity map.

(e) Find all ring-homomorphisms from $\mathbb{Z}_{12}$ to $\mathbb{Z}_{12}$.

17. Let $\det : \mathrm{Mat}_n(\mathbb{R}) \to \mathbb{R}$ stand for the determinant mapping that maps every matrix $M \in \mathrm{Mat}_n(\mathbb{R})$ to its determinant. Is det a ring-homomorphism?

18. Let $f : R \to S$ be a ring-homomorphism. If J is an ideal of S, show that the pre-image $f^{-1}(J) = \{x \in R \mid f(x) \in J\}$ is an ideal of R and that Ker $f \subseteq f^{-1}(J)$.

19. Let $f : R \to S$ be a ring-homomorphism that maps R onto S.

(a) If I is an ideal of R, show that the image $f(I) = \{f(x) \in S \mid x \in I\}$ is an ideal of S.

(b) Let $K = \mathrm{Ker}\, f$. If J is an ideal of S, show that $f^{-1}(J)/K \cong J$.

(c) Show that there is a 1–1 correspondence between the ideals of S and those ideals of R that contain Ker f.

(d) If R has an identity 1_R, show that S has an identity 1_S and that $1_S = f(1_R)$.

20. Let $f : R \to S$ be a ring-isomorphism.

(a) Show that I is an ideal of R if and only if $f(I)$ is an ideal of S.

(b) Show that the correspondence $I \leftrightarrow f(I)$ is a 1–1 correspondence between the ideals of R and the ideals of S.

(c) Show that R has an identity if and only if S has an identity. In this case show that $f(1_R) = 1_S$, where 1_R and 1_S stand for the identities of R and S, respectively.

21. Let $f : R \to S$ be a ring-isomorphism. Show that $f^{-1} : S \to R$ is a ring-isomorphism.

22. Let R and S be rings with identity elements 1_R and 1_S, respectively, and let $f : R \to S$ be a ring-homomorphism such that $f(1_R) = 1_S$.

(a) If u is a unit of R, show that $f(u)$ is a unit of S.

(b) Show that the restriction map $f \mid U(R) : U(R) \to U(S)$ is a group-homomorphism from the group of units of R to the group of units of S.

(c) If $R \cong S$, show that $U(R) \cong U(S)$.

23. Define the function $f : \mathbb{Z}[i] \to \mathbb{Z}_2$ by setting $f(n + mi) = [n] + [m]$ for all $n + mi \in \mathbb{Z}[i]$, where $[n]$ and $[m]$ stand for the congruence classes of n and m modulo 2, respectively.

(a) Show that f is a ring-homomorphism
(b) Show that Ker $f = (1 + i)\mathbb{Z}[i]$ and Im $f = \mathbb{Z}_2$.
(c) Show that $\mathbb{Z}[i]/(1 + i)\mathbb{Z}[i] \cong \mathbb{Z}_2$.
(d) If I is an ideal of $\mathbb{Z}[i]$ such that $(1 + i)\mathbb{Z}[i] \subseteq I$, show that either I $= (1 + i)\mathbb{Z}[i]$ or $I = \mathbb{Z}[i]$.

24. Define the function $f:\mathbb{Z}[i] \to \mathbb{Z}_5$ by setting $f(n + mi) = [n] + [2m]$ for all $n + mi \in \mathbb{Z}[i]$, where $[n]$ and $[2m]$ stand for the congruence classes of n and $2m$ modulo 5, respectively.
   (a) Show that f is a ring-homomorphism.
   (b) Show that Ker $f = (-2 + i)\mathbb{Z}[i]$ and Im $f = \mathbb{Z}_5$.
   (c) Show that $\mathbb{Z}[i]/(-2 + i)\mathbb{Z}[i] \cong \mathbb{Z}_5$.
   (d) If I is an ideal of $\mathbb{Z}[i]$ such that $(-2 + i)\mathbb{Z}[i] \subseteq I$, show that either $I = (-2 + i)\mathbb{Z}[i]$ or $I = \mathbb{Z}[i]$.

25. Find an element $\alpha \in \mathbb{Z}[i]$ such that $\mathbb{Z}[i]/\alpha\mathbb{Z}[i] \cong \mathbb{Z}_{13}$.

26. Is there an element $\alpha \in \mathbb{Z}[i]$ such that $\mathbb{Z}[i]/\alpha\mathbb{Z}[i] \cong \mathbb{Z}_7$?

27. Show that $\mathbb{Z}[\sqrt{2}]/(3 + \sqrt{2})\mathbb{Z}[\sqrt{2}] \cong \mathbb{Z}_7$.

28. Let R be a ring and let Hom(R) stand for the set of ring-homomorphisms from R to R.
   (a) Is Hom(R) a subring of $F(R)$, the ring of functions on R?
   (b) Is Hom(R) an ideal of $F(R)$?
   (c) If R is commutative, is Hom(R) a subring of $F(R)$?
   (d) If R is commutative, is Hom(R) an ideal of $F(R)$?

29. Let G be an additive abelian group and let Hom(G) stand for the set of group-homomorphisms from G to G. If $f, g \in$ Hom(G), define the sum $f + g:G \to G$ and product $fg:G \to G$ by setting $(f + g)(x) = f(x) + g(x)$ and $(fg)(x) = f(g(x))$ for all $x \in G$.
   (a) Show that Hom(G) is a ring under addition and multiplication. Hom(G) is called the *endomorphism ring* of the abelian group G.
   (b) Show that $U($Hom$(G)) =$ Aut(G); that is, the units of the endomorphism ring Hom(G) are precisely the automorphisms of G.
   (c) Let $H \leq G$ and let Hom$_H(G) = \{f \in$ Hom$(G)|f(H) \subseteq H\}$. Show that Hom$_H(G)$ is a subring of Hom(G).

30. Let R and S be rings and let $R \times S$ stand for the direct product of R and S (see Exercise 20, Section 1). Let I be an ideal of R, J an ideal of S, and let $I \times J = \{(x, y) \in R \times S | x \in I, y \in J\}$.
   (a) Show that $I \times J$ is an ideal of $R \times S$.
   (b) Show that $(R \times S)/(I \times J) \cong R/I \times S/J$.
   (c) Write out the addition and multiplication tables for the ring $\mathbb{Z}_2 \times \mathbb{Z}_4$.
      (1) Let $K = \{([0], [0]), ([1], [1]), ([0], [2]), ([1], [3])\}$. Show that K is an ideal of $\mathbb{Z}_2 \times \mathbb{Z}_4$.
      (2) Does $K = I \times J$ for some ideals I of $\mathbb{Z}_2$ and J of $\mathbb{Z}_4$?
      (3) Show that $(\mathbb{Z}_2 \times \mathbb{Z}_4)/K \cong \mathbb{Z}_2$.

31. Let $R = \{(\begin{smallmatrix} a & 0 \\ 0 & b \end{smallmatrix}) \in$ Mat$_2(\mathbb{R})|a, b \in \mathbb{R}\}$ stand for the set of 2×2 real diagonal matrices.
   (a) Show that R is a commutative ring with identity under addition and multiplication of matrices.

(b) Let $\mathbb{R} \times \mathbb{R}$ stand for the direct product of the ring $\mathbb{R}$ with itself. Define the function $f : R \to \mathbb{R} \times \mathbb{R}$ by setting $f(\left(\begin{smallmatrix} a & 0 \\ 0 & b \end{smallmatrix}\right)) = (a, b)$ for all matrices $\left(\begin{smallmatrix} a & 0 \\ 0 & b \end{smallmatrix}\right) \in R$. Show that f is a ring-isomorphism.

(c) Let $I = \{\left(\begin{smallmatrix} a & 0 \\ 0 & 0 \end{smallmatrix}\right) \in R \mid a \in \mathbb{R}\}$ and $J = \{\left(\begin{smallmatrix} 0 & 0 \\ 0 & b \end{smallmatrix}\right) \in R \mid b \in \mathbb{R}\}$. Show that I and J are ideals of R. Describe the ideals $f(I)$ and $f(J)$ of $\mathbb{R} \times \mathbb{R}$ that correspond to I and J under the isomorphism f in part (b).

32. Let $I = \{(x, x) \in \mathbb{R} \times \mathbb{R} \mid x \in \mathbb{R}\}$. Is I an ideal of $\mathbb{R} \times \mathbb{R}$?

33. **Algebra of sets.** Let X be a nonempty set and $\mathscr{P}(X)$ the power set of X, that is, the collection of all subsets of X. In Exercise 18, Section 1, we defined addition and multiplication on $\mathscr{P}(X)$ by setting $A + B = A \cup B - A \cap B$ and $AB = A \cap B$ for all subsets $A, B \in \mathscr{P}(X)$, and asked the reader to show that $\mathscr{P}(X)$ is a commutative ring with identity under these operations. Let 2^X stand for the set of all functions from X into the ring $\mathbb{Z}_2$.

(a) Show that 2^X is a commutative ring with identity under addition and multiplication of functions.

(b) For each subset A of X, the *characteristic function of A* is the function $\varepsilon_A : X \to \mathbb{Z}_2$ defined by setting

$$\varepsilon_A(x) = \begin{cases} [0], & \text{if } x \notin A \\ [1], & \text{if } x \in A \end{cases}$$

for all $x \in X$. Define $f : \mathscr{P}(X) \to 2^X$ by setting $f(A) = \varepsilon_A$ for all subsets $A \in \mathscr{P}(X)$. Show that f is a ring-isomorphism and conclude, therefore, that $\mathscr{P}(X) \cong 2^X$.

34. **Left and right ideals.** Let R be a ring. A nonempty subset I of R is called a *left ideal* of R if $a + b \in I$, $-a \in I$, and $xa \in I$ for all elements $a, b \in I$ and $x \in R$; I is called a *right ideal* of R if $a + b \in I$, $-a \in I$, and $ax \in I$ for all elements $a, b \in I$ and $x \in R$. Left and right ideals are frequently referred to as *one-sided ideals*; the ordinary ideals of R, as defined in this section, are then called *two-sided ideals*.

(a) Let $I = \{\left(\begin{smallmatrix} a & b \\ 0 & 0 \end{smallmatrix}\right) \in \mathrm{Mat}_2(\mathbb{R}) \mid a, b \in \mathbb{R}\}$ and $J = \{\left(\begin{smallmatrix} 0 & 0 \\ a & b \end{smallmatrix}\right) \in \mathrm{Mat}_2(\mathbb{R}) \mid a, b \in \mathbb{R}\}$. Show that I and J are right ideals of the ring $\mathrm{Mat}_2(\mathbb{R})$. Are I and J two-sided ideals of $\mathrm{Mat}_2(\mathbb{R})$?

(b) Let $M = \{\left(\begin{smallmatrix} a & 0 \\ b & 0 \end{smallmatrix}\right) \in \mathrm{Mat}_2(\mathbb{R}) \mid a, b \in \mathbb{R}\}$ and $N = \{\left(\begin{smallmatrix} 0 & a \\ 0 & b \end{smallmatrix}\right) \in \mathrm{Mat}_2(\mathbb{R}) \mid a, b \in \mathbb{R}\}$. Show that M and N are left ideals of the ring $\mathrm{Mat}_2(\mathbb{R})$. Are M and N two-sided ideals of $\mathrm{Mat}_2(\mathbb{R})$?

(c) Show that every right ideal and left ideal of R is a subring of R.

(d) Show that I is a two-sided ideal of R if and only if I is both a right ideal and a left ideal.

(e) If R is commutative, show that all ideals are two-sided ideals.

35. **Nilpotent elements and the nilradical.** Let R be a commutative ring with identity. An element $x \in R$ is said to be *nilpotent* if $x^n = 0$ for some positive integer n. Let $N(R)$ stand for the set of nilpotent elements of R (see Exercise 21, Section 1).

(a) Show that $N(R)$ is an ideal of R. $N(R)$ is called the *nilradical* of R.

(b) Show that the only nilpotent element in the quotient ring $R/N(R)$ is the zero element. Thus, $N(R/N(R)) = \{0\}$.

(c) Find the nilradical $N(\mathbb{Z}_8)$ and show that $\mathbb{Z}_8/N(\mathbb{Z}_8) \cong \mathbb{Z}_2$. Verify by direct calculation that the only nilpotent element in $\mathbb{Z}_8/N(\mathbb{Z}_8)$ is the zero element.

36. **Idempotent elements.** Let R be a ring. An element $x \in R$ is said to be *idempotent* if $x^2 = x$.

(a) Find all idempotent elements of the rings $\mathbb{Z}$, $\mathbb{R}$, $\mathbb{Z}_2$, $\mathbb{Z}_3$, and $\mathbb{Z}_6$.

(b) Let $A = \left(\begin{smallmatrix} 1 & 0 \\ 0 & 0 \end{smallmatrix}\right)$. Show that A is an idempotent matrix in the ring $\text{Mat}_2(\mathbb{R})$.

(c) If every element in R is idempotent, show that R is commutative.

(d) If R is commutative and x and y are idempotent elements of R, show that xy is also idempotent.

(e) Let R be a commutative ring and let c be an idempotent element of R. Define the function $f_c : R \to R$ by setting $f_c(x) = cx$ for all $x \in R$. Show that f_c is a ring-homomorphism of R.

(f) Show that every ring-homomorphism mapping $\mathbb{Z}_n$ to itself has the form f_c for some idempotent $c \in \mathbb{Z}_n$.

(g) Show that there is a 1–1 correspondence between the set of ring-homomorphisms mapping $\mathbb{Z}_n$ to itself and the set of idempotent elements of $\mathbb{Z}_n$.

37. **A geometric interpretation of idempotent elements.** Let V be a finite-dimensional real vector space and let U be a subspace of V. Let W be any complement of U in V; that is, $V = U \oplus W$. The *projection mapping of V onto U in the direction of W* is the function $P_U : V \to V$ defined by setting $P_U(X) = X_U$ for every vector $X \in V$, where $X = X_U + X_W$ is the decomposition of X into unique U- and W-components X_U and X_W, respectively.

(a) Show that the projection mapping P_U is a linear transformation.

(b) Show that P_U is an idempotent element of the ring $L(V)$ of linear transformations on V.

(c) If $n = \dim V$, show that the matrix representation of P_U relative to any coordinate system is an idempotent matrix in the matrix ring $\text{Mat}_n(\mathbb{R})$.

(d) Let $V = \mathbb{R}^2$, the real plane, and let $U = \langle e_1 \rangle_{\mathbb{R}}$ and $W = \langle e_2 \rangle_{\mathbb{R}}$ stand for the x- and y-axes, respectively, where $e_1 = (1, 0)$ and $e_2 = (0, 1)$. In this case the projection function $P_U : \mathbb{R}^2 \to \mathbb{R}^2$ represents the perpendicular projection of the plane $\mathbb{R}^2$ onto the x-axis. Find a formula for $P_U(x, y)$ for any point $(x, y) \in \mathbb{R}^2$, and verify by direct calculation that P_U is an idempotent element of $L(\mathbb{R}^2)$. Find the matrix of P_U relative to the standard coordinate system and verify that it is idempotent.

10

Commutative Rings

In this chapter we turn our attention to commutative rings with identity. Recall that many of the rings we discussed in the previous chapter were commutative rings with identity. The rings of integers, rational numbers, real and complex numbers, for example, are all commutative rings with identity. Commutative rings with identity are especially important because they not only form the basic algebraic model for number systems such as these, but for many different types of finite number systems as well. In this chapter our purpose is to discuss the theory of commutative rings with identity in more detail. Let us emphasize that the concepts and results of this chapter are especially important for our study of fields later in the book.

We begin by first identifying two particular types of commutative rings with identity that play a central role throughout the chapter: integral domains and fields. Integral domains are commutative rings with identity in which the product of nonzero elements is nonzero. The ring of integers, for example, is an integral domain. Fields, on the other hand, are commutative rings with identity in which every nonzero element is a unit. The rational, real, and complex numbers, for example, are all fields. We will also see many examples of finite fields. Fields are especially convenient to work with since, in a field, every nonzero element has a multiplicative inverse and hence it is possible to carry out all four arithmetic operations in any field—addition, subtraction, multiplication, and division. Thus, from an arithmetic point of view, when we work in a field we have the freedom to do what we want, so to speak. Fields also provide the natural algebraic setting in which to discuss the roots of a polynomial; when we discuss the complex roots of a polynomial, for example, we are working in the field of complex numbers. And of course, in linear algebra the scalars that act on vector spaces are always chosen from a field.

We then turn our attention to the ideals of a commutative ring with identity. Recall that in the previous chapter we introduced the concept of an ideal and showed that every element of a commutative ring with identity generates an ideal called a principal ideal. We begin our study of ideals by extending this notion to ideals generated by any collection of elements. Following this, we identify two types of ideals that play a particularly important role in the study of commutative rings with identity: prime ideals and maximal ideals. These ideals are especially important because they extend the notion of prime number from the ring of integers to arbitrary commutative rings with identity. Our goal is to characterize prime and maximal ideals in terms of their quotient rings; specifically, we show that prime ideals are those ideals whose corresponding quotient rings are integral domains, while maximal ideals are those for which the quotient ring is a field. In particular, this gives us a useful technique for constructing fields, namely as quotient rings modulo maximal ideals.

We then conclude the chapter by introducing a new and important family of rings, the polynomial rings. Our goal is to first discuss and illustrate the basic properties of polynomials whose coefficients lie in an arbitrary commutative ring R with identity, and show that the collection $R[X]$ of all such polynomials is itself a commutative ring with identity. We then explore in more detail the properties of polynomials whose coefficients lie in a field, beginning with the division algorithm. The division algorithm for polynomials over a field k is an important tool in the study of the ring $k[X]$. We show, for example, that every ideal of $k[X]$ is a principal ideal, and that if a polynomial is irreducible over k—that is, if it cannot be factored into a product of polynomials of smaller degree—then the principal ideal generated by it is a prime ideal of $k[X]$. Ideals generated by irreducible polynomials, it turns out, are also maximal ideals of $k[X]$ and hence their quotient rings are fields. This, then, gives us an important technique for constructing new fields: simply form the quotient ring of $k[X]$ modulo the ideal generated by an irreducible polynomial. We will use this method to construct and study several such fields, both finite and infinite, and see that many properties of the field depend, to a large extent, upon the irreducible polynomial used to create it.

1. A FEW CONVENTIONS

Throughout this chapter the term *ring* will always mean commutative ring with identity 1, where $1 \neq 0$. Thus, a ring will always contain at least two distinct elements, 0 and 1. If S is a *subring* of a ring R, we will assume that S contains the identity 1. Every subring of a commutative ring with identity is therefore a commutative ring with identity. And finally, if R and S are rings and $f : R \rightarrow S$ is a *ring-homomorphism*, then we assume that $f(1) = 1$; that is, that f maps the identity of R to the identity of S.

2.   INTEGRAL DOMAINS AND FIELDS

As we noted earlier, many of the rings we discussed in the previous chapter, such as the integers, the rational, real and complex numbers, and the Gaussian integers, have important properties in addition to those required by the ring axioms. The ring $\mathbb{Z}$ of integers, for example, is a commutative ring with identity, but also satisfies the additional property that the product of two integers is zero if and only if one or both of the factors is zero, a property also shared by the ring $\mathbb{Q}$ of rational numbers, the ring $\mathbb{R}$ of real numbers, and the ring $\mathbb{C}$ of complex numbers. Moreover, each of the rings $\mathbb{Q}$, $\mathbb{R}$, and $\mathbb{C}$ have the additional property that every nonzero element is a unit. Let us now extend these observations to arbitrary rings.

> **Definition**
>
> Let R be a commutative ring with identity. R is called an *integral domain* if it has the property that whenever $ab = 0$ for some elements $a, b \in R$, either $a = 0$ or $b = 0$. R is called a *field* if every nonzero element in R is a unit of R.

The ring $\mathbb{Z}$ of integers, for example, is an integral domain since, as we noted above, the product of two integers is zero if and only if one or both of the factors is zero. But it is not a field since the only integers that are units are ± 1. The ring $\mathbb{Q}$ of rational numbers, on the other hand, is both an integral domain and a field since every nonzero rational number is a unit of $\mathbb{Q}$. Similarly, the ring $\mathbb{R}$ of real numbers and the ring $\mathbb{C}$ of complex numbers are both integral domains and fields. The ring $\mathbb{Z}_6$ of integers modulo 6, however, is a commutative ring with identity but is neither an integral domain nor a field: it is not an integral domain since $[2][3] = [0]$, for example, but $[2] \neq [0]$ and $[3] \neq [0]$; and it is not a field since $[2]$, for example, is not a unit of $\mathbb{Z}_6$.

Let us mention that an element a of a ring R is frequently called a *zero-divisor* of R if $ab = 0$ for some nonzero element $b \in R$. The zero element is, of course, a zero-divisor of any ring. But there may be nonzero zero-divisors as well. For example, in the ring $\mathbb{Z}_6$ both $[2]$ and $[3]$ are zero-divisors since $[2][3] = [0]$ but neither $[2]$ nor $[3]$ is zero. We may think of an integral domain as a ring that has no zero-divisors other than the zero element.

Now, it is clear that every subring of an integral domain is an integral domain. Thus it follows, for example, that the ring $\mathbb{Z}[i]$ of Gaussian integers is an integral domain since it is a subring of the integral domain $\mathbb{C}$ of complex numbers. But $\mathbb{Z}[i]$ is not a field since the number 2, for example, is not a unit of $\mathbb{Z}[i]$. This example also illustrates the fact that a subring of a field is not necessarily a field, for $\mathbb{C}$ is a field but $\mathbb{Z}[i]$ is not. A subring of a field that is itself a field is called a *subfield* of the field. For example, $\mathbb{Q}$ is a subfield of $\mathbb{R}$, and $\mathbb{R}$ is a subfield of $\mathbb{C}$. Clearly, if R is a subring of a field K, then R is a subfield of K if and only if every nonzero element in R is a unit of R. Finally, if K is a field, then every nonzero element in K is a unit of K and hence the group $U(K)$ of units of K consists of all nonzero elements in K; that is, $U(K) = K^*$, the set of nonzero elements in K. The group K^* is called the *multiplicative group of nonzero elements of* K. Observe that if k is a subfield of K, then k^* is a subgroup of K^*.

PROPOSITION 1. Every field is an integral domain.

Proof. Let K be a field and let a and b be elements in K such that $ab = 0$. To show that K is an integral domain, we must show that either $a = 0$ or $b = 0$. Suppose that $a \neq 0$. Then a is a unit of K and therefore $b = 1b = (a^{-1}a)b = a^{-1}(ab) = a^{-1}0 = 0$. Thus, either a or b is zero and therefore K is an integral domain. ■

PROPOSITION 2. Every finite integral domain is a field.

Proof. Let $R = \{a_1, \ldots, a_n\}$ be a finite integral domain and let $a \neq 0$ be a typical nonzero element in R. We must show that a is a unit of R; that is, $ab = 1$ for some element $b \in R$. To this end, consider the set $aR = \{aa_1, \ldots, aa_n\}$ consisting of all multiples of a by elements in R. Clearly, $aR \subseteq R$. On the other hand, if $aa_i = aa_j$ for some $i \neq j$, then $a(a_i - a_j) = 0$, which is a contradiction since R is an integral domain and $a \neq 0$ and $a_i - a_j \neq 0$. Thus the products $aa_1, \ldots, aa_n$ are n distinct elements of R and therefore $aR = R$. Since $1 \in R$, it now follows that $1 = ab$ for some element $b \in R$ and hence a is a unit of R, as required. Thus, R is a field. ■

These two results show that a field is a special type of integral domain, namely an integral domain in which all nonzero elements are units, and that finite integral domains are the same as finite fields. In the following example we use these results to describe an important collection of finite fields.

EXAMPLE 1. The finite fields $\mathbb{F}_p$

Let n be a positive integer. Let us show that the ring $\mathbb{Z}_n$ of integers modulo n is an integral domain, and hence a field, if and only if n is prime, and then discuss a method for finding the multiplicative inverse of a given element.

(A) If p is prime, then $\mathbb{Z}_p$ is an integral domain. For suppose that $[s][t] = [0]$ for some elements $[s], [t] \in \mathbb{Z}_p$. Then $st \equiv 0 \pmod{p}$ and therefore p divides the product st. It follows that p divides either s or t and hence that $[s] = [0]$ or $[t] = [0]$. Thus $\mathbb{Z}_p$ is an integral domain. Now suppose, conversely, that $\mathbb{Z}_n$ is an integral domain. To show that n is prime, let $n = st$ be any factorization of n into positive integers s, t. Then $[s][t] = [0]$ in $\mathbb{Z}_n$, and hence either $[s] = [0]$ or $[t] = [0]$ since $\mathbb{Z}_n$ is, by assumption, an integral domain. Therefore, n divides either s or t. Since s

and t lie between 1 and n, either $s = n$, in which case $t = 1$, or $s = 1$, in which case $t = n$. Consequently, n has only the trivial factorizations and hence must be prime. We conclude, therefore, that $\mathbb{Z}_n$ is an integral domain if and only if n is prime. Since finite integral domains are the same as fields, it follows that $\mathbb{Z}_n$ is a field if and only if n is prime. If $n = p$ is prime, we denote the field $\mathbb{Z}_p$ by the symbol $\mathbb{F}_p$ and refer to it as the *field of integers modulo p.*

(B) For any prime p, $\mathbb{F}_p = \{[0], [1], \ldots, [p-1]\}$, where addition and multiplication are performed modulo the prime p. For example, $\mathbb{F}_2 = \{[0], [1]\}$ is a finite field containing two elements in which $[1] + [1] = [0]$, while $\mathbb{F}_3 = \{[0], [1], [2]\}$ is a finite field containing three elements in which $[1] + [1] + [1] = [0]$. And finally, the addition and multiplication tables for the field $\mathbb{F}_5 = \{[0], [1], [2], [3], [4]\}$ are shown below where, for convenience, we no longer include the zero element in the multiplication table.

+	[0]	[1]	[2]	[3]	[4]
[0]	[0]	[1]	[2]	[3]	[4]
[1]	[1]	[2]	[3]	[4]	[0]
[2]	[2]	[3]	[4]	[0]	[1]
[3]	[3]	[4]	[0]	[1]	[2]
[4]	[4]	[0]	[1]	[2]	[3]

·	[1]	[2]	[3]	[4]
[1]	[1]	[2]	[3]	[4]
[2]	[2]	[4]	[1]	[3]
[3]	[3]	[1]	[4]	[2]
[4]	[4]	[3]	[2]	[1]

(C) If p is prime, then $\mathbb{F}_p$ is a field and hence each of its nonzero elements has a multiplicative inverse. How does one determine the multiplicative inverse of a given element in $\mathbb{F}_p$? In $\mathbb{F}_3$, for example, it is easy to see that $[2]^{-1} = [2]$, while in $\mathbb{F}_5$ we find from the multiplication table that $[2]^{-1} = [3]$, $[3]^{-1} = [2]$, and $[4]^{-1} = [4]$. But how do we find the multiplicative inverse of $[26]$ in the field $\mathbb{F}_{37}$, for example, where it is impractical to multiply $[26]$ by each element and wait for $[1]$ to appear? Here is a simple method for finding multiplicative inverses in the fields $\mathbb{F}_p$ that uses the division algorithm for integers. Consider the element $[26] \in \mathbb{F}_{37}$. First divide 37 by 26 to get $37 = (1)(26) + 11$. It follows that $[0] = [1][26] + [11]$ in $\mathbb{F}_{37}$ and hence $[26]^{-1} = [-1][11]^{-1}$. Continue in this manner, dividing 37 by each successive remainder, until the remainder 1 appears—it must appear eventually since the remainders are a strictly decreasing sequence of positive integers:

$$37 = (1)(26) + 11, \quad \text{hence } [26]^{-1} = [-1][11]^{-1};$$

$$37 = (3)(11) + 4, \quad \text{hence } [11]^{-1} = [-1][3][4]^{-1};$$

$$37 = (9)(4) + 1, \quad \text{hence } [4]^{-1} = [-1][9].$$

Therefore, $[26]^{-1} = [-1]^3[3][9] = [-27] = [10]$. Indeed, we find by direct calculation that $[26][10] = [260] = [1]$. In the exercises we formalize this algorithm and ask the reader to show that it will give the multiplicative inverse of any nonzero element in the fields $\mathbb{F}_p$ whenever p is prime. In fact, it will work for any unit in any of the rings $\mathbb{Z}_n$.

EXAMPLE 2

Let $\mathbb{Q}[i] = \{a + bi \in \mathbb{C} \,|\, a, b \in \mathbb{Q}\}$ stand for the set of complex numbers whose real and imaginary parts are rational numbers. We claim that $\mathbb{Q}[i]$ is a field under addition and multiplication of complex numbers. To show this, let $x = a + bi$ and $y = c + di$ be typical elements in $\mathbb{Q}[i]$, where $a, b, c, d \in \mathbb{Q}$. Then

$$x + y = (a + c) + (b + d)i,$$

$$-x = (-a) + (-b)i, \quad \text{and}$$

$$xy = (ac - bd) + (bc + ad)i,$$

all of which are elements in $\mathbb{Q}[i]$. It follows that $\mathbb{Q}[i]$ is a subring of the field $\mathbb{C}$ of complex numbers and hence is a commutative ring. Moreover, since $1 = 1 + 0i \in \mathbb{Q}[i]$, $\mathbb{Q}[i]$ is a commutative ring with identity. Finally, to show that $\mathbb{Q}[i]$ is a field it remains to verify that every nonzero element in $\mathbb{Q}[i]$ is a unit. To this end, let $x = a + bi \neq 0$ be a typical nonzero element in $\mathbb{Q}[i]$ and set

$$y = \frac{a}{a^2 + b^2} - \frac{b}{a^2 + b^2} i.$$

Then $y \in \mathbb{Q}[i]$ since not both a and b are zero, and we find by direct calculation that $xy = 1$. Thus, every nonzero element in $\mathbb{Q}[i]$ is a unit and therefore $\mathbb{Q}[i]$ is a field. Note that $\mathbb{Q} \subseteq \mathbb{Q}[i] \subseteq \mathbb{C}$, and that $\mathbb{Q}$ is a subfield of $\mathbb{Q}[i]$ and $\mathbb{Q}[i]$ is a subfield of $\mathbb{C}$.

We showed earlier that every field is an integral domain. But to what extent is an integral domain a field? We know, for example, that the ring of integers is an integral domain but not a field, although every finite integral domain is a field. In general, an integral domain may fail to be a field for only one reason, namely that its nonzero elements are not all units of the ring. We now describe a formal construction that begins with an arbitrary integral domain and enlarges it to a field by adjoining a multiplicative inverse for each of its nonzero elements. This new field is called the quotient field of the integral domain. The construction is important because it shows that every integral domain may be embedded in a field, thus allowing us to carry out division by arbitrary nonzero elements. Since the construction is similar to the process by which the rational numbers are constructed from the set of integers, let us begin by first reviewing the construction of the rational numbers.

Rational numbers, we recall, are obtained from the integers by first defining a relation $\sim$ on the set $\mathbb{Z} \times \mathbb{Z}^*$ as follows: if (n, m) and (s, t) are ordered pairs of integers, where $m \neq 0$ and $t \neq 0$, then $(n, m) \sim (s, t)$ whenever $nt = ms$. The relation $\sim$ is then an equivalence relation on the set $\mathbb{Z} \times \mathbb{Z}^*$, and a rational number is defined to be the equivalence class $[(n, m)]$ of a typical pair (n, m). For convenience, we usually denote this class by the symbol $\dfrac{n}{m}$. Thus, $\dfrac{n}{m} = \dfrac{s}{t}$ if and only if $nt = ms$. If $\dfrac{n}{m}$ and $\dfrac{s}{t}$ are typical rational numbers, we set

$$\frac{n}{m} + \frac{s}{t} = \frac{nt + ms}{mt} \quad \text{and} \quad \frac{n}{m}\frac{s}{t} = \frac{ns}{mt}.$$

One then proves that these operations are well-defined binary operations and that the set $\mathbb{Q}$ of all such rational numbers is a field under these operations. Let us now imitate this construction for an arbitrary integral domain.

Let R be an integral domain and let $R \times R^*$ stand for the set of ordered pairs (a, b) of elements in R, where $b \neq 0$. Define a relation $\sim$ on $R \times R^*$ as follows: if $(a, b), (c, d) \in R \times R^*$, then $(a, b) \sim (c, d)$ means that $ad = bc$.

PROPOSITION 3. The relation $\sim$ is an equivalence relation on the set $R \times R^*$.

Proof. We verify the three requirements for an equivalence relation.

(A) Reflexivity. Let $(a, b) \in R \times R^*$. Then $ab = ba$ since R is commutative and hence $(a, b) \sim (a, b)$. Therefore $\sim$ is reflexive.

(B) Symmetry. Suppose that $(a, b) \sim (c, d)$ for two pairs $(a, b), (c, d) \in R \times R^*$. Then $ad = bc$. It follows that $cb = da$ and therefore $(c, d) \sim (a, b)$. Hence, $\sim$ is symmetric.

(C) Transitivity. Suppose that $(a, b) \sim (c, d)$ and $(c, d) \sim (e, f)$ for pairs (a, b), $(c, d), (e, f) \in R \times R^*$. Then $ad = bc$ and $cf = de$. It follows that $(af)d = (ad)f = (bc)f = b(cf) = b(de) = (be)d$. Therefore, $(af)d = (be)d$ and hence, since R is an integral domain and $d \neq 0$, $af = be$. Hence, $(a, b) \sim (e, f)$ and therefore $\sim$ is transitive. Thus, $\sim$ is an equivalence relation on the set $R \times R^*$. ∎

For convenience, we denote the equivalence class $[(a, b)]$ of a typical pair $(a, b) \in R \times R^*$ by the symbol $\dfrac{a}{b}$. Thus, $\dfrac{a}{b} = \dfrac{c}{d}$ if and only if $ad = bc$. Now, let

$$R \times R^*/\sim \; = \left\{ \frac{a}{b} \,\middle|\, a, b \in R, b \neq 0 \right\}$$

stand for the set of equivalence classes in $R \times R^*$ under the relation $\sim$. We define addition and multiplication of classes as follows: if $\dfrac{a}{b}$ and $\dfrac{c}{d}$ are typical classes in $R \times R^*$, let

$$\frac{a}{b} + \frac{c}{d} = \frac{ad + bc}{bd} \quad \text{and} \quad \frac{a}{b}\frac{c}{d} = \frac{ac}{bd}.$$

PROPOSITION 4. Addition and multiplication are well-defined binary operations on the set $R \times R^*/\sim$ of equivalence classes and $R \times R^*/\sim$ is a field under these operations. The zero element is the class $\dfrac{0}{1}$, and $\dfrac{a}{b} = \dfrac{0}{1}$ if and only if $a = 0$. Moreover, if $\dfrac{a}{b} \neq \dfrac{0}{1}$, then $\left(\dfrac{a}{b}\right)^{-1} = \dfrac{b}{a}$.

Proof. We begin by showing that addition is a well-defined operation on $R \times R^*/\sim$; the proof that multiplication is well-defined is similar and is left as an exercise for the reader.

(A) Let $\dfrac{a}{b}$ and $\dfrac{c}{d}$ be typical classes in $R \times R^*/\sim$ and suppose that $\dfrac{a'}{b'} = \dfrac{a}{b}$ and $\dfrac{c'}{d'} = \dfrac{c}{d}$ for some $\dfrac{a'}{b'}, \dfrac{c'}{d'} \in R \times R^*/\sim$. We must show that

$$\frac{a'}{b'} + \frac{c'}{d'} = \frac{a}{b} + \frac{c}{d}.$$

Now, $a'b = b'a$ since $\dfrac{a'}{b'} = \dfrac{a}{b}$ and $c'd = d'c$ since $\dfrac{c'}{d'} = \dfrac{c}{d}$. Therefore,

$$
\begin{aligned}
(a'd' + b'c')bd &= a'd'bd + b'c'bd \\
&= (a'b)(dd') + (c'd)(bb') \\
&= (b'a)(dd') + (d'c)(bb') \\
&= b'd'ad + b'd'bc \\
&= b'd'(ad + bc).
\end{aligned}
$$

It now follows that

$$\frac{a'd' + b'c'}{b'd'} = \frac{ad + bc}{bd}$$

and therefore

$$\frac{a'}{b'} + \frac{c'}{d'} = \frac{a'd' + b'c'}{b'd'} = \frac{ad + bc}{bd} = \frac{a}{b} + \frac{c}{d}.$$

Hence, addition is a well-defined binary operation on the set $R \times R^*/\sim$.

To complete the proof that the set $R \times R^*/\sim$ is a field under addition and multiplication of classes, we must show that it is an abelian group under addition, that multiplication is associative, commutative, distributes over addition and has an identity, and, finally, that every nonzero element is a unit.

(B) $R \times R^*/\sim$ is an abelian group under addition. For let $\frac{a}{b}, \frac{c}{d}$, and $\frac{e}{f}$ be typical classes in $R \times R^*/\sim$. Then

$$\left(\frac{a}{b} + \frac{c}{d}\right) + \frac{e}{f} = \frac{ad + bc}{bd} + \frac{e}{f} = \frac{(ad + bc)f + (bd)e}{(bd)f}$$

$$= \frac{adf + bcf + bde}{bdf} = \frac{a(df) + b(cf + de)}{b(df)}$$

$$= \frac{a}{b} + \frac{cf + de}{df} = \frac{a}{b} + \left(\frac{c}{d} + \frac{e}{f}\right).$$

Therefore, addition is associative. Similarly, addition is commutative. Now, since

$$\frac{0}{1} + \frac{a}{b} = \frac{0b + 1a}{1b} = \frac{a}{b} = \frac{a}{b} + \frac{0}{1},$$

it follows that the class $\frac{0}{1}$ is the zero element for addition. Moreover, $\frac{a}{b} = \frac{0}{1}$ if and only if $a = 0$. Finally, if $\frac{a}{b}$ is a typical class in $R \times R^*/\sim$, then

$$\frac{a}{b} + \frac{-a}{b} = \frac{ab + b(-a)}{bb} = \frac{0}{1} = \frac{-a}{b} + \frac{a}{b}.$$

Hence, the additive inverse of $\frac{a}{b}$ is the class $\frac{-a}{b}$; that is, $-\frac{a}{b} = \frac{-a}{b}$. It now follows that $R \times R^*/\sim$ is an abelian group under addition of classes.

(C) The proof that multiplication of classes is associative, commutative, distributes over addition, and has the class $\frac{1}{1}$ as identity element is similar to the above and is left as an exercise for the reader.

(D) Finally, let us show that every nonzero element in $R \times R^*/\sim$ is a unit. Let $\frac{a}{b}$ be a typical nonzero element in $R \times R^*/\sim$. Then $a \neq 0$ and hence $\frac{b}{a}$ is an element in $R \times R^*/\sim$. Since

$$\frac{a}{b}\frac{b}{a} = \frac{ab}{ba} = \frac{1}{1},$$

it follows that $\frac{a}{b}$ is a unit whose multiplicative inverse is $\frac{b}{a}$. Therefore, every nonzero element in $R \times R^*/\sim$ is a unit and hence, we conclude that $R \times R^*/\sim$ is a field under addition and multiplication of classes. ∎

Definition

Let R be an integral domain. The field $R \times R^*/\sim$ is called the *quotient field* of R and is denoted by $QF(R)$.

For example, the quotient field of the ring of integers is just the field of rational numbers; that is, $QF(\mathbb{Z}) = \mathbb{Q}$. And, in the same way that an integer $n \in \mathbb{Z}$ is regarded as the fraction $\dfrac{n}{1} \in \mathbb{Q}$, let us now show that every integral domain R may be embedded in its quotient field $QF(R)$.

PROPOSITION 5. Let R be an integral domain, $QF(R)$ its quotient field, and let $R' = \left\{ \dfrac{a}{1} \in QF(R) \,\middle|\, a \in R \right\}$. Then R' is a subring of $QF(R)$ and $R' \cong R$. Moreover, the quotient field $QF(R)$ is the smallest field containing R in the sense that if K is any subfield of $QF(R)$ such that $R' \subseteq K \subseteq QF(R)$, then $K = QF(R)$.

Proof. To show that R' is a subring of $QF(R)$, let $x = \dfrac{a}{1}$ and $y = \dfrac{b}{1}$ be typical elements in R'. Then $x + y = \dfrac{a+b}{1}$, $-x = \dfrac{-a}{1}$, and $xy = \dfrac{ab}{1}$, all of which are elements in R'. Hence, R' is a subring of $QF(R)$. Now, define the function $f : R \to R'$ by setting $f(a) = \dfrac{a}{1}$ for all elements $a \in R$. Then, for all $a, b \in R$,

$$f(a + b) = \frac{a+b}{1} = \frac{a}{1} + \frac{b}{1} = f(a) + f(b)$$

$$f(ab) = \frac{ab}{1} = \frac{a}{1}\frac{b}{1} = f(a)f(b),$$

and therefore f is a ring-homomorphism. Moreover, f is 1–1; for if $f(a) = f(b)$ for some $a, b \in R$, then $\dfrac{a}{1} = \dfrac{b}{1}$ and hence $a = b$. Since f is clearly an onto mapping, we conclude that f is a ring-isomorphism and therefore $R \cong R'$, as required. Finally, to show that $QF(R)$ is the smallest field containing R', suppose that K is any subfield of $QF(R)$ such that $R' \subseteq K \subseteq QF(R)$ and let $\dfrac{a}{b}$ be a typical element in $QF(R)$. Then $\dfrac{a}{1} \in K$ since $\dfrac{a}{1} \in R'$, and $\dfrac{1}{b} \in K$ since $\dfrac{1}{b} = \left(\dfrac{b}{1}\right)^{-1}$ and $\dfrac{b}{1}$ is a nonzero element of the field K. Therefore, $\dfrac{a}{b} \in K$ since $\dfrac{a}{b} = \dfrac{a}{1}\dfrac{1}{b}$. Hence, $K = QF(R)$. ∎

If R is an integral domain and $\dfrac{a}{b}$ is a typical class in the quotient field $QF(R)$, we think of $\dfrac{a}{b}$ informally as a fraction whose numerator and denominator are elements in R, with the denominator not equal to zero. These fractions are added and multiplied in the same manner that we add and multiply ordinary rational numbers. Indeed, as we noted earlier, $QF(\mathbb{Z}) = \mathbb{Q}$. What we have shown is that the construction of rational numbers from integers may in fact be applied to any integral domain. Moreover, in the same way that we think of an integer as a fraction whose denominator is 1, we may think of any element $a \in R$ as the fraction $\dfrac{a}{1}$ in the quotient field $QF(R)$. And finally, we showed that the quotient field of an integral domain is the smallest field containing the domain since its quotient field is in fact generated by the elements in the domain together with their inverses; that is, $\dfrac{a}{b} = \dfrac{a}{1}\dfrac{1}{b}$. It is in this sense that the quotient field $QF(R)$ is obtained from the integral domain R by adjoining the inverse of every nonzero element in R.

EXAMPLE 3

Let $\mathbb{Z}[i]$ stand for the ring of Gaussian integers and let $\mathbb{Q}[i]$ stand for the ring discussed in Example 2 consisting of those complex numbers whose real and imaginary parts are rational. Then $\mathbb{Z}[i]$ is an integral domain and $\mathbb{Q}[i]$ is a field. We claim that $\mathbb{Q}[i] = QF(\mathbb{Z}[i])$.

(A) Clearly $\mathbb{Z}[i] \subseteq \mathbb{Q}[i]$. Now, if $a + bi$ is a typical element in $\mathbb{Q}[i]$ and $a = \dfrac{n}{m}$ and $b = \dfrac{s}{t}$, then

$$a + bi = \frac{n}{m} + \frac{s}{t}i = \frac{nt + msi}{mt}$$

and therefore $a + bi \in QF(\mathbb{Z}[i])$. Hence, $\mathbb{Z}[i] \subseteq \mathbb{Q}[i] \subseteq QF(\mathbb{Z}[i])$. Since $\mathbb{Q}[i]$ is a subfield of $QF(\mathbb{Z}[i])$ containing $\mathbb{Z}[i]$, it follows that $\mathbb{Q}[i] = QF(\mathbb{Z}[i])$ since $QF(\mathbb{Z}[i])$ is the smallest such subfield.

(B) In part (A) we showed that $\mathbb{Q}[i] = QF(\mathbb{Z}[i])$ by using the fact that $QF(\mathbb{Z}[i])$ is the smallest field containing $\mathbb{Z}[i]$. Let us verify this result by direct calculation. A typical element $x \in QF(\mathbb{Z}[i])$ has the form $x = \dfrac{n + mi}{s + ti}$, where $n, m, s, t \in \mathbb{Z}$ and not both s and t are zero. Therefore,

$$x = \frac{n + mi}{s + ti} \cdot \frac{s - ti}{s - ti}$$

$$= \frac{(ns + mt) + (ms - nt)i}{s^2 + t^2}$$

$$= \frac{ns + mt}{s^2 + t^2} + \frac{ms - nt}{s^2 + t^2}i$$

and hence $x \in Q[i]$. Thus, every element in $QF(\mathbb{Z}[i])$ may be written in the form $a + bi$ for some $a, b \in \mathbb{Q}$ and therefore $QF(\mathbb{Z}[i]) = \mathbb{Q}[i]$. For example,

$$\frac{3 + 4i}{5 + 3i} = \frac{3 + 4i}{5 + 3i} \cdot \frac{5 - 3i}{5 - 3i} = \frac{27}{34} + \frac{11}{34}i.$$

Exercises

1. Determine which of the following rings are integral domains, fields, or neither.
   (a) $\mathbb{Z}[\sqrt{2}] = \{n + m\sqrt{2} \in \mathbb{R} \mid n, m \in \mathbb{Z}\}$
   (b) $\mathbb{Q}[\sqrt{2}] = \{a + b\sqrt{2} \in \mathbb{R} \mid a, b \in \mathbb{Q}\}$
   (c) $\left\{ \dfrac{n}{m} \in \mathbb{Q} \mid n, m \in \mathbb{Z}, m \text{ odd} \right\}$
   (d) $\left\{ \begin{pmatrix} a & 0 \\ 0 & b \end{pmatrix} \in \text{Mat}_2(\mathbb{R}) \mid a, b \in \mathbb{R} \right\}$
   (e) $\left\{ \begin{pmatrix} a & 0 \\ 0 & a \end{pmatrix} \in \text{Mat}_2(\mathbb{R}) \mid a \in \mathbb{R} \right\}$
   (f) $\left\{ \begin{pmatrix} a & 0 \\ 0 & 0 \end{pmatrix} \in \text{Mat}_2(\mathbb{R}) \mid a \in \mathbb{R} \right\}$

2. Show that $QF(\mathbb{Z}[\sqrt{2}]) = \mathbb{Q}[\sqrt{2}]$.
3. Let $\mathbb{Z}[\sqrt{3}] = \{n + m\sqrt{3} \in \mathbb{R} \mid n, m \in \mathbb{Z}\}$ and $\mathbb{Q}[\sqrt{3}] = \{a + b\sqrt{3} \in \mathbb{R} \mid a, b \in \mathbb{Q}\}$.
   (a) Show that $\mathbb{Z}[\sqrt{3}]$ is an integral domain.
   (b) Show that $\mathbb{Q}[\sqrt{3}]$ is a field.
   (c) Show that $QF(\mathbb{Z}[\sqrt{3}]) = \mathbb{Q}[\sqrt{3}]$.
   (d) If $x = \dfrac{n + m\sqrt{3}}{s + t\sqrt{3}}$ is a typical element in $QF(\mathbb{Z}[\sqrt{3}])$, find rational numbers a and b such that $x = a + b\sqrt{3}$.
4. Let $\mathbb{Z}[i\sqrt{2}] = \{n + mi\sqrt{2} \in \mathbb{C} \mid n, m \in \mathbb{Z}\}$ and $\mathbb{Q}[i\sqrt{2}] = \{a + bi\sqrt{2} \in \mathbb{C} \mid a, b \in \mathbb{Q}\}$.
   (a) Show that $\mathbb{Z}[i\sqrt{2}]$ is an integral domain.
   (b) Show that $\mathbb{Q}[i\sqrt{2}]$ is a field.
   (c) Show that $QF(\mathbb{Z}[i\sqrt{2}]) = \mathbb{Q}[i\sqrt{2}]$.
   (d) If $x = \dfrac{n + mi\sqrt{2}}{s + ti\sqrt{2}}$ is a typical element in $QF(\mathbb{Z}[i\sqrt{2}])$, find rational numbers a and b such that $x = a + bi\sqrt{2}$.
5. Let $K = \left\{ \begin{pmatrix} a & -b \\ b & a \end{pmatrix} \in \text{Mat}_2(\mathbb{R}) \mid a, b \in \mathbb{R} \right\}$.
   (a) Show that K is a field under addition and multiplication of matrices.
   (b) Show that $K \cong \mathbb{C}$, the field of complex numbers.

6. Let $F(\mathbb{Z})$ stand for the ring of functions on $\mathbb{Z}$ and let $R = \{f \in F(\mathbb{Z}) | f$ is constant$\}$.
   (a) Show that R is an integral domain.
   (b) Show that $QF(R) \cong \mathbb{Q}$, the field of rational numbers.
7. Let $F(\mathbb{R})$ stand for the ring of functions on the field $\mathbb{R}$ of real numbers. Is $F(\mathbb{R})$ an integral domain?
8. Let R be an integral domain.
   (a) If a is any nonzero element in R, show that $ax = ay$ for some elements $x, y \in R$ if and only if $x = y$.
   (b) Let a and b be elements in R with $b \neq 0$. Show that $\dfrac{-a}{b} = \dfrac{a}{-b} = -\dfrac{a}{b}$ in the quotient field $QF(R)$.
   (c) Let a and b be elements in R with $b \neq 0$. Show that $\dfrac{ac}{bc} = \dfrac{a}{b}$ in the quotient field $QF(R)$ for any nonzero element $c \in R$.
9. Let K be a field. Show that $QF(K) \cong K$.
10. Let R be an integral domain that is a subring of a field K. Suppose that for every element $x \in K$ there is some element $a \in R$ such that $ax \in R$. Show that $QF(R) \cong K$.
11. Let R be an integral domain that is a subring of a field K. Let $R^* = \bigcap L$, where the intersection is taken over all subfields L of K such $R \subseteq L$. Show that R^* is a subfield of K and $QF(R) \cong R^*$.
12. Show that the only ideals of a field are the trivial ideals.
13. Show that the only nilpotent element in an integral domain is the zero element (see Exercise 21, Section 1, Chapter 9).
14. Show that the only idempotent elements in an integral domain are 0 and 1 (see Exercise 36, Section 2, Chapter 9).
15. Write out the multiplication table for the multiplicative group $\mathbb{F}_7^*$ of non-zero elements is the field $\mathbb{F}_7$. Determine the inverse and order of each element in $\mathbb{F}_7^*$ and show that $\mathbb{F}_7^*$ is a cyclic group of order 6.
16. Let p be a prime. Show that the only subfield of the finite field $\mathbb{F}_p$ is $\mathbb{F}_p$ itself.
17. Let p be a prime and let $[s]$ be a nonzero element in the field $\mathbb{F}_p$, where $1 \leq s \leq p - 1$.
    (a) Show that there are positive integers $q_1, \ldots, q_n$ and $r_1, \ldots, r_n$ such that

$$p = q_1 s + r_1$$

$$p = q_2 r_1 + r_2$$

$$\vdots$$

$$p = q_n r_{n-1} + r_n,$$

   where $s > r_1 > r_2 > \cdots > r_{n-1} > r_n = 1$.
    (b) Show that $[s]^{-1} = [-1]^n [q_1][q_2] \cdots [q_n]$ in $\mathbb{F}_p$.
    (c) Using the algorithm in parts (a) and (b), find $[32]^{-1}$ in the field $\mathbb{F}_{23}$.
18. Let $2\mathbb{Z}[i]$ stand for the principal ideal of $\mathbb{Z}[i]$ generated by the number 2. Is the quotient ring $\mathbb{Z}[i]/2\mathbb{Z}[i]$ an integral domain?

19. Let R be an integral domain. Is the direct product $R \times R$ an integral domain?

3.   IDEALS IN COMMUTATIVE RINGS

In this section we turn our attention to the ideals of a commutative ring with identity. We begin by first generalizing the notion of a principal ideal, which is an ideal generated by a single element, to ideals generated by any collection of elements in a ring. We then introduce two especially important types of ideals: prime ideals and maximal ideals. Prime ideals and maximal ideals are important because they extend the concept of prime number from the ring of integers to any commutative ring with identity. Our goal is to characterize these ideals in terms of their quotient rings by showing that prime ideals are those ideals for which the quotient ring is an integral domain, while maximal ideals are those for which the quotient ring is a field. Since every field is an integral domain, it follows, in particular, that every maximal ideal is a prime ideal.

Let R be a commutative ring with identity. Recall that an ideal of R is any nonempty subset I with the property that if a and b are arbitrary elements in I, then $a + b \in I$, $-a \in I$, and $xa \in I$ for all $x \in R$. If $a \in R$, the principal ideal of R generated by a is the set $aR = \{ax \in R \mid x \in R\}$ consisting of all multiples of a by elements in R. More generally, let A be any nonempty subset of R and let $AR = \{a_1 x_1 + \cdots + a_n x_n \in R \mid a_1, \ldots, a_n \in A, x_1, \ldots, x_n \in R, n$ a positive integer$\}$ stand for the set of all finite sums of products of the form ax, where $a \in A$ and $x \in R$.

PROPOSITION 1. AR is an ideal of R.

Proof.   Let $a = a_1 x_1 + \cdots + a_n x_n$ and $b = b_1 y_1 + \cdots + b_m y_m$ be typical elements in AR, where $a_1, \ldots, a_n, b_1, \ldots, b_m \in A$ and $x_1, \ldots, x_n, y_1, \ldots, y_m \in R$. Then $a + b = a_1 x_1 + \cdots + b_m y_m$, which is an element of AR, and $-a = a_1(-x_1) + \cdots + a_n(-x_n)$, which is also an element of AR. Moreover, for any $x \in R$, $xa = a_1(xx_1) + \cdots + a_n(xx_n) \in AR$. Therefore, AR is an ideal of R.   ∎

The ideal AR is called the *ideal of R generated by A*. If $A = \{a_1, \ldots, a_n\}$ is a finite subset of R, we write $(a_1, \ldots, a_n)R$ for the ideal generated by A and call it the ideal generated by $a_1, \ldots, a_n$. In particular, principal ideals of R are ideals of R generated by a single element. Note also that $A \subseteq AR$ for any nonempty subset A of R since $a = a1$ for every element $a \in A$. For example, consider the ideal $(6, 8, 10)\mathbb{Z}$ of $\mathbb{Z}$ generated by the integers 6, 8, and 10. By definition, a typical integer in $(6, 8, 10)\mathbb{Z}$ has the form $6n_1 + 8n_2 + 10n_3$, where n_1, n_2, n_3 are arbitrary integers in $\mathbb{Z}$. We claim that $(6, 8, 10)\mathbb{Z} = 2\mathbb{Z}$. For clearly, $(6, 8, 10)\mathbb{Z} \subseteq 2\mathbb{Z}$. On the other hand, $2 \in (6, 8, 10)\mathbb{Z}$ since $2 = (-1)6 + (1)8 + (0)10$ and hence $2\mathbb{Z} \subseteq (6, 8, 10)\mathbb{Z}$. Therefore, $(6, 8, 10)\mathbb{Z}$

$= 2\mathbb{Z}$. Thus, the ideal of $\mathbb{Z}$ generated by 6, 8, and 10 is in fact the principal ideal generated by 2. Observe that the generator 2 is the greatest common divisor of 6, 8, and 10. In the exercises we ask the reader to show, more generally, that if $a_1, \ldots, a_n$ are any nonzero integers and $d = \gcd(a_1, \ldots, a_n)$ is their greatest common divisor, then $(a_1, \ldots, a_n)\mathbb{Z} = d\mathbb{Z}$.

As a second example, consider the ideal $(3 - 2i, 1 + i)\mathbb{Z}[i]$ generated by $3 - 2i$ and $1 + i$ in the ring $\mathbb{Z}[i]$ of Gaussian integers. Since $i(3 - 2i) + (-2)(1 + i) = i$, it follows that $i \in (3 - 2i, 1 + i)\mathbb{Z}[i]$. But i is a unit of the ring $\mathbb{Z}[i]$. Therefore, $(3 - 2i, 1 + i)\mathbb{Z}[i] = \mathbb{Z}[i]$ since the only ideal that contains a unit is the ring itself.

Now, an integer in the ring $\mathbb{Z}$ of integers is said to be prime if its only positive divisors are itself and 1. But this definition makes little sense in an arbitrary ring. In order to discuss the concept of primality within the context of an arbitrary ring, we need a general definition that applies to all rings. To this end, recall that if p is a prime number, then p has the property that whenever it divides the product nm of two integers n and m, then p divides either n or m, a property which in fact characterizes the primality of p. Expressed in terms of ideals of the ring $\mathbb{Z}$, this property states that if $nm \in p\mathbb{Z}$ for some integers $n, m \in \mathbb{Z}$, then either $n \in p\mathbb{Z}$ or $m \in p\mathbb{Z}$. It is this last statement that expresses primality within the more general framework of set membership, free from the divisibility concept in $\mathbb{Z}$, and which serves as our guide for extending the concept of prime number to arbitrary commutative rings with identity.

Definition

Let R be a commutative ring with identity. A *prime ideal* of R is any ideal $P \neq R$ having the property that if $xy \in P$ for some elements $x, y \in R$, then either $x \in P$ or $y \in P$.

In other words, prime ideals are those ideals that contain the product of two elements if and only if they contain at least one of the factors. For example, it is clear from the above remarks that $p\mathbb{Z}$ is a prime ideal of $\mathbb{Z}$ whenever p is a prime number. The zero ideal $\{0\}$ is also a prime ideal of $\mathbb{Z}$ since the product of two integers is zero if and only if one or both of the factors is zero.

In general, one ideal of a ring is "larger" than another ideal if it contains the ideal as a subset. If the ideal is such that there is no proper ideal of the ring that contains it, other than itself, then we say that the ideal is maximal. The following definition expresses this idea more formally.

Definition

Let R be a commutative ring with identity. A *maximal ideal* of R is any ideal $M \neq R$ having the property that if $M \subseteq I \subseteq R$ for some ideal I of R, then either $I = M$ or $I = R$.

Thus, maximal ideals are those ideals of a ring that are not properly contained in any larger ideal except the ring itself and, in that sense, are the "largest" ideals of the ring. Let us emphasize, however, that a ring may have many distinct maximal ideals, and that a given maximal ideal does not necessarily contain every ideal of the ring. For example, we claim that if p is a prime in the ring $\mathbb{Z}$ of integers, then the ideal $p\mathbb{Z}$ is in fact a maximal ideal of $\mathbb{Z}$. Suppose that $p\mathbb{Z} \subseteq I \subseteq \mathbb{Z}$ for some ideal I of $\mathbb{Z}$. Then $I = n\mathbb{Z}$ for some integer $n > 0$ since every ideal of $\mathbb{Z}$ is principal. Since $p \in I$, it follows that n divides p. But p is prime. Therefore, either $n = 1$, in which case $I = \mathbb{Z}$, or $n = p$, in which case $I = p\mathbb{Z}$. Thus, $p\mathbb{Z}$ is not properly contained in any ideal of $\mathbb{Z}$ other than $\mathbb{Z}$ itself and hence is a maximal ideal of $\mathbb{Z}$. The ideals $2\mathbb{Z}$, $3\mathbb{Z}$, and $5\mathbb{Z}$, for example, are all maximal ideals of $\mathbb{Z}$. Note, however, that the zero ideal $\{0\}$ is not a maximal ideal of $\mathbb{Z}$ since, for example, $\{0\} \subsetneq 2\mathbb{Z} \subsetneq \mathbb{Z}$, although it is a prime ideal of $\mathbb{Z}$.

Before we discuss other examples of prime and maximal ideals, let us first characterize these ideals in terms of their quotient rings. Recall that if I is any ideal of a commutative ring R with identity 1, then the quotient ring R/I consists of all cosets $x + I$, where $x \in R$, and R/I is itself a commutative ring with identity $1 + I$ in which addition and multiplication of cosets is defined by $(x + I) + (y + I) = (x + y) + I$ and $(x + I)(y + I) = xy + I$ for all $x + I$, $y + I \in R/I$. The zero element of R/I is the coset I, and $x + I = I$ if and only if $x \in I$.

PROPOSITION 2. Let R be a commutative ring with identity and let I be an ideal of R. Then:

(1) I is a prime ideal of R if and only if the quotient ring R/I is an integral domain;

(2) I is a maximal ideal of R if and only if the quotient ring R/I is a field.

Proof. Suppose that I is a prime ideal of R. Then R/I is a commutative ring with identity. To show that R/I is an integral domain, we must verify that if $ab = 0$ for some $a, b \in R/I$, then either $a = 0$ or $b = 0$. Let $a = x + I$ and $b = y + I$. Then $ab = xy + I$ and hence $ab = 0$ if and only if $xy \in I$. Since I is prime, this means that either $x \in I$, in which case $a = 0$, or $y \in I$, in which case $b = 0$. Thus, R/I is an integral domain. Conversely, if R/I is an integral domain and $xy \in I$ for some $x, y \in R$, then $ab = (x + I)(y + I) = xy + I = 0$, and hence $a = 0$, in which case $x \in I$, or $b = 0$, in which case $y \in I$. Thus, I is a prime ideal of R. It follows, therefore, that I is prime if and only if R/I is an integral domain, which proves statement (1).

Now suppose that I is a maximal ideal of R. Then R/I is a commutative ring with identity. To show that R/I is a field, we must verify that every nonzero element in R/I is a unit. Let $a = x + I$ be a typical nonzero element in R/I. Then $x \notin I$ and hence the ideal $(x, I)R$ generated by x and I properly contains I; that is, $I \subsetneq (x, I)R \subseteq R$. Since I is maximal, $(x, I)R = R$ and therefore $1 = xy + z$ for some $y \in R$ and $z \in I$. Let $b = y + I$. Then $1 + I$ $= (x + I)(y + I) = ab$ and therefore a is a unit of the ring R/I. Thus, every nonzero element in R/I is a unit and hence R/I is a field. Conversely, if R/I is a field, then I is a maximal ideal of R. To show this, suppose that $I \subseteq J \subseteq R$ for some ideal J of R. We must verify that either $J = I$ or $J = R$. Assume $J \neq I$. Then there is some element $x \in J$ such that $x \notin I$. Let $a = x + I$. Then a is a unit of R/I since $a \neq 0$, and therefore $ab = 1$ for some $b \in R/I$. Letting $b = y + I$, it follows that $ab = xy + I = 1 + I$ and hence that $1 = xy + z$ for some $z \in I$. But $x \in J$ and $z \in I \subseteq J$. Therefore, both xy and z lie in J and hence $1 \in J$, which shows that $J = R$, as required. Thus, I is a maximal ideal of R. We conclude, therefore, that I is maximal if and only if R/I is a field and the proof is complete. ■

COROLLARY. Every maximal ideal of a ring is a prime ideal.

Proof. This follows immediately from Proposition 2 since every field is an integral domain. ■

COROLLARY. Let R be a commutative ring with identity. Then the zero ideal $\{0\}$ is a prime ideal of R if and only if R is an integral domain, while $\{0\}$ is a maximal ideal if and only if R is a field.

Proof. These statements follow immediately from Proposition 2 since $R/\{0\} \cong R$. ■

Consider the quotient ring $\mathbb{Z}/n\mathbb{Z}$, for example, where n is a positive integer. Since $\mathbb{Z}/n\mathbb{Z} = \mathbb{Z}_n$, the ring of integers modulo n, and since $\mathbb{Z}_n$ is a field if and only if n is a nonzero prime, it follows that the maximal ideals of $\mathbb{Z}$ are precisely those ideals of the form $p\mathbb{Z}$ for some nonzero prime p. Moreover, since maximal ideals are also prime ideals and since $\{0\}$ is the only other prime ideal of $\mathbb{Z}$, we conclude that the maximal ideals of $\mathbb{Z}$ are just the nonzero prime ideals.

Let us note that a field has only two ideals, namely the zero ideal and the field itself; for if R is a field, then $\{0\}$ is a maximal ideal of R and hence there are no other ideals except $\{0\}$ and R. Of course, this also follows from the fact that every nonzero element in a field is a unit.

Proposition 2 is an especially useful result for several reasons. On the one hand it provides a means, other than the definition, for determining if an ideal I is a prime ideal or maximal ideal of a ring R, or neither: construct the quotient ring R/I and determine if it is an integral domain, a field, or neither. On the other hand, it provides an important method for constructing new fields: form the quotient ring of a commutative ring with identity modulo a maximal ideal. This method for obtaining new fields is an especially important algebraic technique and will be used frequently throughout the rest of the book. Let us now conclude this section by discussing several examples that illustrate these two aspects of Proposition 2.

EXAMPLE 1

Let R be a ring and let $F(R)$ stand for the ring of functions on R. For each element $c \in R$, let $I_c = \{f \in F(R) \mid f(c) = 0\}$ stand for the set of functions on R that vanish at c. Recall that I_c is an ideal of $F(R)$ (see Examples 4 and 9, Section 2, Chapter 9).

(A) We claim that I_c is a prime ideal of $F(R)$ if and only if R is an integral domain, and is a maximal ideal of $F(R)$ if and only if R is a field. To show this, consider the evaluation homomorphism $\varphi_c : F(R) \to R$, where $\varphi_c(f) = f(c)$ for all $f \in F(R)$. Clearly, φ_c is a ring-homomorphism mapping $F(R)$ onto R whose kernel is I_c. Therefore, $F(R)/I_c \cong R$. It follows from Proposition 2, therefore, that I_c is prime if and only if R is an integral domain, while I_c is maximal if and only if R is a field.

(B) For example, the ideal I_0 of functions in $F(\mathbb{R})$ vanishing at 0 is a maximal ideal of $F(\mathbb{R})$ since $F(\mathbb{R})/I_0 \cong \mathbb{R}$, the field of real numbers. Similarly, the ideal I_1 of functions in $F(\mathbb{Q})$ vanishing at 1 is a maximal ideal of $F(\mathbb{Q})$ since $F(\mathbb{Q})/I_1 \cong \mathbb{Q}$, the field of rational numbers. But the ideal I_n of functions in $F(\mathbb{Z})$ vanishing at an arbitrary integer n is not a maximal ideal of $F(\mathbb{Z})$ since $F(\mathbb{Z})/I_n \cong \mathbb{Z}$, which is not a field. However, I_n is a prime ideal of $F(\mathbb{Z})$ since $\mathbb{Z}$ is an integral domain.

(C) We showed in part (B) that the ideal I_n of functions in $F(\mathbb{Z})$ vanishing at the integer n is a prime ideal but not a maximal ideal of $F(\mathbb{Z})$ by using Proposition 2. Let us verify this fact directly from the definition. Suppose that f and g are functions in $F(\mathbb{Z})$ such that $fg \in I_n$. Then $(fg)(n) = f(n)g(n) = 0$. Since $\mathbb{Z}$ is an integral domain, it follows that either $f(n) = 0$, in which case $f \in I_n$, or $g(n) = 0$, in which case $g \in I_n$. Thus, I_n is a prime ideal of $F(\mathbb{Z})$. But I_n is not a maximal ideal of $F(\mathbb{Z})$ since we may construct a larger ideal containing I_n. For example, let $f_2 : \mathbb{Z} \to \mathbb{Z}$ stand for the constant function defined by setting $f_2(m) = 2$ for all $m \in \mathbb{Z}$. Then $I_n \subsetneqq (I_n, f_2)$ and $(I_n, f_2) \neq F(\mathbb{Z})$. For if $(I_n, f_2) = F(\mathbb{Z})$, then $1 = g + hf_2$ for some $g \in I_2$ and $h \in F(\mathbb{Z})$, where 1 stands for the function mapping every integer to 1. This is a contradiction, however, since we find, by evaluating at n, that $1(n) = g(n) + h(n)f_2(n)$, or $1 = 2h(n)$, which is impossible. Hence, $I_n \subsetneqq (I_n, f_2) \subsetneqq F(\mathbb{Z})$ and therefore I_n is not maximal.

EXAMPLE 2. Prime and maximal ideals in $\mathbb{Z}_n$

Recall that if n is a positive integer, the ideals of the ring $\mathbb{Z}_n$ of integers modulo n are principal ideals and hence have the form $[m]\mathbb{Z}_n$, where $[m] \in \mathbb{Z}_n$. We claim that the prime ideals of $\mathbb{Z}_n$ are the same as the maximal ideals, and that a typical ideal $[m]\mathbb{Z}_n$ is maximal if and only if $\gcd(m, n)$ is prime.

(A) Let us begin by showing that the prime ideals of the ring $\mathbb{Z}_n$ are the same as the maximal ideals. Let P be an ideal of $\mathbb{Z}_n$. Then the quotient ring $\mathbb{Z}_n/P$ is a finite ring since $\mathbb{Z}_n$ is finite, and hence $\mathbb{Z}_n/P$ is an integral domain if and only if it is a field. It follows, therefore, that P is prime if and only if P is maximal. Thus, the prime ideals of the ring $\mathbb{Z}_n$ are the same as the maximal ideals.

(B) Now, let $[m]\mathbb{Z}_n$ be a typical ideal in $\mathbb{Z}_n$, where $1 \le m \le n$, and let $d = \gcd(m, n)$. Define the function $f : \mathbb{Z}_n \to \mathbb{Z}_d$ by setting $f([s]) = [s]_d$ for every element $[s] \in \mathbb{Z}_n$, where $[s]_d$ stands for the congruence class of s modulo d. We leave the reader to verify that f is a well-defined ring-homomorphism mapping $\mathbb{Z}_n$ onto $\mathbb{Z}_d$. The kernel of f consists of all elements $[s]$ for which $[s]_d = [0]_d$. In particular, since d divides m, $[m]_d = [0]_d$ and hence $[m] \in \operatorname{Ker} f$. Therefore, $[m]\mathbb{Z}_n \subseteq \operatorname{Ker} f$. On the other hand, $d = \gcd(m, n)$ and hence $d = am + bn$ for some $a, b \in \mathbb{Z}$. Thus, if $[s]_d = [0]_d$, then $s = kd = kam + kbn$ for some integer k and hence $[s] = [k][a][m] \in [m]\mathbb{Z}_n$. It follows that $\operatorname{Ker} f = [m]\mathbb{Z}_n$ and hence, by the fundamental theorem, $\mathbb{Z}_n/[m]\mathbb{Z}_n \cong \mathbb{Z}_d$. Since $\mathbb{Z}_d$ is a field if and only if d is prime, we conclude that $[m]\mathbb{Z}_n$ is a maximal ideal of $\mathbb{Z}_n$, or equivalently, a prime ideal, if and only if d is prime, as required.

(C) Let us illustrate the above results by discussing the ideals of the ring $\mathbb{Z}_{12}$. The ideals of $\mathbb{Z}_{12}$ have the form $[s]\mathbb{Z}_{12}$, where $s = 0, \dots, 11$. We find that:

$[0]\mathbb{Z}_{12} = \{[0]\}$ $[6]\mathbb{Z}_{12} = \{[0], [6]\}$

$[1]\mathbb{Z}_{12} = \mathbb{Z}_{12}$ $[7]\mathbb{Z}_{12} = \mathbb{Z}_{12}$

$[2]\mathbb{Z}_{12} = \{[0], [2], [4], [6], [8], [10]\}$ $[8]\mathbb{Z}_{12} = \{[0], [8], [4]\}$

$[3]\mathbb{Z}_{12} = \{[0], [3], [6], [9]\}$ $[9]\mathbb{Z}_{12} = \{[0], [9], [6], [3]\}$

$[4]\mathbb{Z}_{12} = \{[0], [4], [8]\}$ $[10]\mathbb{Z}_{12} = \{[0], [10], [8], [6], [4], [2]\}$

$[5]\mathbb{Z}_{12} = \mathbb{Z}_{12}$ $[11]\mathbb{Z}_{12} = \mathbb{Z}_{12}$

Thus the ring $\mathbb{Z}_{12}$ has six distinct ideals: $\{[0]\}$, $[2]\mathbb{Z}_{12}$, $[3]\mathbb{Z}_{12}$, $[4]\mathbb{Z}_{12}$, $[6]\mathbb{Z}_{12}$, and $\mathbb{Z}_{12}$, as shown to the right. Since $\gcd(2, 12) = 2$, a prime, $[2]\mathbb{Z}_{12}$ is a maximal ideal of $\mathbb{Z}_{12}$; in this case $\mathbb{Z}_{12}/[2]\mathbb{Z}_{12} \cong \mathbb{Z}_2$, the field of integers modulo 2. Similarly, $\gcd(3, 12) = 3$ and hence $[3]\mathbb{Z}_{12}$ is a maximal ideal of $\mathbb{Z}_{12}$ with $\mathbb{Z}_{12}/[3]\mathbb{Z}_{12} \cong \mathbb{Z}_3$, the field of integers modulo 3. These are the only maximal ideals, and hence prime ideals, of $\mathbb{Z}_{12}$. On the other hand, the ideals $[4]\mathbb{Z}_{12}$ and $[6]\mathbb{Z}_{12}$ are neither maximal nor prime ideals of $\mathbb{Z}_{12}$; in these cases, $\mathbb{Z}_{12}/[4]\mathbb{Z}_{12} \cong \mathbb{Z}_4$ and $\mathbb{Z}_{12}/[6]\mathbb{Z}_{12} \cong \mathbb{Z}_6$, neither of which are integral domains.

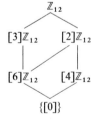

The next three examples illustrate maximal ideals and their quotient fields in the ring $\mathbb{Z}[i]$ of Gaussian integers. In the first two examples we show by direct calculation that the quotient ring $\mathbb{Z}[i]/M$ is a field and conclude that M is a maximal ideal of $\mathbb{Z}[i]$, while in the last example we verify that M is maximal directly from the definition and conclude that the quotient ring is a field. Although our main goal in discussing these examples is to illustrate the relationship between maximal ideals and their quotient fields in $\mathbb{Z}[i]$, an important theme that runs throughout and which the reader should be aware of is that certain prime numbers behave differently when regarded as elements in the ground ring $\mathbb{Z}$ or as elements of the larger ring $\mathbb{Z}[i]$.

EXAMPLE 3

Let $(1 + 2i)\mathbb{Z}[i]$ stand for the principal ideal of $\mathbb{Z}[i]$ generated by $1 + 2i$ and let $K = \mathbb{Z}[i]/(1 + 2i)\mathbb{Z}[i]$ stand for its quotient ring. Let us show by direct calculation that K is a field and hence that $(1 + 2i)\mathbb{Z}[i]$ is a maximal ideal of $\mathbb{Z}[i]$, and then discuss some of the properties of the field K.

(A) We begin by showing that every coset in K has an integer representative. Let $n + mi + (1 + 2i)\mathbb{Z}[i]$ be a typical coset in K. Then

$$n + mi + (1 + 2i)\mathbb{Z}[i] = n + 2m + (1 + 2i)\mathbb{Z}[i]$$

since $(n + mi) - (n + 2m) = m(i - 2) = mi(1 + 2i) \in (1 + 2i)\mathbb{Z}[i]$. Thus, every coset in K has an integer representative and hence may be written in the form $n + (1 + 2i)\mathbb{Z}[i]$ for some integer $n \in \mathbb{Z}$.

(B) Now, we already know that K is a commutative ring with identity. To show that it is a field, it remains to verify that every nonzero element is a unit of K. To this end let $a = n + (1 + 2i)\mathbb{Z}[i]$ be a typical nonzero element in K. We show that a has an inverse as follows. Let $[n]$ stand for the congruence class of n in the field $\mathbb{F}_5$; the number 5 is significant because $5 = (1 + 2i)(1 - 2i) \in (1 + 2i)\mathbb{Z}[i]$. Then $[n] \neq [0]$. For if $[n] = [0]$ in $\mathbb{F}_5$, then $n = 5k$ for some integer k and therefore

$$\begin{aligned} n + (1 + 2i)\mathbb{Z}[i] &= 5k + (1 + 2i)\mathbb{Z}[i] \\ &= (1 + 2i)(1 - 2i)k + (1 + 2i)\mathbb{Z}[i] \\ &= 0 + (1 + 2i)\mathbb{Z}[i]. \end{aligned}$$

In other words $a = 0$ in K, which is a contradiction. Therefore, $[n] \neq [0]$ and hence $[n]$ has an inverse in the field $\mathbb{F}_5$. Let $[n][s] = [1]$, where $[s] \in \mathbb{F}_5$, and let $b = s + (1 + 2i)\mathbb{Z}[i]$. Then $ns = 1 + 5k$ for some integer k and hence

$$\begin{aligned} (n + (1 + 2i)\mathbb{Z}[i])(s + (1 + 2i)\mathbb{Z}[i]) &= ns + (1 + 2i)\mathbb{Z}[i] \\ &= 1 + (1 + 2i)\mathbb{Z}[i] \end{aligned}$$

since $5k \in (1 + 2i)\mathbb{Z}[i]$. Thus, $ab = 1$ in the field K and therefore a is a unit of K. It now follows that K is a field since every nonzero element in K is a unit. Consequently $(1 + 2i)\mathbb{Z}[i]$ is a maximal ideal of $\mathbb{Z}[i]$.

(C) Let us illustrate some calculations in the field K. Let $a = 2 + (1 + 2i)\mathbb{Z}[i]$ and $b = 2 + 3i + (1 + 2i)\mathbb{Z}[i]$. Then, $a + b = 4 + 3i + (1 + 2i)\mathbb{Z}[i]$ and $ab = 4 + 6i + (1 + 2i)\mathbb{Z}[i]$. Now, to find a^{-1} we follow the method described in part (B). In this case $[2]^{-1} = [3]$ in $\mathbb{F}_5$

and hence $a^{-1} = 3 + (1 + 2i)\mathbb{Z}[i]$. Indeed, we find by direct calculation that

$$(2 + (1 + 2i)\mathbb{Z}[i])(3 + (1 + 2i)\mathbb{Z}[i]) = 6 + (1 + 2i)\mathbb{Z}[i];$$

but $6 + (1 + 2i)\mathbb{Z}[i] = 1 + (1 + 2i)\mathbb{Z}[i]$ since $6 - 1 = 5 \in (1 + 2i)\mathbb{Z}[i]$ and therefore

$$(2 + (1 + 2i)\mathbb{Z}[i])(3 + (1 + 2i)\mathbb{Z}[i]) = 1 + (1 + 2i)\mathbb{Z}[i].$$

To find b^{-1}, however, we must first find an integer representative for the coset b. Using part (A), we find that

$$\begin{aligned}
b &= 2 + 3i + (1 + 2i)\mathbb{Z}[i] \\
&= 2 + 2(3) + (1 + 2i)\mathbb{Z}[i] \\
&= 8 + (1 + 2i)\mathbb{Z}[i].
\end{aligned}$$

Since $[8]^{-1} = [2]$ in $\mathbb{F}_5$, it follows that $b^{-1} = 2 + (1 + 2i)\mathbb{Z}[i]$; indeed,

$$\begin{aligned}
(2 + 3i + (1 + 2i)\mathbb{Z}[i])(2 + (1 + 2i)\mathbb{Z}[i]) &= 4 + 6i + (1 + 2i)\mathbb{Z}[i] \\
&= 1 + (1 + 2i)\mathbb{Z}[i]
\end{aligned}$$

since $(4 + 6i) - 1 = 3 + 6i = (1 + 2i)3 \in (1 + 2i)\mathbb{Z}[i]$.

(D) The number 5 has played an important role in the discussion of the field K; indeed, in part (B) we used the fact that $5 = (1 + 2i)(1 - 2i)$ to show that every nonzero element in K has an inverse that can be obtained from the inverse of the corresponding element in $\mathbb{F}_5$. Let us now show that K is, in fact, isomorphic to the field $\mathbb{F}_5$. Define the function $f : \mathbb{F}_5 \to K$ by letting $f([n]) = n + (1 + 2i)\mathbb{Z}[i]$ for every element $[n] \in \mathbb{F}_5$. Then f is well-defined; for if $[n] = [m]$ in $\mathbb{F}_5$, then $n - m$ is a multiple of 5 and hence $n - m \in (1 + 2i)\mathbb{Z}[i]$ since $5 = (1 + 2i)(1 - 2i)$. Therefore, $n + (1 + 2i)\mathbb{Z}[i] = m + (1 + 2i)\mathbb{Z}[i]$, thus showing that f is well-defined. Now, it is easy to see that f is a ring-homomorphism. Furthermore, f maps $\mathbb{F}_5$ onto K since, by part (A), every element in K may be written in the form $n + (1 + 2i)\mathbb{Z}[i]$ for some integer n. Finally, f is 1–1; for its kernel is an ideal of $\mathbb{F}_5$ and hence is either trivial or all of $\mathbb{F}_5$. But $\mathrm{Ker}\, f \neq \mathbb{F}_5$ since f does not map every element in $\mathbb{F}_5$ to the zero element in K. Therefore, $\mathrm{Ker}\, f = \{0\}$ and hence f is 1–1. It now follows that f is an isomorphism and hence that $K \cong \mathbb{F}_5$.

(E) Let us conclude this example by making some observations about the number 5. As an integer, the number 5 is prime and hence it generates a prime ideal $5\mathbb{Z}$ in the ring $\mathbb{Z}$ of integers. But 5 may also be written as a sum of two squares, namely $5 = 1^2 + 2^2$ and hence factors in the ring $\mathbb{Z}[i]$ of Gaussian integers as $5 = (1 + 2i)(1 - 2i)$. Consequently, the ideal $5\mathbb{Z}[i]$ generated by 5 in $\mathbb{Z}[i]$ is no longer a prime ideal of $\mathbb{Z}[i]$ since $5 = (1 + 2i)(1 - 2i) \in 5\mathbb{Z}[i]$, but neither of the factors $1 + 2i$ nor $1 - 2i$ are in $5\mathbb{Z}[i]$ since their real and imaginary parts are not multiples of 5.

It follows, in particular, that the quotient ring $\mathbb{Z}[i]/5\mathbb{Z}[i]$ is not an integral domain; indeed, if we let $a = 1 + 2i + 5\mathbb{Z}[i]$ and $b = 1 - 2i + 5\mathbb{Z}[i]$, then $ab = 0$ in $\mathbb{Z}[i]/5\mathbb{Z}[i]$ but $a \neq 0$ and $b \neq 0$. In the exercises we ask the reader to show that $\mathbb{Z}[i]/5\mathbb{Z}[i] \cong \mathbb{F}_5 \times \mathbb{F}_5$. We also ask the reader to show, more generally, that if p is any prime number that may be written as the sum of two squares in $\mathbb{Z}$, then $p\mathbb{Z}[i]$ is no longer a prime ideal but rather splits as the product of two prime ideals of $\mathbb{Z}[i]$. This concludes the example.

EXAMPLE 4

Let $3\mathbb{Z}[i]$ stand for the principal ideal of $\mathbb{Z}[i]$ generated by the integer 3 and let $K = \mathbb{Z}[i]/3\mathbb{Z}[i]$. Let us show by direct calculation that K is a field and hence that $3\mathbb{Z}[i]$ is a maximal ideal of $\mathbb{Z}[i]$, and then construct the multiplication table for K and discuss its properties in more detail.

(A) We begin by first describing a unique representation for the elements in K. Let $n + mi + 3\mathbb{Z}[i]$ be a typical element in K. Then, by the division algorithm, there are integers q, q', r, r' such that $n - 3q + r$ and $m = 3q' + r'$, where $0 \leq r, r' \leq 2$. Hence,

$$
\begin{aligned}
n + mi + 3\mathbb{Z}[i] &= 3q + r + (3q' + r')i + 3\mathbb{Z}[i] \\
&= r + r'i + 3(q + q'i) + 3\mathbb{Z}[i] \\
&= r + r'i + 3\mathbb{Z}[i].
\end{aligned}
$$

To show that this representation is unique, suppose that $r + r'i + 3\mathbb{Z}[i] = s + s'i + 3\mathbb{Z}[i]$ for some integers r, r', s, s' lying between 0 and 2. Then $r - s + (r' - s')i \in 3\mathbb{Z}[i]$ and therefore $r - s$ and $r' - s'$ must be integer multiples of 3. This is impossible, however, unless $r - s = 0$ and $r' - s' = 0$ since all of these numbers lie between 0 and 2. Therefore, $r = s$ and $r' = s'$. Thus, every element in K may be written uniquely in the form $r + r'i + 3\mathbb{Z}[i]$, where r and r' are integers between 0 and 2. To write a typical element $n + mi + 3\mathbb{Z}[i]$ in K in this form, simply replace n and m by their remainders upon division by 3; for example, $5 + 7i + 3\mathbb{Z}[i] = 2 + i + 3\mathbb{Z}[i]$. Since there are three choices for n and m, namely 0, 1, or 2, it follows that K contains nine elements: $3\mathbb{Z}[i], 1 + 3\mathbb{Z}[i], 2 + 3\mathbb{Z}[i], i + 3\mathbb{Z}[i], 2i + 3\mathbb{Z}[i], 1 + i + 3\mathbb{Z}[i], 2 + i + 3\mathbb{Z}[i], 1 + 2i + 3\mathbb{Z}[i]$, and $2 + 2i + 3\mathbb{Z}[i]$.

(B) Now, K is a commutative ring with identity. To show that it is a field we need only verify that every nonzero element in K is a unit. Here we find by direct calculation, and leave the reader to verify, that the eight nonzero elements in K have the following inverses:

$$
\begin{aligned}
(1 + 3\mathbb{Z}[i])^{-1} &= 1 + 3\mathbb{Z}[i] & (1 + i + 3\mathbb{Z}[i])^{-1} &= 2 + i + 3\mathbb{Z}[i] \\
(2 + 3\mathbb{Z}[i])^{-1} &= 2 + 3\mathbb{Z}[i] & (2 + i + 3\mathbb{Z}[i])^{-1} &= 1 + i + 3\mathbb{Z}[i] \\
(i + 3\mathbb{Z}[i])^{-1} &= 2i + 3\mathbb{Z}[i] & (1 + 2i + 3\mathbb{Z}[i])^{-1} &= 2 + 2i + 3\mathbb{Z}[i] \\
(2i + 3\mathbb{Z}[i])^{-1} &= i + 3\mathbb{Z}[i] & (2 + 2i + 3\mathbb{Z}[i])^{-1} &= 1 + 2i + 3\mathbb{Z}[i].
\end{aligned}
$$

$$K = \mathbb{Z}[i]/3\mathbb{Z}[i] = \{0, 1, 2, a, 2a, 1 + a, 2 + a, 1 + 2a, 2 + 2a\}$$

	1	2	a	$2a$	$1 + a$	$2 + a$	$1 + 2a$	$2 + 2a$
1	1	1	a	$2a$	$1 + a$	$2 + a$	$1 + 2a$	$2 + 2a$
2	2	1	$2a$	a	$2 + 2a$	$1 + 2a$	$2 + a$	$1 + a$
a	a	$2a$	2	1	$2 + a$	$2 + 2a$	$1 + a$	$1 + 2a$
$2a$	$2a$	a	1	2	$1 + 2a$	$1 + a$	$2 + 2a$	$2 + a$
$1 + a$	$1 + a$	$2 + 2a$	$2 + a$	$1 + 2a$	$2a$	1	2	a
$2 + a$	$2 + a$	$1 + a$	$2 + 2a$	$1 + a$	1	a	$2a$	2
$1 + 2a$	$1 + 2a$	$2 + a$	$1 + a$	$2 + 2a$	2	$2a$	a	1
$2 + 2a$	$2 + 2a$	$1 + a$	$1 + 2a$	$2 + a$	a	2	1	$2a$

FIGURE 1. Multiplication table for the quotient field K of Gaussian integers modulo the maximal ideal $3\mathbb{Z}[i]$.

It follows therefore that K is a field containing nine elements and hence that $3\mathbb{Z}[i]$ is a maximal ideal of $\mathbb{Z}[i]$.

(C) Let us discuss the properties of the field K in more detail. For convenience, we denote a typical coset $n + mi + 3\mathbb{Z}[i] \in K$ by the symbol $n + ma$. Then

$$K = \{0, 1, 2, a, 1 + a, 2 + a, 2a, 1 + 2a, 2 + 2a\}.$$

Now, $a^2 = 2$ since $(i + 3\mathbb{Z}[i])^2 = -1 + 3\mathbb{Z}[i] = 2 + 3\mathbb{Z}[i]$. In general, the formulas for addition and multiplication in K are given by:

$$(n + ma) + (s + ta) = (n + s) + (m + t)a$$

$$(n + ma)(s + ta) = (ns + 2mt) + (ms + nt)a$$

for all integers n, m, s, and t. For example, $(1 + 2a)(2 + a) = 2 + a + 4a + 2a^2 = 6 + 5a = 2a$. The table in Figure 1 gives the complete multiplication table for the field K; for convenience, we omit the zero element. Observe that the elements 0, 1, and 2 form a subfield of K isomorphic to $\mathbb{F}_3$. The table in Figure 1 is, of course, the multiplication table for the multiplicative group K^* of nonzero elements in K. Let us show that K^* is in fact a cyclic group of order 8. Consider the element $1 + a \in K^*$. Using the multiplication table in Figure 1, we find that $(1 + a)^2 = 2a$, $(1 + a)^3 = 1 + 2a$, $(1 + a)^4 = 2$, $(1 + a)^5 = 2 + 2a$, $(1 + a)^6 = a$, $(1 + a)^7 = 2 + a$, and $(1 + a)^8 = 1$. Therefore, $1 + a$ has order 8 and hence K^* is a cyclic group of order 8 generated by $1 + a$. The table in the margin summarizes the inverse and order of each element in K^*.

x	$|x|$
1	1
2	2
a	4
$2a$	4
$1 + a$	8
$2 + a$	8
$1 + 2a$	8
$2 + 2a$	8

(D) Finally, let us make an observation about the prime 3. We showed in part (B) that $K = \mathbb{Z}[i]/3\mathbb{Z}[i]$ is a field, from which it follows that $3\mathbb{Z}[i]$

is a maximal ideal of $\mathbb{Z}[i]$. Therefore, $3\mathbb{Z}[i]$ is a prime ideal of $\mathbb{Z}[i]$ since maximal ideals are also prime ideals. Thus, unlike the prime 5 discussed in the previous example, the prime 3 generates a prime ideal in $\mathbb{Z}$ as well as a prime ideal in the larger ring $\mathbb{Z}[i]$. Note also that, unlike the prime 5, the prime 3 cannot be written as the sum of two squares.

EXAMPLE 5

Let us show directly from the definition that the principal ideal $7\mathbb{Z}[i]$ generated by the prime 7 is a maximal ideal of the ring $\mathbb{Z}[i]$ and then discuss the corresponding quotient field $\mathbb{Z}[i]/7\mathbb{Z}[i]$.

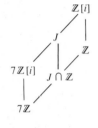

(A) Suppose that J is an ideal of $\mathbb{Z}[i]$ such that $7\mathbb{Z}[i] \subseteq J \subseteq \mathbb{Z}[i]$. To show that $7\mathbb{Z}[i]$ is maximal, we must show that either $J = 7\mathbb{Z}[i]$ or $J = \mathbb{Z}[i]$. Consider the ideal $J \cap \mathbb{Z}$. Clearly, $7\mathbb{Z} \subseteq J \cap \mathbb{Z} \subseteq \mathbb{Z}$. Since $7\mathbb{Z}$ is a maximal ideal of $\mathbb{Z}$, it follows that either $J \cap \mathbb{Z} = 7\mathbb{Z}$ or $J \cap \mathbb{Z} = \mathbb{Z}$. If $J \cap \mathbb{Z} = \mathbb{Z}$, then $1 \in J$; in this case $J = \mathbb{Z}[i]$ and we are done. Assume therefore, that $J \cap \mathbb{Z} = 7\mathbb{Z}$ and let $n + mi$ be a typical element in J. Then $n^2 + m^2 \in J \cap \mathbb{Z}$ since $n^2 + m^2 = (n + mi)(n - mi)$, and therefore $n^2 + m^2 = 7k$ for some integer k. It follows that $[n]^2 + [m]^2 = [0]$ in the field $\mathbb{F}_7$, or equivalently, that $[m]^2 = -[n]^2$. But the table to the left shows that the only solution of this equation in $\mathbb{F}_7$ is the trivial solution $[n] = [m] = [0]$ since the only element in $\mathbb{F}_7$ that is both a square and the negative of a square is $[0]$. Therefore, $n = 7s$ and $m = 7t$ for some integers s and t and hence $n + mi = 7(s + ti)$. Consequently, $n + mi \in 7\mathbb{Z}[i]$ and therefore $J = 7\mathbb{Z}[i]$. Thus, J is either $7\mathbb{Z}[i]$ or $\mathbb{Z}[i]$ and we conclude, therefore, that $7\mathbb{Z}[i]$ is a maximal ideal of $\mathbb{Z}[i]$.

(B) Let $K = \mathbb{Z}[i]/7\mathbb{Z}[i]$. Then it follows from part (A) that K is a field. In the exercises we ask the reader to show that K is a finite field containing 49 elements and that the multiplicative group K^* is a cyclic group of order 48.

(C) Observe, finally, that the prime 7 behaves like the prime 3 discussed in the previous example relative to the rings $\mathbb{Z}$ and $\mathbb{Z}[i]$: both $3\mathbb{Z}$ and $7\mathbb{Z}$ are prime ideals of $\mathbb{Z}$ and, when extended to the larger ring $\mathbb{Z}[i]$, $3\mathbb{Z}[i]$ and $7\mathbb{Z}[i]$ remain prime ideals. The prime 5, however, does not have this property.

$[m]$	$[m]^2$	$-[m]^2$
$[0]$	$[0]$	$[0]$
$[1]$	$[1]$	$[6]$
$[2]$	$[4]$	$[3]$
$[4]$	$[2]$	$[5]$
$[5]$	$[4]$	$[3]$
$[6]$	$[1]$	$[6]$

Exercises

1. Find a single generator for each of the following ideals:
   (a) $(20, 16)\mathbb{Z}$ (c) $(16, 20, 25)\mathbb{Z}$
   (b) $(12, 18, -24)\mathbb{Z}$ (d) $(1 + i, 2)\mathbb{Z}[i]$
2. Let $a_1, \ldots, a_n$ be nonzero integers and let $d = \gcd(a_1, \ldots, a_n)$ stand for their greatest common divisor. Show that $(a_1, \ldots, a_n)\mathbb{Z} = d\mathbb{Z}$.

3. Let $I = (2 + 3i)\mathbb{Z}[i]$ stand for the principal ideal of $\mathbb{Z}[i]$ generated by $2 + 3i$.
   (a) Show that a typical element in I may be written in the form $2n - 3m + (3n + 2m)i$ for some integers n and m.
   (b) Which of the following elements belong to I: $-3 + 2i$, $11 + 10i$, $3 + 11i$, 13, i?

4. Show that:
   (a) $(1 + \sqrt{2})\mathbb{Z}[\sqrt{2}] = (3 + 2\sqrt{2})\mathbb{Z}[\sqrt{2}]$
   (b) $(2 + \sqrt{2})\mathbb{Z}[\sqrt{2}] = \sqrt{2}\mathbb{Z}[\sqrt{2}]$
   (c) $(2 - i, 3)\mathbb{Z}[i] = (1 + 2i)\mathbb{Z}[i]$
   (d) $(2 + 3i)\mathbb{Z}[i] = (3 - 2i)\mathbb{Z}[i]$
   (e) $(4 - 5i, 4 + 4i)\mathbb{Z}[i] = \mathbb{Z}[i]$
   (f) $(2 - 3i)\mathbb{Z}[i] \cap (5 + 6i)\mathbb{Z}[i] = (28 - 3i)\mathbb{Z}[i]$

5. Let $F(\mathbb{Z})$ stand for the ring of functions on $\mathbb{Z}$. For each integer n and prime number p, including $p = 0$, let $P(n;p) = \{f \in F(\mathbb{Z}) \mid f(n) \in p\mathbb{Z}\}$.
   (a) Show that $P(n;p)$ is a prime ideal of $F(\mathbb{Z})$ for all integers n and all primes p.
   (b) If $p \neq 0$, show that $P(n;p)$ is a maximal ideal of $F(\mathbb{Z})$ and that $F(\mathbb{Z})/P(n;p) \cong F_p$.
   (c) Show that $P(n;0)$ is not a maximal ideal of $F(\mathbb{Z})$ and that $F(\mathbb{Z})/P(n;0) \cong \mathbb{Z}$ for all integers n.
   (d) Show that $P(n;0) \subseteq P(n;p)$ for all integers n and all primes p.
   (e) Show that $\bigcap P(n;p) = \{0\}$, where the intersection is over all integers n and all primes p.

6. Let $\mathbb{Q}_2 = \left\{ \dfrac{n}{m} \in \mathbb{Q} \mid m \text{ is odd} \right\}$ stand for the set of rational numbers having odd denominator.
   (a) Show that $\mathbb{Q}_2$ is an integral domain but not a field.
   (b) Let $2\mathbb{Q}_2$ stand for the principal ideal of $\mathbb{Q}_2$ generated by the number 2. Show that $2\mathbb{Q}_2$ is a prime ideal of $\mathbb{Q}_2$.
   (c) Let $x \in \mathbb{Q}_2$. Show that x is a unit of $\mathbb{Q}_2$ and only if $x \notin 2\mathbb{Q}_2$.
   (d) Show that $2\mathbb{Q}_2$ is a maximal ideal of $\mathbb{Q}_2$ and that $\mathbb{Q}_2/2\mathbb{Q}_2 \cong F_2$.
   (e) Show that $2\mathbb{Q}_2$ is the only maximal ideal of $\mathbb{Q}_2$.
   (f) Let I be an ideal of $\mathbb{Q}_2$. Show that $I = 2^s \mathbb{Q}_2$ for some integer $s \geq 0$ and hence conclude that the ideals of $\mathbb{Q}_2$ form a descending chain $\mathbb{Q}_2 \supseteq 2\mathbb{Q}_2 \supseteq 4\mathbb{Q}_2 \supseteq 8\mathbb{Q}_2 \supseteq \dots$.

7. Find the ideals of the ring $\mathbb{Z}_8$ and indicate the inclusion relationships among the ideals by means of a diagram. Show that $[2]\mathbb{Z}_8$ is the only maximal ideal of $\mathbb{Z}_8$ and that $\mathbb{Z}_8/[2]\mathbb{Z}_8 \cong F_2$.

8. Find the ideals of the ring $\mathbb{Z}_{10}$ and indicate the inclusion relationships among the ideals by means of a diagram. Determine which ideals are maximal and find their quotient fields.

9. Let n be a positive integer and p a prime. Show that the ideals of the ring $\mathbb{Z}_{p^n}$ form a descending chain and that $\mathbb{Z}_{p^n}$ has a unique maximal ideal.

10. Let $7\mathbb{Z}[i]$ stand for the principal ideal of $\mathbb{Z}[i]$ generated by the prime 7 and let $K = \mathbb{Z}[i]/7\mathbb{Z}[i]$.
    (a) Show that every element in K may be written uniquely in the form $n + mi + 7\mathbb{Z}[i]$, where n and m are integers between 0 and 6.
    (b) Show that K is a finite field containing 49 elements.
    (c) Let $n + ma$ stand for the typical element $n + mi + 7\mathbb{Z}[i]$ in K, where n and m are integers. Show that $(n + ma) + (s + ta) = (n + s) + (m + t)a$ and $(n + ma)(s + ta) = (ns + 6mt) + (ms + nt)a$ for all integers n, m, s, t.
    (d) Find the order of the elements a, $1 + a$, $2 + a$ in the multiplicative group K^* of nonzero elements in K.
    (g) Show that K^* is a cyclic group of order 48.

11. Let $K = \mathbb{Z}[i]/(2 + 3i)\mathbb{Z}[i]$.
    (a) Show that K is a finite field containing 13 elements.
    (b) Show that $K \cong \mathbb{F}_{13}$.
    (c) Show that the multiplicative group K^* of nonzero elements in K is a cyclic group of order 12.

12. Show that $\mathbb{Z}[i]/5\mathbb{Z}[i] \cong \mathbb{F}_5 \times \mathbb{F}_5$.

13. Let p be a prime number and suppose that $p = n^2 + m^2$ is the sum of squares of two integers n and m.
    (a) Let $\alpha = n + mi$. Show that the principal ideal $\alpha\mathbb{Z}[i]$ is a maximal ideal of the ring $\mathbb{Z}[i]$ and that $\mathbb{Z}[i]/\alpha\mathbb{Z}[i] \cong \mathbb{F}_p$.
    (b) Show that the principal ideal $p\mathbb{Z}[i]$ is neither a prime ideal nor maximal ideal of the ring $\mathbb{Z}[i]$ and that $\mathbb{Z}[i]/p\mathbb{Z}[i] \cong \mathbb{F}_p \times \mathbb{F}_p$. Thus, if p is a prime that may be written as a sum of squares, then $p\mathbb{Z}$ is a prime ideal of $\mathbb{Z}$ but, when extended to $\mathbb{Z}[i]$, $p\mathbb{Z}[i]$ is no longer prime.

14. Let p be a prime number that cannot be written as the sum of squares of two integers. Show that the principal ideal $p\mathbb{Z}[i]$ is a maximal ideal of the ring $\mathbb{Z}[i]$ and that $\mathbb{Z}[i]/p\mathbb{Z}[i]$ is a field containing p^2 elements. Thus, if p cannot be written as a sum of squares, then both $p\mathbb{Z}$ and $p\mathbb{Z}[i]$ are prime ideals.

15. Let P be a nonzero prime ideal of $\mathbb{Z}[i]$.
    (a) Show that $P \cap \mathbb{Z} = p\mathbb{Z}$ for some prime p.
    (b) Show that $\mathbb{Z}[i]/P$ is a finite integral domain.
    (c) Show that P is a maximal ideal of $\mathbb{Z}[i]$.
    (d) Show that the maximal ideals of $\mathbb{Z}[i]$ are the same as the nonzero prime ideals.

16. Let $\mathbb{Z}[\sqrt{2}] = \{n + m\sqrt{2} \in \mathbb{R} \mid n, m \in \mathbb{Z}\}$.
    (a) Show that $\mathbb{Z}[\sqrt{2}]$ is an integral domain but not a field.
    (b) Let $(2 + \sqrt{2})\mathbb{Z}[\sqrt{2}]$ stand for the principal ideal of $\mathbb{Z}[\sqrt{2}]$ generated by the number $2 + \sqrt{2}$. Show that every coset in $\mathbb{Z}[\sqrt{2}]/(2 + \sqrt{2})\mathbb{Z}[\sqrt{2}]$ may be written in the form $n + (2 + \sqrt{2})\mathbb{Z}[\sqrt{2}]$ for some integer n.

(c) Let $K = \mathbb{Z}[\sqrt{2}]/(2 + \sqrt{2})\mathbb{Z}[\sqrt{2}]$. Show that K is a field and that $K \cong \mathbb{F}_2$.

17. Let $K = \mathbb{Z}[\sqrt{2}]/(5 - 3\sqrt{2})\mathbb{Z}[\sqrt{2}]$, where $\mathbb{Z}[\sqrt{2}]$ is the ring discussed in Exercise 16.

 (a) Show that K is a field containing 7 elements and that $K \cong \mathbb{F}_7$.

 (b) Write out the multiplication table for the multiplicative group K^* of nonzero elements in K.

 (c) Show that K^* is a cyclic group of order 6.

18. Let $\mathbb{Z}[\sqrt{2}]$ stand for the ring discussed in Exercise 16. Suppose that p is a prime number that may be written in the form $p = n^2 - 2m^2$ for some integers n and m.

 (a) Show that the principal ideal $p\mathbb{Z}[\sqrt{2}]$ generated by p is neither a prime nor maximal ideal of the ring $\mathbb{Z}[\sqrt{2}]$ and that $\mathbb{Z}[\sqrt{2}]/p\mathbb{Z}[\sqrt{2}] \cong \mathbb{F}_p \times \mathbb{F}_p$.

 (b) Let $\alpha = n + m\sqrt{2}$. Show that the principal ideal $\alpha\mathbb{Z}[\sqrt{2}]$ is a maximal ideal of the ring $\mathbb{Z}[\sqrt{2}]$ and that $\mathbb{Z}[\sqrt{2}]/\alpha\mathbb{Z}[\sqrt{2}] \cong \mathbb{F}_p$.

19. Let $\omega = -\dfrac{1}{2} + \dfrac{1}{2}\sqrt{3}i$ stand for a primitive cube root of unity and let $\mathbb{Z}[\omega] = \{n + m\omega \in \mathbb{C} \,|\, n, m \in \mathbb{Z}\}$.

 (a) Show that $\mathbb{Z}[\omega]$ is an integral domain but not a field.

 (b) Define the function $N : \mathbb{Z}[\omega] \to \mathbb{Z}$ by setting $N(n + m\omega) = (n + m\omega)$ $(n + m\omega^2) = n^2 + m^2 - nm$ for all $n + m\omega \in \mathbb{Z}[\omega]$.

   (1) Show that $N(\alpha\beta) = N(\alpha)N(\beta)$ for all $\alpha, \beta \in \mathbb{Z}[\omega]$.

   (2) Show that α is a unit of $\mathbb{Z}[\omega]$ if and only if $N(\alpha) = 1$.

 (c) Show that the group of units of $\mathbb{Z}[\omega]$ is $U(\mathbb{Z}[\omega]) = \{1, -1, \omega, -\omega, 1 + \omega, -1 - \omega\}$ and that this is a cyclic group of order 6.

20. Let $\mathbb{Z}[\omega]$ stand for the ring discussed in Exercise 19 and let $(2 + 3\omega)\mathbb{Z}[\omega]$ stand for the principal ideal generated by $2 + 3\omega$. Show that $(2 + 3\omega)\mathbb{Z}[\omega]$ is a maximal ideal of $\mathbb{Z}[\omega]$ and that $\mathbb{Z}[\omega]/(2 + 3\omega)\mathbb{Z}[\omega] \cong \mathbb{F}_7$.

21. Let $\mathbb{Z}[\omega]$ stand for the ring discussed in Exercise 19.

 (a) Let $K = \mathbb{Z}[\omega]/2\mathbb{Z}[\omega]$.

   (1) Show that K is a finite field containing four elements.

   (2) Write out the addition and multiplication tables for K.

   (3) Show that the multiplicative group K^* of nonzero elements in K is a cyclic group of order 3.

 (b) Show that the principal ideal $3\mathbb{Z}[\omega]$ is neither a prime ideal nor maximal ideal of $\mathbb{Z}[\omega]$ and that $\mathbb{Z}[\omega]/3\mathbb{Z}[\omega] \cong \mathbb{F}_3 \times \mathbb{F}_3$.

22. Let $\mathbb{Z}[\omega]$ stand for the ring discussed in Exercise 19. Suppose that p is a prime that may be written in the form $p = n^2 + m^2 - nm$ for some integers n and m.

 (a) Show that the principal ideal $p\mathbb{Z}[\omega]$ is neither a prime ideal nor maximal ideal of the ring $\mathbb{Z}[\omega]$ and that $\mathbb{Z}[\omega]/p\mathbb{Z}[\omega] \cong \mathbb{F}_p \times \mathbb{F}_p$.

 (b) Let $\alpha = n + m\omega$.

   (1) Show that n and m are relatively prime and hence that $an + bm = 1$ for some integers a and b.

(2) If $s + t\omega + \alpha\mathbb{Z}[\omega]$ is a typical element in $\mathbb{Z}[\omega]/\alpha\mathbb{Z}[\omega]$, show that $s + t\omega + \alpha\mathbb{Z}[\omega] = N + \mathbb{Z}[\omega]$, where $N = s - t(an + bn - am)$.

(3) Show that the principal ideal $\alpha\mathbb{Z}[\omega]$ is a maximal ideal of the ring $\mathbb{Z}[\omega]$ and that $\mathbb{Z}[\omega]/\alpha\mathbb{Z}[\omega] \cong \mathbb{F}_p$.

23. Let R be a commutative ring with identity and let I be an ideal of R. Show that a coset $x + I \in R/I$ is a unit of the ring R/I if and only if $(I, x) = R$, where (I, x) is the ideal of R generated by I and x.

24. Let R and S be rings and let $f : R \to S$ be a ring-homomorphism.
    (a) If P is a prime ideal of S, show that the pre-image $f^{-1}(P) = \{x \in R \mid f(x) \in P\}$ is a prime ideal of R.
    (b) If f maps R onto S and P is a prime ideal of S, show that $R/f^{-1}(P) \cong S/P$.
    (c) If M is a maximal ideal of S, is $f^{-1}(M)$ necessarily a maximal ideal of R?

25. Let R be a ring and let S be a subring of R.
    (a) If P is a prime ideal of R, show that $P \cap S$ is a prime ideal of S.
    (b) If M is a maximal ideal of R, is $M \cap S$ necessarily a maximal ideal of S?

26. Let R be a ring having the property that for every element $x \in R$, $x^n = x$ for some integer $n > 1$.
    (a) Show that every prime ideal of R is a maximal ideal.
    (b) Show that every Boolean ring has the property mentioned above (see Exercise 19, Section 1, Chapter 9).

27. Let R be an integral domain, P a prime ideal of R, and let $R_p = \left\{ \dfrac{a}{b} \in QF(R) \mid b \notin P \right\}$, where $QF(R)$ stands for the quotient field of R.
    (a) Show that R_p is an integral domain.
    (b) Let PR_p stand for the ideal of R_p generated by P. Show that PR_p is a maximal ideal of R_p.
    (c) If I is a proper ideal of R_p, show that $I \subseteq PR_p$.
    (d) Show that R_p has a unique maximal ideal, namely PR_p.
    (e) Show that $R_{\{0\}} = QF(R)$.

28. **The arithmetic of ideals.** Let R be a commutative ring with identity and let I and J be ideals of R. Let $I + J = \{a + b \in R \mid a \in I, b \in J\}$ and $IJ = \{a_1 b_1 + \cdots + a_n b_n \in R \mid a_1, \ldots, a_n \in I, b_1, \ldots, b_n \in J$, n a positive integer$\}$.
    (a) Show that $I + J$ and IJ are ideals of R. We refer to these ideals as the *sum* and *product* of I and J, respectively.
    (b) Show that $I \subseteq I + J$, $J \subseteq I + J$, and $IJ \subseteq I \cap J$.
    (c) Let I, J, and K be ideals of R. Show that $\{0\} + I = I$, $I + (J + K) = (I + J) + K$, $I + J = J + I$, $RI = I$, $I(JK) = (IJ)K$, $IJ = JI$, $I(J + K) = IJ + IK$, and $(I + J)K = IK + JK$.
    (d) Show that addition of ideals is a binary operation on the set of ideals of R that has an identity element and which satisfies both general associativity and general commutativity.

(e) Show that multiplication of ideals is a binary operation on the set of ideals of R that has an identity element and which satisfies both general associativity and general commutativity.

(f) If n is a positive integer, let $I^n = I \cdots I$ stand for the product of I with itself n times; if $n = 0$, set $I^0 = R$. Show that $I^n I^m = I^{n+m}$ for all nonnegative integers n and m.

(g) Let $n\mathbb{Z}$ and $m\mathbb{Z}$ be ideals of the ring $\mathbb{Z}$ of integers. Show that:

   (1) $n\mathbb{Z} + m\mathbb{Z} = d\mathbb{Z}$, where $d = \gcd(n, m)$ is the greatest common divisor of n and m;

   (2) $n\mathbb{Z} \cap m\mathbb{Z} = t\mathbb{Z}$, where $t = l\,\mathrm{cm}(n, m)$ is the least common multiple of n and m;

   (3) $(n\mathbb{Z})^s = n^s\mathbb{Z}$ for all integers $s > 0$;

   (4) $(n\mathbb{Z})(m\mathbb{Z}) = nm\mathbb{Z}$.

(h) Let $I = n\mathbb{Z}$ be an ideal of $\mathbb{Z}$, where n is a positive integer, and let $n = p_1^{s_1} \cdots p_t^{s_t}$ be the factorization of n into a product of prime numbers $p_1, \ldots, p_t$. Let $P_i = p_i\mathbb{Z}$ for $i = 1, \ldots, t$. Show that $I = P_1^{s_1} \cdots P_t^{s_t}$. Thus, the ideal I factors into a product of prime ideals P_1, $\ldots, P_t$.

(i) Let $I = [2]\mathbb{Z}_8$ stand for the principal ideal of $\mathbb{Z}_8$ generated by $[2]$. Show that $I^3 = \{[0]\}$.

(j) Let $P = (1 + i)\mathbb{Z}[i]$ stand for the principal ideal of $\mathbb{Z}[i]$ generated by $1 + i$.

   (1) Show that $P^2 = 2\mathbb{Z}[i]$.

   (2) Show that P is a prime ideal of $\mathbb{Z}[i]$ but $2\mathbb{Z}[i]$ is not (see Exercise 13). The prime 2 therefore generates the prime ideal $2\mathbb{Z}$ in the ring $\mathbb{Z}$, but in the larger ring $\mathbb{Z}[i]$ it generates the ideal $2\mathbb{Z}[i]$ which is no longer prime but is, instead, the square of a prime ideal.

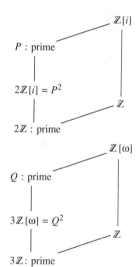

P : prime

$2\mathbb{Z}[i] = P^2$

$2\mathbb{Z}$: prime

$\mathbb{Z}[i]$

$\mathbb{Z}$

(k) Let $Q = (1 - \omega)\mathbb{Z}[\omega]$ stand for the principal ideal of $\mathbb{Z}[\omega]$ generated by $1 - \omega$, where $\mathbb{Z}[\omega]$ is the ring discussed in Exercise 19.

   (1) Show that $Q^2 = 3\mathbb{Z}[\omega]$.

   (2) Show that Q is a prime ideal of $\mathbb{Z}[\omega]$ but $3\mathbb{Z}[\omega]$ is not (see Exercise 21). The prime 3 therefore generates a prime ideal $3\mathbb{Z}$ in the ring $\mathbb{Z}$, but in the larger ring $\mathbb{Z}[\omega]$ it generates the ideal $3\mathbb{Z}[\omega]$ which is no longer prime but is, instead, the square of a prime ideal.

Q : prime

$3\mathbb{Z}[\omega] = Q^2$

$3\mathbb{Z}$: prime

$\mathbb{Z}[\omega]$

$\mathbb{Z}$

29. **Multiplicative norms.** Let R stand for any of the rings $\mathbb{Z}[i]$, $\mathbb{Z}[\sqrt{2}]$ or $\mathbb{Z}[\omega]$ discussed above in Exercises 14, 18, 19, and 22. In these exercises we showed that each of these rings admits a function $N : R \to \mathbb{Z}$ with the property that $N(xy) = N(x)N(y)$ for all $x, y \in R$. Any such function N is called a *multiplicative norm* on R. In the case of $\mathbb{Z}[i]$, $N(n + mi) = n^2 + m^2$; for $\mathbb{Z}[\sqrt{2}]$, $N(n + m\sqrt{2}) = n^2 - 2m^2$; and for $\mathbb{Z}[\omega]$, $N(n + m\omega) = n^2 + m^2 - nm$. Now, if p is a prime number in $\mathbb{Z}$, then $p\mathbb{Z}$ is a maximal ideal of $\mathbb{Z}$ but its extension pR to R may or may not be maximal. Summarize and discuss the results of Exercises 14, 18, 19, and 22 concerning the nature of the ideal pR in terms of the norm mapping N.

4.   POLYNOMIAL RINGS

In this section we introduce a new and exceptionally important family of rings—the polynomial rings. Polynomial rings are important to us for two reasons: first, they have a number of important algebraic properties and are thus a rich source of examples in the study of ring theory; and second, polynomial rings play an important role in the construction of new fields.

We begin by first defining the concept of a polynomial whose coefficients lie in an arbitrary ring R and showing that the collection $R[X]$ of all such polynomials is a ring under addition and multiplication of polynomials. The rest of the section then deals with the algebraic properties of polynomials over a field. We begin by proving the division algorithm, a result that is as important in the study of polynomials as the division algorithm for integers is in the study of integers, and then use the algorithm to derive a number of useful results about roots of polynomials. Following this, we introduce the concept of an irreducible polynomial over a field and show that every nonconstant polynomial over a field may be written in an essentially unique way as the product of irreducible polynomials. This result, the unique factorization theorem for polynomials over a field, shows that irreducible polynomials form the basic building blocks from which all polynomials are constructed. And finally, we conclude the section by discussing the ideals and quotient rings of the ring of polynomials over a field.

Polynomials are familiar to us from elementary algebra; they are expressions of the form $1 + X$ and $2 - 3X + X^2$. As they stand, however, these expressions are meaningless—how do we add the number 1 to the symbol X, for example? The problem, of course, is that these expressions are not actually polynomials but only represent polynomials. Let us begin, therefore, by taking a moment to define the concept of a polynomial whose coefficients lie in an arbitrary commutative ring with identity. We do this by means of infinite sequences.

Definition

Let R be a commutative ring with identity. A *polynomial over* R is an infinite sequence $\{a_i \in R \mid i = 0, 1, \ldots\}$ of elements in R for which all but a finite number of terms in the sequence are zero.

If $\{a_i \in R \mid i = 0, 1, \ldots\}$ is a polynomial over R, then only finitely many of the terms a_i are nonzero and hence there is a largest nonnegative integer n such that $a_n \neq 0$. We then represent the polynomial by means of the formal sum $f(X) = a_0 + a_1 X + \cdots + a_n X^n$ and call this expression a *polynomial in X of degree n with coefficients* $a_0, \ldots, a_n$ in R. The integer n is called the *degree* of the polynomial. On the other hand, if every term $a_i = 0$, we write

$f(X) = 0$ and call this the *zero polynomial*. Let us emphasize that the zero polynomial does not have a degree.

When we represent a polynomial by a formal sum such as $f(X)$, we usually omit those terms whose coefficients are zero. For example, $1 + X^2$, $1 + 0X + X^2$, and $1 + 0X + X^2 + 0X^3$ all represent the same polynomial of degree 2, namely $\{1, 0, 1, 0, 0, 0, \ldots\}$. Moreover, since polynomials are infinite sequences, it is clear that two polynomials are equal if and only if their corresponding coefficients are equal; that is,

$$a_0 + a_1 X + \cdots + a_n X^n = b_0 + b_1 X + \cdots + b_m X^m$$

if and only if $a_i = b_i$ for $i = 0, 1, \ldots$. From now on we will think of polynomials over a ring R as we have always thought of them, namely as expressions of the form $a_0 + a_1 X + \cdots + a_n X^n$, where n is a nonnegative integer and $a_0, \ldots, a_n$ are elements in R.

Let $f(X) = a_0 + a_1 X + \cdots + a_n X^n$ be a nonzero polynomial of degree n over a commutative ring R with identity, where $a_n \neq 0$. We let deg $f(X)$ stand for the degree of $f(X)$; thus, deg $f(X) = n$. The coefficients a_0 and a_n are called the *constant term* and *lead coefficient* of $f(X)$, respectively. A *monic* polynomial is a polynomial whose lead coefficient is the identity element 1 of R. A *constant* polynomial over R is any polynomial of the form $f(X) = c$, where c is any element in R. A *linear*, or *first degree*, polynomial over R is any polynomial having degree 1 and thus has the form $f(X) = a + bX$ for some elements $a, b \in R$, where $b \neq 0$. *Quadratic*, *cubic*, and higher degree polynomials are defined similarly.

Now, let R be a commutative ring with identity and let $R[X]$ stand for the set of polynomials in X with coefficients in the ring R. We define addition and multiplication on $R[X]$ as follows: if $f(X) = a_0 + a_1 X + \cdots + a_n X^n$ and $g(X) = b_0 + b_1 X + \cdots + b_m X^m$ are typical polynomials in $R[X]$, let

$$f(X) + g(X) = c_0 + c_1 X + \cdots + c_s X^s,$$

where $c_i = a_i + b_i$ for $i = 0, 1, \ldots$, and

$$f(X)g(X) = d_0 + d_1 X + \cdots + d_s X^s,$$

where $d_i = a_0 b_i + a_1 b_{i-1} + \cdots + a_i b_0$ for $i = 0, 1, \ldots$. Clearly, the sum and product of two polynomials are also polynomials since all but a finite number of the coefficients c_i and d_i are zero. Thus, addition and multiplication of polynomials are binary operations on the set $R[X]$. We now claim that $R[X]$ is a commutative ring with identity under these operations.

PROPOSITION 1. Let R be a commutative ring with identity. Then $R[X]$ is a commutative ring with identity under addition and multiplication of polynomials. Moreover, if R is an integral domain, then $R[X]$ is an integral domain and $\deg(f(X)g(X)) = \deg f(X) + \deg g(X)$ for all polynomials $f(X)$, $g(X) \in R[X]$.

Proof. We must verify that $R[X]$ is an abelian group under addition of polynomials and that multiplication of polynomials is associative, commutative, distributes over addition, and has an identity.

(A) $R[X]$ is an abelian group under addition. To show this, we begin by showing that addition is associative. Let $f(X) = a_0 + a_1 X + \cdots + a_n X^n$, $g(X) = b_0 + b_1 X + \cdots + b_m X^m$, and $h(X) = c_0 + c_1 X + \cdots + c_p X^p$ be typical polynomials in $R[X]$. Then, for any nonnegative integer i, the ith coefficient of $(f(X) + g(X)) + h(X)$ is $(a_i + b_i) + c_i$. But this is equal to $a_i + (b_i + c_i)$, the ith coefficient of $f(X) + (g(X) + h(X))$, since addition in R is associative. Hence, $(f(X) + g(X)) + h(X) = f(X) + (g(X) + h(X))$ and therefore addition of polynomials is associative. Similarly, addition is commutative. Clearly, the zero element of $R[X]$ is the zero polynomial, and, if $f(X) = a_0 + a_1 X + \cdots + a_n X^n$ is any polynomial in $R[X]$, the additive inverse of $f(X)$ is the polynomial $-f(X) = -a_0 + (-a_1)X + \cdots + (-a_n)X^n$ since $a_i + (-a_i) = 0$ for all nonnegative integers i. It follows, therefore, that $R[X]$ is an abelian group under addition.

(B) The proof that multiplication is associative, commutative, distributes over addition, and has the constant polynomial $f(X) = 1$ as identity, where 1 stands for the identity of R, is similar and is left as an exercise for the reader. Therefore, $R[X]$ is a commutative ring with identity.

(C) Finally, suppose that R is an integral domain and let $f(X) = a_0 + a_1 X + \cdots + a_n X^n$ and $g(X) = b_0 + b_1 X + \cdots + b_m X^m$ be nonzero polynomials in $R[X]$, with $n = \deg f(X)$ and $m = \deg g(X)$. Then $f(X)g(X) = c_0 + c_1 X + \cdots + c_s X^s$, where $c_i = a_0 b_i + \cdots + a_i b_0$ for $i = 0, 1, \ldots$. It follows, in particular, that $c_{n+m} = a_n b_m$, and therefore $c_{n+m} \neq 0$ since R is an integral domain and $a_n \neq 0$ and $b_m \neq 0$. Hence, $f(X)g(X) \neq 0$ and therefore $R[X]$ is an integral domain. Finally, if $s > n + m$, then $c_s = a_0 b_s + \cdots + a_n b_{s-n} + \cdots + a_s b_0$, which is equal to zero since $b_s = \cdots = b_{s-n} = 0$ and $a_{n+1} = \cdots = a_s = 0$. Thus, since $c_{n+m} \neq 0$ but $c_s = 0$ for $s > n + m$, it follows that $\deg(f(X)g(X)) = n + m = \deg f(X) + \deg g(X)$ and the proof is complete. ∎

The ring $R[X]$ is called the *ring of polynomials in X over R*. Let us emphasize that if the coefficient ring R is an integral domain, then $R[X]$ is also an integral domain; in this case the degree of the product of two nonzero polynomials is equal to the sum of their degrees. This is an important point since the product of two nonzero polynomials may, in general, have degree smaller than either of the factors. Let us also note that the subset of constant polynomials is a subring of $R[X]$; for if $f_a(X)$ stands for the constant polynomial $f_a(X) = a$ for each element $a \in R$, then $f_a + f_b = f_{a+b}$, $-f_a = f_{-a}$, and $f_a f_b = f_{ab}$ for all $a, b \in R$, and hence the constant polynomials form a subring of R. In fact, the correspondence $a \leftrightarrow f_a$ between the elements of R and the constant polynomials is a ring-isomorphism between R and the subring of constant polynomials, and it is in this sense that we regard R as a subring of $R[X]$.

EXAMPLE 1

Here are some examples of calculations in various polynomial rings. In doing these calculations, let us emphasize that the coefficient arithmetic is done in the coefficient ring R, while the exponents on X are handled as ordinary integers.

(A) Polynomials in $\mathbb{Z}[X]$. Since the ring $\mathbb{Z}$ of integers is an integral domain, $\mathbb{Z}[X]$ is an integral domain. For example, if $f(X) = X^3 + 2X - 1$ and $g(X) = X^3 + X^2 + 3$, then $f(X)$ and $g(X)$ are monic polynomials of degree 3 and we find that

$$f(X) + g(X) = (X^3 + 2X - 1) + (X^3 + X^2 + 3) = 2X^3 + X^2 + 2X + 2$$

$$f(X) - g(X) = (X^3 + 2X - 1) - (X^3 + X^2 + 3) = -X^2 + 2X - 4$$

$$-f(X) = -(X^3 + 2X - 1) = -X^3 - 2X + 1,$$

$$\begin{aligned} f(X)g(X) &= (X^3 + 2X - 1)(X^3 + X^2 + 3) \\ &= X^3(X^3 + X^2 + 3) + 2X(X^3 + X^2 + 3) - (X^3 + X^2 + 3) \\ &= X^6 + X^5 + 3X^3 + 2X^4 + 2X^3 + 6X - X^3 - X^2 - 3 \\ &= X^6 + X^5 + 2X^4 + 4X^3 - X^2 + 6X - 3 \end{aligned}$$

Notice that in calculating the product $f(X)g(X)$ we have used the fact that multiplication distributes over addition rather than the definition of polynomial multiplication.

(B) Polynomials in $\mathbb{C}[X]$. Since the field $\mathbb{C}$ of complex numbers is an integral domain, $\mathbb{C}[X]$ is also an integral domain. For example, if $f(X) = X - i$ and $g(X) = X + i$, then $f(X)$ and $g(X)$ are monic polynomials of degree 1 in $\mathbb{C}[X]$ and we find that $f(X) + g(X) = 2X$, $f(X) - g(X) = -2i$, a constant polynomial, $-f(X) = -X + i$, and $f(X)g(X) = X^2 - i^2 = X^2 + 1$.

(C) Polynomials in $\mathbb{F}_3[X]$. Here $\mathbb{F}_3 = \{[0], [1], [2]\}$ is the field of integers modulo 3, which is an integral domain, and hence $\mathbb{F}_3[X]$ is an integral domain. For example, if $f(X) = [2]X + [1]$ and $g(X) = X^3 + [2]X + [2]$, then

$$f(X) + g(X) = X^3 + [4]X + [3] = X^3 + X, \quad \text{since } [4] = [1] \text{ and } [3] = [0] \text{ in } \mathbb{F}_3;$$

$$f(X) - g(X) = -X^3 - [1] = [2]X^3 + [2], \quad \text{since } -[1] = [2] \text{ in } \mathbb{F}_3;$$

$$-f(X) = -[2]X - [1] = X + [2]$$

$$\begin{aligned} f(X)g(X) &= [2]X(X^3 + [2]X + [2]) + [1](X^3 + [2]X + [2]) \\ &= [2]X^4 + [4]X^2 + [4]X + X^3 + [2]X + [2] \\ &= [2]X^4 + X^3 + X^2 + [2]. \end{aligned}$$

Note that we have not written the identity element $[1]$ when it occurs as a coefficient, as, for example, in $[1]X$, but write simply X.

(D) Polynomials in $\mathbb{Z}_4[X]$. Here $\mathbb{Z}_4 = \{[0], [1], [2], [3]\}$ is the ring of integers modulo 4 and is not an integral domain. For example, if $f(X) = g(X) = [2]X + [2]$, then

$$f(X) + g(X) = [4]X + [4] = [0],$$

$$f(X) - g(X) = [0],$$

$$-f(X) = -[2]X - [2] = [2]X + [2] = f(X)$$

$$\begin{aligned} f(X)g(X) &= [2]X([2]X + [2]) + [2]([2]X + [2]) \\ &= [4]X^2 + [4]X + [4]X + [4] \\ &= [0]. \end{aligned}$$

Since $f(X)g(X) = 0$ in $\mathbb{Z}_4[X]$ but $f(X) \neq 0$ and $g(X) \neq 0$, it follows that $\mathbb{Z}_4[X]$ is not an integral domain. Note that both $f(X)$ and $g(X)$ have degree 1, but their product has no degree since it is the zero polynomial.

Now, if R is a commutative ring with identity, we may interpret polynomials over R as functions mapping R to R as follows. Let $f(X) = a_0 + a_1 X + \cdots + a_n X^n$ be a polynomial in X over R. Define the function $f(X): R \to R$ by setting $[f(X)](c) = f(c) = a_0 + a_1 c + \cdots + a_n c^n$ for all elements $c \in R$; in other words, replace the symbol X by the element $c \in R$ and calculate the resulting expression in R. Then it follows easily from the definition of addition and multiplication of polynomials, together with the arithmetic properties of rings, that $[f(X) + g(X)](c) = f(c) + g(c)$ and $[f(X)g(X)](c) = f(c)g(c)$ for all polynomials $f(X), g(X) \in R[X]$ and all elements $c \in R$. For example, if $f(X) = X^3 + 2X - 1 \in \mathbb{Z}[X]$, then $f(0) = -1$ and $f(1) = 2$; if $f(X) = X^2 + X + [2] \in \mathbb{Z}_4[X]$, then $f([0]) = [2]$ and $f([1]) = [0]$.

If $f(X)$ is a polynomial in $R[X]$, an element $c \in R$ is called a *root* of $f(X)$ if $f(c) = 0$. For example, 0 is a root of $X^3 + X$ in $\mathbb{Z}$, $[1]$ is a root of $X^2 + X + [2]$ in $\mathbb{Z}_4$, and i is a root of $X^2 + 1$ in $\mathbb{C}$. Observe that the polynomial $X^2 + 1$ does not have a root in the ring $\mathbb{Z}$, but, by viewing it as a polynomial over the larger ring $\mathbb{C}$ of complex numbers, it has two roots, namely, i and $-i$. Observe, also, that the polynomial $f(X) = [2]X^2 + [2]X \in \mathbb{Z}_4[X]$ has every element in $\mathbb{Z}_4$ as a root since $f([0]) = f([1]) = f([2]) = f([3]) = [0]$, but $f(X)$ is not the zero polynomial. Thus, it is possible for a polynomial to be identically zero as a function but not identically zero as a polynomial. The following example illustrates how the functional interpretation of polynomials may be used to define evaluation homomorphisms between rings.

EXAMPLE 2. Evaluation homomorphisms

Let R and S be commutative rings with identity such that R is a subring of S, and let c be any element in S. Then every polynomial over R is also a polynomial over S and hence $f(c)$ is a well-defined element in S for any polynomial $f(X) \in R[X]$. Define the function $\varphi_c : R[X] \to S$ by setting $\varphi_c(f(X)) = f(c)$ for every polynomial $f(X) \in R[X]$. Let us show that φ_c is a ring-homomorphism and then discuss its image and kernel.

(A) To show that φ_c is a ring-homomorphism, let $f(X)$ and $g(X)$ be arbitrary polynomials in $R[X]$. Then

$$\varphi_c(f(X) + g(X)) = [f(X) + g(X)](c)$$
$$= f(c) + g(c)$$
$$= \varphi_c(f(X)) + \varphi_c(g(X)),$$

$$\varphi_c(f(X)g(X)) = [f(X)g(X)](c)$$
$$= f(c)g(c)$$
$$= \varphi_c(f(X)) \, \varphi_c(g(X)),$$

and $\varphi_c(1) = 1$. Therefore, φ_c is a ring-homomorphism. The mapping φ_c is called the *evaluation homomorphism on $R[X]$ at c*. For example, the evaluation homomorphism on $\mathbb{Z}[X]$ at the integer 0 is the function $\varphi_0 : \mathbb{Z}[X] \to \mathbb{Z}$, where $\varphi_0(f(X)) = f(0)$ for all polynomials $f(X) \in \mathbb{Z}[X]$; we find, for example, that $\varphi_0(3 - 2X + X^2) = 3$, $\varphi_0(2X - X^4) = 0$, and $\varphi_0(-1 - X^3) = -1$. On the other hand, $\mathbb{Z}$ is a subring of $\mathbb{Q}$, the ring of rational numbers, and hence we have an evaluation homomorphism $\varphi_{1/2} : \mathbb{Z}[X] \to \mathbb{Q}$, for example, where $\varphi_{1/2}(f(X)) = f(1/2)$ for every polynomial $f(X) \in \mathbb{Z}[X]$; here we find, for example, that $\varphi_{1/2}(1 - 2X) = 0$ and $\varphi_{1/2}(X + X^2) = 3/4$.

(B) Let $R[c]$ stand for the image of the evaluation homomorphism $\varphi_c : R[X] \to S$. Then $R[c] = \{f(c) \in S \mid f(X) \in R[X]\}$ and hence consists of all polynomials $a_0 + a_1 c + \cdots + a_n c^n$ in c with coefficients $a_0, \ldots, a_n$ in R. Clearly, $R[c]$ is a subring of S since it is the image of a ring-homomorphism. $R[c]$ is called the *subring of S generated by R and c*. Let us emphasize that although every element in $R[c]$ may be written in the form $a_0 + a_1 c + \cdots + a_n c^n$ for some $a_0, \ldots, a_n \in R$, the coefficients $a_0, \ldots, a_n$ are not, in general, unique. For example, the subring of $\mathbb{C}$ generated by $\mathbb{Z}$ and i is the ring $\mathbb{Z}[i]$ of Gaussian integers. For the subring generated by $\mathbb{Z}$ and i consists of all polynomial expressions of the form $a_0 + a_1 i + \cdots + a_n i^n$, where $a_0, \ldots, a_n$ are integers. But since $i^2 = -1$, every such element may be written in the form $a + bi$ for some $a, b \in \mathbb{Z}$. Therefore, the subring generated by $\mathbb{Z}$ and i is the ring $\mathbb{Z}[i]$ of Gaussian integers. Note that in this case every element in $\mathbb{Z}[i]$ may in fact be written uniquely in the form $a + bi$. We leave the reader to verify, similarly, that $\mathbb{Z}[\sqrt{2}] = \{a + b\sqrt{2} \in \mathbb{R} \mid a, b \in \mathbb{Z}\}$, where each element is uniquely represented in the form $a + b\sqrt{2}$ for some $a, b \in \mathbb{Z}$, and that $\mathbb{R}[i] = \mathbb{C}$.

(C) Finally, observe that the kernel of the evaluation homomorphism $\varphi_c : R[X] \to S$ is the ideal $I_c = \{f(X) \in R[X] \mid f(c) = 0\}$ consisting of all polynomials in $R[X]$ that have c as a root. Since the image of φ_c is the subring $R[c]$ generated by R and c, it follows from the fundamental theorem for ring-homomorphisms that $R[X]/I_c \cong R[c]$. In particular, $R[c]$ is an integral domain if and only if I_c is a prime ideal of the polynomial ring $R[X]$, while $R[c]$ is a field if and only if I_c is a maximal ideal of $R[X]$. For example, consider the evaluation homomorphism $\varphi_0 : \mathbb{Z}[X] \to \mathbb{Z}$ discussed in part (A). In this case $\varphi_0(a_0 + a_1 X + \cdots + a_n X^n) = a_0$ for a typical polynomial $a_0 + a_1 X + \cdots + a_n X^n \in \mathbb{Z}[X]$. Therefore, Im $\varphi_0 = \mathbb{Z}$ and

$$\text{Ker } \varphi_0 = \{a_1 X + \cdots + a_n X^n \in \mathbb{Z}[X] \mid a_1, \ldots, a_n \in \mathbb{Z}, n \text{ a positive integer}\}.$$

Thus, the kernel of φ_0 consists of all polynomials in $\mathbb{Z}[X]$ whose constant term is zero and hence Ker $\varphi_0 = X\mathbb{Z}[X]$, the principal ideal of $\mathbb{Z}[X]$ generated by X. Therefore, $\mathbb{Z}[X]/X\mathbb{Z}[X] \cong \mathbb{Z}$. Since $\mathbb{Z}$ is an integral domain but not a field, it follows that the ideal $X\mathbb{Z}[X]$ is a prime ideal but not a maximal ideal of $\mathbb{Z}[X]$; we leave the reader to verify, for example, that $X\mathbb{Z}[X] \subsetneqq (2, X)\mathbb{Z}[X] \subsetneqq \mathbb{Z}[X]$.

The polynomials that we have discussed up to this point have involved only one symbol—or variable, when we think of polynomials in terms of functions—namely X. But the ring $R[X]$ is itself a commutative ring with identity and hence may be used as a coefficient ring for polynomials in a new symbol as follows.

EXAMPLE 3. Polynomials in several variables

Let R be a commutative ring with identity. Then $R[X]$ is a commutative ring with identity and hence we may form the ring $R[X][Y]$ of polynomials over $R[X]$ in a new symbol Y. We denote this ring by $R[X, Y]$ and call it the *ring of polynomials in X and Y over R*. A typical element in $R[X, Y]$ has the form

$$f(X, Y) = a_{00} + a_{10} X + a_{01} Y + a_{11} XY + \cdots + a_{nm} X^n Y^m,$$

where $a_{00}, \ldots, a_{nm}$ are elements in R and n and m are nonnegative integers. We do not assign a degree to $f(X, Y)$ but refer, instead, to its *degree in X* and *degree in Y*. For example, $f(X, Y) = 2 + 5X + XY + 3Y^2$ is a polynomial in $\mathbb{Z}[X, Y]$ whose degree in X is 1 and whose degree in Y is 2, while $g(X, Y) = 1 + X^2$ is a polynomial in $\mathbb{Z}[X, Y]$ whose degree in X is 2 and degree in Y is 0. In general, if n is a positive integer and $X_1, \ldots, X_n$ are n distinct symbols, the polynomial ring $R[X_1, \ldots, X_n]$ over R in $X_1, \ldots, X_n$ is defined inductively by setting $R[X_1, \ldots, X_n] = R[X_1, \ldots, X_{n-1}][X_n]$.

As the above examples show, the arithmetic properties of the polynomial ring $R[X]$ depend to a large extent on those of the coefficient ring R. If R is an integral domain, for example, then $R[X]$ is also an integral domain and the product of two nonzero polynomials has degree equal to the sum of their degrees. But it is when the coefficient ring is a field that we can say the most about the polynomials. Let us now turn our attention to the properties of polynomials whose coefficients lie in a field.

Let k be a field. Then k is an integral domain and hence the polynomial ring $k[X]$ is an integral domain in which the product of two nonzero polynomials is nonzero and has degree equal to the sum of their degrees. Moreover, since k is a field, every nonzero element in k is a unit and hence divides any polynomial in $k[X]$. Our first goal is to extend this result to the division algorithm for polynomials over a field by showing that every polynomial in $k[X]$ may be divided by any nonzero polynomial to give a unique quotient and a unique remainder that is either zero or whose degree is smaller than the degree of the divisor. We then use this result to derive a number of important properties of the ring $k[X]$, including the existence and uniqueness of greatest common divisors for polynomials. The division algorithm for polynomials over a field is an exceptionally important result since most of the basic properties of the ring $k[X]$ follow directly from it; indeed, it is as important in the study of polynomials over a field as its counterpart, the division algorithm for integers, is in the study integers. Its proof is simply a matter of using the familiar process of long-division within the framework of induction.

PROPOSITION 2. (DIVISION ALGORITHM FOR POLYNOMIALS OVER A FIELD) Let k be a field and let $f(X)$ and $g(X)$ be polynomials over k with $g(X) \neq 0$. Then there are uniquely determined polynomials $q(X)$ and $r(X)$ in $k[X]$ such that $f(X) = q(X)g(X) + r(X)$, where either $r(X) = 0$ or $\deg r(X) < \deg g(X)$.

Proof. We begin by first taking care of a few special cases. If $g(X)$ is a nonzero constant polynomial, say $g(X) = c \neq 0$, then it is easily verified that $q(X) = \dfrac{1}{c} f(X)$ and $r(X) = 0$ are the unique polynomials satisfying the requirements of the Proposition. On the other hand if $f(X) = 0$, then $q(X) = r(X) = 0$ satisfy the requirements of the Proposition and we are done. For the rest of the proof we assume therefore that $g(X)$ is nonconstant and that $f(X) \neq 0$. In this case $f(X)$ has a degree, say n, and we use induction on n to show that there are polynomials $q(X)$ and $r(X)$ in $k[X]$ having the required properties.

(A) Suppose first that $n = 0$. Then $f(X)$ is a nonzero constant polynomial, say $f(X) = c \neq 0$. Let $q(X) = 0$ and $r(X) = c$. Then $f(X) = q(X)g(X) + r(X)$ and $\deg r(X) < \deg g(X)$, as required.

(B) Now assume as induction hypothesis that $n > 0$ and that every polynomial in $k[X]$ whose degree is less than n may be divided by any nonzero polynomial to give a quotient and remainder having the required properties. If $\deg g(X) > n$, let $q(X) = 0$ and $r(X) = f(X)$. Then $f(X) = q(X)g(X) + r(X)$ and $\deg r(X) = \deg f(X) = n < \deg g(X)$, as required.

On the other hand, if $\deg g(X) \leq n$, let $f(X) = a_0 + \cdots + a_n X^n$ and $g(X) = b_0 + \cdots + b_m X^m$, where $a_n \neq 0$, $b_m \neq 0$, and $m \leq n$. Then

$$f(X) - \frac{a_n}{b_m} X^{n-m} g(X)$$

is a polynomial in $k[X]$ whose degree is less than n and hence, by the induction hypothesis, there are polynomials $Q(X), R(X) \in k[X]$ such that

$$f(X) - \frac{a_n}{b_m} X^{n-m} g(X) = Q(X)g(X) + R(X),$$

where either $R(X) = 0$ or $\deg R(X) < \deg g(X)$. Therefore,

$$f(X) = \left[\frac{a_n}{b_m} X^{n-m} + Q(X) \right] g(X) + R(X).$$

Hence, if we let $q(X) = (a_n/b_m)X^{n-m} + Q(X)$ and $r(X) = R(X)$, then $q(X)$ and $r(X)$ satisfy the requirements of the Proposition and the proof of their existence by induction is therefore complete.

To complete the proof, it remains to show that $q(X)$ and $r(X)$ are unique. Let $f(X)$ and $g(X)$ be polynomials over k, with $g(X) \neq 0$, and suppose there are polynomials $q_1(X), q_2(X), r_1(X),$ and $r_2(X)$ in $k[X]$ such that

$$f(X) = q_1(X)g(X) + r_1(X) = q_2(X)g(X) + r_2(X),$$

where either $r_i(X) = 0$ or $\deg r_i(X) < \deg g(X)$ for $i = 1, 2$. Then $[q_1(X) - q_2(X)] g(X) = r_2(X) - r_1(X)$. If either side of this equation is nonzero, then the degree of the left side is greater than or equal to $\deg g(X)$, while the degree of the right side is less than $\deg g(X)$. Since this is impossible, it follows that both sides of the equation must be zero and therefore $q_1(X) = q_2(X)$ and $r_1(X) = r_2(X)$. Thus, the polynomials $q(X)$ and $r(X)$ are uniquely determined and the proof is complete. ∎

For example, if $f(X) = 2X^3 + X - 1$ and $g(X) = X^2 + 3X + 2$ in the ring $\mathbb{Q}[X]$, then we find, using long-division in $\mathbb{Q}[X]$, that $q(X) = 2X - 6$ and $r(X) = 15X + 11$; thus

$$2X^3 + X - 1 = (2X - 6)(X^2 + 3X + 2) + (15X + 11)$$

in $\mathbb{Q}[X]$. Similarly, if we divide $X^4 + [2]X^3 + X^2 + [2]$ by $[2]X^2 + [1]$ in the ring $\mathbb{F}_3[X]$, as illustrated below, we find that $q(X) = [2]X^2 + X + [1]$ and $r(X) = [2]X + [1]$.

$$
\begin{array}{r}
[2]X^2 + X - [2] \\
[2]X^2 + [1] \overline{\smash{\big)}\; X^4 + [2]X^3 + X^2 + [2]} \\
X^4 + [2]X^2 \\
\hline
[2]X^3 - X^2 \\
[2]X^3 + X \\
\hline
-X^2 - X + [2] \\
-X^2 - [2] \\
\hline
- X + [1]
\end{array}
$$

Hence,

$$X^4 + [2]X^3 + X^2 + [2] = ([2]X^2 + X - [2])([2]X^2 + [1]) + ([2]X + [1])$$

in $\mathbb{F}_3[X]$.

COROLLARY.   Let $f(X)$ be a polynomial over a field k and let $c \in k$. Then $f(X) = q(X)(X - c) + f(c)$ for some polynomial $q(X) \in k[X]$. In particular, c is a root of $f(X)$ if and only if $f(X) = q(X)(X - c)$.

Proof.   By the division algorithm there are polynomials $q(X)$, $r(X) \in k[X]$ such that $f(X) = q(X)(X - c) + r(X)$, where either $r(X) = 0$ or deg $r(X) < 1$. It follows that $r(X)$ is constant and hence, since $f(c) = q(c)0 + r(c), r(X) = f(c)$. Therefore, $f(X) = q(X)(X - c) + f(c)$, as required. It follows, in particular, that c is a root of $f(X)$ if and only if $f(X) = q(X)(X - c)$.   ■

COROLLARY.   Let $f(X)$ be a polynomial over a field k and let $c_1, \ldots, c_n$ be distinct roots of $f(X)$ in k. Then $f(X) = q(X)(X - c_1) \cdots (X - c_n)$ for some polynomial $q(X) \in k[X]$. In particular, if $f(X)$ has degree n, then $f(X)$ has at most n distinct roots in k.

Proof.   This follows immediately from the previous corollary by induction on n. We leave the details as an exercise for the reader.   ■

The last corollary shows that a nonzero polynomial of degree n over a field cannot have more than n distinct roots in the field. If the coefficient ring R is not a field, however, then it is possible for a nonzero polynomial of degree n over R to have more than n distinct roots in R. For example, $f(X) = [2]X^2 + [2]X$ is a nonzero polynomial of degree 2 in $\mathbb{Z}_4[X]$ but it has all four elements in $\mathbb{Z}_4$ as roots since $f([0]) = f([1]) = f([2]) = f([3]) = [0]$.

COROLLARY.   Let k be a field. Then every ideal of the polynomial ring $k[X]$ is a principal ideal.

Proof.   Let I be an ideal of $k[X]$. If $I = \{0\}$, then I is principal and we are done. Hence, assume that $I \neq \{0\}$. In this case I contains some nonzero polynomial of least degree, say $g(X)$. We claim that $I = g(X)k[X]$, the principal ideal of $k[X]$ generated by $g(X)$. For clearly, $g(X)k[X] \subseteq I$. To show the reverse inclusion, let $f(X)$ be a polynomial in I. Then it follows from the division algorithm that there are polynomials $q(X)$, $r(X) \in k[X]$ such that $f(X) = q(X)g(X) + r(X)$, where $r(X) = 0$ or deg $r(X) <$ deg $g(X)$. Therefore, $r(X) = f(X) - q(X)g(X)$ and hence $r(X) \in I$. Now, if $r(X)$ is not zero, then it is a polynomial in I having degree smaller than deg $g(X)$, which is the smallest degree of any nonzero polynomial in I. Thus, we must have that $r(X) = 0$ and consequently $f(X) = q(X)g(X)$. Therefore, $f(X) \in g(X)k[X]$. It now follows that $I = g(X)k[X]$ and the proof is complete.   ■

The fact that all ideals of the polynomial rings $k[X]$ over a field k are principal ideals is an especially important and useful fact. The reader should note the similarity between the proof of this result and the proof that every subgroup of the additive group of $\mathbb{Z}$ is cyclic (Example 2, Section 2, Chapter 2); in each case it is the appropriate division algorithm that plays a crucial role in the proof. Let us now use this result to introduce the concept of a greatest common divisor for polynomials; later, in Chapter 11, we will discuss the concept of a greatest common divisor within the framework of arbitrary integral domains.

Let k be a field and let $f(X)$ and $g(X)$ be polynomials over k, with $g(X) \neq 0$. We say that $g(X)$ *divides* $f(X)$ if $f(X) = q(X)g(X)$ for some polynomial $q(X)$ $\in k[X]$; note that if such a polynomial $q(X)$ exists, then it is necessarily unique since $k[X]$ is an integral domain. A *proper divisor* of $f(X)$ is any nonconstant polynomial that divides $f(X)$ and whose degree is less than $\deg f(X)$; all other divisors are called *trivial divisors*. Thus, the trivial divisors of $f(X)$ are the nonzero constant polynomials and the constant multiples $cf(X)$, where c is any nonzero constant in k. For example, $X + 1$ is a proper divisor of $X^2 - 1$ in $\mathbb{Q}[X]$, while the number 3 and the polynomial $3X^2 - 3$ are both trivial divisors. The polynomial $X^2 + 1$, on the other hand, has no proper divisor in $\mathbb{Q}[X]$; but in the ring $\mathbb{C}[X]$, both $X - i$ and $X + i$ are proper divisors of $X^2 + 1$.

Now, let $f(X)$ and $g(X)$ be nonzero polynomials over the field k and let $(f(X), g(X))k[X]$ stand for the ideal of $k[X]$ generated by $f(X)$ and $g(X)$; recall that a typical element in $(f(X), g(X))k[X]$ has the form $a(X)f(X) + b(X)g(X)$, where $a(X)$ and $b(X)$ are arbitrary polynomials in $k[X]$. Then $(f(X), g(X))k[X] = d(X)k[X]$ for some polynomial $d(X) \in k[X]$ since every ideal of $k[X]$ is principal. Moreover, we may assume that $d(X)$ is monic since it may be divided by its lead coefficient and still generate the same ideal. We now claim that $d(X)$ is unique; that is, that there is a unique monic polynomial $d(X)$ for which $(f(X), g(X))k[X] = d(X)k[X]$. For suppose that $(f(X), g(X))k[X] = d'(X)k[X]$ for some monic polynomial $d'(X) \in k[X]$. Then $d(X)k[X] = d'(X)k[X]$ and hence $d(X)$ and $d'(X)$ divide each other. Since both are monic, it follows that $d'(X) = d(X)$, as required. The unique monic polynomial $d(X)$ for which $(f(X), g(X))k[X] = d(X)k[X]$ is called the *greatest common divisor of* $f(X)$ *and* $g(X)$ and is denoted by $\gcd(f(X), g(X))$. Let us now show that the greatest common divisor of two polynomials is in fact a common divisor, that it may be expressed in terms of the polynomials themselves, and that it is the greatest of the common divisors in the sense that it is divisible by every common divisor.

PROPOSITION 3. Let $f(X)$ and $g(X)$ be nonzero polynomials in $k[X]$ and let $d(X) = \gcd(f(X), g(X))$ stand for their greatest common divisor. Then:

(1) $d(X)$ divides both $f(X)$ and $g(X)$;

(2) $d(X) = a(X)f(X) + b(X)g(X)$ for some polynomials $a(X), b(X) \in k[X]$;

(3) if $d'(X)$ is any polynomial in $k[X]$ that divides both $f(X)$ and $g(X)$, then $d'(X)$ divides $d(X)$.

Proof. Since $(f(X), g(X))k[X] = d(X)k[X]$, both $f(X)$ and $g(X)$ belong to $d(X)k[X]$ and hence $d(X)$ divides both $f(X)$ and $g(X)$, which proves statement (1). Statement (2) follows immediately from the fact that $d(X)$ belongs to $(f(X), g(X))k[X]$. And finally, statement (3) follows directly from (2). ■

The easiest way to find greatest common divisors is, of course, to factor the polynomials and then multiply those factors that the polynomials have in common; for example, $\gcd(X^2 - 1, X^3 - 1) = \gcd((X - 1)(X + 1), (X - 1)(X^2 + X + 1)) = X - 1$, $\gcd(2, 3X + 1) = 1$, and $\gcd(2X + 2, 4X^2 - 4) = X + 1$. But at this point we have not yet proven that such a factorization is possible. Let us now turn to this result.

Definition

Let $f(X)$ be a nonconstant polynomial over a field k. Then $f(X)$ is said to be *reducible over* k if $f(X) = g(X)h(X)$ for some nonconstant polynomials $g(X), h(X) \in k[X]$. Otherwise, $f(X)$ is *irreducible over* k.

In other words, a polynomial is reducible over a field k if it has a proper divisor in $k[X]$, while it is irreducible over k if it has only trivial divisors. For example, $X^2 - 1$ is reducible over the field $\mathbb{Q}$ of rational numbers since both $X + 1$ and $X - 1$ are proper divisors of $X^2 - 1$ in $\mathbb{Q}[X]$; in this case, $X^2 - 1 = (X - 1)(X + 1)$ is a proper factorization of $X^2 - 1$. The polynomial $X^2 + 1$, however, is irreducible over $\mathbb{Q}$ since it has no proper divisors in $\mathbb{Q}[X]$. But $X^2 + 1$ is reducible as a polynomial over the larger field $\mathbb{C}$ of complex numbers since $X^2 + 1 = (X - i)(X + i)$ in $\mathbb{C}[X]$. The polynomial $X^2 + 1$ illustrates the fact that the reducibility or irreducibility of a polynomial depends not only on the polynomial, but also on the underlying field: a polynomial that is irreducible over a given field may, in fact, become reducible when viewed as a polynomial over a larger field. And finally, observe that a linear polynomial $f(X) = aX + b$, $a, b \in k$, $a \neq 0$, over an arbitrary field k is always irreducible over k.

PROPOSITION 4. Let $p(X)$ be an irreducible polynomial over a field k and suppose that $p(X)$ divides the product $f(X)g(X)$ of two polynomials $f(X)$ and $g(X)$ in $k[X]$. Then $p(X)$ divides either $f(X)$ or $g(X)$.

Proof. If $p(X)$ divides $f(X)$, we are done. Assume therefore that $p(X)$ does not divide $f(X)$. Then $\gcd(p(X), f(X)) = 1$ and hence $a(X)p(X) + b(X)g(X) = 1$ for some polynomials $a(X), b(X) \in k[X]$. Multiplying both sides of this equation by $g(X)$, we find that $g(X) = a(X)p(X)g(X) + b(X)f(X)g(X)$. Since $p(X)$ divides both $p(X)$ and $f(X)g(X)$, it follows that $p(X)$ divides $g(X)$.   ∎

COROLLARY.   Let $p(X)$ be an irreducible polynomial over a field k and suppose that $p(X)$ divides the product $f_1(X) \cdots f_n(X)$ of a finite number of polynomials in $k[X]$. Then $p(X)$ divides at least one of the polynomials $f_1(X), \ldots, f_n(X)$.

Proof.   This follows immediately from Proposition 4 by induction on n; we leave the details as an exercise for the reader.   ∎

PROPOSITION 5. (UNIQUE FACTORIZATION OF POLYNOMIALS OVER A FIELD)

Let $f(X)$ be a nonconstant polynomial over a field k. Then there are irreducible polynomials $p_1(X), \ldots, p_s(X)$ over k such that $f(X) = p_1(X) \cdots p_s(X)$, and these irreducible factors are uniquely determined to within constants and rearrangements.

Proof.   Let $n = \deg f(X)$. We prove the statement by induction on n. If $n = 1$, then $f(X)$ is linear and hence is irreducible over k. In this case $f(X)$ is the product of a single irreducible factor which is uniquely determined to within constants. Now assume as induction hypothesis that $n > 1$ and that every polynomial over k having degree less than n may be written as the product of irreducible polynomials that are unique to within constants and rearrangements. If $f(X)$ is irreducible, then $f(X)$ is the product of a single irreducible factor, unique to within constants, and we are done. Otherwise $f(X)$ is reducible over k and hence is the product of two nonconstant polynomials each having degree less than n. By the induction hypothesis, each factor may then be written as the product of irreducible polynomials over k. It follows, therefore, that $f(X)$ is a product of irreducible polynomials over k. To prove the uniqueness of this factorization, suppose that $p_1(X), \ldots, p_s(X)$ and $q_1(X), \ldots, q_t(X)$ are irreducible polynomials over k such that $f(X) = p_1(X) \cdots p_s(X) = q_1(X) \cdots q_t(X)$. Then $p_1(X)$ divides the product $q_1(X) \cdots q_t(X)$ and hence divides one of the factors, say $q_j(X)$. Since $q_j(X)$ is also irreducible, it follows that $q_j(X) = cp_1(X)$ for some nonzero constant $c \in k$. Hence,

$$p_1(X)[p_2(X) \cdots p_s(X)] = cp_1(X)[q_1(X) \cdots q_{j-1}(X)q_{j+1}(X) \cdots q_t(X)],$$

and therefore

$$p_2(X) \cdots p_s(X) = cq_1(X) \cdots q_{j-1}(X)q_{j+1}(X) \cdots q_t(X)$$

since $k[X]$ is an integral domain. But the right and left sides of the last equation represent two factorizations of the same polynomial, whose degree is less than n, into a product of irreducible factors. It follows from the induction hypothesis, therefore, that the factors $p_2(X), \ldots, p_s(X)$ may be rearranged and multiplied by appropriate constants to give the factors $q_1(X), \ldots, q_{j-1}(X)$, $q_{j+1}(X), \ldots, q_t(X)$. Thus, the factors $p_1(X), \ldots, p_s(X)$ are the same, to within constants and rearrangements, as the factors $q_1(X), \ldots, q_t(X)$ and the proof by induction is complete.   ∎

Proposition 5 is the basic structure result for polynomials over a field. It guarantees that every nonconstant polynomial over a field may be factored into the product of irreducible polynomials, thus showing that the irreducible polynomials over the field are the basic building blocks from which all other polynomials are constructed, and also insures that the irreducible factors are essentially unique. The uniqueness of the factorization is useful, for example, in determining all divisors of a given polynomial: first find the irreducible factors of the given polynomial, and then multiply these factors in all possible ways. For example, consider the polynomial $X^4 - 1 \in \mathbb{Q}[X]$. Since $X^4 - 1 = (X - 1)(X + 1)(X^2 + 1)$ in $\mathbb{Q}[X]$, and since each of these factors is irreducible over $\mathbb{Q}$, it follows that the divisors of $X^4 - 1$ in $\mathbb{Q}[X]$ are $1, X - 1,$ $X + 1, \ X^2 + 1, \ (X - 1)(X + 1), \ (X - 1)(X^2 + 1), \ (X + 1)(X^2 + 1),$ and $(X - 1)(X + 1)(X^2 + 1),$ or nonzero constant multiples of these eight polynomials.

As we noted earlier, the unique factorization of polynomials over a field is also useful in finding the greatest common divisor of two nonconstant polynomials; in this case, first find the irreducible monic factors of each polynomial and then multiply the irreducible factors that the two polynomials have in common. For example,

$$\gcd(X^2 + X - 2, X^4 - 1) = \gcd((X + 2)(X - 1), (X - 1)(X + 1)(X^2 + 1)) = X - 1$$

and

$$\gcd((X + 2)(X^2 + 2)(2X - 3), 3(X^2 + 2)(2X - 3)(X^2 + 1)) = (X^2 + 2)\left(X - \frac{3}{2}\right).$$

Using this method to find the greatest common divisor of two polynomials assumes, of course, that we can actually find the irreducible factors of the polynomials, which is not always easy to do. There is also an algorithm for finding the greatest common divisor that does not rely on factoring but uses repeated application of the division algorithm for polynomials, and is similar to the algorithm for finding the greatest common divisor of integers. We discuss this method in more detail in the exercises.

Let us now conclude this section by discussing the prime and maximal ideals of the polynomial ring $k[X]$ and their corresponding quotient rings. Recall that every ideal of $k[X]$ is a principal ideal and hence has the form $f(X)k[X]$ for some polynomial $f(X)$. Under what circumstances is $f(X)k[X]$ a prime ideal or maximal ideal of $k[X]$? In the case of integers, for example, it is the prime numbers that generate the prime ideals, with the nonzero prime ideals being maximal. We claim that a similar result is true for polynomials, with irreducible polynomials taking the place of prime numbers.

PROPOSITION 6. Let k be a field and let $f(X)$ be a polynomial over k. Then:

(1) $f(X)k[X]$ is a prime ideal of $k[X]$ if and only if $f(X)$ is either irreducible over k or $f(X) = 0$;

(2) $f(X)k[X]$ is a maximal ideal of $k[X]$ if and only if $f(X)$ is irreducible over k.

Proof. Suppose first that $f(X)k[X]$ is a prime ideal of $k[X]$, with $f(X) \neq 0$. Then $f(X)$ is irreducible over k. For if $f(X) = g(X)h(X)$ for some nonconstant polynomials $g(X)$ and $h(X)$, then one of the factors $g(X)$ or $h(X)$ belongs to $f(X)k[X]$; but this is impossible since both $g(X)$ and $h(X)$ have degree smaller than deg $f(X)$, while every nonzero polynomial in $f(X)k[X]$ has degree at least equal to deg $f(X)$. Hence, $f(X)$ has no proper factorization over k and is therefore irreducible over k. Conversely, if $f(X) = 0$, then $f(X)k[X] = \{0\}$ and hence is a prime ideal since $k[X]$ is an integral domain. On the other hand, if $f(X)$ is irreducible over k and $g(X)h(X) \in f(X)k[X]$, then $f(X)$ divides the product $g(X)h(X)$; hence $f(X)$ divides either $g(X)$, in which case $g(X) \in f(X)k[X]$, or $f(X)$ divides $h(X)$, in which case $h(X) \in f(X)k[X]$. Thus, $f(X)k[X]$ is a prime ideal and the proof of statement (1) is complete.

To prove statement (2), suppose first that $f(X)k[X]$ is a maximal ideal of $k[X]$. Then $f(X)k[X]$ is a prime ideal and hence $f(X)$ is either zero or irreducible over k. But $f(X) \neq 0$ since the zero ideal is not maximal in $k[X]$. Thus, $f(X)$ is irreducible, as required. Conversely, if $f(X)$ is any irreducible polynomial over k then $f(X)k[X]$ is maximal. To show this, let I be an ideal such that $f(X)k[X] \subseteq I$. Then $I = g(X)k[X]$ for some nonzero polynomial $g(X)$, and $g(X)$ divides $f(X)$ since $f(X) \in I$. Since $f(X)$ is irreducible, $g(X)$ is either a nonzero constant, in which case $I = k[X]$, or $g(X)$ is a constant multiple of $f(X)$, in which case $I = f(X)k[X]$. Thus, $f(X)k[X]$ is not properly contained in any larger ideal of $k[X]$ and hence is maximal. ∎

Thus, in answer to our previous question, the maximal ideals of $k[X]$ are those principal ideals that are generated by an irreducible polynomial, while the prime ideals are the maximal ideals together with the zero ideal. It follows, in particular, that if $f(X)$ is irreducible over k, then the quotient ring $k[X]/f(X)k[X]$ is a field. Proposition 6 therefore provides an important method for constructing new fields: simply form the quotient ring of $k[X]$ modulo the principal ideal generated by any polynomial that is irreducible over k. Consider the quotient ring $K = \mathbb{R}[X]/(X^2 + 1)\mathbb{R}[X]$, for example. Since $X^2 + 1$ is irreducible over $\mathbb{R}$, K is a field. We claim, in fact, that $K \cong \mathbb{C}$, the field of complex numbers. For if $\varphi_i : \mathbb{R}[X] \to \mathbb{C}$ is the evaluation homomorphism on $\mathbb{R}[X]$ at i, where $\varphi_i(f(X)) = f(i)$ for all $f(X) \in \mathbb{R}[X]$, then $X^2 + 1 \in \text{Ker } \varphi_i$ and hence $(X^2 + 1)\mathbb{R}[X] \subseteq \text{Ker } \varphi_i$. But $(X^2 + 1)\mathbb{R}[X]$ is a maximal ideal of $\mathbb{R}[X]$. Hence $\text{Ker } \varphi_i = (X^2 + 1)\mathbb{R}[X]$. Since $\text{Im } \varphi_i = \mathbb{C}$, we conclude that $K = \mathbb{R}[X]/(X^2 + 1)\mathbb{R}[X] \cong \mathbb{C}$. Note that under this isomorphism a typical coset $f(X) + (X^2 + 1)\mathbb{R}[X]$ corresponds to the complex number $f(i)$.

EXAMPLE 4

Let $K = \mathbb{Q}[X]/(X^2 - 2)\mathbb{Q}[X]$. Let us show that K is a field and that $K \cong \mathbb{Q}[\sqrt{2}]$, the subring of $\mathbb{R}$ generated by $\mathbb{Q}$ and $\sqrt{2}$.

(A) The polynomial $X^2 - 2$ is irreducible over $\mathbb{Q}$. For if $X^2 - 2$ could be written as the product of two nonconstant polynomials in $\mathbb{Q}[X]$, both factors would be linear and hence $X^2 - 2$ would have roots in the field $\mathbb{Q}$ of rational numbers. Since $X^2 - 2$ has no rational roots, it follows that it is irreducible over $\mathbb{Q}$.

(B) It follows from part (A) that $(X^2 - 2)\mathbb{Q}[X]$ is a maximal ideal of $\mathbb{Q}[X]$ and hence that the quotient ring K is a field. Now, to show that $K \cong \mathbb{Q}[\sqrt{2}]$, consider the evaluation homomorphism $\varphi_{\sqrt{2}} : \mathbb{Q}[X] \to \mathbb{R}$, where $\varphi_{\sqrt{2}}(f(X)) = f(\sqrt{2})$ for every polynomial $f(X) \in \mathbb{Q}[X]$. Clearly $(X^2 - 2)\mathbb{Q}[X] \subseteq \operatorname{Ker} \varphi_{\sqrt{2}}$. Since $(X^2 - 2)\mathbb{Q}[X]$ is a maximal ideal of $\mathbb{Q}[X]$, it follows, as above, that $\operatorname{Ker} \varphi_{\sqrt{2}} = (X^2 - 2)\mathbb{Q}[X]$. Finally, since $\operatorname{Im} \varphi_{\sqrt{2}} = \mathbb{Q}[\sqrt{2}]$ by definition, we conclude that $K = \mathbb{Q}[X]/(X^2 - 2)\mathbb{Q}[X] \cong \mathbb{Q}[\sqrt{2}]$. This shows, in particular, that the subring $\mathbb{Q}[\sqrt{2}]$ generated by $\mathbb{Q}$ and $\sqrt{2}$ is, in fact, a field.

EXAMPLE 5

Let $K = \mathbb{Q}[X]/(X^2 + 1)\mathbb{Q}[X]$. Then K is a field since $X^2 + 1$ is irreducible over the field $\mathbb{Q}$ of rational numbers. We claim that $K \cong \mathbb{Q}[i]$, the subfield of $\mathbb{C}$ generated by $\mathbb{Q}$ and i discussed in Example 2, Section 2. For, as above, the kernel of the evaluation homomorphism $\varphi_i : \mathbb{Q}[X] \to \mathbb{C}$ is the ideal $(X^2 + 1)\mathbb{Q}[X]$ and its image is, by definition, $\mathbb{Q}[i]$. Hence $K \cong \mathbb{Q}[i]$.

EXAMPLE 6

Let $K = \mathbb{F}_2[X]/(X^2 + X + 1)\mathbb{F}_2[X]$. Let us show that the quotient ring K is a finite field containing four elements, construct its addition and multiplication tables, and then show that the multiplicative group K^* of nonzero elements is a cyclic group of order three.

(A) Let $f(X) = X^2 + X + 1 \in \mathbb{F}_2[X]$. Then $f(X)$ is irreducible over $\mathbb{F}_2$. For if $f(X)$ could be written as the product of two nonconstant polynomials, each factor would be linear and hence $f(X)$ would have a root in $\mathbb{F}_2$. But $f(0) = 1$ and $f(1) = 1$. Therefore, $f(X)$ has no root in $\mathbb{F}_2$ and hence is irreducible over $\mathbb{F}_2$.

(B) Since $X^2 + X + 1$ is irreducible over $\mathbb{F}_2$, the quotient ring K is a field. We now claim that every element in K may be written uniquely in the form $aX + b + (X^2 + X + 1)\mathbb{F}_2[X]$, where $a, b \in \mathbb{F}_2$. For let $f(X) + (X^2 + X + 1)\mathbb{F}_2[X]$ be a typical element in K. Then by the division algorithm there are polynomials $q(X), r(X) \in \mathbb{F}_2[X]$ such that $f(X) = q(X)(X^2 + X + 1) + r(X)$, where $r(X) = 0$ or deg $r(X) < 2$. Thus, $r(X) = aX + b$ for some $a, b \in \mathbb{F}_2$ and hence

$$f(X) + (X^2 + X + 1)\mathbb{F}_2[X] = aX + b + (X^2 + X + 1)\mathbb{F}_2[X].$$

To show that the coefficients a and b are uniquely determined suppose that

$$aX + b + (X^2 + X + 1)\mathbb{F}_2[X] = cX + d + (X^2 + X + 1)\mathbb{F}_2[X]$$

for some $a, b, c, d \in \mathbb{F}_2$. Then $(a - c)X + (b - d)$ is a polynomial in the ideal $(X^2 + X + 1)\mathbb{F}_2[X]$ having degree less than 2 and hence must be the zero polynomial. Therefore, $a - c = b - d = 0$, or equivalently, $a = c$ and $b = d$. We conclude, therefore, that every element in K may be written uniquely in the form $aX + b + (X^2 + X + 1)\mathbb{F}_2[X]$.

(C) Let $ax + b$ stand for the typical element $aX + b + (X^2 + X + 1)\mathbb{F}_2[X]$ in K. Then, since there are only two choices for a and b, namely, 0 or 1, it follows that $K = \{0, 1, x, 1 + x\}$. Thus, K is a finite field containing four elements. We find, for example, that $x + x = (1 + 1)x = 0x = 0$ and

$$
\begin{aligned}
x^2 &= (X + (X^2 + X + 1)\mathbb{F}_2[X])^2 \\
&= X^2 + (X^2 + X + 1)\mathbb{F}_2[X] \\
&= X + 1 + (X^2 + X + 1)\mathbb{F}_2[x] \\
&= x + 1.
\end{aligned}
$$

Figure 1 shows the complete addition and multiplication tables for K. Note that $x^2 + x + 1 = x + 1 + x + 1 = 0$, which shows that the element $x \in K$ is a root of the polynomial $X^2 + X + 1$. Thus, the polynomial $X^2 + X + 1$ has no root in the field $\mathbb{F}_2$ but, when regarded as a polynomial over the larger field K, it has the element x as a root.

$$K = \mathbb{F}_2[X]/(X^2 + X + 1)\mathbb{F}_2[X] = \{0, 1, x, 1 + x\}$$

+	0	1	x	$1 + x$
0	0	1	x	$1 + x$
1	1	0	$1 + x$	x
x	x	$1 + x$	0	1
$1 + x$	$1 + x$	x	1	0

$\cdot$	1	x	$1 + x$
1	1	x	$1 + x$
x	x	$1 + x$	1
$1 + x$	$1 + x$	1	x

FIGURE 1.   Addition and multiplication tables for the field K.

(D) Finally, let K^* stand for the multiplicative group of nonzero elements in K. Then $K^* = \{1, x, 1 + x\}$. Since $x^2 = 1 + x$, $K^* = \{1, x, x^2\}$ and therefore K^* is a cyclic group of order three.

EXAMPLE 7

The quotient ring $\mathbb{Q}[X]/(X^2 - 1)\mathbb{Q}[X]$ is not a field since the polynomial $X^2 - 1$ is reducible over the field $\mathbb{Q}$ of rational numbers. Let us show that

$$\mathbb{Q}[X]/(X^2 - 1)\mathbb{Q}[X] \cong \mathbb{Q} \oplus \mathbb{Q}.$$

(A) Recall that the direct sum $\mathbb{Q} \oplus \mathbb{Q}$ is the ring of ordered pairs (a, b), where $a, b \in \mathbb{Q}$, and that addition and multiplication are defined componentwise: $(a, b) + (c, d) = (a + c, b + d)$ and $(a, b)(c, d) = (ac, bd)$ (see Exercise 20, Section 1, Chapter 9). Define the function $\varphi : \mathbb{Q}[X] \to \mathbb{Q} \oplus \mathbb{Q}$ by setting $\varphi(f(X)) = (f(1), f(-1))$ for all polynomials $f(X) \in \mathbb{Q}[X]$. Then φ is a ring-homomorphism; for if $f(X)$ and $g(X)$ are typical polynomials in $\mathbb{Q}[X]$, then

$$\begin{aligned} \varphi(f(X) + g(X)) &= (f(1) + g(1), f(-1) + g(-1)) \\ &= (f(1), f(-1)) + (g(1), g(-1)) \\ &= \varphi(f(X)) + \varphi(g(X)) \end{aligned}$$

and, similarly, $\varphi(f(X)g(X)) = \varphi(f(X))\varphi(g(X))$. Moreover, φ maps $\mathbb{Q}[X]$ onto $\mathbb{Q} \oplus \mathbb{Q}$; for if (a, b) is a typical ordered pair in $\mathbb{Q} \oplus \mathbb{Q}$ and we let

$$f(X) = \frac{1}{2}(a - b)X + \frac{1}{2}(a + b),$$

then $f(1) = a$ and $f(-1) = b$ and therefore $\varphi(f(X)) = (a, b)$. Hence, φ is a ring-homomorphism mapping $\mathbb{Q}[X]$ onto $\mathbb{Q} \oplus \mathbb{Q}$. Finally, we claim that $\operatorname{Ker} \varphi = (X^2 - 1)\mathbb{Q}[X]$. For clearly $(X^2 - 1)\mathbb{Q}[X] \subseteq \operatorname{Ker} \varphi$ since $\varphi(X^2 - 1) = (0, 0)$. On the other hand, if $f(X) \in \operatorname{Ker} \varphi$, then $\varphi(f(X)) = (f(1), f(-1)) = (0, 0)$. Therefore, $f(1) = 0$ and $f(-1) = 0$ and hence both $X - 1$ and $X + 1$ divide $f(X)$. It follows that $f(X) = q(X)(X^2 - 1)$ for some polynomial $q(X) \in \mathbb{Q}[X]$ and therefore $f(X) \in (X^2 - 1)\mathbb{Q}[X]$. Hence, $\operatorname{Ker} \varphi = (X^2 - 1)\mathbb{Q}[X]$. Since φ is a ring-homomorphism mapping $\mathbb{Q}[X]$ onto $\mathbb{Q} \oplus \mathbb{Q}$ whose kernel is $(X^2 - 1)\mathbb{Q}[X]$, it follows that $\mathbb{Q}[X]/(X^2 - 1)\mathbb{Q}[X] \cong \mathbb{Q} \oplus \mathbb{Q}$.

(B) We showed in part (A) that $\mathbb{Q}[X]/(X^2 - 1)\mathbb{Q}[X]$ is not a field since $X^2 - 1$ is not irreducible over $\mathbb{Q}$. Alternatively, it is clear that $\mathbb{Q} \oplus \mathbb{Q}$ is not a field since the pair $(1, 0)$, for example, is not a unit of $\mathbb{Q} \oplus \mathbb{Q}$. In fact, $\mathbb{Q} \oplus \mathbb{Q}$ is not even an integral domain since the pairs $(1, 0)$ and $(0, 1)$, for example, are nonzero elements in $\mathbb{Q} \oplus \mathbb{Q}$ but $(1, 0)(0, 1) = (0, 0)$. This agrees with the fact that $(X^2 - 1)\mathbb{Q}[X]$ is not a prime ideal of $\mathbb{Q}[X]$.

Throughout this section, we have discussed the ring $R[X]$ of polynomials whose coefficients lie in an arbitrary commutative ring R with identity. If the coefficient ring is a field k, then $k[X]$ is an integral domain and every ideal is principal. In this case the maximal ideals are generated by an irreducible polynomial $f(X)$ over k, in which case the quotient ring $k[X]/f(X)k[X]$ is a field. This then provides us with a new and important source of fields. We will study these fields in more detail in Chapter 12. The algebraic properties of the ring $k[X]$ are very similar to those of the ring $\mathbb{Z}$ of integers. Indeed, from an algebraic point of view $k[X]$ and $\mathbb{Z}$ are very similar: each is an integral domain, and each contains special elements—irreducible polynomials in $k[X]$, prime numbers in $\mathbb{Z}$—with the property that every nontrivial element of the ring may be written in an essentially unique way as the product of these special elements. These rings are examples of a general class of rings called unique factorization domains, which are the basic algebraic model for rings in which unique factorization is always possible. We discuss these rings and their properties in more detail in the next chapter.

Finally, let us mention that many times throughout this section we discussed polynomials that were irreducible over the field $\mathbb{Q}$ of rational numbers and observed that they became reducible when regarded as polynomials over the larger field $\mathbb{C}$ of complex numbers. This is no coincidence, of course, since the fundamental theorem of algebra shows that every nonconstant polynomial over $\mathbb{C}$ has a root in $\mathbb{C}$ and hence must factor into a product of linear factors in $\mathbb{C}[X]$. Thus, any nonlinear polynomial over $\mathbb{Q}$ must not only be reducible when viewed as a polynomial over the larger field $\mathbb{C}$, but must factor completely into the product of linear factors. The fundamental theorem of algebra is not an easy theorem to prove since it involves both algebraic as well as analytic properties of the real and complex numbers. It was first proved in 1797 by Carl Friedrich Gauss. Since then there have been many other proofs, some of them strictly analytic in nature, others more algebraic. In Chapter 14 we will give a proof of the fundamental theorem that combines ideas from both group theory and field theory, and which is due to the twentieth-century mathematician Emil Artin (1898–1962).

Exercises

1. Calculate the following sums and products in the indicated polynomial ring:
   (a) $(3X^2 - 2X + 4) + (X^3 + 2X^2 - X + 3)$, in $\mathbb{F}_5[X]$
   (b) $(2X + 1)(4X^2 - X + 3)$, in $\mathbb{F}_7[X]$
   (c) $(X + 1)^2$, in $\mathbb{F}_2[X]$
   (d) $(2X - 3)(3X^2 + 1)$, in $\mathbb{F}_5[X]$
   (e) $(3X^2 + 2X + 4)(2X^2 + X + 1)$, in $\mathbb{Z}_6[X]$
   (f) $(4X + 3)^2$, in $\mathbb{Z}_8[X]$
2. For the polynomials $f(X)$ and $g(X)$ listed below, find polynomials $q(X)$

and $r(X)$ such that $f(X) = q(X)g(X) + r(X)$, where either $r(X) = 0$ or $\deg r(X) < \deg g(X)$.

(a) $f(X) = X^3 + 2X^2 + X - 1$, $g(X) = 2X + 3$ in $\mathbb{Q}[X]$

(b) $f(X) = 2X^3 + X^2 + 3X + 1$, $g(X) = 3X^2 + X + 2$ in $\mathbb{F}_5[X]$

3. Show that $X^3 + 1$ is a reducible polynomial over the field $\mathbb{F}_3$. Find its irreducible factors in $\mathbb{F}_3[X]$ and list all of its divisors to within constant multiples.

4. Show that $X^4 + X^2 + 1$ is a reducible polynomial over the field $\mathbb{Q}$. Find its irreducible factors in $\mathbb{Q}[X]$ and list all of its divisors to within constant multiples.

5. Show that $X^3 + X + 1$ is irreducible over the field $\mathbb{Q}$.

6. Show that $X^4 + X^3 + X^2 + X + 1$ is irreducible over the field $\mathbb{Q}$.

7. Determine which of the following polynomials are irreducible over the indicated field.

(a) $X^2 - 2$ over $\mathbb{Q}$ (e) $2X^2 + 3X + 4$ over $\mathbb{F}_5$

(b) $X^2 - 2$ over $\mathbb{R}$ (f) $4X^2 + 3X + 3$ over $\mathbb{F}_{11}$

(c) $X^2 + X + 3$ over $\mathbb{Q}$ (g) $X^4 + 1$ over $\mathbb{F}_2$

(d) $3X^2 + X - 5$ over $\mathbb{Q}$

8. Find a single generator for the following ideals:

(a) $(X^2 + 3X + 2, X^2 + 5X + 6)\mathbb{Q}[X]$

(b) $(X^3 + X^2, X^3 + 2X^2 + X)\mathbb{Q}[X]$

(c) $(X^2 + 1, X^3 + 1, X^4 + 1)\mathbb{F}_2[X]$

(d) $(3X^2 - 6, X^3 - X^2 - 2X + 2)\mathbb{R}[X]$

(e) $(X^2 + X - 2, 3X^2 - 3X - 6)\mathbb{Q}[X]$

9. Is $2X^3 - 5X^2 + 5X - 3 \in (2X^2 - X + 3)\mathbb{Q}[X]$?

10. Let k be a field and let $f(X)$ be a polynomial over k having degree 2 or 3. Show that $f(X)$ is irreducible over k if and only if $f(X)$ does not have a root in k.

11. Let k be a field in which $1 + 1 \neq 0$ and let $q(X) = aX^2 + bX + c$ stand for the general quadratic polynomial over k, where $a \neq 0$. The *discriminant* of $q(X)$ is the element Disc $q(X) = b^2 - 4ac$ in k. Show that $q(X)$ is reducible over k if and only if Disc $q(X) = d^2$ for some element $d \in k$.

12. Let $K = \mathbb{Q}[X]/(X^3 - 2)\mathbb{Q}[X]$.

(a) Show that K is a field.

(b) Show that $K \cong \mathbb{Q}[\sqrt[3]{2}]$, the subring of $\mathbb{R}$ generated by $\mathbb{Q}$ and $\sqrt[3]{2}$.

(c) Show that every element in $\mathbb{Q}[\sqrt[3]{2}]$ may be written uniquely in the form $a + b\sqrt[3]{2} + c\sqrt[3]{4}$ for some $a, b, c \in \mathbb{Q}$.

13. Let $\omega = \frac{1}{2}(-1 + i\sqrt{3})$ stand for a primitive cube-root of unity and let $\mathbb{Q}[\omega]$ stand for the subring of $\mathbb{C}$ generated by $\mathbb{Q}$ and ω.

(a) Show that $\omega^2 + \omega + 1 = 0$.

(b) Show that $\mathbb{Q}[\omega]$ is a field and that $\mathbb{Q}[\omega] \cong \mathbb{Q}[X]/(X^2 + X + 1)\mathbb{Q}[X]$.

(c) Show that every element in $\mathbb{Q}[\omega]$ may be written uniquely in the form $a + b\omega$, where $a, b \in \mathbb{Q}$.

(d) Write the following elements of $\mathbb{Q}[\omega]$ in the form $a + b\omega$ for some $a, b \in \mathbb{Q}$:

$$(2 - 3\omega)(1 + \omega), \quad 3 + 5\omega - 2\omega^2 + \omega^5, \quad \frac{2 - \omega}{3 + 2\omega}.$$

(e) Is the polynomial $X^2 + X + 1$ irreducible as a polynomial over the field $\mathbb{Q}[\omega]$?

14. Let $K = \mathbb{F}_3[X]/(X^3 + X^2 + 2)\mathbb{F}_3[X]$.
   (a) Show that K is a finite field containing 27 elements.
   (b) Let $a = 1 + X + (X^3 + X^2 + 2)\mathbb{F}_3[X] \in K$. Find a polynomial $f(X) \in \mathbb{F}_3[X]$ such that $a^{-1} = f(X) + (X^3 + X^2 + 2)\mathbb{F}_3[X]$, where $\deg f(X) \leq 2$.

15. Let k be a field and let $R = k[X]/X^2 k[X]$.
   (a) Is R an integral domain?
   (b) Is R a field?
   (c) Let $x = X + X^2 k[X]$. Show that every element in R may be written uniquely in the form $ax + b$ for some elements $a, b \in k$.
   (d) If $ax + b$ and $cx + d$ are typical elements in R, show that $(ax + b)(cx + d) = (bc + ad)x + bd$.
   (e) Show that an element $ax + b \in R$ is a unit of R if and only if $b \neq 0$.
   (f) Show that R has exactly three ideals: $\{0\}$, xR, and R.
   (g) Show that every prime ideal of R is maximal.

16. Let k be a field and let $c \in k$. Show that $k[X]/(X - c)k[X] \cong k$.

17. Let k be a field and let $c_1, \ldots, c_n$ be a finite number of distinct elements in k. Show that $k[X]/(X - c_1) \cdots (X - c_n)k[X] \cong k \oplus \cdots \oplus k$, the direct sum of k with itself n times.

18. Let $f(X)$ be a polynomial over a field k.
   (a) Show that $f(X)$ may be written uniquely in the form $c \, p_1(X)^{s_1} \cdots p_n(X)^{s_n}$, where $c \in k$, $s_1, \ldots, s_n$ are nonnegative integers, and $p_1(X), \ldots, p_n(X)$ are distinct monic irreducible polynomials over k.
   (b) Show that $f(X)$ has $(s_1 + 1)(s_2 + 1) \cdots (s_n + 1)$ divisors to within constants.

19. Let K be a field and let $f(X) \in K[X]$ be a polynomial whose coefficients lie in a subfield k of K. If $f(X)$ is irreducible over K, show that $f(X)$ is irreducible over k.

20. Let k be a field. Two nonzero polynomials $f(X)$ and $g(X)$ over k are said to be *relatively prime* if $\gcd(f(X), g(X)) = 1$.
   (a) Show that $f(X)$ and $g(X)$ are relatively prime if and only if $a(X)f(X) + b(X)g(X) = 1$ for some polynomials $a(X), b(X) \in k[X]$.
   (b) Let $p(X)$ be an irreducible polynomial over k. Show that $p(X)$ and $f(X)$ are relatively prime if and only if $p(X)$ does not divide $f(X)$.
   (c) Let K be a field containing k as a subfield and let $f(X)$ and $g(X)$ be nonzero polynomials over k. If $f(X)$ and $g(X)$ are relatively prime over k, show that $f(X)$ and $g(X)$ are relatively prime over K.

(d) Let K be a field containing k as a subfield and let $f(X)$ and $g(X)$ be nonzero polynomials over k. If $f(X)$ and $g(X)$ have a proper divisor in common over K, show that they also have a proper divisor in common over k.

21. Let k be a field and let $f(X)$ and $g(X)$ be nonzero polynomials over k.

(a) Show that there are polynomials $q_1(X), \ldots, q_{n+1}(X)$ and $r_1(X), \ldots,$ $r_{n+1}(X)$ in $k[X]$ such that

$$f(X) = q_1(X)g(X) + r_1(X)$$
$$g(X) = q_2(X)r_1(X) + r_2(X)$$
$$r_1(X) = q_3(X)r_2(X) + r_3(X)$$
$$\vdots$$
$$r_{n-1}(X) = q_{n+1}(X)r_n(X),$$

where $\deg r_n(X) < \deg r_{n-1}(X) < \cdots < \deg r_1(X) < \deg g(X)$.

(b) Show that $\gcd(f(X), g(X)) = \dfrac{1}{c}r_n(X)$, where c is the lead coefficient of $r_n(X)$.

(c) Find $\gcd(X^3 - 2X^2 - X + 2, X^4 + X^2 - 2)$ in $\mathbb{Q}[X]$.

22. Let $p(X)$ be an irreducible polynomial over a field k and let $f(X) + p(X)k[X]$ be a nonzero element of the field $k[X]/p(X)k[X]$.

(a) Show that there are polynomials $q_1(X), \ldots, q_n(X), r_1(X), \ldots, r_n(X)$ in $k[X]$ such that

$$p(X) = q_1(X)f(X) + r_1(X)$$
$$p(X) = q_2(X)r_1(X) + r_2(X)$$
$$\vdots$$
$$p(X) = q_n(X)r_{n-1}(X) + r_n(X),$$

where $\deg f > \deg r_1 > \deg r_2 > \cdots > \deg r_{n-1} > \deg r_n$ and $r_n(X) = c$, a nonzero constant.

(b) Show that

$$(f(X) + p(X)k[X])^{-1}$$
$$= (-1)^n(c^{-1} + p(X)k[X])(q_1(X) + p(X)k[X]) \cdots (q_n(X) + p(X)k[X])$$

in $k[X]/p(X)k[X]$.

23. In the ring $\mathbb{Z}_4[X]$, does the polynomial $f(X) = X$ divide the polynomial $g(X) = (X + 2)^2$?

24. **Polynomial functions.** Let R be a commutative ring with identity. A *polynomial function* on R is any function $\alpha: R \to R$ for which there is some polynomial $f(X) \in R[X]$ such that $\alpha(c) = f(c)$ for all elements $c \in R$. Let $PF(R)$ stand for the set of polynomial functions on R.

(a) Show that $PF(R)$ is a subring of the ring $F(R)$ of functions on R.

(b) Show that $PF(R)$ contains a subring isomorphic to R.

(c) Let $\sin: \mathbb{R} \to \mathbb{R}$ stand for the sine function on the field $\mathbb{R}$ of real numbers. Show that $\sin$ is not a polynomial function on $\mathbb{R}$.

(d) Let $||: \mathbb{R} \to \mathbb{R}$ stand for the absolute value function on the field $\mathbb{R}$. Show that $||$ is not a polynomial function on $\mathbb{R}$.

(e) Let $R = \mathbb{F}_2$, the field of integers modulo 2, and let $f(X) = X$ and $g(X) = X^2$. Show that $f(c) = g(c)$ for all elements $c \in \mathbb{F}_2$. Thus, $f(X)$ and $g(X)$ define the same polynomial function on $\mathbb{F}_2$. Does $f(X) = g(X)$ as polynomials in $\mathbb{F}_2[X]$?

(f) Is $PF(R) \cong R[X]$?

25. **Subrings generated by a finite number of elements.** Let S be a commutative ring with identity, R a subring of S, and let $a_1, \ldots, a_n$ be any finite number of elements in S. Define $R[a_1, \ldots, a_n]$ inductively by setting $R[a_1, \ldots, a_n] = R[a_1, \ldots, a_{n-1}][a_n]$.

(a) Show that $R[a_1, \ldots, a_n]$ is a subring of S and that every element in $R[a_1, \ldots, a_n]$ may be written as a polynomial in $a_1, \ldots, a_n$ with coefficients in the ring R. $R[a_1, \ldots, a_n]$ is called the *subring of S generated by R and $a_1, \ldots, a_n$*.

(b) Let $\mathbb{Z}[\sqrt{2}, i]$ stand for the subring of $\mathbb{C}$ generated by $\mathbb{Z}$, $\sqrt{2}$ and i. Show that every element in $\mathbb{Z}[\sqrt{2}, i]$ may be written uniquely in the form $a + b\sqrt{2} + ci + di\sqrt{2}$ for some integers a, b, c, d. Is $\mathbb{Z}[\sqrt{2}, i]$ a field?

(c) Let $\mathbb{Q}[\sqrt{2}, \sqrt{3}]$ stand for the subring of $\mathbb{R}$ generated by $\mathbb{Q}$, $\sqrt{2}$ and $\sqrt{3}$. Show that every element in $\mathbb{Q}[\sqrt{2}, \sqrt{3}]$ may be written uniquely in the form $a + b\sqrt{2} + c\sqrt{3} + d\sqrt{6}$, where $a, b, c, d \in \mathbb{Q}$.

(d) Show that $\mathbb{Q}[\sqrt{2}, \sqrt{3}]$ is a field.

(e) Show that $\mathbb{Q}[\sqrt{2}, \sqrt{3}] = \mathbb{Q}[\sqrt{2} + \sqrt{3}]$.

26. **Transcendental numbers.** A real number c is said to be *transcendental over $\mathbb{Q}$* if it is not a root of a nonzero polynomial over $\mathbb{Q}$. The real numbers e and π, for example, are transcendental over $\mathbb{Q}$. If $c \in \mathbb{R}$ is transcendental over $\mathbb{Q}$, show that $\mathbb{Q}[c] \cong \mathbb{Q}[X]$.

27. Let k be a field and let $k[X, Y]$ stand for the ring of polynomials in X and Y over k.

(a) Show that $Xk[X, Y]$ is a prime ideal of $k[X, Y]$.

(b) Show that $Xk[X, Y]$ is not a maximal ideal of $k[X, Y]$.

(c) Show that $k[X, Y]/Xk[X, Y] \cong k[Y]$.

28. Show that $(X + Y, X - Y)\mathbb{Q}[X, Y] = (X, Y)\mathbb{Q}[X, Y]$.

29. **Formal power series.** Let R be a commutative ring with identity. A *power series over R* is an infinite sequence $\{a_n \in R \,|\, n = 0, 1, \ldots\}$ of elements in R. We represent a power series by the infinite formal sum $a_0 + a_1 X + \cdots + a_n X^n + \cdots$ in X and call this expression a *formal power series in X over R*. If $f(X) = a_0 + a_1 X + \cdots$ and $g(X) = b_0 + b_1 X + \cdots$ are formal power series in X over R, let

$$f(X) + g(X) = c_0 + c_1 X + \cdots, \quad \text{where } c_n = a_n + b_n \text{ for } n = 0, 1, \ldots,$$

$$f(X)g(X) = d_0 + d_1 X + \cdots, \quad \text{where } d_n = a_n b_0 + \cdots + a_0 b_n \text{ for } n = 0, 1, \ldots.$$

Let $R[[X]]$ stand for the set of formal power series in X over R.

(a) Show that $R[[X]]$ is a commutative ring with identity under addition and multiplication of formal power series. $R[[X]]$ is called the *ring of formal power series in X over R*.

(b) Show that the polynomial ring $R[X]$ is a subring of $R[[X]]$.

(c) Show that $R[[X]]$ is an integral domain if and only if R is an integral domain.

(d) Let k be a field.

   (1) Show that $(1 - X)^{-1} = 1 + X + X^2 + \cdots + X^n + \cdots$ in the ring $k[[X]]$.

   (2) Let $f(X) \in k[[X]]$. Show that $f(X)$ is a unit of $k[[X]]$ if and only if the constant term of $f(X)$ is not zero.

   (3) Show that the ideal $Xk[[X]]$ is a maximal ideal of $k[[X]]$.

   (4) Show that $k[[X]]/Xk[[X]] \cong k$.

   (5) Show that $Xk[[X]]$ is the only maximal ideal of $k[[X]]$.

   (6) Show that $\{0\}$ and $Xk[[X]]$ are the only prime ideals of $k[[X]]$.

──── **11** ────

Unique Factorization Domains

The ring of integers and the ring of polynomials over a field are examples of integral domains in which every nonzero element except a unit may be written as the product of special elements: in the case of integers, the fundamental theorem of arithmetic shows that every integer greater than 1 is the product of uniquely determined primes, while for polynomials, we showed in the previous chapter that every nonconstant polynomial over a field is the product of irreducible polynomials that are uniquely determined to within constant multiples. These results are important because they identify special elements in the rings—prime numbers and irreducible polynomials—from which all other elements are constructed and which therefore form the basic building blocks of the rings. In this chapter our goal is to extend these ideas to arbitrary integral domains. We begin by first defining the concept of irreducibility for elements of an arbitrary integral domain. Integral domains in which every element other than a unit may be factored into a product of irreducible elements in an essentially unique way are then called unique factorization domains and form the basic algebraic model in which to discuss factorization. In Section 1 we discuss the basic properties of unique factorization domains and characterize them among integral domains in general, while Section 2 deals with the properties of polynomials over a unique factorization domain.

1. BASIC PROPERTIES OF UNIQUE FACTORIZATION DOMAINS

We begin with some basic terminology. Let R be an integral domain and let a be any nonzero element in R. By an *associate* of a, we mean any element of the form ua, where u is a unit of R; that is, any unit multiple of a. If b is any nonzero element in R, we say that b *divides* a, and write $b|a$, if $a = bc$ for some element $c \in R$. Note that the concept of a divisor is well-defined in an integral domain; for if $bc = bc'$ for some elements c, $c' \in R$, then $c' = c$. Every unit of R is clearly a divisor of a, and every associate of a is a divisor of a; these divisors are called *improper divisors* of a. All other divisors, if any, are called *proper divisors* of a. Finally, the element a is said to be *irreducible* if it has no proper divisors and is not a unit of R.

In the ring $\mathbb{Z}$ of integers, for example, the units are ± 1 and hence the associates of an arbitrary integer n are $\pm n$. Since the prime numbers are the only numbers that do not have any proper divisors, the irreducible elements in $\mathbb{Z}$ are numbers of the form $\pm p$, where p is any prime. The situation for polynomials over a field is similar: if k is a field, then every nonzero element in k is a unit and hence the associates of any arbitrary polynomial $f(X)$ have the form $cf(X)$, where c is any nonzero element in k. The irreducible elements in $k[X]$ have the form $cp(X)$, where $p(X)$ is any irreducible polynomial over k.

Definition

A *unique factorization domain* is an integral domain in which every nonzero element other than a unit may be written as the product of a finite number of irreducible elements that are uniquely determined to within order and associates.

The expression "unique to within order and associates" means that if $p_1 \cdots p_n$ and $q_1 \cdots q_m$ are two factorizations of an element in R into a product of irreducible elements $p_1, \ldots, p_n$ and $q_1, \ldots, q_m$, then $n = m$ and $q_1, \ldots, q_n$ are just rearranged associates of $p_1, \ldots, p_n$; that is, there are units $u_1, \ldots, u_n$ of R such that $q_1 = u_1 p_{i_1}, \ldots, q_n = u_n p_{i_n}$. For example, $(2)(-3)$ and $(3)(-2)$ are two factorizations of the number -6 in the ring $\mathbb{Z}$, but they are the same to within order and associates.

It is clear from our discussion above that the ring of integers and the ring of polynomials over a field are examples of unique factorization domains. In fact, any field is a unique factorization domain; for if k is a field, then every nonzero element in k is a unit and hence the requirement that every nonzero element in k other than a unit be a product of irreducible elements is vacuously true. Later in this section we will show that the ring $\mathbb{Z}[i]$ of Gaussian integers is also a unique factorization domain. The following example shows, however, that not every integral domain is a unique factorization domain.

EXAMPLE 1. An integral domain that is not a unique factorization domain

Let $\mathbb{Z}[\sqrt{-5}] = \{n + m\sqrt{-5} \in \mathbb{C} \mid n, m \in \mathbb{Z}\}$ stand for the subring of the field of complex numbers generated by $\mathbb{Z}$ and $\sqrt{-5}$. Then $\mathbb{Z}[\sqrt{-5}]$ is clearly an integral domain. We claim, however, that it is not a unique factorization domain. To show this, observe that $(2)(3)$ and $(1 + \sqrt{-5})(1 - \sqrt{-5})$ are two factorizations of the number 6 in $\mathbb{Z}[\sqrt{-5}]$. Moreover, each of the numbers $2, 3, 1 + \sqrt{-5}$ and $1 - \sqrt{-5}$ are irreducible as elements of the ring $\mathbb{Z}[\sqrt{-5}]$. For consider the number 2. If $n + m\sqrt{-5}$ is a divisor of 2 in $\mathbb{Z}[\sqrt{-5}]$, then $2 = (n + m\sqrt{-5})(s + t\sqrt{-5})$ for some integers s and t, and therefore

$$2(n - m\sqrt{-5}) = (n^2 + 5m^2)(s + t\sqrt{-5}).$$

It follows that the integer $n^2 + 5m^2$ divides both $2n$ and $2m$. But this is possible if and only if $m = 0$ and $n = \pm 1$ or $n = \pm 2$, in which case the divisor $n + m\sqrt{-5}$ is either ± 1 or ± 2 and hence is either a unit or an associate of 2. Therefore, 2 is irreducible. Similarly, the numbers $3, 1 + \sqrt{-5}$, and $1 - \sqrt{-5}$ are irreducible elements in $\mathbb{Z}[\sqrt{-5}]$. Thus, $(2)(3)$ and $(1 + \sqrt{-5})(1 - \sqrt{-5})$ are two factorizations of the number 6 into the product of irreducible elements of $\mathbb{Z}[i]$. Finally, $1 + \sqrt{-5}$ cannot be an associate of either 2 or 3 since neither 2 nor 3 divides 1. Therefore, the two factorizations are not the same to within associates and hence $\mathbb{Z}[\sqrt{-5}]$ is not a unique factorization domain.

Later in this section we will discuss conditions under which an arbitrary integral domain is a unique factorization domain. For the moment, however, let us turn our attention to the basic properties of unique factorization domains.

PROPOSITION 1. Let R be a unique factorization domain. Then:

(1) every nonzero element in R has only a finite number of divisors to within associates;

(2) every nonzero element in R other than a unit is divisible by some irreducible element; and

(3) if p is an irreducible element in R and $p \mid ab$ for some elements $a, b \in R$, then $p \mid a$ or $p \mid b$.

Proof.   Let a be any nonzero element in R other than a unit and let $a = p_1 \cdots p_n$ be its factorization into irreducible elements $p_1, \ldots, p_n$. Since this factorization is unique to within order and associates, every divisor of a must have the form $up_{i_1} \cdots p_{i_s}$, where u is a unit of R and $s \leq n$. Thus, a has at most a finite number of divisors to within order and associates, which proves statement (1). Moreover, $p_1 \mid a$, which proves statement (2). And finally, suppose that p is irreducible and that $p \mid ab$ for some $a, b \in R$. If either $a = 0$ or $b = 0$, then p divides a or b and we are done. On the other hand if a is a unit of R, then $ac = 1$ for some $c \in R$; in this case, $b = abc$ and therefore $p \mid b$. Similarly, $p \mid a$ if b is a unit of R. Assume therefore that neither a nor b are units of R and let $a = p_1 \cdots p_n$ and $b = q_1 \cdots q_m$ be their factorizations into irreducible elements. Then $ab = p_1 \cdots p_n q_1 \cdots q_m$ is a factorization of ab into a product of irreducible elements. Since such a factorization is unique to within order and associates and since $p|ab$, it follows that p must be an associate of some p_i or q_j and hence divide either a or b, as required. ∎

Recall that in the ring of integers and the ring of polynomials over a field k any finite collection of nonzero elements has a greatest common divisor which can be found by using the unique factorization property of the rings: simply factor each element into its irreducible factors and then multiply the common irreducible factors. Let us now show that greatest common divisors exist in any unique factorization domain and can be found the same way.

Definition

Let R be an integral domain and let $a_1, \ldots, a_n$ be a finite collection of nonzero elements in R. A *greatest common divisor* of $a_1, \ldots, a_n$ is any element $d \in R$ that satisfies the following two conditions:

(1) $d|a_1, \ldots, d|a_n$; and
(2) if $d' \in R$ divides each of the elements $a_1, \ldots, a_n$, then $d'|d$.

PROPOSITION 2. Let R be a unique factorization domain. Then every finite collection of nonzero elements in R has a greatest common divisor and such an element is unique to within associates.

Proof.   Let $a_1, \ldots, a_n$ be a finite collection of nonzero elements in R. Since each of these elements is either a unit or the product of irreducible elements, we may write

$$a_1 = u_1 p_1^{s_{11}} \cdots p_m^{s_{1m}}$$
$$\vdots$$
$$a_n = u_n p_1^{s_{n1}} \cdots p_m^{s_{nm}},$$

where $u_1, \ldots, u_n$ are units of R, $p_1, \ldots, p_m$ are irreducible elements of R no two of which are associates, and the s_{ij} are nonnegative integers. Let $d = p_1^{t_1} \cdots p_m^{t_m}$, where t_j is the minimum of $s_{1j}, \ldots, s_{nj}$ for $j = 1, \ldots, m$. Then d is a greatest common divisor of $a_1, \ldots, a_n$. For clearly, $d|a_i$ for $i = 1, \ldots, n$. On the other hand, if $d' \in R$ is such that $d'|a_i$ for all i, then, because of unique factorization, $d' = u p_1^{w_1} \cdots p_m^{w_m}$ for some unit u and some integers $w_j \le t_j$, and therefore $d'|d$. Hence, d is a greatest common divisor of $a_1, \ldots, a_n$. Finally, to show that d is unique to within associates, suppose that d^* is any greatest common divisor of $a_1, \ldots, a_n$. Then d^* divides each of the elements $a_1, \ldots, a_n$ and hence divides d by condition (2). Similarly, d divides d^*. Therefore, $d^* = ud$ and $d = vd^*$ for some $u, v \in R$, from which it follows that $uv = 1$. Hence, u and v are units of R and therefore d and d^* are associates, as required. ∎

If $a_1, \ldots, a_n$ are nonzero elements of a unique factorization domain, we denote their greatest common divisor by $\gcd(a_1, \ldots, a_n)$. Let us emphasize that, in general, such an element is determined only to within associates, although in certain cases, such as integers and polynomials, it is possible to choose a unique greatest common divisor. Nevertheless, it is clear that if some greatest common divisor of $a_1, \ldots, a_n$ is a unit, then every greatest common divisor is a unit; in this case we say that $a_1, \ldots, a_n$ are *relatively prime*. In the ring of integers, for example, every collection of integers has two greatest common divisors, one positive, one negative, and thus we may always choose the unique greatest common divisor that is positive; for example, the greatest common divisors of 4 and 6 are ± 2 and we set $\gcd(4, 6) = 2$. Similarly, in the ring of polynomials over a field the greatest common divisors of a collection of polynomials are all constant multiples of a single monic polynomial, which we therefore choose as the greatest common divisor; for example, in $\mathbb{Q}[X]$ we set $\gcd(2X^2 - 2, 3X + 3) = X + 1$.

Let us also mention that if $a_1, \ldots, a_n$ are a finite number of elements in a unique factorization domain, then their greatest common divisor cannot always be written in the form $x_1 a_1 + \cdots + x_n a_n$ for some elements $x_1, \ldots, x_n \in R$. Although this is possible in the rings $\mathbb{Z}$ and $k[X]$, it is not possible in general. We leave the reader to verify that if $d = \gcd(a_1, \ldots, a_n)$, then $d = x_1 a_1 + \cdots + x_n a_n$ for some $x_1, \ldots, x_n \in R$ if and only if $(a_1, \ldots, a_n)R = dR$. Thus, in particular, if all ideals of R are principal—as they are in $\mathbb{Z}$ and $k[X]$—then greatest common divisors may always be expressed in terms of the elements themselves.

Now, we showed in Example 1 that the ring $\mathbb{Z}[\sqrt{-5}]$ is an integral domain but not a unique factorization domain. To characterize those integral domains that are unique factorization domains, let us first make a general observation about factorization in an integral domain. Let R be an arbitrary integral domain and let a be any nonzero element in R other than a unit. If a is irreducible, then a is its own factorization and we are done. But if a is not irreducible, then $a = bc$ for some elements $b, c \in R$ that are neither units nor associates of a. If b and c are irreducible, then we have factored the element a into a product of irreducible elements in R; otherwise, continue this process with each of the factors b and c. Unfortunately, the process may continue indefinitely, thus generating an infinite chain of factors, unless there is some type of finiteness condition on the ring R to guarantee that the process eventually stops with a factorization of a as a product of irreducible factors. As for the uniqueness of such a factorization, all that is needed is property (3) of Proposition 1. Let us now show that these two requirements—no infinite chains of divisibility and property (3) of Proposition 1—are in fact sufficient to characterize unique factorization domains.

PROPOSITION 3. Let R be an integral domain. Then R is a unique factorization domain if and only if the following two conditions are satisfied:

(1) there is no infinite sequence $d_1, d_2, \ldots$ of elements in R such that d_{i+1} is a proper divisor of d_i for all $i \geq 1$;

(2) if p is an irreducible element in R and $p \mid ab$ for some elements $a, b \in R$, then $p \mid a$ or $p \mid b$.

Proof. If R is a unique factorization domain, then, by Proposition 1, R satisfies conditions (1) and (2). Now suppose, conversely, that R is an integral domain satisfying the two conditions and let a be any nonzero element in R other than a unit. Define a sequence $d_1, d_2, \ldots$ of elements in R as follows. Set $d_1 = a$. If d_n has been defined and is irreducible, set $d_{n+1} = 1$; otherwise d_n is reducible and we set $d_n = p_n d_{n+1}$, where p_n is any irreducible element that divides d_n. Then $d_{i+1} | d_i$ for all $i \geq 1$. Since the sequence $d_1, d_2, \ldots$ cannot be infinite by assumption, it follows that $d_n = 1$ for some n and hence $a = p_1 \cdots p_n$. Thus, every nonzero element in R other than a unit may be written as the product of irreducible elements. To prove the uniqueness of such a factorization, suppose that $p_1 \cdots p_n = q_1 \cdots q_m$ are two factorizations of an element into irreducible elements $p_1, \ldots, p_n$ and $q_1, \ldots, q_m$ of R. Then $q_1 | p_1 \cdots p_n$, and hence by condition (2) and induction it follows that $q_1 | p_{i_1}$ for some p_{i_1}. Since p_{i_1} is irreducible, $q_1 = u_1 p_{i_1}$ for some unit $u_1 \in R$. Therefore $p_1 \cdots p_{i_1 - 1} p_{i_1 + 1} \cdots p_n = u_1 q_2 \cdots q_m$. It now follows by induction that $n = m$ and that $q_2, \ldots, q_m$ are rearranged associates of $p_1, \ldots, p_{i_1 - 1}$, $p_{i_1 + 1}, \ldots, p_n$. Therefore, factorization in R is unique to within order and associates and hence R is a unique factorization domain. ∎

Let us now use the criteria in Proposition 3 to discuss two particularly important types of unique factorization domains: principal ideal domains and Euclidean domains.

Definition

A *principal ideal domain* is an integral domain in which every ideal is principal.

Both the ring $\mathbb{Z}$ of integers and the ring $k[X]$ of polynomials over a field k are principal ideal domains. Moreover, every field is a principal ideal domain since a field has only two ideals, the zero ideal and the field itself, both of which are principal.

PROPOSITION 4. Every principal ideal domain is a unique factorization domain.

Proof. Let R be a principal ideal domain and suppose that $d_1, d_2, \ldots$ is an infinite sequence of elements in R such that $d_{i+1} | d_i$ for all $i \geq 1$. Consider the ideals $d_1 R, d_2 R, \ldots$ generated by these elements. Since $d_{i+1} | d_i$, $d_i R \subseteq d_{i+1} R$ for all $i \geq 1$. Thus, we have an ascending chain of ideals $d_1 R \subseteq d_2 R \subseteq \ldots$ of R. Let $I = \bigcup d_i R$ stand for the union of these ideals. Then it is easily verified that I is in fact an ideal of R and hence, since every ideal of R is principal, $I = aR$ for some element $a \in I$. But $a \in d_n R$ for some integer n. Therefore, $I \subseteq d_n R \subseteq I$ and hence $I = d_n R$. Finally, since $d_{n+1} \in I$, it follows that $d_n | d_{n+1}$ and therefore d_n and d_{n+1} must be associates since $d_{n+1} | d_n$. Thus, the ring R contains no infinite chain $d_1, d_2, \ldots$ of elements in which

d_{i+1} is a proper divisor of d_i for all i, which verifies condition (1) of Proposition 3. To verify condition (2), let p be an irreducible element in R and suppose that $p|ab$ for some $a, b \in R$. If $p|a$, we are done. Otherwise p does not divide a and we consider the ideal $(a, p)R$ generated by a and p. Since every ideal of R is principal, $(a, p)R = dR$ for some $d \in R$ and hence $d|a$ and $d|p$. Since p is irreducible and does not divide a, it follows that d is a unit and hence $(a, p)R = R$. Therefore, $xa + yp = 1$ for some $x, y \in R$ and hence $b = x(ab) + (by)p$, from which it follows that $p|b$, as required. Thus condition (2) is satisfied and therefore R is a unique factorization domain. ∎

Since every principal ideal domain is a unique factorization domain, it now follows that greatest common divisors exist in any principal ideal domain. Indeed, if $a_1, \ldots, a_n$ are elements of a principal ideal domain R, then $(a_1, \ldots, a_n)R = dR$ if and only if d is the greatest common divisor of $a_1, \ldots, a_n$. Moreover, in this case $d = x_1 a_1 + \cdots + x_n a_n$ for some elements $x_1, \ldots, x_n \in R$. Thus, greatest common divisors not only exist in principal ideal domains but in fact may be expressed in terms of the elements themselves.

Now recall that an ideal P of a ring R is a prime ideal if, whenever $ab \in P$, either $a \in P$ or $b \in P$, while an ideal is maximal if it is not contained in any ideal of R other than itself and R. What can be said about the prime ideals and maximal ideals of a principal ideal domain? In the ring of integers, for example, the prime ideals have the form $\{0\}$ or $p\mathbb{Z}$, p a prime, while the maximal ideals are the nonzero prime ideals. The following result shows that this is essentially the case for any principal ideal domain.

PROPOSITION 5. Let R be a principal ideal domain. Then the prime ideals of R have the form pR, where p is either zero or an irreducible element in R, while the maximal ideals of R are the nonzero prime ideals.

Proof.   If p is zero or an irreducible element in R, we leave the reader to show that the ideal pR is a prime ideal of R. Now suppose, conversely, that $P = pR$ is a prime ideal of R. Then p is irreducible. For let a be any divisor of p. Then $p = ab$ for some $b \in R$. It follows that $ab \in P$ and hence either $a \in P$ or $b \in P$. If $a \in P$, then $p|a$ and therefore a and p are associates since $a|p$; if $b \in P$, then $p|b$ and it follows that a is a unit. Therefore, p has no proper divisors and hence is irreducible, as required. Thus, the prime ideals of R are precisely those ideals of the form pR, where p is either zero or an irreducible element in R. We leave the proof that the maximal ideals are the nonzero prime ideals as an exercise for the reader. ∎

One of the most useful tools for working with integers and polynomials over a field is the division algorithm, or Euclidean algorithm as it is frequently called. The division algorithm provides a way of expressing one element in the ring as a multiple of another element, with an error term, or remainder, whose "size" may be measured and bounded in some way. The "size" of a polynomial, for example, is measured by its degree, and the division algorithm

shows that any polynomial may be written as a multiple of a second polynomial with a remainder that is either zero or whose degree is smaller than
that of the second polynomial. In general, the "size" of an element in an
integral domain is measured by means of a valuation function on the ring,
and any integral domain together with such a function is called a Euclidean
domain.

Definition

A *Euclidean domain* is an integral domain R together with a function φ,
called the *valuation function*, that associates to every nonzero element $a \in R$
a nonnegative integer $\varphi(a)$ satisfying the following two conditions:

(1) $\varphi(a) \leq \varphi(ab)$ for all nonzero elements $a, b \in R$;

(2) for any nonzero elements $a, b \in R$, there are elements $q, r \in R$ such that
$b = qa + r$, where either $r = 0$ or $\varphi(r) < \varphi(a)$.

For example, the ring of integers together with the absolute value function
is a Euclidean domain; in this case the valuation function $\varphi(n) = |n|$ for every
nonzero integer n. Similarly, the ring $k[X]$ of polynomials over a field k
together with the degree function is a Euclidean domain; here, the valuation
function $\varphi(f(X)) = \deg f(X)$ for every nonzero polynomial $f(X) \in k[X]$.
Since the division algorithm for integers and polynomials is the main reason
these rings are principal ideal domains, the following result is not surprising.

PROPOSITION 6. Every Euclidean domain is a principal ideal domain.

Proof.   Let R be a Euclidean domain with valuation function φ and let I
be an ideal of R. If $I = \{0\}$, then I is principal and we are done. Otherwise
there is some nonzero element $a \in I$ whose valuation is minimal among all
nonzero elements in I; that is, $\varphi(a) \leq \varphi(x)$ for all nonzero elements $x \in I$. We
claim that $I = aR$, the principal ideal generated by a. For clearly, $aR \subseteq I$.
On the other hand, if x is any nonzero element in I, then $x = qa + r$ for
some elements $q, r \in R$, where either $r = 0$ or $\varphi(r) < \varphi(a)$. If $r \neq 0$, then
$r = x - qa$ is an element in I and $\varphi(r) < \varphi(a)$, which contradicts the definition
of a. Therefore, $r = 0$ and hence $x = qa \in aR$. Thus, $I = aR$ and hence every
ideal of R is principal, as required.   ■

COROLLARY.   Every Euclidean domain is a unique factorization
domain.

Proof.   Immediate from Propositions 4 and 6.   ■

Determining the units of a ring is, in general, a difficult problem. For
Euclidean domains, however, the following result shows that the units are
easily described in terms of the valuation function.

PROPOSITION 7. Let R be a Euclidean domain with valuation function φ. Then a nonzero element $a \in R$ is a unit of R if and only if $\varphi(a) = \varphi(1)$.

Proof.   If a is a unit of R, then $ab = 1$ for some element $b \in R$ and hence $\varphi(1) \leq \varphi(1a) = \varphi(a) \leq \varphi(ab) = \varphi(1)$. Therefore, $\varphi(a) = \varphi(1)$. Conversely, suppose that $\varphi(a) = \varphi(1)$ for some nonzero element $a \in R$. Then $1 = qa + r$ for some elements $q, r \in R$, where either $r = 0$ or $\varphi(r) < \varphi(a)$. If $r \neq 0$, then $\varphi(r) < \varphi(a) = \varphi(1) \leq \varphi(1r) = \varphi(r)$, which is a contradiction. Therefore, $r = 0$. Hence, $1 = qa$ and therefore a is a unit of R.   ∎

In the ring of integers, for example, $|n| = 1$ if and only if $n = \pm 1$, and ± 1 are indeed the units of $\mathbb{Z}$. Similarly, in the ring $k[X]$ of polynomials over a field k, $\deg f(X) = \deg(1) = 0$ if and only if $f(X)$ is a nonzero constant polynomial, which agrees with the fact that the nonzero constant polynomials are the units of $k[X]$.

The existence of a valuation function on an integral domain is an especially useful method for showing that the ring is a unique factorization domain. Let us now conclude our discussion of unique factorization domains by using this method to show that the ring of Gaussian integers is a unique factorization domain.

EXAMPLE 2. The ring $\mathbb{Z}[i]$ of Gaussian integers

Let $\mathbb{Z}[i] = \{n + mi \in \mathbb{C} \mid n, m \in \mathbb{Z}\}$ stand for the ring of Gaussian integers. Define the function $N : \mathbb{Z}[i] \to \mathbb{Z}$ by setting $N(n + mi) = n^2 + m^2$ for all Gaussian integers $n + mi \in \mathbb{Z}[i]$. Let us show that $\mathbb{Z}[i]$ is a Euclidean domain with valuation function N and consequently that $\mathbb{Z}[i]$ is a principal ideal domain and unique factorization domain, and then discuss the units, prime ideals, and maximal ideals of $\mathbb{Z}[i]$.

(A) We begin by showing that N is a valuation function on $\mathbb{Z}[i]$. Let $a = n + mi$ be a typical Gaussian integer and let $\bar{a} = n - mi$ be its complex conjugate. Then $N(a) = n^2 + m^2 = a\bar{a}$. It follows that if a and b are nonzero elements in $\mathbb{Z}[i]$, then $N(a) > 0$ and $N(ab) = ab\bar{a}\bar{b} = a\bar{a}b\bar{b} = N(a)N(b) \geq N(a)$. Now, let $a, b \in \mathbb{Z}[i]$, with $a \neq 0$. Then there are rational numbers $s, t \in \mathbb{Q}$ such that $b/a = s + ti$. Let n and m be integers such that $|s - n| \leq 1/2$ and $|t - m| \leq 1/2$, and let $q = n + mi$ and $r = a[(s - n) + (t - m)i]$. Then

$$qa + r = (n + mi)a + a[(s - n) + (t - m)i]$$
$$= as + ati = b.$$

Therefore, $r = b - qa$ and hence $r \in \mathbb{Z}[i]$. Moreover, since $r = a[(s - n) + (t - m)i]$,

$$N(r) = N(a)N([(s - n) + (t - m)i]) = N(a)[(s - n)^2 + (t - m)^2]$$
$$\leq N(a)(1/4 + 1/4)$$
$$< N(a).$$

Hence, $b = qa + r$, where either $r = 0$ or $N(r) < N(a)$ and therefore $\mathbb{Z}[i]$ is a Euclidean domain with valuation function N.

(B) Since $\mathbb{Z}[i]$ is a Euclidean domain with valuation function N, it follows that the units of $\mathbb{Z}[i]$ are those Gaussian integers $n + mi$ for which $N(n + mi) = n^2 + m^2 = 1$. Since the only integer solutions to this equation are $n = \pm 1$, $m = 0$, and $n = 0$, $m = \pm 1$, it follows that ± 1 and $\pm i$ are the only units of $\mathbb{Z}[i]$. Thus, the group of units $U(\mathbb{Z}[i]) = \{1, -1, i, -i\}$, which is a cyclic group of order 4. It follows, in particular, that every element $a \in \mathbb{Z}[i]$ has four associates, namely $\pm a$ and $\pm ia$. For example, the elements $2 + 3i$ and $-3 + 2i$ are associates since $-3 + 2i = i(2 + 3i)$.

(C) Since $\mathbb{Z}[i]$ is a Euclidean domain, it is also a principal ideal domain and hence a unique factorization domain. Thus, every Gaussian integer may be written as the product of irreducible elements that are unique to within order and associates. But it is not immediately clear which Gaussian integers are irreducible. Here is where the valuation function N is useful. We claim that if a is a Gaussian integer such that $N(a) = p$, a prime, then a is an irreducible element of $\mathbb{Z}[i]$. For if $a = bc$ for some $b, c \in \mathbb{Z}[i]$, then $p = N(a) = N(bc) = N(b)N(c)$. Therefore, $N(b) = 1$ or $N(c) = 1$ and hence either b or c is a unit. Thus, a has no proper factorization and is therefore irreducible. For example, $1 + i$ is irreducible since $N(1 + i) = 2$, a prime. Similarly, $2 - i$ and $2 + 3i$ are irreducible since $N(2 - i) = 5$ and $N(2 + 3i) = 13$. Let us emphasize, however, that not every irreducible element has prime valuation; for example, it is not difficult to show that the number 3 is irreducible as an element of $\mathbb{Z}[i]$, but $N(3) = 9$.

(D) Finally, since $\mathbb{Z}[i]$ is a principal ideal domain, the prime ideals of $\mathbb{Z}[i]$ have the form $p\mathbb{Z}[i]$, where $p = 0$ or an irreducible element in $\mathbb{Z}[i]$, with the maximal ideals being the nonzero prime ideals. For example, since $1 + i$ is irreducible, the ideal $(1 + i)\mathbb{Z}[i]$ is both a prime ideal and maximal ideal of $\mathbb{Z}[i]$. Similarly, both $(2 - i)\mathbb{Z}[i]$ and $(2 + 3i)\mathbb{Z}[i]$ are prime as well as maximal ideals of $\mathbb{Z}[i]$. Note also that since $(1 + i)(1 - i) = 2$, the number 2 is no longer prime when regarded as a Gaussian integer but rather splits into the product of $1 + i$ and $1 - i$; in terms of ideals, this shows that if $P = (1 + i)\mathbb{Z}[i]$ and $Q = (1 - i)\mathbb{Z}[i]$, then P and Q are prime ideals of $\mathbb{Z}[i]$ and $2\mathbb{Z}[i] = PQ$.

The valuation function N on the ring of Gaussian integers is an example of a multiplicative valuation function, that is, a valuation function with the property that $N(ab) = N(a)N(b)$ for all $a, b \in \mathbb{Z}[i]$, and it is for this reason that an element $a \in \mathbb{Z}[i]$ is irreducible if $N(a) = p$, for some prime p. In the exercises we ask the reader to show, in general, that if R is a Euclidean domain with a multiplicative valuation function φ and $\varphi(a) = p$, a prime, for some element $a \in R$, then a is irreducible. Thus, multiplicative valuation functions are especially useful for determining both the units of a Euclidean domain as well as irreducible elements.

Finally, let us mention that although Euclidean domains and principal ideal domains are unique factorization domains, not every unique factorization domain is principal. In the next section we will see that the ring $k[X, Y]$ of polynomials in two variables over a field k is a unique factorization domain but is neither principal nor a Euclidean domain. It is not as easy, however, to give an example of a principal ideal domain that is not a Euclidean domain. The ring $\mathbb{Z}[\frac{1}{2}(1 + \sqrt{-19})]$ is such a ring. (For a detailed discussion of this ring, we refer the reader to the article "A Principal Ideal Domain That Is Not a Euclidean Domain," by Oscar A. Cámpoli, in *The American Mathematical Monthly*, Vol. 95, No. 9, 1988.)

Exercises

1. Let R be an integral domain and let a, b, and c be nonzero elements in R.
   (a) Show that a and b are associates if and only if $a|b$ and $b|a$.
   (b) If $a|b$ and $b|c$, show that $a|c$.
   (c) If $a|b$, show that $ca|cb$.
2. Show that the ring $\mathbb{Z}[\sqrt{-6}]$ is not a unique factorization domain.
3. Show that the ring $\mathbb{Z}[\sqrt{-3}]$ is not a unique factorization domain.
4. Show that $1 + \sqrt{-5}$ is an irreducible element in the ring $\mathbb{Z}[\sqrt{-5}]$ but the ideal $(1 + \sqrt{-5})\mathbb{Z}[\sqrt{-5}]$ is not a prime ideal of $\mathbb{Z}[\sqrt{-5}]$.
5. Determine which of the following elements in $\mathbb{Z}[i]$ are irreducible; if the element is not irreducible, factor it into a product of irreducible elements.
   (a) $2 + i$ (b) $3 - 2i$ (c) $3 + 6i$ (d) $2 + 6i$
6. Show that a Gaussian integer $a \in \mathbb{Z}[i]$ is irreducible if and only if its complex conjugate $\bar{a}$ is irreducible.
7. Find the following greatest common divisors in $\mathbb{Z}$:
   (a) $\gcd(-35, 63)$ (b) $\gcd(60, -10, 21)$ (c) $\gcd(182, 143, 156)$
8. Find the following greatest common divisors in $\mathbb{Z}[i]$:
   (a) $\gcd(1 + i, 2 - 3i)$ (b) $\gcd(i, 2 + 5i)$ (c) $\gcd(-1 + 9i, 5 - i)$
9. Let R be an integral domain. Show that R is a unique factorization domain if and only if it satisfies the following two conditions:
   (1) every finite collection of nonzero elements in R has a greatest common divisor;
   (2) if p is an irreducible element in R and $p|ab$ for some $a, b \in R$, then either $p|a$ or $p|b$.
10. Let R be a unique factorization domain and let R^* stand for the set of nonzero elements in R. Define a relation $\sim$ on R^* as follows: if $a, b \in R^*$, then $a \sim b$ means that $b = ua$ for some unit $u \in R$. Show that $\sim$ is an equivalence relation on R^* and that the equivalence class of an element $a \in R^*$ is the set of associates of a.

11. Let R be a unique factorization domain. If p is an irreducible element in R and $p | a_1 \cdots a_n$ for some elements $a_1, \ldots, a_n \in R$, show that p divides at least one of the factors $a_1, \ldots, a_n$.

12. Let R be a principal ideal domain and let a and b be nonzero elements in R. Show that a and b are relatively prime if and only if $xa + yb = 1$ for some $x, y \in R$.

13. Let R be a Euclidean domain with valuation function φ and let a and b be nonzero elements in R.
   (a) If $a | b$, show that $\varphi(a) \le \varphi(b)$.
   (b) If a and b are associates, show that $\varphi(a) = \varphi(b)$.

14. Let $\mathbb{Z}[\omega] = \{n + m\omega \in \mathbb{C} \,|\, n, m \in \mathbb{Z}\}$ stand for the subring of the field $\mathbb{C}$ of complex numbers generated by $\mathbb{Z}$ and ω, where $\omega = \dfrac{1}{2}(-1 + i\sqrt{3})$ is a primitive cube root of unity. Define the function $N : \mathbb{Z}[\omega] \to \mathbb{Z}$ by setting $N(n + m\omega) = n^2 + m^2 - nm$ for all elements $n + m\omega \in \mathbb{Z}[\omega]$.
   (a) Show that $\mathbb{Z}[\omega]$ is a Euclidean domain with valuation function N.
   (b) Show that the units of $\mathbb{Z}[\omega]$ are ± 1, $\pm\omega$, and $\pm(1 + \omega)$.
   (c) If $N(a) = p$, a prime, for some element $a \in \mathbb{Z}[\omega]$, show that a is irreducible.
   (d) Determine which of the following elements are irreducible in $\mathbb{Z}[\omega]$:
       $2 + \omega$, $3 + 2\omega$, 2, 3, $4 + 5\omega$.
   (e) Find $\gcd(1 + 2\omega, -2 + \omega)$.

15. Show that the ring $\mathbb{Z}[\sqrt{-2}]$ is a Euclidean domain.

16. Show that the ring $\mathbb{Z}[\sqrt{2}]$ is a Euclidean domain.

17. Let $\mathbb{Q}_2 = \{n/m \in \mathbb{Q} \,|\, n, m \in \mathbb{Z}, m \neq 0, m \text{ odd}\}$ stand for the set of rational numbers having odd denominator.
   (a) Show that $\mathbb{Q}_2$ is an integral domain.
   (b) Show that every nonzero element in $\mathbb{Q}_2$ may be written uniquely in the form $2^s \dfrac{n}{m}$, where s is a nonnegative integer and n and m are odd integers that are relatively prime.
   (c) Show that the units of $\mathbb{Q}_2$ are those elements of the form n/m, where n and m are odd integers that are relatively prime.
   (d) Let $a = 2^s \dfrac{n}{m}$ and $b = 2^t \dfrac{u}{v}$ be typical elements in $\mathbb{Q}_2$. Show that $a | b$ in $\mathbb{Q}_2$ if and only if $s \le t$.
   (e) Define the function $\varphi : \mathbb{Q}_2^* \to \mathbb{Z}$ by setting $\varphi\left(2^s \dfrac{n}{m}\right) = 2^s$ for every nonzero element $2^s \dfrac{n}{m} \in \mathbb{Q}_2$. Show that $\mathbb{Q}_2$ is a Euclidean domain with valuation function φ.
   (f) Show that the number 2 is irreducible in $\mathbb{Q}_2$ and that every irreducible element in $\mathbb{Q}_2$ is an associate of 2.
   (g) Show that the ideal $2\mathbb{Q}_2$ generated by 2 is both a prime ideal and maximal ideal of $\mathbb{Q}_2$ and is the only such ideal.
   (h) Verify that a nonzero element $a \in \mathbb{Q}_2$ is a unit if and only if $\varphi(a) = \varphi(1)$.

18. Let R be a Euclidean domain with valuation function φ and suppose that $\varphi(ab) = \varphi(a)\varphi(b)$ for all nonzero elements $a, b \in R$.
    (a) Show that $\varphi(1) = 1$.
    (b) Show that a nonzero element $a \in R$ is a unit of R if and only if $\varphi(a) = 1$.
    (c) If $\varphi(a) = p$, a prime, for some nonzero element $a \in R$, show that a is an irreducible element in R.
19. Let K be a field and let K^* stand for the set of nonzero elements in K. Define the function $\varphi : K^* \to \mathbb{Z}$ by setting $\varphi(a) = 1$ for every nonzero element $a \in K^*$. Show that K is a Euclidean domain with valuation function φ. Thus, every field is a Euclidean domain under the trivial valuation function.
20. (a) Is every subring of a unique factorization domain a unique factorization domain?
    (b) Is every quotient ring of a unique factorization domain a unique factorization domain?
21. (a) Is every subring of a principal ideal domain a principal ideal domain?
    (b) Is every quotient ring of a principal ideal domain a principal ideal domain?

2. POLYNOMIALS OVER A UNIQUE FACTORIZATION DOMAIN

In the previous chapter we discussed the basic properties of polynomials over a field. In this section we turn our attention to polynomials whose coefficients lie in a unique factorization domain. The situation here is somewhat different from the case where the coefficients lie in a field since the nonzero elements of an arbitrary unique factorization domain are not necessarily units and hence division by nonzero elements is not always possible; indeed, the coefficients of such a polynomial may themselves factor. We begin, therefore, by first discussing the concept of an irreducible polynomial over a unique factorization domain R and indicating the relationship between irreducibility over R and irreducibility over the quotient field $QF(R)$. The main goal of the section is to then show that if R is any unique factorization domain, the ring $R[X]$ of polynomials over R is also a unique factorization domain, thus insuring that polynomials over R do indeed factor uniquely into a product of irreducible polynomials to within order and associates.

Let us begin by recalling that if R is a unique factorization domain, then R is an integral domain and hence the ring $R[X]$ of polynomials over R is also an integral domain.

Definition

Let R be a unique factorization domain. A nonzero polynomial $f(X)$ over R is said to be *irreducible over* R if $f(X)$ is irreducible as an element of the integral domain $R[X]$.

Thus, a polynomial $f(X)$ over a unique factorization domain R is irreducible over R if its only divisors in $R[X]$ are its associates and the units of R. The associates of $f(X)$, we recall, are polynomials of the form $uf(X)$, where u is any unit of R. For example, the polynomial $2X + 3 \in \mathbb{Z}[X]$ is irreducible over $\mathbb{Z}$ since its only divisors in $\mathbb{Z}[X]$ are the units ± 1 and the associates $\pm(2X + 3)$. The polynomial $2X + 4 \in \mathbb{Z}[X]$, on the other hand, is not irreducible over $\mathbb{Z}$ since $2X + 4 = 2(X + 2)$ is a proper factorization in $\mathbb{Z}[X]$; the factors 2 and $X + 2$ are, in fact, the irreducible factors of $2X + 4$ in $\mathbb{Z}[X]$. Note, however, that $2X + 4$ is irreducible as a polynomial over the field $\mathbb{Q}$ of rational numbers.

The polynomial $2X + 4 \in \mathbb{Z}[X]$ illustrates the fact that the coefficients of a polynomial may have a nontrivial common divisor which, when factored out, gives a proper factorization of the polynomial. In discussing irreducibility it is therefore important to first remove the greatest common divisor from among the coefficients before factoring the polynomial itself; this is possible since greatest common divisors always exist in a unique factorization domain. We refer to the greatest common divisor of the coefficients as the content of the polynomial.

Definition

Let R be a unique factorization domain and let $f(X)$ be a nonzero polynomial over R. The *content* of $f(X)$ is the greatest common divisor of the nonzero coefficients of $f(X)$ and is denoted by $C(f)$. The polynomial $f(X)$ is said to be *primitive over* R if its content is a unit of R.

If $f(X)$ is a nonzero polynomial over a unique factorization domain R, then we may factor out its content $C(f)$ and write $f(X) = C(f)f^*(X)$, where $f^*(X)$ is primitive over R. For example, if $f(X) = 2X + 3 \in \mathbb{Z}[X]$, then $C(f) = 1$ and hence $f(X)$ is primitive over $\mathbb{Z}$. But for $f(X) = 2X + 4$, $C(f) = 2$ and therefore $2X + 4$ is not primitive over $\mathbb{Z}$; in this case $f(X) = 2(X + 2)$, $f^*(X) = X + 2$, and $X + 2$ is, indeed, primitive over $\mathbb{Z}$.

PROPOSITION 1. (GAUSS) Let R be a unique factorization domain and let $f(X)$ and $g(X)$ be nonzero polynomials over R. Then $C(fg) = C(f)C(g)$.

Proof. Let $f(X) = C(f)f^*(X)$ and $g(X) = C(g)g^*(X)$, where $f^*(X)$ and $g^*(X)$ are primitive. Then $f(X)g(X) = C(f)C(g)f^*(X)g^*(X)$, and hence to show that $C(fg) = C(f)C(g)$ it remains only to show that the product $f^*(X)g^*(X)$ is primitive. To this end let

$$f^*(X) = a_0 + a_1 X + \cdots + a_n X^n$$

$$g^*(X) = b_0 + b_1 X + \cdots + b_m X^m,$$

where $a_0, \ldots, a_n, b_0, \ldots, b_m \in R$. Now, if the product $f^*(X)g^*(X)$ is not primitive then there is some irreducible element $p \in R$ that divides each of its coefficients. But p does not divide all coefficients of $f^*(X)$ since $f^*(X)$ is primitive; let a_i be the first coefficient of $f*(X)$ not divisible by p. Similarly, let b_j be the first coefficient of $g*(X)$ not divisible by p. Then the coefficient of X^{i+j} in the product $f*(X)g*(X)$ is

$$a_0 b_{i+j} + \cdots + a_{i-1}b_{j+1} + a_i b_j + a_{i+1}b_{j-1} + \cdots + a_{i+j}b_0.$$

By assumption, p divides this coefficient. But p also divides each term in this sum except $a_i b_j$, which is impossible. Hence, no such irreducible element p exists and therefore the product $f*(X)g*(X)$ is primitive. It follows, then, that $C(fg) = C(f)C(g)$. ∎

COROLLARY.   The product of primitive polynomials is primitive.

Proof.   Let $f(X)$ and $g(X)$ be primitive polynomials over a unique factorization domain R. Then $C(f)$ and $C(g)$ are units of R and therefore $C(fg)$ is a unit of R since $C(fg) = C(f)C(g)$. Hence, $f(X)g(X)$ is primitive over R, as required. ∎

COROLLARY.   Every proper divisor of a primitive polynomial is a primitive polynomial of strictly smaller degree.

Proof.   Let $f(X)$ be a primitive polynomial over a unique factorization domain R and let $d(X)$ be a proper divisor of $f(X)$ in $R[X]$. Then $d(X)$ is nonconstant since $f(X)$ is primitive, and $f(X) = q(X)d(X)$ for some polynomial $q(X) \in R[X]$. Thus, $C(f) = C(q)C(d)$. Since $C(f)$ is a unit of R, it follows that both $C(q)$ and $C(d)$ are units of R and hence, in particular, that $d(X)$ is primitive. Finally, if $f(X)$ and $d(X)$ have the same degree, then $q(X)$ is constant and therefore a unit of R. In this case $d(X)$ is an associate of $f(X)$, which contradicts the fact that $d(X)$ is a proper divisor of $f(X)$. Thus, $d(X)$ is a primitive polynomial whose degree is strictly smaller than that of $f(X)$ and the proof is complete. ∎

To summarize, we have noted that factorization of polynomials over a unique factorization domain is different from factorization over a field since nonzero constants may be proper divisors of a given polynomial; the constant 2, for example, is a proper divisor of $2X + 4$ in $\mathbb{Z}[X]$. But for primitive polynomials, proper divisors are nonconstant, primitive, and have strictly smaller degree, as one expects, and hence factoring primitive polynomials over a unique factorization domain is essentially the same as factoring over a field. We emphasize, however, that this is only true for primitive polynomials.

There is an important relationship between the irreducibility of a primitive polynomial over a unique factorization domain R and its irreducibility over the quotient field $QF(R)$. Recall that if R is a unique factorization domain, the quotient field $QF(R)$ consists of all formal fractions $\dfrac{a}{b}$, where $a, b \in R$, $b \neq 0$; for example, $QF(\mathbb{Z}) = \mathbb{Q}$, the field of rational numbers. Now, we noted earlier that the polynomial $2X + 4 \in \mathbb{Z}[X]$, for example, is reducible over $\mathbb{Z}$

since it has the proper factorization $2X + 4 = 2(X + 2)$, but, when viewed over the quotient field $\mathbb{Q}$, it is irreducible. Thus, it is possible for a polynomial over a unique factorization domain R to be reducible over R but irreducible over the field $QF(R)$. We now claim that this cannot happen for primitive polynomials; that is, a primitive polynomial over R is irreducible over the quotient field $QF(R)$ if and only if it was already irreducible over R. Thus, for primitive polynomials, irreducibility over R is the same as irreducibility over the quotient field of R.

PROPOSITION 2. Let R be a unique factorization domain. Then a primitive polynomial over R is irreducible over the quotient field $QF(R)$ if and only if it is irreducible over R.

Proof. Let $f(X)$ be a primitive polynomial over R. Suppose first that $f(X)$ is irreducible over $QF(R)$. If $f(X)$ factors over R, then $f(X) = g(X)h(X)$ for some polynomials $g(X), h(X) \in R[X]$ which, by the previous corollaries, are primitive and of strictly smaller degree than $f(X)$. This factorization is therefore a proper factorization of $f(X)$ over the field $QF(R)$, which contradicts its irreducibility over $QF(R)$. Hence, $f(X)$ is irreducible over R.

Now suppose, conversely, that $f(X)$ is irreducible over R. To show that $f(X)$ is irreducible over $QF(R)$ we assume, to the contrary, that $f(X) = g(X)h(X)$ is a proper factorization of $f(X)$ in $QF(R)[X]$ and work for a contradiction. Since the coefficients of $g(X)$ and $h(X)$ are formal fractions whose denominators lie in R, there are elements $a, b \in R$ such that $ag(X), bh(X) \in R[X]$. Therefore,

$$abf(X) = [ag(X)][bh(X)] = C(ag)C(bh)[ag(X)]^*[bh(X)]^*.$$

But $C(ag)C(bh) = C(abgh) = abC(f)$, and $C(f) = u$ for some unit $u \in R$ since $f(X)$ is primitive. Therefore, $abf(X) = abu[ag(X)]^*[bh(X)]^*$ and hence $f(X) = u[ag(X)]^*[bh(X)]^*$, which is a proper factorization of $f(X)$ in $R[X]$, a contradiction since $f(X)$ is assumed to be irreducible over R. It follows that $f(X)$ has no proper factorization over $QF(R)$ and hence is irreducible over $QF(R)$. ∎

COROLLARY. Let $f(X)$ be a nonzero polynomial over a unique factorization domain R. Then $f(X)$ is irreducible over R if and only if $f(X) = p$ for some irreducible element $p \in R$ or $f(X)$ is primitive and irreducible over the quotient field $QF(R)$.

Proof. Clearly, if $f(X) = p$ for some irreducible element $p \in R$ or is primitive and irreducible over $QF(R)$, then $f(X)$ is irreducible over R. Conversely, suppose that $f(X)$ is any irreducible polynomial over R. If $f(X)$ is constant, then $f(X) = p$ for some element $p \in R$ and p is clearly irreducible in R. On the other hand if $f(X)$ is nonconstant, then $f(X) = C(f)f^*(X)$ is a factorization of $f(X)$ over R and hence $C(f)$ must be a unit of R. Therefore, $f(X)$ is primitive and irreducible over R and hence, by Proposition 2, is irreducible over $QF(R)$. ∎

This corollary gives us a complete description of the irreducible polynomials over a unique factorization domain R; they are the constant polynomials of the form $f(X) = p$, where p is any irreducible element in R, and the primitive polynomials over R that are irreducible over the quotient field $QF(R)$. For example, the constant polynomials 2, 3, and 5 are irreducible over $\mathbb{Z}$ since these constants are prime, while the polynomials $2X + 3$ and $X^2 + X + 1$ are irreducible over $\mathbb{Z}$ since they are primitive and irreducible over the quotient field $\mathbb{Q}$. Note, in particular, that a monic polynomial over R is irreducible over R if and only if it is irreducible over $QF(R)$; for example, a monic polynomial in $\mathbb{Z}[X]$ is irreducible over $\mathbb{Z}$ if and only if it is irreducible over $\mathbb{Q}$. Finally, note that Proposition 2 is equivalent to the statement that a primitive polynomial over a unique factorization domain R has a proper factorization over the quotient field $QF(R)$ if and only if it has a proper factorization over R itself.

The above results relate the irreducibility of a polynomial over R to its irreducibility over $QF(R)$. But determining whether or not a polynomial over R is or is not irreducible over $QF(R)$ is not always a simple matter. The following result is a useful numerical criterion for deciding when a polynomial over a unique factorization domain is irreducible over the quotient field of the domain.

PROPOSITION 3.   (EISENSTEIN IRREDUCIBILITY CRITERION) Let R be a unique factorization domain and let $f(X) = a_0 + a_1 X + \cdots + a_n X^n$ be a polynomial over R. Suppose there is an irreducible element $p \in R$ such that $p \mid a_0, \ldots, p \mid a_{n-1}$, but $p \nmid a_n$ and $p^2 \nmid a_0$. Then $f(X)$ is irreducible over the quotient field $QF(R)$.

Proof.   Let $f(X) = C(f)f^*(X)$, where $f^*(X)$ is primitive. Then $p \nmid C(f)$ and hence $f^*(X)$ is a primitive polynomial over R whose coefficients satisfy the same numerical criterion as those of $f(X)$. Since $f(X)$ is irreducible over $QF(R)$ if and only if $f^*(X)$ is irreducible over $QF(R)$, we may assume that $f(X)$ is itself primitive. We now claim that $f(X)$ is irreducible over R. For suppose, to the contrary, that $f(X)$ has a proper factorization in $R[X]$, say $f(X) = g(X)h(X)$ for some polynomials $g(X), h(X) \in R[X]$, and let

$$g(X) = b_0 + b_1 X + \cdots + b_s X^s$$

$$h(X) = c_0 + c_1 X + \cdots + c_t X^t,$$

where $b_0, \ldots, b_s, c_0, \ldots, c_t \in R$. Then $b_0 c_0 = a_0$. Since $p \mid a_0$ but $p^2 \nmid a_0$, it follows that p divides either b_0 or c_0, but not both. Let $p \mid b_0$, $p \nmid c_0$. Then, since $b_s c_t = a_n$ and $p \nmid a_n$, it follows that p does not divide b_s and hence there is a smallest positive integer m such that $p \nmid b_m$. Now consider the coefficient a_m of X^m in $f(X)$. Since $f(X) = g(X)h(X)$, $a_m = c_0 b_m + c_1 b_{m-1} + \cdots$. By assumption, $p \mid a_m$. But p also divides each of the terms $c_1 b_{m-1}, \ldots$. Therefore, p must divide $c_0 b_m$, which is impossible since p divides neither c_0 nor b_m. It follows that $f(X)$ has no proper factorization in $R[X]$ and hence is irreducible over R. Since $f(X)$ is primitive, we conclude that $f(X)$ is irreducible over $QF(R)$, as required.   ■

EXAMPLE 1

Let $f(X) = 2X^3 + 5X^2 + 10X + 5 \in \mathbb{Z}[X]$. Then $f(X)$ satisfies the Eisenstein criterion for the prime $p = 5$ and hence is irreducible over $\mathbb{Q}$. Moreover, since $f(X)$ is primitive it is also irreducible over $\mathbb{Z}$.

EXAMPLE 2

Let $f(X) = 4X^4 + 12X^3 + 18X^2 + 6 \in \mathbb{Z}[X]$. Then $f(X)$ satisfies the Eisenstein criterion for the prime $p = 3$ and hence is irreducible over $\mathbb{Q}$. But in this case $f(X)$ is not primitive and hence is not irreducible over $\mathbb{Z}$. In fact, $f(X) = 2(2X^4 + 6X^3 + 9X^2 + 3)$ is a proper factorization of $f(X)$ in $\mathbb{Z}[X]$; moreover, each of these factors is irreducible over $\mathbb{Z}$: 2 is prime and hence irreducible over $\mathbb{Z}$, while $2X^4 + 6X^3 + 9X^2 + 3$ is irreducible over $\mathbb{Q}$ since it satisfies the Eisenstein criterion for $p = 3$, and hence is irreducible over $\mathbb{Z}$ since it is primitive.

EXAMPLE 3. Cyclotomic polynomials of degree $p - 1$, p a prime

Let p be a prime and let $f_p(X) = X^{p-1} + X^{p-2} + \cdots + X + 1 \in \mathbb{Z}[X]$. We claim that $f_p(X)$ is irreducible over both $\mathbb{Z}$ and $\mathbb{Q}$.

(A) Observe that we cannot use the Eisenstein criterion directly since there is no prime that satisfies the requirements on the coefficients of $f_p(X)$. We therefore modify $f_p(X)$ by letting $g(X) = f_p(X + 1)$. Then $f_p(X)$ is irreducible over $\mathbb{Q}$ if and only if $g(X)$ is irreducible over $\mathbb{Q}$. Now, to show that $g(X)$ is irreducible over $\mathbb{Q}$, note that

$$f_p(X) = \frac{X^p - 1}{X - 1}$$

and therefore

$$g(X) = \frac{(X + 1)^p - 1}{X}$$

$$= X^{p-1} + \binom{p}{1}X^{p-2} + \cdots + \binom{p}{p-1},$$

where $\binom{p}{s} = \dfrac{p!}{s!(p - s)!}$ stands for the binomial coefficient for $s = 1, \ldots, p$.
Now, if $1 \leq s \leq p - 1$, then $s!(p - s)!\binom{p}{s} = p!$. Since p does not divide either $s!$ or $(p - s)!$ but does divide $p!$, it follows that $p \mid \binom{p}{s}$. Moreover, p does not divide the coefficient of X^{p-1} in $g(X)$ and p^2 does not divide $\binom{p}{p-1}$. Therefore, $g(X)$ satisfies the Eisenstein criterion for the prime p and hence is irreducible over $\mathbb{Q}$. Thus, $f_p(X)$ is irreducible over $\mathbb{Q}$ and, since $f_p(X)$ is monic, it is also irreducible over $\mathbb{Z}$.

(B) The polynomial $f_p(X)$ is called the *cyclotomic polynomial of degree $p - 1$*. For example, the cyclotomic polynomials corresponding to the primes $p = 2$, 3, and 5 are $X + 1$, $X^2 + X + 1$, and $X^4 + X^3 + X^2 + X + 1$, respectively. These polynomials are irreducible over both $\mathbb{Z}$ and $\mathbb{Q}$. We use the term "cyclotomic" to describe these polynomials because they are related to complex roots of unity and the geometric problem of dividing a circle into equal arcs. Recall that if p is prime, the pth roots of unity are the complex roots of the equation $X^p = 1$ and have the form $1, z_p, \ldots, z_p^{p-1}$, where z_p is any primitive pth root of unity (Example 4, Section 4, Chapter 2). Geometrically, these roots divide the unit complex circle into p equal arcs. Removing the rational root 1 leaves us with the $p - 1$ roots $z_p, \ldots, z_p^{p-1}$, which are the roots of the polynomial $f_p(X)$, all of which are primitive pth roots of unity. When we discuss these polynomials in more detail in Chapter 14, and later in Chapter 16 when we discuss constructibility by straight-edge and compass, we will see that the irreducibility of the polynomials $f_p(X)$ shows that the division of the circle into p equal arcs is possible by means of straight-edge and compass if and only if p is a prime of the form $2^n + 1$.

We now conclude this section by showing that the ring of polynomials over a unique factorization domain is itself a unique factorization domain.

PROPOSITION 4. Let R be a unique factorization domain. Then the ring $R[X]$ of polynomials over R is a unique factorization domain.

Proof.   We must show that every nonzero polynomial in $R[X]$ may be written as the product of irreducible polynomials over R and that such a factorization is unique to within order and associates. Let us begin by showing that such a factorization is always possible. Let $f(X)$ be a nonzero polynomial in $R[X]$. Since R is a unique factorization domain, there are irreducible elements $p_1, \ldots, p_n$ in R such that $C(f) = p_1 \cdots p_n$. Hence, $f(X) = C(f)f^*(X) = p_1 \cdots p_n f^*(X)$, where $f^*(X)$ is primitive. Now, if $f^*(X)$ is irreducible over R we are done. Otherwise, $f^*(X)$ has a proper divisor which is primitive and of strictly smaller degree. It then follows easily by induction that $f^*(X)$ may be written as the product of irreducible polynomials over R, and hence so may $f(X)$. To complete the proof, it remains to show that such a factorization is unique to within order and associates. Suppose that

$$p_1 \cdots p_n f_1(X) \cdots f_s(X) = q_1 \cdots q_m g_1(X) \cdots g_t(X)$$

are two factorizations of a nonzero polynomial in $R[X]$, where $p_1, \ldots, p_n$, $q_1, \ldots, q_m$ are irreducible elements in R, and $f_1(X), \ldots, f_s(X), g_1(X), \ldots, g_t(X)$ are irreducible primitive polynomials in $R[X]$. Taking the content of both sides of the equation, it follows that $p_1 \cdots p_n = q_1 \cdots q_m$, and therefore the p_i and q_j are the same to within order and associates since R is a unique factorization domain. Therefore, $f_1(X) \cdots f_s(X) = g_1(X) \cdots g_t(X)$ to within a unit factor. Now, the $f_i(X)$ and $g_j(X)$ are irreducible over the quotient field $QF(R)$ since they are primitive and irreducible over R. Hence, they are the same to within order and constant multiples since factorization in $QF(R)[X]$

is unique. To show that they are in fact associates in $R[X]$, let $g_j(X) = cf_i(X)$ for some $c \in QF(R)$. Then $dc \in R$ for some nonzero element $d \in R$ and hence $dg_j(X) = dcf_i(X)$. Therefore, $dC(g_j) = dcC(f_i)$ and hence, since $C(f_i)$ is a unit in R, it follows that $c \in R$ and is a unit in R. Thus, $f_i(X)$ and $g_j(X)$ are associates in $R[X]$. Therefore, factorization in $R[X]$ is unique to within order and associates and hence $R[X]$ is a unique factorization domain. ∎

COROLLARY. If R is a unique factorization domain, then the ring $R[X_1, \ldots, X_n]$ of polynomials over R in a finite number of variables $X_1, \ldots, X_n$ is also a unique factorization domain. In particular, if k is a field, then the ring $k[X_1, \ldots, X_n]$ of polynomials over k is a unique factorization domain.

Proof. This follows immediately from Proposition 4 by induction. We leave the details as an exercise for the reader. ∎

For example, in $\mathbb{Z}[X]$ we find that $6X + 12 = 2 \cdot 3 \cdot (X + 2)$ is the factorization of $6X + 12$ into irreducible polynomials over $\mathbb{Z}$. Similarly,

$$10X^2 + 35X + 30 = 5(2X^2 + 7X + 6) = 5(2X + 3)(X + 2)$$

$$4X^3 - 4 = 2^2(X^3 - 1) = 2^2(X - 1)(X^2 + X + 1)$$

$$4X^4 + 12X^2 + 6X = 2X(2X^3 + 6X + 3).$$

Note that the factor $X^2 + X + 1$ is irreducible over $\mathbb{Z}$ since it is the cyclotomic polynomial for the prime 3, while $2X^3 + 6X + 3$ is irreducible over $\mathbb{Z}$ since it is primitive and irreducible over $\mathbb{Q}$ by Eisenstein's criterion with $p = 3$.

Since $R[X]$ is a unique factorization domain whenever R is, it follows that any finite collection of nonzero polynomials over R have a greatest common divisor, that it may be found by factoring the polynomials into irreducible factors and multiplying the common irreducible factors, and that it is unique to within associates. For example, in $\mathbb{Z}[X]$ we find that

$$\gcd(6(X^3 - X^2 + X - 1), 4X^4 + 4X^2) = \gcd(6(X - 1)(X^2 + 1), 4X^2(X^2 + 1))$$
$$= 2(X^2 + 1).$$

In conclusion, observe that the polynomial ring $k[X, Y]$ is a unique factorization domain for any field k. We claim, however, that $k[X, Y]$ is not a principal ideal domain. For consider the ideal $(X, Y)k[X, Y]$ generated by X and Y. If $(X, Y)k[X, Y] = f(X, Y)k[X, Y]$ for some $f(X, Y) \in k[X, Y]$, then $X = f(X, Y)g(X, Y)$ and $Y = f(X, Y)h(X, Y)$ for some $g(X, Y), h(X, Y) \in k[X, Y]$. But then $f(X, Y)$ contains no X-terms and no Y-terms and is therefore constant. Hence, $(X, Y)k[X, Y] = k[X, Y]$, from which it follows that $a(X, Y)X + b(X, Y)Y = 1$ for some $a(X, Y), b(X, Y) \in k[X, Y]$. But this is impossible; for if we let $X = Y = 0$, it follows that $0 = 1$, a contradiction. We conclude, therefore, that no such polynomial $f(X, Y)$ exists and hence

$(X, Y)k[X, Y]$ is not a principal ideal. The ring $k[X, Y]$ is therefore an example of a unique factorization domain that is not a principal ideal domain.

Exercises

1. Show that the following polynomials in $\mathbb{Z}[X]$ are irreducible over $\mathbb{Z}$:
   (a) $4X^2 + 9$ (d) $2X^5 + 9X^4 + 3X^3 + 6X^2 + 3X + 3$
   (b) $3X^2 + 2X + 1$ (e) $5X^6 - 3X + 12$
   (c) $2X^3 + 5$

2. Factor the following polynomials in $\mathbb{Z}[X]$ into a product of irreducible polynomials over $\mathbb{Z}$:
   (a) $6X^3 - 48$ (d) $-4X^4 + 16$
   (b) $3X^2 + 6X + 3$ (e) $12X^4 - 12$
   (c) $15X + 45$

3. Determine which of the following polynomials in $\mathbb{Z}[i][X]$ are irreducible over the ring $\mathbb{Z}[i]$ of Gaussian integers; if the polynomial is not irreducible, factor it into a product of irreducible polynomials over $\mathbb{Z}[i]$.
   (a) $X^2 + 1$ (e) $X^2 - 2i$
   (b) $2X + 4$ (f) $X^3 + (-1 + i)X^2 + (-1 + i)X + 1 + i$
   (c) $X^2 - 2$ (g) $X^4 + 2X^2 + 2$
   (d) $5X^3 + 5$

4. Let R be a unique factorization domain and let K stand for the quotient field of R. If $f(X)$ and $g(X)$ are polynomials in $R[X]$ such that $g(X)$ divides $f(X)$ in $K[X]$, is it true that $g(X)$ divides $f(X)$ in $R[X]$?

5. Let R be a unique factorization domain and let K stand for the quotient field of R. Let $f_1(X), \ldots, f_n(X)$ be a finite number of nonzero polynomials in $R[X]$ and let $D(X)$ stand for their greatest common divisor in $R[X]$. If d is the lead coefficient of $D(X)$, show that $\dfrac{1}{d} D(X)$ is the greatest common divisor of $f_1(X), \ldots, f_n(X)$ in $K[X]$.

6. Let $(2, X)\mathbb{Z}[X]$ stand for the ideal of $\mathbb{Z}[X]$ generated by 2 and X. Show that $(2, X)\mathbb{Z}[X]$ is not a principal ideal of $\mathbb{Z}[X]$.

7. **The Eisenstein criterion for polynomials.** Let k be a field and let $f(X, Y)$ be a nonzero polynomial in $k[X, Y]$. Write $f(X, Y) = a_0(X) + a_1(X)Y + \cdots + a_n(X)Y^n$, where $a_0(X), \ldots, a_n(X) \in k[X]$. Suppose there is an irreducible polynomial $p(X) \in k[X]$ such that $p(X) | a_0(X), \ldots, p(X) | a_{n-1}(X)$, but $p(X) \nmid a_n(X)$ and $p(X)^2 \nmid a_0(X)$. Show that $f(X, Y)$ is irreducible in $k[X, Y]$.

8. Let $f(X, Y) = X^2 + Y^2 - 1 \in \mathbb{Q}[X, Y]$. Show that $f(X, Y)$ is an irreducible polynomial in $\mathbb{Q}[X, Y]$.

9. Let $I = (X^2 + XY - 2Y^2, X^3 - Y^3)\mathbb{Q}[X, Y]$ stand for the ideal of $\mathbb{Q}[X, Y]$ generated by the polynomials $X^2 + XY - 2Y^2$ and $X^3 - Y^3$. Show that I is not a principal ideal of $\mathbb{Q}[X, Y]$.

THREE

BASIC FIELD THEORY

12

Fields

In the third and final part of this book we turn our attention to fields, an algebraic structure in which all four basic arithmetic operations—addition, subtraction, multiplication, and division by nonzero elements—may be carried out, and in which these operations satisfy the standard algebraic properties. Fields provide the basic algebraic model for studying polynomials, particularly the roots of polynomials, and it is this theme that will dominate our study of field theory. Indeed, the concept of a field evolved along with that of a group out of early nineteenth-century attempts to solve algebraic equations by providing the appropriate algebraic framework in which to discuss their solutions. The mutual interaction between field theory and group theory soon became the basis for one of the most useful and important algebraic theories in mathematics—the Galois theory of fields. Today, Galois theory not only provides the necessary mathematical tools for dealing with such classical problems as the solvability of algebraic equations and the impossibility of certain geometric constructions, but, when applied to finite fields, plays an important role in such contemporary areas of mathematics as combinatorics and coding theory.

The reader will recall that we introduced the concept of a field in Chapter 10 as a special type of commutative ring with identity. In this chapter our goal is to discuss and illustrate the basic concepts of field theory in more detail: fields and subfields, characteristic of a field, field-isomorphisms, extension fields and the degree of a field extension, and, finally, algebraic extensions. We begin by first reviewing the basic properties of fields and the different methods for constructing fields. Then, in Section 2, we adopt a new point of view: rather than discuss subfields of a given field, we concentrate instead on those fields that contain the given field as a subfield. These larger fields are called extension fields of the given field and are important because they form vector spaces over the given field. This new point of view—that extension fields form vector spaces over a given subfield—therefore provides a way to use the concepts and results of linear algebra in the study of fields. For example, if K/k is an extension of fields—that is, if k is a subfield of K—then K is a vector space over k and, as such, has a dimension which is called the degree of the extension and denoted by $[K:k]$. If this number is finite, we say that K is a finite extension of k. In this case the elements in K can be written in terms of a fixed basis using elements in k as coefficients.

We conclude the chapter by studying an especially important type of field extension that occurs when the elements of the extension field are roots of polynomials over the ground field. These extensions are called algebraic extensions. In the third section we use elementary results from linear algebra to discuss the basic properties of algebraic extensions. We will show, for example, that every finite extension of fields is an algebraic extension, and, moreover, has the form $k[X]/f(X)k[X]$, where $f(X)$ is irreducible over k, that the degree of the extension is equal to the degree of $f(X)$, and that the basic algebraic properties of this field come directly from the polynomial $f(X)$.

1. FIELDS AND THEIR ARITHMETIC PROPERTIES

Let us begin by reviewing some basic terminology and results from Chapter 10.

Definition

A *field* is a commutative ring with identity in which every nonzero element is a unit.

In other words, a field is a nonempty set K together with two binary operations, addition and multiplication, such that addition is associative, commutative, has a unique identity, denoted by 0, and every element $x \in K$ has a unique additive inverse, denoted by $-x$; likewise, multiplication is associative, commutative, has a unique identity, denoted by 1, and every nonzero element $x \in K$ has a unique multiplicative inverse, denoted by x^{-1}, or $1/x$ as we will write from now on; and finally, multiplication distributes over addition. Informally, we think of a field as an algebraic structure in which it is possible to carry out the four basic arithmetic operations— addition, subtraction, multiplication, and division by nonzero elements— and in which these operations satisfy the standard algebraic properties.

For example, the set $\mathbb{Q}$ of rational numbers, the set $\mathbb{R}$ of real numbers, and the set $\mathbb{C}$ of complex numbers are all fields under addition and multiplication of numbers. Recall, in particular, that if $a + bi$ and $c + di$ are typical complex numbers in the field $\mathbb{C}$, then

$$(a + bi) + (c + di) = (a + c) + (b + d)i$$

$$(a + bi)(c + di) = (ac - bd) + (ad + bc)i,$$

and, if $a + bi \neq 0$,

$$\frac{1}{a + bi} = \frac{a}{a^2 + b^2} + \frac{-b}{a^2 + b^2} i.$$

It follows from the definition that if K is a field, the set K^* of nonzero elements in K is an abelian group under multiplication; K^* is called the *multiplicative group of nonzero elements of K*. A *subfield* of K is any nonempty subset that inherits both addition and multiplication and is a field under these operations. Clearly, if k is a nonempty subset of K, then k is a subfield of K if and only if four conditions are satisfied:

(1) if $x, y \in k$, then $x + y \in k$;
(2) if $x, y \in k$, then $xy \in k$;
(3) if $x \in k$, then $-x \in k$;
(4) if $x \in k$, $x \neq 0$, then $\frac{1}{x} \in k$.

In other words, k is a subfield of K if it is closed under sums, products, and inverses, both additive and multiplicative. The field $\mathbb{Q}$ of rational numbers is a subfield of the field $\mathbb{R}$ of real numbers, for example, and $\mathbb{R}$ is a subfield of $\mathbb{C}$. The ring $\mathbb{Z}$ of integers, however, is not a subfield of $\mathbb{Q}$ since it is not closed under multiplicative inverses.

And finally, recall that a *field-isomorphism* from a field K to a field K' is any 1–1 and onto function $f: K \to K'$ such that $f(x + y) = f(x) + f(y)$ and $f(xy) = f(x)f(y)$ for all elements $x, y \in K$. If $f: K \to K'$ is a field-isomorphism, then $f(0) = 0$, $f(1) = 1$, $f(-x) = -f(x)$ for all $x \in K$, and $f(1/x) = \frac{1}{f(x)}$ for all nonzero $x \in K$. A field-isomorphism from a field K onto itself is called a *field-automorphism*, or simply *automorphism*, of K. For example, the complex conjugation mapping $f: \mathbb{C} \to \mathbb{C}$ that maps every complex number $a + bi$ to its complex conjugate $f(a + bi) = a - bi$ is a field-isomorphism mapping $\mathbb{C}$ onto $\mathbb{C}$ and hence is an automorphism of $\mathbb{C}$.

EXAMPLE 1

Let $\mathbb{Q}[i] = \{a + bi \in \mathbb{C} \mid a, b \in \mathbb{Q}\}$ stand for the set of complex numbers whose real and imaginary parts are rational numbers. Then $\mathbb{Q}[i]$ is a subfield of $\mathbb{C}$ and hence, in particular, is a field under addition and multiplication of complex numbers.

EXAMPLE 2. The field $\mathbb{F}_p$ of integers modulo a prime p

Let p be a prime and let $\mathbb{F}_p = \{[0], [1], \ldots, [p-1]\}$ stand for the set of congruence classes of integers modulo p. Then $\mathbb{F}_p$ is a field under addition and multiplication of classes defined by setting $[s] + [t] = [s + t]$ and $[s][t] = [st]$ for all $[s], [t] \in \mathbb{F}_p$. For convenience, we write $0, 1, \ldots, p - 1$ for the elements in $\mathbb{F}_p$. Recall that if $s \neq 0$, the inverse $1/s$ may be found either by trial and error or by using the division algorithm. In $\mathbb{F}_3$, for example, we find easily that $\frac{1}{2} = 2$, while in $\mathbb{F}_{37}$ we find, using the division algorithm, that $1/26 = 10$ (Example 1, Section 2, Chapter 10).

There are two important methods for constructing new fields: forming the quotient field of an integral domain, and forming the quotient ring of a commutative ring with identity modulo a maximal ideal. The following examples review and illustrate these methods.

EXAMPLE 3.   The quotient field of an integral domain

Let R be an integral domain. Then the quotient field of R is the set $QF(R)$ $= \left\{ \frac{a}{b} \,\middle|\, a, b \in R, b \neq 0 \right\}$ consisting of all formal fractions of elements in R with nonzero denominator. For example, $QF(\mathbb{Z}) = \mathbb{Q}$ and $QF(\mathbb{Z}[i]) = \mathbb{Q}[i]$. Recall that if $\frac{a}{b}$ and $\frac{c}{d}$ are typical elements in $QF(R)$, then $\frac{a}{b} = \frac{c}{d}$ if and only if $ad = bc$, and that addition and multiplication in $QF(R)$ are defined by the usual algebraic formulas: $\frac{a}{b} + \frac{c}{d} = \frac{ad + bc}{bd}$ and $\frac{a}{b} \cdot \frac{c}{d} = \frac{ac}{bd}$. Moreover, if $a \neq 0$, then $\left(\frac{a}{b} \right)^{-1} = \frac{b}{a}$.

EXAMPLE 4.   Fields of rational functions

Let k be a field. Then the ring $k[X]$ of polynomials in X over k is an integral domain and hence we may form its quotient field $QF(k[X])$. Let $k(X) = QF(k[X])$. The field $k(X)$ is called the *field of rational functions in X over k*. A typical element in $k(X)$ has the form $\frac{f(X)}{g(X)}$, where $f(X)$ and $g(X)$ are polynomials over k with $g(X) \neq 0$. If $\frac{f(X)}{g(X)}$ and $\frac{s(X)}{t(X)}$ are typical elements in $k(X)$, then

$$\frac{f(X)}{g(X)} = \frac{s(X)}{t(X)} \quad \text{if and only if} \quad f(X)t(X) = g(X)s(X).$$

For example, in $\mathbb{Q}(X)$,

$$\frac{X^2 - 1}{X^2 - X} = \frac{X + 1}{X}.$$

A rational function over a field k defines a function on those elements in k at which it has a nonvanishing denominator as follows. Let $r(X)$ be a rational function over k and let $c \in k$. Then, if $r(X) = \dfrac{f(X)}{g(X)}$ for some polynomials $f(X)$ and $g(X)$, and $g(c) \neq 0$, we set $r(c) = \dfrac{f(c)}{g(c)}$; if $r(X)$ has no such representation, $r(c)$ is undefined. Let us illustrate these ideas with two particular examples.

(A) The field $\mathbb{Q}(X)$ of rational functions over $\mathbb{Q}$. In this case

$$\mathbb{Q}(X) = \left\{ \frac{f(X)}{g(X)} \,\middle|\, f(X), g(X) \in \mathbb{Q}[X], g(X) \neq 0 \right\}.$$

For example, let

$$u(X) = \frac{X + 1}{X^2 - 1} \quad \text{and} \quad v(X) = \frac{X}{X + 1}.$$

Then $u(X)$ and $v(X)$ are typical elements in $\mathbb{Q}(X)$. Moreover, since

$$\frac{X + 1}{X^2 - 1} = \frac{1}{X - 1}$$

in $\mathbb{Q}(X)$, it follows that

$$u(X) = \frac{1}{X - 1}.$$

Therefore, $u(X)$ defines a function at all numbers $c \in \mathbb{Q}$ except where $c = 1$; if $c \neq 1$, $u(c) = \dfrac{1}{c - 1}$. Similarly, $v(X)$ defines a function at all numbers $c \in \mathbb{Q}$ except $c = -1$; if $c \neq -1$, $v(c) = \dfrac{c}{c + 1}$. Finally, we find that

$$u(X) + v(X) = \frac{1}{X - 1} + \frac{X}{X + 1}$$

$$= \frac{X^2 + 1}{X^2 - 1}, \quad \text{which is defined at all numbers in } \mathbb{Q} \text{ except } \pm 1;$$

$$u(X)v(X) = \frac{X}{X^2 - 1}, \quad \text{which is defined at all numbers in } \mathbb{Q} \text{ except } \pm 1;$$

$$\frac{1}{u(X)} = X - 1, \quad \text{which is defined at all numbers in } \mathbb{Q};$$

$$\frac{1}{v(X)} = \frac{X + 1}{X}, \quad \text{which is defined at all numbers in } \mathbb{Q} \text{ except } 0.$$

(B) The field $\mathbb{F}_3(X)$ of rational functions over $\mathbb{F}_3$. In this case

$$\mathbb{F}_3(X) = \left\{ \frac{f(X)}{g(X)} \,\middle|\, f(X), g(X) \in \mathbb{F}_3[X], g(X) \neq 0 \right\}.$$

For example, let

$$u(X) = \frac{X^2 + X}{2X^2 + 1} \quad \text{and} \quad v(X) = \frac{X + 1}{X + 2}.$$

Then $u(X)$ and $v(X)$ are typical elements in $\mathbb{F}_3(X)$. Observe that the denominator of $u(X)$ is $2X^2 + 1$ and that $2X^2 + 1 = 1 - X^2$ in $\mathbb{F}_3[X]$. It follows, therefore, that

$$u(X) = \frac{X^2 + X}{1 - X^2} = \frac{X(X + 1)}{(1 - X)(1 + X)}$$

$$= \frac{X}{1 - X} = -\frac{X}{X - 1} = \frac{2X}{X - 1}.$$

Therefore, $u(X)$ defines a function at all numbers $c \in \mathbb{F}_3$ except $c = 1$; we find that $u(0) = 0$ and $u(2) = 1$. Similarly, $v(X)$ defines a function at all numbers except 1; in this case, $v(0) = 2$ and $v(2) = 0$. Finally,

$$u(X) + v(X) = \frac{2X}{X - 1} + \frac{X + 1}{X + 2}$$

$$= \frac{2X}{X + 2} + \frac{X + 1}{X + 2}$$

$$= \frac{1}{X + 2}, \qquad \text{which is defined at all numbers in } \mathbb{F}_3 \text{ except 1;}$$

$$u(X)v(X) = \frac{2X(X + 1)}{(X + 2)^2}, \quad \text{which is defined at all numbers in } \mathbb{F}_3 \text{ except 1;}$$

$$\frac{1}{u(X)} = \frac{X - 1}{2X}, \qquad \text{which is defined at all numbers in } \mathbb{F}_3 \text{ except 0;}$$

$$\frac{1}{v(X)} = \frac{X + 2}{X + 1}, \qquad \text{which is defined at all numbers in } \mathbb{F}_3 \text{ except 2.}$$

The table here summarizes these functions and their values on $\mathbb{F}_3$: an asterisk (*) indicates that the function is not defined at the particular value.

Function	0	1	2
$u(X)$	0	*	1
$v(X)$	2	*	0
$u(X) + v(X)$	2	*	1
$u(X)v(X)$	0	*	0
$1/u(X)$	*	0	1
$1/v(X)$	2	0	*

EXAMPLE 5. Quotient rings modulo maximal ideals

Let R be a commutative ring with identity and let M be a maximal ideal of R. Then the quotient ring R/M is a field under addition and multiplication of cosets defined by setting

$$(x + M) + (y + M) = (x + y) + M$$

$$(x + M)(y + M) = xy + M$$

for all cosets $x + M$, $y + M \in R/M$. For example, if p is a prime, then the principal ideal $p\mathbb{Z}$ is a maximal ideal of the ring $\mathbb{Z}$ of integers and the quotient ring $\mathbb{Z}/p\mathbb{Z} = \mathbb{F}_p$, the field of integers modulo p.

EXAMPLE 6. Fields of polynomials modulo an irreducible polynomial

Let $f(X)$ be an irreducible polynomial over a field k. Then the ideal $f(X)k[X]$ generated by $f(X)$ is a maximal ideal of $k[X]$ and hence the quotient ring $K = k[X]/f(X)k[X]$ is a field. We refer to K as the *field of polynomials over k modulo $f(X)$*. Recall that if $g(X) + f(X)k[X]$ is a typical element in K, then $g(X) + f(X)k[X] = r(X) + f(X)k[X]$, where $r(X)$ is the remainder of $g(X)$ upon division by $f(X)$. Since $r(X)$ is either zero or has degree smaller than that of $f(X)$, it follows that every element in K may be written uniquely in the form $a_0 + a_1 X + \cdots + a_{n-1}X^{n-1} + f(X)k[X]$, where $n = \deg f(X)$ and $a_0, \ldots, a_{n-1} \in k$. Now,

$$a_0 + a_1 X + \cdots + a_{n-1}X^{n-1} + f(X)k[X]$$
$$= (a_0 + f(X)k[X]) + (a_1 + f(X)k[X])(X + f(X)k[X])$$
$$+ \cdots + (a_{n-1} + f(X)k[X])(X + f(X)k[X])^{n-1}.$$

Hence, if we let $x = X + f(X)k[X]$ and if we write a_i in place of the coset $a_i + f(X)k[X]$ for $i = 0, \ldots, n - 1$, then every element in K may be written uniquely in the form $a_0 + a_1 x + \cdots + a_{n-1}x^{n-1}$, where $a_0, \ldots, a_{n-1} \in k$. Observe that if $a_0 + a_1 x + \cdots + a_{n-1}x^{n-1}$ is a typical nonzero element in K, then, since K is a field, there are uniquely determined elements $b_0, \ldots, b_{n-1} \in k$ such that

$$\frac{1}{a_0 + a_1 x + \cdots + a_{n-1}x^{n-1}} = b_0 + b_1 x + \cdots + b_{n-1}x^{n-1}.$$

Let us illustrate these remarks by discussing two particular fields of polynomials modulo an irreducible polynomial.

(A) Let $K = \mathbb{R}[X]/(X^2 + 1)\mathbb{R}[X]$. Then K is a field since $X^2 + 1$ is irreducible over $\mathbb{R}$. In this case $\deg(X^2 + 1) = 2$ and hence a typical element in K may be written uniquely in the form $a + bx$, where $a, b \in \mathbb{R}$ and where $x = X + (X^2 + 1)\mathbb{R}[X]$. Observe that $x^2 + 1 = 0$ in K since

$X^2 + 1 + (X^2 + 1)\mathbb{R}[X] = 0 + (X^2 + 1)\mathbb{R}[X]$. If $a + bx$ and $c + dx$ are typical elements in K, we find that

$$(a + bx) + (c + dx) = (a + c) + (b + d)x$$

$$(a + bx)(c + dx) = (ac - bd) + (ad + bc)x.$$

If, in addition, $a + bx \neq 0$, then

$$\frac{1}{a + bx} = \frac{a}{a^2 + b^2} + \frac{-b}{a^2 + b^2} x.$$

Finally, we leave the reader to verify that the function $f : K \to \mathbb{C}$ defined by setting $f(a + bx) = a + bi$ for all elements $a + bx \in K$ is a field-isomorphism, and therefore $K \cong \mathbb{C}$, the field of complex numbers.

(B) Let $K = \mathbb{F}_2[X]/(X^2 + X + 1)\mathbb{F}_2[X]$. Then K is a field since $X^2 + X + 1$ is irreducible over $\mathbb{F}_2$. In this case a typical element in K may be written uniquely in the form $a + bx$, where $a, b \in \mathbb{F}_2$ and $x = X + (X^2 + X + 1)\mathbb{F}_2[X]$. It follows that K contains only four elements, namely 0, 1, x, and $1 + x$. Moreover, $x^2 + x + 1 = 0$ in K and therefore $x^2 = 1 + x$. Thus, if $a + bx$ and $c + dx$ are typical elements in K, we find that

$$(a + bx) + (c + dx) = (a + c) + (b + d)x$$

and

$$(a + bx)(c + dx) = ac + bcx + adx + bdx^2$$
$$= ac + bcx + adx + bd(1 + x)$$
$$= (ac + bd) + (ad + bd + bc)x.$$

If, in addition, $a + bx \neq 0$, then

$$\frac{1}{a + bx} = \frac{1}{a + bx} \cdot \frac{a + b + bx}{a + b + bx}$$

$$= \frac{a + b}{a^2 + ab + b^2} + \frac{b}{a^2 + ab + b^2} x.$$

The multiplying factor $a + b + bx$ may be obtained as follows. Given $a + bx \neq 0$, we want to find an element $c + dx$ such that the product $(a + bx)(c + dx)$ has no x term. The x term of this product is $ad + bd + bc$, or $(a + b)d + bc$. Setting $d = b$, it follows that $c = a + b$ and hence $c + dx = a + b + bx$. The complete addition and multiplication tables for K are shown in Example 6, Section 4, Chapter 10. This concludes the example.

As the above examples show, the arithmetic properties of one field may differ significantly from those of another field. Nowhere are these differences more apparent than in the additive property of the identity element 1. For example, in the number fields $\mathbb{Q}$, $\mathbb{R}$, and $\mathbb{C}$, the number 1 when added to itself any finite number of times is never zero. But in the field $\mathbb{F}_p$ of integers modulo a prime p, the number 1 when added to itself p times is zero:

$$\underbrace{1 + \cdots + 1}_{p} = 0.$$

With this observation in mind, we make the following definition.

Definition

Let K be a field. The *characteristic* of K is the smallest positive integer n such that

$$\underbrace{1 + \cdots + 1}_{n} = 0,$$

where 1 stands for the identity of K. If no such integer exists, the characteristic of K is defined to be zero. We denote the characteristic of K by char(K).

For example, char($\mathbb{Q}$) = char($\mathbb{R}$) = char($\mathbb{C}$) = 0, but char($\mathbb{F}_2$) = 2, char($\mathbb{F}_3$) = 3, and, in general, char($\mathbb{F}_p$) = p for all primes p. The characteristic of a field is an elementary yet important way to distinguish between fields. In the rest of this section our goal is to show, first, that the characteristic of any field is either zero or a prime number, and second, that every field of characteristic zero contains a subfield isomorphic to $\mathbb{Q}$ while every field of prime characteristic p contains a subfield isomorphic to $\mathbb{F}_p$. Before we do this, however, let us first recall some notation. Let 1 stand for the identity element of a field K and let n be an integer. If n is positive, then $n1 = 1 + \cdots + 1$ stands for the sum of 1 with itself n times; if n is negative, $n1 = -1 + \cdots + -1 = (-n)(-1)$, the sum of -1 with itself $-n$ times. Using this notation, the characteristic of K is then either zero or the smallest positive integer n such that $n1 = 0$. Finally, recall that $(n1)(m1) = (nm)1$ and $n1 + m1 = (n + m)1$ for all integers n and m.

PROPOSITION 1. The characteristic of a field is either zero or a prime number.

Proof.   Let K be a field, $n = $ char(K), and assume that $n \neq 0$. If $n = st$ is any factorization of n as a product of integers s and t, where $1 \leq s, t \leq n$, then $(s1)(t1) = (st)1 = n1 = 0$. Therefore, $s1 = 0$ or $t1 = 0$ since K is a field. Since n is the smallest positive integer with the property that $n1 = 0$, it follows that either $s = n$, in which case $t = 1$, or $t = n$, in which case $s = 1$. Thus, the number n has only the trivial factorizations and is therefore prime. ∎

PROPOSITION 2. Let K be a field of prime characteristic p. Then $(x + y)^p = x^p + y^p$ for all elements $x, y \in K$.

Proof. Let $x, y \in K$. Then, since K is a commutative ring with identity,

$$(x + y)^p = x^p + \binom{p}{1} x^{p-1} y + \cdots + y^p,$$

where $\binom{p}{s} = \dfrac{p!}{s!(p - s)!}$ stands for the binomial coefficient for $s = 1, \ldots, p$.

Now, for $s = 1, \ldots, p - 1$, $s!(p - s)! \binom{p}{s} = p!$ and therefore p divides either

$s!$, $(p - s)!$, or $\binom{p}{s}$. Since p does not divide $s!$ or $(p - s)!$, it follows that p

divides $\binom{p}{s}$ and hence $\binom{p}{s} x^{p-s} = 0$ in K. Therefore, $(x + y)^p = x^p + y^p$. ■

PROPOSITION 3. Let K be a field.

(1) If char$(K) = 0$, then K contains a subfield k such that $k \cong \mathbb{Q}$.
(2) If char$(K) = p$, a prime, then K contains a subfield k such that $k \cong \mathbb{F}_p$.

Proof. We claim that the subfield generated by the identity element 1 has the required property. To show this, suppose first that char$(K) = 0$ and let

$$k = \left\{ \frac{n1}{m1} \in K \,\middle|\, n, m \in \mathbb{Z}, m \neq 0 \right\}.$$

Observe that if m is a nonzero integer, then $m1 \neq 0$ since char$(K) = 0$ and therefore $\dfrac{n1}{m1}$ is in fact an element of K. We claim that k is a subfield of K and that $k \cong \mathbb{Q}$, the field of rational numbers. To show this, let $x = \dfrac{n1}{m1}$ and $y = \dfrac{s1}{t1}$ be typical elements in k. Then

$$x + y = \frac{(n1)(t1) + (m1)(s1)}{(m1)(t1)} = \frac{(nt + ms)1}{(mt)1} \in k,$$

$$xy = \frac{(ns)1}{(mt)1} \in k,$$

$$-x = \frac{(-n)1}{m1} \in k.$$

If, in addition, $x \neq 0$, then $n \neq 0$ and hence $\dfrac{1}{x} = \dfrac{m1}{n1} \in k$. Thus, k is a subfield of K. Finally, we leave the reader to verify that the function $f : \mathbb{Q} \to k$ defined by setting $f\left(\dfrac{n}{m}\right) = \dfrac{n1}{m1}$ for all rational numbers $\dfrac{n}{m} \in \mathbb{Q}$ is a field-isomorphism. Therefore, $k \cong \mathbb{Q}$, which completes the proof of statement (1). To prove statement (2), suppose that char$(K) = p$, a prime, and let $k = \{n1 \in K \,|\, n \in \mathbb{Z}\}$.

Then k is a subfield of K. For if $x = n1$ and $y = m1$ are typical elements in k, then $x + y = (n + m)1$, $xy = (nm)1$, and $-x = (-n)1$, all of which are elements in k. If, in addition, $x \neq 0$, then p does not divide n. Therefore, p and n are relatively prime and hence there are integers a and b such that $an + bp = 1$. It follows that $1 = (an)1 + (bp)1 = (a1)(n1) = (a1)x$ since $p1 = 0$ in K, and therefore $\dfrac{1}{x} = a1 \in k$. Thus, k is a subfield of K. Finally, define the mapping $f : \mathbb{F}_p \to k$ by setting $f([n]) = n1$ for all congruence classes $[n] \in \mathbb{F}_p$; observe that f is well-defined since, if $[n] = [m]$, then $n = m + sp$ for some integer s and therefore $n1 = m1$. We leave the reader to verify that f is a 1–1 and onto function that preserves both addition and multiplication and hence is a field-isomorphism. Therefore, $k \cong \mathbb{F}_p$, which proves statement (2) and completes the proof. ∎

It follows from Proposition 3 that the unique subfield of a given field generated by the identity element is isomorphic to either $\mathbb{Q}$ or $\mathbb{F}_p$ depending upon the characteristic of the field. We refer to this subfield as the *prime subfield* of the given field. Clearly, the prime subfield is the smallest subfield of the given field. For example, the prime subfield of the fields $\mathbb{Q}$, $\mathbb{R}$, and $\mathbb{C}$ is the field $\mathbb{Q}$ of rational numbers, while the prime subfield of the field $\mathbb{F}_3(X)$ of rational functions over $\mathbb{F}_3$ discussed in Example 4 is the field $\mathbb{F}_3$ of integers modulo 3. The table here lists the prime subfield of some of the fields discussed in this section.

Field	Prime subfield
$\mathbb{Q}, \mathbb{R}, \mathbb{C}$	$\mathbb{Q}$
$\mathbb{Q}(X)$	$\mathbb{Q}$
$\mathbb{F}_3(X)$	$\mathbb{F}_3$
$\mathbb{R}[X]/(X^2 + 1)\mathbb{R}[X]$	$\mathbb{Q}$
$\mathbb{F}_2[X]/(X^2 + X + 1)\mathbb{F}_2[X]$	$\mathbb{F}_2$

Exercises

1. Let $\mathbb{Q}[\sqrt{2}] = \{a + b\sqrt{2} \in \mathbb{R} \mid a, b \in \mathbb{Q}\}$ stand for the subring of $\mathbb{R}$ generated by $\mathbb{Q}$ and $\sqrt{2}$.
   (a) Show that $\mathbb{Q}[\sqrt{2}]$ is a field under addition and multiplication of real numbers.
   (b) Determine the characteristic and prime subfield of $\mathbb{Q}[\sqrt{2}]$.
   (c) Does $\mathbb{Q}[\sqrt{2}]$ contain an element whose square is $3 - 2\sqrt{2}$?
   (d) Does $\mathbb{Q}[\sqrt{2}]$ contain an element whose square is 3?
   (e) Find all elements in $\mathbb{Q}[\sqrt{2}]$ whose square is 2.
   (f) Write the element $\dfrac{2 + 3\sqrt{2}}{1 + \sqrt{2}}$ in the form $a + b\sqrt{2}$ for some $a, b \in \mathbb{Q}$.

(g) Define the function $f:\mathbb{Q}[\sqrt{2}] \to \mathbb{Q}[\sqrt{2}]$ by setting $f(a + b\sqrt{2})$ $= a - b\sqrt{2}$ for all elements $a + b\sqrt{2} \in \mathbb{Q}[\sqrt{2}]$. Show that f is an automorphism of $\mathbb{Q}[\sqrt{2}]$.

(h) Let $g:\mathbb{Q}[\sqrt{2}] \to \mathbb{Q}[\sqrt{2}]$ be an automorphism of $\mathbb{Q}[\sqrt{2}]$.
   (1) Show that $g(a + b\sqrt{2}) = a + bg(\sqrt{2})$ for every element $a + b\sqrt{2}$ $\in \mathbb{Q}[\sqrt{2}]$.
   (2) Show that $g(\sqrt{2}) = \pm\sqrt{2}$.

(i) Show that the identity map and the function f defined in part (g) are the only automorphisms of $\mathbb{Q}[\sqrt{2}]$.

2. Let $\mathbb{Q}[\sqrt{3}] = \{a + b\sqrt{3} \in \mathbb{R} \,|\, a, b \in \mathbb{Q}\}$ stand for the subring of $\mathbb{R}$ generated by $\mathbb{Q}$ and $\sqrt{3}$.

(a) Show that $\mathbb{Q}[\sqrt{3}]$ is a field under addition and multiplication of real numbers.

(b) Determine the characteristic and prime subfield of $\mathbb{Q}[\sqrt{3}]$.

(c) Does $\mathbb{Q}[\sqrt{3}]$ contain an element whose square is $4 - 2\sqrt{3}$?

(d) Does $\mathbb{Q}[\sqrt{3}]$ contain an element whose square is 2?

(e) Find all elements in $\mathbb{Q}[\sqrt{3}]$ whose square is 3.

(f) Write the element $\dfrac{\sqrt{3}}{5 - 2\sqrt{3}}$ in the form $a + b\sqrt{3}$ for some $a, b \in \mathbb{Q}$.

(g) Define the function $f:\mathbb{Q}[\sqrt{3}] \to \mathbb{Q}[\sqrt{3}]$ by setting $f(a + b\sqrt{3})$ $= a - b\sqrt{3}$ for all elements $a + b\sqrt{3} \in \mathbb{Q}[\sqrt{3}]$. Show that f is an automorphism of $\mathbb{Q}[\sqrt{3}]$.

(h) Let $g:\mathbb{Q}[\sqrt{3}] \to \mathbb{Q}[\sqrt{3}]$ be an automorphism of $\mathbb{Q}[\sqrt{3}]$.
   (1) Show that $g(a + b\sqrt{3}) = a + bg(\sqrt{3})$ for every element $a + b\sqrt{3} \in \mathbb{Q}[\sqrt{3}]$.
   (2) Show that $g(\sqrt{3}) = \pm\sqrt{3}$.

(i) Show that the identity map and the function f defined in part (g) are the only automorphisms of $\mathbb{Q}[\sqrt{3}]$.

3. Let $\mathbb{Q}[\sqrt{2}]$ and $\mathbb{Q}[\sqrt{3}]$ stand for the fields discussed in Exercises 1 and 2.

(a) Define the function $f:\mathbb{Q}[\sqrt{2}] \to \mathbb{Q}[\sqrt{3}]$ by setting $f(a + b\sqrt{2})$ $= a + b\sqrt{3}$ for all elements $a + b\sqrt{2} \in \mathbb{Q}[\sqrt{2}]$. Is f a field-isomorphism?

(b) Are the fields $\mathbb{Q}[\sqrt{2}]$ and $\mathbb{Q}[\sqrt{3}]$ isomorphic?

4. Let $\mathbb{Q}[\sqrt{2}, \sqrt{3}]$ stand for the subring of $\mathbb{R}$ generated by $\mathbb{Q}$, $\sqrt{2}$, and $\sqrt{3}$.

(a) Show that $\mathbb{Q}[\sqrt{2}, \sqrt{3}] = \{a + b\sqrt{2} + c\sqrt{3} + d\sqrt{6} \in \mathbb{R} \,|\, a, b, c, d \in \mathbb{Q}\}$.

(b) Show that $\mathbb{Q}[\sqrt{2}, \sqrt{3}]$ is a field under addition and multiplication of real numbers.

(c) Determine the characteristic and prime subfield of $\mathbb{Q}[\sqrt{2}, \sqrt{3}]$.

(d) Write the element $\dfrac{2 - \sqrt{2} + \sqrt{3} + 2\sqrt{6}}{1 + \sqrt{2} - 2\sqrt{3} + \sqrt{6}}$ in the form $a + b\sqrt{2}$ $+ c\sqrt{3} + d\sqrt{6}$ for some $a, b, c, d \in \mathbb{Q}$.

(e) Does the field $\mathbb{Q}[\sqrt{2}, \sqrt{3}]$ contain an element whose square is $2 - \sqrt{3}$?

(f) Show that both $\mathbb{Q}[\sqrt{2}]$ and $\mathbb{Q}[\sqrt{3}]$ are subfields of $\mathbb{Q}[\sqrt{2}, \sqrt{3}]$.

(g) Does $\mathbb{Q}[\sqrt{2}] \cap \mathbb{Q}[\sqrt{3}] = \mathbb{Q}$?

5. Let $\mathbb{Q}[\sqrt[3]{2}]$ stand for the subring of $\mathbb{R}$ generated by $\mathbb{Q}$ and $\sqrt[3]{2}$.

(a) Show that $\mathbb{Q}[\sqrt[3]{2}] = \{a + b\sqrt[3]{2} + c\sqrt[3]{4} \in \mathbb{R} \,|\, a, b, c \in \mathbb{Q}\}$.

(b) Show that $\mathbb{Q}[\sqrt[3]{2}]$ is a field under addition and multiplication of real numbers.

(c) Determine the characteristic and prime subfield of $\mathbb{Q}[\sqrt[3]{2}]$.

(d) Define the function $f : \mathbb{Q}[\sqrt[3]{2}] \to \mathbb{Q}[\sqrt[3]{2}]$ by setting $f(a + b\sqrt[3]{2} + c\sqrt[3]{4}) = a + c\sqrt[3]{2} + b\sqrt[3]{4}$ for all elements $a + b\sqrt[3]{2} + c\sqrt[3]{4} \in \mathbb{Q}[\sqrt[3]{2}]$. Is f an automorphism of $\mathbb{Q}[\sqrt[3]{2}]$?

(e) Let $S = \{a + b\sqrt[3]{2} \in \mathbb{Q}[\sqrt[3]{2}] \,|\, a, b \in \mathbb{Q}\}$. Is S a subfield of $\mathbb{Q}[\sqrt[3]{2}]$?

6. Let $K = \{\left(\begin{smallmatrix} a & -b \\ b & a \end{smallmatrix}\right) \in \mathrm{Mat}_2(\mathbb{R}) \,|\, a, b \in \mathbb{R}\}$.

(a) Show that K is a field under addition and multiplication of matrices.

(b) Determine the characteristic and prime subfield of K.

(c) Show that K contains an element J such that $J^2 = -I_2$, where $I_2 = \left(\begin{smallmatrix} 1 & 0 \\ 0 & 1 \end{smallmatrix}\right)$ stands for the 2×2 identity matrix.

(d) Show that every element in K may be written uniquely in the form $aI_2 + bJ$ for some real numbers a and b, where J is the matrix in part (c).

(e) Show that $K \cong \mathbb{C}$, the field of complex numbers.

(f) Define the function $f : K \to K$ by setting $f(M) = M^T$ for every matrix $M \in K$, where M^T stands for the transpose of M. Show that f is an automorphism of K.

7. Let $K = \{\left(\begin{smallmatrix} a & -b \\ b & a \end{smallmatrix}\right) \in \mathrm{Mat}_2(\mathbb{F}_3) \,|\, a, b \in \mathbb{F}_3\}$.

(a) Show that K contains an element J such that $J^2 = -I_2$, where $I_2 = \left(\begin{smallmatrix} 1 & 0 \\ 0 & 1 \end{smallmatrix}\right)$ stands for the 2×2 identity matrix.

(b) Show that every element in K may be written uniquely in the form $aI_2 + bJ$, where J stands for the matrix in part (a) and where $a, b \in \mathbb{F}_3$.

(c) Show that K contains nine elements. Assuming that matrices in K are added and multiplied in the usual way, write out the addition and multiplication tables for K and verify that K is a field under addition and multiplication of matrices.

(d) Find the inverse of each nonzero element in K and verify that the multiplicative group K^* is a cyclic group of order 8.

(e) Define the function $f : K \to K$ by setting $f(M) = M^T$ for every matrix $M \in K$, where M^T stands for the transpose of M. Show that f is an automorphism of K.

8. Let $\mathbb{F}_7(X)$ stand for the field of rational functions in X over $\mathbb{F}_7$.

(a) Determine the characteristic and prime subfield of $\mathbb{F}_7(X)$.

(b) Let

$$u(X) = \frac{2X^2 + 3}{X^2 + 3} \quad \text{and} \quad v(X) = \frac{3X + 2}{4X + 5}.$$

(1) Find $u(X) + v(X)$, $u(X) - v(X)$, $u(X)v(X)$, $\dfrac{1}{u(X)}$, and $\dfrac{1}{v(X)}$.

(2) Evaluate $u(X)$ and $v(X)$ at all numbers in $\mathbb{F}_7$ at which they are defined.

9. Let $K = \mathbb{F}_3[X]/(X^2 + X + 2)\mathbb{F}_3[X]$.

(a) Show that K is a field.

(b) Show that every element in K may be written uniquely in the form $a + bx$ for some elements $a, b \in \mathbb{F}_3$, where $x = X + (X^2 + X + 2)\mathbb{F}_3[X]$.

(c) If $a + bx$ and $c + dx$ are typical elements in K, show that

$$(a + bx) + (c + dx) = (a + c) + (b + d)x$$

and

$$(a + bx)(c + dx) = (ac + bd) + (bc + ad + 2bd)x.$$

(d) Show that K is a finite field containing nine elements and write out the addition and multiplication tables for K.

(e) Show that the multiplicative group K^* of nonzero elements in K is a cyclic group of order 8.

(f) Define the function $f : K \to K$ by setting $f(a + bx) = (a + 2b) + 2bx$ for all elements $a + bx \in K$. Show that f is an automorphism of K.

10. Let k be a field and let $k(X)$ stand for the field of rational functions in X over k. Define the function $f : k(X) \to k(X)$ by setting $f(r(X)) = r(X^2)$ for every rational function $r(X) \in k(X)$. Is f an automorphism of $k(X)$?

11. At what numbers in $\mathbb{F}_3$ is the rational function $r(X) = \dfrac{1}{X^3 + 2X} \in \mathbb{F}_3(X)$ defined?

12. Let K and K' be isomorphic fields.

(a) Show that $\text{char}(K) = \text{char}(K')$.

(b) Show that $K^* \cong K'^*$.

13. Let K be a field and let c be an element in K. If n is a positive integer, show that there are at most n elements $x \in K$ such that $x^n = c$.

14. Let K be a field of characteristic $p \neq 0$. Show that $px = 0$ for every element $x \in K$.

15. Show that the characteristic of a finite field is a nonzero prime number.

16. Let K be a field of characteristic $p \neq 0$ and let $K^p = \{x^p \in K \mid x \in K\}$ stand for the set of pth powers of elements in K.

(a) Show that K^p is a subfield of K.

(b) If $x^p = 1$ for some element $x \in K$, show that $x = 1$.

(c) Define the function $f_p : K \to K^p$ by setting $f_p(x) = x^p$ for all elements $x \in K$. Show that f_p is a field-isomorphism.

(d) If K is a finite field, show that $K^p = K$.

(e) If K is a finite field of characteristic p, then it follows from part (d) that every element in K has a unique pth root; that is, if $x \in K$, there

is a unique element $y \in K$ such that $y^p = x$. Verify by direct calculation that $\mathbb{F}_5^5 = \mathbb{F}_5$ and determine the unique 5th root of each element
in $\mathbb{F}_5$.

(f) If K is a finite field, then it follows from parts (c) and (d) that the
function $f_p : K \to K$ is an automorphism of K. Show that the function
f defined in part (f) of Exercise 9 is the automorphism f_3.

17. Let K be a field and let $K^{*2} = \{x^2 \in K^* \mid x \in K^*\}$ stand for the set of
squares of nonzero elements in K.

(a) Show that K^{*2} is a subgroup of the multiplicative group K^*.

(b) Define the function $f : K^* \to K^{*2}$ by setting $f(x) = x^2$ for all elements
$x \in K^*$. Show that f is a group-homomorphism mapping K^* onto
K^{*2}.

(c) If K is a finite field with $\text{char}(K) \neq 2$, show that the index $[K^* : K^{*2}]$
$= 2$.

(d) If K is a finite field with $\text{char}(K) = 2$, show that $K^{*2} = K^*$; in this
case, every element in K is a square. Thus, in a finite field of characteristic 2, every element has a unique square root.

2. EXTENSION FIELDS

In the previous section we reviewed the definition and basic properties of
fields and subfields, and some of the techniques for constructing fields. In
this section we adopt a new point of view in our study of fields: instead of
discussing subfields of a given field, we concentrate instead on fields that
contain the given field as a subfield. When looked at from this point of view,
these larger fields are called extension fields of the given field. The field $\mathbb{R}$
of real numbers, for example, is a subfield of the field $\mathbb{C}$ of complex numbers;
from the new point of view, however, we think of $\mathbb{C}$ as an extension field of
$\mathbb{R}$. The main reason for adopting this point of view, as we will show shortly,
is that every field may then be interpreted as a vector space over any of its
subfields. This, then, provides a way to use concepts and results from linear
algebra in the study of fields.

Definition

Let k be a field. An *extension field* of k is any field K that contains k as a
subfield. We write K/k to indicate that K is an extension field of k and say
that K/k is an *extension of fields*.

As we mentioned earlier, the goal of this section is to first interpret extension fields as vector spaces over the subfield, or ground field as it is usually
called, and then use concepts and results from linear algebra to study extension fields. Since the language and terminology of linear algebra will play

an important role throughout our study of fields, let us begin by first reviewing some terminology and basic facts about vector spaces over an arbitrary
field.

Let k be a field. A *vector space over k* is an abelian group V whose elements
are called *vectors*—the sum of two vectors $A, B \in V$ being written $A + B$—
together with an operation of k on V that assigns to each pair of elements
$c \in k$ and $A \in V$ a unique vector $cA \in V$ such that the following four conditions are satisfied for all elements $c, d \in k$ and all vectors $A, B \in V$:

(1) $c(dA) = (cd)A$;
(2) $(c + d)A = cA + dA$;
(3) $c(A + B) = cA + cB$;
(4) $1A = A$, where 1 stands for the identity of k.

The elements in k are called *scalars* and the operation of k on V is called
scalar multiplication. The zero element of V is called the *zero vector* and is
denoted by 0. For example, the set $\mathbb{Q}^2 = \mathbb{Q} \times \mathbb{Q}$ of ordered pairs of rational
numbers is a vector space over the field $\mathbb{Q}$; in this case vector addition is
defined by setting $(a, b) + (c, d) = (a + c, b + d)$ for all vectors $(a, b), (c, d)$
$\in \mathbb{Q}^2$, and scalar multiplication is defined by setting $c(a, b) = (ca, cb)$ for all
scalars $c \in \mathbb{Q}$ and all vectors $(a, b) \in \mathbb{Q}^2$.

Let V be a vector space over a field k and let $A_1, \ldots, A_n$ be a finite number
of vectors in V. A *linear combination* of $A_1, \ldots, A_n$ is any vector of the form
$c_1 A_1 + \cdots + c_n A_n$, where $c_1, \ldots, c_n \in k$. The vectors $A_1, \ldots, A_n$ are said to
be *linearly independent* if the only scalars $c_1, \ldots, c_n \in k$ such that $c_1 A_1 + \cdots$
$+ c_n A_n = 0$ are $c_1 = \cdots = c_n = 0$; otherwise, they are *linearly dependent*.
Thus, $A_1, \ldots, A_n$ are linearly dependent if there are scalars $c_1, \ldots, c_n \in k$,
not all of which are zero, such that $c_1 A_1 + \cdots + c_n A_n = 0$. The vectors
$A_1, \ldots, A_n$ are said to *span* V if every vector in V may be written as a linear
combination of $A_1, \ldots, A_n$. If V contains a finite number of vectors that are
linearly independent over k and span V, we call such a collection of vectors
a *basis* for V and say that V is a *finite-dimensional vector space over k*. Clearly,
the vectors $A_1, \ldots, A_n$ form a basis for V if and only if every vector in V
may be written uniquely in the form $c_1 A_1 + \cdots + c_n A_n$ for some scalars
$c_1, \ldots, c_n \in k$. Finally, recall that if V is a finite-dimensional vector space
over a field k, then all bases for V contain the same number of vectors; this
common number of vectors in any basis is called the *dimension* of V and is
denoted by $\dim_k V$.

Now, let K/k be an extension of fields. Then K is a vector space over the
subfield k. In this case the elements in K are the vectors, which form an
abelian group under addition since K is a field, while the elements in k are
the scalars. Moreover, if $c \in k$ and $x \in K$, then the scalar product cx is just
the ordinary product of c and x in the field K. Scalar multiplication clearly
satisfies the four requirements mentioned earlier since multiplication in the
field K is associative, distributes over addition, and has the element 1 as
identity. Thus, K is a vector space over k. For example, the field $\mathbb{C}$ of complex

numbers is an extension field of $\mathbb{R}$, which we indicate by saying that $\mathbb{C}/\mathbb{R}$ is an extension of fields, in which the real numbers act as scalars on the complex numbers. Similarly, $\mathbb{R}/\mathbb{Q}$ is an extension of fields in which the field $\mathbb{Q}$ of rational numbers acts as the field of scalars on the real numbers, and $\mathbb{C}/\mathbb{Q}$ is an extension of fields with $\mathbb{Q}$ as the field of scalars.

If K/k is an extension of fields, then every basis for K as a vector space over k contains the same number of elements, which may or may not be finite, and this common number is the dimension of K as a vector space over k. With this in mind, we make the following definition.

Definition

Let K/k be an extension of fields. The dimension of K as a vector space over k is called the *degree* of the extension K/k and is denoted by $[K:k]$. If $[K:k]$ is finite, we say that K is a *finite extension* of k.

Consider the extension $\mathbb{C}/\mathbb{R}$, for example. Since every complex number may be written uniquely in the form $a + bi$ for some scalars $a, b \in \mathbb{R}$, it follows that the numbers 1 and i form a basis for $\mathbb{C}$ as a vector space over the field $\mathbb{R}$ of real numbers. Therefore, $[\mathbb{C}:\mathbb{R}] = 2$ and hence $\mathbb{C}$ is a finite extension of $\mathbb{R}$ of degree 2. Similarly, the field $\mathbb{Q}[i]$ of complex numbers whose real and imaginary parts are rational is a finite extension of $\mathbb{Q}$ having the numbers 1 and i as a basis over $\mathbb{Q}$ and hence $[\mathbb{Q}[i]:\mathbb{Q}] = 2$. Finally, let us note that the field $\mathbb{R}$ of real numbers is an extension field of $\mathbb{Q}$ but it is not a finite extension. For if $[\mathbb{R}:\mathbb{Q}] = n$ for some positive integer n, then $\mathbb{R}$ is an n-dimensional vector space over $\mathbb{Q}$ and hence is isomorphic to the vector space $\mathbb{Q}^n$ of n-tuples over $\mathbb{Q}$. Since $\mathbb{Q}^n$ is a countable set, it follows that the set $\mathbb{R}$ of real numbers is also countable, a contradiction. Thus, no such integer n exists and therefore the degree $[\mathbb{R}:\mathbb{Q}]$ is infinite. The extension $\mathbb{R}/\mathbb{Q}$ is therefore an example of a field extension having infinite degree.

Here are some more examples that illustrate the concept of field extensions and their degrees in more detail.

EXAMPLE 1

Let $\mathbb{Q}[\sqrt{2}] = \{a + b\sqrt{2} \in \mathbb{R} \,|\, a, b \in \mathbb{Q}\}$ stand for the subring of $\mathbb{R}$ generated by $\mathbb{Q}$ and $\sqrt{2}$. We claim that $\mathbb{Q}[\sqrt{2}]$ is a field under addition and multiplication of real numbers and that $[\mathbb{Q}[\sqrt{2}]:\mathbb{Q}] = 2$.

(A) To show that $\mathbb{Q}[\sqrt{2}]$ is a field, let $x = a + b\sqrt{2}$ and $y = c + d\sqrt{2}$ be typical elements in $\mathbb{Q}[\sqrt{2}]$. Then

$$x + y = (a + c) + (b + d)\sqrt{2} \in \mathbb{Q}[\sqrt{2}],$$

$$xy = (ac + 2bd) + (ad + bc)\sqrt{2} \in \mathbb{Q}[\sqrt{2}],$$

$$-x = (-a) + (-b)\sqrt{2} \in \mathbb{Q}[\sqrt{2}].$$

If, in addition, $x \neq 0$, then

$$\frac{1}{x} = \frac{a}{a^2 - 2b^2} + \frac{-b}{a^2 - 2b^2} \sqrt{2} \in \mathbb{Q}[\sqrt{2}].$$

Note that if $x \neq 0$, then $a^2 - 2b^2 \neq 0$ since 2 is not the square of a rational number, and therefore the above expression is indeed an element in $\mathbb{Q}[\sqrt{2}]$. It now follows that $\mathbb{Q}[\sqrt{2}]$ is a subfield of $\mathbb{R}$ and hence is a field under addition and multiplication of real numbers.

(B) Since $\mathbb{Q}[\sqrt{2}]$ is a field by part (A) and contains $\mathbb{Q}$ as a subfield, it follows that $\mathbb{Q}[\sqrt{2}]$ is an extension field of $\mathbb{Q}$. Moreover, the numbers 1 and $\sqrt{2}$ form a basis for $\mathbb{Q}[\sqrt{2}]$ as a vector space over $\mathbb{Q}$; for clearly, 1 and $\sqrt{2}$ span $\mathbb{Q}[\sqrt{2}]$ since every element in $\mathbb{Q}[\sqrt{2}]$ may be written in the form $a + b\sqrt{2}$ for some $a, b \in \mathbb{Q}$, and are linearly independent over $\mathbb{Q}$ since, if $a + b\sqrt{2} = 0$ for some nonzero $a, b \in \mathbb{Q}$, then $\sqrt{2} \in \mathbb{Q}$, which is impossible. Thus, $\{1, \sqrt{2}\}$ is a basis for $\mathbb{Q}[\sqrt{2}]$ as a vector space over $\mathbb{Q}$ and therefore $[\mathbb{Q}[\sqrt{2}]:\mathbb{Q}] = 2$.

EXAMPLE 2

Let $\mathbb{Q}[\sqrt{2}][\sqrt{3}]$ stand for the subring of $\mathbb{R}$ generated by $\mathbb{Q}[\sqrt{2}]$ and $\sqrt{3}$. Let us show that $\mathbb{Q}[\sqrt{2}][\sqrt{3}]$ is a field extension of both $\mathbb{Q}$ and $\mathbb{Q}[\sqrt{2}]$ and that $[\mathbb{Q}[\sqrt{2}][\sqrt{3}]:\mathbb{Q}[\sqrt{2}]] = 2$ and $[\mathbb{Q}[\sqrt{2}][\sqrt{3}]:\mathbb{Q}] = 4$.

(A) Let us begin by first observing that since $(\sqrt{3})^2 = 3$, every element in $\mathbb{Q}[\sqrt{2}][\sqrt{3}]$ may be written in the form $A + B\sqrt{3}$ for some $A, B \in \mathbb{Q}[\sqrt{2}]$ and hence

$$\mathbb{Q}[\sqrt{2}][\sqrt{3}] = \{A + B\sqrt{3} \in \mathbb{R} \mid A, B \in \mathbb{Q}[\sqrt{2}]\}.$$

Now, to show that $\mathbb{Q}[\sqrt{2}][\sqrt{3}]$ is a field, let $x = A + B\sqrt{3}$ and $y = C + D\sqrt{3}$ be typical elements in $\mathbb{Q}[\sqrt{2}][\sqrt{3}]$, where $A, B, C, D \in \mathbb{Q}[\sqrt{2}]$. Then it follows, as in Example 1, that $x + y$, xy, and $-x$ are elements in $\mathbb{Q}[\sqrt{2}][\sqrt{3}]$. Now, if $x \neq 0$, we claim that

$$\frac{1}{x} = \frac{A}{A^2 - 3B^2} + \frac{-B}{A^2 - 3B^2} \sqrt{3}.$$

To show this, we must first verify that $A^2 - 3B^2 \neq 0$, or equivalently, that 3 is not the square of an element in $\mathbb{Q}[\sqrt{2}]$. Suppose, to the contrary, that

$$3 = (s + t\sqrt{2})^2 = (s^2 + 2t^2) + 2st\sqrt{2}$$

for some element $s + t\sqrt{2} \in \mathbb{Q}[\sqrt{2}]$. Then $s^2 + 2t^2 = 3$ and $2st = 0$ since the numbers 1 and $\sqrt{2}$ are linearly independent over $\mathbb{Q}$. Therefore, either $s = 0$, in which case $\sqrt{\dfrac{3}{2}} \in \mathbb{Q}$, a contradiction, or $t = 0$, in which case $\sqrt{3} \in \mathbb{Q}$, also a contradiction. Thus, 3 is not the square of any number in $\mathbb{Q}[\sqrt{2}]$ and hence $A^2 - 3B^2 \neq 0$ if $x \neq 0$. We now find by direct calculation that

$$x\left(\frac{A}{A^2 - 3B^2} + \frac{-B}{A^2 - 3B^2}\sqrt{3}\right) = 1$$

and hence

$$\frac{1}{x} = \frac{A}{A^2 - 3B^2} + \frac{-B}{A^2 - 3B^2}\sqrt{3} \in \mathbb{Q}[\sqrt{2}][\sqrt{3}],$$

as required. Therefore, $\mathbb{Q}[\sqrt{2}][\sqrt{3}]$ is a subfield of $\mathbb{R}$ and hence is a field under addition and multiplication of real numbers.

(B) It follows from part (A) that $\mathbb{Q}[\sqrt{2}][\sqrt{3}]$ is an extension field of $\mathbb{Q}[\sqrt{2}]$ since it is a field and contains $\mathbb{Q}[\sqrt{2}]$ as a subfield. Moreover, we leave the reader to verify, as in Example 1, that the numbers 1 and $\sqrt{3}$ form a basis for $\mathbb{Q}[\sqrt{2}][\sqrt{3}]$ as a vector space over $\mathbb{Q}[\sqrt{2}]$. Hence, $[\mathbb{Q}[\sqrt{2}][\sqrt{3}]:\mathbb{Q}[\sqrt{2}]] = 2$.

(C) Now, the field $\mathbb{Q}[\sqrt{2}][\sqrt{3}]$ is also an extension field of $\mathbb{Q}$. Let us show, in conclusion, that the numbers 1, $\sqrt{2}$, $\sqrt{3}$, and $\sqrt{6}$ form a basis for $\mathbb{Q}[\sqrt{2}][\sqrt{3}]$ as a vector space over $\mathbb{Q}$ and hence that $[\mathbb{Q}[\sqrt{2}][\sqrt{3}]:\mathbb{Q}] = 4$. For let $x = A + B\sqrt{3}$ be a typical element in $\mathbb{Q}[\sqrt{2}][\sqrt{3}]$, where $A, B \in \mathbb{Q}[\sqrt{2}]$. Then $A = a + a'\sqrt{2}$ and $B = b + b'\sqrt{2}$ for some $a, a', b, b' \in \mathbb{Q}$ and therefore

$$x = (a + a'\sqrt{2}) + (b + b'\sqrt{2})\sqrt{3}$$
$$= a + a'\sqrt{2} + b\sqrt{3} + b'\sqrt{6},$$

which shows that the numbers 1, $\sqrt{2}$, $\sqrt{3}$, and $\sqrt{6}$ span $\mathbb{Q}[\sqrt{2}][\sqrt{3}]$ as a vector space over $\mathbb{Q}$. Moreover, these numbers are linearly independent over $\mathbb{Q}$; for if $a + a'\sqrt{2} + b\sqrt{3} + b'\sqrt{6} = 0$ for some $a, a', b, b' \in \mathbb{Q}$, then $a + a'\sqrt{2} = b + b'\sqrt{2} = 0$ since 1 and $\sqrt{3}$ are linearly independent over $\mathbb{Q}[\sqrt{2}]$, and therefore $a = a' = b = b' = 0$ since 1 and $\sqrt{2}$ are linearly independent over $\mathbb{Q}$. We conclude, therefore, that the numbers 1, $\sqrt{2}$, $\sqrt{3}$, and $\sqrt{6}$ form a basis for $\mathbb{Q}[\sqrt{2}][\sqrt{3}]$ as a vector space over $\mathbb{Q}$ and hence $[\mathbb{Q}[\sqrt{2}][\sqrt{3}]:\mathbb{Q}] = 4$.

EXAMPLE 3. Fields of polynomials modulo an irreducible polynomial

Let k be a field, $f(X)$ an irreducible polynomial over k, and let $K = k[X]/f(X)k[X]$ stand for the field of polynomials over k modulo $f(X)$. Then every element in K may be written uniquely in the form $a_0 + a_1x + \cdots + a_{n-1}x^{n-1}$, where $a_0, \ldots, a_{n-1} \in k$, $n = \deg f(X)$, and $x = X + f(X)k[X]$ (Example 6, Section 1). Therefore, K is an extension field of k. Moreover, the elements

1, $x, \ldots, x^{n-1}$ form a basis for K as a vector space over k and hence K is a finite extension of k with $[K:k] = n$. Here are some examples of extension fields of this form.

(A) Let $K = \mathbb{R}[X]/(X^2 + 1)\mathbb{R}[X]$. Then $[K:\mathbb{R}] = \deg(X^2 + 1) = 2$ and every element in K may be written uniquely in the form $a + bx$, where $a, b \in \mathbb{R}$, $x = X + (X^2 + 1)\mathbb{R}[X]$, and $x^2 = -1$. In this case the function that maps a typical element $a + bx$ in K to the complex number $a + bi$ is a field-isomorphism and hence $K \cong \mathbb{C}$.

(B) Let $K = \mathbb{Q}[X]/(X^2 - 2)\mathbb{Q}[X]$. Then $[K:\mathbb{Q}] = \deg(X^2 - 2) = 2$ and every element in K may be written uniquely in the form $a + bx$, where $a, b \in \mathbb{Q}$, $x = X + (X^2 - 2)\mathbb{Q}[X]$, and $x^2 = 2$. In this case the function that maps a typical element $a + bx$ in K to the real number $a + b\sqrt{2}$ is a field-isomorphism and hence $K \cong \mathbb{Q}[\sqrt{2}]$. Observe that since the elements 1 and x form a basis for K as a vector space over $\mathbb{Q}$, their images 1 and $\sqrt{2}$ form a basis for $\mathbb{Q}[\sqrt{2}]$ over $\mathbb{Q}$ and are therefore linearly independent over $\mathbb{Q}$, a result that we obtained by direct calculation in Example 1.

(C) Let $K = \mathbb{F}_2[X]/(X^2 + X + 1)\mathbb{F}_2[X]$ stand for the field discussed in Example 6(B), Section 1. Then K is a finite extension field of $\mathbb{F}_2$ with $[K:\mathbb{F}_2] = 2$, and every element in K may be written uniquely in the form $a + bx$, where $a, b \in \mathbb{F}_2$, $x = X + (X^2 + X + 1)\mathbb{F}_2[X]$, and $x^2 = 1 + x$.

Let us note that the isomorphisms $\mathbb{R}[X]/(X^2 + 1)\mathbb{R}[X] \cong \mathbb{C}$ and $\mathbb{Q}[X]/(X^2 - 2)\mathbb{Q}[X] \cong \mathbb{Q}[\sqrt{2}]$ discussed above in parts (A) and (B) of Example 3 may also be obtained directly from the fundamental theorem for ring-homomorphisms by using the evaluation homomorphisms $\varphi_i: \mathbb{R}[X] \to \mathbb{C}$ and $\varphi_{\sqrt{2}}: \mathbb{Q}[X] \to \mathbb{Q}[\sqrt{2}]$; in the first case, the kernel of φ_i is $(X^2 + 1)\mathbb{R}[X]$ and hence $\mathbb{R}[X]/(X^2 + 1)\mathbb{R}[X] \cong \mathbb{C}$, while in the second case the kernel is $(X^2 - 2)\mathbb{Q}[X]$ and hence $\mathbb{Q}[X]/(X^2 - 2)\mathbb{Q}[X] \cong \mathbb{Q}[\sqrt{2}]$. The following example illustrates how this type of isomorphism may be used to obtain information about subfields of $\mathbb{R}$.

EXAMPLE 4

Let $K = \mathbb{Q}[X]/(X^3 - 2)\mathbb{Q}[X]$. Then K is a finite extension field of $\mathbb{Q}$ with $[K:\mathbb{Q}] = 3$ since $X^3 - 2$ is an irreducible polynomial of degree 3 over $\mathbb{Q}$. Moreover, every element in K may be written uniquely in the form $a + bx + cx^2$, where $a, b, c \in \mathbb{Q}$, and $x^3 = 2$. Now, the kernel of the evaluation homomorphism

$$\varphi_{\sqrt[3]{2}}: \mathbb{Q}[X] \to \mathbb{Q}[\sqrt[3]{2}]$$

$$f(X) \mapsto f(\sqrt[3]{2})$$

is $(X^3 - 2)\mathbb{Q}[X]$ and hence $K \cong \mathbb{Q}[\sqrt[3]{2}]$. It follows, in particular, that the subring $\mathbb{Q}[\sqrt[3]{2}]$ of $\mathbb{R}$ generated by $\mathbb{Q}$ and $\sqrt[3]{2}$ is a subfield of $\mathbb{R}$ and that

$[\mathbb{Q}[\sqrt[3]{2}]:\mathbb{Q}] = 3$. Let us use these facts to show that the numbers 1, $\sqrt[3]{2}$, and $\sqrt[3]{4}$ form a basis for $\mathbb{Q}[\sqrt[3]{2}]$ as a vector space over $\mathbb{Q}$, and then use the isomorphism $K \cong \mathbb{Q}[\sqrt[3]{2}]$ to find the inverse of an element in $\mathbb{Q}[\sqrt[3]{2}]$.

(A) Under the isomorphism $K \cong \mathbb{Q}[\sqrt[3]{2}]$, the image of a typical element $a + bx + cx^2$ in K is $a + b\sqrt[3]{2} + c(\sqrt[3]{2})^2$. Since the elements 1, x, x^2 form a basis for K as a vector space over $\mathbb{Q}$, it follows that their images 1, $\sqrt[3]{2}$ and $\sqrt[3]{4}$ form a basis for $\mathbb{Q}[\sqrt[3]{2}]$ as a vector space over $\mathbb{Q}$. Thus, the real numbers 1, $\sqrt[3]{2}$, and $\sqrt[3]{4}$ are linearly independent over $\mathbb{Q}$, every element in $\mathbb{Q}[\sqrt[3]{2}]$ may be written uniquely in the form $a + b\sqrt[3]{2} + c(\sqrt[3]{2})^2$ for some $a, b, c \in \mathbb{Q}$, and $[\mathbb{Q}[\sqrt[3]{2}]:\mathbb{Q}] = 3$.

(B) Let us use the isomorphism $K \cong \mathbb{Q}[\sqrt[3]{2}]$ to write the element

$$c = \frac{1}{1 + \sqrt[3]{2} + \sqrt[3]{4}}$$

in the form $a + b\sqrt[3]{2} + c\sqrt[3]{4}$, where $a, b, c \in \mathbb{Q}$. This is possible, of course, since $\mathbb{Q}[\sqrt[3]{2}]$ is a field, but it is not immediately clear how to do so. We proceed as follows. The element $1 + \sqrt[3]{2} + \sqrt[3]{4}$ in $\mathbb{Q}[\sqrt[3]{2}]$ corresponds to the element $1 + x + x^2$ in the field K. Consider the corresponding polynomial $1 + X + X^2$ in $\mathbb{Q}[X]$.

$$\mathbb{Q}[X] \xrightarrow{\varphi_{\sqrt[3]{2}}} K \xrightarrow{\cong} \mathbb{Q}[\sqrt[3]{2}]$$

$$1 + X + X^2 \mapsto 1 + x + x^2 \mapsto 1 + \sqrt[3]{2} + \sqrt[3]{4}$$

Since $\gcd(1 + X + X^2, X^3 - 2) = 1$, there are polynomials $a(X)$, $b(X) \in \mathbb{Q}[X]$ such that $a(X)(1 + X + X^2) + b(X)(X^3 - 2) = 1$. Using the division algorithm, for example, we find that $(X - 1)(1 + X + X^2) + (-1)(X^3 - 2) = 1$. Therefore, $(x - 1)(1 + x + x^2) = 1$ in the field K and hence, taking images under the isomorphism, $(\sqrt[3]{2} - 1)(1 + \sqrt[3]{2} + \sqrt[3]{4}) = 1$ in $\mathbb{Q}[\sqrt[3]{2}]$. Therefore,

$$c = \frac{1}{1 + \sqrt[3]{2} + \sqrt[3]{4}} = \sqrt[3]{2} - 1.$$

Many times throughout this chapter we have illustrated results in field theory by using finite fields and in each case have observed that the multiplicative group of nonzero elements of the field form a cyclic group. As our final example, let us use the idea of extension fields together with basic results from group theory to show that this is indeed the case for every finite field.

EXAMPLE 5.   Finite fields

Let K be a finite field of prime characteristic p and let k stand for the prime subfield of K. Then K is a finite extension field of k. We claim that if $[K:k] = n$, then K is a finite field containing p^n elements and the multiplicative group K^* of nonzero elements in K is a cyclic group of order $p^n - 1$.

(A) Let $A_1, \ldots, A_n$ be a basis for K as a vector space over k. Then every element in K may be written uniquely in the form $c_1 A_1 + \cdots + c_n A_n$, where $c_1, \ldots, c_n \in k$. Hence, since $k \cong \mathbb{F}_p$, there are p choices for each coefficient c_i and therefore K contains p^n elements.

(B) Now, to show that K^* is a cyclic group first observe that it is a finite group of order $p^n - 1$ and hence the order of each of its elements divides $p^n - 1$. Let d be any positive divisor of $p^n - 1$ and let N_d stand for the number of elements of order d in K^*. Then N_d is either 0 or $\varphi(d)$, where φ stands for the Euler phi-function. For if there is an element x of order d in K^*, then the $\varphi(d)$ generators of the cyclic subgroup (x) all have order d; if there were any other such elements not in (x), the polynomial $X^d - 1 \in K[X]$ would have more than the d elements in (x) as roots, which is impossible. Thus, N_d is either 0 or $\varphi(d)$. Now, we may count the elements in K^* by simply adding the numbers N_d since the order of every element in K^* is a divisor of $p^n - 1$. Thus, $\Sigma N_d = |K^*| = p^n - 1$, where the summation is over all divisors d of $p^n - 1$. On the other hand, we recall from Section 4, Chapter 2, that $\Sigma \varphi(d) = p^n - 1$. It follows, therefore, that $\Sigma (\varphi(d) - N_d) = 0$. But $\varphi(d) - N_d \geq 0$ for all divisors d since N_d is either 0 or $\varphi(d)$. Therefore, $N_d - \varphi(d) = 0$, or $N_d = \varphi(d)$, for all d, and hence, in particular, $N_{p^n-1} \neq 0$. Thus, K^* contains some element of order $p^n - 1$ and hence is a cyclic group.

(C) Let us illustrate the preceding results with three examples. First, let $K = \mathbb{F}_2[X]/(X^2 + X + 1)\mathbb{F}_2[X]$. Then K is a finite extension of $\mathbb{F}_2$ and $[K:\mathbb{F}_2] = 2$. Therefore, K is a finite field containing 2^2, or 4, elements, and K^* is a cyclic group of order 3. In this case every element in K may be written uniquely in the form $a + bx$, where $a, b \in \mathbb{F}_2$, and hence $K = \{0, 1, x, 1 + x\}$, where $x = X + (X^2 + X + 1)\mathbb{F}_2[X]$. Since $x^2 = 1 + x$, $x^3 = x + x^2 = 1$ and therefore $|x| = 3$. Similarly, we find that $|1 + x| = 3$. Thus,

$$K^* = (x) = (1 + x) = \{1, x, 1 + x\} \cong \mathbb{Z}_3.$$

In this case both x and $1 + x$ generate the cyclic group K^*.

(D) As a second example, let $K = \mathbb{F}_3[X]/(X^3 + X^2 + 2)\mathbb{F}_3[X]$. The polynomial $X^3 + X^2 + 2$ is irreducible over $\mathbb{F}_3$ since it has no root in $\mathbb{F}_3$ and hence K is a finite extension of $\mathbb{F}_3$ with $[K:\mathbb{F}_3] = 3$. Therefore, K contains 3^3, or 27, elements, and the multiplicative group K^* is a cyclic group of order 26. In this case every element in K may be written uniquely in the form $a + bx + cx^2$, where $a, b, c \in \mathbb{F}_3$, and hence

$$K = \{0, 1, 2, x, 2x, x^2, 2x^2, 1 + x, \ldots, 2 + 2x + 2x^2\}.$$

Using the fact that $x^3 = -x^2 - 2 = 1 + 2x^2$, we find that $|x| = 13$ and $|2 + x| = 26$. Hence, $2 + x$ generates the cyclic group K^*.

(E) As a third and final example, let p be a prime and consider the field $\mathbb{F}_p$ of integers modulo p. Since $\mathbb{F}_p$ is a finite field containing p elements, the multiplicative group $\mathbb{F}_p^*$ is a cyclic group of order $p - 1$. Hence, there is some element $c \in \mathbb{F}_p$ such that

$$\mathbb{F}_p^* = \{1, \ldots, p - 1\} = \{c, c^2, \ldots, c^{p-1}\}.$$

For example, in $\mathbb{F}_5$ we find that $|2| = 4$; hence $\mathbb{F}_5^* = \{1, 2, 3, 4\}$ $= \{2, 2^2, 2^3, 2^4\}$. Similarly, in $\mathbb{F}_7$, $|3| = 6$ and hence $\mathbb{F}_7^* = \{1, 2, 3, 4, 5, 6\}$ $= \{3, 3^2, 3^3, 3^4, 3^5, 3^6\}$.

These examples illustrate many different types of field extensions. Example 5, in particular, is especially important since it shows that every finite field must contain a prime power number of elements and that its multiplicative group of nonzero elements is always a cyclic group. In Chapter 13 we will show that there is, in fact, a finite field containing p^n elements for any prime power p^n.

Let us now discuss the degree of a field extension in more detail. In Example 2 we showed that $\mathbb{Q}[\sqrt{2}][\sqrt{3}]$ is an extension of $\mathbb{Q}[\sqrt{2}]$ of degree 2 and $\mathbb{Q}[\sqrt{2}]$ is an extension of $\mathbb{Q}$ of degree 2, and then used the bases for these extensions to construct a basis for $\mathbb{Q}[\sqrt{2}][\sqrt{3}]$ over $\mathbb{Q}$, showing that it is an extension of degree 4 of $\mathbb{Q}$. In general, let $k \subseteq K \subseteq L$ be fields, with k a subfield of K and K a subfield of L. Then L/K, K/k, and L/k are extensions of fields. In the following result, due to Dedekind, we use the bases for K/k and L/K to construct a basis for L/k and show that $[L:k]$ is the product of the two degrees $[L:K]$ and $[K:k]$.

PROPOSITION 1. (DEDEKIND) Let $k \subseteq K \subseteq L$ be finite extensions of fields. Then L is a finite extension of k and $[L:k] = [L:K][K:k]$.

Proof. Let $[L:K] = n$, $[K:k] = m$, and let $A_1, \ldots, A_n \in L$ be a basis for L as a vector space over K and $B_1, \ldots, B_m \in K$ a basis for K as a vector space over k. We claim that the nm products $A_1 B_1, \ldots, A_n B_m$ in L form a basis for L as a vector space over k.

(A) Let us first show that the products $A_1 B_1, \ldots, A_n B_m$ span L over k. Let $x \in L$. Then $x = c_1 A_1 + \cdots + c_n A_n$ for some scalars $c_1, \ldots, c_n \in K$. Since $B_1, \ldots, B_m$ form a basis for K as a vector space over k, it follows that for each coefficient c_i, $i = 1, \ldots, n$, there are scalars $d_{i1}, \ldots, d_{im} \in k$ such that $c_i = d_{i1} B_1 + \cdots + d_{im} B_m$. Therefore,

$$\begin{aligned}
x &= c_1 A_1 + \cdots + c_n A_n \\
&= (d_{11} B_1 + \cdots + d_{1m} B_m) A_1 + \cdots + (d_{n1} B_1 + \cdots + d_{nm} B_m) A_n \\
&= d_{11} A_1 B_1 + \cdots + d_{nm} A_n B_m.
\end{aligned}$$

Hence, every element in L may be written as a linear combination of $A_1 B_1, \ldots, A_n B_m$ with coefficients in k and therefore the products $A_1 B_1, \ldots, A_n B_m$ span L over k.

(B) $A_1B_1, \ldots, A_nB_m$ are linearly independent over k. For suppose that $d_{11}A_1B_1 + \cdots + d_{nm}A_nB_m = 0$ for some scalars $d_{11}, \ldots, d_{nm} \in k$. Then

$$(d_{11}B_1 + \cdots + d_{1m}B_m)A_1 + \cdots + (d_{n1}B_1 + \cdots + d_{nm}B_m)A_n = 0,$$

and therefore

$$d_{11}B_1 + \cdots + d_{1m}B_m = 0$$
$$\vdots$$
$$d_{n1}B_1 + \cdots + d_{nm}B_m = 0$$

since $A_1, \ldots, A_n$ are linearly independent over K. Since $B_1, \ldots, B_m$ are linearly independent over k and since the coefficients d_{ij} are elements in k, it now follows that $d_{11} = \cdots = d_{nm} = 0$. Hence, the products A_1B_1, $\ldots, A_nB_m$ are linearly independent over k.

It follows from parts (A) and (B) that the nm products $A_1B_1, \ldots, A_nB_m$ form a basis for L as a vector space over k and therefore L is a finite extension of k with $[L:k] = nm = [L:K][K:k]$. ■

COROLLARY. Let $k \subseteq K \subseteq L$ be finite extensions of fields. Then $[K:k]$ divides $[L:k]$.

Proof. Immediate from Proposition 1. ■

COROLLARY. Let $k \subseteq K$ be a finite extension of fields such that $[K:k]$ is a prime number. Then the only subfields of K that contain k are k and K.

Proof. We leave this as an exercise for the reader. ■

If K/k is an extension of fields, any subfield k' such that $k \subseteq k' \subseteq K$ is called an *intermediate subfield* of the extension K/k. The above corollaries show that for finite extensions, the degree of an intermediate subfield must divide the over-all degree of the extension. Consequently an extension K/k of prime degree has no intermediate subfields except the ground field k and the extension K itself. More generally, a *tower of fields* is any chain of fields $k_1 \subseteq k_2 \subseteq \cdots \subseteq k_n$, where k_i is a subfield of k_{i+1} for $i = 1, \ldots, n-1$. It follows easily from Proposition 1 that if $k_1 \subseteq k_2 \subseteq \cdots \subseteq k_n$ is a tower of fields and k_{i+1} is a finite extension of k_i for $i = 1, \ldots, n-1$, then k_n is a finite extension of k_1 and $[k_n:k_1] = [k_n:k_{n-1}] \cdots [k_2:k_1]$. Thus, in a tower of fields degree is multiplicative.

For example, since $[\mathbb{C}:\mathbb{R}] = 2$, the field $\mathbb{C}$ of complex numbers is an extension of $\mathbb{R}$ of prime degree and hence there are no subfields of $\mathbb{C}$ that contain the real numbers other than $\mathbb{R}$ and $\mathbb{C}$ themselves. As a second example, consider the field $\mathbb{Q}[\sqrt{2}][\sqrt{3}]$ discussed in Example 1. Since $[\mathbb{Q}[\sqrt{2}][\sqrt{3}]: \mathbb{Q}[\sqrt{2}]] = 2$, a prime, it follows that there are no subfields of $\mathbb{Q}[\sqrt{2}][\sqrt{3}]$

containing $\mathbb{Q}[\sqrt{2}]$ except $\mathbb{Q}[\sqrt{2}][\sqrt{3}]$ and $\mathbb{Q}[\sqrt{2}]$. Moreover, since $\mathbb{Q} \subseteq \mathbb{Q}[\sqrt{2}] \subseteq \mathbb{Q}[\sqrt{2}][\sqrt{3}]$ is a tower of fields and since degree is multiplicative, it follows that

$$[\mathbb{Q}[\sqrt{2}][\sqrt{3}]:\mathbb{Q}] = [\mathbb{Q}[\sqrt{2}][\sqrt{3}]:\mathbb{Q}[\sqrt{2}]][\mathbb{Q}[\sqrt{2}]:\mathbb{Q}] = 2 \cdot 2 = 4,$$

a result that we obtained by direct calculation in Example 2. In this case the numbers 1 and $\sqrt{2}$ form a basis for $\mathbb{Q}[\sqrt{2}]$ as a vector space over $\mathbb{Q}$, while 1 and $\sqrt{3}$ form a basis for $\mathbb{Q}[\sqrt{2}][\sqrt{3}]$ as a vector space over $\mathbb{Q}[\sqrt{2}]$. Hence, by Dedekind's theorem, the four products 1, $\sqrt{2}$, $\sqrt{3}$, and $\sqrt{6}$ form a basis for $\mathbb{Q}[\sqrt{2}][\sqrt{3}]$ as a vector space over $\mathbb{Q}$, a result that we also obtained by direct calculation in Example 1. Finally, observe that $\mathbb{Q}[\sqrt{3}]$ is also an intermediate subfield of the extension $\mathbb{Q}[\sqrt{2}][\sqrt{3}]/\mathbb{Q}$ and that $[\mathbb{Q}[\sqrt{2}][\sqrt{3}]:\mathbb{Q}[\sqrt{3}]]$ $= \dfrac{4}{2} = 2$, a prime; thus, the only subfields of $\mathbb{Q}[\sqrt{2}][\sqrt{3}]$ that contain $\mathbb{Q}[\sqrt{3}]$ are $\mathbb{Q}[\sqrt{2}][\sqrt{3}]$ and $\mathbb{Q}[\sqrt{3}]$.

An especially important method for obtaining intermediate subfields of an extension K/k is to first choose an element c in the extension field K and then form the intermediate subfield $k(c)$ generated by the ground field and the chosen element. We used this idea, for example, to form the fields $\mathbb{Q}[i]$ and $\mathbb{Q}[\sqrt{2}]$. Let us now discuss this process in more detail.

PROPOSITION 2. Let K/k be an extension of fields, $c \in K$, and let

$$k(c) = \left\{ \frac{f(c)}{g(c)} \in K \,\middle|\, f(X), g(X) \in k[X], g(c) \neq 0 \right\}$$

stand for the set of rational expressions in c over k. Then $k(c)$ is an intermediate subfield of the extension K/k and is the smallest subfield of K that contains both k and c.

Proof. To show that $k(c)$ is a subfield of K, let $x = \dfrac{f(c)}{g(c)}$ and $y = \dfrac{p(c)}{q(c)}$ be typical elements in $k(c)$, where $f(X)$, $g(X)$, $p(X)$, and $q(X)$ are polynomials over k with $g(c) \neq 0$ and $q(c) \neq 0$. Then

$$\begin{aligned} x + y &= \frac{f(c)q(c) + g(c)p(c)}{g(c)q(c)} \\ &= \frac{[f(X)q(X) + g(X)p(X)](c)}{[g(X)q(X)](c)} \in k(c), \end{aligned}$$

and, similarly, $xy \in k(c)$ and $-x \in k(c)$. If, in addition, $x \neq 0$, then $f(c) \neq 0$ and hence

$$\frac{1}{x} = \frac{g(c)}{f(c)} \in k(c).$$

It follows, therefore, that $k(c)$ is a subfield of K. Hence, since k is clearly a subfield of $k(c)$, $k(c)$ is an intermediate subfield of the extension K/k. Finally, to show that $k(c)$ is the smallest subfield of K that contains both k and c, let k' be any subfield of K containing both k and c. Then $f(c)$ and $g(c)$ are elements in k' for all polynomials $f(X)$, $g(X) \in k[X]$ and hence the quotient $\dfrac{f(c)}{g(c)}$ is also an element in k' whenever $g(c) \neq 0$. Therefore, $k(c) \subseteq k'$ and hence $k(c)$ is the smallest subfield of K containing both k and c. ∎

Definition

Let K/k be an extension of fields and let $c \in K$. The intermediate subfield $k(c)$ is called the subfield of K *generated by k and c*. The extension K is a *simple extension* of k if $K = k(c)$ for some element $c \in K$; in this case any element $c \in K$ such that $K = k(c)$ is called a *primitive element* for the extension K/k.

In other words, a simple extension of fields is any field extension that is generated over the ground field by adjoining a single element. For example, $\mathbb{C} = \mathbb{R}(i)$ since every complex number may be written in the form $a + bi$ for some $a, b \in \mathbb{R}$, and therefore the field of complex numbers is a simple extension of the real numbers with i as a primitive element. Observe also that $\mathbb{C} = \mathbb{R}(-i)$, which shows that a simple extension may, in general, have more than one primitive element.

Now, let K/k be an extension of fields, $c \in K$, and let $k[c]$ stand for the ring of polynomials in c over k. Then $k[c] \subseteq k(c)$. In many cases $k[c]$ is itself a field and hence $k[c] = k(c)$ since $k(c)$ is the smallest subfield of K containing both c and k. In this case every element in $k(c)$, that is, every rational expression in c over k, may be written as a polynomial in c. For example, the ring $\mathbb{Q}[i]$ of complex numbers whose real and imaginary parts are rational is a field and hence $\mathbb{Q}[i] = \mathbb{Q}(i)$; in this case every quotient of complex numbers may be written as a complex number whose real and imaginary parts are rational numbers. Similarly, $\mathbb{Q}[\sqrt{2}] = \mathbb{Q}(\sqrt{2})$.

If $k[c] = k(c)$ for some element $c \neq 0$, then the element $1/c$, in particular, may be written as a polynomial in c over k, say $1/c = a_0 + a_1 c + \cdots + a_n c^n$, where $a_0, \ldots, a_n \in k$, and hence c is a root of the polynomial $a_n X^{n+1} + \cdots + a_0 X - 1$. Thus, if $k[c] = k(c)$, then the element c is a root of some polynomial over k. Let us now identify those elements of an extension field that are roots of polynomials over the ground field.

Definition

Let K/k be an extension of fields and let $c \in K$. The element c is said to be *algebraic over k* if it is a root of some nonconstant polynomial over k. Otherwise c is said to be *transcendental over k*.

The complex number i, for example, is algebraic over the field $\mathbb{R}$ of real numbers since it is a root of the polynomial $X^2 + 1 \in \mathbb{R}[X]$. Similarly, the real number $\sqrt{2}$ is algebraic over the field $\mathbb{Q}$ of rational numbers since it is a root of $X^2 - 2 \in \mathbb{Q}[X]$. But the polynomial X, as an element of the field $\mathbb{Q}(X)$ of rational functions in X over $\mathbb{Q}$, is not algebraic over $\mathbb{Q}$ since it is not the root of any nonconstant polynomial in $\mathbb{Q}[X]$. The polynomial X is therefore transcendental over $\mathbb{Q}$. The real numbers e and π are also transcendental over $\mathbb{Q}$, although this is not so obvious. The fact that e is transcendental over $\mathbb{Q}$ was first proved in 1873 by the French mathematician Charles Hermite, while π was shown to be transcendental by the German mathematician C. L. F. Lindemann in 1882. Let us emphasize that the property of being algebraic or transcendental is a relative concept since it depends upon the ground field; the real number π, for example, is transcendental over the subfield $\mathbb{Q}$, but it is algebraic over the subfield $\mathbb{Q}(\pi)$ generated by $\mathbb{Q}$ and π since it is a root of the polynomial $X - \pi \in \mathbb{Q}(\pi)[X]$.

PROPOSITION 3. Let K/k be an extension of fields and let $c \in K$ be algebraic over k. Then there is a unique monic irreducible polynomial in $k[X]$ having c as a root.

Proof. By definition, c is a root of some nonconstant polynomial in $k[X]$ and consequently there is a monic polynomial $f(X)$ over k of least positive degree having c as a root. We claim that $f(X)$ is unique and irreducible over k. To show uniqueness, we need only observe that if $g(X)$ is any monic polynomial over k of least degree having c as a root, then c is also a root of $f(X) - g(X)$, and hence, if $f(X) - g(X) \neq 0$, then dividing out its leading coefficient gives a monic polynomial over k of smaller degree having c as a root, which is a contradiction. Therefore, $f(X)$ is unique. Finally, $f(X)$ is irreducible over k since otherwise it has a monic factor of smaller degree having c as a root, which contradicts the definition of $f(X)$. Thus, c is a root of a unique monic irreducible polynomial over k. ∎

Definition

Let K/k be an extension of fields and let $c \in K$ be algebraic over k. The unique monic irreducible polynomial over k having c as a root is called the *irreducible polynomial of c over k* and is denoted by $\mathrm{Irr}(c;k)$

For example, $\mathrm{Irr}(i;\mathbb{R}) = X^2 + 1$, $\mathrm{Irr}(\sqrt{2};\mathbb{Q}) = X^2 - 2$, $\mathrm{Irr}(\sqrt{2};\mathbb{Q}(\sqrt{2}))$ $= X - \sqrt{2}$, and $\mathrm{Irr}(\sqrt[3]{2};\mathbb{Q}) = X^3 - 2$.

Now, if K/k is an extension of fields and $c \in K$ is algebraic over k, then there are many polynomials over k having c as a root in addition to $\mathrm{Irr}(c;k)$. The complex number i, for example, is a root of $X^4 - 1$ even though $\mathrm{Irr}(i;\mathbb{R})$ $= X^2 + 1$. Observe, however, that $X^4 - 1$ is a multiple of $X^2 + 1$. Let us now show, in general, that the only polynomials over k that have c as a root are precisely the multiples of $\mathrm{Irr}(c;k)$.

PROPOSITION 4. Let K/k be an extension of fields and let c be an element in K that is algebraic over k. If $g(X)$ is a polynomial over k, then $g(c) = 0$ if and only if $\text{Irr}(c;k)$ divides $g(X)$.

Proof. If $\text{Irr}(c;k)$ divides $g(X)$, then clearly $g(c) = 0$. Conversely, suppose that $g(c) = 0$. By the division algorithm, there are polynomials $q(X)$, $r(X) \in k[X]$ such that $g(X) = q(X)\text{Irr}(c;k) + r(X)$, where either $r(X) = 0$ or $\deg r(X) < \deg \text{Irr}(c;k)$, and $r(c) = 0$ since $g(c) = 0$. Hence, if $r(X) \neq 0$, then $r(X)$ is a polynomial over k having c as a root whose degree is smaller than the degree of $\text{Irr}(c;k)$, which contradicts the definition of $\text{Irr}(c;k)$. Therefore, $r(X) = 0$ and hence $\text{Irr}(c;k)$ divides $g(X)$. ∎

Thus, if an element $c \in K$ is algebraic over k, then the polynomials in $k[X]$ that have c as a root are precisely the multiples of the irreducible polynomial of c over k. This result also provides a useful way to find $\text{Irr}(c;k)$: if $g(X)$ is any polynomial over k such that $g(c) = 0$, then $\text{Irr}(c;k)$ must occur as an irreducible factor of $g(X)$. Thus, we need only factor $g(X)$ into its irreducible monic factors and find that factor having c as a root. In particular, if $g(X)$ is an irreducible monic polynomial over k such that $g(c) = 0$, then $\text{Irr}(c;k) = g(X)$. For example, $\sqrt[3]{2}$ is a root of the polynomial $X^3 - 2 \in \mathbb{Q}[X]$; since $X^3 - 2$ is monic and irreducible over $\mathbb{Q}$, it follows that $\text{Irr}(\sqrt[3]{2};\mathbb{Q}) = X^3 - 2$.

We noted earlier that if K/k is an extension of fields and $k[c] = k(c)$ for some element $c \in K$, then c is algebraic over k. To complete our discussion of algebraic elements, let us show now that $k[c] = k(c)$ if and only if c is an algebraic element over k, and then use this fact to give a complete description of the simple extension $k(c)$ when the element c is algebraic over k.

PROPOSITION 5. Let K/k be an extension of fields and let $c \in K$ be algebraic over k. Then:

(1) $k(c) = k[c] \cong k[X]/\text{Irr}(c;k)k[X]$;

(2) the elements $1, c, \dots, c^{n-1}$ form a basis for $k(c)$ as a vector space over k, where $n = \deg \text{Irr}(c;k)$;

(3) $[k(c):k] = \deg \text{Irr}(c;k)$.

Proof. Let $\varphi_c : k[X] \to K$ stand for the evaluation homomorphism defined by setting $\varphi_c(f(X)) = f(c)$ for all polynomials $f(X) \in k[X]$. Then $\text{Im } \varphi_c = k[c]$ by definition, and $\text{Ker } \varphi_c = \{g(X) \in k[X] \mid g(c) = 0\} = \text{Irr}(c;k)k[X]$ by Proposition 4. Hence, by the fundamental theorem of ring-homomorphism, $k[c] \cong k[X]/\text{Irr}(c;k)k[X]$. In particular, since $\text{Irr}(c;k)$ is irreducible over k, $k[X]/\text{Irr}(c;k)k[X]$ is a field, namely the field of polynomials modulo $\text{Irr}(c;k)$, and therefore $k[c]$ is a field and hence $k[c] = k(c)$. Therefore, $k(c) = k[c] \cong k[X]/\text{Irr}(c;k)k[X]$. Finally, recall that the elements $1, x, \dots, x^{n-1}$ form a basis for $k[X]/\text{Irr}(c;k)k[X]$ as a vector space over k, where $n = \deg \text{Irr}(c;k)$ and $x = X + \text{Irr}(c;k)k[X]$. Since $\varphi_c(X) = c$, the element x corresponds to the element c under the above isomorphism and therefore the elements

$1, c, \ldots, c^{n-1}$ form a basis for $k(c)$ as a vector space over k and $[k(c):k] = n$, as required.   ∎

If an element $c \in K$ is algebraic over k, then it follows from Proposition 5 that $k(c) = k[c]$ and therefore every rational expression in c, that is, an expression of the form

$$\frac{a_0 + a_1 c + \cdots + a_s c^s}{b_0 + b_1 + \cdots + b_t c^t},$$

where $b_0 + b_1 + \cdots + b_t c^t \neq 0$, may in fact be written in the form $d_0 + d_1 c + \cdots + d_m c^m$ for some elements $d_0, d_1, \ldots, d_m \in k$. As we saw in Example 4, however, it is not always easy to find such a representation.

To illustrate these results, consider the complex number i. Since i is algebraic over $\mathbb{R}$ and $\mathrm{Irr}(i;\mathbb{R}) = X^2 + 1$, it follows that $\mathbb{R}(i) = \mathbb{R}[i] \cong \mathbb{R}[X]/(X^2 + 1)\mathbb{R}[X]$, $[\mathbb{R}(i):\mathbb{R}] = \deg(X^2 + 1) = 2$, and the numbers 1 and i form a basis for $\mathbb{R}(i)$ as a vector space over $\mathbb{R}$. If

$$x = \frac{a + bi}{c + di}$$

is a typical rational expression in i, where $c + di \neq 0$, then

$$x = \frac{a + bi}{c + di}\frac{c - di}{c - di} = \frac{ac + bd}{c^2 + d^2} + \frac{bc - ad}{c^2 + d^2} i,$$

which is a representation of x as a polynomial in i over $\mathbb{R}$. Similarly, the real number $\sqrt{2}$ is algebraic over $\mathbb{R}$ with $\mathrm{Irr}(\sqrt{2};\mathbb{Q}) = X^2 - 2$. Therefore, $\mathbb{Q}(\sqrt{2}) = \mathbb{Q}[\sqrt{2}] \cong \mathbb{Q}[X]/(X^2 - 2)\mathbb{Q}[X]$, $[\mathbb{Q}(\sqrt{2}):\mathbb{Q}] = \deg(X^2 - 2) = 2$, and the numbers 1 and $\sqrt{2}$ form a basis for $\mathbb{Q}(\sqrt{2})$ as a vector space over $\mathbb{Q}$; in this case

$$x = \frac{a + b\sqrt{2}}{c + d\sqrt{2}} = \frac{a + b\sqrt{2}}{c + d\sqrt{2}}\frac{c - d\sqrt{2}}{c - d\sqrt{2}} = \frac{ac - 2bd}{c^2 - 2d^2} + \frac{bc - ad}{c^2 - 2d^2}\sqrt{2}$$

is the representation of a typical rational expression in $\sqrt{2}$ as a polynomial in $\sqrt{2}$.

EXAMPLE 6

Let us show that the number $\sqrt{2} + \sqrt{3}$ is algebraic over $\mathbb{Q}$, find its irreducible polynomial over $\mathbb{Q}$, and then discuss the simple extension $\mathbb{Q}(\sqrt{2} + \sqrt{3})$.

(A) To show that $\sqrt{2} + \sqrt{3}$ is algebraic over $\mathbb{Q}$, let

$$f(X) = (X - \sqrt{2} - \sqrt{3})(X - \sqrt{2} + \sqrt{3})(X + \sqrt{2} - \sqrt{3})(X + \sqrt{2} + \sqrt{3}).$$

Then we find by direct calculation that $f(X) = X^4 - 10X^2 + 1$ and hence $f(X) \in \mathbb{Q}[X]$. Since $\sqrt{2} + \sqrt{3}$ is obviously a root of $f(X)$, it follows that $\sqrt{2} + \sqrt{3}$ is algebraic over $\mathbb{Q}$. Now, we claim that $f(X)$ is in fact irreducible over $\mathbb{Q}$ and hence is the irreducible polynomial of $\sqrt{2} + \sqrt{3}$ over $\mathbb{Q}$. To show this, first observe that $f(X)$ cannot be written as the product of a first degree polynomial and a third degree polynomial in $\mathbb{Q}[X]$ since none of its roots lie in $\mathbb{Q}$. On the other hand, if $f(X)$ is the product of two second degree polynomials in $\mathbb{Q}[X]$, then two of the linear factors $X - \sqrt{2} - \sqrt{3}$, $X - \sqrt{2} + \sqrt{3}$, $X + \sqrt{2} - \sqrt{3}$, and $X + \sqrt{2} + \sqrt{3}$ must have a product lying in $\mathbb{Q}[X]$. But we find by direct calculation that this is not the case; for example,

$$(X - \sqrt{2} - \sqrt{3})(X - \sqrt{2} + \sqrt{3}) = X^2 - 2\sqrt{2}X - 1 \notin \mathbb{Q}[X].$$

Therefore, $f(X)$ is a monic irreducible polynomial over $\mathbb{Q}$ having $\sqrt{2} + \sqrt{3}$ as a root and hence $\mathrm{Irr}(\sqrt{2} + \sqrt{3}; \mathbb{Q}) = f(x) = X^4 - 10X^2 + 1$.

(B) It now follows that

$$\mathbb{Q}(\sqrt{2} + \sqrt{3}) = \mathbb{Q}[\sqrt{2} + \sqrt{3}] \cong \mathbb{Q}[X]/(X^4 - 10X^2 + 1)\mathbb{Q}[X],$$

$$[\mathbb{Q}(\sqrt{2} + \sqrt{3}):\mathbb{Q}] = \deg(X^4 - 10X^2 + 1) = 4,$$

and that the numbers 1, $\sqrt{2} + \sqrt{3}$, $(\sqrt{2} + \sqrt{3})^2$, and $(\sqrt{2} + \sqrt{3})^3$ form a basis for $\mathbb{Q}(\sqrt{2} + \sqrt{3})$ as a vector space over $\mathbb{Q}$. Expanding these numbers, we find that

$$(\sqrt{2} + \sqrt{3})^2 = 5 + 2\sqrt{6}$$

$$(\sqrt{2} + \sqrt{3})^3 = 11\sqrt{2} + 9\sqrt{3}.$$

Thus, in particular, the numbers $\sqrt{2}$, $\sqrt{3}$, and $\sqrt{6}$ are elements of the field $\mathbb{Q}(\sqrt{2} + \sqrt{3})$ since

$$\sqrt{2} = -\frac{9}{2}(\sqrt{2} + \sqrt{3}) + \frac{1}{2}(\sqrt{2} + \sqrt{3})^3,$$

$$\sqrt{3} = \frac{11}{2}(\sqrt{2} + \sqrt{3}) - \frac{1}{2}(\sqrt{2} + \sqrt{3})^3,$$

$$\sqrt{6} = \sqrt{2}\sqrt{3}.$$

(C) Finally, let $\mathbb{Q}(\sqrt{2})(\sqrt{3})$ stand for the subfield of $\mathbb{Q}(\sqrt{2} + \sqrt{3})$ generated by $\mathbb{Q}(\sqrt{2})$ and $\sqrt{3}$. Then $\mathbb{Q}(\sqrt{2})(\sqrt{3}) = \mathbb{Q}(\sqrt{2} + \sqrt{3})$; for clearly, $\mathbb{Q}(\sqrt{2})(\sqrt{3}) \subseteq \mathbb{Q}(\sqrt{2} + \sqrt{3})$ since both $\sqrt{2}$ and $\sqrt{3}$ are elements in $\mathbb{Q}(\sqrt{2} + \sqrt{3})$, while $\mathbb{Q}(\sqrt{2} + \sqrt{3}) \subseteq \mathbb{Q}(\sqrt{2})(\sqrt{3})$ since $\sqrt{2} + \sqrt{3} \in \mathbb{Q}(\sqrt{2})(\sqrt{3})$. Since the numbers 1, $\sqrt{2}$, $\sqrt{3}$, and $\sqrt{6}$ form a basis for $\mathbb{Q}(\sqrt{2})(\sqrt{3})$ as a vector space over $\mathbb{Q}$, it now follows that every element in $\mathbb{Q}(\sqrt{2} + \sqrt{3})$ may be written uniquely in the form $a + b\sqrt{2} + c\sqrt{3} + d\sqrt{6}$ for some a, b, c, $d \in \mathbb{Q}$.

EXAMPLE 7

Let $\omega = \cos 120° + i \sin 120° = -\frac{1}{2} + \frac{1}{2}i\sqrt{3}$ stand for a primitive cube root of unity. Let us show that ω is algebraic over $\mathbb{Q}$, discuss the simple extension

$\mathbb{Q}(\omega)$ of $\mathbb{Q}$ generated by ω, and then conclude by showing that every element in $\mathbb{Q}(\omega)$ is algebraic over $\mathbb{Q}$.

(A) We claim first that ω is algebraic over $\mathbb{Q}$ and that $\text{Irr}(\omega;\mathbb{Q}) = X^2 + X + 1$. To show that ω is algebraic over $\mathbb{Q}$, we need only observe that ω is a root of the polynomial $X^3 - 1 \in \mathbb{Q}[X]$. The polynomial $X^3 - 1$ is not irreducible over $\mathbb{Q}$, however, but factors into the product $(X - 1)(X^2 + X + 1)$. Since each of these factors is irreducible over $\mathbb{Q}$ and since ω is not a root of $X - 1$, it must therefore be a root of $X^2 + X + 1$. Therefore, $\text{Irr}(\omega;\mathbb{Q}) = X^2 + X + 1$ and hence, in particular, $\omega^2 + \omega + 1 = 0$.

(B) It follows from part (A) that

$$\mathbb{Q}(\omega) = \mathbb{Q}[\omega] \cong \mathbb{Q}[X]/(X^2 + X + 1)\mathbb{Q}[X],$$

$$[\mathbb{Q}(\omega):\mathbb{Q}] = 2,$$

and that the numbers 1 and ω form a basis for $\mathbb{Q}(\omega)$ as a vector space over $\mathbb{Q}$. Consequently, every element in $\mathbb{Q}(\omega)$ may be written uniquely in the form $a + b\omega$ for some $a, b \in \mathbb{Q}$. Let us illustrate some typical calculations in $\mathbb{Q}(\omega)$. Using the fact that $\omega^2 = -1 - \omega$, we find, for example, that

$$3 + 5\omega - 2\omega^2 + \omega^5 = 3 + 5\omega - 2(-1 - \omega) + \omega^3\omega^2$$
$$= 3 + 5\omega - 2(-1 - \omega) + 1(-1 - \omega)$$
$$= 4 + 6\omega$$

and

$$\frac{2 - \omega}{3 + 2\omega} = \frac{2 - \omega}{3 + 2\omega}\frac{3 + 2\omega^2}{3 + 2\omega^2} = \frac{6 - 3\omega + 4\omega^2 - 2\omega^3}{9 + 6\omega + 6\omega^2 + 4\omega^3} = -\omega.$$

(C) We claim next that the field $\mathbb{Q}(\omega)$ contains a square root of ω, that is, an element whose square is ω. For, since $\omega^3 = 1$, it follows that $(\omega^2)^2 = \omega^4 = \omega$. Therefore, ω^2, or $-1 - \omega$, is an element in $\mathbb{Q}(\omega)$ whose square is ω. In fact, the field $\mathbb{Q}(\omega)$ contains two elements whose squares are ω, namely $\pm(1 + \omega)$.

(D) Finally, let us show that not only ω but every element in $\mathbb{Q}(\omega)$ is algebraic over $\mathbb{Q}$. Let $a + b\omega$ be a typical element in $\mathbb{Q}(\omega)$, where $a, b \in \mathbb{Q}$, and let $f(X) = (X - a)^2 + b(X - a) + b^2$. Then $f(X)$ is a polynomial over $\mathbb{Q}$ and we find that $f(a + b\omega) = b^2\omega^2 + b^2\omega + b^2 = b^2(\omega^2 + \omega + 1) = 0$. Therefore, $a + b\omega$ is a root of $f(X)$ and hence is algebraic over $\mathbb{Q}$. The polynomial $f(X)$ may or may not be irreducible over $\mathbb{Q}$, however, and hence is not necessarily the irreducible polynomial of $a + b\omega$. Since $f(X)$ is a second degree polynomial, we may use its discriminant to decide if it is irreducible over $\mathbb{Q}$. Thus, expanding $f(X)$ we find that $f(X) = X^2 + (b - 2a)X + (a^2 - ab + b^2)$ and therefore

Discriminant $f(X) = (b - 2a)^2 - 4(a^2 - ab + b^2) = -3b^2$.

Since $-3b^2$ is negative whenever $b \neq 0$, it follows that $f(X)$ is irreducible over $\mathbb{Q}$ if and only if $b \neq 0$. Thus, if $b \neq 0$, $\mathrm{Irr}(a + b\omega; \mathbb{Q}) = (X - a)^2 + b(X - a) + b^2$; but if $b = 0$, then $a + b\omega = a$ and hence $\mathrm{Irr}(a + b\omega; \mathbb{Q}) = X - a$. The table here summarizes these results. For example, we find that

$$\mathrm{Irr}(1 + \omega; \mathbb{Q}) = (X - 1)^2 + (X - 1) + 1 = X^2 - X + 1$$

and

$$\mathrm{Irr}(2 - 3\omega; \mathbb{Q}) = (X - 2)^2 - 3(X - 2) + 9 = X^2 - 7X + 19.$$

$a + b\omega \in \mathbb{Q}(\omega)$	$\mathrm{Irr}(a + b\omega : \mathbb{Q})$
$b \neq 0$	$(X - a)^2 + b(X - a) + b^2$
$b = 0$	$X - a$

This concludes the example.

Up to this point we have been discussing field extensions generated by a single element, that is, extensions of the form $k(c)$, where c is an element of some extension field of k. We showed that if the element c is algebraic over k, then $k(c)$ is a finite extension of the ground field k, $[k(c):k] = \deg \mathrm{Irr}(c;k)$, and $k(c) = k[c]$, which means that every element in $k(c)$ may be written as a polynomial in c. If c is not algebraic over k, then $k(c)$ consists of all rational expressions in c over k and is not a finite extension field of k. In general, we may generate an intermediate subfield by using any collection of elements in the field, not just a single element. These extensions are called multiple extensions of the ground field. Let us now conclude our study of extension fields by discussing multiple field extensions. Our goal is to first define the intermediate subfield generated over the ground field k by any collection of elements and then show that multiple extensions generated by adjoining a finite number of elements may in fact be obtained through a sequence of simple extensions by adjoining each element separately.

To begin, recall that if K/k is an extension of fields and $c \in K$, then the simple extension $k(c)$ is the smallest subfield of K containing both k and c. Thus, $k(c)$ is the intersection of all subfields of K containing both k and c. In general, if S is any nonempty subset of K, let $k(S)$ stand for the intersection of all subfields of K that contain both k and S.

PROPOSITION 6. Let K/k be an extension of fields and let S be a non-empty subset of K. Then $k(S)$ is a subfield of K containing both k and S.

Proof. Let $x, y \in k(S)$. To show that $x + y$, xy, $-x$, and $\dfrac{1}{x}$, for $x \neq 0$, are also elements in $k(S)$, we must show that these four elements belong to every subfield of K that contains both k and S. Suppose that k' is such a subfield of K. Then $k(S) \subseteq k'$ and hence $x, y \in k'$. Since k' is a field, $x + y$, xy, $-x$, and $\dfrac{1}{x}$, for $x \neq 0$, are also elements in k'. Thus, these four elements lie in every subfield of K that contains both k and S and hence, by definition, lie in $k(S)$. Therefore, $k(S)$ is a subfield of K. Finally, it is clear that $k(S)$ contains both k and S and the proof is complete. ∎

If S is a nonempty subset of an extension field K of k, the subfield $k(S)$ is called the subfield of K *generated by k and S*. If $S = \{c_1, \ldots, c_n\}$ is a finite set of elements in K, we denote this subfield by $k(c_1, \ldots, c_n)$.

PROPOSITION 7. Let K/k be an extension of fields and let $c_1, \ldots, c_n$ be a finite number of elements in K. Then $k(c_1, \ldots, c_n) = k(c_1, \ldots, c_{n-1})(c_n)$.

Proof.   Since $k(c_1, \ldots, c_{n-1})(c_n)$ is a particular subfield of K that contains both k and $c_1, \ldots, c_n$, it follows that $k(c_1, \ldots, c_n) \subseteq k(c_1, \ldots, c_{n-1})(c_n)$. On the other hand, $k(c_1, \ldots, c_n)$ is a field containing both the subfield $k(c_1, \ldots, c_{n-1})$ and the element c_n. Therefore, $k(c_1, \ldots, c_{n-1})(c_n) \subseteq k(c_1, \ldots, c_n)$. Hence, $k(c_1, \ldots, c_n) = k(c_1, \ldots, c_{n-1})(c_n)$. ∎

Proposition 7 is important because it shows that a field extension generated by a finite number of elements can always be obtained through a sequence of simple extensions: if $c_1, \ldots, c_n$ are a finite number of elements in K, then

$$k \subseteq k(c_1) \subseteq k(c_1)(c_2) = k(c_1, c_2) \subseteq \cdots \subseteq k(c_1, \ldots, c_{n-1})(c_n)$$
$$= k(c_1, \ldots, c_n)$$

is a tower of fields in which each field is a simple extension of the previous field. It follows, therefore, from the multiplicativity of degree that if $k(c_1, \ldots, c_n)$ is a finite extension of k, then

$$[k(c_1, \ldots, c_n):k] = [k(c_1, \ldots, c_{n-1})(c_n):k(c_1, \ldots, c_{n-1})] \cdots [k(c_1):k].$$

If, in addition, each element c_i is algebraic over k, then c_i is also algebraic over the subfield $k(c_1, \ldots, c_{i-1})$ and therefore

$$[k(c_1, \ldots, c_n):k] = \deg \mathrm{Irr}(c_n;k(c_1, \ldots, c_{n-1})) \cdots \deg \mathrm{Irr}(c_1;k).$$

Let us also observe that in a multiple extension $k(c_1, \ldots, c_n)$, the order in which the elements $c_1, \ldots, c_n$ are adjoined does not matter since the definition does not depend upon their ordering; for example, $k(c_1)(c_2) = k(c_2)(c_1) = k(c_1, c_2)$. Finally, let us emphasize that although the multiple extension $k(c_1, \ldots, c_n)$ is generated by the n elements $c_1, \ldots, c_n$, it may in fact be generated by a single element and hence be a simple extension of k; in other words, there may be an element $c \in k(c_1, \ldots, c_n)$ such that $k(c_1, \ldots, c_n) = k(c)$. In this case the element c is a primitive element for the extension $k(c_1, \ldots, c_n)$. Although not every multiple extension has a primitive element, many extensions do. In Chapter 14 we will prove a criterion for deciding precisely when a field extension has a primitive element.

For example, consider the multiple extension $\mathbb{Q}(\sqrt{2}, \sqrt{3})$ discussed in Example 2. The field $\mathbb{Q}(\sqrt{2}, \sqrt{3})$ contains two towers of subfields, as illustrated in Figure 1. Using the tower on the left, for example, we find that

$$
\begin{aligned}
[\mathbb{Q}(\sqrt{2}, \sqrt{3}):\mathbb{Q}] &= [\mathbb{Q}(\sqrt{2})(\sqrt{3}):\mathbb{Q}(\sqrt{2})][\mathbb{Q}(\sqrt{2}):\mathbb{Q}] \\
&= \deg \text{Irr}(\sqrt{3};\mathbb{Q}(\sqrt{2})) \deg \text{Irr}(\sqrt{2};\mathbb{Q}) \\
&= \deg(X^2 - 3) \deg(X^2 - 2) \\
&= 4,
\end{aligned}
$$

a result that we obtained by direct calculation in Example 2. Finally, recall from Example 6 that $\mathbb{Q}(\sqrt{2}, \sqrt{3}) = \mathbb{Q}(\sqrt{2} + \sqrt{3})$. Therefore, $\sqrt{2} + \sqrt{3}$ is a primitive element for the extension $\mathbb{Q}(\sqrt{2}, \sqrt{3})$ and hence $\mathbb{Q}(\sqrt{2}, \sqrt{3})$ is in fact a simple extension of $\mathbb{Q}$.

Throughout this section we have looked at fields from a new perspective, namely as extension fields. From this point of view, every field K is a vector space over any one of its subfields k and its dimension as a vector space over k is the degree of the extension. Like elements of a group, which may be used to generate a subgroup, and elements of a ring, which may be used to generate an ideal, we have shown that any collection of elements of K generate an intermediate subfield of the extension K/k. In general, the subfield $k(c)$ generated by a single element $c \in K$ consists of all rational expressions in c, unless c is algebraic over the ground field k, in which case every rational expression in c may in fact be written as a polynomial in c. In the next section we will study the nature of algebraic elements in more detail and show, in particular, that if c is algebraic over k, then every polynomial in c, and hence every element in $k(c)$, is algebraic over k.

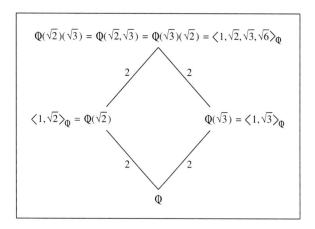

FIGURE 1 Intermediate subfields of the extension $\mathbb{Q}(\sqrt{2} + \sqrt{3})/\mathbb{Q}$.

Exercises

1. Let $c = i\sqrt[3]{2} \in \mathbb{C}$.
   (a) Show that the complex number c is algebraic over $\mathbb{Q}$ and find $\mathrm{Irr}(c;\mathbb{Q})$.
   (b) Find $[\mathbb{Q}(c):\mathbb{Q}]$ and a basis for $\mathbb{Q}(c)$ as a vector space over $\mathbb{Q}$.
   (c) Show that i and $\sqrt[3]{2}$ are elements of the field $\mathbb{Q}(c)$.
   (d) Show that $\mathbb{Q}(i, \sqrt[3]{2}) = \mathbb{Q}(c)$ and that the numbers 1, $\sqrt[3]{2}$, $\sqrt[3]{4}$, i, $i\sqrt[3]{2}$, $i\sqrt[3]{4}$ form a basis for $\mathbb{Q}(c)$ as a vector space over $\mathbb{Q}$.
   (e) Find rational numbers $a, b, c, d, e, f \in \mathbb{Q}$ such that

   $$\frac{1}{i\sqrt[3]{2}} = a + b\sqrt[3]{2} + c\sqrt[3]{4} + di + ei\sqrt[3]{2} + fi\sqrt[3]{4}.$$

   (f) Find $[\mathbb{Q}(c):\mathbb{Q}(i)]$ and $\mathrm{Irr}(c;\mathbb{Q}(i))$.
   (g) Find $[\mathbb{Q}(c):\mathbb{Q}(\sqrt[3]{2})]$ and $\mathrm{Irr}(c;\mathbb{Q}(i\sqrt[3]{2}))$.
2. Let $c = \sqrt{2} + \sqrt{-2} \in \mathbb{C}$.
   (a) Show that the complex number c is algebraic over $\mathbb{Q}$ and find $\mathrm{Irr}(c;\mathbb{Q})$.
   (b) Find $[\mathbb{Q}(c):\mathbb{Q}]$ and find a basis for $\mathbb{Q}(c)$ as a vector space over $\mathbb{Q}$.
   (c) Show that i and $\sqrt{2}$ are elements in $\mathbb{Q}(c)$.
   (d) Show that $\mathbb{Q}(i, \sqrt{2}) = \mathbb{Q}(c)$ and that the numbers 1, $\sqrt{2}$, i, $i\sqrt{2}$ form a basis for $\mathbb{Q}(c)$ as a vector space over $\mathbb{Q}$.
   (e) Find rational numbers $a, b, c, d \in \mathbb{Q}$ such that

   $$\frac{1}{\sqrt{2} + \sqrt{-2}} = a + b\sqrt{2} + ci + di\sqrt{2}.$$

   (f) Find $[\mathbb{Q}(c):\mathbb{Q}(\sqrt{2})]$ and $\mathrm{Irr}(c;\mathbb{Q}(\sqrt{2}))$.
   (g) Find $[\mathbb{Q}(c):\mathbb{Q}(i)]$ and $\mathrm{Irr}(c;\mathbb{Q}(i))$.
   (h) Show that $\sqrt{i} \in \mathbb{Q}(c)$.
   (i) Show that $\mathbb{Q}(c) = \mathbb{Q}(\sqrt{i})$.
   (j) Find $\mathrm{Irr}(\sqrt{i};\mathbb{Q})$, $\mathrm{Irr}(\sqrt{i};\mathbb{Q}(i))$, and $\mathrm{Irr}(\sqrt{i};\mathbb{Q}(\sqrt{2}))$.
3. Let ω stand for a primitive cube root of unity.
   (a) Find $[\mathbb{Q}(\sqrt{2}, \omega):\mathbb{Q}]$.
   (b) Show that $\mathbb{Q}(\sqrt{2}, \omega) = \mathbb{Q}(\omega\sqrt{2})$.
   (c) How many complex roots of the polynomial $X^6 - 8$ are contained in the field $\mathbb{Q}(\sqrt{2}, \omega)$?
4. Let $c = \sqrt{1 + 4i} \in \mathbb{C}$.
   (a) Show that the complex number c is algebraic over $\mathbb{Q}$ and find $\mathrm{Irr}(c;\mathbb{Q})$.
   (b) Find $[\mathbb{Q}(c):\mathbb{Q}]$ and find a basis for $\mathbb{Q}(c)$ as a vector space over $\mathbb{Q}$.
   (c) Show that $\mathbb{Q}(i) \subseteq \mathbb{Q}(c)$.
   (d) Find $[\mathbb{Q}(c):\mathbb{Q}(i)]$ and $\mathrm{Irr}(c;\mathbb{Q}(i))$.
5. Let $K = \mathbb{F}_3[X]/(2X^2 + X + 1)\mathbb{F}_3[X]$.
   (a) Show that K is a field and find $[K:\mathbb{F}_3]$.
   (b) Show that K contains nine elements and write out the addition and multiplication tables for K.

(c) Find the order of each nonzero element in K and show that the multiplicative group K^* of nonzero elements in K is a cyclic group of order 8.

6. Let $K = \mathbb{F}_5[X]/(X^3 + 2X + 1)\mathbb{F}_5[X]$.
   (a) Show that K is a field and find $[K:\mathbb{F}_5]$.
   (b) Let $x = X + (X^3 + 2X + 1)\mathbb{F}_5[X]$.
       (1) Show that the element $2x \in K$ is algebraic over $\mathbb{F}_5$ and find $\mathrm{Irr}(2x; \mathbb{F}_5)$.
       (2) Show that the element $x + x^2 \in K$ is algebraic over $\mathbb{F}_5$ and find $\mathrm{Irr}(x + x^2; \mathbb{F}_5)$.

7. Let $K = \mathbb{F}_5[X]/(X^4 + 2)\mathbb{F}_5[X]$.
   (a) Show that K is a field and find $[K:\mathbb{F}_5]$.
   (b) Let $x = X + (X^4 + 2)\mathbb{F}_5[X]$. Show that the element $1 + x^2 \in K$ is algebraic over $\mathbb{F}_5$ and find $\mathrm{Irr}(1 + x^2; \mathbb{F}_5)$.

8. Find the following irreducible polynomials:
   (a) $\mathrm{Irr}(1 + \sqrt{2}; \mathbb{Q})$
   (b) $\mathrm{Irr}(2i - 1; \mathbb{Q})$ and $\mathrm{Irr}(2i - 1; \mathbb{C})$
   (c) $\mathrm{Irr}(\sqrt{2} + 2i; \mathbb{Q})$, $\mathrm{Irr}(\sqrt{2} + 2i; \mathbb{R})$, and $\mathrm{Irr}(\sqrt{2} + 2i; \mathbb{Q}(i))$
   (d) $\mathrm{Irr}(\sqrt[3]{4}; \mathbb{R})$
   (e) $\mathrm{Irr}(\sqrt[3]{2}; \mathbb{Q}(\sqrt[3]{4}))$
   (f) $\mathrm{Irr}(i\sqrt[3]{2}; \mathbb{R})$
   (g) $\mathrm{Irr}(\sqrt{i}; \mathbb{R})$

9. Find the following degrees:
   (a) $[\mathbb{Q}(\sqrt{2}, \sqrt[3]{2}):\mathbb{Q}]$
   (b) $[\mathbb{Q}(\sqrt{2} + \sqrt[3]{2}):\mathbb{Q}]$
   (c) $[\mathbb{Q}(\sqrt[6]{1728}):\mathbb{Q}]$
   (d) $[\mathbb{Q}(i, \sqrt{1 + i}):\mathbb{Q}]$
   (e) $[\mathbb{Q}(\sqrt[4]{2i}):\mathbb{Q}(i)]$
   (f) $[\mathbb{Q}(\sqrt{6}, i\sqrt{2}, i\sqrt{3}):\mathbb{Q}]$

10. Find $[\mathbb{R}(\sqrt[4]{-5}:\mathbb{R}]$ and $\mathrm{Irr}(\sqrt[4]{-5}; \mathbb{R})$.

11. Show that every complex number is algebraic over the field $\mathbb{R}$ of real numbers. If $a + bi$ is a typical complex number, find $\mathrm{Irr}(a + bi; \mathbb{R})$.

12. Let K/k be an extension of fields. Show that $[K:k] = 1$ if and only if $K = k$.

13. Let k be a field and let $k(X)$ stand for the field of rational functions in X over k. Show that the only elements in $k(X)$ that are algebraic over k are the elements in k itself.

14. Let K/k be an extension of fields and let $c \in K$ be transcendental over k. Show that $k[c] \cong k[X]$, the ring of polynomials in X over k, and $k(c) \cong k(X)$, the field of rational functions in X over k.

15. Let $\mathbb{Q}(X)$ stand for the field of rational functions in X over $\mathbb{Q}$.
    (a) Show that $[\mathbb{Q}(X):\mathbb{Q}(X^2)] = 2$.
    (b) Show that the polynomial $1 + X \in \mathbb{Q}(X)$ is algebraic over the subfield $\mathbb{Q}(X^2)$ and find $\mathrm{Irr}(1 + X; \mathbb{Q}(X^2))$.
    (c) Show that the polynomial $X^3 + X + 1 \in \mathbb{Q}(X)$ is algebraic over the subfield $\mathbb{Q}(X^2)$ and find $\mathrm{Irr}(X^3 + X + 1; \mathbb{Q}(X^2))$.

16. Let K/k be an extension of fields and let $c \in K$. Show that the subring $k[c]$ generated by k and c is a field if and only if c is algebraic over k.

17. Let K/k be a finite extension of fields and suppose that $[K:k]$ is a prime number. Show that K is a simple extension of k and that any element in K not in k is a primitive element for the extension.

18. Let K/k be a finite extension of fields and suppose that $[K:k]$ is an odd number. Show that K contains no new square root; that is, if $a \in k$ is such that $a = x^2$ for some element $x \in K$, then $a = y^2$ for some element $y \in k$.

19. Let K/k be an extension of fields and suppose that c and d are elements in K that are algebraic over k. Let $n = \deg \text{Irr}(c;k)$ and $m = \deg \text{Irr}(d;k)$.
    (a) If n and m are relatively prime, show that $[k(c, d):k] = nm$.
    (b) If n and m are not relatively prime, is it necessarily true that $[k(c, d):k] = nm$?

20. Let K/k be an extension of fields and let $c \in K$ be algebraic over k. Let $f(X) = \text{Irr}(c;k)$.
    (a) If $a \in k$, show that the element $c + a$ is algebraic over k and that $\text{Irr}(c + a;k) = f(X - a)$.
    (b) If $a \in k$, $a \neq 0$, show that the element $\dfrac{c}{a}$ is algebraic over k and that $\text{Irr}\left(\dfrac{c}{a};k\right) = \dfrac{1}{a^n} f(aX)$, where $n = \deg f(X)$.
    (c) Is it true that $\text{Irr}(c + a;k) = f(X - a)$ for every element $a \in K$?

3. ALGEBRAIC EXTENSIONS

In the previous section we introduced the concept of an algebraic element of a field extension: if K/k is an extension of fields, an element $c \in K$ is algebraic over k if c is a root of some nonconstant polynomial over k. We then showed that if an element $c \in K$ is algebraic over the ground field k, every rational expression in c can be written as a polynomial in c. In this section our goal is to study the properties of algebraic elements in more detail and discuss, in particular, those extensions in which every element is algebraic over the ground field. The fact that extension fields can be viewed as vector spaces over the ground field will continue to play an important role throughout our discussion.

Definition

Let K/k be an extension of fields. Then K is called an *algebraic extension* of k if every element in K is algebraic over k.

The field $\mathbb{C}$ of complex numbers, for example, is an algebraic extension of $\mathbb{R}$; for if $c = a + bi$ is a typical complex number, then

$$f(X) = [X - (a + bi)][X - (a - bi)] = X^2 - 2aX + (a^2 + b^2)$$

is a polynomial over $\mathbb{R}$ having c as a root. Similarly, we showed in Example 7 of the previous section that every element in the field $\mathbb{Q}(\omega)$ is algebraic over $\mathbb{Q}$, where ω is a primitive cube root of unity, and therefore $\mathbb{Q}(\omega)$ is an algebraic extension of $\mathbb{Q}$. Let us note, however, that neither of the fields $\mathbb{C}$ or $\mathbb{R}$ is an algebraic extension of $\mathbb{Q}$ since both of these fields contain an element—the number π, for example—that is not algebraic over $\mathbb{Q}$. And finally, let us observe that if K/k is an algebraic extension of fields and k' is any intermediate subfield, then K is an algebraic extension of k' and k' is an algebraic extension of k. Later in this section we will show that the converse of this statement is also true, namely if both K/k' and k'/k are algebraic extensions, then K is in fact algebraic over k.

Before we discuss other examples of algebraic extensions, let us first prove an especially useful and important criterion for deciding when an extension of fields is algebraic over the ground field.

PROPOSITION 1. Let K/k be an extension of fields. If $[K:k]$ is finite, then K is an algebraic extension of k.

Proof. Let $[K:k] = n$ and let $c \in K$. Then K is an n-dimensional vector space over k and therefore the $n + 1$ elements $1, c, \ldots, c^n$ are linearly dependent over k. Thus, there are scalars $a_0, \ldots, a_n \in k$, not all zero, such that $a_0 + a_1 c + \cdots + a_n c^n = 0$. Let $f(X) = a_0 + a_1 X + \cdots + a_n X^n$. Then $f(X)$ is a nonconstant polynomial over k having c as a root and hence c is algebraic over k. Thus, every element in K is algebraic over k and therefore K is an algebraic extension of k. ∎

This result is exceptionally important since it shows that every finite extension of fields is an algebraic extension and, as such, provides a simple method for deciding when all the elements of an extension field are algebraic over the ground field without having to find a particular polynomial having the given element as a root. Finding such a polynomial is not always a simple matter. Later in this section we discuss several methods for finding a polynomial having a given element as root. For the moment, however, let us discuss some of the consequences of Proposition 1.

COROLLARY. Let K/k be an extension of fields and let $c \in K$. Then the subfield $k(c)$ generated by k and c is an algebraic extension of k if and only if c is algebraic over k.

Proof. If $k(c)$ is an algebraic extension of k, then the element c is algebraic over k by definition. Conversely, if c is algebraic over k, then $k(c)$ is a finite extension of k since $[k(c):k] = \deg \mathrm{Irr}(c;k)$, and hence, by Proposition 1, is an algebraic extension of k. ∎

COROLLARY. Let K/k be an extension of fields and let $c, d \in K$ be elements that are algebraic over k. Then $c \pm d$, cd, and c/d, for $d \neq 0$, are also algebraic over k.

Proof. Observe that the elements $c \pm d$, cd, and c/d, for $d \neq 0$, are all elements of the subfield $k(c, d)$ generated by k, c, and d. Now, c is algebraic over k and hence $[k(c):k]$ is finite. Furthermore, since d is algebraic over k it is also algebraic over $k(c)$ and hence $[k(c)(d):k(c)]$ is finite. Therefore, $k(c, d)$ is a finite extension of k since $[k(c, d):k] = [k(c)(d):k(c)][k(c):k]$ and hence, by Proposition 1, is an algebraic extension of k. Thus, $c \pm d$, cd, and c/d, for $d \neq 0$, are algebraic over k, as required. ∎

A simple extension $k(c)$ is therefore an algebraic extension if and only if it is generated by an element c that is algebraic over k; in this case $k(c) = k[c]$, which means that every rational expression in c may be written as a polynomial in c. Moreover, every such expression is algebraic over k. For example, each of the numbers $\sqrt{2}$, $\sqrt{3}$, $\sqrt[3]{2}$, and i are algebraic over $\mathbb{Q}$; therefore the numbers $\sqrt{2} + \sqrt{3}$, $\sqrt[3]{2} - i$, and $i\sqrt[3]{2}$, for example, are also algebraic over $\mathbb{Q}$.

Let us now extend these results to any finite number of algebraic elements.

PROPOSITION 2. Let K/k be an extension of fields and let $c_1, \ldots, c_n$ be any finite number of elements in K. Then the subfield $k(c_1, \ldots, c_n)$ is an algebraic extension of k if and only if its generators $c_1, \ldots, c_n$ are algebraic over k.

Proof. If $k(c_1, \ldots, c_n)$ is an algebraic extension of k, then $c_1, \ldots, c_n$ are algebraic over k by definition. Conversely, if $c_1, \ldots, c_n$ are algebraic over k, then each c_i is algebraic over the subfield $k(c_1, \ldots, c_{i-1})$ and hence $k(c_1, \ldots, c_{i-1}) \times (c_i)$ is a finite extension of $k(c_1, \ldots, c_{i-1})$. Thus, we have a tower of fields $k \subseteq k(c_1) \subseteq \ldots \subseteq k(c_1, \ldots, c_n)$ in which each field is a finite extension of the previous field. It follows therefore that $k(c_1, \ldots, c_n)$ is a finite extension and hence an algebraic extension of k. ∎

COROLLARY. If K/k is an extension of fields and $c_1, \ldots, c_n$ are any finite number of elements in K that are algebraic over k, then any arithmetic combination of $c_1, \ldots, c_n$ is algebraic over k.

Proof. If $c_1, \ldots, c_n$ are algebraic over k, then $k(c_1, \ldots, c_n)$ is an algebraic extension of k and hence any arithmetic combination of $c_1, \ldots, c_n$, as an element of this field, is algebraic over k. ∎

Finally, we claim that any field obtained through a sequence of algebraic extensions is itself an algebraic extension of the ground field.

PROPOSITION 3. Let $k \subseteq K \subseteq L$ be fields such that L is an algebraic extension of K and K is an algebraic extension of k. Then L is an algebraic extension of k.

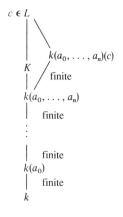

$c \in L$

$k(a_0, \ldots, a_n)(c)$

K

finite

$k(a_0, \ldots, a_n)$

finite

finite

$k(a_0)$

finite

k

Proof.   We must show that every element of L is algebraic over k. Let $c \in L$. Then by assumption c is algebraic over the intermediate subfield K and hence there is some nonzero polynomial $f(X) \in K[X]$ such that $f(c) = 0$. Let $f(X) = a_0 + a_1 X + \cdots + a_n X^n$, where $a_0, \ldots, a_n \in K$. Then c is clearly algebraic over the subfield $k(a_0, \ldots, a_n)$ generated by the coefficients $a_0, \ldots, a_n$ and hence $k(a_0, \ldots, a_n)(c)$ is a finite extension of $k(a_0, \ldots, a_n)$. Furthermore, since each coefficient a_i lies in K, a_i is algebraic over k and hence algebraic over $k(a_0, \ldots, a_{i-1})$. Therefore, $k(a_0, \ldots, a_{i-1})(a_i)$ is a finite extension of $k(a_0, \ldots, a_{i-1})$ for each i. Consequently, we have a tower of fields, $k \subseteq k(a_0) \subseteq \cdots \subseteq k(a_0, \ldots, a_n) \subseteq k(a_0, \ldots, a_n, c)$, in which each field is a finite extension of the previous field. Thus, $k(a_0, \ldots, a_n, c)$ is a finite extension of k and hence an algebraic extension of k. It follows that c, as an element of this field, is algebraic over the ground field k. Thus, L is an algebraic extension of k and the proof is complete.   ∎

COROLLARY.   Let $k_1 \subseteq k_2 \subseteq \cdots \subseteq k_n$ be a tower of fields in which each field is an algebraic extension of the previous field. Then k_n is an algebraic extension of k_1.

Proof.   This follows immediately from Proposition 3 by induction. We leave the details as an exercise for the reader.   ∎

Let us make a few comments about these results. Proposition 1 is, of course, the basic result. It shows that every finite extension of fields is an algebraic extension and, as such, provides a criterion—the finiteness of the degree—for deciding when an extension of fields is algebraic. It then follows that a simple extension $k(c)$ is an algebraic extension of k if and only if it is generated by an element c that is algebraic over the ground field k. In this case every rational expression in c may be written as a polynomial in c since $k(c) = k[c]$, and every polynomial in c is algebraic over k. More generally, a finitely generated extension $k(c_1, \ldots, c_n)$ is an algebraic extension of k if and only if the generators $c_1, \ldots, c_n$ are algebraic over k, which shows that any arithmetic combination of algebraic elements is algebraic. And finally, any field obtained through a sequence of algebraic extensions is itself an algebraic extension of the ground field.

We emphasize that these results, powerful as they are, are existential in nature: they describe whether or not an element of an extension field is algebraic over the ground field in terms of the finiteness of the extension, not by specifying a particular polynomial having the given element as a root. Let us now look at some more examples of algebraic extensions and illustrate, in particular, several ways to find such a polynomial.

EXAMPLE 1

Consider the field $\mathbb{Q}(\sqrt[3]{2})$. Since $\sqrt[3]{2}$ is algebraic over $\mathbb{Q}$, $\mathbb{Q}(\sqrt[3]{2})$ is an algebraic extension of $\mathbb{Q}$. Therefore, every element in $\mathbb{Q}(\sqrt[3]{2})$ is algebraic over $\mathbb{Q}$ and hence is a root of some irreducible polynomial over $\mathbb{Q}$. Let us find $\mathrm{Irr}(1 + \sqrt[3]{2} + \sqrt[3]{4}; \mathbb{Q})$.

(A) We begin by first finding a monic polynomial over $\mathbb{Q}$ having the number $1 + \sqrt[3]{2} + \sqrt[3]{4}$ as a root. Let $c = 1 + \sqrt[3]{2} + \sqrt[3]{4}$. Then $c - 1 = \sqrt[3]{2}(1 + \sqrt[3]{2})$ and hence

$$(c - 1)^3 = 2(1 + 3\sqrt[3]{2} + 3\sqrt[3]{4} + 2)$$

$$= 6 + 6(\sqrt[3]{2} + \sqrt[3]{4})$$

$$= 6 + 6(c - 1) = 6c.$$

Expanding the left side of this equation and collecting terms, we find that $c^3 - 3c^2 - 3c - 1 = 0$. Thus, c is a root of the monic polynomial $X^3 - 3X^2 - 3X - 1$.

(B) We now claim that the polynomial $X^3 - 3X^2 - 3X - 1$ is irreducible over $\mathbb{Q}$ and hence is the irreducible polynomial of $1 + \sqrt[3]{2} + \sqrt[3]{4}$. To show this, let $f(X) = X^3 - 3X^2 - 3X - 1$. If $f(X)$ is not irreducible over $\mathbb{Q}$, then it has some rational number as a root, say $\dfrac{n}{m}$, where n and m are relatively prime. In this case

$$\left(\frac{n}{m}\right)^3 - 3\left(\frac{n}{m}\right)^2 - 3\left(\frac{n}{m}\right) - 1 = 0,$$

and hence $n^3 - 3n^2m - 3nm^2 - m^3 = 0$. It follows that if p is a prime divisor of n, then p divides m, and conversely, if p divides m, then p divides n. Since n and m are relatively prime, it now follows that $\dfrac{n}{m} = \pm 1$.

But neither ± 1 are roots of $f(X)$. Therefore, $f(X)$ has no rational root and hence is irreducible over $\mathbb{Q}$. Thus, $\mathrm{Irr}(1 + \sqrt[3]{2} + \sqrt[3]{4}; \mathbb{Q}) = X^3 - 3X^2 - 3X - 1$.

EXAMPLE 2

The field $\mathbb{Q}(\sqrt{2}, i)$ is an algebraic extension of $\mathbb{Q}$ since its generators $\sqrt{2}$ and i are algebraic over $\mathbb{Q}$. Moreover, $\mathbb{Q}(\sqrt{2}, i)$ is a 4-dimensional vector space over $\mathbb{Q}$ since $[\mathbb{Q}(\sqrt{2}, i):\mathbb{Q}] = [\mathbb{Q}(\sqrt{2})(i):\mathbb{Q}(\sqrt{2})][\mathbb{Q}(\sqrt{2}):\mathbb{Q}] = 4$, and hence any five of its elements are linearly dependent over $\mathbb{Q}$. Let us describe the linear dependence relations among the five numbers 1, $\sqrt{2} + i$, $(\sqrt{2} + i)^2$, $(\sqrt{2} + i)^3$, and $(\sqrt{2} + i)^4$, and then use these relations to find $\mathrm{Irr}(\sqrt{2} + i; \mathbb{Q})$.

(A) Let $c = \sqrt{2} + i$ and let a_0, a_1, a_2, a_3 and a_4 be rational numbers such that

$$a_0 + a_1 c + a_2 c^2 + a_3 c^3 + a_4 c^4 = 0.$$

Now, we find by direct calculation that

$$c^2 = 1 + 2i\sqrt{2},$$
$$c^3 = -\sqrt{2} + 5i,$$
$$c^4 = -7 + 4i\sqrt{2},$$

and therefore

$$a_0 + a_1 c + a_2 c^2 + a_3 c^3 + a_4 c^4$$
$$= (a_0 + a_2 - 7a_4) + (a_1 - a_3)\sqrt{2} + (a_1 + 5a_3)i + (2a_2 + 4a_4)i\sqrt{2}.$$

Since the numbers 1, $\sqrt{2}$, i and $i\sqrt{2}$ form a basis for $\mathbb{Q}(\sqrt{2}, i)$ as a vector space over $\mathbb{Q}$, it follows that $a_0 + a_1 c + a_2 c^2 + a_3 c^3 + a_4 c^4 = 0$ if and only if

$$
\begin{aligned}
a_0 \quad + \ a_2 \qquad\quad - 7a_4 &= 0 \\
a_1 \qquad\ - \ a_3 \qquad &= 0 \\
a_1 \qquad + 5a_3 \qquad &= 0 \\
2a_2 \qquad + 4a_4 &= 0.
\end{aligned}
$$

Solving this system of homogeneous equations, we find that $a_0 = 9t$, $a_1 = 0$, $a_2 = -2t$, $a_3 = 0$, and $a_4 = t$, where $t \in \mathbb{Q}$. Thus, all linear dependence relations among the numbers $1, c, c^2, c^3, c^4$ have the form $(9t)1 + (-2t)c^2 + (t)c^4 = 0$, where t is any rational number.

(B) Now, every dependence relation

$$a_0 + a_1 c + a_2 c^2 + a_3 c^3 + a_4 c^4 = 0$$

obtained in part (A) has associated with it the polynomial $a_0 + a_1 X + a_2 X^2 + a_3 X^3 + a_4 X^4$ in $\mathbb{Q}[X]$ having c as a root, and conversely, every such polynomial determines such a dependence relation. Since the most general dependence relation has the form $(9t)1 + (-2t)c^2 + (t)c^4 = 0$, where $t \in \mathbb{Q}$, it follows that the monic polynomial of least degree in $\mathbb{Q}[X]$ having c as a root occurs when $t = 1$ and is $X^4 - 2X^2 + 9$. Therefore, $\mathrm{Irr}(\sqrt{2} + i; \mathbb{Q}) = X^4 - 2X^2 + 9$.

EXAMPLE 3. Finite fields

Let us show that every finite field is an algebraic extension of its prime subfield and then describe a method for finding the irreducible polynomial of any element in the field over the prime subfield.

(A) Let K be a finite field and let k stand for the prime subfield of k. Then K is finite and hence is a finite extension of k. It follows that K is an algebraic extension of k and hence that every element in K is algebraic over the prime subfield k.

(B) To find the irreducible polynomial of an element in K, let char $(K) = p$, a prime, and let $[K:k] = n$. We claim that the irreducible polynomial over k of any element in K must be a divisor of the polynomial $X^{p^n} - X$. To show this, recall that the multiplicative group K^* of nonzero elements in K is a finite group of order $p^n - 1$ and hence the order of every element in K^* divides $p^n - 1$ by Lagrange's theorem. Therefore $c^{p^n - 1} = 1$ for all elements $c \in K^*$, or equivalently, $c^{p^n} - c = 0$ for all $c \in K$. It follows therefore that every element in K is a root of the polynomial $X^{p^n} - X \in k[X]$ and hence $\mathrm{Irr}(c;k)$ divides $X^{p^n} - X$ for every element $c \in K$, as required. Thus, we may find the irreducible polynomial of each element in K by first factoring the polynomial $X^{p^n} - X$ into its irreducible factors over k and then finding the roots of each factor.

(C) Let us illustrate the above results using the field $K = \mathbb{F}_3[X]/(X^2 + 1)\mathbb{F}_3[X]$. Since $X^2 + 1$ is irreducible over $\mathbb{F}_3$ and has degree 2, K is a finite field containing 3^2, or 9, elements, and every element may be written uniquely in the form $a + bx$, where $a, b \in \mathbb{F}_3$ and $x = X + (X^2 + 1)\mathbb{F}_3[X]$. Therefore,

$$K = \{0, 1, 2, x, 2x, 1 + x, 1 + 2x, 2 + x, 2 + 2x\}.$$

Now, since K contains 9 elements, every element is a root of the polynomial $X^9 - X \in \mathbb{F}_3[X]$. To factor this polynomial into irreducible factors over $\mathbb{F}_3$, we proceed as follows:

$$
\begin{aligned}
X^9 - X &= X(X^8 - 1) \\
&= X(X^4 - 1)(X^4 + 1) \\
&= X(X - 1)(X + 1)(X^2 + 1)(X^4 + 1).
\end{aligned}
$$

The factors X, $X - 1$, $X + 1$ and $X^2 + 1$ are clearly irreducible over $\mathbb{F}_3$, while $X^4 + 1 = (X^2 + X + 2)(X^2 + 2X + 2)$, and both of these factors are irreducible over $\mathbb{F}_3$ since they have no root in $\mathbb{F}_3$. Hence,

$$X^9 - X = X(X - 1)(X + 1)(X^2 + 1)(X^2 + X + 2)(X^2 + 2X + 2)$$

is the factorization of $X^9 - X$ into irreducible polynomials over $\mathbb{F}_3$. Now, every element in K is a root of exactly one of these factors. For example, since $x^2 = -1 = 2$ in the field K, x is a root of $X^2 + 1$. The element $1 + x$, on the other hand, is clearly not a root of any of the first four factors; but substituting it into the remaining two factors, we find that

$$(1 + x)^2 + (1 + x) + 2 = 6 + 3x = 0,$$

$$(1 + x)^2 + 2(1 + x) + 2 = x + 1 \neq 0.$$

$c \in K$	$\mathrm{Irr}(c; \mathbb{F}_p)$
0	X
1	$X - 1$
2	$X - 2$
x	$X^2 + 1$
$2x$	$X^2 + 1$
$1 + x$	$X^2 + X + 2$
$1 + 2x$	$X^2 + X + 2$
$2 + x$	$X^2 + 2X + 2$
$2 + 2x$	$X^2 + 2X + 2$

Therefore, $1 + x$ is a root of the irreducible monic polynomial $X^2 + X + 2$ and hence $\mathrm{Irr}(1 + x; \mathbb{F}_3) = X^2 + X + 2$. The irreducible polynomials for the remaining elements are found similarly and are summarized in the marginal table.

Let us conclude this section with a few remarks about the algebraic elements in an arbitrary field extension. Let K/k be an extension of fields and let

$$\mathrm{Alg}(K/k) = \{c \in K \,|\, c \text{ is algebraic over } k\}$$

stand for the set of elements in K that are algebraic over k. Then $k \subseteq \mathrm{Alg}(K/k)$. Moreover, $\mathrm{Alg}(K/k)$ is closed under addition, subtraction, multiplication and division by nonzero elements; for if $c, d \in \mathrm{Alg}(K/k)$, then c and d are algebraic over k and hence, as we showed earlier in this section, $c \pm d$, cd, and $\dfrac{c}{d}$, for $d \neq 0$, are also algebraic over k and hence elements of $\mathrm{Alg}(K/k)$.

Therefore, $\mathrm{Alg}(K/k)$ is a field and hence an intermediate subfield of the extension K/k. We call $\mathrm{Alg}(K/k)$ the *algebraic closure of k in K*.

In general, if K is an algebraic extension of k, then $\mathrm{Alg}(K/k) = K$; otherwise $\mathrm{Alg}(K/k)$ is a proper intermediate subfield of the extension K/k. For example, $\mathrm{Alg}(\mathbb{C}/\mathbb{R}) = \mathbb{C}$ since every complex number is algebraic over $\mathbb{R}$, and, similarly, $\mathrm{Alg}(\mathbb{Q}(\sqrt{2})/\mathbb{Q}) = \mathbb{Q}(\sqrt{2})$. On the other hand $\mathbb{Q} \subseteq \mathrm{Alg}(\mathbb{R}/\mathbb{Q}) \subseteq \mathbb{R}$, but $\mathrm{Alg}(\mathbb{R}/\mathbb{Q}) \neq \mathbb{R}$ since there are real numbers—π, for example—that are not algebraic over $\mathbb{Q}$, although $\sqrt{2} + \sqrt{3}$ is a typical element in $\mathrm{Alg}(\mathbb{R}/\mathbb{Q})$. Similarly, $\mathrm{Alg}(\mathbb{C}/\mathbb{Q}) \neq \mathbb{C}$. Let us also mention, and leave the reader to verify as an exercise, that the field $\mathrm{Alg}(\mathbb{R}/\mathbb{Q})$ of real numbers that are algebraic over $\mathbb{Q}$ is countable since polynomials over $\mathbb{Q}$ may be enumerated by degree, and there are only a countable number of polynomials of each degree each of which has only a finite number of roots in $\mathbb{R}$.

The algebraic closure of a field k in an arbitrary extension K is important for several reasons. First, it is an algebraic extension of k since each of its elements is algebraic over k, and, as such, provides a framework in which to study the algebraic elements of an arbitrary extension since we may then apply the results of this section to the extension $\mathrm{Alg}(K/k)/k$. And second, any element in K that is algebraic over $\mathrm{Alg}(K/k)$ is in fact an element of $\mathrm{Alg}(K/k)$; for if $c \in K$ is algebraic over $\mathrm{Alg}(K/k)$, then $k \subseteq \mathrm{Alg}(K/k) \subseteq \mathrm{Alg}(K/k)(c)$ is a tower of algebraic extensions. Therefore, $\mathrm{Alg}(K/k)(c)$ is algebraic over the ground field k and hence so is the element c, which shows that $c \in \mathrm{Alg}(K/k)$. Thus, all elements in K that are algebraic over $\mathrm{Alg}(K/k)$ already lie in $\mathrm{Alg}(K/k)$ and it is in this sense that the subfield $\mathrm{Alg}(K/k)$ is the algebraic "closure" of k in K.

Exercises

1. Let $K = \mathbb{Q}(i, \sqrt{3})$ and let $c = i + \sqrt{3} \in K$.
   (a) Show that K is an algebraic extension of $\mathbb{Q}$ and find $\mathrm{Irr}(c; \mathbb{Q})$.
   (b) Show that K is an algebraic extension of $\mathbb{Q}(i)$ and find $\mathrm{Irr}(c; \mathbb{Q}(i))$.
   (c) Show that K is an algebraic extension of $\mathbb{Q}(\sqrt{3})$ and find $\mathrm{Irr}(c; \mathbb{Q}(\sqrt{3}))$.

2. Let $K = \mathbb{Q}(\sqrt{2})$.
   (a) Show that K is an algebraic extension of $\mathbb{Q}$.
   (b) If $c = a + b\sqrt{2}$ is a typical element in K, where $a, b \in \mathbb{Q}$, find $\operatorname{Irr}(c;\mathbb{Q})$.

3. Let $K = \mathbb{Q}(\sqrt{i})$ and $c = \sqrt{i} + \sqrt{-i}$.
   (a) Show that K is an algebraic extension of $\mathbb{Q}$.
   (b) Show that $c \in K$ and find $\operatorname{Irr}(c;\mathbb{Q})$.

4. Let $K = \mathbb{F}_3[X]/(2X^2 + X + 1)\mathbb{F}_3[X]$.
   (a) Show that K is a finite field containing nine elements and that every element in K may be written uniquely in the form $a + bx$, where $a, b \in \mathbb{F}_3$ and $x = X + (2X^2 + X + 1)\mathbb{F}_3[X]$.
   (b) Show that K is an algebraic extension of $\mathbb{F}_3$.
   (c) Find the irreducible polynomial of each element in K over $\mathbb{F}_3$.
   (d) Show that $K = \mathbb{F}_3(y)$ for every element $y \in K$ such that $y \notin \mathbb{F}_3$.
   (e) Show that the multiplicative group K^* of nonzero elements in K is a cyclic group of order 8 and find all generators of K^*.

5. Let $K = \mathbb{F}_2[X]/(X^4 + X + 1)\mathbb{F}_2[X]$.
   (a) Show that K is a finite field containing 16 elements and that every element in K may be written uniquely in the form $a + bx + cx^2 + dx^3$, where $a, b, c, d \in \mathbb{F}_2$ and $x = X + (X^4 + X + 1)\mathbb{F}_2[X]$.
   (b) Show that K is an algebraic extension of $\mathbb{F}_2$.
   (c) Find the irreducible polynomial of each element in K over $\mathbb{F}_2$.
   (d) Let $y = x + x^2 \in K$ and let $\mathbb{F}_2(y)$ stand for the subfield of K generated by $\mathbb{F}_2$ and y.
      (1) Show that $\mathbb{F}_2(y)$ contains four elements.
      (2) Find $[K:\mathbb{F}_2(y)]$ and $[\mathbb{F}_2(y):\mathbb{F}_2]$.
      (3) Show that K is an algebraic extension of $\mathbb{F}_2(y)$ and $\mathbb{F}_2(y)$ is an algebraic extension of $\mathbb{F}_2$.
      (4) Find the irreducible polynomial of each element in K over $\mathbb{F}_2(y)$.
      (5) Show that $\mathbb{F}_2(y) \cong \mathbb{F}_2[X]/(X^2 + X + 1)\mathbb{F}_2[X]$.

6. Let K/k be an algebraic extension of fields and let k' be any intermediate subfield of the extension. Show that K is an algebraic extension of k' and k' is an algebraic extension of k.

7. Let K/k be a finite extension of fields and let $[K:k] = n$.
   (a) Show that $\deg \operatorname{Irr}(c;k)$ divides n for every element $c \in K$.
   (b) Show that an element $c \in K$ is a primitive element for the extension K/k if and only if $\deg \operatorname{Irr}(c;k) = n$.

8. Let K/k be an extension of fields and let $c \in K$ be algebraic over k. If $f(X)$ is any polynomial in $k[X]$, show that the element $f(c) \in K$ is also algebraic over k.

9. Show that an extension of fields K/k is a finite extension if and only if $K = k(c_1, \ldots, c_n)$ for some elements $c_1, \ldots, c_n \in K$ that are algebraic over k.

10. For each positive integer n, let $K_n = \mathbb{Q}(\sqrt{2}, \sqrt[3]{2}, \ldots, \sqrt[n]{2})$ stand for the subfield of $\mathbb{R}$ generated by $\mathbb{Q}$ and the numbers $\sqrt{2}, \sqrt[3]{2}, \ldots, \sqrt[n]{2}$. Let $K = \bigcup K_n$ stand for the union of these subfields.
   (a) Show that K is a subfield of $\mathbb{R}$ containing $\mathbb{Q}$.

(b) Show that $K \neq \mathbb{R}$.

(c) Show that K is an algebraic extension of $\mathbb{Q}$.

(d) Show that K is not a finite extension of $\mathbb{Q}$.

11. Let K/k be an extension of fields and let $\mathrm{Alg}(K/k)$ stand for the algebraic closure of k in K.

(a) Show that $\mathrm{Alg}(K/\mathrm{Alg}(K/k)) = \mathrm{Alg}(K/k)$.

(b) Show that every element in K that is not in $\mathrm{Alg}(K/k)$ is transcendental over k.

(c) Show that $[K:\mathrm{Alg}(K/k)]$ cannot be finite unless $\mathrm{Alg}(K/k) = K$, in which case $[K:\mathrm{Alg}(K/k)] = 1$.

12. Let $\mathbb{Q}(X)$ stand for the field of rational functions in X over $\mathbb{Q}$.

(a) Show that $\mathrm{Alg}(\mathbb{Q}(X)/\mathbb{Q}) = \mathbb{Q}$.

(b) Show that $[\mathbb{Q}(X):\mathbb{Q}]$ is infinite.

(c) Show that $\mathbb{Q}(X)$ is an algebraic extension of the subfield $\mathbb{Q}(X^2)$.

(d) Let $c = \dfrac{X}{X+1}$. Find $\mathrm{Irr}(c;\mathbb{Q}(X^2))$.

13. **Finding irreducible polynomials using the null space of a matrix.** Let K be a finite algebraic extension of a field k with $[K:k] = n$ and let $A_1, \ldots, A_n$ be a basis for K as a vector space over k. Let $c \in K$. Then each of the powers $1, c, \ldots, c^n$ may be written as a linear combination of $A_1, \ldots, A_n$. Let

$$1 = a_{10}A_1 + \cdots + a_{n0}A_n$$

$$c = a_{11}A_1 + \cdots + a_{n1}A_n$$

$$\vdots$$

$$c^n = a_{1n}A_1 + \cdots + a_{nn}A_n,$$

where $a_{10}, \ldots, a_{nn} \in k$, and let

$$M(c) = \begin{pmatrix} a_{10} & \cdots & a_{1n} \\ a_{20} & \cdots & a_{2n} \\ & \vdots & \\ a_{n0} & \cdots & a_{nn} \end{pmatrix}$$

stand for the $n \times (n+1)$ matrix whose (i,j)th entry is a_{ij}, $1 \leq i \leq n$, $0 \leq j \leq n$. Then *null space* of $M(c)$ is the set $M(c)^0$ consisting of all $(n+1)$-tuples $(c_0, \ldots, c_n) \in k^{n+1}$ such that

$$M(c) \begin{pmatrix} c_0 \\ \vdots \\ c_n \end{pmatrix} = \begin{pmatrix} 0 \\ \vdots \\ 0 \end{pmatrix}.$$

(a) Show that $M(c)^0$ is a subspace of the vector space k^{n+1}.

(b) Show that $\dim_k M(c)^0 > 1$ and hence that $M(c)^0$ is nontrivial.

(c) If $A = (c_0, \ldots, c_n)$ is any nonzero vector in $M(c)^0$, the *length* of A is the number $l(A) = i$, where $c_i \neq 0$ but $c_j = 0$ for $j > i$; if $l(A) = i$ and $c_i = 1$, A is said to be *normalized*. Show that $M(c)^0$ contains a unique normalized vector of minimal length.

(d) Let $(c_0, \ldots, c_n)$ be the unique normalized vector of minimal length in $M(c)^0$. Show that $\text{Irr}(c;k) = c_0 + c_1 X + \cdots + c_n X^n$.

(e) Show that $[k(c):k] = \text{rank } M(c)$, the rank of the matrix $M(c)$.

(f) Show that c is a primitive element for the extension K/k if and only if the n rows of $M(c)$ are linearly independent over k.

(g) Let $K = \mathbb{Q}(\sqrt{2}, \sqrt{3})$, $k = \mathbb{Q}$, and choose the numbers 1, $\sqrt{2}, \sqrt{3}$, and $\sqrt{6}$ as basis for $\mathbb{Q}(\sqrt{2}, \sqrt{3})$ over $\mathbb{Q}$. Let $c = \sqrt{2} + \sqrt{3}$.

   (1) Calculate the powers $1, c, c^2, c^3$, and c^4, and verify that

$$M(\sqrt{2} + \sqrt{3}) = \begin{pmatrix} 1 & 0 & 5 & 0 & 49 \\ 0 & 1 & 0 & 8 & 0 \\ 0 & 1 & 0 & 9 & 0 \\ 0 & 0 & 2 & 0 & 20 \end{pmatrix}.$$

   (2) Find the reduced echelon form of the matrix $M(\sqrt{2} + \sqrt{3})$ and verify that

$$M(\sqrt{2} + \sqrt{3})^0 = \{t(1, 0, -10, 0, 1) \in \mathbb{Q}^5 \mid t \in \mathbb{Q}\}.$$

   (3) Using the result in part (2), show that the unique normalized vector of minimal length in $M(\sqrt{2} + \sqrt{3})^0$ is $(1, 0, -10, 0, 1)$ and conclude, therefore, that $\text{Irr}(\sqrt{2} + \sqrt{3}; \mathbb{Q}) = X^4 - 10X^2 + 1$.

(h) Using the method described above, find $\text{Irr}(i\sqrt{3} - \sqrt{2}; \mathbb{Q})$.

14. **Finding irreducible polynomials using the characteristic polynomial of a matrix.** Let K be a finite algebraic extension of a field k with $[K:k] = n$ and let $c \in K$. Define the function $L_c : K \to K$ by setting $L_c(x) = cx$ for every element $x \in K$.

(a) If $c \neq 0$, show that L_c is a nonsingular linear transformation of K as a vector space over k.

(b) Let M_c stand for the matrix of L_c relative to any coordinate system and let $f(X) = \det(M_c - XI_n)$ stand for the characteristic polynomial of M_c, where I_n is the $n \times n$ identity matrix and det means determinant. The Hamilton-Cayley theorem in linear algebra states that **every matrix satisfies its characteristic polynomial; that is, $f(M_c) = O$,** the $n \times n$ zero matrix. Using this result, show that $f(c) = 0$.

(c) If c is a nonzero element in K, show that $\text{Irr}(c;k)$ must occur as an irreducible factor of the characteristic polynomial $f(X)$ of M_c. Thus, to find $\text{Irr}(c;k)$ we need only determine which irreducible factor of $f(X)$ has c as a root.

(d) Using the method described in part (c), find $\text{Irr}(2\sqrt{3} - 1; \mathbb{Q})$.

(e) Let $a + b\sqrt[3]{2} + c\sqrt[3]{4}$ be a typical element in the field $\mathbb{Q}(\sqrt[3]{2})$. If not both b and c are zero, show that

$$\text{Irr}(a + b\sqrt[3]{2} + c\sqrt[3]{4}; \mathbb{Q}) = \det \begin{pmatrix} a - x & 2c & 2b \\ b & a - x & 2c \\ c & b & a - x \end{pmatrix}.$$

13

Roots of Polynomials

As we have seen throughout the previous chapters, polynomials play an especially important role in the study of fields since they may be used to construct extension fields of a given field. Fields, on the other hand, provide the basic algebraic framework in which to discuss the properties of polynomials. This mutual interaction between polynomials and fields therefore provides an appropriate setting in which to discuss the roots of polynomials: first, use polynomials to construct fields that contain roots of the polynomials, and then use properties of the field to discuss the nature and properties of the roots.

In this chapter our goal is to prove two basic facts about polynomials and their roots: first, that every nonconstant polynomial over a field k factors completely, or splits, into a product of linear factors over some extension field of k and hence that the extension field contains a complete set of roots for the polynomial; and second, that the subfield generated by these roots is uniquely determined to within isomorphism. Any field over which a polynomial splits into linear factors and which is generated by the roots of the polynomial is called a splitting field for the polynomial. The goal of the chapter is to show, therefore, that splitting fields exist and are unique to within isomorphism. Splitting fields are important because, as we noted above, they provide the basic algebraic model in which to discuss the roots of a given polynomial. The fact that they exist and are unique to within isomorphism guarantees that we may always find an appropriate extension of the ground field containing the roots of a given polynomial, and that the algebraic properties of the roots are the same regardless of which splitting field is used.

The results in this chapter have a number of important consequences. First, since the algebraic properties of the roots of a polynomial are independent of any particular splitting field, it follows that we may assign a multiplicity to each root by simply counting the number of times the corresponding linear factor occurs in any splitting of the polynomial. Thus, by counting each root according to its multiplicity, the total number of roots is always equal to the degree of the polynomial.

As a second application of splitting fields, we discuss the existence and uniqueness of finite fields. Recall that we have already shown that every finite field contains a prime power number of elements. In this chapter we will show, conversely, that if p^n is any given prime power, then there is in fact a finite field containing p^n elements and all such fields are isomorphic. Thus, finite fields containing any prime power number of elements not only exist but are unique to within isomorphism.

And finally, we conclude the chapter by combining our results on splitting fields and multiplicity to prove the primitive element theorem, which shows that every finite extension of a field of characteristic zero must have a primitive element and hence be a simple extension of the ground field. In the next chapter we will prove a more general criterion for the existence of primitive elements that is independent of the characteristic of the field.

1. SPLITTING FIELDS

In this section our goal is to prove the existence of splitting fields for polynomials; that is, if $f(X)$ is a nonconstant polynomial over a field k, then there is a finite extension field of k that contains a complete set of roots of $f(X)$ and which is generated by those roots. We begin by first defining what it means for a polynomial to split over a given field.

Definition

Let $f(X)$ be a nonconstant polynomial over a field k. We say that $f(X)$ *splits over* k if $f(X) = a(X - c_1) \cdots (X - c_n)$ for some elements $a, c_1, \ldots, c_n \in k$.

Thus, a polynomial splits over a field k if it may be written as the product of linear factors in $k[X]$. The polynomial $X^2 + 3X + 2 \in \mathbb{Q}[X]$ splits over $\mathbb{Q}$, for example, since $X^2 + 3X + 2 = (X + 1)(X + 2)$ in $\mathbb{Q}[X]$. But $X^2 + 1 \in \mathbb{Q}[X]$ does not split over $\mathbb{Q}$ since it is irreducible over $\mathbb{Q}$. Observe, however, that $X^2 + 1$ splits over the extension field $\mathbb{C}$ of complex numbers since $X^2 + 1 = (X - i)(X + i)$ in $\mathbb{C}[X]$. In fact, $X^2 + 1$ already splits over the subfield $\mathbb{Q}(i)$, which is a finite extension of $\mathbb{Q}$. Clearly, any linear polynomial over a field k splits over k, while an irreducible polynomial over k does not. And finally, observe that a polynomial may be reducible over a field without necessarily splitting over the field; the polynomial $X^3 + X \in \mathbb{Q}[X]$, for example, is reducible over $\mathbb{Q}$ since $X^3 + X = X(X^2 + 1)$, but it does not split over $\mathbb{Q}$ since the factor $X^2 + 1$ is irreducible over $\mathbb{Q}$.

Now, the reader may have observed that when we constructed the field of polynomials modulo an irreducible polynomial in the previous chapter the new field contained a root of the original polynomial; the field $K = \mathbb{F}_2[X]/(X^2 + X + 1)\mathbb{F}_2[X]$, for example, contains the element $x = X + (X^2 + X + 1)\mathbb{F}_2[X]$, which is a root of the polynomial $X^2 + X + 1 = 0$ since $x^2 + x + 1 = 0$ in K. In order to show that any nonconstant polynomial over a field k splits over some extension of k, we need only imitate this procedure by first constructing a finite extension field of k that contains at least one root of $f(X)$, and then repeat this construction, using induction, to obtain a tower of fields each containing more roots of $f(X)$, until we arrive, finally, at a finite extension of k over which $f(X)$ splits. We begin this construction by first proving the following fundamental result, due to Kronecker, which shows that there is in fact a finite extension of k containing at least one root of $f(X)$.

PROPOSITION 1.   (KRONECKER) Let $f(X)$ be a nonconstant polynomial of degree n over a field k. Then there is an extension field K of k such that $f(X)$ has a root in K and $[K:k] \leq n$.

Proof.   Let $p(X)$ be an irreducible factor of $f(X)$ over k and let $K = k[X]/p(X)k[X]$ stand for the field of polynomials over k modulo $p(X)$. Then K is an extension field of k with $[K:k] = \deg p(X) \leq n$. Now, let $c = X + p(X)k[X]$. Then $p(c) = 0$ in K and hence $f(c) = 0$ since $f(X)$ is a multiple of $p(X)$. Thus, K is a finite extension field of k containing a root of $f(X)$ and $[K:k] \leq n$, as required.   ∎

PROPOSITION 2. Let $f(X)$ be a nonconstant polynomial of degree n over a field k. Then there is an extension field K of k such that $f(X)$ splits over K and $[K:k] \leq n!$.

Proof.   We prove this statement by induction on n. If $n = 1$, $f(X)$ is linear and hence k itself satisfies all requirements. Now assume that $n > 1$ and that every nonconstant polynomial of degree $s < n$ over an arbitrary field k' splits over some extension K' of k', where $[K':k'] \leq s!$. Let $f(X)$ be a polynomial of degree n over k. By Proposition 1 there is an extension field k' of k such that k' contains a root c of $f(X)$ and $[k':k] \leq n$. Therefore, $f(X) = q(X)(X - c)$ for some polynomial $q(X) \in k'[X]$. Clearly, $q(X)$ is a nonconstant polynomial of degree $n - 1$ and hence, by the induction hypothesis, $q(X)$ splits over some extension field K of k', where $[K:k'] \leq (n - 1)!$. It follows that $f(X)$ splits over K and that $[K:k] = [K:k'][k':k] \leq n(n-1)! = n!$. Thus, the field K satisfies all requirements and the proof by induction is complete. ∎

Definition

Let $f(X)$ be a nonconstant polynomial over a field k. A *splitting field* for $f(X)$ over k is any extension field K of k such that $f(X)$ splits over K but does not split over any proper subfield of K containing k.

We think of a splitting field for a given polynomial as a minimal extension of the ground field over which the given polynomial splits. Such a field is necessarily generated by the roots of the polynomial. For if $f(X)$ is a non-constant polynomial of degree n over a field k and K is any extension of k over which $f(X)$ splits, then $f(X) = a(X - c_1) \cdots (X - c_n)$ for some elements $a, c_1, \ldots, c_n \in K$, and $c_1, \ldots, c_n$ are a complete set of roots for $f(X)$ in K since a polynomial of degree n has at most n roots in any given field. Since $f(X)$ splits over the subfield $k(c_1, \ldots, c_n)$ and $k(c_1, \ldots, c_n)$ is the smallest subfield of K containing k over which $f(X)$ splits, it follows that $k(c_1, \ldots, c_n)$ is a splitting field for $f(X)$ over k. Thus, splitting fields are generated over the ground field by the roots of the polynomial. It is clear therefore that we may obtain a splitting field for a given polynomial by first finding, or constructing, a field over which the polynomial splits—such fields exist by Proposition 2—and then forming the subfield generated by the roots of the polynomial. Since there are usually many choices for a field over which a polynomial splits, there are usually many different splitting fields for the poly-nomial. Moreover, since any polynomial of degree n over a field k splits over some extension K having degree at most $n!$ over k, it follows that $f(X)$ has a splitting field whose degree over k is at most $n!$. In many cases, however, there is a splitting field of significantly smaller degree.

For example, the polynomial $X^2 + 1 \in \mathbb{Q}[X]$ splits over the field $\mathbb{C}$ of complex numbers and has i and $-i$ as roots. Therefore, the extension $\mathbb{Q}(i)$ is a splitting field for $X^2 + 1$ over $\mathbb{Q}$, and $X^2 + 1 = (X - i)(X + i)$ is the splitting of $X^2 + 1$ in $\mathbb{Q}(i)[X]$. In this case $[\mathbb{Q}(i):\mathbb{Q}] = 2 = 2!$. Observe that $\mathbb{Q}(i)$ is also a splitting field for the polynomial $X^3 + X$ over $\mathbb{Q}$ since $X^3 + X = X(X - i)(X + i)$; but in this case $[\mathbb{Q}(i):\mathbb{Q}] = 2 < 3!$. Similarly, $\mathbb{Q}(\sqrt{2})$ is a splitting field for $X^2 - 2$ over $\mathbb{Q}$.

In the following examples we illustrate splitting fields in more detail. The first few examples discuss roots of polynomials within the framework of the field of complex numbers, a field over which every nonconstant polynomial splits, while the last two examples illustrate how to construct splitting fields when there is no underlying field from which to select the roots. As we discuss these examples, let us keep in mind that the degree of a splitting field over the ground field may be significantly smaller than the guaranteed upper bound of $n!$.

EXAMPLE 1

Let $f(X) = X^4 - 1 \in \mathbb{Q}[X]$. The roots of $f(X)$ in the field $\mathbb{C}$ are 1, -1, i, and $-i$. Therefore, $\mathbb{Q}(i)$ is a splitting field for $f(X)$ over $\mathbb{Q}$ and $f(X) = (X - 1)(X + 1)(X - i)(X + i)$ is the splitting of $f(X)$ over $\mathbb{Q}(i)$. In this case $[\mathbb{Q}(i):\mathbb{Q}] = 2 < 4!$.

EXAMPLE 2

Let $f(X) = X^4 + 1 \in \mathbb{Q}[X]$. The roots of $f(X)$ in the field $\mathbb{C}$ are $\sqrt{i}$, $-\sqrt{i}$, $\sqrt{-i}$, and $-\sqrt{-i}$. Now, it follows from DeMoivre's theorem that

$$\sqrt{i} = \left(\cos\frac{\pi}{2} + i\sin\frac{\pi}{2}\right)^{1/2} = \cos\frac{\pi}{4} + i\sin\frac{\pi}{4} = \tfrac{1}{2}\sqrt{2}(1 + i).$$

Similarly, $\sqrt{-i} = \tfrac{1}{2}\sqrt{2}(1 - i)$. Thus, the roots of $f(X)$ are $\pm\tfrac{1}{2}\sqrt{2}(1 + i)$ and $\pm\tfrac{1}{2}\sqrt{2}(1 - i)$, and therefore the extension $\mathbb{Q}(\sqrt{2}, i)$ is a splitting field for $f(X)$ over $\mathbb{Q}$. In this case $[\mathbb{Q}(\sqrt{2}, i):\mathbb{Q}] = 4 < 4!$, and

$$f(X) = [X - \tfrac{1}{2}\sqrt{2}(1 + i)][X + \tfrac{1}{2}\sqrt{2}(1 + i)][X - \tfrac{1}{2}\sqrt{2}(1 - i)][X + \tfrac{1}{2}\sqrt{2}(1 - i)]$$

is the splitting of $f(X)$ over $\mathbb{Q}(\sqrt{2}, i)$.

EXAMPLE 3

Let $f(X) = X^3 - 1 \in \mathbb{Q}[X]$. Then the roots of $f(X)$ in the field $\mathbb{C}$ are 1, ω, and ω^2, where ω stands for a primitive cube root of unity, and hence the field $\mathbb{Q}(\omega)$ is a splitting field for $f(X)$ over $\mathbb{Q}$. In this case $[\mathbb{Q}(\omega):\mathbb{Q}] = 2 < 3!$, and $f(X) = (X - 1)(X - \omega)(X - \omega^2)$ is the splitting of $f(X)$ over $\mathbb{Q}(\omega)$.

EXAMPLE 4

Let $f(X) = X^3 - 2 \in \mathbb{Q}[X]$. Then the roots of $f(X)$ in the field $\mathbb{C}$ are $\sqrt[3]{2}$, $\omega\sqrt[3]{2}$, and $\omega^2\sqrt[3]{2}$, where ω stands for a primitive cube root of unity, and hence the field $\mathbb{Q}(\omega, \sqrt[3]{2})$ is a splitting field for $f(X)$ over $\mathbb{Q}$. In this case we find that $[\mathbb{Q}(\omega, \sqrt[3]{2}):\mathbb{Q}] = 6 = 3!$, and $f(X) = (X - \sqrt[3]{2})(X - \omega\sqrt[3]{2})(X - \omega^2\sqrt[3]{2})$ over $\mathbb{Q}(\omega, \sqrt[3]{2})$.

EXAMPLE 5

Let $f(X) = X^4 + X^2 + 1 \in \mathbb{Q}[X]$. Then $f(X) = (X^2 + 1)^2 - X^2 = (X^2 + 1 - X)(X^2 + 1 + X)$. Now, the roots of $X^2 + X + 1$ in the field $\mathbb{C}$ are ω and ω^2, where ω stands for a primitive cube root of unity, and the roots of $X^2 - X + 1$ are $-\omega$ and $-\omega^2$. Hence, the roots of $f(X)$ in $\mathbb{C}$ are $\pm\omega$ and $\pm\omega^2$ and therefore the field $\mathbb{Q}(\omega)$ is a splitting field for $f(X)$ over $\mathbb{Q}$. In this case $f(X) = (X - \omega)(X + \omega)(X - \omega^2)(X + \omega^2)$ is the splitting of $f(X)$ over $\mathbb{Q}(\omega)$ and $[\mathbb{Q}(\omega):\mathbb{Q}] = 2 < 4!$.

$f(X) \in \mathbb{Q}[X]$	Roots of $f(X)$ in $\mathbb{C}$	Splitting field K	$[K:\mathbb{Q}]$	$n!$
$X^2 - 2$	$\pm\sqrt{2}$	$\mathbb{Q}(\sqrt{2})$	2	2
$X^2 + 1$	$\pm i$	$\mathbb{Q}(i)$	2	2
$X^2 + X + 1$	ω, ω^2	$\mathbb{Q}(\omega)$	2	2
$X^3 - 1$	$1, \omega, \omega^2$	$\mathbb{Q}(\omega)$	2	6
$X^3 - 2$	$\sqrt[3]{2}, \omega\sqrt[3]{2}, \omega^2\sqrt[3]{2}$	$\mathbb{Q}(\omega,\sqrt[3]{2})$	6	6
$X^4 - 1$	$\pm 1, \pm i$	$\mathbb{Q}(i)$	2	24
$X^4 + 1$	$\pm\frac{1}{2}\sqrt{2}(1+i), \pm\frac{1}{2}\sqrt{2}(1-i)$	$\mathbb{Q}(i, \sqrt{2})$	4	24
$X^4 + X^2 + 1$	$\pm\omega, \pm\omega^2$	$\mathbb{Q}(\omega)$	2	24

FIGURE 1. Splitting fields and their degrees for selected polynomials over the field $\mathbb{Q}$ of rational numbers.

Figure 1 summarizes the results in the preceding five examples. In each of these examples we found a splitting field for the given polynomial by working within the field of complex numbers. The next two examples illustrate the construction of a splitting field for a given polynomial using only the polynomial and the ground field.

EXAMPLE 6

Let $f(X) = X^2 + X + 1 \in \mathbb{F}_2[X]$. In this case there is no obvious extension of $\mathbb{F}_2$ that contains a root of $f(X)$. Let us therefore use the method described in Proposition 1 to construct a splitting field for $f(X)$ over $\mathbb{F}_2$.

(A) Since $X^2 + X + 1$ is irreducible over $\mathbb{F}_2$, let $K = \mathbb{F}_2[X]/(X^2 + X + 1)\mathbb{F}_2[X]$ stand for the field of polynomials over $\mathbb{F}_2$ modulo $X^2 + X + 1$. Then K is an extension field of $\mathbb{F}_2$ with $[K:\mathbb{F}_2] = 2$, and every element in K may be written uniquely in the form $a + bc$, where $a, b \in \mathbb{F}_2$ and $c = X + (X^2 + X + 1)\mathbb{F}_2[X]$; in particular, $K = \mathbb{F}_2(c)$.

(B) The element $c \in K$ is a root of $f(X)$ since $f(c) = c^2 + c + 1 = 0$ in K. Moreover, we find by long-division in $K[X]$ that $f(X) = (X - c)(X + 1 + c)$. Therefore, c and $1 + c$ are the roots of $f(X)$ in K and hence $f(X)$ splits over K. Finally, since $K = \mathbb{F}_2(c)$, K is generated over $\mathbb{F}_2$ by the roots of $f(X)$ and hence is a splitting field for $f(X)$ over $\mathbb{F}_2$. In this case $[K:\mathbb{F}_2] = 2 = 2!$.

EXAMPLE 7

Let $f(X) = X^3 - 1 \in \mathbb{Q}[X]$. We showed in Example 3 that the field $\mathbb{Q}(\omega)$ is a splitting field for $f(X)$ over $\mathbb{Q}$, where ω stands for a primitive cube root of unity. Let us use the method of Proposition 1 to construct a different splitting field for $f(X)$ over $\mathbb{Q}$ and then show that it is in fact isomorphic to $\mathbb{Q}(\omega)$.

(A) We begin by factoring $f(X)$ in $\mathbb{Q}[X]$: $f(X) = (X - 1)(X^2 + X + 1)$, where $X^2 + X + 1$ is irreducible over $\mathbb{Q}$. Let $K = \mathbb{Q}[X]/(X^2 + X + 1)\mathbb{Q}[X]$. Then K is an extension field of $\mathbb{Q}$ with $[K:\mathbb{Q}] = 2$, and every element in K may be written uniquely in the form $a + bc$, where $a, b \in \mathbb{Q}$ and $c = X + (X^2 + X + 1)\mathbb{Q}[X]$. The element c is a root of $f(X)$ in K since $c^2 + c + 1 = 0$ in K, and we find by long-division in $K[X]$ that $X^2 + X + 1 = (X - c)(X + 1 + c)$. Therefore,

$$f(X) = X^3 - 1 = (X - 1)(X - c)(X + 1 + c)$$

and hence $f(X)$ splits over K and has $1, c$, and $-1 - c$ as roots. Finally, since $K = \mathbb{Q}(c)$, K is generated by these roots and hence is a splitting field for $f(X)$ over $\mathbb{Q}$.

(B) Let us now show that the two splitting fields $\mathbb{Q}(\omega)$ and K are in fact isomorphic. Define the function $\sigma: K \to \mathbb{Q}(\omega)$ by setting $\sigma(a + bc) = a + b\omega$ for all elements $a + bc \in K$. Then, if $a + bc$ and $a' + b'c$ are typical elements in K, we find that

$$\sigma((a + bc) + (a' + b'c)) = \sigma((a + a') + (b + b')c)$$
$$= (a + a') + (b + b')\omega$$
$$= (a + b\omega) + (a' + b'\omega)$$
$$= \sigma(a + bc) + \sigma(a' + b'c)$$

and

$$\sigma((a + bc)(a' + b'c)) = \sigma(aa' + (ab' + ba')c + bb'c^2)$$
$$= \sigma(aa' + (ab' + ba')c + bb'(-1 - c))$$
$$= \sigma((aa' - bb') + (ab' + ba' - bb')c)$$
$$= (aa' - bb') + (ab' + ba' - bb')\omega$$
$$= aa' + (ab' + ba')\omega + bb'\omega^2$$
$$= (a + b\omega)(a' + b'\omega)$$
$$= \sigma(a + bc)\sigma(a' + b'c).$$

Therefore, σ preserves the arithmetic on the fields K and $\mathbb{Q}(\omega)$. Since σ is clearly 1–1 and onto, we conclude that σ is a field-isomorphism and hence that $K \cong \mathbb{Q}(\omega)$. Observe that under the isomorphism σ, the roots $1, c$, and $-1 - c$ of $f(X)$ in K are mapped to the roots $1, \omega$, and ω^2 in $\mathbb{Q}(\omega)$.

$$K \overset{\sigma}{\to} \mathbb{Q}(\omega)$$

$$1 \mapsto 1$$

$$c \mapsto \omega$$

$$-1 - c \mapsto -1 - \omega$$

Let us now use our results on splitting fields to discuss the existence of finite fields. Recall that in the previous chapter we showed that every finite field must contain a prime power number of elements; for if K is a finite field of prime characteristic p and $[K:k] = n$, where k stands for the prime subfield of K, then K is an n-dimensional vector space over k and hence contains p^n elements. We now claim that there is, in fact, a finite field containing p^n elements for any prime power p^n.

EXAMPLE 8. Existence of finite fields

Let n be a positive integer, p a prime, and let K stand for a splitting field of the polynomial $f(X) = X^{p^n} - X$ over $\mathbb{F}_p$. We claim that K is a finite field containing p^n elements and that these elements are precisely the roots of $f(X)$.

(A) Let us begin by first showing that the splitting field K consists of precisely the roots of $f(X)$. To do this, let $k' = \{x \in K \mid f(x) = 0\}$ stand for the set of roots of $f(X)$ in K and let $x, y \in k'$. Then $x^{p^n} = x$ and $y^{p^n} = y$, and therefore

$$(x + y)^{p^n} = x^{p^n} + y^{p^n} = x + y,$$

$$(xy)^{p^n} = x^{p^n} y^{p^n} = xy,$$

$$(-x)^{p^n} = -x^{p^n} = -x.$$

If, in addition, $x \neq 0$, then

$$\left(\frac{1}{x}\right)^{p^n} = \frac{1}{x^{p^n}} = \frac{1}{x}.$$

Thus, $x + y$, xy, $-x$, and $\dfrac{1}{x}$, for $x \neq 0$, are roots of $f(X)$ and hence are elements in k'. Therefore, k' is a subfield of K. Moreover, $\mathbb{F}_p \subseteq k'$ since, by Fermat's Little Theorem, $a^p = a$ for every element $a \in \mathbb{F}_p$. Therefore, k' is an intermediate subfield of the extension $K/\mathbb{F}_p$ that contains the roots of $f(X)$ and hence $k' = K$ since K is the smallest subfield with this property. Thus, K consists precisely of the roots of $f(X)$, as required.

(B) Now, to show that K contains p^n elements, we need only verify that the polynomial $f(X)$ has p^n distinct roots since the elements in K are, by part (A), the roots of $f(X)$. To this end, let $f(X) = (X - c_1) \cdots (X - c_m)$ be the splitting of $f(X)$ in $K[X]$, where $m = p^n$. Then, if $X - c_i$ is a typical factor of $f(X)$, $c_i^{p^n} = c_i$ and therefore

$$(X - c_i)^{p^n} - (X - c_i) = X^{p^n} - c_i^{p^n} - X + c_i = X^{p^n} - X.$$

Hence,

$$\frac{X^{p^n} - X}{X - c_i} = (X - c_i)^{p^n - 1} - 1.$$

Now, if two of the linear factors in the splitting of $f(X)$ are the same, say $X - c_i = X - c_j$, where $i \neq j$, then it follows from this equation that c_j is a root of the left side but not a root of the right side, which is impossible. Consequently the factors $X - c_1, \ldots, X - c_m$ must be distinct. Therefore, $f(X)$ has p^n distinct roots and hence K contains p^n elements since it consists of just these roots.

(C) For example, if K is a splitting field for the polynomial $X^{25} - X$ over $\mathbb{F}_5$, then K contains 25 elements and $[K:\mathbb{F}_5] = 2$; if K is a splitting field for $X^{27} - X$ over $\mathbb{F}_3$, then K contains 27 elements and $[K:\mathbb{F}_3] = 3$; and, if K is a splitting field for $X^{16} - X$ over $\mathbb{F}_2$, then K contains 16 elements and $[K:\mathbb{F}_2] = 4$.

(D) Finally, let us construct a splitting field for the polynomial $X^9 - X$ over $\mathbb{F}_3$ and verify by direct calculation that the elements of the field are indeed the nine roots of $X^9 - X$. We begin by first finding the irreducible factors of $X^9 - X$ over $\mathbb{F}_3$:

$$X^9 - X = X(X^8 - 1) = X(X - 1)(X + 1)(X^2 + 1)(X^2 + X - 1)(X^2 - X - 1).$$

We leave the reader to verify that each of these factors is indeed irreducible over $\mathbb{F}_3$. Now, to find an extension field of $\mathbb{F}_3$ containing a root of $X^9 - X$, first choose an irreducible factor, say $X^2 + 1$, and let $K = \mathbb{F}_3[X]/(X^2 + 1)\mathbb{F}_3[X]$. Then K is a field containing 3^2, or 9 elements, and $[K:\mathbb{F}_3] = 2$. Moreover, since every element in K may be written uniquely in the form $a + bc$, where $a, b \in \mathbb{F}_3$ and $c = X + (X^2 + 1)\mathbb{F}_3[X]$, we find that

$$K = \{0, 1, 2, c, 2c, 1 + c, 2 + c, 1 + 2c, 2 + 2c\}.$$

Now, it is clear that 0 is the root of X, 1 the root of $X - 1$, and 2 the root of $X + 1$; furthermore, since $c^2 = -1$ in K, it follows easily that c and $2c$ are the roots of $X^2 + 1$, while $1 + c$ and $1 + 2c$ are the roots of $X^2 + X - 1$, and $2 + c$ and $2 + 2c$ are the roots of $X^2 - X - 1$. Thus, the nine elements in K are precisely the roots of $X^9 - X$, which agrees with the result in part (A). Moreover, since K is generated over $\mathbb{F}_3$ by these roots, K is indeed a splitting field for $X^9 - X$ over $\mathbb{F}_3$. Finally, observe that the splitting of $X^9 - X$ over the field K is given by the equation

$$X^9 - X = X(X - 1)(X + 1)$$
$$\times \underbrace{(X - c)(X - 2c)}_{X^2 + 1}\underbrace{(X - 1 - c)(X - 1 - 2c)}_{X^2 + X - 1}\underbrace{(X - 2 - c)(X - 2 - 2c)}_{X^2 - X - 1}.$$

Let us close this section with a few general remarks about splitting fields. As we noted in the introduction, splitting fields give us an algebraic framework in which to discuss the roots of a given polynomial over a given field: if $f(X)$ is a nonconstant polynomial of degree n over a field k, there is an extension field K of degree at most $n!$ over k that contains a complete set of roots of $f(X)$ and which is generated over k by these roots. The field K thus provides an appropriate context in which to discuss the nature and properties of the roots of $f(X)$. But splitting fields, while they contain a complete set of roots for a given polynomial over a given field, do not, in general, contain a complete set of roots for every polynomial whose coefficients lie in the field. In contrast to this, a field K is said to be *algebraically closed* if every nonconstant polynomial over K splits over K, or equivalently, has a complete set of roots in K. Thus, while most fields must be extended in order to split a given polynomial, algebraically closed fields are sufficiently large so as to contain a complete set of roots for any nonconstant polynomial whose coefficients lie in the field. The field $\mathbb{C}$ of complex numbers, for example, is an algebraically closed field since, by the fundamental theorem of algebra—which we prove in the next chapter—every nonconstant polynomial over $\mathbb{C}$ has a root in $\mathbb{C}$ and hence splits over $\mathbb{C}$. More generally, using Zorn's lemma it is possible to show that any field k can be extended to an algebraically closed field K. But unfortunately an algebraically closed extension of k is not necessarily algebraic over k; the field $\mathbb{C}$ of complex numbers, for example, is an algebraically closed extension of the field $\mathbb{Q}$ of rational numbers but it is not algebraic over $\mathbb{Q}$. Nevertheless, if K is an algebraically closed extension of a field k, it turns out that the algebraic closure $\mathrm{Alg}(K/k)$ of k in K is both algebraically closed and algebraic over k. Any such extension of k that is both algebraically closed and algebraic over k is called an *algebraic closure* of k. It then follows, using Zorn's lemma, that all algebraic closures of a given field are isomorphic. Thus, algebraic closures of any given field not only exist, but are unique to within isomorphism.

Exercises

1. Let $f(X) = X^2 + 4 \in \mathbb{Q}[X]$. Find a splitting field K for $f(X)$ over $\mathbb{Q}$ and find $[K:\mathbb{Q}]$.
2. Let $f(X) = X^2 + 3X + 1 \in \mathbb{Q}[X]$. Find a splitting field K for $f(X)$ over $\mathbb{Q}$ and find $[K:\mathbb{Q}]$.
3. Let $f(X) = X^3 - 8 \in \mathbb{Q}[X]$. Find a splitting field K for $f(X)$ over $\mathbb{Q}$ and find $[K:\mathbb{Q}]$.
4. Let $f(X) = X^6 - 1 \in \mathbb{Q}[X]$. Find a splitting field K for $f(X)$ over $\mathbb{Q}$ and find $[K:\mathbb{Q}]$.
5. Let $f(X) = X^2 - \sqrt{2}X + 1 \in \mathbb{Q}(\sqrt{2})[X]$. Find a splitting field K for $f(X)$ over $\mathbb{Q}(\sqrt{2})$ and find $[K:\mathbb{Q}(\sqrt{2})]$.
6. Let $f(X) = X^2 - \pi \in \mathbb{R}[X]$. Find a splitting field K for $f(X)$ over $\mathbb{R}$ and find $[K:\mathbb{R}]$.
7. Let $f(X) = X^2 + 2X + 2 \in \mathbb{F}_3[X]$. Construct a splitting field K for $f(X)$ over $\mathbb{F}_3$ and find $[K:\mathbb{F}_3]$.
8. Let $f(X) = X^4 + X + 1 \in \mathbb{F}_2[X]$. Construct a splitting field K for $f(X)$ over $\mathbb{F}_2$ and find $[K:\mathbb{F}_2]$.

9. Let $f(X) = X^5 - X^4 + X - 1 \in \mathbb{F}_5[X]$. Construct a splitting field K for $f(X)$ over $\mathbb{F}_5$ and find $[K:\mathbb{F}_5]$.

10. Let $f(X) = X^4 - 10X^2 + 1 \in \mathbb{Q}[X]$. Show that $\mathbb{Q}(\sqrt{2} + \sqrt{3})$ is a splitting field for $f(X)$ over $\mathbb{Q}$.

11. Let $f(X) = X^4 - 2X^2 + 17 \in \mathbb{Q}[X]$ and let $K = \mathbb{Q}(\sqrt{1 + 4i}, \sqrt{1 - 4i})$.
    (a) Show that K is a splitting field for $f(X)$ over $\mathbb{Q}$.
    (b) Find $[K:\mathbb{Q}]$.
    (c) Show that $\sqrt{17} \in K$.
    (d) Is $\mathbb{Q}(\sqrt{1 + 4i})$ a splitting field for $f(X)$ over $\mathbb{Q}$?
    (e) Is K a splitting field for $X^2 - 17$ over $\mathbb{Q}$?

12. Let $f(X) = X^3 - 2X^2 - X \in \mathbb{Q}[X]$.
    (a) Show that $\mathbb{Q}(\sqrt{2})$ is a splitting field for $f(X)$ over $\mathbb{Q}$.
    (b) Let $K = \mathbb{Q}[X]/(X^2 - 2X - 1)\mathbb{Q}[X]$. Show that K is a splitting field for $f(X)$ over $\mathbb{Q}$.
    (c) Define the function $\sigma: K \to \mathbb{Q}(\sqrt{2})$ by setting $\sigma(a + bc) = a - b + b\sqrt{2}$ for all elements $a + bc \in K$, where $a, b \in \mathbb{Q}$ and $c = X + (X^2 - 2X - 1)\mathbb{Q}[X]$. Show that σ is a field-isomorphism and conclude that $K \cong \mathbb{Q}(\sqrt{2})$.

13. (a) Is the field $\mathbb{C}$ of complex numbers a splitting field for some polynomial over $\mathbb{R}$?
    (b) Is the field $\mathbb{C}$ of complex numbers a splitting field for some polynomial over $\mathbb{Q}$?

14. Let k be a field and let K be a splitting field for some nonconstant polynomial over k. Show that K is a finite algebraic extension of k.

15. Show that the field $\mathbb{Q}(\sqrt[3]{2})$ is a finite algebraic extension of $\mathbb{Q}$ but is not a splitting field for any polynomial over $\mathbb{Q}$.

16. Let $k \subseteq k' \subseteq K$ be a tower of fields such that K is a splitting field for some nonconstant polynomial $f(X)$ over k. Show that K is a splitting field for $f(X)$ over k'.

17. Let $f(X)$ be a nonconstant polynomial over a field k and let K be a finite extension of k over which $f(X)$ splits. If $[K:k]$ is prime, show that either K is a splitting field for $f(X)$ over k or else $f(X)$ already splits over k.

18. Let n be a positive integer and let $z_n = \cos\dfrac{2\pi}{n} + i \sin\dfrac{2\pi}{n}$ stand for a primitive nth root of unity in the field $\mathbb{C}$ of complex numbers. Show that the field $\mathbb{Q}(z_n)$ is a splitting field for the polynomial $f(X) = X^n - 1$ over $\mathbb{Q}$.

19. Let $f_1(X), \ldots, f_n(X)$ be a finite number of nonconstant polynomials over an arbitrary field k. Show that there is a finite extension field K of k such that each of the polynomials $f_1(X), \ldots, f_n(X)$ splits over K.

20. Let K be a field. Show that the following four statements are equivalent:
    (a) K is algebraically closed;
    (b) Every irreducible polynomial over K is linear;
    (c) The only algebraic extension of K is K itself;
    (d) The only finite extension of K is K itself.

21. Let $k = \{c_1, \ldots, c_n\}$ be a finite field and let $f(X) = (X - c_1) \cdots (X - c_n)$ $+ 1 \in k[X]$. Show that $f(X)$ does not have a root in k. Using this fact, show that an algebraically closed field cannot be finite.

22. Let k be a field and suppose that K is an extension field of k that is algebraically closed. Let $\tilde{k} = \text{Alg}(K/k) = \{c \in K \mid c \text{ is algebraic over } k\}$ stand for the algebraic closure of k in K; that is, those elements in K that are algebraic over k.

(a) Show that $\tilde{k}$ is an algebraic extension of k.

(b) Show that $\tilde{k}$ is an algebraically closed field.

(c) Show that $\tilde{k}$ is the smallest algebraically closed extension field of k that is contained in K; that is, if k' is an algebraically closed field such that $k \subseteq k' \subseteq K$, then $\tilde{k} \subseteq k'$.

(d) Show that $\tilde{k} = \bigcap k'$, where the intersection is over all intermediate subfields k' of the extension K/k that are algebraically closed. The field $\tilde{k}$ is called an *algebraic closure* of k; it is an algebraic extension of k that is algebraically closed.

(e) Does $\tilde{\mathbb{R}} = \mathbb{C}$?

(f) Does $\tilde{\mathbb{Q}} = \mathbb{C}$?

2. THE ISOMORPHISM THEOREM FOR SPLITTING FIELDS

In the previous section we showed that every nonconstant polynomial over an arbitrary ground field k has a splitting field; that is, an extension of k containing a complete set of roots of the polynomial and which is generated over k by these roots. As we saw, splitting fields are not unique since there is usually more than one choice for an extension field over which the polynomial splits. But the next best thing is true, namely that all splitting fields for a given polynomial over a given field are in fact isomorphic. This result, known as the isomorphism theorem for splitting fields, is important because it guarantees that any particular splitting field for a given polynomial contains all the essential information about the roots of the polynomial. Thus, to study the algebraic properties of the roots of a polynomial we need only construct a single splitting field for the polynomial; whatever properties exist in such a field, they are the same in all such fields. We conclude the section by discussing two particularly important consequences of the isomorphism theorem; first, root multiplicities, which refers to the number of times a linear factor corresponding to a root occurs in any splitting of the polynomial; and second, the primitive element theorem, which shows that every finite extension of a field of characteristic zero has a primitive element and hence is a simple extension of the ground field.

Before we begin, let us first take a moment to describe in general terms how the isomorphism theorem is proved. Splitting fields, we recall, are generated by the roots of a polynomial and hence may be obtained through a finite sequence of simple algebraic extensions. Specifically, if $f(X)$ is a non-constant polynomial over an arbitrary field k and $c_1, \ldots, c_n$ are the roots of $f(X)$ in some splitting field K over k, then

$$k \subseteq k(c_1) \subseteq k(c_1, c_2) \subseteq \cdots \subseteq k(c_1, \ldots, c_n) = K$$

is a finite sequence of simple algebraic extensions beginning with the ground field k and ending with the splitting field K. To show that two such splitting fields K and K' are isomorphic, we first show that by choosing the sequence of roots properly, any isomorphism σ_i between corresponding intermediate subfields may be extended to an isomorphism σ_{i+1} of the simple extensions at the next level in the tower of fields, as illustrated in the diagram.

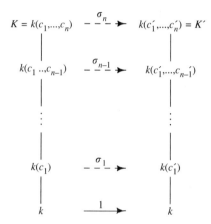

Then, beginning with the identity isomorphism 1 on the ground field k and extending it a finite number of times, we arrive, finally, at an isomorphism of the splitting fields themselves. In order to apply this procedure, however, we must first determine conditions under which a given isomorphism may be extended. Let us begin, therefore, by first discussing extensions of field-isomorphisms.

Definition

Let $\sigma : k \to k'$ be an isomorphism of fields k and k', and let K and K' be extension fields of k and k', respectively, such that K and K' are isomorphic. A field-isomorphism $\sigma^* : K \to K'$ is called an *extension of* σ if $\sigma^*(a) = \sigma(a)$ for every element $a \in k$.

In other words, a field-isomorphism $\sigma^* : K \to K'$ is an extension of a field-isomorphism $\sigma : k \to k'$ if the restriction of σ^* to k is just σ; that is, $\sigma^* | k = \sigma$. For example, the complex conjugation mapping $\sigma^* : \mathbb{C} \to \mathbb{C}$ defined by setting $\sigma^*(a + bi) = a - bi$ for all complex numbers $a + bi$ is a field-isomorphism from the field $\mathbb{C}$ onto itself, and σ^* is an extension of the identity isomorphism $1: \mathbb{R} \to \mathbb{R}$ on the subfield $\mathbb{R}$ of real numbers since $\sigma^*(a) = 1(a) = a$ for all real numbers $a \in \mathbb{R}$. In general, an isomorphism $\sigma^*: K \to K'$ is an extension of the identity isomorphism $1: k \to k$ if and only if σ^* fixes every element in k.

Now, let $\sigma: k \to k'$ be an isomorphism of fields. If $f(X) = a_0 + a_1 X + \cdots + a_n X^n$ is a typical polynomial over k, let $f_\sigma(X) = \sigma(a_0) + \sigma(a_1)X + \cdots + \sigma(a_n)X^n$ stand for the polynomial over k' whose coefficients are the images under σ of the corresponding coefficients of $f(X)$. We call $f_\sigma(X)$ the *polynomial corresponding to $f(X)$ under σ*, and leave the reader to verify that if $f(X)$ and $g(X)$ are arbitrary polynomials over k, then

$$(f + g)_\sigma(X) = f_\sigma(X) + g_\sigma(X),$$

$$(f - g)_\sigma(X) = f_\sigma(X) - g_\sigma(X),$$

$$(-f)_\sigma(X) = -f_\sigma(X),$$

$$(fg)_\sigma(X) = f_\sigma(X)g_\sigma(X).$$

Furthermore, since σ is an isomorphism, every polynomial over k' corresponds under σ to some polynomial over k, and a polynomial $f(X)$ is irreducible over k if and only if the corresponding polynomial $f_\sigma(X)$ is irreducible over k'. It follows, therefore, that the mapping $f(X) \mapsto f_\sigma(X)$ that maps an arbitrary polynomial $f(X)$ over k to its corresponding polynomial $f_\sigma(X)$ over k' is a ring-isomorphism from the ring $k[X]$ of polynomials over k onto the ring $k'[X]$ of polynomials over k'.

PROPOSITION 1. Let $\sigma: k \to k'$ be an isomorphism of fields and let K and K' be extension fields of k and k', respectively. If σ extends to a field-isomorphism $\sigma^*: K \to K'$ and c is any element in K that is algebraic over k, then the image $\sigma^*(c)$ is algebraic over k' and $\text{Irr}(\sigma^*(c); k') = \text{Irr}(c; k)_\sigma$.

Proof. Let $c \in K$ be algebraic over k and let $f(X) = \text{Irr}(c; k) = X^n + a_{n-1}X^{n-1} + \cdots + a_0$, where $a_0, \ldots, a_{n-1} \in k$. Then $f_\sigma(X) = X^n + \sigma(a_{n-1})X^{n-1} + \cdots + \sigma(a_0)$ and is a monic irreducible polynomial over k'. Now, if σ extends to an isomorphism $\sigma^*: K \to K'$, then $\sigma(a_i) = \sigma^*(a_i)$ for each coefficient a_i of $f(X)$ and therefore

$$f_\sigma(\sigma^*(c)) = [\sigma^*(c)]^n + \sigma(a_{n-1})[\sigma^*(c)]^{n-1} + \cdots + \sigma(a_0)$$
$$= [\sigma^*(c)]^n + \sigma^*(a_{n-1})[\sigma^*(c)]^{n-1} + \cdots + \sigma^*(a_0)$$
$$= \sigma^*(c^n + a_{n-1}c^{n-1} + \cdots + a_0)$$
$$= \sigma^*(f(c))$$
$$= 0.$$

Therefore, $\sigma^*(c)$ is a root of $f_\sigma(X)$ and hence is algebraic over k'. Moreover, since $f_\sigma(X)$ is monic and irreducible over k', we conclude that $\text{Irr}(\sigma^*(c); k') = f_\sigma(X) = \text{Irr}(c; k)_\sigma$. ∎

It follows from Proposition 1 that if a field-isomorphism $\sigma: k \to k'$ extends to an isomorphism $\sigma^*: K \to K'$ and c is any element in K algebraic over k, then the image $\sigma^*(c)$ must be a root of the polynomial $\text{Irr}(c;k)_\sigma$. Thus, to construct such an extension σ^* we are limited in our choice of images for $\sigma^*(c)$ to those elements in K' that are roots of $\text{Irr}(c;k)_\sigma$. Let us now show that every such choice does in fact define an extension of σ.

PROPOSITION 2. Let $\sigma: k \to k'$ be an isomorphism of fields and let K and K' be extension fields of k and k', respectively. Let $c \in K$ and $c' \in K'$ be elements that are algebraic over k and k', respectively, such that c' is a root of the polynomial $\text{Irr}(c;k)_\sigma$. Define the function $\sigma_{c,c'}^*: k(c) \to k'(c')$ by setting

$$\sigma_{c,c'}^*(a_0 + a_1 c + \cdots + a_{n-1}c^{n-1}) = \sigma(a_0) + \sigma(a_1)c' + \cdots + \sigma(a_{n-1})c'^{n-1}$$

for every element $a_0 + a_1 c + \cdots + a_{n-1}c^{n-1} \in k(c)$, where $n = \deg \text{Irr}(c;k)$ and $a_0, \ldots, a_{n-1} \in k$. Then $\sigma_{c,c'}^*$ is a field-isomorphism that extends σ to $k(c)$ and maps c onto c'.

Proof.   Recall first that since $c \in K$ is algebraic over k, $k(c) = k[c]$ and hence every element in $k(c)$ may be written as a polynomial in c with coefficients in k, and similarly, every element in $k'(c')$ may be written as a polynomial in c' with coefficients in k'. Using these facts, let us now show that $\sigma_{c,c'}^*$ is well-defined and has the required properties.

(A) We claim first that $\sigma_{c,c'}^*$ is well-defined and that $\sigma_{c,c'}^*(f(c)) = f_\sigma(c')$ for all polynomials $f(c) \in k(c)$. For suppose that $f(c) = g(c)$ for some polynomials $f(X), g(X) \in k[X]$. Then $\text{Irr}(c;k)$ divides $f(X) - g(X)$ and therefore $\text{Irr}(c;k)_\sigma$ divides $f_\sigma(X) - g_\sigma(X)$. But $\text{Irr}(c;k)_\sigma = \text{Irr}(c';k')$ since c' is a root of the monic irreducible polynomial $\text{Irr}(c;k)_\sigma$. Therefore, $\text{Irr}(c';k')$ divides $f_\sigma(X) - g_\sigma(X)$ and hence $f_\sigma(c') = g_\sigma(c')$. Thus, $\sigma_{c,c'}^*$ is well-defined and $\sigma_{c,c'}^*(f(c)) = f_\sigma(c')$ for all polynomials $f(c) \in k(c)$.

(B) $\sigma_{c,c'}^*$ is 1–1 and onto. For clearly $\sigma_{c,c'}^*$ is an onto mapping since every element in $k'(c')$ is a polynomial in c'. Moreover, by an argument similar to the one in part (A), it follows that if $f_\sigma(c') = g_\sigma(c')$ for some $f(X), g(X) \in k[X]$, then $f(c) = g(c)$. Hence, $\sigma_{c,c'}^*$ is a 1–1 mapping.

(C) $\sigma_{c,c'}^*$ preserves the arithmetic on $k(c)$. For let $f(c)$ and $g(c)$ be typical elements in $k(c)$. Then

$$\begin{aligned}
\sigma_{c,c'}^*(f(c) + g(c)) &= (f + g)_\sigma(c') \\
&= f_\sigma(c') + g_\sigma(c') \\
&= \sigma_{c,c'}^*(f(c)) + \sigma_{c,c'}^*(g(c)),
\end{aligned}$$

and, similarly,

$$\sigma_{c,c'}^*(f(c)g(c)) = \sigma_{c,c'}^*(f(c))\sigma_{c,c'}^*(g(c)).$$

Therefore $\sigma_{c,c'}^*$ preserves addition and multiplication on $k(c)$.

It follows that $\sigma_{c,c'}^*$ is a field-isomorphism mapping $k(c)$ onto $k'(c')$. Finally, since $\sigma_{c,c'}^*(a_0) = \sigma(a_0)$ for every element $a_0 \in k$, $\sigma_{c,c'}^*$ extends the isomorphism σ, and maps c onto c' since $\sigma_{c,c'}^*(c) = c'$. ∎

Proposition 2 is especially important since it describes explicitly how to construct field-isomorphisms. It is particularly useful in constructing automorphisms of a given field and will play an important role in our discussion of the Galois theory of fields in the next chapter. Before we illustrate this result, let us first use it to prove the main result of this section.

PROPOSITION 3.   (ISOMORPHISM THEOREM FOR SPLITTING FIELDS) Let $\sigma:k \to k'$ be an isomorphism of fields. If K is a splitting field for a nonconstant polynomial over k and K' is a splitting field for the corresponding polynomial over k', then K and K' are isomorphic fields and there is an isomorphism $\sigma^*:K \to K'$ that extends σ.

Proof.   Let K be a splitting field for a nonconstant polynomial $f(X)$ over k and let $n = [K:k]$. We prove the statement by induction on n. If $n = 1$, then $K = k$ and hence $f(X)$ has all its roots in k. It follows that k' is a splitting field for the corresponding polynomial $f_\sigma(X)$ and hence the isomorphism σ itself has the required property. Now, let $n > 1$ and assume the statement is true for any isomorphism of fields $k^* \to k'^*$ and any extension K^*/k^* for which K^* is a splitting field over k^* such that $[K^*:k^*] < n$. Let c be a root of $f(X)$ in K; since $n > 1$, we may choose c not in k. Then $\mathrm{Irr}(c;k)$ divides $f(X)$ and therefore $\mathrm{Irr}(c;k)_\sigma$ divides $f_\sigma(X)$. Now, if K' is any splitting field for $f_\sigma(X)$ over k', then K' contains a root c' of $\mathrm{Irr}(c;k)_\sigma$ and hence, by Proposition 2, the isomorphism $\sigma:k \to k'$ extends to the isomorphism $\sigma_{c,c'}^*:k(c) \to k'(c')$. Finally, the isomorphism $\sigma_{c,c'}^*$ extends, by the induction hypothesis, to an isomorphism $\sigma^*:K \to K'$ since K and K' are splitting fields for $f(X)$ and $f_\sigma(X)$ over $k(c)$ and $k'(c')$, respectively, and $[K:k(c)] < n$. Therefore, σ^* is an extension of σ, as required, and the proof by induction is complete. ∎

COROLLARY.   Let $f(X)$ be a nonconstant polynomial over a field k. Then all splitting fields for $f(X)$ over k are isomorphic.

Proof.   This follows immediately from the isomorphism theorem by choosing $\sigma = 1$, the identity isomorphism on k. ∎

COROLLARY.   Let k be a field and let K be a splitting field over k. If c and c' are elements in K that have the same irreducible polynomial over k, then there is an automorphism of K that maps c onto c' and fixes all elements in k.

Proof.   Since c and c' are roots of the same irreducible polynomial over k, it follows from Proposition 2 that the identity map on k extends to the isomorphism $\sigma_{c,c'}^*:k(c) \to k(c')$, which maps c onto c' and fixes every element in

k. Since K is a splitting field over both $k(c)$ and $k(c')$, this isomorphism extends to an isomorphism $\sigma^*:K \to K$. Thus, σ^* is an automorphism of K that maps c onto c' and fixes every element in k, as required. ∎

These results now show that all splitting fields for a given polynomial over a given field are isomorphic, or simply that splitting fields are unique to within isomorphism, and hence that the algebraic properties of the roots of a polynomial are the same regardless of which splitting field is used. For example, consider the field $\mathbb{Q}(\sqrt{2})$. $\mathbb{Q}(\sqrt{2})$ is a splitting field for the polynomial $X^2 - 2$ over $\mathbb{Q}$ and $\mathrm{Irr}(\sqrt{2};\mathbb{Q}) = X^2 - 2$. Therefore, every automorphism of $\mathbb{Q}(\sqrt{2})$ that fixes $\mathbb{Q}$ maps $\sqrt{2}$ to either $\sqrt{2}$ or $-\sqrt{2}$. By Proposition 2, there are two such automorphisms: the automorphism

$$\sigma^*_{\sqrt{2},\sqrt{2}}:\mathbb{Q}(\sqrt{2}) \to \mathbb{Q}(\sqrt{2}),$$

which maps a typical element $a + b\sqrt{2}$ to $a + b\sqrt{2}$ and hence is the identity automorphism, and the automorphism

$$\sigma^*_{\sqrt{2},-\sqrt{2}}:\mathbb{Q}(\sqrt{2}) \to \mathbb{Q}(\sqrt{2}),$$

which maps $a + b\sqrt{2}$ to $a - b\sqrt{2}$. We claim, in fact, that every automorphism of $\mathbb{Q}(\sqrt{2})$ fixes $\mathbb{Q}$ and hence is one of these two automorphisms; for if σ is an automorphism of $\mathbb{Q}(\sqrt{2})$, then $\sigma(1) = 1$ and therefore $\sigma\left(\dfrac{n}{m}\right) = \dfrac{n}{m}$ for all rational numbers $\dfrac{n}{m} \in \mathbb{Q}$. Thus, every automorphism of $\mathbb{Q}(\sqrt{2})$ fixes the field $\mathbb{Q}$ and hence is either the identity automorphism or the automorphism $\sigma^*_{\sqrt{2},-\sqrt{2}}$.

EXAMPLE 1

Let $f(X) = X^3 - 2 \in \mathbb{Q}[X]$. Then the roots of $f(X)$ are $\sqrt[3]{2}$, $\omega\sqrt[3]{2}$, and $\omega^2\sqrt[3]{2}$, where ω stands for a primitive cube root of unity, and hence the field $\mathbb{Q}(\omega, \sqrt[3]{2})$ is a splitting field for $f(X)$ over $\mathbb{Q}$. Let us construct another splitting field for $f(X)$ over $\mathbb{Q}$ and then construct an isomorphism between these two fields.

(A) Let $K = \mathbb{Q}[X]/(X^3 - 2)\mathbb{Q}[X]$. Then K is a finite extension field of $\mathbb{Q}$ with $[K:\mathbb{Q}] = 3$ since $X^3 - 2$ is irreducible over $\mathbb{Q}$, and every element in K may be written uniquely in the form $a + bx + cx^2$, where $a, b, c \in \mathbb{Q}$ and $x = X + (X^3 - 2)\mathbb{Q}[X]$. Moreover, the element $x \in K$ is a root of $X^3 - 2$ and we find, by long-division in $K[X]$, that $X^3 - 2 = (X - x)$ $\times (X^2 + xX + x^2)$. Now, we leave the reader to verify that the polynomial $X^2 + xX + x^2$ is irreducible over K. Thus, in order to split the polynomial $X^3 - 2$ we must first extend K to a larger field L that splits $X^2 + xX + x^2$. To this end, let $L = K[X]/(X^2 + xX + x^2)K[X]$. Then L is a finite extension field of K with $[L:K] = 2$, and every element in L may be written uniquely in the form $A + By$, where $A, B \in K$ and $y = X + (X^2 + xX + x^2)L[X]$. Then y is a root of $X^2 + xX + x^2$ in L

and we find that $X^2 + xX + x^2 = (X - y)(X + x + y)$. Therefore, $f(X) = (X - x)(X - y)(X + x + y)$ in $L[X]$ and hence $f(X)$ splits over L and has x, y, and $-x - y$ as roots. Since $L = K(y) = \mathbb{Q}(x, y)$, L is generated over $\mathbb{Q}$ by these roots and is therefore a splitting field for $f(X)$ over $\mathbb{Q}$.

(B) Let us now construct a field-isomorphism between the splitting fields L and $\mathbb{Q}(\omega, \sqrt[3]{2})$. To do this we use Proposition 2 twice, first to extend the identity isomorphism $1 : \mathbb{Q} \to \mathbb{Q}$ to an isomorphism $\sigma : K \to \mathbb{Q}(\sqrt[3]{2})$, and then again to extend σ to an isomorphism $\sigma^* : L \to \mathbb{Q}(\omega, \sqrt[3]{2})$, as illustrated in the diagram.

$$
\begin{array}{ccc}
L = K(y) & \xrightarrow{\sigma^*} & \mathbb{Q}(\sqrt[3]{2})(\omega\sqrt[3]{2}) \\
| & & | \\
K = \mathbb{Q}(x) & \xrightarrow{\sigma} & \mathbb{Q}(\sqrt[3]{2}) \\
| & & | \\
\mathbb{Q} & \xrightarrow{1} & \mathbb{Q}
\end{array}
$$

Let us begin by constructing σ. Since $\mathrm{Irr}(x; \mathbb{Q}) = X^3 - 2$ and since the roots of this polynomial in $\mathbb{Q}(\omega, \sqrt[3]{2})$ are $\sqrt[3]{2}$, $\omega\sqrt[3]{2}$, $\omega^2\sqrt[3]{2}$, it follows that these three elements are the only possible images for $\sigma(x)$. Let us choose $\sqrt[3]{2}$ for the image of x and let

$$
\sigma = \sigma^*_{x, \sqrt[3]{2}} : K \to \mathbb{Q}(\sqrt[3]{2}).
$$

Then σ is a field-isomorphism mapping x onto $\sqrt[3]{2}$ and hence $\sigma(a + bx + cx^2) = a + b\sqrt[3]{2} + c\sqrt[3]{4}$ for every element $a + bx + cx^2 \in L$. Now, to extend σ to an isomorphism from L onto $\mathbb{Q}(\omega, \sqrt[3]{2})$, first write $L = K(y)$ and recall that $\mathrm{Irr}(y; K) = X^2 + xX + x^2$. Then the polynomial over $\mathbb{Q}(\sqrt[3]{2})$ corresponding to $\mathrm{Irr}(y; K)$ under σ is $\mathrm{Irr}(y; K)_\sigma = X^2 + \sqrt[3]{2}X + \sqrt[3]{4}$, whose roots in $\mathbb{Q}(\omega, \sqrt[3]{2})$ are $\omega\sqrt[3]{2}$ and $\omega^2\sqrt[3]{2}$. Let us choose $\omega\sqrt[3]{2}$ for the image of y and let

$$
\sigma^* = \sigma^*_{y, \omega\sqrt[3]{2}} : L \to \mathbb{Q}(\omega, \sqrt[3]{2}).
$$

Then σ^* is a field-isomorphism mapping y onto $\omega\sqrt[3]{2}$ and $\sigma^*(A + By) = \sigma(A) + \sigma(B)\omega\sqrt[3]{2}$ for every element $A + By \in L$, where $A, B \in K$. Thus, σ^* is an isomorphism between the two splitting fields and, under this isomorphism, $\sigma^*(x) = \sqrt[3]{2}$ and $\sigma^*(y) = \omega\sqrt[3]{2}$. If $a + bx + cx^2 + dy + exy + fx^2y$ is a typical element in L, where $a, b, c, d, e, f \in \mathbb{Q}$ we find that

$$
\sigma^*(a + bx + cx^2 + dy + exy + fx^2y)
$$
$$
= a + b\sqrt[3]{2} + c\sqrt[3]{4} + d\omega\sqrt[3]{2} + e\omega\sqrt[3]{4} + 2f\omega.
$$

(C) It is clear from the discussion in part (B) that there are several possible isomorphisms mapping K onto $\mathbb{Q}(\omega, \sqrt[3]{2})$ depending upon the choice of images for $\sigma^*(x)$ and $\sigma^*(y)$: for $\sigma^*(x)$ we may choose either $\sqrt[3]{2}$, $\omega\sqrt[3]{2}$, or $\omega^2\sqrt[3]{2}$, while for $\sigma^*(y)$ we may choose either $\omega\sqrt[3]{2}$ or $\omega^2\sqrt[3]{2}$. Thus, there are six possible isomorphisms mapping the splitting field K onto the splitting field $\mathbb{Q}(\omega, \sqrt[3]{2})$.

EXAMPLE 2. Uniqueness of finite fields

In the previous section we showed that there is a finite field containing any given prime power number of elements. Let us now use the isomorphism theorem for splitting fields to show that all finite fields containing the same number of elements are in fact isomorphic.

(A) Suppose that K and K' are finite fields containing p^n elements, where p is prime and n is a positive integer, and let k and k' stand for the prime subfields of K and K', respectively. Then k and k' are isomorphic since each is isomorphic to the field $\mathbb{F}_p$ of integers modulo p. Since K and K' are splitting fields for the polynomial $X^{p^n} - X$ over k and k', it now follows from the isomorphism theorem that K and K' are isomorphic.

(B) It follows from part (A) that finite fields containing p^n elements not only exist but are unique to within isomorphism. Let us illustrate this result by constructing an explicit isomorphism between two finite fields each containing nine elements. Let

$$K = \mathbb{F}_3[X]/(X^2 + 1)\mathbb{F}_3[X]$$

$$K' = \mathbb{F}_3[X]/(X^2 + X + 1)\mathbb{F}_3[X].$$

Then K and K' are finite fields, and each contains nine elements. Every element in K may be written uniquely in the form $a + bc$, where $a, b \in \mathbb{F}_3$ and $c = X + (X^2 + 1)\mathbb{F}_3[X]$, and hence

$$K = \{0, 1, 2, c, 2c, 1 + c, 2 + c, 1 + 2c, 2 + 2c\}.$$

Similarly, every element in K' may be written uniquely in the form $a + bc'$, where $a, b \in \mathbb{F}_3$ and $c' = X + (X^2 + X + 1)\mathbb{F}_3[X]$, and therefore

$$K' = \{0, 1, 2, c', 2c', 1 + c', 2 + c', 1 + 2c', 2 + 2c'\}.$$

Now, to construct an isomorphism mapping K onto K' first observe that $K = \mathbb{F}_3(c)$ and $\text{Irr}(c; \mathbb{F}_3) = X^2 + 1$. By inspection, the only roots of this polynomial in K' are $1 + 2c'$ and $2 + c'$. Choosing $2 + c'$ for the image of c, it then follows that the mapping $\sigma^*_{c,2+c'}: K \to K'$ is a field-isomorphism from K onto K' that maps c onto $2 + c'$. If $a + bc$ is a typical element in K, we find that

$$\sigma^*_{c,2+c'}(a + bc) = a + b(2 + c') = (a + 2b) + 2c',$$

as shown in the marginal table. This concludes the example.

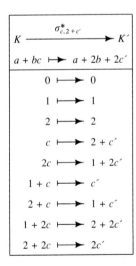

$$\begin{array}{ccc}
& \sigma^*_{c,2+c'} & \\
K & \longrightarrow & K' \\
a + bc & \longmapsto & a + 2b + 2c' \\
0 & \longmapsto & 0 \\
1 & \longmapsto & 1 \\
2 & \longmapsto & 2 \\
c & \longmapsto & 2 + c' \\
2c & \longmapsto & 1 + 2c' \\
1 + c & \longmapsto & c' \\
2 + c & \longmapsto & 1 + c' \\
1 + 2c & \longmapsto & 2 + 2c' \\
2 + 2c & \longmapsto & 2c'
\end{array}$$

Let us now turn our attention to some of the consequences of the isomorphism theorem for splitting fields. If K and K' are splitting fields for a non-constant polynomial $f(X)$, then, as we have shown, there is an isomorphism $\sigma: K \to K'$ that fixes every element in k. Therefore, $K[X] \cong K'[X]$. It follows that if $f(X) = a(X - c_1) \cdots (X - c_n)$ is the splitting of $f(X)$ in $K[X]$, then $f(X) = a(X - \sigma(c_1)) \cdots (X - \sigma(c_n))$ is the splitting of $f(X)$ in $K'[X]$. Thus, if $c_1, \ldots, c_n$ are the roots of $f(X)$ in K, then $\sigma(c_1), \ldots, \sigma(c_n)$ are its roots in

K'. Moreover, if a linear factor $X - c_i$ occurs s times in the splitting of $f(X)$ over K, then the corresponding linear factor $X - \sigma(c_i)$ occurs s times in the splitting of $f(X)$ over K'. In other words, the number of times a linear factor occurs in a splitting of $f(X)$ is independent of the particular splitting field in which we are working. With this observation in mind, we make the following definition.

Definition

Let $f(X)$ be a nonconstant polynomial over a field k and let c be a root of $f(X)$ in any splitting field K. The *multiplicity* of c is the largest positive integer s such that $(X - c)^s$ divides $f(X)$ in $K[X]$. A root whose multiplicity is 1 is called a *simple root* of $f(X)$, while any root of higher multiplicity is called a *multiple root*.

For example, let $f(X) = X^5 + 2X^3 + X \in \mathbb{Q}[X]$. Then $f(X) = X(X^2 + 1)^2 = X(X - i)^2(X + i)^2$ over the splitting field $\mathbb{Q}(i)$. Therefore, 0 is a simple root of $f(X)$, while both i and $-i$ are multiple roots of multiplicity 2. By counting each of these roots according to their multiplicities, we find that $f(X)$ has five roots, which agrees with the degree of $f(X)$. Similarly, if $f(X) = X^2 + 1 \in \mathbb{F}_2[X]$, then $f(X) = (X + 1)^2$ in $\mathbb{F}_2[X]$ and therefore $f(X)$ has only one root, namely 1, but it is a multiple root of multiplicity 2. In general, if the roots of a polynomial are counted according to their multiplicity, then a polynomial of degree n has exactly n roots in any splitting field.

There is a simple and useful method involving formal derivatives for deciding when a polynomial has a multiple root without actually knowing the roots of the polynomial. To describe this method, let k be an arbitrary field and let $f(X) = a_0 + a_1 X + a_2 X^2 + \cdots + a_n X^n$ be a typical polynomial over k. The *formal derivative* of $f(X)$ is the polynomial $f'(X) = a_1 + 2a_2 X + \cdots + na_n X^{n-1}$. We leave the reader to verify that if $f(X)$ and $g(X)$ are any polynomials over k, then $[f(X)g(X)]' = f'(X)g(X) + f(X)g'(X)$. This "product rule" follows easily by comparing coefficients of corresponding powers of X on both sides of the equation.

PROPOSITION 4. Let $f(X)$ be a nonconstant polynomial over a field k and let c be a root of $f(X)$ in any splitting field. Then c is a multiple root of $f(X)$ if and only if $f'(c) = 0$; that is, if and only if c is a root of the derivative $f'(X)$.

Proof. Let K be a splitting field for $f(X)$ over k containing the root c. Then $f(X) = (X - c)q(X)$ for some polynomial $q(X) \in K[X]$. Therefore, $f'(X) = (X - c)q'(X) + q(X)$ and hence $f'(c) = 0$ if and only if $q(c) = 0$. But $q(c) = 0$ if and only if c is a multiple root of $f(X)$. Thus, c is a multiple root of $f(X)$ if and only if $f'(c) = 0$. ∎

PROPOSITION 5. Let $f(X)$ be a nonconstant polynomial over a field k and let $d(X) = \gcd(f(X), f'(X))$ stand for the greatest common divisor of $f(X)$ and $f'(X)$ over k. Then the multiple roots of $f(X)$ are precisely the roots of $d(X)$.

Proof. Let K be a field over which both $f(X)$ and $f'(X)$ split. Then $d(X)$ also splits over K and every root of $d(X)$ in K is clearly a common root of $f(X)$ and $f'(X)$. Thus, by Proposition 4, every root of $d(X)$ is a multiple root of $f(X)$. Conversely, if c is a multiple root of $f(X)$ in K, then $f(c) = f'(c) = 0$ and therefore $d(c) = 0$ since $d(X)$ may be written in the form $a(X)f(X) + b(X)f'(X)$ for some polynomials $a(X), b(X) \in K[X]$. It follows, therefore, that the multiple roots of $f(X)$ are precisely the roots of $d(X)$. ∎

COROLLARY. Let $f(X)$ be a nonconstant polynomial over a field k. Then every root of $f(X)$ is a simple root if and only if $f(X)$ and $f'(X)$ are relatively prime.

Proof. This follows immediately from Proposition 5. ∎

Thus, the formal derivative of a polynomial provides a convenient way to determine whether or not the polynomial has a multiple root without actually knowing the roots of the polynomial. For example, consider the polynomial $f(X) = X^5 + 2X^3 + X$ in $\mathbb{Q}[X]$ that we discussed earlier in this section. In this case $f'(X) = 5X^4 + 6X^2 + 1$, and we find by long division in $\mathbb{Q}[X]$ that

$$d(X) = \gcd(X^5 + 2X^3 + X, 5X^4 + 6X^2 + 1) = X^2 + 1.$$

Since $X^2 + 1$ has two roots in the field $\mathbb{C}$ of complex numbers, it follows that $f(X)$ has exactly two multiple roots, namely, i and $-i$; indeed, we find that $f(i) = f'(i) = 0$ and $f(-i) = f'(-i) = 0$, which agrees with Proposition 4. Finally, observe that 0 is a root of $f(X)$ but not of $f'(X)$ and hence is a simple root of $f(X)$, which agrees with our earlier result. Or consider the polynomial $f(X) = X^{p^n} - X \in \mathbb{F}_p[X]$ that we discussed in connection with finite fields, where p^n is an arbitrary prime power. In this case $f'(X) = p^n X^{p^n - 1} - 1 = -1$. Therefore, $f(X)$ and $f'(X)$ are relatively prime and hence $f(X)$ has p^n simple roots—a result that we verified explicitly in Example 8, Section 1.

Now, if $f(X)$ is a polynomial over a field k and $f'(X) = 0$, then every root of $f(X)$ is a root of $f'(X)$ and hence all roots of $f(X)$ are multiple roots. The following result describes all such polynomials, depending upon the characteristic of the ground field k.

PROPOSITION 6. Let $f(X)$ be a polynomial over a field k such that $f'(X) = 0$.

(1) If $\text{char}(k) = 0$, then $f(X)$ is constant; that is, $f(X) = a$ for some element $a \in k$.

(2) If char$(k) = p$, a prime, then $f(X)$ may be written as a polynomial in X^p; that is, $f(X) = g(X^p)$ for some polynomial $g(X) \in k[X]$.

Proof. Let $f(X) = a_0 + a_1 X + \cdots + a_n X^n$. Then $f'(X) = a_1 + 2a_2 + \cdots + na_n X^{n-1}$. Since $f'(X) = 0$, it follows that $a_1 = 2a_2 = \cdots = na_n = 0$. Now, if char$(k) = 0$, then $a_1 = a_2 = \cdots = a_n = 0$ and hence $f(X) = a_0$, a constant polynomial. But if char$(k) \neq 0$, then every coefficient a_s for which p does not divide s must be zero while the remaining coefficients are arbitrary. In this case $f(X)$ may be written as a polynomial in X^p, as required. ■

COROLLARY. An irreducible polynomial over a field of characteristic zero has only simple roots.

Proof. Let $f(X)$ be an irreducible polynomial over a field of characteristic zero. Then $f(X)$ is nonconstant and hence $f'(X) \neq 0$ by Proposition 6. Therefore, $\gcd(f(X), f'(X)) = 1$ since $f(X)$ is irreducible and hence $f(X)$ has only simple roots. ■

While irreducible polynomials over fields of characteristic zero cannot have multiple roots, the following example shows that this is no longer true over fields of nonzero characteristic.

EXAMPLE 3. An irreducible polynomial having a multiple root

Let $\mathbb{F}_2(t)$ stand for the field of rational functions in an indeterminant t over $\mathbb{F}_2$; recall that a typical element in $\mathbb{F}_2(t)$ has the form $\dfrac{a(t)}{b(t)}$, where $a(t)$, $b(t)$ are polynomials in t over $\mathbb{F}_2$ and $b(t) \neq 0$. Now, let $f(X) = X^2 - t \in \mathbb{F}_2(t)[X]$. We claim that $f(X)$ is an irreducible polynomial over $\mathbb{F}_2(t)$ that has a multiple root.

(A) Let us first show that $f(X)$ is irreducible over $\mathbb{F}_2(t)$. Suppose to the contrary that $f(X)$ is reducible. Then $f(X)$ has a root in the field $\mathbb{F}_2(t)$ and hence there are polynomials $a(t)$, $b(t) \in \mathbb{F}_2[t]$ such that

$$\left[\frac{a(t)}{b(t)}\right]^2 = t,$$

where $b(t) \neq 0$. Therefore, $a(t)^2 = tb(t)^2$. But this equation is impossible since every power of t occurring in $a(t)^2$ is an even power—since we are in characteristic 2, the square of a sum is the sum of the squares—while every power of t in $tb(t)^2$ is an odd power. Consequently no such polynomials exist and we conclude, therefore, that $f(X)$ is irreducible over $\mathbb{F}_2(t)$.

(B) To show that $f(X)$ has a multiple root we need only observe that $f'(X) = 2X = 0$ in $\mathbb{F}_2(t)[X]$, from which it follows that every root of $f(X)$ is a multiple root. Indeed, since $\deg f(X) = 2$, $f(X)$ has a single root of multiplicity 2.

(C) Let us conclude this example by constructing a splitting field for $f(X)$ over $\mathbb{F}_2(t)$ and verifying, explicitly, that $f(X)$ has a single root of multiplicity 2. Let $K = \mathbb{F}_2(t)[X]/(X^2 - t)\mathbb{F}_2(t)[X]$. Since $X^2 - t$ is irreducible over $\mathbb{F}_2(t)$, K is a finite field extension of $\mathbb{F}_2(t)$ with $[K:\mathbb{F}_2(t)] = 2$, and every element in K may be written uniquely in the form $a + bc$, where $a, b \in \mathbb{F}_2(t)$ and $c = X + (X^2 - t)\mathbb{F}_2(t)[X]$. Since $c^2 = t$, it follows that c is a root of $f(X)$ in K. Therefore, $f(X) = X^2 - t = (X - c)^2$ is the splitting of $f(X)$ in $K[X]$ and hence c is a root of $f(X)$ of multiplicity 2.

We now conclude this section by combining our results on splitting fields and multiple roots to prove the primitive element theorem, which shows that every finite extension of a field of characteristic zero has a primitive element and hence is a simple extension of the ground field. In the next chapter we will prove a more general form of the primitive element theorem that is independent of the characteristic and which characterizes precisely when an extension of fields has a primitive element.

PROPOSITION 7. (PRIMITIVE ELEMENT THEOREM) Every finite extension of a field of characteristic zero has a primitive element; that is, if K is a finite extension of a field k of characteristic zero, then there is some element $c \in K$ such that $K = k(c)$.

Proof.   Let $[K:k] = n$. We prove the statement by induction on n. If $n = 1$, then $K = k$ and hence any element in k is a primitive element for the extension. Now suppose that $n > 1$ and assume that any finite extension of fields K'/k' has a primitive element whenever $[K':k'] < n$ and $\operatorname{char}(k') = 0$. Since $[K:k] > 1$, there is some element $a \in K$ such that $a \notin k$. Therefore, $[K:k(a)] < n$ and hence by the induction assumption there is some element $b \in K$ such that $K = k(a, b)$. Thus, to complete the proof it remains to show that $k(a, b) = k(c)$ for some element $c \in k(a, b)$. Let us first construct such an element c and then verify that it has the required property.

(A) We begin by observing that since K is a finite extension of k, K is an algebraic extension of k and hence a and b are algebraic over k. Let $f(X) = \operatorname{Irr}(a;k)$ and $g(X) = \operatorname{Irr}(b;k)$. Furthermore, let K' be any extension of K over which both $f(X)$ and $g(X)$ split, and let $a_1, \ldots, a_n$ and $b_1, \ldots, b_m$ stand for the roots of $f(X)$ and $g(X)$, respectively, in K', where $a_1 = a$ and $b_1 = b$. Now, for each pair of roots a_i, b_j, where $b_j \neq b_1$, there is exactly one element $u_{ij} \in K'$ such that $a + bu_{ij} = a_i + b_j u_{ij}$. But the field k is infinite since $\operatorname{char}(k) = 0$. Thus, we may choose an element $u \in k$ such that $a + bu$ is not one of the elements $a_i + b_j u$ for all $i, j \neq 1$. Let $c = a + bu$.

(B) We now claim that $k(a, b) = k(c)$. For, since $c = a + bu$, where $u \in k$, it follows that $k(c) \subseteq k(a, b)$. To show the reverse inclusion, let us first show that $b \in k(c)$. To this end, let $d(X) = \gcd(f(c - uX), g(X))$ in $K[X]$. Since

$f(c - bu) = f(a) = 0 = g(b)$, b is a root of both $f(c - uX)$ and $g(X)$. Therefore, b is a root of $d(X)$ and hence $X - b$ divides $d(X)$. The element b is in fact the only root of $d(X)$. For the roots of $d(X)$ are among the b_j since $d(X)$ divides $g(X)$. But if b_j is a root of $d(X)$, then $c - ub_j$ is a root of $f(X)$ and hence $c - ub_j = a_i$ for some i. In this case $c = a_i + ub_j$, which contradicts the choice of c unless $b_j = b$. Thus, the only root of $d(X)$ is b and therefore $d(X) = (X - b)^s$ for some positive integer s. It follows that $(X - b)^s$ divides $g(X)$. But $g(X)$ is an irreducible polynomial over a field of characteristic zero and hence cannot have a multiple root. Therefore, $s = 1$ and hence $d(X) = \gcd(f(c - uX), g(X)) = X - b$. Thus, the polynomials $f(c - uX)$ and $g(X)$ are not relatively prime over the field K and consequently are not relatively prime over the subfield $k(c)$. Therefore, $\gcd(f(c - uX), g(X)) = X - b$ in $k(c)[X]$ and hence $b \in k(c)$, as required. Finally, since $a = c - bu$ and c, b and u are elements in $k(c)$, we conclude that $a \in k(c)$. It now follows that $k(a, b) \subseteq k(c)$ and hence $k(a, b) = k(c)$. The element c is therefore a primitive element for the extension K/k and the proof by induction is complete. ∎

Let us use the method described in the proof of the primitive element theorem to find a primitive element for the extension $\mathbb{Q}(\sqrt{2}, \sqrt{3}, \sqrt{5})$ of $\mathbb{Q}$. We begin by writing $\mathbb{Q}(\sqrt{2}, \sqrt{3}, \sqrt{5}) = \mathbb{Q}(\sqrt{2})(\sqrt{3}, \sqrt{5})$. Now,

$$\text{Irr}(\sqrt{3}; \mathbb{Q}(\sqrt{2})) = X^2 - 3$$

$$\text{Irr}(\sqrt{5}; \mathbb{Q}(\sqrt{2})) = X^2 - 5,$$

and the roots of these polynomials are, respectively,

$$a = a_1 = \sqrt{3}, \quad a_2 = -\sqrt{3},$$

$$b = b_1 = \sqrt{5}, \quad b_2 = -\sqrt{5}.$$

If we set $u = 1$, then $c = a + bu = \sqrt{3} + \sqrt{5}$ is not one of the elements $a_1 + b_2 u = \sqrt{3} - \sqrt{5}$ or $a_2 + b_2 u = -\sqrt{3} - \sqrt{5}$, and therefore $\mathbb{Q}(\sqrt{2})(\sqrt{3}, \sqrt{5}) = \mathbb{Q}(\sqrt{2}, \sqrt{3} + \sqrt{5})$. Repeating this procedure again, we find, for example, that $\mathbb{Q}(\sqrt{2}, \sqrt{3} + \sqrt{5}) = \mathbb{Q}(\sqrt{2} + \sqrt{3} + \sqrt{5})$. Hence, $\mathbb{Q}(\sqrt{2}, \sqrt{3}, \sqrt{5}) = \mathbb{Q}(\sqrt{2} + \sqrt{3} + \sqrt{5})$ and therefore $\sqrt{2} + \sqrt{3} + \sqrt{5}$ is a primitive element for the extension $\mathbb{Q}(\sqrt{2}, \sqrt{3}, \sqrt{5})/\mathbb{Q}$.

Exercises

1. Find all automorphisms of the field $\mathbb{Q}(\sqrt{3})$.
2. Find all automorphisms of the field $\mathbb{Q}(\omega)$, where ω is a primitive cube root of unity.
3. Let σ stand for the automorphism of the field $\mathbb{Q}(\sqrt{2})$ that maps a typical element $a + b\sqrt{2}$ to $a - b\sqrt{2}$. Find all automorphisms σ^* of the extension field $\mathbb{Q}(\sqrt{2}, i)$ that extend σ.

4. Let $K = \mathbb{Q}(\sqrt{2}, i)$ and $K' = \mathbb{Q}[X]/(X^4 + 1)\mathbb{Q}[X]$.
   (a) Show that K and K' are splitting fields for $X^4 + 1$ over $\mathbb{Q}$.
   (b) Construct an isomorphism $\sigma: K \to K'$.
5. Let $K = \mathbb{Q}(\sqrt{3}, \omega)$ and $K' = \mathbb{Q}(\sqrt{3})[X]/(X^2 - \sqrt{3}X + 3)\mathbb{Q}(\sqrt{3})[X]$, where ω stands for a primitive cube root of unity.
   (a) Show that K and K' are splitting fields for $X^6 - 27$ over $\mathbb{Q}$.
   (b) Construct an isomorphism $\sigma: K \to K'$.
6. Let $K = \mathbb{F}_2[X]/(X^3 + X^2 + 1)\mathbb{F}_2[X]$ and $K' = \mathbb{F}_2[X]/(X^3 + X + 1)\mathbb{F}_2[X]$.
   (a) Show that K and K' are finite fields containing eight elements.
   (b) Show that K and K' are splitting fields for the polynomial $X^3 + X + 1$ over $\mathbb{F}_2$.
   (c) Construct an isomorphism $\sigma: K \to K'$.
   (d) Verify that K and K' are splitting fields for the polynomial $X^8 - X$ over $\mathbb{F}_2$.
   (e) Factor the polynomial $X^8 - X$ into irreducible factors over $\mathbb{F}_2$.
   (f) Factor the polynomial $X^8 - X$ into irreducible factors over both K and K'.
7. Let K/k be a finite extension of fields having an element $c \in K$ as a primitive element. Show that every root of $\mathrm{Irr}(c; k)$ in K is also a primitive element for the extension K/k.
8. Let K/k be a finite extension of fields having a primitive element $c \in K$ and let $f(X) = \mathrm{Irr}(c; k)$.
   (a) Let k' be an intermediate subfield of K/k and let $g(X) = \mathrm{Irr}(c; k')$.
      (1) Show that $g(X)$ divides $f(X)$ in $k'[X]$.
      (2) If $g(X) = a_0 + a_1 X + \cdots + a_{n-1}X^{n-1} + X^n$, where $a_0, \ldots, a_{n-1} \in k'$, show that $k' = k(a_0, \ldots, a_{n-1})$.
   (b) Show that the extension K/k has only a finite number of intermediate subfields.
   (c) Show that the only intermediate subfields of the extension $\mathbb{Q}(\sqrt{2}, \sqrt{3})/\mathbb{Q}$ are $\mathbb{Q}$, $\mathbb{Q}(\sqrt{2})$, $\mathbb{Q}(\sqrt{3})$, $\mathbb{Q}(\sqrt{6})$, and $\mathbb{Q}(\sqrt{2}, \sqrt{3})$.
9. Let k be a field of characteristic zero and let K be any finite extension field of k. Show that the extension K/k has only a finite number of intermediate subfields.
10. Let $\mathbb{Q}(\pi)$ stand for the simple extension of $\mathbb{Q}$ generated by the real number π. Show that the extension $\mathbb{Q}(\pi)/\mathbb{Q}$ has an infinite number of intermediate subfields.
11. Let $f(X)$ be a nonconstant polynomial over a field k of characteristic zero and let K be a splitting field for $f(X)$ over k.
   (a) Show that the extension K/k has a primitive element.
   (b) Let c be a primitive element for the extension K/k.
      (1) If $a_1, \ldots, a_n$ stand for the roots of $f(X)$ in K, show that there are polynomials $g_1(X), \ldots, g_n(X)$ in $k[X]$ such that $g_1(c) = a_1, \ldots, g_n(c) = a_n$. Thus, the roots of $f(X)$ may be expressed as polynomials in the primitive element c.

(2) Let b stand for any root of $\mathrm{Irr}(c;k)$ in any extension field of k. Show that $g_1(b), \ldots, g_n(b)$ are roots of $f(X)$, where $g_1(X), \ldots, g_n(X)$ are the polynomials in part (1).

(c) Let $f(X) = X^4 - 5X^2 + 6 \in \mathbb{Q}[X]$.

(1) Show that $\mathbb{Q}(\sqrt{2}, \sqrt{3})$ is a splitting field for $f(X)$ over $\mathbb{Q}$.

(2) The element $c = \sqrt{2} + \sqrt{3}$ is a primitive element for the extension $\mathbb{Q}(\sqrt{2}, \sqrt{3})$ over $\mathbb{Q}$. Find the roots a_1, a_2, a_3, a_4 of $f(X)$ in $\mathbb{Q}(\sqrt{2}, \sqrt{3})$, and find polynomials $g_1(X), g_2(X), g_3(X)$, and $g_4(X)$ in $\mathbb{Q}[X]$ such that $a_i = g_i(\sqrt{2} + \sqrt{3})$ for $i = 1, 2, 3, 4$.

(3) Show that $\sqrt{2} - \sqrt{3}$ is a root of $\mathrm{Irr}(\sqrt{2} + \sqrt{3};\mathbb{Q})$ and verify by direct calculation that $g_i(\sqrt{2} - \sqrt{3})$ is a root of $f(X)$ for $i = 1, 2, 3, 4$, where the $g_i(X)$ are the polynomials in part (2).

12. **A field extension that does not have a primitive element.** Let $\mathbb{F}_2[s, t]$ stand for the ring of polynomials in two indeterminants s and t over the field $\mathbb{F}_2$, and let $k = \mathbb{F}_2(s, t) = QF(\mathbb{F}_2[s, t])$ stand for the quotient field of $\mathbb{F}_2[s, t]$; a typical element in k is a rational function in s and t with coefficients in $\mathbb{F}_2$. Let $f(X) = (X^2 - s)(X^2 - t) \in k[X]$ and let K be a splitting field for $f(X)$ over k.

(a) Show that there are elements $a, b \in K$ such that $K = k(a, b)$, where $\mathrm{Irr}(a;k) = X^2 - s$ and $\mathrm{Irr}(b;k) = X^2 - t$.

(b) Show that $[K:k] = 4$ and that the elements 1, a, b, and ab form a basis for K as a vector space over k.

(c) Show that the extension K/k does not have a primitive element.

13. Determine which of the following polynomials have multiple roots:

(a) $f(X) = X^3 - X^2 - X + 1 \in \mathbb{Q}[X]$

(b) $f(X) = 3X^2 + 12X - 4 \in \mathbb{Q}[X]$

(c) $f(X) = X^3 - (3 + \sqrt{2})X^2 + (1 + 2\sqrt{2})X + (1 + \sqrt{2}) \in \mathbb{Q}(\sqrt{2})[X]$

(d) $f(X) = X^4 + 2X^3 + X + 2 \in \mathbb{F}_3[X]$

(e) $f(X) = 2X^4 - 4X^3 - X^2 + 3X + 3 \in \mathbb{F}_5[X]$

14. Let $f(X)$ be a nonconstant polynomial of degree n over a field k of characteristic zero and let K be a splitting field for $f(X)$ over k. For each positive integer s, define the sth formal derivative $f^{(s)}(X)$ of $f(X)$ inductively by setting $f^{(s)}(X) = [f^{(s-1)}(X)]'$. Let $c \in K$.

(a) Show that $f(X) = f(c) + f'(c)(X - c) + \dfrac{f''(c)}{2!}(X - c)^2 + \cdots$

$$+ \frac{f^{(n)}(c)}{n!}(X - c)^n \text{ in } K[X].$$

(b) Show that c is a root of $f(X)$ of multiplicity s if and only if $f(c) = f'(c) = \cdots = f^{(s-1)}(c) = 0$, but $f^{(s)}(c) \neq 0$.

(c) Show that c is a root of $f(X)$ of multiplicity s if and only if $f(X) = (X - c)^s q(X)$ for some polynomial $q(X) \in K[X]$ such that $q(c) \neq 0$.

(d) Let $f(X) = X^5 + 2X^3 + X \in \mathbb{Q}[X]$.

(1) Show that i is a root of $f(X)$ of multiplicity 2.

(2) Calculate $f(i), f'(i), \ldots, f^{(5)}(i)$ and write $f(X)$ as a polynomial in $X - i$ over $\mathbb{Q}(i)$.

(3) Find a polynomial $q(X) \in \mathbb{Q}(i)[X]$ such that $f(X) = (X - i)^2 q(X)$, where $q(i) \neq 0$.

15. Let $f(X) \in \mathbb{F}_p[X]$ and suppose that $f'(X) = 0$.

(a) Show that $f(X) = g(X)^p$ for some polynomial $g(X) \in \mathbb{F}_p[X]$.

(b) Show that the multiplicity of every root of $f(X)$ is a multiple of p.

(c) Let $f(X) = 2X^{10} - 3X^5 + 4 \in \mathbb{F}_5[X]$. Find a polynomial $g(X) \in \mathbb{F}_5[X]$ such that $f(X) = g(X)^5$, and find the distinct roots of $f(X)$ and their multiplicities.

14

The Galois Theory of Fields

Up to this point in our study of field theory, we have concentrated on the construction and properties of specific fields. We have used polynomials to construct new fields and then used these fields to study the roots of the polynomials. In this chapter we turn our attention to a more general and consequently more powerful point of view. Our aim, briefly, is to associate a group with every field extension and then discuss the relationship between the properties of the group and the structure of the field. Let us take a moment to describe these ideas and results in more detail.

Let K/k be an extension of fields and let G stand for the set of k-automorphisms of K, that is, those automorphisms of K that leave every element in k fixed. Then G is a group under function composition called the automorphism group of the extension K/k. Our main goal in the first section is to show that if K/k is a finite extension of fields, then its automorphism group G is a finite group whose order is at most equal to the degree of the extension.

Now, if H is a subgroup of G, then the collection K^H of elements in K that are fixed by every automorphism in H is a subfield of K containing k and hence is an intermediate subfield of the extension K/k. Thus, we obtain a correspondence $H \mapsto K^H$ between the subgroups of G and the intermediate subfields of K/k which is called the Galois correspondence for the extension K/k. It is this correspondence and its properties that lie at the core of the Galois theory of fields. We will show that under the Galois correspondence, $|H| = [K:K^H]$ for every subgroup H of G. It follows, then, that $|G| = [K:K^G]$ and hence that the number of k-automorphisms of K is equal to the degree of K over the fixed field K^G. In general, the fixed field K^G is larger than the ground field k; that is, there are, in general, elements in K that are fixed by every k-automorphism other than those in k. In fact, since $|G| = [K:K^G]$,

$K^G = k$ if and only if $|G| = [K:k]$. Therefore, the number of k-automorphisms of the extension K/k is its maximum number, namely, $[K:k]$, if and only if $K^G = k$, that is, when the elements in the ground field k are the only elements fixed by every such automorphism. An extension of fields having this property is called a normal extension. In the second section of the chapter we characterize and discuss the basic properties of normal field extensions. Our main goal here is to show that a finite extension K/k is a normal extension if and only if K is the splitting field for some polynomial over k all of whose irreducible factors have only simple roots. It then follows, for example, that all splitting fields over a field of characteristic zero are normal extensions since, as we saw in the previous chapter, an irreducible polynomial over such a field has only simple roots.

In the third and final section of the chapter we bring together our results on automorphism groups and normal field extensions in order to establish the basic properties of the Galois correspondence for normal extensions. We begin by showing that if K/k is a normal extension of fields with automorphism group G and H is any subgroup of G, then not only is the fixed field K^H an intermediate subfield of the extension K/k, but in fact every intermediate subfield of K/k has the form K^H for some subgroup H of G. It follows, then, that the Galois correspondence $H \mapsto K^H$ is a 1–1 order-reversing correspondence between the subgroups of G and the intermediate subfields of K/k. Moreover, under this correspondence a subgroup H is a normal subgroup of G if and only if the corresponding subfield K^H is a normal extension of k, and, in this case, the quotient group G/H is isomorphic to the automorphism group of the extension K^H/k. These results are known collectively as the fundamental theorem of Galois theory and are exceptionally important since they show that the Galois correspondence for normal field extensions provides a mirror that faithfully reflects the intermediate subfield structure of the extension K/k onto the subgroup structure of its automorphism group G. As such, the Galois correspondence provides a way to transfer questions set within the framework of field theory into questions about finite groups.

We will illustrate these ideas and results by discussing the Galois theory of finite fields, cyclotomic extensions and roots of unity, symmetric polynomials, and then conclude by using the Galois theory together with the Sylow theory to give a proof, due to Emil Artin, of the fundamental theorem of algebra. In the supplementary chapters that follow we use the Galois theory to solve a number of classical problems that may be modeled within the framework of field theory, from Euclidean constructibility, to the solvability of algebraic equations by means of radicals—a problem which led, in fact, to the development of Galois theory. With its mutual interaction between field theory and group theory, it is no exaggeration to say that the Galois theory of fields is one of the most important and useful algebraic theories in mathematics today.

1. THE AUTOMORPHISM GROUP OF A FIELD EXTENSION

Let K be a field. We begin by recalling that an automorphism of K is any function $\sigma: K \to K$ that satisfies the following three conditions:

(1) σ is 1–1 and onto;
(2) $\sigma(x + y) = \sigma(x) + \sigma(y)$ for all elements $x, y \in K$;
(3) $\sigma(xy) = \sigma(x)\sigma(y)$ for all elements $x, y \in K$.

The identity map $1: K \to K$, for example, is an automorphism of any field K; it is, in fact, the only automorphism of the field $\mathbb{Q}$ of rational numbers and the field $\mathbb{F}_p$ of integers modulo a prime p since these fields are generated by the identity element 1 and $\sigma(1) = 1$. On the other hand, the conjugation mapping $\sigma: \mathbb{C} \to \mathbb{C}$ that maps an arbitrary complex number $a + bi$ to its complex conjugate $a - bi$ is a nontrivial automorphism of the field $\mathbb{C}$ of complex numbers.

Definition

Let K/k be an extension of fields. A *k-automorphism* of K is any automorphism $\sigma: K \to K$ such that $\sigma(a) = a$ for every element $a \in k$. We let $G(K/k)$ stand for the set of k-automorphisms of K.

Thus, a k-automorphism of an extension of fields K/k is any automorphism of K that fixes every element in the ground field k. The identity automorphism $1: K \to K$ is a k-automorphism of K, for example, and the conjugation automorphism $\sigma: \mathbb{C} \to \mathbb{C}$ is an $\mathbb{R}$-automorphism of $\mathbb{C}$ since $\sigma(a) = a$ for every real number $a \in \mathbb{R}$. Let us also note that if k is the prime subfield of a field K, then every automorphism of K fixes every element in k and hence is a k-automorphism of K; in this case $G(K/k)$ is the set of all automorphisms of the field K.

PROPOSITION 1. Let K/k be an extension of fields. Then the set $G(K/k)$ of k-automorphisms of K is a group under function composition. Moreover, if K is a finite simple extension of degree n over k, then $G(K/k)$ is a finite group and $|G(K/k)| \leq n$.

Proof. As we noted earlier, the identity automorphism $1: K \to K$ is a k-automorphism of K and hence is an element of $G(K/k)$. Now, let σ and τ be k-automorphisms of K. Then the composition $\sigma\tau$ and inverse σ^{-1} are also automorphisms of K. Moreover, $\sigma(a) = \tau(a) = a$ for all elements $a \in k$ and therefore $\sigma\tau(a) = \sigma(\tau(a)) = \sigma(a) = a$ and $\sigma^{-1}(a) = a$. Hence, $\sigma\tau$ and σ^{-1} are k-automorphisms of K and are therefore elements of $G(K/k)$. It follows that $G(K/k)$ is a group under function composition. Suppose, finally, that K is a finite simple extension of degree n over k and let $K = k(c)$, where $c \in K$. Then c is algebraic over k. Let $f(X) = \operatorname{Irr}(c;k) = a_0 + a_1 X + \cdots + a_{n-1} X^{n-1} + X^n$, where $a_0, \ldots, a_{n-1} \in k$. Now, if σ is any k-automorphism of K, then $\sigma(c)$ is a root of $f(X)$ since

$$f(\sigma(c)) = a_0 + a_1\sigma(c) + \cdots + a_{n-1}\sigma(c)^{n-1} + \sigma(c)^n$$
$$= \sigma(a_0) + \sigma(a_1)\sigma(c) + \cdots + \sigma(a_{n-1})\sigma(c^{n-1}) + \sigma(c^n)$$
$$= \sigma(a_0 + a_1 c + \cdots + a_{n-1}c^{n-1} + c^n)$$
$$= 0.$$

Moreover, σ is completely determined by the image $\sigma(c)$; for if τ is any k-automorphism of K such that $\tau(c) = \sigma(c)$ and $x = b_0 + b_1 c + \cdots + b_{n-1}c^{n-1}$ is a typical element in K, where $b_0, \ldots, b_{n-1} \in k$, then

$$\tau(x) = \tau(b_0) + \tau(b_1)\tau(c) + \cdots + \tau(b_{n-1})\tau(c)^{n-1}$$
$$= b_0 + b_1\tau(c) + \cdots + b_{n-1}\tau(c)^{n-1}$$
$$= b_0 + b_1\sigma(c) + \cdots + b_{n-1}\sigma(c)^{n-1}$$
$$= \sigma(x).$$

Therefore, $\tau = \sigma$ and hence every k-automorphism $\sigma \in G(K/k)$ is uniquely determined by the image $\sigma(c)$. Since $\sigma(c)$ is a root of $f(X)$ and since $f(X)$ has at most n roots in K, it now follows that there are at most n k-automorphisms of K. Thus, $|G(K/k)| \leq n$ and the proof is complete. ∎

The group $G(K/k)$ of k-automorphisms of K is called the *automorphism group* of the extension K/k. Let us emphasize that the field K may have many automorphisms in addition to those in its automorphism group since $G(K/k)$ contains only those automorphisms that fix every element in k. But if k is the prime subfield of K, then, as we noted earlier, every automorphism of K is a k-automorphism and hence $G(K/k)$ is the group of all automorphisms of K.

Now, if $K = k(c)$ is a finite simple extension of degree n over k, then, as we showed in Proposition 1, the automorphism group $G(K/k)$ is a finite group whose order is at most n. In this case a k-automorphism σ of $k(c)$ is completely determined by the image $\sigma(c)$ and $\sigma(c)$ must be a root of the polynomial $\operatorname{Irr}(c;k)$ in $k(c)$. We claim, in fact, that there is exactly one k-automorphism of $k(c)$ corresponding to each root of $\operatorname{Irr}(c;k)$ in $k(c)$. For recall from the previous chapter that if c' is any root of $\operatorname{Irr}(c;k)$ in $k(c)$, then there is at least one automorphism of $k(c)$ that maps c onto c' and fixes every element in k, namely the extension $\sigma_{c,c'}^*$ of the identity automorphism on k. Since this automorphism is completely determined by the mapping $c \mapsto c'$, it follows that there is exactly one k-automorphism of $k(c)$ corresponding to each root of $\operatorname{Irr}(c;k)$ in $k(c)$ and hence $|G(K/k)| \leq n$, where $n = \deg \operatorname{Irr}(c;k) = [K:k]$. In particular, if $\operatorname{Irr}(c;k)$ splits over $k(c)$, then $|G(K/k)| = n$, the degree of the extension. Later in this section we will show that $|G(K/k)| \leq [K:k]$ for any finite extension of fields K/k, whether it is a simple extension or not, and, in the next section, we discuss in detail those extensions for which $|G(K/k)| = [K:k]$.

Consider the field $\mathbb{C}$ of complex numbers. Since $\mathbb{C} = \mathbb{R}(i)$, $\mathbb{C}$ is a finite simple extension of degree 2 over $\mathbb{R}$ and hence has at most two $\mathbb{R}$-automorphisms. But we already know two distinct $\mathbb{R}$-automorphisms of $\mathbb{C}$, namely the identity automorphism $1:\mathbb{C} \to \mathbb{C}$ and the conjugation automorphism $\sigma:\mathbb{C} \to \mathbb{C}$. Therefore, $G(\mathbb{C}/\mathbb{R}) = \{1, \sigma\}$ and hence $G(\mathbb{C}/\mathbb{R})$ is a cyclic group of order 2.

Before we discuss some other examples of automorphism groups, let us first introduce the concept of the fixed field of a group of automorphisms. If K/k is an extension of fields, any subgroup of the automorphism group $G(K/k)$ is called a *group of k-automorphisms of K*. If H is a group of k-automorphisms of K, let $K^H = \{x \in K \mid \sigma(x) = x \text{ for all } \sigma \in H\}$ stand for the set of elements in K that are fixed by every k-automorphism in H.

PROPOSITION 2. Let K/k be an extension of fields and let H be a group of k-automorphisms of K. Then K^H is an intermediate subfield of the extension K/k.

Proof.   First observe that $k \subseteq K^H$ since every automorphism in H fixes every element in k. Therefore, K^H is a nonempty subset of K. Now, let x and y be typical elements in K^H and let $\sigma \in H$. Then $\sigma(x) = x$ and $\sigma(y) = y$, and therefore

$$\sigma(x + y) = \sigma(x) + \sigma(y) = x + y,$$

$$\sigma(xy) = \sigma(x)\sigma(y) = xy,$$

$$\sigma(-x) = -\sigma(x) = -x,$$

and, if $x \neq 0$,

$$\sigma\left(\frac{1}{x}\right) = \frac{1}{\sigma(x)} = \frac{1}{x}.$$

It follows that $x + y$, xy, $-x$, and $\dfrac{1}{x}$, if $x \neq 0$, are fixed by every automorphism in H and hence are elements in K^H. Therefore, K^H is an intermediate subfield of the extension K/k. ∎

The subfield K^H is called the *fixed field* of the group H of k-automorphisms. In particular, if $G = G(K/k)$ is the automorphism group of the extension K/k, then K^G is the fixed field of the entire automorphism group $G(K/k)$ and consists of those elements in K that are fixed by every k-automorphism of K. Let us emphasize that although every element in the ground field k is fixed by every k-automorphism of K, and hence $k \subseteq K^G$, the fixed field K^G is, in general, larger than k; that is, there may be elements in K in addition to those in k that are fixed by every k-automorphism of K. The following examples illustrate these ideas in more detail. As we discuss these examples, let us pay particular attention to the relationship between the order of the automorphism group $G(K/k)$, the degree of the extension K/k, and to whether or not the fixed field $K^{G(K/k)}$ is equal to the ground field k.

EXAMPLE 1

Let $\mathbb{Q}(i)$ stand for the field of complex numbers whose real and imaginary parts are rational and let $G = G(\mathbb{Q}(i)/\mathbb{Q})$. Let us find the automorphisms in the group G and the fixed field $\mathbb{Q}(i)^G$.

(A) If σ is a $\mathbb{Q}$-automorphism of $\mathbb{Q}(i)$, then $\sigma(i)$ must be a root of the polynomial $X^2 + 1$ since $\mathrm{Irr}(i;\mathbb{Q}) = X^2 + 1$. Hence, $\sigma(i) = i$ or $\sigma(i) = -i$. Moreover, each of these choices does in fact define a $\mathbb{Q}$-automorphism of $\mathbb{Q}(i):\sigma(i) = i$ defines the identity automorphism $1:\mathbb{Q}(i) \to \mathbb{Q}(i)$, while $\sigma(i) = -i$ defines the conjugation automorphism $\sigma:\mathbb{Q}(i) \to \mathbb{Q}(i)$ that maps a typical element $a + bi$ to its conjugate $a - bi$. Therefore, $G(\mathbb{Q}(i)/\mathbb{Q}) = \{1, \sigma\}$ and hence is a cyclic group of order 2. In this case $|G(\mathbb{Q}(i)/\mathbb{Q})| = [\mathbb{Q}(i):\mathbb{Q}] = 2$.

(B) We claim that the fixed field $\mathbb{Q}(i)^G = \mathbb{Q}$. For suppose that $x = a + bi$ is an element in $\mathbb{Q}(i)$ fixed by every $\mathbb{Q}$-automorphism of $\mathbb{Q}(i)$. Then $\sigma(x) = x$, where σ is the conjugation automorphism, and hence $a - bi = a + bi$. It follows that $b = 0$ and a is arbitrary. Therefore, $\mathbb{Q}(i)^G = \{a + bi \in \mathbb{Q}(i) \,|\, b = 0\} = \mathbb{Q}$. Thus, the only elements in $\mathbb{Q}(i)$ that are fixed by every $\mathbb{Q}$-automorphism of $\mathbb{Q}(i)$ are the elements in the ground field $\mathbb{Q}$.

EXAMPLE 2

Consider the field $\mathbb{Q}(\sqrt{2})$. Since $\mathrm{Irr}(\sqrt{2};\mathbb{Q}) = X^2 - 2$ and since this polynomial has two roots in $\mathbb{Q}(\sqrt{2})$, namely $\sqrt{2}$ and $-\sqrt{2}$, it follows that $\mathbb{Q}(\sqrt{2})$ has two $\mathbb{Q}$-automorphisms: the identity automorphism 1, and the automorphism σ mapping a typical element $a + b\sqrt{2}$ to $a - b\sqrt{2}$. Therefore, $G(\mathbb{Q}(\sqrt{2})/\mathbb{Q}) = \{1, \sigma\}$ and is a cyclic group of order 2, and $|G(\mathbb{Q}(\sqrt{2})/\mathbb{Q})| = [\mathbb{Q}(\sqrt{2}):\mathbb{Q}] = 2$. Moreover, we find as in Example 1 that the fixed field $\mathbb{Q}(\sqrt{2})^{G(\mathbb{Q}(\sqrt{2})/\mathbb{Q})} = \mathbb{Q}$.

EXAMPLE 3

Consider the field $\mathbb{Q}(\omega)$, where ω stands for a primitive cube root of unity. Since $\mathrm{Irr}(\omega;\mathbb{Q}) = X^2 + X + 1$, whose roots in $\mathbb{Q}(\omega)$ are ω and ω^2, it follows that the field $\mathbb{Q}(\omega)$ has two $\mathbb{Q}$-automorphisms: the identity automorphism 1, and the automorphism σ mapping a typical element $a + b\omega$ to $a + b\omega^2$. Hence, $G(\mathbb{Q}(\omega)/\mathbb{Q}) = \{1, \sigma\}$ and is a cyclic group of order 2, $|G(\mathbb{Q}(\omega)/\mathbb{Q})| = [\mathbb{Q}(\omega):\mathbb{Q}] = 2$, and, as before, $\mathbb{Q}(\omega)^{G(\mathbb{Q}(\omega)/\mathbb{Q})} = \mathbb{Q}$.

Figure 1 summarizes the three preceding examples. Observe that in each case the order of the automorphism group is equal to the degree of the extension, which is its largest possible value, and that the fixed field of the automorphism group is just the ground field $\mathbb{Q}$, which is the smallest possible subfield.

$G(\mathbb{Q}(i)/\mathbb{Q}) = \{1, \sigma\}$	$G(\mathbb{Q}(\sqrt{2})/\mathbb{Q}) = \{1, \sigma\}$	$G(\mathbb{Q}(\omega)/\mathbb{Q}) = \{1, \sigma\}$
$1 : a + bi \mapsto a + bi$	$1 : a + b\sqrt{2} \mapsto a + b\sqrt{2}$	$1 : a + b\omega \mapsto a + b\omega$
$\sigma : a + bi \mapsto a - bi$	$\sigma : a + b\sqrt{2} \mapsto a - b\sqrt{2}$	$\sigma : a + b\omega \mapsto a + b\omega^2$
$|G(\mathbb{Q}(i)/\mathbb{Q})| = 2$	$|G(\mathbb{Q}(\sqrt{2})/\mathbb{Q})| = 2$	$|G(\mathbb{Q}(\omega)/\mathbb{Q})| = 2$
$[\mathbb{Q}(i):\mathbb{Q}] = 2$	$[\mathbb{Q}(\sqrt{2}):\mathbb{Q}] = 2$	$[\mathbb{Q}(\omega):\mathbb{Q}] = 2$
$\mathbb{Q}(i)^{G(\mathbb{Q}(i)/\mathbb{Q})} = \mathbb{Q}$	$\mathbb{Q}(\sqrt{2})^{G(\mathbb{Q}(\sqrt{2})/\mathbb{Q})} = \mathbb{Q}$	$\mathbb{Q}(\omega)^{G(\mathbb{Q}(\omega)/\mathbb{Q})} = \mathbb{Q}$

FIGURE 1. The automorphism groups of selected field extensions, their orders and fixed fields.

EXAMPLE 4

Let $K = \mathbb{Q}(\sqrt[3]{2})$. We claim that the automorphism group $G(K/\mathbb{Q})$ is trivial and that the fixed field $K^{G(K/\mathbb{Q})} = K$. For if σ is a $\mathbb{Q}$-automorphism of K, then $\sigma(\sqrt[3]{2})$ must be a root of the polynomial $X^3 - 2$ in K since $\text{Irr}(\sqrt[3]{2}; \mathbb{Q}) = X^3 - 2$. But the roots of $X^3 - 2$ are $\sqrt[3]{2}$, $\omega \sqrt[3]{2}$, and $\omega^2 \sqrt[3]{2}$, and of these only $\sqrt[3]{2}$ lies in K. Therefore, $\sigma(\sqrt[3]{2}) = \sqrt[3]{2}$ and hence σ is the identity automorphism. Thus, $G(K/\mathbb{Q}) = \{1\}$, the trivial group, and hence the fixed field $K^{G(K/\mathbb{Q})} = K$ since the identity automorphism fixes all elements in K. In this case $|G(K/\mathbb{Q})| = 1 < [K:\mathbb{Q}] = 3$.

EXAMPLE 5

Let $K = \mathbb{Q}(\omega, \sqrt[3]{2})$, where ω stands for a primitive cube root of unity, and let $G = G(K/\mathbb{Q})$. Let us determine the automorphism group G, show that it is isomorphic to the symmetric group $\text{Sym}(3)$, and then find the fixed subfield of K corresponding to each subgroup of G.

(A) To construct the automorphisms of K, we extend the two automorphisms of $\mathbb{Q}(\omega)$ up to K. Let $1 : \mathbb{Q}(\omega) \to \mathbb{Q}(\omega)$ stand for the identity automorphism on the subfield $\mathbb{Q}(\omega)$. Since K is a splitting field for $X^3 - 2$ over $\mathbb{Q}(\omega)$, the automorphism 1 extends to a $\mathbb{Q}$-automorphism $\sigma : K \to K$ that maps the root $\sqrt[3]{2}$ onto the root $\omega \sqrt[3]{2}$. The map σ is therefore a $\mathbb{Q}$-automorphism of K that maps ω onto ω and $\sqrt[3]{2}$ onto $\omega \sqrt[3]{2}$, and is completely determined by these two images since $\sqrt[3]{2}$ and $\omega \sqrt[3]{2}$ generate K over $\mathbb{Q}$. We indicate this by writing

$$\sigma : \begin{array}{l} \sqrt[3]{2} \to \omega \sqrt[3]{2} \\ \omega \to \omega \end{array}.$$

Similarly, by extending the automorphism of $\mathbb{Q}(\omega)$ that maps ω onto ω^2, we obtain a $\mathbb{Q}$-automorphism τ of K such that

$$\tau : \begin{array}{l} \sqrt[3]{2} \to \sqrt[3]{2} \\ \omega \to \omega^2 \end{array}.$$

Now, since G contains σ and τ and is a group under function composition, it must also contain the six automorphisms 1, σ, σ^2, τ, $\sigma\tau$, and $\tau\sigma$. We find that:

$$1:\begin{matrix} \sqrt[3]{2} \to \sqrt[3]{2} \\ \omega \to \omega \end{matrix} \qquad \sigma:\begin{matrix} \sqrt[3]{2} \to \omega\sqrt[3]{2} \\ \omega \to \omega \end{matrix} \qquad \sigma^2:\begin{matrix} \sqrt[3]{2} \to \omega^2\sqrt[3]{2} \\ \omega \to \omega \end{matrix}$$

$$\tau:\begin{matrix} \sqrt[3]{2} \to \sqrt[3]{2} \\ \omega \to \omega^2 \end{matrix} \qquad \sigma\tau:\begin{matrix} \sqrt[3]{2} \to \omega\sqrt[3]{2} \\ \omega \to \omega^2 \end{matrix} \qquad \tau\sigma:\begin{matrix} \sqrt[3]{2} \to \omega^2\sqrt[3]{2} \\ \omega \to \omega^2 \end{matrix}$$

These automorphisms represent six distinct $\mathbb{Q}$-automorphisms of K. To show that there are no other such automorphisms, we reason as follows. Let α be any $\mathbb{Q}$-automorphism of K. Then $\alpha(\sqrt[3]{2})$ must be a root of $X^3 - 2$, for which there are three possibilities, and $\alpha(\omega)$ must be a root of $X^2 + X + 1$, for which there are two possibilities. Thus, there are at most six possibilities for α since α is completely determined by the images of $\sqrt[3]{2}$ and ω, and hence α is one of the six automorphisms listed above. We conclude, therefore, that $G = \{1, \sigma, \sigma^2, \tau, \sigma\tau, \tau\sigma\}$. In this case $|G| = [K:\mathbb{Q}] = 6$.

(B) To show that $G \cong \text{Sym}(3)$, first observe that every automorphism in G permutes the three roots $\sqrt[3]{2}$, $\omega\sqrt[3]{2}$, and $\omega^2\sqrt[3]{2}$ of $X^3 - 2$ and is uniquely determined by its permutation of these roots. The group G may therefore be regarded as a subgroup of $\text{Sym}(\{\sqrt[3]{2}, \omega\sqrt[3]{2}, \omega^2\sqrt[3]{2}\})$. Since both of these groups have order 6, it follows that $G \cong \text{Sym}(\{\sqrt[3]{2}, \omega\sqrt[3]{2}, \omega^2\sqrt[3]{2}\}) \cong \text{Sym}(3)$.

(C) Since G is isomorphic to the symmetric group $\text{Sym}(3)$, G has six subgroups: (1), (σ), (τ), $(\sigma\tau)$, $(\tau\sigma)$, and G. Let us conclude this example by finding the fixed subfield of K corresponding to each of these subgroups. Let $H = (\sigma) = \{1, \sigma, \sigma^2\}$. Then an element $x \in K$ is fixed by every automorphism in H if and only if $\sigma(x) = x$. Now, if

$$x = a + b\sqrt[3]{2} + c\sqrt[3]{4} + d\omega + e\omega\sqrt[3]{2} + f\omega\sqrt[3]{4}$$

is a typical element in K, where $a, b, c, d, e, f \in \mathbb{Q}$, we find that

$$\sigma(x) = a + b(\omega\sqrt[3]{2}) + c(\omega^2\sqrt[3]{4}) + d(\omega) + e(\omega^2\sqrt[3]{2}) + f(\sqrt[3]{4})$$
$$= a + (-e)\sqrt[3]{2} + (-c + f)\sqrt[3]{4} + d\omega + (b - e)\omega\sqrt[3]{2} + (-c)\omega\sqrt[3]{4},$$

since $\omega^2 = -1 - \omega$. Hence, $\sigma(x) = x$ if and only if $b = -e$, $c = -c + f$, $e = b - e$, and $f = -c$, from which it follows that $b = c = e = f = 0$, with a and d arbitrary. Therefore, $K^H = \{a + d\omega \in K \mid a, d \in \mathbb{Q}\} = \mathbb{Q}(\omega)$. Similarly, we find that

$$K^G = \mathbb{Q} \qquad\qquad K^{(\sigma\tau)} = \mathbb{Q}(\omega^2\sqrt[3]{2})$$

$$K^{(\sigma)} = \mathbb{Q}(\omega) \qquad\qquad K^{(\tau\sigma)} = \mathbb{Q}(\omega\sqrt[3]{2})$$

$$K^{(\tau)} = \mathbb{Q}(\sqrt[3]{2}) \qquad\qquad K^{(1)} = K.$$

This concludes the example.

In the following example we combine our knowledge of finite fields and our results on automorphism groups to determine the automorphism group of a finite field.

EXAMPLE 6. The automorphism group of a finite field

Let K be a finite field of prime characteristic p and let $[K:k] = n$, where k stands for the prime subfield of K. Define the function $\sigma_p : K \to K$ by setting $\sigma_p(x) = x^p$ for every element $x \in K$. We claim that σ_p is a k-automorphism of K, that $|\sigma_p| = n$, and that the automorphism group $G(K/k)$ is a cyclic group of order n generated by σ_p.

(A) Let us begin by showing that σ_p is a k-automorphism of K. Let x and y be typical elements in K. Then $\sigma_p(x) = x^p$ and $\sigma_p(y) = y^p$, and therefore

$$\sigma_p(x + y) = (x + y)^p = x^p + y^p = \sigma_p(x) + \sigma_p(y)$$

$$\sigma_p(xy) = (xy)^p = x^p y^p = \sigma_p(x)\sigma_p(y).$$

Thus, σ_p preserves addition and multiplication on K. Now, if $\sigma_p(x) = 0$ for some element $x \in K$, then $x = 0$; hence σ_p is 1–1. Moreover, since every 1–1 mapping of a finite set into itself is necessarily onto, it follows that σ_p is also an onto mapping. Therefore, σ_p is an automorphism of K. Finally, σ_p fixes every element in the ground field k since k is the prime subfield of K. Thus, σ_p is a k-automorphism of K. The automorphism σ_p is called the *Frobenius automorphism of K*.

(B) We show next that $|\sigma_p| = n$. To do this, recall that the finite field K is a splitting field for the polynomial $X^{p^n} - X$ over k and that the elements in K are the roots of this polynomial. It follows that $\sigma_p^n(x) = x^{p^n} = x$ for every element $x \in K$ and hence $\sigma_p^n = 1$, the identity automorphism on K. Therefore, $|\sigma_p| \le n$. On the other hand, if $s = |\sigma_p|$, then $\sigma_p^s = 1$ and hence $\sigma_p^s(x) = x^{p^s} = x$ for all elements $x \in K$. But this means that all p^n elements in K are roots of the polynomial $X^{p^s} - X$, which is impossible unless $s \ge n$. Therefore, $|\sigma_p| = n$.

(C) Finally, to show that $G(K/k) = (\sigma_p)$, the cyclic subgroup generated by σ_p, observe that $(\sigma_p) \le G(K/k)$. Therefore, $n \le |G(K/k)|$ since $|\sigma_p| = n$. On the other hand, K is a simple extension of k since the multiplicative group K^* of nonzero elements in K is a cyclic group and any generator for K^* is clearly a primitive element for K. Therefore, $|G(K/k)| \le [K:k] = n$. It now follows that $|G(K/k)| = n$ and therefore $G(K/k) = (\sigma_p) = \{1, \sigma_p, \dots, \sigma_p^{n-1}\}$. Thus, the automorphism group of a finite field over its prime field is a cyclic group of order n generated by the Frobenius automorphism, where n is the degree of the extension.

(D) Let us illustrate these results by finding the automorphism group of the extension $K/\mathbb{F}_3$, where

$$K = \mathbb{F}_3[X]/(X^2 + 1)\mathbb{F}_3[X].$$

In this case $[K:\mathbb{F}_3] = 2$ and hence K is a finite field containing nine elements. We find that

$$K = \{0, 1, 2, c, 2c, 1 + c, 2 + c, 1 + 2c, 2 + 2c\},$$

where $c = X + (X^2 + 1)\mathbb{F}_3[X]$ and $c^2 = 2$. The Frobenius automorphism of the field K is the mapping $\sigma_3 : K \to K$ defined by $\sigma_3(x) = x^3$ for all elements $x \in K$. Since $[K:\mathbb{F}_3] = 2, |\sigma_3| = 2$ and hence the automorphism group $G(K/\mathbb{F}_3) = \{1, \sigma_3\}$ is a cyclic group of order 2. We find, for example, that $\sigma_3(1 + c) = (1 + c)^3 = 1 + c^3 = 1 + 2c$, and, in general, $\sigma_3(a + bc) = a + bc^3 = a + 2bc$ for every element $a + bc \in K$. The table here summarizes the automorphisms in the group $G(K/\mathbb{F}_3)$ and the image of each element in K under these automorphisms.

$1 : K \to K$ $a + bc \mapsto a + bc$	$\sigma_3 : K \to K$ $a + bc \mapsto a + 2bc$
$0 \mapsto 0$	$0 \mapsto 0$
$1 \mapsto 1$	$1 \mapsto 1$
$2 \mapsto 2$	$2 \mapsto 2$
$c \mapsto c$	$c \mapsto 2c$
$2c \mapsto 2c$	$2c \mapsto c$
$1 + c \mapsto 1 + c$	$1 + c \mapsto 1 + 2c$
$2 + c \mapsto 2 + c$	$2 + c \mapsto 2 + 2c$
$1 + 2c \mapsto 1 + 2c$	$1 + 2c \mapsto 1 + c$
$2 + 2c \mapsto 2 + 2c$	$2 + 2c \mapsto 2 + c$

(E) In this example we have shown that the automorphism group of a finite field K over its prime subfield k is a cyclic group generated by the Frobenius automorphism of K. Let us mention, in conclusion, that if k' is any subfield of K, not necessarily the prime subfield, then the automorphism group $G(K/k')$ is a subgroup of $G(K/k)$ and hence is also a cyclic group. In the exercises we indicate how the Frobenius automorphism on K may be used to determine the group $G(K/k')$.

The preceding examples illustrated automorphism groups for many different types of field extensions. In each case, however, the extension field K was a finite simple extension of the ground field k and hence $|G(K/k)| \le [K:k]$. Let us now conclude this section by discussing the relationship between the degree of an arbitrary finite extension of fields K/k and the order of its automorphism group $G(K/k)$. Our goal is to first show that $|G(K/k)| \le [K:k]$, whether K is a simple extension of k or not, and then show that $|H| = [K:K^H]$ for any group H of k-automorphisms of K. This, then, gives us an exact relationship between the number of automorphisms in any group H of k-automorphisms and the fixed field of H. Most of these results are due to Emil Artin and do not require the existence of a primitive element, as was the case in Proposition 1, but instead use some basic facts about systems of homogeneous linear equations. Let us begin, therefore, by first recalling a few basic facts about homogeneous systems of equations.

A system of n homogeneous linear equations in m unknowns $X_1, \ldots, X_m$ over a field k is any linear system of the form

$$a_{11}X_1 + \cdots + a_{1m}X_m = 0$$
$$\vdots$$
$$a_{n1}X_1 + \cdots + a_{nm}X_m = 0,$$

where $a_{11}, \ldots, a_{nm} \in k$. Such a system always has the trivial solution $X_1 = \cdots = X_m = 0$; any other solution is called a nontrivial solution. The basic fact that we need concerning these systems is that if $n < m$, then the system must have a nontrivial solution; that is, any system of homogeneous linear equations having more unknowns than equations must have at least one solution in which not every unknown is equal to zero. Using this fact, let us now prove our main results. We begin by first showing that any finite collection of field automorphisms must be linearly independent.

PROPOSITION 3. (ARTIN) Let K/k be an extension of fields and let $\sigma_1, \ldots, \sigma_n$ be any finite number of distinct k-automorphisms of K. Then $\sigma_1, \ldots, \sigma_n$ are linearly independent over K; that is, if $c_1, \ldots, c_n$ are scalars in K such that $c_1\sigma_1(x) + \cdots + c_n\sigma_n(x) = 0$ for all elements $x \in K$, then $c_1 = \cdots = c_n = 0$.

Proof. Suppose the statement is false. Then there is a least positive number of k-automorphisms $\sigma_1, \ldots, \sigma_n$ of K and scalars $c_1, \ldots, c_n$ in K, none of which is zero, such that $c_1\sigma_1(x) + \cdots + c_n\sigma_n(x) = 0$ for all elements $x \in K$. Now, $n \neq 1$; for if $n = 1$, then $c_1\sigma_1(x) = 0$ for all $x \in K$ and hence, in particular, $c_1 = c_1\sigma_1(1) = 0$, which contradicts the fact that $c_1 \neq 0$. Hence, $n > 1$. In this case there is an automorphism $\sigma_2 \neq \sigma_1$ and an element $y \in K$ such that $\sigma_2(y) \neq \sigma_1(y)$. Now, let x be any element in K. Then by assumption

$$c_1\sigma_1(y)\sigma_1(x) + \cdots + c_n\sigma_n(y)\sigma_n(x) = c_1\sigma_1(yx) + \cdots + c_n\sigma_n(yx) = 0.$$

On the other hand,

$$c_1\sigma_1(y)\sigma_1(x) + \cdots + c_n\sigma_1(y)\sigma_n(x) = \sigma_1(y)[c_1\sigma_1(x) + \cdots + c_n\sigma_n(x)] = 0.$$

Subtracting these equations, we find that

$$c_2[\sigma_2(y) - \sigma_1(y)]\sigma_2(x) + \cdots + c_n[\sigma_n(y) - \sigma_1(y)]\sigma_n(x) = 0$$

for all elements $x \in K$. But the coefficient $c_2[\sigma_2(y) - \sigma_1(y)] \neq 0$, which means that this last equation is a nontrivial dependence relation involving fewer than n automorphisms of K. Since this contradicts the minimality of n, it follows that there is no such integer n and hence that the statement is true for any finite number of k-automorphisms of K. ∎

PROPOSITION 4.   (ARTIN) Let K/k be a finite extension of fields. Then the automorphism group $G(K/k)$ is a finite group and $|G(K/k)| \leq [K:k]$.

Proof.   Let $[K:k] = n$ and let $A_1, \ldots, A_n$ be a basis for K as a vector space over k. We must show that the group $G(K/k)$ contains at most n distinct k-automorphisms. Suppose, to the contrary, that we may find m distinct automorphisms $\sigma_1, \ldots, \sigma_m$ in $G(K/k)$ for some integer $m > n$, and consider the following system of n homogeneous linear equations in m unknowns:

$$\sigma_1(A_1)X_1 + \cdots + \sigma_m(A_1)X_m = 0$$
$$\vdots$$
$$\sigma_1(A_n)X_1 + \cdots + \sigma_m(A_n)X_m = 0.$$

Since $m > n$, this system has a nontrivial solution, say $X_1 = c_1, \ldots, X_m = c_m$, where $c_1, \ldots, c_m \in K$ are not all zero. We now claim that $c_1\sigma_1(x) + \cdots + c_m\sigma_m(x) = 0$ for every element $x \in K$. For let $x = a_1A_1 + \cdots + a_nA_n$ be a typical element in K, where $a_1, \ldots, a_n \in k$. Then

$$c_1\sigma_1(x) + \cdots + c_m\sigma_m(x) = c_1[a_1\sigma_1(A_1) + \cdots + a_n\sigma_1(A_n)]$$
$$+ \cdots + c_m[a_1\sigma_m(A_1) + \cdots + a_n\sigma_m(A_n)]$$
$$= a_1[c_1\sigma_1(A_1) + \cdots + c_m\sigma_m(A_1)]$$
$$+ \cdots + a_n[c_1\sigma_1(A_n) + \cdots + c_m\sigma_m(A_n)]$$
$$= 0.$$

Since not all of the c_i are zero, it follows that $\sigma_1, \ldots, \sigma_m$ are linearly dependent over K, which contradicts Proposition 3. Thus, there cannot be more than n distinct automorphisms in the group $G(K/k)$ and consequently $G(K/k)$ is a finite group whose order is at most n; that is, $|G(K/k)| \leq [K:k]$, as required.   ■

PROPOSITION 5.   (ARTIN) Let K/k be a finite extension of fields and let H be any group of k-automorphisms of K. Then $|H| = [K:K^H]$.

Proof.   First observe that H is a subgroup of $G(K/K^H)$ and hence $|H| \leq |G(K/K^H)|$. On the other hand K is a finite extension of the subfield K^H and hence, by Proposition 4, $|G(K/K^H)| \leq [K:K^H]$. Therefore, $|H| \leq |G(K/K^H)| \leq [K:K^H]$ and hence, in particular, H is a finite group. To show that $|H| = [K:K^H]$ and thus complete the proof, let $H = \{\sigma_1, \ldots, \sigma_m\}$, $[K:K^H] = n$, and let $A_1, \ldots, A_n$ be a basis for K as a vector space over the fixed field K^H. Then $m \leq n$. Now suppose that $m < n$ and consider the system (S) of m homogeneous linear equations in n unknowns over the field K:

$$\sigma_1(A_1)X_1 + \cdots + \sigma_1(A_n)X_n = 0$$
$$(S) \qquad \vdots$$
$$\sigma_m(A_1)X_1 + \cdots + \sigma_m(A_n)X_n = 0.$$

Since $m < n$, this system has a nontrivial solution, say $X_1 = c_1, \ldots, X_n = c_n$, where $c_1, \ldots, c_n \in K$; for convenience, we may assume that $c_1 \neq 0$. We now claim that system (S) in fact has a solution $c_1^*, \ldots, c_n^*$ in which each element c_i^* lies in the subfield K^H. To construct such a solution, first note that $\sigma_1, \ldots, \sigma_m$ are linearly independent over K since they represent m distinct automorphisms of K. It follows that the expression $\sigma_1(x) + \cdots + \sigma_m(x)$ is not zero for some element $x \in K$. Let $x_0 \in K$ be such that $\sigma_1(x_0) + \cdots + \sigma_m(x_0) \neq 0$. Let $y_0 = x_0/c_1$ and set $c_i^* = \sigma_1(y_0 c_i) + \cdots + \sigma_m(y_0 c_i)$ for $i = 1, \ldots, n$. We claim that $X_1 = c_1^*, \ldots, X_n = c_n^*$ is a nontrivial solution of system (S) and that each of the elements c_i^* lie in K^H.

(A) Let us first show that $X_1 = c_1^*, \ldots, X_n = c_n^*$ is a nontrivial solution of system (S). Substituting $c_1^*, \ldots, c_n^*$ into the ith equation of the system, we find that

$$
\begin{aligned}
\sigma_i(A_1)c_1^* + \cdots + \sigma_i(A_n)c_n^* &= \sigma_i(A_1)[\sigma_1(y_0 c_1) + \cdots + \sigma_m(y_0 c_1)] + \cdots \\
&\quad + \sigma_i(A_n)[\sigma_1(y_0 c_n) + \cdots + \sigma_m(y_0 c_n)] \\
&= \sigma_1(y_0)[\sigma_i(A_1)\sigma_1(c_1) + \cdots + \sigma_i(A_n)\sigma_1(c_n)] + \cdots \\
&\quad + \sigma_m(y_0)[\sigma_i(A_1)\sigma_m(c_1) + \cdots + \sigma_i(A_n)\sigma_m(c_n)] \\
&= \sigma_1(y_0)\sigma_1[\sigma_1^{-1}\sigma_i(A_1)c_1 + \cdots + \sigma_1^{-1}\sigma_i(A_n)c_n] + \cdots \\
&\quad + \sigma_m(y_0)\sigma_m[\sigma_m^{-1}\sigma_i(A_1)c_1 + \cdots + \sigma_m^{-1}\sigma_i(A_n)c_n].
\end{aligned}
$$

But since H is a group containing the elements $\sigma_1, \ldots, \sigma_m$, left multiplication of σ_i by $\sigma_1^{-1}, \ldots, \sigma_m^{-1}$ simply permutes the elements in H; that is, $\{\sigma_1^{-1}\sigma_i, \ldots, \sigma_m^{-1}\sigma_i\} = \{\sigma_{i_1}, \ldots, \sigma_{i_m}\} = H$. Therefore, the last expression is equal to

$$
\begin{aligned}
\sigma_1(y_0)\sigma_1[\sigma_{i_1}(A_1)c_1 + \cdots + \sigma_{i_1}(A_n)c_n] \\
+ \cdots + \sigma_m(y_0)\sigma_m[\sigma_{i_m}(A_1)c_1 + \cdots + \sigma_{i_m}(A_n)c_n],
\end{aligned}
$$

which is equal to 0 since $X_1 = c_1, \ldots, X_n = c_n$ is a solution of system (S). It follows, therefore, that $X_1 = c_1^*, \ldots, X_n = c_n^*$ is a solution of the system. Finally, since

$$
\begin{aligned}
X_1 = c_1^* &= \sigma_1(y_0 c_1) + \cdots + \sigma_m(y_0 c_1) \\
&= \sigma_1(x_0) + \cdots + \sigma_m(x_0) \\
&\neq 0,
\end{aligned}
$$

the solution $X_1 = c_1^*, \ldots, X_n = c_n^*$ is nontrivial.

(B) To show that $c_1^*, \ldots, c_n^*$ lie in the fixed field K^H, let $\sigma \in H$. Then for $i = 1, \ldots, n$, $\sigma(c_i^*) = \sigma\sigma_1(y_0 c_i) + \cdots + \sigma\sigma_m(y_0 c_i)$. As in part (A), left multiplication of the elements in H by σ simply permutes the elements. Hence, $\sigma(c_i^*) = \sigma_1(y_0 c_i) + \cdots + \sigma_m(y_0 c_i) = c_i^*$. Therefore, each of the elements $c_1^*, \ldots, c_n^*$ is fixed by every automorphism in H and hence lie in the fixed field K^H, as required.

Now, since $X_1 = c_1^*, \ldots, X_n = c_n^*$ is a nontrivial solution of system (S), $\sigma_i(A_1)c_1^* + \cdots + \sigma_i(A_n)c_n^* = 0$ for $i = 1, \ldots, m$. But for some i, $\sigma_i = 1$, the identity automorphism, and hence $c_1^* A_1 + \cdots + c_n^* A_n = 0$. Since the scalars $c_1^*, \ldots, c_n^*$ lie in K^H and are not all zero, this contradicts the fact that

$A_1, \ldots, A_n$ are linearly independent over K^H. Therefore, $m = n$, that is, $|H| = [K:K^H]$, and the proof is complete. ∎

COROLLARY. Let K/k be a finite extension of fields and let H be any group of k-automorphisms of K. Then $|H|$ divides $[K:k]$.

Proof. Immediate from Proposition 5. ∎

COROLLARY. Let K/k be a finite extension of fields such that $[K:k] = p$, a prime. Then the automorphism group $G(K/k)$ is either trivial or is a cyclic group of order p.

Proof. Let $G = G(K/k)$. Then $|G|$ divides $[K:k]$. Since $[K:k] = p$, $|G|$ is either 1 or p and therefore G is either trivial or is a cyclic group of order p. ∎

These results, when taken as a whole, give us exceptionally important relationships between groups of automorphisms and their fixed fields. They show, for example, that if K/k is any finite extension of fields with automorphism group G, then not only is $|G| \leq [K:k]$, but $|G| = [K:K^G]$. Thus, the extension K/k has only a finite number of k-automorphisms and the number of such automorphisms is equal to the degree of K over the fixed subfield K^G. More generally, if H is any group of k-automorphisms of K, then $|H| = [K:K^H]$. Consequently, if the fixed field K^H is known, then $|H| = [K:K^H]$, while if $|H|$ is known, then

$$[K^H:k] = \frac{[K:k]}{[K:K^H]} = \frac{1}{|H|}[K:k].$$

Thus, whenever we have a group of k-automorphisms of an extension field K, the number of such automorphisms is equal to the degree of K over the fixed subfield of the group, while the degree of the fixed subfield over the ground field k is equal to the degree $[K:k]$ of the extension divided by the number of such automorphisms.

For example, let $K = \mathbb{Q}(\omega, \sqrt[3]{2})$ stand for the field discussed in Example 5 and let σ stand for the $\mathbb{Q}$-automorphism of K that maps $\sqrt[3]{2}$ to $\omega\sqrt[3]{2}$ and ω to ω. Let $H = \{1, \sigma, \sigma^2\}$ stand for the cyclic subgroup generated by σ. Then $|H| = 3$ and hence $[K^H:\mathbb{Q}] = \frac{1}{3}[K:\mathbb{Q}] = \frac{6}{3} = 2$. Since ω is fixed by σ, $\omega \in K^H$ and hence $K^H = \mathbb{Q}(\omega)$, a result that we obtained by direct calculation in Example 5. We find, similarly, that

$$K^G = \mathbb{Q}; \qquad\qquad |G| = [K:\mathbb{Q}] = 6$$
$$K^{(\tau)} = \mathbb{Q}(\sqrt[3]{2}); \qquad |(\tau)| = [K:\mathbb{Q}(\sqrt[3]{2})] = 2$$
$$K^{(\sigma\tau)} = \mathbb{Q}(\omega\sqrt[3]{2}); \qquad |(\sigma\tau)| = [K:\mathbb{Q}(\omega\sqrt[3]{2})] = 2$$
$$K^{(\tau\sigma)} = \mathbb{Q}(\omega^2\sqrt[3]{2}); \qquad |(\tau\sigma)| = [K:\mathbb{Q}(\omega^2\sqrt[3]{2})] = 2$$
$$K^{(1)} = K; \qquad\qquad |(1)| = [K:K] = 1.$$

As mentioned in the introduction, our main goal in this chapter is to associate a group of automorphisms with every field extension and then discuss the relationship between the group and the structure of the field extension. We began in this section by associating the automorphism group $G(K/k)$ with any extension of fields K/k: $G(K/k)$ consists of all k-automorphisms of K, that is, all automorphisms of K that fix every element in the ground field k. If H is any group of k-automorphisms of K, then H is a subgroup of $G(K/k)$ and has associated with it its fixed field K^H, which is the intermediate subfield of K/k consisting of those elements in K that are fixed by every automorphism in H. Most of our results in this first section have been concerned with the order of the automorphism group $G(K/k)$ and its relationship to the degree $[K:k]$ of the extension. We proved two basic results: first, if K/k is any finite extension of fields, then its automorphism group $G(K/k)$ is a finite group whose order is at most equal to the degree of the extension, that is, $|G(K/k)| \le [K:k]$; and second, if H is any group of k-automorphisms of K, then $|H| = [K:K^H]$. It follows, in particular, that if $G = G(K/k)$, then $|G| = [K:K^G]$ and hence $|G| = [K:k]$ if and only if $K^G = k$. In general, the fixed field K^G contains but is not necessarily equal to the ground field k, although Examples 1–6 indicate that in many cases these two fields are in fact the same. In the next section we will discuss and characterize those extensions K/k, called normal extensions, for which the fixed field K^G is equal to the ground field k. Then, in the third and final section we establish the fundamental relationship between the intermediate subfields of a normal extension and the subgroups of its automorphism group.

Exercises

1. Find the automorphism group $G(\mathbb{Q}(\sqrt{3})/\mathbb{Q})$ and show that $G(\mathbb{Q}(\sqrt{3})/\mathbb{Q})$ $\cong \mathbb{Z}_2$.

2. Let $K = \mathbb{Q}(\sqrt{2}, \sqrt{3})$ and let $G = G(K/\mathbb{Q})$.
   (a) Find G and show that $G \cong \mathbb{Z}_2 \times \mathbb{Z}_2$.
   (b) Let σ stand for the $\mathbb{Q}$-automorphism of K that maps $\sqrt{2}$ to $-\sqrt{2}$ and $\sqrt{3}$ to $\sqrt{3}$. Find the fixed field $K^{(\sigma)}$ corresponding to the cyclic subgroup generated by σ and verify that $|(\sigma)| = [K:K^{(\sigma)}]$.
   (c) Show that G has five subgroups and find the fixed subfield of K corresponding to each subgroup.
   (d) Find $G(K/\mathbb{Q}(\sqrt{6}))$.
   (e) Find $G(K/\mathbb{Q}(\sqrt{2} + \sqrt{3}))$.

3. Let $K = \mathbb{Q}(\sqrt{2}, \sqrt{-2})$ and let $G = G(K/\mathbb{Q})$.
   (a) Find G and show that $G \cong \mathbb{Z}_2 \times \mathbb{Z}_2$.
   (b) Find all subgroups of G and find the fixed subfield of K corresponding to each subgroup.
   (c) Find $G(K/\mathbb{Q}(i))$.
   (d) Find $G(K/\mathbb{Q}(\sqrt{2} + \sqrt{-2}))$.

4. Let $K = \mathbb{Q}(\sqrt[4]{2}, i)$ and let $G = G(K/\mathbb{Q})$.
   (a) Find G and show that $G \cong D_4$, the group of the square.
   (b) Show that there is a $\mathbb{Q}$-automorphism σ of K that maps $\sqrt[4]{2}$ onto $i\sqrt[4]{2}$ and i onto i, and find the fixed subfield $K^{(\sigma)}$ corresponding to the cyclic subgroup generated by σ.
   (c) Show that there is a $\mathbb{Q}$-automorphism τ of K that maps $\sqrt[4]{2}$ onto $\sqrt[4]{2}$ and i into $-i$, and find the fixed subfield $K^{(\tau)}$ corresponding to the cyclic subgroup generated by τ.
   (d) Let $H = \{1, \sigma^2, \tau, \sigma^2\tau\}$, where σ and τ are the automorphisms defined in parts (b) and (c).
      (1) Show that H is a group of $\mathbb{Q}$-automorphisms of K.
      (2) Find the fixed field K^H.
      (3) Verify that $|H| = [K:K^H]$.
   (e) Show that $\mathbb{Q}(i\sqrt{2})$ is a subfield of K and find $G(K/\mathbb{Q}(i\sqrt{2}))$.
5. Let $K = \mathbb{F}_3[X]/(X^2 + 2X + 2)\mathbb{F}_3[X]$. Show that K is a finite field containing nine elements and find the automorphism group $G(K/\mathbb{F}_3)$. Find the image of each element in K under the automorphisms in $G(K/\mathbb{F}_3)$.
6. Let $K = \mathbb{F}_2[X]/(X^3 + X^2 + 1)\mathbb{F}_2[X]$. Show that K is a finite field containing eight elements and find the automorphism group $G(K/\mathbb{F}_2)$. Find the image of each element in K under the automorphisms in $G(K/\mathbb{F}_2)$.
7. Find $G(\mathbb{Q}(\sqrt{1 + 4i})/\mathbb{Q})$.
8. Find $G(\mathbb{Q}(i, \sqrt[3]{2})/\mathbb{Q})$.
9. Let K/k be an extension of fields and let k' be an intermediate subfield of the extension.
   (a) Show that $G(K/k') \leq G(K/k)$.
   (b) Show that $K^{G(K/k)} \subseteq K^{G(K/k')}$.
10. Let K/k be an extension of fields and let k' be an intermediate subfield of the extension. Let σ be a k-automorphism of K and let $\sigma(k') = \{\sigma(x) \in K \mid x \in k'\}$ stand for the image of k' under σ.
   (a) Show that $\sigma(k')$ is an intermediate subfield of K/k.
   (b) Show that $G(K/\sigma(k')) = \sigma G(K/k')\sigma^{-1}$.
11. Let $K = \mathbb{Q}(\omega, \sqrt[3]{2})$, where ω stands for a primitive cube root of unity.
   (a) Find $G(K/\mathbb{Q}(\sqrt[3]{2}))$, $G(K/\mathbb{Q}(\omega, \sqrt[3]{2}))$, and $G(K/\mathbb{Q}(\omega^2, \sqrt[3]{2}))$.
   (b) Let σ stand for the $\mathbb{Q}$-automorphism of K that maps $\sqrt[3]{2}$ onto $\omega\sqrt[3]{2}$ and ω onto ω. Show that $G(K/\mathbb{Q}(\omega\sqrt[3]{2})) = \sigma G(K/\mathbb{Q}(\sqrt[3]{2}))\sigma^{-1}$ and $G(K/\mathbb{Q}(\omega^2\sqrt[3]{2})) = \sigma^2 G(K/\mathbb{Q}(\sqrt[3]{2}))\sigma^{-2}$.
12. **The automorphism group of the field of real numbers.** Let σ be a $\mathbb{Q}$-automorphism of the field $\mathbb{R}$ of real numbers.
   (a) If $a \in \mathbb{R}$, $a \geq 0$, show that $\sigma(a) \geq 0$.
   (b) If $a, b \in \mathbb{R}$ with $a \leq b$, show that $\sigma(a) \leq \sigma(b)$.
   (c) Using the fact that there is some rational number between any two distinct real numbers, show that $\sigma(a) = a$ for every real number $a \in \mathbb{R}$.

(d) Show that $G(\mathbb{R}/\mathbb{Q}) = \{1\}$ and conclude that the only automorphism of the field $\mathbb{R}$ of real numbers is the identity automorphism.

13. Let $\mathbb{Q}(X)$ stand for the field of rational functions in X over $\mathbb{Q}$. For every rational number $a \in \mathbb{Q}$, define the function $\sigma_a : \mathbb{Q}(X) \to \mathbb{Q}(X)$ by setting

$$\sigma_a\left(\frac{f(X)}{g(X)}\right) = \frac{f(X + a)}{g(X + a)} \text{ for all rational functions } \frac{f(X)}{g(X)} \in \mathbb{Q}(X).$$

   (a) Show that σ_a is a $\mathbb{Q}$-automorphism of $\mathbb{Q}(X)$ for every rational number $a \in \mathbb{Q}$.

   (b) Let $H_\mathbb{Q} = \{\sigma_a \in G(\mathbb{Q}(X)/\mathbb{Q}) \,|\, a \in \mathbb{Q}\}$. Show that $H_\mathbb{Q}$ is a subgroup of $G(\mathbb{Q}(X)/\mathbb{Q})$.

   (c) Show that $H_\mathbb{Q} \cong \mathbb{Q}$, the additive group of rational numbers.

   (d) Define the function $\tau : \mathbb{Q}(X) \to \mathbb{Q}(X)$ by setting $\tau\left(\dfrac{f(X)}{g(X)}\right) = \dfrac{f(-X)}{g(-X)}$ for all rational functions $\dfrac{f(X)}{g(X)} \in \mathbb{Q}(X)$. Show that τ is a $\mathbb{Q}$-automorphism of $\mathbb{Q}(X)$.

   (e) Does $G(\mathbb{Q}(X)/\mathbb{Q}) = H_\mathbb{Q}$?

14. **The automorphism group of a finite field over an arbitrary intermediate subfield.** Let K be a finite field of characteristic p and let k be a subfield of K with $[K:k] = n$. Let $\sigma_p : K \to K$ stand for the Frobenius automorphism of K.

   (a) If $[k:\mathbb{F}_p] = m$, show that σ_p^m fixes every element in k.

   (b) Show that $G(K/k) = (\sigma_p^m) = \{1, \sigma_p^m, \ldots, \sigma_p^{m(n-1)}\}$. Thus, the automorphism group of a finite field over an arbitrary intermediate subfield is the cyclic group generated by σ_p^m, where m is the degree of the extension.

   (c) Show that $|G(K/k)| = [K:k]$.

   (d) Let $K = \mathbb{F}_2[X]/(X^4 + X + 1)\mathbb{F}_2[X]$.

   　　(1) Show that K is a finite field containing 16 elements.

   　　(2) Let $k = \{x \in K \,|\, x^4 = x\}$. Show that k is an intermediate subfield of the extension $K/\mathbb{F}_2$ and find $[K:k]$ and $[k:\mathbb{F}_2]$.

   　　(3) Find $G(K/k)$.

15. Let K/k be an extension of fields and let K^* stand for the multiplicative group of nonzero elements in K.

   (a) Show that $G(K/k)$ is a subgroup of the automorphism group $\text{Aut}(K^*)$ of K^*.

   (b) Let K be a finite field containing p^n elements, where p is prime, and let $n = [K:\mathbb{F}_p]$.

   　　(1) Show that $|\text{Aut}(K^*)| = \varphi(p^n - 1)$, where φ stands for the Euler phi-function.

   　　(2) Show that n divides $\varphi(p^n - 1)$.

16. **Matrix representations of automorphisms.** Let K/k be a finite extension of fields.

    (a) Show that every k-automorphism of K is a nonsingular linear transformation of K as a vector space over k.

    (b) Let σ be a k-automorphism of K and let $K^{(\sigma)}$ stand for the fixed subfield corresponding to the cyclic subgroup (σ). Show that $K^{(\sigma)}$ $= \mathrm{Ker}(\sigma - 1)$, the kernel of the linear transformation $\sigma - 1$.

    (c) In view of part (a), every k-automorphism of the extension K/k may be represented by a matrix relative to any given basis for K as a vector space over k.

        (1) Find the matrix representation of each $\mathbb{Q}$-automorphism of $\mathbb{Q}(\omega)$ relative to the basis $\{1, \omega\}$.

        (2) Find the matrix representation of each $\mathbb{Q}$-automorphism of $\mathbb{Q}(\sqrt{2}, \sqrt{3})$ relative to the basis $\{1, \sqrt{2}, \sqrt{3}, \sqrt{6}\}$.

        (3) Let $K = \mathbb{F}_2[X]/(X^3 + X^2 + 1)\mathbb{F}_2[X]$ stand for the field discussed in Exercise 6. Find the matrix representation of each $\mathbb{F}_2$-automorphism of K relative to the basis $\{1, c, c^2\}$, where $c = X + (X^3 + X^2 + 1)\mathbb{F}_2[X]$.

    (d) Let $[K:k] = n$ and let $\{A_1, \ldots, A_n\}$ be a basis for K as a vector space over k. Define the function $f : G(K/k) \to \mathrm{Mat}_n^*(k)$ by letting $f(\sigma)$ be the matrix representation of σ relative to the basis $\{A_1, \ldots, A_n\}$ for each automorphism $\sigma \in G(K/k)$. Show that f is a 1–1 group-homomorphism from the automorphism group $G(K/k)$ into the group $\mathrm{Mat}_n^*(k)$ of $n \times n$ invertible matrices with entries in k.

17. **Permutation representations of automorphisms of splitting fields.** Let K/k be a finite extension of fields and suppose that K is a splitting field for some polynomial $f(X)$ over k. Let $c_1, \ldots, c_n$ stand for the roots of $f(X)$ in K.

    (a) Show that every k-automorphism of K permutes the roots of $f(X)$.

    (b) Let σ be a k-automorphism of K and let π_σ stand for the permutation of the roots $c_1, \ldots, c_n$ that maps c_i onto $\sigma(c_i)$ for $i = 1, \ldots, n$. Define the function $\pi : G(K/k) \to \mathrm{Sym}(\{c_1, \ldots, c_n\})$ by setting $\pi(\sigma) = \pi_\sigma$ for every k-automorphism $\sigma \in G(K/k)$. Show that π is a 1–1 group-homomorphism from the automorphism group $G(K/k)$ into the symmetric group on the roots of $f(X)$.

    (c) The field $\mathbb{Q}(\sqrt{2}, \sqrt{3})$ is a splitting field for the polynomial $f(X) = X^4 - 10X^2 + 1$ over $\mathbb{Q}$. Find the permutation of the roots of $f(X)$ corresponding to each automorphism in $G(\mathbb{Q}(\sqrt{2}, \sqrt{3})/\mathbb{Q})$.

    (d) Let $K = \mathbb{F}_2[X]/(X^3 + X^2 + 1)\mathbb{F}_2[X]$ stand for the field discussed in Exercise 6. Show that K is a splitting field for the polynomial $f(X) = X^3 + X^2 + 1$ over $\mathbb{F}_2$ and find the permutation of the roots of $f(X)$ corresponding to each automorphism in $G(K/\mathbb{F}_2)$.

    (e) Is the homomorphism π defined in part (b) necessarily onto; that is, is every permutation of the roots of a polynomial $f(X)$ necessarily represented by some automorphism of its splitting field?

    (f) Show that a polynomial $f(X)$ over a field K is irreducible over k if and only if for any pair of roots c_i and c_j of $f(X)$, there is some k-automorphism σ of its splitting field K such that $\sigma(c_i) = c_j$.

2.   NORMAL EXTENSIONS

We showed in the previous section that if K/k is a finite extension of fields with automorphism group G, then the fixed field K^G of all elements in K fixed by every k-automorphism of K is an intermediate subfield of the extension K/k. In many cases, however, the elements in the ground field k are the only elements in K that are fixed by every k-automorphism, in which case we say that K is a normal extension of k. In this section our goal is to discuss and characterize normal field extensions, first in terms of the degree of the extension, then in terms of the irreducible polynomials of elements in K, and finally in terms of splitting fields. We then conclude the section by discussing roots of unity and those fields, called cyclotomic fields, that are generated from the rational numbers by adjoining roots of unity.

> **Definition**
>
> Let k be a field. A *normal extension* of k is any finite extension K of k with the property that the only elements in K fixed by every k-automorphism are the elements in the ground field k.

Thus, if K/k is a finite extension of fields with automorphism group G, then K is a normal extension of k if $K^G = k$. The fields $\mathbb{Q}(\sqrt{2})$, $\mathbb{Q}(i)$ and $\mathbb{Q}(\omega)$, for example, are normal extensions of $\mathbb{Q}$ since in each case the automorphism group consists of two automorphisms whose fixed field is $\mathbb{Q}$. The field $\mathbb{Q}(\sqrt[3]{2})$, however, is not a normal extension of $\mathbb{Q}$ since its automorphism group is trivial and hence fixes all elements in $\mathbb{Q}(\sqrt[3]{2})$.

Now recall that if K/k is a finite extension of fields with automorphism group G, then $|G| = [K:K^G] \le [K:k]$ and hence the number of k-automorphisms of K is at most equal to the degree of the extension. Let us now show that normal extensions of k are precisely those extensions that admit the maximum number of k-automorphisms.

PROPOSITION 1. Let K/k be a finite extension of fields with automorphism group G. Then K is a normal extension of k if and only if $|G| = [K:k]$.

Proof.   Since $|G| = [K:K^G]$, it follows that $|G| = [K:k]$ if and only if $K^G = k$, that is, if and only if K is a normal extension of k.   ■

For example, let $K = \mathbb{Q}(\omega, \sqrt[3]{2})$ stand for the field generated by ω and $\sqrt[3]{2}$, where ω is a primitive cube root of unity. Then K is a finite extension of $\mathbb{Q}$ and $[\mathbb{Q}(\omega, \sqrt[3]{2}):\mathbb{Q}] = 6$. In this case the automorphism group $G(\mathbb{Q}(\omega, \sqrt[3]{2})/\mathbb{Q})$ consists of six automorphisms, namely 1, σ, σ^2, τ, $\sigma\tau$, and $\tau\sigma$, where

$$\sigma: \begin{matrix} \sqrt[3]{2} \to \omega\sqrt[3]{2} \\ \omega \to \omega \end{matrix} \quad \text{and} \quad \tau: \begin{matrix} \sqrt[3]{2} \to \sqrt[3]{2} \\ \omega \to \omega^2 \end{matrix}.$$

Therefore, $|G(\mathbb{Q}(\omega, \sqrt[3]{2})/\mathbb{Q})| = [\mathbb{Q}(\omega, \sqrt[3]{2}):\mathbb{Q}] = 6$ and hence $\mathbb{Q}(\omega, \sqrt[3]{2})$ is a normal extension of $\mathbb{Q}$. Observe that since every $\mathbb{Q}$-automorphism must map ω to either ω or ω^2, and $\sqrt[3]{2}$ to either $\sqrt[3]{2}$, $\omega \sqrt[3]{2}$, or $\omega^2 \sqrt[3]{2}$, there are at most six possible $\mathbb{Q}$-automorphisms of $\mathbb{Q}(\omega, \sqrt[3]{2})$ and all six possibilities do in fact occur.

Now observe that if K is a normal extension of a field k, then K is algebraic over k since it is a finite extension of k and hence every element $c \in K$ is a root of a unique monic irreducible polynomial over k, namely its irreducible polynomial $\mathrm{Irr}(c;k)$. Moreover, if σ is any k-automorphism of K, $\sigma(c)$ is also a root of $\mathrm{Irr}(c;k)$; for if $f(X)$ is any polynomial over k having c as a root, then $\sigma(c)$ is a root of $f(X)$ since $f(\sigma(c)) = \sigma(f(c)) = 0$. We say that an element $c' \in K$ is *conjugate* to c if $c' = \sigma(c)$ for some k-automorphism σ of K. It is easily verified that conjugation is an equivalence relation on the field K and hence partitions the field into disjoint subsets called conjugacy classes; the conjugates of an element are simply its images under all k-automorphisms. In view of our earlier remarks, all conjugates of c are roots of the irreducible polynomial $\mathrm{Irr}(c;k)$. Let us now show that, for normal extensions, the distinct conjugates of c are in fact the only roots of $\mathrm{Irr}(c;k)$ and hence completely determine the irreducible polynomial.

PROPOSITION 2. Let K/k be a normal extension of fields and let $c \in K$. If $c_1, \ldots, c_n$ are the distinct conjugates of c in K, then $\mathrm{Irr}(c;k) = (X - c_1) \cdots (X - c_n)$.

Proof. Let $f(X) = (X - c_1) \cdots (X - c_n)$. We first claim that the coefficients of $f(X)$ lie in the ground field k. For let $S_t(c_1, \ldots, c_n)$ stand for the coefficient of X^t in $f(X)$ for $t = 0, 1, \ldots$. To show that $S_t(c_1, \ldots, c_n) \in k$, we need only verify that $S_t(c_1, \ldots, c_n)$ is fixed by every k-automorphism of K since K is a normal extension of k. To this end, let σ be a k-automorphism of K. Then $\sigma S_t(c_1, \ldots, c_n) = S_t(\sigma(c_1), \ldots, \sigma(c_n))$ since $S_t(c_1, \ldots, c_n)$ is a polynomial in $c_1, \ldots, c_n$ with coefficients in k. But $S_t(\sigma(c_1), \ldots, \sigma(c_n)) = S_t(c_1, \ldots, c_n)$ since σ just permutes the conjugates $c_1, \ldots, c_n$, and the product of the factors $X - c_1, \ldots, X - c_n$ is the same regardless of how the factors are rearranged. Therefore, $\sigma S_t(c_1, \ldots, c_n) = S_t(c_1, \ldots, c_n)$ and hence $S_t(c_1, \ldots, c_n) \in k$ since it is fixed by every k-automorphism of K. Consequently, $f(X)$ is a polynomial over k. Now, since c is a root of $f(X)$, $\mathrm{Irr}(c;k)$ divides $f(X)$. On the other hand, $f(X)$ divides $\mathrm{Irr}(c;k)$ since every conjugate of c is also a root of $\mathrm{Irr}(c;k)$. Therefore, $\mathrm{Irr}(c;k) = f(X) = (X - c_1) \cdots (X - c_n)$, as required, and the proof is complete. ∎

COROLLARY. Let K/k be a normal extension of fields. Then two elements in K are conjugate if and only if they have the same irreducible polynomial over k.

Proof.   If c and c' are elements in K, then, by Proposition 2, $\mathrm{Irr}(c;k) = \mathrm{Irr}(c';k)$ if and only if the conjugates of c are the same as those of c', which is the case if and only if c and c' are conjugate.   ∎

Proposition 2 is an especially useful result for normal extensions since it gives us an explicit description of the irreducible polynomial of any element over the ground field. For example, the field $\mathbb{Q}(\sqrt{2})$ is a normal extension of $\mathbb{Q}$ and $G(\mathbb{Q}(\sqrt{2})/\mathbb{Q}) = \{1, \sigma\}$, where 1 stands for the identity automorphism and σ the automorphism that maps a typical element $a + b\sqrt{2}$ to $a - b\sqrt{2}$. It follows that the conjugates of the element $2 + 3\sqrt{2}$, for example, are $2 + 3\sqrt{2}$ and $2 - 3\sqrt{2}$, and hence

$$\mathrm{Irr}(2 + 3\sqrt{2}; \mathbb{Q}) = [X - (2 + 3\sqrt{2})][X - (2 - 3\sqrt{2})] = X^2 - 4X - 14.$$

Similarly, $\mathbb{Q}(i)$ and $\mathbb{Q}(\omega)$ are normal extensions of $\mathbb{Q}$ and we find, for example, that

$$\mathrm{Irr}(3 + 2i; \mathbb{Q}) = [X - (3 + 2i)][X - (3 - 2i)] = X^2 - 6X + 13$$

$$\mathrm{Irr}(4 - 3\omega; \mathbb{Q}) = [X - (4 - 3\omega)][X - (4 - 3\omega^2)] = X^2 - 9X + 25.$$

EXAMPLE 1. Finite fields

Let us show that every finite field is a normal extension of its prime subfield and then use the above results, together with the Frobenius automorphism, to describe the irreducible polynomial of each element in the field over the prime subfield.

(A) Let K be a finite field of prime characteristic p, k its prime subfield, and let $[K:k] = n$. Recall that the automorphism group $G(K/k)$ is a cyclic group of order n generated by the Frobenius automorphism $\sigma_p : K \to K$, where $\sigma_p(x) = x^p$ for all elements $x \in K$. Therefore, $|G(K/k)| = [K:k] = n$ and hence K is a normal extension of k. Now, since $G(K/k) = \{1, \sigma_p, \ldots, \sigma_p^{n-1}\}$, it follows that if c is a typical element in K, then the conjugates of c are the powers $c, c^p, c^{p^2}, \ldots, c^{p^{n-1}}$, although these powers are not, in general, distinct. If $c, c^p, \ldots, c^{p^s}$ are the distinct conjugates of c, then, by Proposition 2,

$$\mathrm{Irr}(c;k) = (X - c)(X - c^p) \cdots (X - c^{p^s}).$$

(B) For example, let $K = \mathbb{F}_3[X]/(X^2 + 1)\mathbb{F}_3[X]$. Then K is a finite extension of $\mathbb{F}_3$ of degree 2 and hence is a finite field containing 3^2, or 9, elements. We find that $K = \{0, 1, 2, c, 2c, 1 + c, 2 + c, 1 + 2c, 2c\}$, where $c = X + (X^2 + 1)\mathbb{F}_3[X]$ and $c^2 = 2$. In this case $G(K/\mathbb{F}_3) = \{1, \sigma_3\}$, where $\sigma_3 : K \to K$ is the Frobenius automorphism defined by $\sigma_3(x) = x^3$ for all $x \in K$. The conjugates of the element c, for example, are c and $\sigma_3(c) = c^3 = c^2 c = 2c$, and hence

$$\mathrm{Irr}(c; \mathbb{F}_3) = (X - c)(X - 2c) = X^2 - 3cX + 2c^2 = X^2 + 1.$$

Similarly, the conjugates of $1 + c$ are $1 + c$ and $\sigma_3(1 + c) = 1 + c^3 = 1 + 2c$; therefore

$$\text{Irr}(1 + c; \mathbb{F}_3) = [X - (1 + c)][X - (1 + 2c)] = X^2 - 2X + 2.$$

The conjugates and irreducible polynomials of the remaining elements in K are found similarly and are summarized in the table here. Observe

$x \in K$	Conjugates of x	$\text{Irr}(x; F_3)$
0	0	X
1	1	$X - 1$
2	2	$X - 2$
c	$c, 2c$	$X^2 + 1$
$2c$	$2c, c$	$X^2 + 1$
$1 + c$	$1 + c, 1 + 2c$	$X^2 - 2X + 2$
$1 + 2c$	$1 + 2c, 1 + c$	$X^2 - 2X + 2$
$2 + c$	$2 + c, 2 + 2c$	$X^2 - X + 2$
$2 + 2c$	$2 + 2c, 2 + c$	$X^2 - X + 2$

that conjugate elements in K have the same irreducible polynomial over $\mathbb{F}_3$, which agrees with the above corollary. Finally, let us note that the decomposition of K into its conjugacy classes,

$$K = \{0\} \cup \{1\} \cup \{2\} \cup \{c, 2c\} \cup \{1 + c, 1 + 2c\} \cup \{2 + c, 2 + 2c\},$$

corresponds to the factorization of $X^9 - X$ into irreducible polynomials over $\mathbb{F}_3$:

$$X^9 - X = X(X - 1)(X - 2)(X^2 + 1)(X^2 - 2X + 2)(X^2 - X + 2);$$

under this correspondence, each irreducible factor corresponds to its set of conjugate roots in K. This concludes the example.

Now, if K/k is a normal extension of fields, then, as we have shown, the irreducible polynomial over k of any element in K has the form $(X - c_1) \cdots (X - c_n)$, where $c_1, \ldots, c_n$ are the distinct conjugates of the element, and hence the polynomial splits over K and has simple roots. Let us now show, as our second description of normal extensions, that this property in fact characterizes normal extensions.

PROPOSITION 3. Let K/k be a finite extension of fields. Then K is a normal extension of k if and only if the irreducible polynomial over k of every element in K splits over K and has only simple roots.

Proof. As we noted above, every normal extension of fields has the required property. Now suppose, conversely, that K/k is a finite extension of fields, say $K = k(c_1, \ldots, c_n)$, where $c_1, \ldots, c_n \in K$, and that the irreducible polynomials of elements in K split over K and have only simple roots. Then K is a splitting field for some polynomial over k; for if $f_i(X) = \mathrm{Irr}(c_i; k)$ for $i = 1, \ldots, n$, then, by assumption, each of the polynomials $f_1(X), \ldots, f_n(X)$ splits over K and hence so does their product $f_1(X) \cdots f_n(X)$. Therefore, K is a splitting field for $f_1(X) \cdots f_n(X)$ since it is generated by the roots $c_1, \ldots,$ c_n. Now, to show that K is normal over k, let c be an element in K fixed by every k-automorphism of K. If c' is any root of $\mathrm{Irr}(c; k)$, then c and c' have the same irreducible polynomial over k and hence there is some k-automorphism of K mapping c onto c' since K is a splitting field. But c is fixed by all k-automorphisms of K. Therefore, $c' = c$. Thus, c is the only root of $\mathrm{Irr}(c; k)$ and hence, since the roots are simple, $\mathrm{Irr}(c; k) = X - c$. Therefore, $c \in k$ and hence K is a normal extension of k. ∎

Proposition 3 is an important characterization of normal field extensions although it is difficult to use in practice since it requires rather detailed information about the irreducible polynomial of every element in the field. Nevertheless, it suggests a relationship between normal extensions and splitting fields for polynomials all of whose irreducible factors have only simple roots; indeed, we showed in the proof that every normal extension is a splitting field for such a polynomial. Let us now conclude our descriptions of normal extensions by showing that they are in fact splitting fields for precisely this type of polynomial. This, then, provides a simple and effective method for obtaining normal field extensions.

PROPOSITION 4. Let K/k be a finite extension of fields. Then K is a normal extension of k if and only if K is a splitting field for some polynomial over k all of whose irreducible factors over k have only simple roots.

Proof. If K is a normal extension of k, then, as we showed above, K is a splitting field for some polynomial over k all of whose irreducible factors have only simple roots. Conversely, suppose that K is a splitting field for some polynomial $f(X)$ over k all of whose irreducible factors have only simple roots. Let $[K:k] = n$. We show that K is a normal extension of k by induction on n. If $n = 1$, then $K = k$ and hence K is a normal extension of k. Now, let $n > 1$ and assume as induction hypothesis that if K'/k' is any finite extension of fields of degree less than n such that K' is a splitting field over k' for some polynomial all of whose irreducible factors have only simple roots, then K' is a normal extension of k'. Let c be an element in K fixed by every k-automorphism of K. Since $n > 1$, $f(X)$ has an irreducible factor $p(X)$ of degree $s > 1$ which, by assumption, splits over K and has only simple roots. Let $c_1, \ldots, c_s$ stand for the roots of $p(X)$ in K. Then K is a splitting field for $f(X)$ over the subfield $k(c_1)$ and, since $[K:k(c_1)] < n$, it follows from the induction hypothesis that K is a normal extension of $k(c_1)$. Therefore, $c \in k(c_1)$ since c is fixed by every $k(c_1)$-automorphism of K. Thus, we may write

$$c = a_0 + a_1 c_1 + \cdots + a_{s-1} c_1^{s-1}$$

for some elements $a_0, \ldots, a_{s-1} \in k$. But for each root c_i of $p(X)$, there is a k-automorphism σ_i of K that maps c_1 onto c_i since K is a splitting field over k. Therefore,

$$c = \sigma_i(c) = a_0 + a_1 c_i + \cdots + a_{s-1} c_1^{s-1}$$

for $i = 1, \ldots, s$. It follows that $c_1, \ldots, c_s$ are s distinct roots of the polynomial $(a_0 - c) + a_1 X + \cdots + a_{s-1} X^{s-1}$, which is impossible unless all coefficients are zero. Therefore, $a_0 - c = 0$ and hence $c \in k$. Thus, the only elements in K fixed by every k-automorphism of K are the elements in the ground field k and therefore K is a normal extension of k, as required.   ∎

COROLLARY.   Let K/k be a normal extension of fields and let k' be an intermediate subfield of the extension. Then K is a normal extension of k'.

Proof.   Since K is a normal extension of k, K is a splitting field for some polynomial $f(X)$ over k all of whose irreducible factors have only simple roots. But then K is also a splitting field for $f(X)$ over k' and hence is a normal extension of k'.   ∎

Proposition 4 provides an especially convenient way to obtain normal extensions of a given field k: simply choose a splitting field for any polynomial over k all of whose irreducible factors have only simple roots. For example, since every irreducible polynomial over a field of characteristic zero has only simple roots, any splitting field of characteristic zero is a normal extension of the ground field. The field $\mathbb{Q}(\omega)$, for example, is a normal extension of $\mathbb{Q}$ since it is a splitting field for $X^3 - 1$ over $\mathbb{Q}$, and $\mathbb{Q}(\omega, \sqrt[3]{2})$ is a normal extension of $\mathbb{Q}$ since it is a splitting field for $X^3 - 2$ over $\mathbb{Q}$. Similarly, $\mathbb{Q}(i, \sqrt{2})$ is a normal extension of $\mathbb{Q}$ since it is a splitting field for $X^4 - 4$ over $\mathbb{Q}$. Moreover, the corollary shows that if K is a normal extension of a field k, then K is normal over any intermediate subfield. Observe, however, that an intermediate subfield is not necessarily normal over k; the field $\mathbb{Q}(\omega, \sqrt[3]{2})$ is a normal extension of $\mathbb{Q}$ and hence is normal over the subfield $\mathbb{Q}(\sqrt[3]{2})$, but $\mathbb{Q}(\sqrt[3]{2})$ is not a normal extension of $\mathbb{Q}$.

Finally, let us note that a tower of normal extensions is not necessarily normal over the ground field; that is, if $k_1 \subseteq k_2 \subseteq \cdots \subseteq k_n$ is a tower of fields in which each field is normal over the previous field, then it is not necessarily true that k_n is normal over k_1. For example, $\mathbb{Q} \subseteq \mathbb{Q}(\sqrt{2}) \subseteq \mathbb{Q}(\sqrt[4]{2})$ is a tower of fields in which each field is a normal extension of the previous field, but

$\mathbb{Q}(\sqrt[4]{2})$ is not normal over $\mathbb{Q}$ since $\mathrm{Irr}(\sqrt[4]{2};\mathbb{Q}) = X^4 - 2$ and this polynomial does not split over $\mathbb{Q}(\sqrt[4]{2})$.

A polynomial over a field k all of whose irreducible factors have only simple roots is called a *separable polynomial* over k. Thus, Proposition 4 shows that normal extensions are simply splitting fields for separable polynomials. Let us note that while irreducible polynomials in characteristic zero must be separable, this is no longer true in fields of nonzero characteristic. We will discuss the concept of separability in more detail in the exercises and show, in particular, that any finite extension of fields, regardless of characteristic, can be "separated" into a separable part and a nonseparable part.

Let us now conclude this section by using our results on normal field extensions to discuss complex roots of unity and the fields they generate. These fields are called cyclotomic extensions and are especially important in field theory because they provide the basic algebraic model for studying the constructibility of regular n-gons in the plane. In addition, they also provide a rich source of examples for illustrating many of the concepts in field theory—the fields $\mathbb{Q}(\omega)$ and $\mathbb{Q}(i)$ that we studied earlier, for example, are cyclotomic extensions. Our goal in the following example is to discuss and illustrate the basic algebraic properties of cyclotomic extensions.

EXAMPLE 2. Roots of unity and cyclotomic extensions

Let n be a positive integer. Then the nth roots of unity are the complex roots of the polynomial $X^n - 1$. Recall that these roots form a cyclic group of order n under multiplication and that the $\varphi(n)$ generators of this group are called the primitive nth roots of unity. Thus, if z_n is a primitive nth root of unity, the roots of $X^n - 1$ are $1, z_n, \ldots, z_n^{n-1}$ and hence the field $\mathbb{Q}(z_n)$ generated by z_n is a splitting field for $X^n - 1$ over $\mathbb{Q}$. It follows, therefore, that $\mathbb{Q}(z_n)$ is a normal extension of the field $\mathbb{Q}$ of rational numbers. Any field of the form $\mathbb{Q}(z_n)$ generated by a primitive root of unity is called a *cyclotomic extension* of $\mathbb{Q}$. Our goal in this example is to first determine the irreducible polynomial of z_n over $\mathbb{Q}$—which is called the nth cyclotomic polynomial over $\mathbb{Q}$ and is itself an important polynomial—and use it to show that $[\mathbb{Q}(z_n):\mathbb{Q}] = \varphi(n)$. We then determine the automorphism group $G(\mathbb{Q}(z_n)/\mathbb{Q})$ and show that $G(\mathbb{Q}(z_n)/\mathbb{Q}) \cong U(\mathbb{Z}_n)$, the multiplicative group of units in the ring $\mathbb{Z}_n$ of integers modulo n.

(A) Let us begin by recalling that the ring $\mathbb{Z}[X]$ is a unique factorization domain; that is, every nonconstant polynomial with integer coefficients may be factored into a product of irreducible polynomials over $\mathbb{Z}$ that are unique to within order and constants. Now, let z_n be a primitive nth root of unity and let $p(X)$ be the irreducible factor of $X^n - 1$ in $\mathbb{Z}[X]$ that has z_n as a root. Then $X^n - 1 = p(X)q(X)$ for some polynomial $q(X)$ $\in \mathbb{Z}[X]$, and both $p(X)$ and $q(X)$ are monic. In particular, since $p(X)$ is a monic irreducible polynomial over $\mathbb{Z}$, it is also irreducible over $\mathbb{Q}$

and hence is the irreducible polynomial of z_n over $\mathbb{Q}$; that is, $p(X) = \mathrm{Irr}(z_n; \mathbb{Q})$. Moreover, since $\mathbb{Q}(z_n)$ is normal over $\mathbb{Q}$, the roots of $p(X)$ are the conjugates of z_n. Since every conjugate of a primitive root is clearly a primitive root, it follows that $p(X)$ has only primitive nth roots of unity as roots. We now claim that every primitive nth root of unity is a root of $p(X)$. To show this, let us first observe that it is sufficient to show that if $p(w) = 0$ for some primitive root w, then $p(w^p) = 0$ for every prime p not dividing n. For if w' is any primitive root, then $w' = z_n^s$ for some positive integer s relatively prime to n and hence there is a sequence of primitive roots $z_n, z_n^p, z_n^{pq}, \ldots, z_n^s = w'$, where $p, q, \ldots$ are prime divisors of s. It follows that if $p(X)$ vanishes on z_n and vanishes on any of these roots whenever it vanishes on the previous root, then $p(w') = 0$. We assume therefore that $p(w) = 0$ for some primitive root w and let p be any prime not dividing n, and work to show that $p(w^p) = 0$. Suppose to the contrary that $p(w^p) \neq 0$. Then $q(w^p) = 0$ and hence w is a root of $q(X^p)$. Therefore, $p(X)$ divides $q(X^p)$, say $q(X^p) = p(X)r(X)$, where $r(X) \in \mathbb{Q}[X]$. Since both $q(X^p)$ and $p(X)$ are monic and have integer coefficients, it follows by long-division that $r(X)$ is also monic and has integer coefficients. Now, reduce the coefficients in the equation $q(X^p) = p(X)r(X)$ modulo p to get the equation $\bar{q}(X^p) = \bar{p}(X)\bar{r}(X)$ in $\mathbb{F}_p[X]$. But $\bar{q}(X^p) = \bar{q}(X)^p$ in $\mathbb{F}_p[X]$. Hence, $\bar{q}(X)^p = \bar{p}(X)\bar{r}(X)$ and therefore $\bar{p}(X)$ and $\bar{q}(X)$ must have some root in common in some extension field of $\mathbb{F}_p$ over which both polynomials split. Since $X^n - 1 = \bar{p}(X)\bar{q}(X)$, it follows that $X^n - 1$ has a multiple root, which is impossible since its derivative is $\bar{n}X^{n-1}$ and $\bar{n} \neq 0$. Therefore, $p(w^p) = 0$ and hence every primitive nth root of unity is a root of $p(X)$, as required. It follows, therefore, that the roots of $p(X)$ are precisely the primitive nth roots of unity.

(B) Let $w_1, \ldots, w_m$ stand for the primitive nth roots of unity, where $m = \varphi(n)$, and let $\Phi_n(X) = (X - w_1) \cdots (X - w_m)$. Then it follows from part (A) that $\mathrm{Irr}(z_n; \mathbb{Q}) = \Phi_n(X)$ and that $\Phi_n(X)$ is an irreducible polynomial over $\mathbb{Q}$ having integer coefficients. The polynomial $\Phi_n(X)$ is called the *nth cyclotomic polynomial over* $\mathbb{Q}$. Although it appears difficult to determine these polynomials, there is in fact a simple recursive procedure for finding them. We first write

$$X^n - 1 = (X - 1)(X - z_n) \cdots (X - z_n^{n-1}).$$

Now observe that each of the roots $1, z_n, \ldots, z_n^{n-1}$ is a primitive dth root of unity for some divisor d of n, and, conversely, every primitive dth root occurs in this factorization for any such divisor d. Since the product of the factors corresponding to the primitive dth roots is $\Phi_d(X)$, it follows that we may arrange the factors to obtain the formula

$$X^n - 1 = \prod \Phi_d(X),$$

where the product is over all positive divisors d or n. Thus, if we know the cyclotomic polynomials $\Phi_d(X)$ for all divisors $d < n$, then $\Phi_n(X)$ is simply $X^n - 1$ divided by their product. This product formula therefore provides a recursive method for finding the cyclotomic polynomials. For example, we find directly from the definition that $\Phi_1(X) = X - 1$ and $\Phi_2(X) = X + 1$. Therefore,

$$\Phi_3(X) = \frac{X^3 - 1}{\Phi_1(X)} = \frac{X^3 - 1}{X - 1} = X^2 + X + 1.$$

Similarly,

$$\Phi_4(X) = \frac{X^4 - 1}{\Phi_1(X)\Phi_2(X)} = X^2 + 1,$$

$$\Phi_5(X) = \frac{X^5 - 1}{\Phi_1(X)} = X^4 + X^3 + X^2 + X + 1,$$

$$\Phi_6(X) = \frac{X^6 - 1}{\Phi_1(X)\Phi_2(X)\Phi_3(X)} = X^2 - X + 1.$$

Finally, observe that the product formula $X^n - 1 = \Pi\, \Phi_d(X)$ also gives the factorization of $X^n - 1$ into the product of irreducible polynomials in $\mathbb{Q}[X]$; for example, $X^6 - 1 = \Phi_1(X)\Phi_2(X)\Phi_3(X)\Phi_6(X) = (X - 1)(X + 1)(X^2 + X + 1)(X^2 - X + 1)$.

(C) Let us now use the cyclotomic polynomials to discuss the extension $\mathbb{Q}(z_n)/\mathbb{Q}$. Since $\mathrm{Irr}(z_n;\mathbb{Q}) = \Phi_n(X)$ and $\deg \Phi_n(X) = \varphi(n)$, it follows that

$$[\mathbb{Q}(z_n):\mathbb{Q}] = \deg \mathrm{Irr}(z_n;\mathbb{Q}) = \varphi(n).$$

Moreover, the conjugates of z_n are the roots of $\Phi_n(X)$ and hence are the primitive nth roots of unity. Since the primitive roots have the form z_n^s, where s and n are relatively prime, it follows that for each integer s relatively prime to n, there is a $\mathbb{Q}$-automorphism $\sigma_s:\mathbb{Q}(z_n) \to \mathbb{Q}(z_n)$ such that $\sigma_s(z_n) = z_n^s$. Since $\sigma_s = \sigma_t$ if and only if $s \equiv t \pmod n$, the mappings σ_s account for $\varphi(n)$ $\mathbb{Q}$-automorphisms of $\mathbb{Q}(z_n)$. But there are only $\varphi(n)$ such automorphisms since $|G(\mathbb{Q}(z_n)/\mathbb{Q})| = [\mathbb{Q}(z_n):\mathbb{Q}] = \varphi(n)$. Therefore,

$$G(\mathbb{Q}(z_n)/\mathbb{Q}) = \{\sigma_s \,|\, 1 \le s \le n,\ \gcd(s, n) = 1\}.$$

We leave the reader to verify that $\sigma_s\sigma_t = \sigma_{st}$ for all $\sigma_s, \sigma_t \in G(\mathbb{Q}(z_n)/\mathbb{Q})$. Finally, let us recall that the multiplicative group $U(\mathbb{Z}_n)$ of units in the ring $\mathbb{Z}_n$ consists of all congruence classes $[s] \in \mathbb{Z}_n$ for which $\gcd(s, n) = 1$. Now, define the mapping $f: U(\mathbb{Z}_n) \to G(\mathbb{Q}(z_n)/\mathbb{Q})$ by setting $f([s]) = \sigma_s$ for every class $[s] \in U(\mathbb{Z}_n)$. Then it follows easily that f is a group-isomorphism and hence $G(\mathbb{Q}(z_n)/\mathbb{Q}) \cong U(\mathbb{Z}_n)$.

(D) Let us conclude this example by discussing in detail the cyclotomic extension $\mathbb{Q}(z_5)$, where z_5 stands for a primitive 5th root of unity. In this case

$$\mathrm{Irr}(z_5;\mathbb{Q}) = \Phi_5(X) = X^4 + X^3 + X^2 + X + 1,$$

$$[\mathbb{Q}(z_5):\mathbb{Q}] = \varphi(5) = 4.$$

It follows therefore that every element in $\mathbb{Q}(z_5)$ may be written uniquely in the form $a + bz_5 + cz_5^2 + dz_5^3$, where $a, b, c, d \in \mathbb{Q}$. Moreover, since z_5 is a root of $\Phi_5(X)$, $z_5^4 + z_5^3 + z_5^2 + z_5 + 1 = 0$ and hence $z_5^4 = -1 - z_5 - z_5^2 - z_5^3$. Thus, we find, for example, that

$$(1 - z_5^3)(z_5 - z_5^2) = z_5 - z_5^4 - z_5^2 + z_5^5$$
$$= z_5 - (-1 - z_5 - z_5^2 - z_5^3) - z_5^2 + 1$$
$$= 2 + 2z_5 + z_5^3.$$

Similarly, since z_5 is a root of $X^5 - 1$, $z_5^5 = 1$ and hence

$$\frac{1}{z_5} = z_5^4, \quad \frac{1}{z_5^2} = z_5^3, \quad \frac{1}{z_5^3} = z_5^2, \quad \frac{1}{z_5^4} = z_5.$$

Now, $G(\mathbb{Q}(z_5)/\mathbb{Q}) = \{1, \sigma_2, \sigma_3, \sigma_4\}$, where

$$1: \mathbb{Q}(z_5) \to \mathbb{Q}(z_5), \quad 1(z_5) = z_5,$$

$$\sigma_2: \mathbb{Q}(z_5) \to \mathbb{Q}(z_5), \quad \sigma_2(z_5) = z_5^2,$$

$$\sigma_3: \mathbb{Q}(z_5) \to \mathbb{Q}(z_5), \quad \sigma_3(z_5) = z_5^3,$$

$$\sigma_4: \mathbb{Q}(z_5) \to \mathbb{Q}(z_5), \quad \sigma_4(z_5) = z_5^4.$$

For example, $\sigma_2\sigma_4(z_5) = \sigma_2(z_5^4) = z_5^8 = z_5^3 = \sigma_3(z_5)$ and therefore $\sigma_2\sigma_4 = \sigma_3$, which agrees with the fact that $2 \cdot 4 \equiv 3 \pmod 5$. The figure here gives the complete multiplication table for the automorphism group $G(\mathbb{Q}(z_5)/\mathbb{Q})$ and the group $U(\mathbb{Z}_5)$ of units of $\mathbb{Z}_5$, and illustrates the isomorphism between these two groups.

	1	σ_2	σ_3	σ_4		[1]	[2]	[3]	[4]
1	1	σ_2	σ_3	σ_4	[1]	[1]	[2]	[3]	[4]
σ_2	σ_2	σ_4	1	σ_3	[2]	[2]	[4]	[1]	[3]
σ_3	σ_3	1	σ_4	σ_2	[3]	[3]	[1]	[4]	[2]
σ_4	σ_4	σ_3	σ_2	1	[4]	[4]	[3]	[2]	[1]

$$G(\mathbb{Q}(z_5)/\mathbb{Q}) \qquad \cong \qquad U(\mathbb{Z})$$

Clearly, $|\sigma_2| = |\sigma_3| = 4$ and $|\sigma_4| = 2$. Therefore, $G(\mathbb{Q}(z_5)/\mathbb{Q})$ is a cyclic group of order 4 generated by either σ_2 or σ_3. For example, if $c = z_5 + 1/z_5 = z_5 + z_5^4$, then

$$1(c) = z_5 + z_5^4 = c$$

$$\sigma_2(c) = z_5^2 + z_5^8 = z_5^2 + z_5^3 = -1 - c$$

$$\sigma_3(c) = z_5^3 + z_5^{12} = z_5^3 + z_5^2 = -1 - c$$

$$\sigma_4(c) = z_5^4 + z_5^{16} = z_5^4 + z_5 = c.$$

The element c therefore has two distinct conjugates, c and $-1 - c$, and hence $\operatorname{Irr}(c;\mathbb{Q}) = (X - c)(X + 1 + c) = X^2 + X - c - c^2 = X^2 + X - (z_5 + z_5^4 + z_5^2 + 2 + z_5^3) = X^2 + X - 1$. It follows therefore that

$$[\mathbb{Q}(c):\mathbb{Q}] = \deg \operatorname{Irr}(c;\mathbb{Q}) = 2$$

$$[\mathbb{Q}(z_5):\mathbb{Q}(c)] = \frac{4}{2} = 2.$$

Note that since $\mathbb{Q}(z_5)$ is a normal extension of $\mathbb{Q}$, $\operatorname{Irr}(c;\mathbb{Q})$ must split over $\mathbb{Q}(z_5)$; it does, of course, since its roots are c and $-1 - c$. In fact, $\operatorname{Irr}(c;\mathbb{Q})$ already splits over $\mathbb{Q}(c)$. Therefore, $\mathbb{Q}(c)$ is a normal extension of $\mathbb{Q}$ since it is a splitting field for $\operatorname{Irr}(c;\mathbb{Q})$ and hence has two $\mathbb{Q}$-automorphisms since $[\mathbb{Q}(c):\mathbb{Q}] = 2$. Moreover, these automorphisms must map the element c to a root of $\operatorname{Irr}(c;\mathbb{Q})$ and hence to either c or $-1 - c$. Thus, $G(\mathbb{Q}(c)/\mathbb{Q}) = \{1, \tau\}$, where $1:\mathbb{Q}(c) \to \mathbb{Q}(c)$ is the identity automorphism of $\mathbb{Q}(c)$ and $\tau:\mathbb{Q}(c) \to \mathbb{Q}(c)$ is the automorphism defined by $\tau(a + bc) = a + b(-1 - c)$ for all $a + bc \in \mathbb{Q}(c)$. Note that the two automorphisms of $\mathbb{Q}(c)$ are just the restriction of the automorphisms of $\mathbb{Q}(z_5)$ to $\mathbb{Q}(c)$: both 1 and σ_4 restrict to the identity automorphism 1, while σ_2 and σ_3 restrict to τ, as illustrated here.

Finally, let us find the automorphism group $G(\mathbb{Q}(z_5)/\mathbb{Q}(c))$ and then use it to find the conjugates and irreducible polynomial of z_5 over the intermediate subfield $\mathbb{Q}(c)$. By definition, the group $G(\mathbb{Q}(z_5)/\mathbb{Q}(c))$ consists of those automorphisms of $\mathbb{Q}(z_5)$ that fix the element c. By inspection, there are two such automorphisms: 1 and σ_4. Hence, $G(\mathbb{Q}(z_5)/\mathbb{Q}(c)) = \{1, \sigma_4\}$. Thus, the conjugates of z_5 over the intermediate subfield $\mathbb{Q}(c)$ are $1(z_5) = z_5$ and $\sigma_4(z_5) = z_5^4$, and therefore

$$\operatorname{Irr}(z_5;\mathbb{Q}(c)) = (X - z_5)(X - z_5^4) = X^2 - cX + 1.$$

Thus, in particular, $[\mathbb{Q}(z_5):\mathbb{Q}(c)] = 2 = |G(\mathbb{Q}(z_5)/\mathbb{Q}(c))|$, which agrees with the fact that $\mathbb{Q}(z_5)$ is a normal extension of the intermediate subfield $\mathbb{Q}(c)$. This concludes the example.

Exercises

1. Let $K = \mathbb{Q}(\sqrt{2}, \sqrt{3})$.
   (a) Show that K is a normal extension of $\mathbb{Q}$ and verify that $|G(K/\mathbb{Q})| = [K:\mathbb{Q}]$.
   (b) Find the conjugates of $\sqrt{2} + \sqrt{3}$ and $\mathrm{Irr}(\sqrt{2} + \sqrt{3}; \mathbb{Q})$.
   (c) Find the conjugates of $\sqrt{6} - \sqrt{2}$ and $\mathrm{Irr}(\sqrt{6} - \sqrt{2}; \mathbb{Q})$.
   (d) Find the conjugates of $3 - \sqrt{6}$ and $\mathrm{Irr}(3 - \sqrt{6}; \mathbb{Q})$.
   (e) Show that K is a normal extension of the intermediate subfield $\mathbb{Q}(\sqrt{6})$ and find the automorphism group $G(K/\mathbb{Q}(\sqrt{6}))$.
   (f) Find the conjugates of $\sqrt{2} + \sqrt{3}$ relative to the extension $K/\mathbb{Q}(\sqrt{6})$ and find $\mathrm{Irr}(\sqrt{2} + \sqrt{3}; \mathbb{Q}(\sqrt{6}))$.

2. Let $K = \mathbb{Q}(\sqrt{2}, \sqrt{-2})$.
   (a) Show that K is a normal extension of $\mathbb{Q}$ and verify that $|G(K/\mathbb{Q})| = [K:\mathbb{Q}]$.
   (b) Find the conjugates of $\sqrt{2} + \sqrt{-2}$ and $\mathrm{Irr}(\sqrt{2} + \sqrt{-2}; \mathbb{Q})$.
   (c) Find the conjugates of $i + \sqrt{2}$ and $\mathrm{Irr}(i + \sqrt{2}; \mathbb{Q})$.
   (d) Find the conjugates of $i\sqrt{2}$ and $\mathrm{Irr}(i\sqrt{2}; \mathbb{Q})$.
   (e) Show that K is a normal extension of the intermediate subfield $\mathbb{Q}(i)$ and find the automorphism group $G(K/\mathbb{Q}(i))$.
   (f) Find the conjugates of $\sqrt{2} + \sqrt{-2}$ relative to the extension $K/\mathbb{Q}(i)$ and find $\mathrm{Irr}(\sqrt{2} + \sqrt{-2}; \mathbb{Q}(i))$.

3. Let $K = \mathbb{Q}(\sqrt[4]{2}, i)$.
   (a) Show that K is a normal extension of $\mathbb{Q}$ and verify that $|G(K/\mathbb{Q})| = [K:\mathbb{Q}]$.
   (b) Find the conjugates of $i\sqrt[4]{2}$ and $\mathrm{Irr}(i\sqrt[4]{2}; \mathbb{Q})$.
   (c) Find the conjugates of $\sqrt{2} + \sqrt[4]{2}$ and $\mathrm{Irr}(\sqrt{2} + \sqrt[4]{2}; \mathbb{Q})$.
   (d) Find the conjugates of $i + \sqrt[4]{2}$ and $\mathrm{Irr}(i + \sqrt[4]{2}; \mathbb{Q})$.
   (e) Show that K is a normal extension of the intermediate subfield $\mathbb{Q}(\sqrt{2})$ and find the automorphism group $G(K/\mathbb{Q}(\sqrt{2}))$.
   (f) Show that $\sqrt{i} \in K$ and find $\mathrm{Irr}(\sqrt{i}; (\mathbb{Q}(\sqrt{2}))$.

4. Let $K = \mathbb{F}_2[X]/(X^3 + X + 1)\mathbb{F}_2[X]$.
   (a) Show that K is a finite field containing eight elements.
   (b) Find the automorphism group $G(K/\mathbb{F}_2)$ and verify that $|G(K/\mathbb{F}_2)| = [K:\mathbb{F}_2]$.
   (c) Find the conjugates and irreducible polynomial over $\mathbb{F}_2$ of each element in K.
   (d) Factor the polynomial $X^8 - X$ into irreducible polynomials over $\mathbb{F}_2$ of each element in K.

5. Let K/k be a finite extension of fields and let $c \in K$. Show that the subfield $k(c)$ generated by c is a normal extension of k if and only if $\mathrm{Irr}(c;k)$ splits over $k(c)$ and has only simple roots; that is, if and only if the roots of $\mathrm{Irr}(c;k)$ are simple roots and may be expressed arithmetically in terms of a single root.

6. Show that every subfield of $\mathbb{Q}(\sqrt{2}, \sqrt{3})$ is a normal extension of $\mathbb{Q}$.

7. Show that every subfield of $\mathbb{Q}(\sqrt{2}, \sqrt{-2})$ is a normal extension of $\mathbb{Q}$.
8. Is every subfield of $\mathbb{Q}(\sqrt[4]{2}, i)$ a normal extension of $\mathbb{Q}$?
9. Let K/k be a finite extension of fields and let k_1 and k_2 be intermediate subfields that are normal extensions of k.
   (a) Show that the subfield $k_1 \cap k_2$ is a normal extension of k.
   (b) Show that the subfield of K generated by k_1 and k_2 is a normal extension of k.
10. Let z_6 stand for a primitive 6th root of unity.
    (a) Find $\text{Irr}(z_6; \mathbb{Q})$ and $[\mathbb{Q}(z_6):\mathbb{Q}]$.
    (b) Find the automorphism group $G(\mathbb{Q}(z_6)/\mathbb{Q})$.
    (c) Find the conjugates of $1 - z_6$ and $\text{Irr}(1 - z_6; \mathbb{Q})$.
    (d) Find the conjugates of $z_6 + \dfrac{1}{z_6}$ and $\text{Irr}\left(z_6 + \dfrac{1}{z_6}; \mathbb{Q}\right)$.
    (e) Let $a + bz_6$ be a typical element in $\mathbb{Q}(z_6)$, where $a, b \in \mathbb{Q}$.
        (1) Show that $(a + bz_6)\left(a + \dfrac{b}{z_6}\right)$ is a rational number.
        (2) If $a + bz_6 \neq 0$, show that

$$\frac{1}{a + bz_6} = \frac{a + b}{a^2 + ab + b^2} - \frac{b}{a^2 + ab + b^2} z_6.$$

11. Let n be a positive integer and let z stand for any nth root of unity.
    (a) Show that $\dfrac{1}{z} = \bar{z}$, the complex conjugate of z.
    (b) Show that $z + \dfrac{1}{z}$ is a real number.
    (c) If $z \neq 1$, show that $1 + z + \cdots + z^{n-1} = 0$.
12. Let n be a positive integer and let z_n be a primitive nth root of unity. Let s be a positive integer, $d = \gcd(n, s)$, and $q = \dfrac{n}{d}$.
    (a) Show that z_n^s is a primitive qth root of unity.
    (b) Show that $\mathbb{Q}(z_n^s) = \mathbb{Q}(z_q)$, where z_q is any primitive qth root of unity.
13. Find the following irreducible polynomials over the indicated fields:
    (a) $\text{Irr}(z_{12}^4; \mathbb{Q})$ (d) $\text{Irr}(z_{20}^8; \mathbb{Q})$
    (b) $\text{Irr}(z_{10}^2; \mathbb{Q})$ (e) $\text{Irr}(z_{20}^6; \mathbb{Q})$
    (c) $\text{Irr}(z_{12}^{10}; \mathbb{Q})$
14. (a) Find the cyclotomic polynomials $\Phi_8(X)$, $\Phi_9(X)$, and $\Phi_{10}(X)$.
    (b) Factor the polynomials $X^8 - 1$, $X^9 - 1$, and $X^{10} - 1$ into the product of irreducible polynomials over $\mathbb{Q}$.
15. Let z_{12} stand for a primitive 12th root of unity.
    (a) Find $\text{Irr}(z_{12}; \mathbb{Q})$ and $[\mathbb{Q}(z_{12}):\mathbb{Q}]$.
    (b) Find the automorphism group $G(\mathbb{Q}(z_{12})/\mathbb{Q})$ and show that $G(\mathbb{Q}(z_{12})/\mathbb{Q}) \cong \mathbb{Z}_2 \times \mathbb{Z}_2$.
    (c) Factor the polynomial $X^{12} - 1$ into the product of irreducible polynomials over $\mathbb{Q}$.

    (d) Let $c = z_{12} + z_{12}^5$.

      (1) Find the conjugates of c, $\mathrm{Irr}(c;\mathbb{Q})$, and $[\mathbb{Q}(c):\mathbb{Q}]$.

      (2) Show that $\mathbb{Q}(z_{12})$ is a normal extension of the intermediate subfield $\mathbb{Q}(c)$ and find $[\mathbb{Q}(z_{12}):\mathbb{Q}(c)]$.

      (3) Find the automorphism group $G(\mathbb{Q}(z_{12})/(\mathbb{Q}(c))$ and verify that $|G(\mathbb{Q}(z_{12})/\mathbb{Q}(c))| = [\mathbb{Q}(z_{12}):\mathbb{Q}(c)]$.

      (4) Find the conjugates of z_{12} relative to the extension $\mathbb{Q}(z_{12})/\mathbb{Q}(c)$ and find $\mathrm{Irr}(z_{12};\mathbb{Q}(c))$.

      (5) Is $\mathbb{Q}(c)$ a normal extension of $\mathbb{Q}$?

      (6) Factor the polynomial $X^{12} - 1$ into the product of irreducible polynomials over $\mathbb{Q}(c)$.

16. Let p be a prime and n a positive integer.

    (a) Show that $\Phi_{p^n}(X) = \Phi_p(X^{p^{n-1}})$.

    (b) If p does not divide n, show that $\Phi_{pn}(X) = \dfrac{\Phi_n(X^p)}{\Phi_n(X)}$.

17. Let $\mathbb{F}_2(t)$ stand for the field of rational functions in t over $\mathbb{F}_2$ and let $f(X) = X^2 - t \in \mathbb{F}_2(t)[X]$. Let K stand for a splitting field for $f(X)$ over $\mathbb{F}_2(t)$.

    (a) Find $[K:\mathbb{F}_2(t)]$.

    (b) Find $G(K/\mathbb{F}_2(t))$.

    (c) Show that K is not a normal extension of $\mathbb{F}_2(t)$.

    (d) Show that $\mathrm{Irr}(c;\mathbb{F}_2(t))$ splits over K for every element $c \in K$.

    (e) Find an element $c \in K$ such that $\mathrm{Irr}(c;\mathbb{F}_2(t))$ does not have simple roots.

18. Let $f(X)$ be an irreducible polynomial over an arbitrary field k and suppose that $f(X)$ has a simple root in some extension field of k over which $f(X)$ splits. Show that every root of $f(X)$ is a simple root.

19. Let $f(X)$ be a nonconstant polynomial over an arbitrary field k. Show that there are unique polynomials $g(X)$ and $h(X)$ over k such that $f(X) = g(X)h(X)$, where $g(X)$ is separable over k—that is, its irreducible factors have only simple roots—and the irreducible factors of $h(X)$ have only multiple roots.

20. **Separable extensions.** Let K/k be a finite extension of fields. A polynomial over k is said to be *separable* over k if its irreducible factors over k have only simple roots, and an element $c \in K$ is *separable* over k if its irreducible polynomial is a separable polynomial over k. The extension K is a *separable extension* of k if every element in K is separable over k.

    (a) If K is a normal extension of k, show that K is a separable extension of k.

    (b) If L is a separable extension of K and K is a separable extension of k, show that L is a separable extension of k.

    (c) Let $c \in K$. Show that c is separable over k if and only if the subfield $k(c)$ is a separable extension of k.

    (d) Show that K is a separable extension of k if and only if K is generated by a finite number of elements that are separable over k.

(e) Let $K_{sep} = \{c \in K \,|\, c$ is separable over $k\}$. Show that K_{sep} is an intermediate subfield of the extension K/k.

(f) Show that the only elements in K that are separable over the subfield K_{sep} are precisely the elements in K_{sep}.

(g) If $\text{char}(k) = 0$, show that $K_{sep} = K$. Thus, every finite extension of a field of characteristic zero is a separable extension.

(h) If k is finite, show that $K_{sep} = K$. Thus, every finite field is a separable extension of its prime subfield.

(i) Find an example of a separable extension of fields that is not a normal extension.

21. **The trace function of a normal extension.** Let K/k be a normal extension of fields with automorphism group G. For each element $c \in K$, let

$$\text{Tr}_{K/k}(c) = \sum \sigma(c),$$

where the summation is over all automorphisms $\sigma \in G$.

(a) Show that $\sigma(\text{Tr}_{K/k}(c)) = \text{Tr}_{K/k}(c)$ for every automorphism $\sigma \in G$ and every element $c \in K$.

(b) Show that $\text{Tr}_{K/k}(c) \in k$ for every element $c \in K$.

(c) If $c \in K$, the element $\text{Tr}_{K/k}(c) \in k$ is called the *trace of c*. The function $\text{Tr}_{K/k} : K \to k$ that maps every element $c \in K$ to its trace $\text{Tr}_{K/k}(c)$ is called the *trace function* of the extension K/k. Show that the trace function $\text{Tr}_{K/k}$ is a k-linear transformation of K as a vector space over k; that is, show that $\text{Tr}_{K/k}(ax + by) = a\,\text{Tr}_{K/k}(x) + b\,\text{Tr}_{K/k}(y)$ for all scalars $a, b \in k$ and all elements $x, y \in K$.

(d) If $a \in k$, show that $\text{Tr}_{K/k}(a) = na$, where $n = [K:k]$.

(e) Let $a + b\sqrt{2}$ be a typical element in $\mathbb{Q}(\sqrt{2})$. Find $\text{Tr}_{\mathbb{Q}(\sqrt{2})/\mathbb{Q}}(a + b\sqrt{2})$.

(f) Let $a + bi$ be a typical element in $\mathbb{Q}(i)$. Find $\text{Tr}_{\mathbb{Q}(i)/\mathbb{Q}}(a + bi)$.

(g) Let $K = \mathbb{Q}(\omega, \sqrt[3]{2})$. Find $\text{Tr}_{K/\mathbb{Q}}(\omega \sqrt[3]{2})$.

(h) Let $K = \mathbb{F}_3[X]/(X^2 + 1)\mathbb{F}_3[X]$ stand for the field discussed in Example 1 of this section. Find $\text{Tr}_{K/\mathbb{F}_3}(c)$ for each element $c \in K$.

(i) Show that the trace function $\text{Tr}_{K/k}$ maps K onto k.

22. **The norm function of a normal extension.** Let K/k be a normal extension of fields with automorphism group G. For each element $c \in K$, let

$$N_{K/k}(c) = \prod \sigma(c),$$

where the product is over all automorphisms $\sigma \in G$.

(a) Show that $\sigma(N_{K/k}(c)) = N_{K/k}(c)$ for every automorphism $\sigma \in G$ and every element $c \in K$.

(b) Show that $N_{K/k}(c) \in k$ for every element $c \in K$.

(c) If $c \in K$, the element $N_{K/k}(c) \in k$ is called the *norm of c*. The function $N_{K/k} : K \to k$ that maps every element $c \in K$ to its norm $N_{K/k}(c)$ is called the *norm function* of the extension K/k. Show that the norm function $N_{K/k}$ is a multiplicative function: that is, show that $N_{K/k}(xy) = N_{K/k}(x)N_{K/k}(y)$ for all elements $x, y \in K$.

(d) If $a \in k$, show that $N_{K/k}(a) = a^n$, where $n = [K:k]$.

(e) Let $a + b\sqrt{2}$ be a typical element in $\mathbb{Q}(\sqrt{2})$. Find $N_{\mathbb{Q}(\sqrt{2})/\mathbb{Q}}(a + b\sqrt{2})$.

(f) Let $a + bi$ be a typical element in $\mathbb{Q}(i)$. Find $N_{\mathbb{Q}(i)/\mathbb{Q}}(a + bi)$.

(g) Let $K = \mathbb{Q}(\omega, \sqrt[3]{2})$. Find $N_{K/\mathbb{Q}}(\omega \sqrt[3]{2})$.

(h) Let $a + b\omega$ be a typical element in $\mathbb{Q}(\omega)$. Show that $N_{\mathbb{Q}(\omega)/\mathbb{Q}}(a + b\omega)$ $= a^2 - ab + b^2$.

(i) Let $a + bz_6$ be a typical element in $\mathbb{Q}(z_6)$, where z_6 stands for a primitive 6th root of unity. Show that $N_{\mathbb{Q}(z_6)/\mathbb{Q}}(a + bz_6) = a^2 + ab + b^2$.

(j) Let $K = \mathbb{F}_3[X]/(X^2 + 1)\mathbb{F}_3[X]$ stand for the field discussed in Example 1 of this section. Find $N_{K/\mathbb{F}_3}(c)$ for each element $c \in K$.

(k) Show that the norm function defines a group-homomorphism $N_{K/k}$: $K^* \to k^*$ from the multiplicative group of nonzero elements in K into the multiplicative group of nonzero elements in k.

23. Let K/k be a normal extension of fields and suppose that $K = k(c)$ for some element $c \in K$. Let $\mathrm{Irr}(c;k) = X^n + a_1 X^{n-1} + \cdots + a_n$, where $a_1, \ldots, a_n \in k$. Show that $\mathrm{Tr}_{K/k}(c) = -a_1$ and $N_{K/k}(c) = (-1)^n a_n$.

3. THE FUNDAMENTAL THEOREM OF GALOIS THEORY

Recall that if K/k is an arbitrary extension of fields with automorphism group G, then every subgroup H of G has associated with it the fixed field K^H. Thus, we have a correspondence $H \mapsto K^H$ between the subgroups of G and the intermediate subfields of K/k. In this section our goal is to bring together our results on automorphism groups and normal field extensions to show that if K/k is a normal extension of fields, then the correspondence $H \mapsto K^H$, called the Galois correspondence, has two basic properties: first, it is a 1–1 correspondence between the subgroups of G and intermediate subfields of K/k that reverses the inclusion relation on subgroups and subfields; and second, H is a normal subgroup of G if and only if the corresponding subfield K^H is a normal extension of k, in which case the automorphism group $G(K^H/k)$ is isomorphic to the quotient group G/H. These results, known collectively as the fundamental theorem of Galois theory, are exceptionally important since they show that the Galois correspondence for normal field extensions provides a mirror that faithfully reflects the intermediate subfield structure of the extension onto the subgroup structure of its automorphism group. We will illustrate these ideas and results by discussing the Galois theory of finite fields, cyclotomic extensions, symmetric polynomials, and the existence of primitive elements for finite extensions, and then conclude the chapter by using the Galois theory together with the Sylow theory to give a proof, due to Emil Artin, of the fundamental theorem of algebra. As we noted at the beginning of the chapter, it is no exaggeration to say that the Galois theory of fields, with its mutual interaction between field theory and group theory, is one of the most important and useful algebraic theories in mathematics today.

Let us begin by first proving a preliminary result about groups of automorphisms of an arbitrary finite extension of fields. Recall that if K/k is a finite extension of fields and H is any group of k-automorphisms of K, then $|H| = [K:K^H]$. We claim that K is in fact a normal extension of the fixed field K^H and that the group of the extension K/K^H is precisely H. It follows, then, that every group of automorphisms of K is completely determined by its fixed field.

PROPOSITION 1. Let K/k be a finite extension of fields and let H be a group of k-automorphisms of K. Then K is a normal extension of the fixed field K^H and $G(K/K^H) = H$.

Proof. Clearly, $H \leq G(K/K^H)$. Moreover, $K^{G(K/K^H)} = K^H$ since an element in K is fixed by every automorphism in H if and only if it is fixed by every automorphism in $G(K/K^H)$. Thus, the fixed field of the group $G(K/K^H)$ is K^H and hence K is a normal extension of K^H. Therefore, $[K:K^H] = |G(K/K^H)|$. Finally, since $H \leq G(K/K^H)$ and $|H| = [K:K^H] = |G(K/K^H)|$, it now follows that $H = G(K/K^H)$. ∎

COROLLARY. Let K/k be a finite extension of fields with automorphism group G and let H_1 and H_2 be subgroups of G. Then $H_1 \leq H_2$ if and only if $K^{H_2} \subseteq K^{H_1}$.

Proof. If $H_1 \leq H_2$, then any element in K fixed by every automorphism in H_2 is clearly fixed by every automorphism in H_1 and hence $K^{H_2} \subseteq K^{H_1}$. Conversely, if $K^{H_2} \subseteq K^{H_1}$, then $G(K/K^{H_1}) \leq G(K/K^{H_2})$ and hence, by Proposition 1, $H_1 = G(K/K^{H_1}) \leq G(K/K^{H_2}) = H_2$. ∎

COROLLARY. Let K/k be a finite extension of fields. Then distinct groups of k-automorphisms of K have distinct fixed fields.

Proof. If H_1 and H_2 are groups of k-automorphisms of K, then, by the previous corollary, $H_1 = H_2$ if and only if $K^{H_1} = K^{H_2}$. Thus, distinct groups of k-automorphisms have distinct fixed fields. ∎

It is clear from these results that if K/k is any finite extension of fields with automorphism group G and H is any group of k-automorphisms of K, then H is the automorphism group of the extension K/K^H and hence is completely determined by its fixed field K^H. Moreover, if H_1 and H_2 are two such groups of k-automorphisms, then $H_1 \leq H_2$ if and only if $K^{H_2} \subseteq K^{H_1}$. It follows, therefore, that the mapping $H \mapsto K^H$ from subgroups of G to intermediate subfields of K/k is a 1–1 mapping that reverses the inclusion relation on subgroups and subfields. However, the mapping is not onto, in general, since there may be intermediate subfields of the extension that are not the fixed field of some group of k-automorphisms; the field $\mathbb{Q}$, for example, is an intermediate subfield of the extension $\mathbb{Q}(\sqrt[3]{2})/\mathbb{Q}$, but it is not the fixed field of a group of $\mathbb{Q}$-automorphisms of $\mathbb{Q}(\sqrt[3]{2})$ since there is only one such automorphism, the identity, whose fixed field is $\mathbb{Q}(\sqrt[3]{2})$, not $\mathbb{Q}$. But for normal extensions the situation is different: in this case we claim that every intermediate subfield is the fixed field of some group of k-automorphisms and hence the mapping $H \mapsto K^H$ is a 1–1 correspondence.

PROPOSITION 2. Let K/k be a normal extension of fields with automorphism group G. Then every intermediate subfield of K/k is the fixed field of some group of k-automorphisms of K and hence the mapping $H \mapsto K^H$ is a 1–1 order-reversing correspondence between the subgroups of G and the intermediate subfields of K/k; under this correspondence, $[G:H] = [K^H:k]$ for every subgroup H of G.

Proof. Let k' be an intermediate subfield of the extension K/k and let $H = G(K/k')$. Since K is normal over k, K is also normal over k' and hence $K^H = k'$. Thus, k' is the fixed field of the group H. It now follows, as we noted above, that the correspondence $H \mapsto K^K$ is a 1–1 order-reversing corresponding. Finally, if H is any subgroup of G, then $|H| = [K:K^H]$ and $|G| = [K:k]$, and therefore

$$[G:H] = \frac{|G|}{|H|} = \frac{[K:k]}{[K:K^H]} = [K^H:k],$$

as required. ■

If K/k is a normal extension of fields with automorphism group G, the group G is called the *Galois group* of the extension and the correspondence $H \mapsto K^H$ between the subgroups of G and intermediate subfields of K/k is called the *Galois correspondence* for the extension. Under this correspondence, a subgroup H corresponds to its fixed field K^H and $[G:H] = [K:K^H]$, as illustrated in Figure 1, while an intermediate subfield k' corresponds to the subgroup $G(K/k')$ of automorphisms of K that leave k' fixed.

For example, let $K = \mathbb{Q}(\omega, \sqrt[3]{2})$ stand for the field discussed in Example 5, Section 1. Then K is a normal extension of $\mathbb{Q}$ and has six $\mathbb{Q}$-automorphisms since $[K:\mathbb{Q}] = 6$. If $G = G(K/\mathbb{Q})$ stands for the Galois group of $K/\mathbb{Q}$, recall that $G = \{1, \sigma, \sigma^2, \tau, \sigma\tau, \tau\sigma\} \cong \text{Sym}(3)$, where

$$\sigma: \begin{matrix} \sqrt[3]{2} \to \omega\sqrt[3]{2} \\ \omega \to \omega \end{matrix} \quad \text{and} \quad \tau: \begin{matrix} \sqrt[3]{2} \to \sqrt[3]{2} \\ \omega \to \omega^2 \end{matrix}.$$

In this case G has six subgroups, namely (1), (σ), (τ), $(\sigma\tau)$, $(\tau\sigma)$, and G, and hence by the Galois correspondence the extension $K/\mathbb{Q}$ has six intermediate subfields. Choosing the subgroup $H = (\sigma)$, for example, we find that $K^H = \mathbb{Q}(\omega)$; thus, the subgroup (σ) corresponds under the Galois correspondence to the intermediate subfield $\mathbb{Q}(\omega)$, as illustrated in Figure 2. We find,

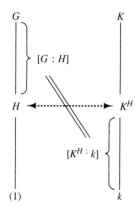

FIGURE 1. The Galois correspondence between a subgroup H and its fixed field K^H.

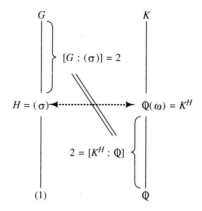

FIGURE 2. The Galois correspondence between the subgroup (σ) and its fixed field $\mathbb{Q}(\omega)$ for the extension $\mathbb{Q}(\omega, \sqrt[3]{2})$.

similarly, that

$$G \leftrightarrow K^G = \mathbb{Q}; \qquad\qquad [G:G] = 1 = [\mathbb{Q}:\mathbb{Q}]$$

$$(\tau) \leftrightarrow K^{(\tau)} = \mathbb{Q}(\sqrt[3]{2}); \qquad [G:(\tau)] = 3 = [K^{(\tau)}:\mathbb{Q}]$$

$$(\sigma\tau) \leftrightarrow K^{(\sigma\tau)} = \mathbb{Q}(\omega^2 \sqrt[3]{2}); \quad [G:(\sigma\tau)] = 3 = [K^{(\sigma\tau)}:\mathbb{Q}]$$

$$(\tau\sigma) \leftrightarrow K^{(\tau\sigma)} = \mathbb{Q}(\omega \sqrt[3]{2}); \quad [G:(\tau\sigma)] = 3 = [K^{(\tau\sigma)}:\mathbb{Q}]$$

$$(1) \leftrightarrow K^{(1)} = K; \qquad\qquad [G:(1)] = 6 = [K^{(1)}:\mathbb{Q}].$$

Figure 3 summarizes the Galois correspondence between the six subgroups of G and the six intermediate subfields of the extension $K/\mathbb{Q}$. In particular, the six intermediate subfields of the extension $K/\mathbb{Q}$ are $\mathbb{Q}$, $\mathbb{Q}(\omega)$, $\mathbb{Q}(\sqrt[3]{2})$, $\mathbb{Q}(\omega \sqrt[3]{2})$, $\mathbb{Q}(\omega^2 \sqrt[3]{2})$, and K. Let us use the Galois correspondence to determine which of these subfields is equal to $\mathbb{Q}(\sqrt[3]{4})$. Let $k' = \mathbb{Q}(\sqrt[3]{4})$. Then k' corresponds to the subgroup $G(K/k')$ under the Galois correspondence. Now, an automorphism of K fixes k' if and only if it fixes $\sqrt[3]{4}$. By inspection, the only such automorphisms are 1 and τ. Hence, $G(K/k') = \{1, \tau\} = (\tau)$. Therefore, $\mathbb{Q}(\sqrt[3]{4}) = \mathbb{Q}(\sqrt[3]{2})$ since the subgroup corresponding to (τ) is $\mathbb{Q}(\sqrt[3]{2})$.

We continue our discussion of the Galois correspondence for normal extensions by studying the intermediate subfields that correspond to conjugate subgroups of the automorphism group. Recall first that if H is a subgroup of an arbitrary group G and $\sigma \in G$, then the subset $\sigma H \sigma^{-1} = \{\sigma h \sigma^{-1} \in G \,|\, h \in H\}$ is a subgroup of G. Two subgroups H_1 and H_2 of G are said to be conjugate if $H_2 = \sigma H_1 \sigma^{-1}$ for some element $\sigma \in G$, and H is a normal subgroup of G if $\sigma H \sigma^{-1} = H$ for every element $\sigma \in G$.

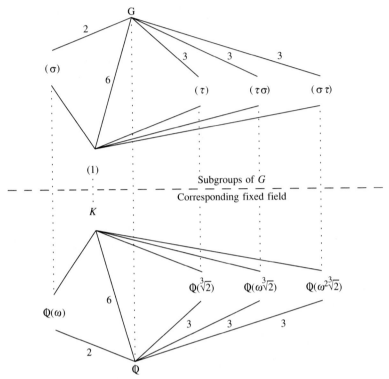

FIGURE 3. The Galois correspondence for the extension $\mathbb{Q}(\omega, \sqrt[3]{2})/\mathbb{Q}$.

Now, let K/k be an extension of fields and let k' be an intermediate subfield of the extension. If σ is a k-automorphism of K, then it follows easily that the image $\sigma(k') = \{\sigma(x) \in K \mid x \in k'\}$ is also an intermediate subfield of the extension K/k. We say that two intermediate subfields k_1 and k_2 of K/k are *conjugate* if $k_2 = \sigma(k_1)$ for some k-automorphism σ of K, and leave the reader to verify that conjugation is an equivalence relation on the collection of intermediate subfields of K/k. Let us now show that every conjugate of a fixed field is a fixed field, and that, under the Galois correspondence, two intermediate subfields are conjugate if and only if their corresponding subgroups are conjugate.

PROPOSITION 3. Let K/k be an extension of fields and let K^H be the fixed subfield of a group H of k-automorphisms of K. Then $\sigma(K^H) = K^{\sigma H \sigma^{-1}}$ for every k-automorphism σ of K.

Proof. Let σ be a k-automorphism of K. To show that $\sigma(K^H) \subseteq K^{\sigma H \sigma^{-1}}$, let $x \in K^H$. Then $(\sigma\tau\sigma^{-1})(\sigma(x)) = \sigma\tau(x) = \sigma(x)$ for every element $\tau \in H$ and hence $\sigma(x) \in K^{\sigma H \sigma^{-1}}$. Therefore, $\sigma(K^H) \subseteq K^{\sigma H \sigma^{-1}}$. Conversely, if $x \in K^{\sigma H \sigma^{-1}}$, then $\sigma\tau\sigma^{-1}(x) = x$ for every element $\tau \in H$ and hence $\tau\sigma^{-1}(x) = \sigma^{-1}(x)$. Therefore,

$\sigma^{-1}(x) \in K^H$ and hence $x \in \sigma(K^H)$. Thus, $K^{\sigma H \sigma^{-1}} \subseteq \sigma(K^H)$ and we conclude, therefore, that $\sigma(K^H) = K^{\sigma H \sigma^{-1}}$.   ∎

PROPOSITION 4. Let K/k be a normal extension of fields and let k_1 and k_2 be intermediate subfields of the extension. Then $k_2 = \sigma(k_1)$ for some k-automorphism σ of K if and only if $G(K/k_2) = \sigma G(K/k_1)\sigma^{-1}$.

Proof.   Let $H_1 = G(K/k_1)$ and $H_2 = G(K/k_2)$ stand for the subgroups of $G(K/k)$ corresponding to k_1 and k_2, respectively, under the Galois correspondence. Then $k_1 = K^{H_1}$ and $k_2 = K^{H_2}$. Hence, by Proposition 3, $k_2 = \sigma(k_1)$ for some $\sigma \in G(K/k)$ if and only if $K^{H_2} = \sigma(K^{H_1}) = K^{\sigma H_1 \sigma^{-1}}$, or equivalently, $H_2 = \sigma H_1 \sigma^{-1}$ since the Galois correspondence is 1–1.   ∎

Proposition 4 shows that under the Galois correspondence, intermediate subfields of a normal extension are conjugate if and only if their corresponding subgroups are conjugate subgroups of the Galois group of the extension. It follows, in particular, that if k' is an intermediate subfield of a normal extension K/k, then $\sigma(k') = k'$ for every k-automorphism σ of K if and only if $\sigma G(K/k')\sigma^{-1} = G(K/k')$; that is, k' is self-conjugate if and only if its corresponding subgroup $G(K/k')$ is a normal subgroup of the Galois group $G(K/k)$. We now claim that this is the case if and only if k' is a normal extension of k and that, in this case, the automorphisms of k' may be obtained by restricting the automorphisms of K to k'.

PROPOSITION 5. Let K/k be a normal extension of fields and let k' be an intermediate subfield of the extension. Then k' is a normal extension of k if and only if the corresponding subgroup $G(K/k')$ is a normal subgroup of $G(K/k)$; in this case, we have that:

(1) every k-automorphism of K restricts to a k-automorphism of k' and every k-automorphism of k' is the restriction of such an automorphism;

(2) $G(k'/k) \cong G(K/k)/G(K/k')$.

Proof.   If $G(K/k')$ is a normal subgroup of $G(K/k)$, then, as we noted above, $\sigma(k') = k'$ for every automorphism $\sigma \in G(K/k)$. In this case every k-automorphism of K restricts to a k-automorphism of k', and hence, if an element $x \in k'$ is fixed by every k-automorphism of k', then x is also fixed by every k-automorphism of K and therefore lies in the ground field k since K is normal over k. Thus, if $G(K/k')$ is a normal subgroup of $G(K/k)$, then k' is a normal extension of k.

Now suppose, conversely, that k' is a normal extension of k. Then k' is a splitting field for some polynomial $f(X)$ over k and hence $k' = k(c_1, \ldots, c_n)$, where $c_1, \ldots, c_n$ are the roots of $f(X)$ in k'. Now, if σ is any k-automorphism of K, then $\sigma(k') = k'$ since the conjugates $\sigma(c_1), \ldots, \sigma(c_n)$ generate $\sigma(k')$ and are also roots of $f(X)$. Therefore, every k-automorphism of K restricts to a k-automorphism of k'. Furthermore, every k-automorphism of k' may be obtained in this way; for if τ is any k-automorphism of k', then τ extends to a

k-automorphism σ of K since K is a splitting field over k', and τ is the restriction of σ to k'. Thus, if k' is a normal extension of k, then every k-automorphism of K restricts to a k-automorphism of k' and every k-automorphism of k' is the restriction of such an automorphism. To complete the proof, it remains to show that the corresponding subgroup $G(K/k')$ is a normal subgroup of $G(K/k)$ and that $G(k'/k) \cong G(K/k)/G(K/k')$. To this end, define the function $r: G(K/k) \to G(k'/k)$ by setting $r(\sigma) = \sigma|k'$, the restriction of σ to k'. Then r is a group-homomorphism mapping $G(K/k)$ onto $G(k'/k)$. Moreover, $\operatorname{Ker} r = \{\sigma \in G(K/k) | \sigma|k' = 1\} = G(K/k')$. It follows, therefore, that $G(K/k')$ is a normal subgroup of $G(K/k)$ and that $G(K/k)/G(K/k') \cong G(k'/k)$, and the proof is complete. ■

Let us take a moment to summarize these results. If K/k is a normal extension of fields with automorphism group G, then there is a 1–1 correspondence $H \mapsto K^H$, the Galois correspondence, between subgroups of G and intermediate subfields of K/k. Under this correspondence, an intermediate subfield k' is the fixed field of a subgroup H, that is, $k' = K^H$, if and only if $H = G(K/k')$, the subgroup of automorphisms of K fixing k'; in this case, $[G:H] = [K^H:k]$. Moreover, under this correspondence conjugate subgroups of G correspond to conjugate intermediate subfields of K/k. It follows, therefore, that H is a normal subgroup of G if and only if the corresponding subfield k' is self-conjugate, which, by Proposition 5, is the case if and only if k' is a normal extension of k. In this case, the k-automorphisms of k' are obtained by simply restricting the k-automorphisms of K to k', a statement that is summarized by the isomorphism $G(k'/k) \cong G/H$.

Proposition 5 concludes our discussion of the basic properties of the Galois correspondence. Let us collect these results together into a single statement—the fundamental theorem of Galois theory—prove a few corollaries, and then illustrate these ideas with some examples.

THE FUNDAMENTAL THEOREM OF GALOIS THEORY. Let K/k be a normal extension of fields with automorphism group G. Then the Galois correspondence $H \mapsto K^H$ is a 1–1 order-reversing correspondence between the subgroups of G and intermediate subfields of K/k. If H is a subgroup of G and k' an intermediate subfield of K/k, then, under this correspondence, we have that:

(1) $k' = K^H$ if and only if $H = G(K/k')$;

(2) $[G:H] = [K^H:k]$;

(3) k' is a normal extension of k if and only if the corresponding subgroup H is a normal subgroup of G; in this case the k-automorphisms of k' are the restrictions of the k-automorphisms of K to k' and $G(k'/k) \cong G/H$.

COROLLARY.   A normal extension of fields has only a finite number of intermediate subfields.

Proof.   Let K/k be a normal extension of fields. Then the intermediate sub-fields of K/k are in 1–1 correspondence with the subgroups of $G(K/k)$ and hence only a finite number of them since $G(K/k)$ is a finite group.   ■

COROLLARY.   If K/k is a normal extension of fields whose automorphism group is abelian, then every intermediate subfield of K/k is a normal extension of k.

Proof.   If the automorphism group $G(K/k)$ is abelian, then every subgroup of $G(K/k)$ is normal and hence by the fundamental theorem every intermediate subfield of K/k is a normal extension of k.   ■

For example, consider the field $K = \mathbb{Q}(\omega, \sqrt[3]{2})$ that we discussed earlier in this section. Since the automorphism σ maps $\sqrt[3]{2}$ onto $\omega\sqrt[3]{2}$, it follows that $\sigma(\mathbb{Q}(\sqrt[3]{2})) = \mathbb{Q}(\omega\sqrt[3]{2})$ and hence $\sigma G(K/\mathbb{Q}(\sqrt[3]{2}))\sigma^{-1} = G(K/\mathbb{Q}(\omega\sqrt[3]{2}))$. Indeed, since $G(K/\mathbb{Q}(\sqrt[3]{2})) = \{1, \tau\}$ and $G(K/\mathbb{Q}(\omega\sqrt[3]{2})) = \{1, \tau\sigma\}$, we find by direct calculation that

$$\sigma G(K/\mathbb{Q}(\sqrt[3]{2}))\sigma^{-1} = \sigma\{1, \tau\}\sigma^{-1} = \{1, \tau\sigma\} = G(K/\mathbb{Q}(\omega\sqrt[3]{2})).$$

Similarly, $\sigma(\mathbb{Q}(\omega\sqrt[3]{2})) = \mathbb{Q}(\omega^2\sqrt[3]{2})$ and   hence $\sigma G(K/\mathbb{Q}(\omega\sqrt[3]{2}))\sigma^{-1}$ $= G(K/\mathbb{Q}(\omega^2\sqrt[3]{2}))$. It follows that $\mathbb{Q}(\sqrt[3]{2})$, $\mathbb{Q}(\omega\sqrt[3]{2})$ and $\mathbb{Q}(\omega^2\sqrt[3]{2})$ are conjugate subfields of K and hence that their corresponding subgroups $\{1, \tau\}$, $\{1, \tau\sigma\}$ and $\{1, \sigma\tau\}$ are conjugate subgroups of the Galois group $G(K/\mathbb{Q})$. Observe, however, that none of these subfields is a normal extension of $\mathbb{Q}$ since the polynomial $X^3 - 2$, for example, has a root in each subfield but does not split over any one of them. The subfield $\mathbb{Q}(\omega)$, on the other hand, is a normal extension of $\mathbb{Q}$ since it is a splitting field for $X^3 - 1$ over $\mathbb{Q}$. In this case the subgroup of $G(K/\mathbb{Q})$ corresponding to $\mathbb{Q}(\omega)$ is $G(K/\mathbb{Q}(\omega)) = (\sigma)$, which is, indeed, a normal subgroup of $G(K/\mathbb{Q})$. Consequently, the $\mathbb{Q}$-automorphisms of $\mathbb{Q}(\omega)$ may be obtained by restricting those of K to $\mathbb{Q}(\omega)$. Let us verify this fact by direct calculation, recalling first that the extension $\mathbb{Q}(\omega)/\mathbb{Q}$ has two automorphisms—the identity automorphism 1, and the automorphism α that maps ω to ω^2. Restricting the six automorphisms of K to the subfield $\mathbb{Q}(\omega)$, we find that

$$1\,|\,\mathbb{Q}(\omega) = 1, \quad \sigma\,|\,\mathbb{Q}(\omega) = 1, \quad \sigma^2\,|\,\mathbb{Q}(\omega) = 1$$

$$\tau\,|\,\mathbb{Q}(\omega) = \alpha, \quad \sigma\tau\,|\,\mathbb{Q}(\omega) = \alpha, \quad \tau\sigma\,|\,\mathbb{Q}(\omega) = \alpha.$$

Thus, every automorphism of K does indeed restrict to an automorphism of $\mathbb{Q}(\omega)$, and every automorphism of $\mathbb{Q}(\omega)$ is in fact the restriction of such an automorphism. Finally, for the quotient group $G(K/\mathbb{Q})/G(K/\mathbb{Q}(\omega))$ we find that

$$G(K/\mathbb{Q})/G(K/\mathbb{Q}(\omega)) = \{(\sigma), \tau(\sigma)\} \cong \{1, \alpha\} = G(\mathbb{Q}(\omega)/\mathbb{Q}),$$

which agrees with the fundamental theorem.

EXAMPLE 1

Let $K = \mathbb{Q}(\sqrt{2}, \sqrt{3})$ and $G = G(K/\mathbb{Q})$. Then K is a splitting field for the polynomial $(X^2 - 2)(X^2 - 3)$ over $\mathbb{Q}$ and hence is a normal extension of $\mathbb{Q}$. Let us find the $\mathbb{Q}$-automorphisms of K and then discuss the Galois correspondence between the intermediate subfields of $K/\mathbb{Q}$ and the subgroups of G.

(A) A $\mathbb{Q}$-automorphism of K maps $\sqrt{2}$ to either $\pm\sqrt{2}$ and $\sqrt{3}$ to either $\pm\sqrt{3}$. The field K therefore has four $\mathbb{Q}$-automorphisms, as summarized in the table here.

$1: K \to K$	$\sigma: K \to K$	$\tau: K \to K$	$\sigma\tau: K \to K$
$\sqrt{2} \mapsto \sqrt{2}$	$\sqrt{2} \mapsto \sqrt{2}$	$\sqrt{2} \mapsto -\sqrt{2}$	$\sqrt{2} \mapsto -\sqrt{2}$
$\sqrt{3} \mapsto \sqrt{3}$	$\sqrt{3} \mapsto -\sqrt{3}$	$\sqrt{3} \mapsto \sqrt{3}$	$\sqrt{3} \mapsto -\sqrt{3}$

Clearly, $\sigma^2 = \tau^2 = (\sigma\tau)^2 = 1$. Therefore, G is a group of order four in which every element has order 2 and hence $G \cong \mathbb{Z}_2 \times \mathbb{Z}_2$. Now, the group G has five subgroups: (1), (σ), (τ), $(\sigma\tau)$, and G. For the fixed subfields corresponding to these subgroups, we find that

$$K^{(1)} = K; \qquad [K^{(1)}:\mathbb{Q}] = [G:(1)] = 4$$

$$K^{(\sigma)} = \mathbb{Q}(\sqrt{2}); \quad [K^{(\sigma)}:\mathbb{Q}] = [G:(\sigma)] = 2$$

$$K^{(\tau)} = \mathbb{Q}(\sqrt{3}); \quad [K^{(\tau)}:\mathbb{Q}] = [G:(\tau)] = 2$$

$$K^{(\sigma\tau)} = \mathbb{Q}(\sqrt{6}); \quad [K^{(\sigma\tau)}:\mathbb{Q}] = [G:(\sigma\tau)] = 2$$

$$K^G = \mathbb{Q}; \qquad [K^G:\mathbb{Q}] = [G:G] = 1.$$

The diagram in Figure 4 illustrates the Galois correspondence for the extension $K/\mathbb{Q}$.

(B) It follows from the Galois correspondence that the extension $K/\mathbb{Q}$ has five intermediate subfields: $\mathbb{Q}$, $\mathbb{Q}(\sqrt{2})$, $\mathbb{Q}(\sqrt{3})$, $\mathbb{Q}(\sqrt{6})$, and K. Moreover, since G is abelian, each of these subfields is a normal extension of $\mathbb{Q}$ and hence their automorphisms may be obtained by restricting the automorphisms of K to the subfield. Let us illustrate this by using the subfield $\mathbb{Q}(\sqrt{2})$. Recall first that the extension $\mathbb{Q}(\sqrt{2})/\mathbb{Q}$ has two automorphisms: the identity automorphism 1, and the automorphism α that maps $\sqrt{2}$ to $-\sqrt{2}$. Now, restricting the four automorphisms of K to $\mathbb{Q}(\sqrt{2})$, we find that:

$$1 \,|\, \mathbb{Q}(\sqrt{2}) = 1 \qquad \tau \,|\, \mathbb{Q}(\sqrt{2}) = \alpha$$

$$\sigma \,|\, \mathbb{Q}(\sqrt{2}) = 1 \qquad \sigma\tau \,|\, \mathbb{Q}(\sqrt{2}) = \alpha.$$

Thus, the automorphisms of K restrict to give the automorphisms of $\mathbb{Q}(\sqrt{2})$. Finally, the subgroup of G corresponding to $\mathbb{Q}(\sqrt{2})$ is $G(K/\mathbb{Q}(\sqrt{2}))$ $= (\sigma)$ and is a normal subgroup of G, which agrees with the fact that

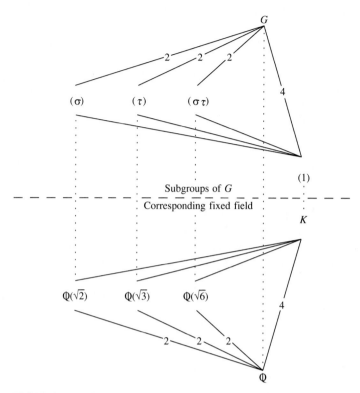

FIGURE 4. The Galois correspondence for the extension $\mathbb{Q}(\sqrt{2}, \sqrt{3})/\mathbb{Q}$.

$\mathbb{Q}(\sqrt{2})$ is a normal extension of $\mathbb{Q}$, while for the quotient group $G/G(K/\mathbb{Q}(\sqrt{2}))$, we find that

$$G/G(K/\mathbb{Q}(\sqrt{2})) = G/(\sigma) = \{(\sigma), \tau(\sigma)\} \cong \{1, \alpha\} = G(\mathbb{Q}(\sqrt{2})/\mathbb{Q}).$$

(C) Let us conclude this example by using the Galois correspondence for the extension $K/\mathbb{Q}$ to show that $\mathbb{Q}(\sqrt{6} - \sqrt{2}) = K$. Under the Galois correspondence, the subfield $\mathbb{Q}(\sqrt{6} - \sqrt{2})$ corresponds to the subgroup $G(K/\mathbb{Q}(\sqrt{6} - \sqrt{2}))$, which consists of those automorphisms of K that leave $\sqrt{6} - \sqrt{2}$ fixed. Since

$$1(\sqrt{6} - \sqrt{2}) = \sqrt{6} - \sqrt{2}, \qquad \tau(\sqrt{6} - \sqrt{2}) = -\sqrt{6} - \sqrt{2}$$
$$\sigma(\sqrt{6} - \sqrt{2}) = -\sqrt{6} + \sqrt{2}, \qquad \sigma\tau(\sqrt{6} - \sqrt{2}) = \sqrt{6} + \sqrt{2}.$$

only the identity automorphism fixes $\sqrt{6} - \sqrt{2}$ and therefore $G(K/\mathbb{Q}(\sqrt{6} - \sqrt{2})) = (1)$. Hence, by the fundamental theorem $\mathbb{Q}(\sqrt{6} - \sqrt{2}) = K^{(1)} = K$. In particular, $\sqrt{6} - \sqrt{2}$ is a primitive element for the extension $K/\mathbb{Q}$.

EXAMPLE 2

Let us show how the Galois theory may be used to obtain information about groups of automorphisms and their fixed fields without resorting to explicit calculation. Let $K = \mathbb{Q}(i, \sqrt[4]{2})$ and $G = G(K/\mathbb{Q})$. Then K is a splitting field for the polynomial $X^4 - 2$ over $\mathbb{Q}$ and hence is a normal extension of $\mathbb{Q}$.

(A) Let $H = G(K/\mathbb{Q}(i\sqrt{2}))$ stand for the subgroup of G corresponding to the intermediate subfield $\mathbb{Q}(i\sqrt{2})$. We claim that H is a group of order 4. For $K^H = \mathbb{Q}(i\sqrt{2})$ and hence, by the fundamental theorem, $[G:H] = [K^H:\mathbb{Q}] = [\mathbb{Q}(i\sqrt{2}):\mathbb{Q}] = 2$. Thus, since $|G| = [K:\mathbb{Q}] = 8$, it follows that $|H| = 4$, as required.

(B) Now, let σ stand for the $\mathbb{Q}$-automorphism of K that maps i to i and $\sqrt[4]{2}$ to $-i\sqrt[4]{2}$. Then we claim that the fixed field $K^{(\sigma)} = \mathbb{Q}(i)$. For clearly, $\mathbb{Q}(i) \subseteq K^{(\sigma)}$ since σ fixes i. On the other hand, $|\sigma| = 4$ and therefore $[K^{(\sigma)}:\mathbb{Q}] = [G:(\sigma)] = 8/4 = 2$. Hence, since $\mathbb{Q} \subseteq \mathbb{Q}(i) \subseteq K^{(\sigma)}$ and both $\mathbb{Q}(i)$ and $K^{(\sigma)}$ have degree 2 over $\mathbb{Q}$, we conclude that $K^{(\sigma)} = \mathbb{Q}(i)$, as required.

EXAMPLE 3. The Galois correspondence for cyclotomic extensions

Recall from the previous section that a cyclotomic extension of $\mathbb{Q}$ is any extension of the form $\mathbb{Q}(z_n)$, where z_n is a primitive nth root of unity and n is a positive integer. The field $\mathbb{Q}(z_n)$ is a normal extension of $\mathbb{Q}$ since it is a splitting field for the polynomial $X^n - 1$ over $\mathbb{Q}$. Let us discuss the Galois correspondence for the extension $\mathbb{Q}(z_n)/\mathbb{Q}$.

(A) We begin by recalling that the automorphism group $G(\mathbb{Q}(z_n)/\mathbb{Q}$ consists of the mappings $\sigma_s: \mathbb{Q}(z_n) \to \mathbb{Q}(z_n)$, where s is any positive integer relatively prime to n and $\sigma_s(z_n) = z_n^s$, and that $G(\mathbb{Q}(z_n)/\mathbb{Q}) \cong U(\mathbb{Z}_n)$, the multiplicative group of units of the ring $\mathbb{Z}_n$. Since the group $G(\mathbb{Q}(z_n)/\mathbb{Q})$ is abelian, all of its subgroups are normal and hence, by the Galois theory, every intermediate subfield of $\mathbb{Q}(z_n)/\mathbb{Q}$ is a normal extension of $\mathbb{Q}$ whose automorphisms may be obtained by restricting the mappings σ_s to the subfield.

(B) For example, consider the cyclotomic extension $\mathbb{Q}(z_5)$, where z_5 stands for a primitive 5th root of unity. In this case $[\mathbb{Q}(z_5):\mathbb{Q}] = \varphi(5) = 4$ and $G = G(\mathbb{Q}(z_5)/\mathbb{Q}) = \{1, \sigma_2, \sigma_3, \sigma_4\}$, where $1(z_5) = z_5$, $\sigma_2(z_5) = z_5^2$, $\sigma_3(z_5) = z_5^3$, and $\sigma_4(z_5) = z_5^4$. Then $|\sigma_2| = |\sigma_3| = 4$ and $|\sigma_4| = 2$ and hence the automorphism group G is a cyclic group of order 4 generated by either σ_2 or σ_3. G therefore has three subgroups: (1), $(\sigma_4) = \{1, \sigma_4\}$, and G. Now, the fixed subfields of $\mathbb{Q}(z_5)$ corresponding to (1) and G are $\mathbb{Q}(z_5)^{(1)} = \mathbb{Q}(z_5)$ and $\mathbb{Q}(z_5)^G = \mathbb{Q}$. To find the fixed field $\mathbb{Q}(z_5)^{(\sigma_4)}$ corresponding to (σ_4), first note that

$$[\mathbb{Q}(z_5)^{(\sigma_4)}:\mathbb{Q}] = [G:(\sigma_4)] = 4/2 = 2.$$

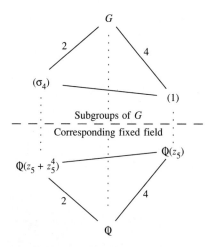

FIGURE 5. The Galois correspondence for the cyclotomic extension $\mathbb{Q}(z_5)/\mathbb{Q}$.

Thus, to find $\mathbb{Q}(z_5)^{(\sigma_4)}$ we need only find a single element in $\mathbb{Q}(z_5)$ not in $\mathbb{Q}$ that is fixed by σ_4; any such element generates the fixed field of (σ_4). For example, we find, either by trial and error or by the method described in Example 5, Section 1, that $\sigma_4(z_5 + z_5^4) = z_5^4 + z_5^{16} = z_5 + z_5^4$. Therefore, $\mathbb{Q}(z_5)^{(\sigma_4)} = \mathbb{Q}(z_5 + z_5^4)$. Hence, the cyclotomic extension $\mathbb{Q}(z_5)/\mathbb{Q}$ has three intermediate subfields, namely $\mathbb{Q}$, $\mathbb{Q}(z_5 + z_5^4)$, and $\mathbb{Q}(z_5)$, which correspond under the Galois correspondence to the subgroups G, (σ_4), and (1), respectively, as shown in Figure 5.

(C) Let us discuss the intermediate subfield $\mathbb{Q}(z_5 + z_5^4)$ and its automorphisms in more detail. As we showed above, $\mathbb{Q}(z_5 + z_5^4)$ is a normal extension of $\mathbb{Q}$ whose corresponding subgroup is (σ_4). Hence, by the fundamental theorem, $G(\mathbb{Q}(z_5 + z_5^4)/\mathbb{Q}) \cong G/(\sigma_4)$. Let us find the automorphisms of $\mathbb{Q}(z_5 + z_5^4)$ and verify this isomorphism directly. Since $\mathbb{Q}(z_5 + z_5^4)$ is a normal extension of $\mathbb{Q}$, its $\mathbb{Q}$-automorphisms may be found by restricting the four automorphisms of $\mathbb{Q}(z_5)$ to $\mathbb{Q}(z_5 + z_5^4)$. Doing so, we find that:

$$1\big|\mathbb{Q}(z_5 + z_5^4): z_5 + z_5^4 \mapsto z_5 + z_5^4$$

$$\sigma_2\big|\mathbb{Q}(z_5 + z_5^4): z_5 + z_5^4 \mapsto z_5^2 + z_5^3$$

$$\sigma_3\big|\mathbb{Q}(z_5 + z_5^4): z_5 + z_5^4 \mapsto z_5^3 + z_5^2$$

$$\sigma_4\big|\mathbb{Q}(z_5 + z_5^4): z_5 + z_5^4 \mapsto z_5^4 + z_5.$$

It follows that $\mathbb{Q}(z_5 + z_5^4)$ has two $\mathbb{Q}$-automorphisms: the identity automorphism 1, and the automorphism α that maps $z_5 + z_5^4$ to $z_5^2 + z_5^3$. Therefore, $G(\mathbb{Q}(z_5 + z_5^4)/\mathbb{Q}) = \{1, \alpha\}$ and hence

$$G/(\sigma_4) = \{(\sigma_4), \sigma_2(\sigma_4)\} \cong \{1, \alpha\} = G(\mathbb{Q}(z_5 + z_5^4)/\mathbb{Q}),$$

which agrees with the result obtained using the fundamental theorem. Now observe that the images of $z_5 + z_5^4$ under the automorphisms of K are simply the conjugates of $z_5 + z_5^4$. Thus, $z_5 + z_5^4$ has two distinct conjugates and hence

$$\text{Irr}(z_5 + z_5^4; \mathbb{Q}) = [X - (z_5 + z_5^4)][X - (z_5^2 + z_5^3)]$$
$$= X^2 - (z_5 + z_5^2 + z_5^3 + z_5^4)X + (z_5^3 + z_5^4 + z_5^6 + z_5^7)$$
$$= X^2 + X - 1.$$

It follows, in particular, that $[\mathbb{Q}(z_5 + z_5^4):\mathbb{Q}] = 2$, which agrees with our result obtained in part (B) using the Galois correspondence. Finally, observe that $z_5 + z_5^4$ is, in fact, a real number since it is equal to its complex conjugate:

$$\overline{z_5 + z_5^4} = \bar{z}_5 + \bar{z}_5^4 = z_5^4 + z_5;$$

indeed, $z_5 + z_5^4$ is a root of the equation $X^2 + X - 1 = 0$ and hence is either

$$\frac{-1 + \sqrt{5}}{2} \quad \text{or} \quad \frac{-1 - \sqrt{5}}{2},$$

depending upon the choice of z_5. In either case, $\mathbb{Q}(z_5 + z_5^4) = \mathbb{Q}(\sqrt{5})$. If we choose

$$z_5 + z_5^4 = \frac{-1 + \sqrt{5}}{2},$$

then

$$z_5^2 + z_5^3 = \frac{-1 - \sqrt{5}}{2}.$$

In this case $\sqrt{5} = 1 + 2(z_5 + z_5^4)$ and hence $\alpha(\sqrt{5}) = 1 + 2(z_5^2 + z_5^3)$ $= -\sqrt{5}$. Thus, the automorphism α is simply the automorphism of $\mathbb{Q}(z_5 + z_5^4)$ that maps $\sqrt{5}$ to $-\sqrt{5}$.

EXAMPLE 4. The Galois correspondence for finite fields

Let K be a finite field of prime characteristic p and let $[K:k] = n$, where k stands for the prime subfield of K. Then K is a splitting field for the polynomial $X^{p^n} - X$ over k and hence is a normal extension of k since the irreducible factors of this polynomial have only simple roots. Let us use the Galois correspondence and our knowledge of finite fields to show that the extension K/k has precisely one subfield of degree d over k for each positive divisor d of n, and that every such subfield is a normal extension of k.

(A) Recall first that the automorphism group $G(K/k)$ is a cyclic group of order n generated by the Frobenius automorphism $\sigma_p: K \to K$, where $\sigma_p(x) = x^p$ for every element $x \in K$, and that a cyclic group of order n

is abelian and has a unique subgroup of index d for each divisor d of n. It then follows, by the Galois correspondence, that the extension K/k has exactly one intermediate subfield of degree d over k, and every such subfield is a normal extension of k. More precisely, let d be a positive divisor of n and let K_d be the intermediate subfield of K/k corresponding to the cyclic subgroup (σ_p^d) under the Galois correspondence; K_d consists of those elements in K that are fixed by the automorphism σ_p^d. Since $|\sigma_p^d| = n/d$, it follows that $[K_d:k] = [G(K/k):(\sigma_p^d)] = d$. Thus, K_d is the unique intermediate subfield of K/k having degree d over k, and is a normal extension of k since the corresponding subgroup (σ_p^d) is a normal subgroup of $G(K/k)$.

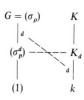

(1)

(B) We may describe the subfields K_d more explicitly as follows. Recall that the multiplicative group K^* of nonzero elements in K is a cyclic group of order $p^n - 1$. Let c stand for a generator of K^*. For each positive divisor d of n, let $s(d) = (p^n - 1)/(p^d - 1)$; note that $s(d)$ is an integer since K_d^* is a subgroup of K^* and hence its order, $p^d - 1$, divides that of K^*. We claim that $K_d = k(c^{s(d)})$. To show this, first observe that $s(d)p^d \equiv s(d) \bmod(p^n - 1)$ since $s(d) = (p^n - 1)/(p^d - 1)$. Therefore, $c^{s(d)p^d} = c^{s(d)}$ and hence $\sigma_p^d(c^{s(d)}) = c^{s(d)p^d} = c^{s(d)}$. It follows, then, that $c^{s(d)} \in K_d$ since K_d is the fixed field of σ_p^d, and hence $k(c^{s(d)}) \subseteq K_d$. Now, the distinct conjugates of $c^{s(d)}$ are

$$1(c^{s(d)}) = c^{s(d)}$$

$$\sigma_p(c^{s(d)}) = c^{s(d)p}$$

$$\vdots$$

$$\sigma_p^{d-1}(c^{s(d)}) = c^{s(d)p^{d-1}}.$$

Therefore, $\mathrm{Irr}(c^{s(d)}:k) = (X - c^{s(d)}) \cdots (X - c^{s(d)p^{d-1}})$ and hence $[k(c^{s(d)}):k] = \deg \mathrm{Irr}(c^{s(d)}:k) = d$. Since $k(c^{s(d)}) \subseteq K_d$ and since $[K_d:k] = d$, it now follows that $K_d = k(c^{s(d)})$. Thus, the unique subfield K_d of K having degree d over k is generated by the element $c^{s(d)}$, where c is any element in K of order $p^n - 1$ and $s(d) = (p^n - 1)/(p^d - 1)$.

(C) Let us illustrate these results with a particular example. Let K stand for a splitting field of the polynomial $X^{64} - X$ over $\mathbb{F}_2$. Then K is a finite field containing 64 elements and is a normal extension of degree 6 over $\mathbb{F}_2$. Let $G = G(K/\mathbb{F}_2)$. Then $G = (\sigma_2) = \{1, \sigma_2, \sigma_2^2, \sigma_2^3, \sigma_2^4, \sigma_2^5\}$, where $\sigma_2 : K \to K$ is the Frobenius automorphism of K, $\sigma_2(x) = x^2$ for all $x \in K$. In this case the possible divisors of 6 are $d = 1, 2, 3$, and 6, and the subgroups (σ_2^d) and corresponding subfields K_d are as follows:

$$d = 1: (\sigma_2) = G, \qquad K_1 = K^{(\sigma_2)} = \mathbb{F}_2, \quad [K_1:\mathbb{F}_2] = [G:(\sigma_2)] = 1$$

$$d = 2: (\sigma_2^2) = \{1, \sigma_2^2, \sigma_2^4\}, \quad K_2 = K^{(\sigma_2^2)}, \qquad [K_2:\mathbb{F}_2] = [G:(\sigma_2^2)] = 2$$

$$d = 3: (\sigma_2^3) = \{1, \sigma_2^3\}, \qquad K_3 = K^{(\sigma_2^3)}, \qquad [K_3:\mathbb{F}_2] = [G:(\sigma_2^3)] = 3$$

$$d = 6: (\sigma_2^6) = \{1\}, \qquad K_6 = K^{(1)} = K, \quad [K_6:\mathbb{F}_2] = [G:(\sigma_2^6)] = 6.$$

Thus, the extension $K/\mathbb{F}_2$ has four intermediate subfields, namely $\mathbb{F}_2$, K_2, K_3, and K, whose corresponding subgroups are (σ_2), (σ_2^2), (σ_2^3), and (1), respectively. Figure 6 summarizes the Galois correspondence for the extension $K/\mathbb{F}_2$. Let us describe these subfields in more detail. In this case the multiplicative group K^* is a cyclic group of order 63. Let c be a generator for K^*. Then for $d = 1, 2, 3$, and 6, we find that

$$s(1) = 63/(2^1 - 1) = 63, \quad \text{and hence } K_1 = \mathbb{F}_2(c^{63}) = \mathbb{F}_2;$$

$$s(2) = 63/(2^2 - 1) = 21, \quad \text{and hence } K_2 = \mathbb{F}_2(c^{21});$$

$$s(3) = 63/(2^3 - 1) = 9, \quad \text{and hence } K_3 = \mathbb{F}_2(c^{9});$$

$$s(6) = 63/(2^6 - 1) = 1, \quad \text{and hence } K_6 = \mathbb{F}_2(c^{1}) = K.$$

Finally, consider the subfield K_3. Since K_3 is a normal extension of $\mathbb{F}_2$, its $\mathbb{F}_2$-automorphisms may be found by restricting the six $\mathbb{F}_2$-automorphisms

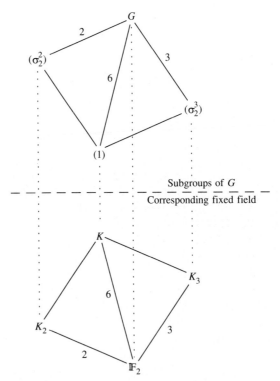

FIGURE 6. The Galois correspondence for the extension $K/\mathbb{F}_2$, where K is a finite field containing 64 elements.

of K to K_3. To this end, let $\alpha = \sigma_2 \vert K_3$. Then:

$$1 \vert K_3 = 1 : c^9 \mapsto c^9$$

$$\sigma_2 \vert K_3 = \alpha : c^9 \mapsto c^{18}$$

$$\sigma_2^2 \vert K_3 = \alpha^2 : c^9 \mapsto c^{36}$$

$$\sigma_2^3 \vert K_3 = \alpha^3 : c^9 \mapsto c^{72} = c^9; \qquad \alpha^3 = 1$$

$$\sigma_2^4 \vert K_3 = \alpha^4 : c^9 \mapsto c^{144} = c^{18}; \quad \alpha^4 = \alpha$$

$$\sigma_2^5 \vert K_3 = \alpha^5 : c^9 \mapsto c^{288} = c^{36}; \quad \alpha^5 = \alpha^2.$$

Thus, the field K_3 has three $\mathbb{F}_2$-automorphisms, namely the identity automorphism 1 and the automorphisms α and α^2, and therefore $G(K_3/\mathbb{F}_2)$ $= \{1, \alpha, \alpha^2\} \cong \mathbb{Z}_3$. Indeed, since $[K_3 : \mathbb{F}_2] = 3$, K_3 is a finite field containing 2^3, or 8, elements, and hence its automorphism group is a cyclic group of order 3 generated by the Frobenius automorphism that maps every element $x \in K_3$ to x^2. The mapping α above is just the Frobenius automorphism on K_3.

EXAMPLE 5. The Galois group of a separable polynomial

Let $f(X)$ be a separable polynomial over a field k—that is, a polynomial all of whose irreducible factors over k have only simple roots—and let K be a splitting field for $f(X)$ over k. Then K is a normal extension of k. In this case the automorphism group $G(K/k)$ is called the *Galois group of $f(X)$*. The Galois group of a separable polynomial has two basic properties: first, the order of the Galois group of $f(X)$ is equal to $[K : k]$, the degree of the splitting field over k; and second, the automorphisms in the Galois group permute the roots of the polynomial $f(X)$ since every k-automorphism of K maps roots of $f(X)$ to roots. Here are some examples of separable polynomials and their Galois groups.

(A) Let $f(X) = X^2 + 1 \in \mathbb{Q}[X]$. Then $f(X)$ is a separable polynomial over $\mathbb{Q}$ and has the field $\mathbb{Q}(i)$ as a splitting field over $\mathbb{Q}$. Hence, the Galois group of $f(X)$ is the group $G(\mathbb{Q}(i)/\mathbb{Q}) = \{1, \sigma\}$, where 1 stands for the identity automorphism and σ the automorphism that maps i to $-i$. As permutations, the identity automorphism fixes the roots of $f(X)$ while σ interchanges the roots:

$$1 \leftrightarrow \begin{pmatrix} i & -i \\ i & -i \end{pmatrix}, \qquad \sigma \leftrightarrow \begin{pmatrix} i & -i \\ -i & i \end{pmatrix}.$$

(B) Let $f(X) = X^3 - 1 \in \mathbb{Q}[X]$. Then $f(X)$ is a separable polynomial over $\mathbb{Q}$ and has the field $\mathbb{Q}(\omega)$ as a splitting field over $\mathbb{Q}$, where ω is a primitive cube root of unity. Hence, the Galois group of $f(X)$ is the group $G(\mathbb{Q}(\omega)/\mathbb{Q}) = \{1, \sigma\}$, where 1 stands for the identity automorphism and

σ the automorphism that maps ω to ω^2. As permutations, the identity automorphism fixes the three roots of $f(X)$ while σ fixes the root 1 and interchanges ω and ω^2:

$$1 \leftrightarrow \begin{pmatrix} 1 & \omega & \omega^2 \\ 1 & \omega & \omega^2 \end{pmatrix}, \qquad \sigma \leftrightarrow \begin{pmatrix} 1 & \omega & \omega^2 \\ 1 & \omega^2 & \omega \end{pmatrix}.$$

(C) Let $f(X) = X^3 - 2 \in \mathbb{Q}[X]$. Then $f(X)$ is a separable polynomial over $\mathbb{Q}$ and its Galois group is $G(\mathbb{Q}(\omega, \sqrt[3]{2})/\mathbb{Q}) = \{1, \sigma, \sigma^2, \tau, \sigma\tau, \tau\sigma\}$. If we let $a = \sqrt[3]{2}$, $b = \omega\sqrt[3]{2}$, and $c = \omega^2\sqrt[3]{2}$ stand for the three roots of $f(X)$, then the six automorphisms in the Galois group permute the roots of $f(X)$ as follows:

$$1 \leftrightarrow \begin{pmatrix} a & b & c \\ a & b & c \end{pmatrix} \qquad \tau \leftrightarrow \begin{pmatrix} a & b & c \\ a & c & b \end{pmatrix}$$

$$\sigma \leftrightarrow \begin{pmatrix} a & b & c \\ b & c & a \end{pmatrix} \qquad \sigma\tau \leftrightarrow \begin{pmatrix} a & b & c \\ b & a & c \end{pmatrix}$$

$$\sigma^2 \leftrightarrow \begin{pmatrix} a & b & c \\ c & a & b \end{pmatrix} \qquad \tau\sigma \leftrightarrow \begin{pmatrix} a & b & c \\ c & b & a \end{pmatrix}.$$

Note that in this case all of the six possible permutations of the roots of $f(X)$ are in fact represented by some automorphism in the Galois group of $f(X)$.

(D) Let $f(X) = X^n - 1 \in \mathbb{Q}[X]$, where n is a positive integer. Then $f(X)$ is a separable polynomial over $\mathbb{Q}$ whose splitting field is the cyclotomic extension $\mathbb{Q}(z_n)$ of $\mathbb{Q}$, where z_n is a primitive nth root of unity. Thus, the Galois group of $f(X)$ is the group $G(\mathbb{Q}(z_n)/\mathbb{Q}) = \{\sigma_s \mid 1 \leq s \leq n, \gcd(s, n) = 1\}$, where σ_s is the automorphism of $\mathbb{Q}(z_n)$ that maps z_n onto z_n^s. In this case the roots of $f(X)$ are the powers $1, z_n, \ldots, z_n^{n-1}$, and the permutation of these roots corresponding to a typical automorphism σ_s is

$$\sigma_s \leftrightarrow \begin{pmatrix} 1 & z_n & \cdots & z_n^{n-1} \\ 1 & z_n^s & \cdots & z_n^{s(n-1)} \end{pmatrix}.$$

For example, the Galois group of the polynomial $f(X) = X^6 - 1$ is $G(\mathbb{Q}(z_6)/\mathbb{Q}) = \{1, \sigma_5\}$, where $\sigma_5(z_6) = z_6^5$. In this case the identity automorphism fixes the six roots of $f(X)$, while σ_5 fixes the roots 1 and z_6^3 but interchanges the pairs z_6, z_6^5 and z_6^2, z_6^4:

$$1 \leftrightarrow \begin{pmatrix} 1 & z_6 & z_6^2 & z_6^3 & z_6^4 & z_6^5 \\ 1 & z_6 & z_6^2 & z_6^3 & z_6^4 & z_6^5 \end{pmatrix}, \qquad \sigma_5 \leftrightarrow \begin{pmatrix} 1 & z_6 & z_6^2 & z_6^3 & z_6^4 & z_6^5 \\ 1 & z_6^5 & z_6^4 & z_6^3 & z_6^2 & z_6 \end{pmatrix}.$$

(E) Let $f(X) = X^{p^n} - X \in \mathbb{F}_p[X]$, where p is a prime and n a positive integer. Then $f(X)$ is a separable polynomial over $\mathbb{F}_p$ and its splitting field K is a finite field containing the p^n distinct roots of $f(X)$ and has degree n over $\mathbb{F}_p$. Thus, the Galois group of $f(X)$ is the cyclic group (σ_p) of order

n generated by the Frobenius automorphism σ_p of K. For example, if $f(X) = X^4 - X \in \mathbb{F}_2[X]$, then $K = \mathbb{F}_2[X]/(X^2 + X + 1)\mathbb{F}_2[X] = \{0, 1, c, 1 + c\}$ is a splitting field for $f(X)$ over $\mathbb{F}_2$, where $c = X + (X^2 + X + 1)\mathbb{F}_2[X]$, and the Galois group of $f(X)$ is $G(K/\mathbb{F}_2) = \{1, \sigma_2\}$, where $\sigma_2(x) = x^2$ for all $x \in K$; the identity automorphism fixes the roots of $f(X)$, while σ_2 fixes 0 and 1 but interchanges c and $1 + c$:

$$1 \leftrightarrow \begin{pmatrix} 0 & 1 & c & 1+c \\ 0 & 1 & c & 1+c \end{pmatrix} \qquad \sigma_2 \leftrightarrow \begin{pmatrix} 0 & 1 & c & 1+c \\ 0 & 1 & 1+c & c \end{pmatrix}.$$

(F) Finally, observe that the automorphism group of any normal extension of fields is in fact the Galois group of some separable polynomial since every such field is the splitting field for some separable polynomial over the ground field. The Galois theory of fields plays an important role in the study of polynomials and their roots since, if K is a splitting field for a separable polynomial $f(X)$ over a field k, then K may be obtained through a sequence of fields $k \subseteq k_1 \subseteq \cdots \subseteq k_n = K$, where k_{i+1} is obtained from k_i by adjoining a single root of $f(X)$, and hence there is a corresponding tower of subgroups in the Galois group of the polynomial. It is possible, therefore, to transfer questions about the roots of polynomials over into questions about finite groups and we will use these ideas in the next chapter to study the solvability of algebraic equations by means of radicals.

In the following example we discuss symmetric polynomials, one of the classic applications of the Galois theory of fields. Symmetric polynomials are polynomials that remain unchanged when the variables are permuted. They are especially important because the coefficients of any polynomial may be expressed in terms of symmetric polynomials of the roots. Our goal in this example is to introduce the reader to symmetric polynomials and symmetric functions and then use the Galois theory of fields to discuss their basic properties.

EXAMPLE 6. Symmetric functions

Let n be a positive integer and let $\mathbb{Q}[X_1, \ldots, X_n]$ stand for the ring of polynomials in n distinct variables $X_1, \ldots, X_n$ over the field $\mathbb{Q}$ of rational numbers. A *symmetric polynomial* in $X_1, \ldots, X_n$ is any polynomial in $\mathbb{Q}[X_1, \ldots, X_n]$ that remains unchanged by any permutation of the variables $X_1, \ldots, X_n$. The polynomial $X_1 + X_2$, for example, is a symmetric polynomial in two variables. More generally, a *symmetric rational function* is any rational function in the function field $\mathbb{Q}(X_1, \ldots, X_n)$ that remains unchanged by any permutation of the variables $X_1, \ldots, X_n$. In this example our goal is to discuss three basic properties of symmetric polynomials: first, we define a special class of symmetric polynomials called the elementary symmetric polynomials, denoted by $S_1, \ldots, S_n$, show that the coefficients of any polynomial may be expressed as elementary symmetric polynomials in the roots of the given polynomial, and then show that every symmetric rational function may in fact be expressed as a rational function in the elementary symmetric polynomials $S_1, \ldots, S_n$; second, we discuss the general polynomial of degree n over $\mathbb{Q}$, whose coefficients are the elementary symmetric polynomials, and show that its Galois group is isomorphic to the symmetric group Sym(n); and finally, we use the Galois theory to show that every finite group is isomorphic to the Galois group of the general polynomial over an appropriately chosen ground field.

(A) We begin by defining the elementary symmetric polynomials. Let t stand for a variable different from $X_1, \ldots, X_n$ and let $f(t) = (t - X_1) \cdots (t - X_n)$. Then $f(t)$ is a polynomial in t over the field $\mathbb{Q}(X_1, \ldots, X_n)$ and the coefficients of $f(t)$ are polynomials in $X_1, \ldots, X_n$. Let $S_1(X_1, \ldots, X_n), \ldots, S_n(X_1, \ldots, X_n)$ stand for the coefficients of $f(t)$ when $f(t)$ is expanded in powers of t; that is, let

$$f(t) = X^n + (-1)S_1(X_1, \ldots, X_n)t^{n-1} + \cdots + (-1)^n S_n(X_1, \ldots, X_n).$$

The polynomials $S_1(X_1, \ldots, X_n), \ldots, S_n(X_1, \ldots, X_n)$ are called the *elementary symmetric polynomials* in the n variables $X_1, \ldots, X_n$. Note that they are, in fact, symmetric polynomials; for if σ is any permutation of the numbers $1, \ldots, n$, then the coefficients of $f(t)$ remain unchanged when the factors $t - X_1, \ldots, t - X_n$ are permuted in any manner and therefore $S_i(X_{\sigma(1)}, \ldots, X_{\sigma(n)}) = S_i(X_1, \ldots, X_n)$ for $i = 1, \ldots, n$. The symmetric polynomials may be calculated from the following formulas, which are easily verified by induction on n:

$$S_1(X_1, \ldots, X_n) = X_1 + X_2 + \cdots + X_n$$

$$S_2(X_1, \ldots, X_n) = X_1 X_2 + \cdots + X_1 X_n + X_2 X_3 + \cdots + X_2 X_n + \cdots + X_{n-1} X_n$$

$$\vdots$$

$$S_n(X_1, \ldots, X_n) = X_1 X_2 \cdots X_n.$$

In general, $S_k(X_1, \ldots, X_n)$ is the sum of all k-fold products of $X_1, \ldots, X_n$. For example, if $n = 3$, the elementary symmetric polynomials are

$$S_1(X_1, X_2, X_3) = X_1 + X_2 + X_3$$

$$S_2(X_1, X_2, X_3) = X_1 X_2 + X_1 X_3 + X_2 X_3$$

$$S_3(X_1, X_2, X_3) = X_1 X_2 X_3.$$

(B) The elementary symmetric polynomials provide a convenient way to express the coefficients of any polynomial in terms of its roots. For suppose that $g(X) = X^n + a_1 X^{n-1} + \cdots + a_{n-1} X + a_n$ is any nonconstant polynomial in $\mathbb{Q}[X]$, and let $c_1, \ldots, c_n$ stand for the roots of $g(X)$ in some splitting field. Then $g(X) = (X - c_1) \cdots (X - c_n)$ and hence

$$a_1 = -S_1(c_1, \ldots, c_n)$$

$$\vdots$$

$$a_n = (-1)^n S_n(c_1, \ldots, c_n).$$

For example, the roots of the polynomial $g(X) = X^3 - 3X^2 + 2X$ are 0, 1 and 2; therefore

$$a_1 = -S_1(0, 1, 2) = -(0 + 1 + 2) = -3$$

$$a_2 = (-1)^2 S_2(0, 1, 2) = (0)(1) + (0)(2) + (1)(2) = 2$$

$$a_3 = (-1)^3 S_3(0, 1, 2) = -(0)(1)(2) = 0.$$

(C) Let us now discuss the relationship between the elementary symmetric polynomials and symmetric functions in general. For convenience, let $S_1, \ldots, S_n$ stand for the elementary symmetric polynomials in the n variables $X_1, \ldots, X_n$. Then $S_1, \ldots, S_n$ remain unchanged by any permutation of the variables and hence every rational expression in $S_1, \ldots, S_n$ is also unchanged by any such permutation. Thus, every rational expression in $S_1, \ldots, S_n$ is a symmetric rational function; for example, the expression

$$\frac{S_1 + S_2}{S_1} = \frac{X_1 + X_2 + X_3 + X_1 X_2 + X_1 X_3 + X_2 X_3}{X_1 X_2 X_3}$$

is a symmetric rational function in X_1, X_2, and X_3. We now claim that every symmetric rational function has this form, that is, may be written as a rational expression in the elementary symmetric polynomials. To show this, let $\mathbb{Q}(S_1, \ldots, S_n)$ stand for the subfield of $\mathbb{Q}(X_1, \ldots, X_n)$ generated by $S_1, \ldots, S_n$ and let G stand for the automorphism group of the extension $\mathbb{Q}(X_1, \ldots, X_n)/\mathbb{Q}(S_1, \ldots, S_n)$. Then $\mathbb{Q}(X_1, \ldots, X_n)$ is a normal extension of $\mathbb{Q}(S_1, \ldots, S_n)$ since it is a splitting field for the polynomial $f(t) = (t - X_1) \cdots (t - X_n)$ over $\mathbb{Q}(S_1, \ldots, S_n)$. Therefore, $\mathbb{Q}(X_1, \ldots, X_n)^G = \mathbb{Q}(S_1, \ldots, S_n)$. Now, G is the Galois group of $f(t)$ over $\mathbb{Q}(S_1, \ldots, S_n)$ and hence every automorphism in G permutes the n roots of $f(t)$ and is completely determined by such a permutation. Therefore, $G \leq \operatorname{Sym}(n)$. On the other hand, if σ is any permutation of the numbers $1, \ldots, n$, then σ defines a $\mathbb{Q}$-automorphism of $\mathbb{Q}(X_1, \ldots, X_n)$ that maps a typical function $r(X_1, \ldots, X_n)$ to $r(X_{\sigma(1)}, \ldots, X_{\sigma(n)})$ and, as such, leaves the subfield $\mathbb{Q}(S_1, \ldots, S_n)$ fixed. Therefore, $G = \operatorname{Sym}(n)$ and hence

$$\mathbb{Q}(X_1, \ldots, X_n)^{\operatorname{Sym}(n)} = \mathbb{Q}(S_1, \ldots, S_n).$$

We conclude therefore that every symmetric rational function in n variables may be written as a rational expression in the elementary symmetric polynomials $S_1, \ldots, S_n$. For example, the expression

$$\frac{1}{X_1} + \frac{1}{X_2} + \frac{1}{X_3}$$

is a symmetric rational function in X_1, X_2, and X_3, and we find that

$$\frac{1}{X_1} + \frac{1}{X_2} + \frac{1}{X_3} = \frac{X_2 X_3 + X_1 X_3 + X_1 X_2}{X_1 X_2 X_3} = \frac{S_2}{S_3}.$$

Similarly, $X_1^2 + X_2^2 + X_3^2$ is a symmetric polynomial in X_1, X_2, and X_3, and we find that

$$X_1^2 + X_2^2 + X_3^2 = (X_1 + X_2 + X_3)^2 - 2(X_1X_2 + X_1X_3 + X_2X_3)$$
$$= S_1^2 - 2S_2.$$

(D) The polynomial $f(t) = t^n + a_1 t^{n-1} + \cdots + a_{n-1}t + a_n$ discussed in part (A), where $a_i = (-1)^i S_i(X_1, \ldots, X_n)$ for $i = 1, \ldots, n$, is called the *general polynomial of degree n over* $\mathbb{Q}$. The coefficients of $f(t)$ lie in the field $\mathbb{Q}(S_1, \ldots, S_n)$, and $\mathbb{Q}(X_1, \ldots, X_n)$ is a splitting field for $f(t)$ over $\mathbb{Q}(S_1, \ldots, S_n)$. Moreover, we showed in part (C) that if G is the Galois group of $f(t)$, then $G \cong \mathrm{Sym}(n)$. Thus, the Galois group of the general polynomial of degree n over $\mathbb{Q}$ is isomorphic to the symmetric group $\mathrm{Sym}(n)$.

(E) Let us conclude this example by using the Galois theory of fields, together with the above results, to show that every finite group is in fact isomorphic to the Galois group of the general polynomial over an appropriately chosen ground field. Recall first that if H is any finite group of order n, then H is isomorphic to a subgroup H^* of $\mathrm{Sym}(n)$. Let

$$k^* = \mathbb{Q}(X_1, \ldots, X_n)^{H^*}$$

stand for the fixed field corresponding to H^* under the Galois correspondence for the extension $\mathbb{Q}(X_1, \ldots, X_n)/\mathbb{Q}(S_1, \ldots, S_n)$. Then $\mathbb{Q}(X_1, \ldots, X_n)$ is a normal extension of k^* and $G(\mathbb{Q}(X_1, \ldots, X_n)/k^*) = H^*$. Therefore, H^* is the Galois group of the general polynomial of degree n over the subfield k^*. Thus, every finite group is isomorphic to the Galois group of the general polynomial over an appropriately chosen ground field. But the ground field may not be the field $\mathbb{Q}$ of rational numbers; indeed, it is not known if every finite group is isomorphic to the Galois group of some polynomial over the field $\mathbb{Q}$ of rational numbers.

For our final example, let us use the Galois theory of fields to prove what is perhaps the most celebrated theorem in mathematics, the fundamental theorem of algebra. This theorem shows that every nonconstant polynomial over the field $\mathbb{C}$ of complex numbers has a complex root, and was first proved in a completely satisfactory manner in 1797 by Carl Friedrich Gauss. Since then there have been many proofs of the theorem using various techniques, most of which rely on analytic properties of functions of a complex variable. In the following example we present a proof due to Emil Artin that assumes only two basic facts about real numbers and uses the Galois theory of fields together with some basic properties of Sylow subgroups of a finite group.

EXAMPLE 7. The fundamental theorem of algebra

The fundamental theorem of algebra states that every nonconstant polynomial over the field $\mathbb{C}$ of complex numbers has a complex root. To prove the theorem, let us first recall two basic facts about real numbers: first, the field of real numbers is an ordered field in which every positive number has a square root; and second, every polynomial of odd degree with real coefficients has a real root. With these two assumptions, let us now prove the fundamental theorem of algebra.

(A) Let $f(X)$ be a nonconstant polynomial over the field $\mathbb{C}$ of complex numbers and let K be a splitting field for $f(X)$ over $\mathbb{C}$. Then K is a finite extension of the field $\mathbb{R}$ of real numbers and hence $K = \mathbb{R}(c_1, \ldots, c_n)$ for some elements $c_1, \ldots, c_n \in K$. Let $f_i(X) = \mathrm{Irr}(c_i; \mathbb{R})$ stand for the irreducible polynomial of c_i over $\mathbb{R}$ for $1 \leq i \leq n$, and let L be a splitting field for the product $f_1(X) \cdots f_n(X)$ over $\mathbb{R}$. Then $\mathbb{R} \subseteq \mathbb{C} \subseteq K \subseteq L$ and L is normal over $\mathbb{R}$.

(B) Let $G = G(L/\mathbb{R})$. We claim that the automorphism group G is a 2-group. To show this, first observe that $G(L/\mathbb{C}) \lhd G$ since the corresponding field $\mathbb{C}$ is a normal extension of $\mathbb{R}$. Therefore, $G/G(L/\mathbb{C}) \cong G(\mathbb{C}/\mathbb{R}) \cong \mathbb{Z}_2$ and hence the prime 2 divides $|G|$. It follows that G has a nontrivial Sylow 2-subgroup G_2. Let L^{G_2} stand for the corresponding fixed field. Then $[L^{G_2}:\mathbb{R}] = [G:G_2]$ and hence $[L^{G_2}:\mathbb{R}]$ is an odd number. But $L^{G_2} = \mathbb{R}(c)$ for some element $c \in L^{G_2}$ since, in characteristic zero, every finite extension has a primitive element. Therefore, $\mathrm{Irr}(c; \mathbb{R})$ is a nonconstant polynomial of odd degree over $\mathbb{R}$ and hence has a root in $\mathbb{R}$. Since the polynomial is irreducible over $\mathbb{R}$, this is impossible unless it has degree 1, in which case $L^{G_2} = \mathbb{R}$. Therefore, $G = G_2$ and hence G is a 2-group.

(C) We now claim that the automorphism group $G(L/\mathbb{C})$ is trivial. For suppose to the contrary that $G(L/\mathbb{C}) \neq (1)$. Then $G(L/\mathbb{C})$ is a 2-group since it is a subgroup of G which, by part (A), is a 2-group, and therefore $G(L/\mathbb{C})$ has a subgroup H of index 2. Let L^H stand for the corresponding fixed field in L. Then $L^H = \mathbb{C}(x)$ for some element $x \in L^H$, and $\mathrm{Irr}(x; \mathbb{C})$ is an irreducible polynomial of degree 2 over $\mathbb{C}$ since $\deg \mathrm{Irr}(x; \mathbb{C}) = [\mathbb{C}(x):\mathbb{C}]$ and $[G(L/\mathbb{C}):H] = 2$. But the field $\mathbb{C}$ of complex numbers contains the square root of all its elements. Therefore, $\mathrm{Irr}(x; \mathbb{C})$ must have degree 1, which is a contradiction, and consequently no such subgroup H exists. Therefore, $G(L/\mathbb{C})$ is trivial, as required.

(D) It follows from parts (A), (B) and (C) that the field L is a normal extension of $\mathbb{C}$ whose automorphism group $G(L/\mathbb{C})$ is trivial. Hence, by the Galois theory of fields, $L = \mathbb{C}$. Therefore, $K = \mathbb{C}$ and hence $f(X)$ has all its roots in $\mathbb{C}$, as required, and the proof of the fundamental theorem of algebra is complete.

The preceding examples have illustrated the Galois correspondence and fundamental theorem of Galois theory in a variety of settings. In many of these examples the automorphism group of the extension is an abelian group, which is an important point since it guarantees that every subgroup is normal and hence that every intermediate subfield of the extension is a normal extension of the ground field whose automorphisms may therefore be obtained by restriction of the automorphisms of the larger field. In general, a normal extension of fields K/k is called an *abelian extension* if its auto-morphism group $G(K/k)$ is an abelian group. Cyclotomic extensions are abe-lian extensions of $\mathbb{Q}$, for example, and any finite field is an abelian extension of its prime subfield. The field $\mathbb{Q}(\omega, \sqrt[3]{2})$, on the other hand, is a nonabelian extension of $\mathbb{Q}$ since its automorphism group is isomorphic to the nonabelian group $\mathrm{Sym}(3)$. There is an important theorem concerning abelian extensions due to the 19th century mathematician Leopold Kronecker (1823–1891) which states that every abelian extension of $\mathbb{Q}$ is isomorphic to a subfield of some cyclotomic extension. Thus, to within isomorphism, cyclotomic exten-sions and their subfields account for all abelian extensions of $\mathbb{Q}$.

Let us now conclude this section by discussing the existence of primitive elements for field extensions. Recall that if K/k is an extension of fields, an element $c \in K$ is called a primitive element for the extension if $K = k(c)$. In Chapter 13 we proved the primitive element theorem for fields of characteristic zero, which shows that every finite extension of a field of characteristic zero has a primitive element. Our goal at present is to prove a more general form of the primitive element theorem, due to Steinitz, which gives a criterion for the existence of primitive elements in any finite extension, regardless of the characteristic, in terms of the intermediate subfields of the extension. Using this result, together with the Galois theory, we then show that every normal extension of fields has a primitive element and hence must be a simple extension of the ground field.

PROPOSITION 6.   (STEINITZ) A finite extension of fields has a primitive element if and only if the extension has only a finite number of in-termediate subfields.

Proof.   Let K/k be a finite extension of fields. Suppose first that K/k has a primitive element, say $K = k(c)$ for some element $c \in K$, and let $f(X) = \mathrm{Irr}(c;k)$. We claim that every intermediate subfield of K/k is generated over k by the coefficients of some divisor of $f(X)$. For suppose that k' is an inter-mediate subfield of K/k and let

$$g(X) = \mathrm{Irr}(c;k') = X^n + a_{n-1}X^{n-1} + \cdots + a_1 X + a_0,$$

where $a_0, \ldots, a_{n-1} \in k'$. Then $g(X)$ divides $f(X)$ and $k(a_0, \ldots, a_{n-1}) \subseteq k'$. Now, the element c generates K over k' and has $g(X)$ as its irreducible poly-nomial over k'. Hence, $[K:k'] = \deg g(X)$. On the other hand, c also gen-erates K over the subfield $k(a_0, \ldots, a_{n-1})$ and has $g(X)$ as its irreducible polynomial over $k(a_0, \ldots, a_{n-1})$. Hence, $[K:k(a_0, \ldots, a_{n-1})] = \deg g(X)$. Therefore, $[K:k'] = [K:k(a_0, \ldots, a_{n-1})]$ and hence, since $k(a_0, \ldots, a_{n-1}) \subseteq k'$, it follows that $k' = k(a_0, \ldots, a_{n-1})$. Thus, every intermediate subfield of K/k is generated over k by the coefficients of some divisor of $f(X)$. Since $f(X)$ has only a finite number of divisors to within constant multiples, we conclude that the extension K/k has only a finite number of intermediate subfields.

Now suppose, conversely, that the extension K/k has only a finite number of intermediate subfields. To show that K/k has a primitive element, first suppose that the ground field k is finite. Then K is also finite and hence its multiplicative group K^* is cyclic. In this case any generator of K^* is a primitive element for the extension K/k and we are done. Assume, therefore, that the ground field k is infinite. Now, since the extension K/k has only a finite number of intermediate subfields, there is some element $c^* \in K$ such that $[k(c^*):k]$ is maximum; that is, $[k(c):k] \leq [k(c^*):k]$ for all elements $c \in K$. We claim that $K = k(c^*)$. For let c be any element in K and consider those intermediate subfields of the form $k(c^* + ac)$, where $a \in k$. Since k is infinite by assumption and since there are only a finite number of intermediate sub-fields, it follows that there are elements $a, a' \in k$, with $a \neq a'$, such that $k(c^* + ac) = k(c^* + a'c)$. Therefore, both $c^* + ac$ and $c^* + a'c$ lie in $k(c^* + ac)$ and hence $c \in k(c^* + ac)$. Consequently, $c^* \in k(c^* + ac)$ and hence $k(c^*) \subseteq k(c^* + ac)$. But $[k(c^*):k]$ is the maximum degree of any simple extension of k within K. Therefore, $k(c^*) = k(c^* + ac)$ and hence $c \in k(c^*)$. But c was an arbitrary element in K. Thus, $K = k(c^*)$, as required, and hence c^* is a primitive element for the extension K/k. ∎

COROLLARY. Every normal extension of fields has a primitive element.

Proof. Let K/k be a normal extension of fields. Then the automorphism group $G(K/k)$ is a finite group and hence, by the Galois correspondence, the extension K/k has only a finite number of intermediate subfields. Thus, by the Steinitz theorem, K/k has a primitive element. ∎

Exercises

1. Let $K = \mathbb{Q}(i, \sqrt{2})$ and $G = G(K/\mathbb{Q})$.
   (a) Find the automorphisms in G and show that $G \cong \mathbb{Z}_2 \times \mathbb{Z}_2$.
   (b) Show that G has five subgroups and find the intermediate subfield of $K/\mathbb{Q}$ corresponding to each subgroup.
   (c) Draw a diagram illustrating the Galois correspondence for the extension $K/\mathbb{Q}$.
   (d) Show that every intermediate subfield of $K/\mathbb{Q}$ is a normal extension of $\mathbb{Q}$.
   (e) Let $k' = \mathbb{Q}(1 + i\sqrt{2})$.
      (1) Find the subgroup H of G corresponding to k' and verify that $|H| = [K:K^H]$ and $[G:H] = [K^H:\mathbb{Q}]$.
      (2) Find $G(k'/\mathbb{Q})$.
      (3) Show that every $\mathbb{Q}$-automorphism of K restricts to a $\mathbb{Q}$-automorphism of k'.
      (4) Show that every $\mathbb{Q}$-automorphism of k' is the restriction of some $\mathbb{Q}$-automorphism of K and verify that $G(k'/\mathbb{Q}) \cong G/H$.
2. Let $K = \mathbb{Q}(i, \sqrt[4]{2})$ and $G = G(K/\mathbb{Q})$.

(a) Find the automorphisms in G and show that $G \cong D_4$, the group of the square.

(b) Show that G has ten subgroups and find the intermediate subfield corresponding to each subgroup.

(c) Draw a diagram illustrating the Galois correspondence for the extension $K/\mathbb{Q}$.

(d) Determine which intermediate subfields of $K/\mathbb{Q}$ are normal extensions of $\mathbb{Q}$, and find all conjugates of those subfields that are not normal extensions of $\mathbb{Q}$.

(e) Let $k' = \mathbb{Q}(\sqrt{2} + \sqrt[4]{2})$.

(1) Find the subgroup H of G corresponding to k' and verify that $|H| = [K:K^H]$ and $[G:H] = [K^H:\mathbb{Q}]$.

(2) Find $G(k'/\mathbb{Q})$.

(3) Find a $\mathbb{Q}$-automorphism of K that does not restrict to a $\mathbb{Q}$-automorphism of k'.

(f) Let $k' = \mathbb{Q}(i\sqrt{2})$.

(1) Find the subgroup H of G corresponding to k' and verify that $|H| = [K:K^H]$ and $[G:H] = [K^H:\mathbb{Q}]$.

(2) Find $G(k'/\mathbb{Q})$.

(3) Show that k' is a normal extension of $\mathbb{Q}$ and verify that every $\mathbb{Q}$-automorphism of K restricts to a $\mathbb{Q}$-automorphism of k'.

(4) Show that every $\mathbb{Q}$-automorphism of k' is the restriction of some $\mathbb{Q}$-automorphism of K and verify that $G(k'/\mathbb{Q}) \cong G/H$.

3. Let $K = \mathbb{F}_2[X]/(X^4 + X + 1)\mathbb{F}_2[X]$ and $G = G(K/\mathbb{F}_2)$.

(a) Show that K is a finite field containing 16 elements and that $[K:\mathbb{F}_2] = 4$.

(b) Find the automorphisms in G and show that $G \cong \mathbb{Z}_4$, the cyclic group of order 4.

(c) Show that G has three subgroups and find the intermediate subfield corresponding to each subgroup.

(d) Draw a diagram illustrating the Galois correspondence for the extension $K/\mathbb{F}_2$.

(e) Show that every intermediate subfield of $K/\mathbb{F}_2$ is a normal extension of $\mathbb{F}_2$.

(f) Let $k' = \mathbb{F}_2(c + c^2)$, where $c = X + (X^4 + X + 1)\mathbb{F}_2[X] \in K$.

(1) Find the subgroup H of G corresponding to k' and verify that $|H| = [K:K^H]$ and $[G:H] = [K^H:\mathbb{F}_2]$.

(2) Find $G(k'/\mathbb{F}_2)$.

(3) Show that every $\mathbb{F}_2$-automorphism of K restricts to an $\mathbb{F}_2$-automorphism of k'.

(4) Show that every $\mathbb{F}_2$-automorphism of k' is the restriction of some $\mathbb{F}_2$-automorphism of K and verify that $G(k'/\mathbb{F}_2) \cong G/H$.

(g) Let c stand for the element of K defined in part (f).

(1) Show that $K = \{0, 1, c, c^2, \ldots, c^{14}\}$.

     (2) Find the subgroup of G corresponding to each of the intermediate subfields $F_2(1), F_2(c), \ldots, F_2(c^{14})$.

     (3) Using the Galois correspondence, determine which of the subfields $F_2(1), F_2(c), \ldots, F_2(c^{14})$ are the same.

(h) How many primitive elements does the extension K/F_2 have?

4. Let $K = Q(i, \omega)$, where ω stands for a primitive cube root of unity, and let $G = G(K/Q)$.

(a) Show that K is a normal extension of Q.

(b) Find the automorphisms in G and show that $G \cong Z_2 \times Z_2$.

(c) Show that $\sqrt{3} \in K$.

(d) Find the subgroup H of G corresponding to the intermediate subfield $Q(\sqrt{3})$. Show that:

     (1) H is a normal subgroup of G;

     (2) $Q(\sqrt{3})$ is a normal extension of Q;

     (3) $G(Q(\sqrt{3})/Q) \cong G/H$.

5. Let K/k be a normal extension of fields and let $c \in K$.

(a) Show that the number of conjugates of c in K is equal to $[k(c):k]$.

(b) If σ and τ are k-automorphisms of K, show that $\sigma(c) = \tau(c)$ if and only if $\sigma G(K/k(c)) = \tau G(K/k(c))$.

(c) Show that $[k(c):k] = [G(K/k):G(K/k(c))]$.

6. Let K/k be a normal extension of fields. Show that an element $c \in K$ is a primitive element for the extension K/k if and only if the identity automorphism is the only k-automorphism of K that fixes c.

7. Let K/k be a normal extension of fields. Show that every intermediate subfield of K/k is a simple extension of k.

8. Find the Galois group of the following polynomials. In each case write out the permutation of the roots corresponding to each automorphism in the Galois group.

(a) $f(X) = X^2 + 2 \in Q[X]$

(b) $f(X) = (X^2 + 1)(X^2 - 2) \in Q[X]$

(c) $f(X) = X^4 - 2 \in Q[X]$

(d) $f(X) = X^4 + 16 \in Q[X]$

(e) $f(X) = X^2 - 2\sqrt{2}X - 1 \in Q(\sqrt{2})[X]$

(f) $f(X) = X^4 + X + 1 \in F_2[X]$

(g) $f(X) = X^3 + 2X^2 + X + 2 \in F_3[X]$

9. Let k be a field such that $\mathrm{char}(k) \neq 2$ and let $f(X) = aX^2 + bX + c$, where $a, b, c \in k$. Let G_f stand for the Galois group of $f(X)$ over k and let $\Delta^2 = b^2 - 4ac$.

(a) Show that G_f is isomorphic to a subgroup of $\mathrm{Sym}(2)$.

(b) Show that $G_f = (1)$ if and only if $\Delta \in k$, while $G_f \cong \mathrm{Sym}(2)$ if and only if $\Delta \notin k$.

10. Let $f(X)$ be a nonconstant polynomial over a field k and let K and K' be splitting fields for $f(X)$ over k. Show that $G(K/k) \cong G(K'/k)$. Thus, the Galois group of a polynomial is unique to within isomorphism.

11. Let n be a positive integer, k a subfield of the field $\mathbb{C}$ of complex numbers that contains a primitive nth root of unity, and let $a \in k$. Show that the Galois group of the polynomial $X^n - a$ over k is abelian.

12. Let $f(X)$ be a nonconstant polynomial over a field k and suppose that $f(X) = g(X)h(X)$ for some polynomials $g(X), h(X) \in k[X]$ that have no roots in common. Let G_f, G_g, and G_h stand for the Galois groups of $f(X), g(X)$, and $h(X)$, respectively. Show that $G_f \cong G_g \times G_h$.

13. Let K be a normal extension of the field $\mathbb{Q}$ of rational numbers. If σ is a $\mathbb{Q}$-automorphism of K, then σ is a nonsingular linear transformation of K as a vector space over $\mathbb{Q}$ and hence has a nonzero determinant. Define the function $\det : G(K/\mathbb{Q}) \to \mathbb{Q}^*$ by letting $\det(\sigma)$ stand for the determinant of σ for every $\mathbb{Q}$-automorphism $\sigma \in G(K/\mathbb{Q})$.

    (a) Show that the mapping det is a group-homomorphism that maps the Galois group $G(K/\mathbb{Q})$ into the multiplicative group $\mathbb{Q}^*$ of nonzero rational numbers.

    (b) Show that $\det(\sigma) = \pm 1$ for every $\mathbb{Q}$-automorphism $\sigma \in G(K/\mathbb{Q})$.

14. **The normal closure of an intermediate subfield.** Let K/k be a normal extension of fields and let L be an intermediate subfield of K/k. Let $G = G(K/k)$ and $H = G(K/L)$.

    (a) Let $N = \cap\, \sigma H \sigma^{-1}$, where the intersection is over all elements $\sigma \in G$. Show that N is a normal subgroup of G and that it is the largest subgroup of H that is normal in G.

    (b) Show that there is a unique intermediate subfield $\bar{L}$ of the extension K/k that satisfies the following three conditions:

    (1) $k \subseteq L \subseteq \bar{L} \subseteq K$

    (2) $\bar{L}$ is a normal extension of k

    (3) $\bar{L}$ is the smallest normal extension of k in K that contains L. The intermediate subfield $\bar{L}$ is called the *normal closure* of L in K.

    (c) Let $c \in L$ be a primitive element for the extension L/k. Show that $\bar{L}$ is generated over k by the conjugates of c in K.

15. **The discriminant of a normal extension.** Let K/k be a normal extension of fields and let $A_1, \ldots, A_n$ be a basis for K as a vector space over k, where $n = [K:k]$. Let $G(K/k) = \{\sigma_1, \ldots, \sigma_n\}$ and let

$$M(A_1, \ldots, A_n) = \begin{pmatrix} \sigma_1(A_1) \cdots \sigma_n(A_1) \\ \vdots \\ \sigma_1(A_n) \cdots \sigma_n(A_n) \end{pmatrix}.$$

    (a) Show that the matrix $M(A_1, \ldots, A_n)$ is nonsingular.

    (b) Show that $\sigma(\det M(A_1, \ldots, A_n)) = \pm \det M(A_1, \ldots, A_n)$ for every k-automorphism σ of K.

    (c) Let $\Delta(A_1, \ldots, A_n) = [\det M(A_1, \ldots, A_n)]^2$. Show that $\Delta(A_1, \ldots, A_n)$ is a nonzero element in k. $\Delta(A_1, \ldots, A_n)$ is called the *discriminant* of the extension K/k relative to the basis $A_1, \ldots, A_n$.

    (d) Let $B_1, \ldots, B_n$ be any basis for K as a vector space over k and let P stand for the transition matrix from the basis $A_1, \ldots, A_n$ to the basis $B_1, \ldots, B_n$ defined by

$$\begin{pmatrix} B_1 \\ \vdots \\ B_n \end{pmatrix} = P \begin{pmatrix} A_1 \\ \vdots \\ A_n \end{pmatrix}.$$

Show that $\Delta(B_1, \ldots, B_n) = (\det P)^2 \Delta(A_1, \ldots, A_n)$.

(e) Let k^* stand for the multiplicative group of nonzero elements in k and $k^{*2} = \{x^2 \in k^* \mid x \in k^*\}$ the subgroup of squares in k^*. Show that the coset $\Delta(A_1, \ldots, A_n)k^{*2} \in k^*/k^{*2}$ is independent of the basis $A_1, \ldots, A_n$. Thus, the discriminant $\Delta(A_1, \ldots, A_n)$ of the extension K/k is uniquely determined modulo squares in k^*.

(f) Let $G(K/k)_\Delta = \{\sigma \in G(K/k) \mid \sigma(\det M(A_1, \ldots, A_n)) = \det M(A_1, \ldots, A_n)\}$. Show that either:

(1) $G(K/k)_\Delta = G(K/k)$, in which case $\det M(A_1, \ldots, A_n) \in k$; or,

(2) $[G(K/k):G(K/k)_\Delta] = 2$, in which case $G(K/k)_\Delta \lhd G(K/k)$.

(g) Show that $k(\sqrt{\Delta(A_1, \ldots, A_n)})$ is the intermediate subfield of K/k corresponding to the subgroup $G(K/k)_\Delta$ under the Galois correspondence for K/k.

(h) Calculate the discriminant of the following normal extensions relative to the indicated basis and find the subgroup $G(K/k)_\Delta$ and corresponding subfield $k(\sqrt{\Delta(A_1, \ldots, A_n)})$:

(1) $\mathbb{Q}(\omega)/\mathbb{Q}$, basis $= \{1, \omega\}$;

(2) $\mathbb{Q}(i)/\mathbb{Q}$, basis $= \{1, i\}$;

(3) $\mathbb{Q}(i)/\mathbb{Q}$, basis $= \{1 + i, 1 - i\}$;

(4) $\mathbb{Q}(\sqrt{2}, \sqrt{3})$, basis $= \{1, \sqrt{2}, \sqrt{3}, \sqrt{6}\}$.

16. Let K/k be a normal extension of fields. Suppose there is a tower of intermediate subfields $k = K_0 \subseteq K_1 \subseteq \cdots \subseteq K_n = K$ in the extension K/k such that K_i is a normal extension of K_{i-1} for $i = 1, \ldots, n$.

(a) Show that the automorphism group $G(K/k)$ contains a chain of subgroups $(1) = H_n \leq H_{n-1} \leq \cdots \leq H_0 = G(K/k)$ such that $H_i \lhd H_{i-1}$ and $H_{i-1}/H_i \cong G(K_i/K_{i-1})$ for $i = 1, \ldots, n$.

(b) Find the chain of subgroups in $G(\mathbb{Q}(i, \sqrt[4]{2})/\mathbb{Q})$ corresponding to the tower of subfields $\mathbb{Q} \subseteq \mathbb{Q}(\sqrt{2}) \subseteq \mathbb{Q}(i, \sqrt{2}) \subseteq \mathbb{Q}(i, \sqrt[4]{2})$ of the extension $\mathbb{Q}(i, \sqrt[4]{2})/\mathbb{Q}$.

17. Let K/k be a normal extension of fields. Suppose that the automorphism group $G(K/k)$ contains a chain of subgroups $(1) = H_0 \leq H_1 \leq \cdots \leq H_n = G(K/k)$ such that $H_{i-1} \lhd H_i$ for $i = 1, \ldots, n$.

(a) Show that the extension K/k contains a tower of intermediate subfields $k = K_n \subseteq K_{n-1} \subseteq \cdots \subseteq K_0 = K$ such that K_{i-1} is a normal extension of K_i and $G(K_{i-1}/K_i) \cong H_i/H_{i-1}$ for $i = 1, \ldots, n$.

(b) Find the tower of intermediate subfields of the extension $\mathbb{Q}(\sqrt{2}, \sqrt{3})$ corresponding to the chain of subgroups $(1) \leq \{1, \sigma\} \leq G(\mathbb{Q}(\sqrt{2}, \sqrt{3})/\mathbb{Q})$ where σ stands for the automorphism of $\mathbb{Q}(\sqrt{2}, \sqrt{3})$ that maps $\sqrt{2}$ to $-\sqrt{2}$ and $\sqrt{3}$ to $-\sqrt{3}$.

15

Solvability by Radicals

The quadratic formula is a well-known formula for finding the roots of any second-degree equation whose coefficients lie in the field of complex numbers: if a, b and c are complex numbers with $a \neq 0$, then the roots of the equation $ax^2 + bx + c = 0$ are given by the formulas

$$x_1 = \frac{-b + \sqrt{b^2 - 4ac}}{2a} \quad \text{and} \quad x_2 = \frac{-b - \sqrt{b^2 - 4ac}}{2a}.$$

These formulas not only find the roots of the equation, they also show that the roots may be expressed in terms of the coefficients using addition, subtraction, multiplication and division, together with extraction of square roots. Moreover, the Renaissance mathematicians Cardan and Tartaglia found similar formulas for the roots of any third or fourth degree polynomial— formulas that once again express the roots in terms of the four arithmetic operations on the coefficients together with root extractions. But to what extent is it always possible to express the roots of a polynomial in this manner? The fundamental theorem of algebra, for example, guarantees that every nonconstant polynomial of degree n over the field of complex numbers has exactly n complex roots, but says nothing about how these roots may be expressed.

It was not until the development of group theory and field theory in the early nineteenth century that the mathematical tools became available to answer this question; field theory provided the framework in which to discuss the roots, while group theory provided the tools to analyze the nature of the solutions. Indeed, it was the solvability of algebraic equations by means of radicals that led to the evolution of group theory and, with it, the realization that not every algebraic equation is solvable by radicals. In 1826 the young Norwegian mathematician Niels Henrik Abel showed that the general quintic is not solvable by radicals. Then in 1832 Galois answered the entire solvability question by creating a general theory for dealing with the solvability of any algebraic equation regardless of its degree. By doing so, he showed that the general algebraic equation of degree n is not solvable by radicals whenever $n \geq 5$.

In this chapter our goal is to use the Galois theory of fields to construct an algebraic model for the solvability of polynomials by means of radicals. When a polynomial is solvable by radicals, its roots can be expressed in terms of the coefficients using the four arithmetic operations together with root extraction, and hence may be obtained through a sequence of field extensions, each field obtained from the previous one by adjoining an appropriate root. Using the Galois correspondence, it turns out that this is possible if and only if the Galois group of the polynomial contains a chain of subgroups each normal in its successor and having an abelian quotient group. A finite group with this property is called a solvable group. Thus, the solvability of a polynomial by means of radicals leads to the concept of a solvable group, and it is this concept that provides the answer to the solvability question: a polynomial is solvable by radicals if and only if its Galois group is a solvable group. After we discuss the general properties of solvable groups, and the solvability of the symmetric groups in particular, we then use the solvability criterion to show that the general polynomial of degree n over the field of rational numbers is not solvable by radicals when $n \geq 5$. It follows, therefore, that there is no general formula for expressing the roots a polynomial of degree n in terms of arithmetic operations on the coefficients and root extractions when $n \geq 5$, although there are particular equations of any given degree whose roots may be so expressed.

Throughout this chapter we restrict our attention to the field of complex numbers and its subfields. Let us begin by recalling that for any positive integer n, the nth roots of unity are the n complex roots of the polynomial $X^n - 1$ and that these roots form a cyclic group of order n generated by any primitive nth root of unity z_n. In general, if a is any complex number, the roots of the polynomial $X^n - a$ have the form $w, wz_n, \ldots, wz_n^{n-1}$, where w is any complex number such that $w^n = a$, and are the nth roots of a. For example, if ω is a primitive cube root of unity, then the cube roots of 2 are $\sqrt[3]{2}, \omega\sqrt[3]{2}$ and $\omega^2\sqrt[3]{2}$.

Definition

Let k be a field. A *radical extension* of k is any extension field generated by an nth root of some element in k. A nonconstant polynomial $f(X)$ over k is said to be *solvable by radicals* if there is a tower of fields $k = k_1 \subseteq k_2 \subseteq \cdots \subseteq k_n$ such that k_{i+1} is a radical extension of k_i for $i = 1, \ldots, n-1$ and $f(X)$ splits over k_n; in this case, any such tower of fields is called a *tower of radicals* for $f(X)$ over k.

Thus, if K is a radical extension of k, then $K = k(w)$ for some element $w \in K$, where $w^n \in k$ for some positive integer n; the fields $\mathbb{Q}(\sqrt{2})$, $\mathbb{Q}(\sqrt[3]{2})$, $\mathbb{Q}(i)$ and $\mathbb{Q}(\omega)$, for example, are radical extensions of $\mathbb{Q}$, while $\mathbb{Q}(\sqrt[4]{2})$ is a radical extension of both $\mathbb{Q}$ and $\mathbb{Q}(\sqrt{2})$, and $\mathbb{Q}(\omega, \sqrt[3]{2})$ is a radical extension of both $\mathbb{Q}(\sqrt[3]{2})$ and $\mathbb{Q}(\omega)$. It follows that if $f(X)$ is a polynomial over k that is solvable by radicals and $k = k_1 \subseteq k_2 \subseteq \cdots \subseteq k_n$ is a tower of radicals for $f(X)$ over k, then $k_{i+1} = k_i(w_{i+1})$, where w_{i+1} is an n_{i+1}th root of some element in k_i for $i = 1, \ldots, n-1$, and hence $k_n = k(w_2, \ldots, w_n)$. Hence, since $f(X)$ splits over k_n, it follows that the roots of $f(X)$ may be expressed in terms of the coefficients by formulas that involve only addition, subtraction, multiplication and division, together with root extraction. The polynomial $f(X) = X^4 - 1$, for example, is solvable by radicals since its roots are ± 1 and $\pm i$; in this case $\mathbb{Q} \subseteq \mathbb{Q}(i)$ is a tower of radicals for $f(X)$ over $\mathbb{Q}$. Similarly, the polynomial $g(X) = X^3 - 2$ is solvable by radicals since its roots are $\sqrt[3]{2}$, $\omega\sqrt[3]{2}$ and $\omega^2\sqrt[3]{2}$, where ω is a primitive cube root of unity, and $\mathbb{Q} \subseteq \mathbb{Q}(\omega) \subseteq \mathbb{Q}(\omega, \sqrt[3]{2})$ is a tower of radicals for $g(X)$ over $\mathbb{Q}$.

EXAMPLE 1. The general quadratic polynomial is solvable by radicals

Let $q(X) = aX^2 + bX + c$ stand for the general quadratic polynomial over the field $\mathbb{Q}$ of rational numbers. Then the roots of $q(X)$ are

$$x_1 = \frac{-b + \sqrt{b^2 - 4ac}}{2a} \quad \text{and} \quad x_2 = \frac{-b - \sqrt{b^2 - 4ac}}{2a}.$$

Therefore, $q(X)$ is solvable by radicals, and, if $k = \mathbb{Q}(a, b, c)$ stands for the field of coefficients of $q(X)$, then $k \subseteq k(\sqrt{b^2 - 4ac})$ is a tower of radicals for $q(X)$ over k.

EXAMPLE 2. The general cubic polynomial is solvable by radicals

Let $f(X) = aX^3 + bX^2 + cX + d$ stand for the general cubic polynomial over $\mathbb{Q}$. Let us derive Cardan's formulas for the roots of $f(X)$ and thus show that the general cubic polynomial is solvable by radicals.

(A) To obtain the roots of $f(X)$, we first transform $f(X)$ into a sixth degree polynomial whose roots are easy to find. Let $X = Y - b/3a$. Substituting this expression into the equation $aX^3 + bX^2 + cX + d = 0$, we find that $Y^3 + pY + q = 0$, where

$$p = \frac{c}{a} - \frac{b^2}{3a^2}$$

$$q = \frac{4b^3}{27a^3} - \frac{bc}{3a^2} + \frac{d}{a}.$$

Now, let $Y = Z - p/3Z$. Substituting this expression into $Y^3 + pY + q = 0$ and simplifying, we obtain the equation

$$Z^6 + qZ^3 - \frac{p^3}{27} = 0,$$

which is easily solved since it is a quadratic in Z^3. Applying the quadratic formula, we find that

$$Z^3 = \frac{-q + \sqrt{q^2 + \frac{4p^3}{27}}}{2} \quad \text{or} \quad Z^3 = \frac{-q - \sqrt{q^2 + \frac{4p^3}{27}}}{2}.$$

Thus, if we let L and M stand for these two expressions, respectively, then the roots of the sixth degree equation are $\sqrt[3]{L}$, $\omega \sqrt[3]{L}$, $\omega^2 \sqrt[3]{L}$, and $\sqrt[3]{M}$, $\omega \sqrt[3]{M}$ and $\omega^2 \sqrt[3]{M}$. Substituting these roots back, first to find Y and then X, and using the fact that $LM = -\frac{2p}{3}$, which is easily verified, and the fact that $\omega^2 = -1 - \omega$, we obtain, finally, the three roots of $f(X)$:

$$x_1 = -\frac{b}{3a} + \sqrt[3]{L} + \sqrt[3]{M}$$

$$x_2 = -\frac{b}{3a} + \omega \sqrt[3]{L} + \omega^2 \sqrt[3]{M}$$

$$x_3 = -\frac{b}{3a} + \omega^2 \sqrt[3]{L} + \omega \sqrt[3]{M}.$$

These formulas are *Cardan's formulas* for the roots of a cubic polynomial and show, clearly, that every cubic polynomial is solvable by radicals. If we let $k = \mathbb{Q}(a, b, c, d)$ stand for the field of coefficients of $f(X)$, then

$$k \subseteq k(\omega) \subseteq k(\omega)\left(\sqrt{q^2 + \frac{4p^3}{27}}\right) \subseteq k\left(\omega, \sqrt{q^2 + \frac{4p^3}{27}}\right)(\sqrt[3]{L})$$

is a tower of radicals for $f(X)$ over k.

(B) Let us illustrate Cardan's formulas by using them to find the roots of the polynomial $f(X) = X^3 + 3X^2 + 6X + 5$. In this case $p = 3$ and $q = 3$, and therefore

$$L = \frac{-3 + \sqrt{13}}{2} \quad \text{and} \quad M = \frac{-3 - \sqrt{13}}{2}.$$

Using Cardan's formulas, it follows that the roots of $f(X)$ are

$$x_1 = -1 + \sqrt[3]{\frac{-3 + \sqrt{13}}{2}} + \sqrt[3]{\frac{-3 - \sqrt{13}}{2}}$$

$$x_2 = -1 + \omega \sqrt[3]{\frac{-3 + \sqrt{13}}{2}} + \omega^2 \sqrt[3]{\frac{-3 - \sqrt{13}}{2}}$$

$$x_3 = -1 + \omega^2 \sqrt[3]{\frac{-3 + \sqrt{13}}{2}} + \omega \sqrt[3]{\frac{-3 - \sqrt{13}}{2}}.$$

In this case,

$$\mathbb{Q} \subseteq \mathbb{Q}(\omega) \subseteq \mathbb{Q}(\omega)(\sqrt{13}) \subseteq \mathbb{Q}(\omega, \sqrt{13})\left(\sqrt[3]{\frac{-3 + \sqrt{13}}{2}}\right)$$

is a tower of radicals for $f(X)$ over $\mathbb{Q}$: extending from $\mathbb{Q}$ to $\mathbb{Q}(\omega)$ gives us a primitive cube root of unity to work with, from $\mathbb{Q}(\omega)$ to $\mathbb{Q}(\omega)(\sqrt{13})$ splits the resolvent $Z^6 + 3Z^3 - 1$ into the product $(Z^3 - L)(Z^3 - M)$, which, over the field $\mathbb{Q}(\omega, \sqrt{13})\left(\sqrt[3]{\frac{-3 + \sqrt{13}}{2}}\right)$, splits to give the roots x_1, x_2, and x_3.

Example 2 illustrates one method for finding the roots of a cubic equation; using a sequence of algebraic transformations, the original equation is transformed into new equations, called resolvents, which are then easily solved and whose roots yield the roots of the original cubic. In the exercises we indicate a similar procedure for finding the roots of the general quartic, or fourth degree equation.

Let us now discuss the basic properties of radical extensions and their automorphism groups.

PROPOSITION 1. Let $K = k(w)$ be a radical extension of a field k, where $w^s \in k$, and assume that k contains a primitive sth root of unity. Then K is a normal extension of k and its automorphism group $G(K/k)$ is abelian.

Proof. Let z_s stand for a primitive sth root of unity in k. Then the roots of the polynomial $X^s - w^s$ are $w, wz_s, \ldots, wz_s^{s-1}$. Since these roots lie in K and generate K over k, it follows that K is a splitting field for $X^s - w^s$ over k and hence is a normal extension of k. Now, if σ is a k-automorphism of K, then $\sigma(w)$ is also an sth root of a w^s and hence $\sigma(w) = wz_s^i$ for some integer $i \geq 0$, and σ is uniquely determined by the image $\sigma(w)$. Hence, if τ is any k-automorphism of K and $\tau(w) = wz_s^j$, then $(\sigma\tau)(w) = \sigma(wz_s^j) = wz_s^{i+j} = (\tau\sigma)(w)$, and therefore $\sigma\tau = \tau\sigma$. Therefore, the automorphism group $G(K/k)$ is abelian. ∎

PROPOSITION 2. Let K be a normal extension of a field k with $[K:k]$ $= n$, and assume that k contains a primitive nth root of unity. If the automorphism group $G(K/k)$ is cyclic, then K is a radical extension of k.

Proof. Let z_n be a primitive nth root of unity in k and let σ be a generator for the automorphism group $G(K/k)$. Then $|G(K/k)| = [K:k] = n$ since K is normal over k and hence $G(K/k) = \{1, \sigma, \ldots, \sigma^{n-1}\}$. Now, it follows from Artin's theorem that the sum $x + z_n\sigma(x) + z_n^2\sigma^2(x) + \cdots + z_n^{n-1}\sigma^{n-1}(x)$ is not identically zero for every element $x \in K$. Let $x_0 \in K$ be such that

$$y = x_0 + z_n\sigma(x_0) + \cdots + z_n^{n-1}\sigma^{n-1}(x_0) \neq 0.$$

Then, since $z_n \in k$ by assumption, $\sigma(z_n) = z_n$ and therefore

$$z_n\sigma(y) = z_n\sigma(x_0) + z_n^2\sigma^2(x_0) + \cdots + z_n^{n-1}\sigma^{n-1}(x_0) + x_0 = y.$$

Hence, $\sigma(y) = yz_n^{-1}$ and therefore $\sigma^s(y) = yz_n^{-s}$ for all integers $s \geq 0$. It follows that the element y has at least n distinct conjugates in the field K and therefore $[k(y):k] \geq n$. But $[K:k] = n$ and $k(y) \subseteq K$. Therefore, $k(y) = K$. Finally, $\sigma(y^n) = (yz_n^{-1})^n = y^n$ and hence y^n is fixed by every automorphism in $G(K/k)$. Therefore, $y^n \in k$ since K is a normal extension of k. Hence, K is a radical extension of k generated by an nth root of an element in k. ∎

It follows from Proposition 1 that every radical extension is a normal extension and has an abelian automorphism group, provided the ground field has an appropriate root of unity. Proposition 2, on the other hand, shows that every normal extension with cyclic automorphism group is in fact a radical extension, assuming, again, that the ground field contains the appropriate root of unity. For example, $\mathbb{Q}(\omega, \sqrt[3]{2})$ is a radical extension of $\mathbb{Q}(\omega)$ in which the ground field $\mathbb{Q}(\omega)$ contains a primitive cube root of unity; in this case the automorphism group $G(\mathbb{Q}(\omega, \sqrt[3]{2})/\mathbb{Q}(\omega)) = \{1, \sigma, \sigma^2\}$ is the abelian group of order 3 generated by the automorphism $\sigma: \sqrt[3]{2} \mapsto \omega\sqrt[3]{2}$. Observe, however, that $\mathbb{Q}(\sqrt[3]{2})$ is a radical extension of $\mathbb{Q}$ whose automorphism group is trivial and hence abelian, but is not a normal extension of $\mathbb{Q}$; in this case the ground field $\mathbb{Q}$ does not contain a primitive cube root of unity.

Now, if $f(X)$ is a polynomial over a field k that is solvable by radicals, then there is a tower of radicals $k = k_1 \subseteq k_2 \subseteq \cdots \subseteq k_n$ for $f(X)$ over k. Although k_{i+1} is a radical extension of k_i for every i, we cannot use Proposition 1 to obtain information about the automorphism groups $G(k_{i+1}/k_i)$ since the ground fields k_i do not necessarily contain the appropriate roots of unity. Indeed, we cannot use the Galois theory of fields to study the extensions since the fields are not necessarily normal over k. Let us now show that by adjoining appropriate roots, any tower of radicals may in fact be extended to a tower of radicals that has these properties.

PROPOSITION 3. Let $k = k_1 \subseteq k_2 \subseteq \cdots \subseteq k_n$ be a tower of radical extensions. Then there is a tower of radical extensions $k = K_1 \subseteq K_2 \subseteq \cdots \subseteq K_m$ having the following properties:

(1) each field k_i is a subfield of some K_j;
(2) K_j is a normal extension of k for $j = 1, \ldots, m$;
(3) the automorphism groups $G(K_{j+1}/K_j)$ are abelian for $j = 1, \ldots, m - 1$.

Proof. We prove the statement by induction on the length n of the tower. If $n = 1$, then $K_1 = k$ satisfies all requirements. Assume therefore that $n > 1$ and that every tower of radicals having fewer than n fields may be extended to a larger tower that satisfies the three requirements. Let $k = k_1 \subseteq k_2 \subseteq \cdots \subseteq k_n$ be a tower of radical extensions having n intermediate subfields. Then $k = k_1 \subseteq k_2 \subseteq \cdots \subseteq k_{n-1}$ is such a tower of length $n - 1$ and hence by the induction hypothesis there is a tower of radical extensions $k = K_1 \subseteq K_2 \subseteq \cdots \subseteq K_m$ such that each of the fields $k_1, \ldots, k_{n-1}$ is a subfield of some K_j, K_j is normal over k, and the automorphism groups $G(K_{j+1}/K_j)$ are abelian. Assume $k_{n-1} \subseteq K_m$. Now, k_n is a radical extension of k_{n-1} and hence $k_n = k_{n-1}(w)$ for some element $w \in k_n$, where $w^s = a \in k_{n-1}$. Let z_s stand for a primitive sth root of unity. Since K_m is normal over k, it is the splitting field for some polynomial $f(X)$ over k. Therefore, $K_m(z_s)$ is a splitting field for $f(X)(X^s - 1)$ over k and hence is a normal extension of k. Moreover, the automorphism group $G(K_m(z_s)/K_m)$ is abelian since it is a subgroup of $G(\mathbb{Q}(z_s)/\mathbb{Q})$, which is abelian. Now, let $G(K_m/k) = \{\sigma_1, \ldots, \sigma_t\}$ and let $w_1, \ldots, w_t$ stand for sth roots of the conjugates $\sigma_1(a), \ldots, \sigma_t(a)$, respectively. Then, for $j = 1, \ldots, t - 1$, $K_m(z_s, w_1, \ldots, w_j)(w_{j+1})$ is a radical extension of $K_m(z_s, w_1, \ldots, w_j)$ and, by Proposition 1, has an abelian automorphism group. Finally, we claim that $K_m(z_s, w_1, \ldots, w_t)$ is a normal extension of k. For let $g(X) = (X^s - \sigma_1(a)) \cdots (X^s - \sigma_t(a))$. Then the coefficients of $g(X)$ are symmetric polynomials in the conjugates of a and hence are fixed by every automorphism in $G(K_m/k)$. Therefore, the coefficients lie in the ground field k since K_m is normal over k, and hence $g(X)$ is a polynomial over k. Since

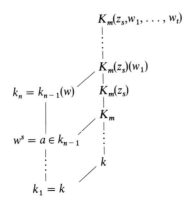

the roots of $g(X)$ are $w_1, \ldots, w_1 z_s^{s-1}, \ldots, w_t, \ldots, w_t z_s^{s-1}$, it follows that $K_m(z_s, w_1, \ldots, w_t)$ is a splitting field for $f(X)(X^s - \sigma_1(a)) \cdots (X^s - \sigma_t(a))$ over k and hence is a normal extension of k. Thus, the tower of radical extensions

$$k = K_1 \subseteq K_2 \subseteq \cdots \subseteq K_m \subseteq K_m(z_s)$$
$$\subseteq K_m(z_s)(w_1) \subseteq \cdots \subseteq K_m(z_s, w_1, \ldots, w_t)$$

satisfies the three requirements and the proof by induction is complete. ∎

It now follows that if a polynomial $f(X)$ over a field k is solvable by radicals, then any tower of radicals for $f(X)$ over k may be enlarged to a tower of radicals of the form $k = K_1 \subseteq K_2 \subseteq \cdots \subseteq K_m$, where each field K_{j+1} is a normal extension of k, each automorphism group $G(K_{j+1}/K_j)$ is abelian, and where K_m contains a splitting field for $f(X)$ over k. Consequently, the Galois theory applies to the extension K_m/k and shows that the automorphism group $G(K_m/k)$ contains a corresponding chain of subgroups,

$$(1) \le \cdots \le G(K_m/K_{j+1}) \le G(K_m/K_j) \le \cdots \le G(K_m/k),$$

in which $G(K_m/K_{j+1}) \lhd G(K_m/K_j)$ and $G(K_m/K_j)/G(K_m/K_{j+1}) \cong G(K_{j+1}/K_j)$ for $1 \le j \le m - 1$, as illustrated here.

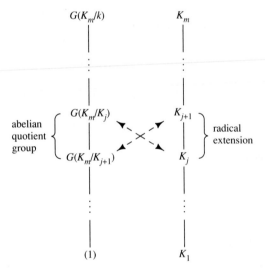

Moreover, since K_m contains a splitting field K for $f(X)$ over k, K is normal over k and hence, by the Galois theory, $G(K/k) \cong G(K_m/k)/G(K_m/K)$. Thus, when a polynomial is solvable by radicals its Galois group is the quotient group of some group that has a chain of subgroups each normal in its successor and having an abelian quotient group.

For example, consider the polynomial $f(X) = X^3 - 2 \in \mathbb{Q}[X]$. In this case $\mathbb{Q} \subseteq \mathbb{Q}(\omega) \subseteq \mathbb{Q}(\omega, \sqrt[3]{2})$ is a tower of radicals for $f(X)$ over $\mathbb{Q}$, and

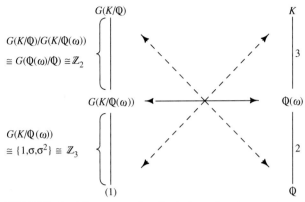

FIGURE 1. The tower of radicals for $X^3 - 2$ over $\mathbb{Q}$ and the corresponding chain of subgroups in $G(K/\mathbb{Q})$ under the Galois correspondence.

$\mathbb{Q}(\omega, \sqrt[3]{2})$ is a normal extension of $\mathbb{Q}$ since it is a splitting field for $f(X)$ over $\mathbb{Q}$. Figure 1 illustrates the tower of radicals for $f(X)$ and the corresponding chain of subgroups in $G(\mathbb{Q}(\omega, \sqrt[3]{2})/\mathbb{Q})$ and their quotient groups.

Before we can complete the proof of the solvability criterion, we first need to discuss the basic properties of groups having a chain of normal subgroups for which the corresponding quotient groups are abelian. Let us begin by first extending these ideas to finite groups in general, discuss their basic properties, and then use the results to prove the main solvability criterion.

Definition

Let G be a finite group. G is said to be *solvable* if there is a chain of subgroups $(1) = H_1 \leq H_2 \leq \cdots \leq H_n = G$ such that $H_i \triangleleft H_{i+1}$ and H_{i+1}/H_i is abelian for $1 \leq i \leq n - 1$.

Clearly, every abelian group is solvable. The group D_4 of the square is also solvable; in this case $D_4 = \{1, r, r^2, r^3, s, sr, sr^2, sr^3\}$ and we find that

$$\{1\} \triangleleft \{1, r, r^2, r^3\} \triangleleft D_4$$

$$\{1\} \triangleleft \{1, r^2\} \triangleleft \{1, r, r^2, r^3\} \triangleleft D_4$$

are two chains of subgroups that satisfy the solvability requirement. In general, every p-group is solvable; for if G is a p-group of order p^n, p a prime, then G contains a normal subgroup H of index p and hence the quotient group G/H is a cyclic group of order p and therefore abelian. Continuing in this manner, we obtain a chain of subgroups of G that satisfy the requirements for solvability. Thus, every p-group is solvable.

PROPOSITION 4. Let G be a finite group. If G is solvable, then every subgroup and every quotient group of G is solvable.

Proof. Suppose that G is solvable and let $(1) = H_1 \leq H_2 \leq \cdots \leq H_n = G$ be a chain of subgroups of G such that $H_i \triangleleft H_{i+1}$ and H_{i+1}/H_i is abelian for $1 \leq i \leq n - 1$. Let K be any subgroup of G and let $K_i = K \cap H_i$ for $1 \leq i \leq n - 1$. For each i, the function $f : K_{i+1} \to H_{i+1}/H_i$ defined by setting $f(x) = xH_i$ for all $x \in K_{i+1}$ is a group-homomorphism mapping K_{i+1} into the quotient group H_{i+1}/H_i. Since $f(x) = 1$ if and only if $x \in K \cap H_i = K_i$, it follows that $\operatorname{Ker} f = K_i$. Therefore, $K_i \triangleleft K_{i+1}$ and K_{i+1}/K_i is isomorphic to a subgroup of H_{i+1}/H_i. But H_{i+1}/H_i is abelian. Thus, K_{i+1}/K_i is also abelian and hence the chain $(1) = K_1 \leq K_2 \leq \cdots \leq K_n = K$ satisfies the solvability requirement for K. Now suppose that K is a normal subgroup of G and consider the the quotient group G/K. Let $\bar{H}_i$ stand for the image of H_i in G/K. Then $(\bar{1}) = \bar{H}_1 \leq \bar{H}_2 \leq \cdots \leq \bar{H}_n = G/K$ is a chain of subgroups in G/K and $\bar{H}_i \triangleleft \bar{H}_{i+1}$ for all i. Moreover, for each i, the function $f : H_{i+1}/H_i \to \bar{H}_{i+1}/\bar{H}_i$ defined by setting $f(xH_i) = xK\bar{H}_i$ for all $xH_i \in H_{i+1}/H_i$ is a group-homomorphism mapping H_{i+1}/H_i onto $\bar{H}_{i+1}/\bar{H}_i$. Since H_{i+1}/H_i is abelian, it follows that the quotient $\bar{H}_{i+1}/\bar{H}_i$ is also abelian. Hence, the chain $(\bar{1}) = \bar{H}_1 \leq \bar{H}_2 \leq \cdots \leq \bar{H}_n$ satisfies the requirements for solvability and therefore G/K is a solvable group. ∎

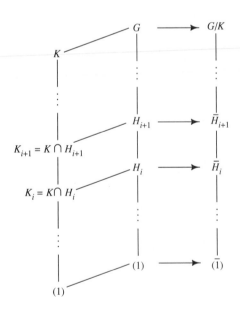

PROPOSITION 5. Let G be a finite group. If N is a normal subgroup of G such that both N and G/N are solvable, then G is solvable.

Proof.   If both N and G/N are solvable, then any chain of subgroups for G/N may be pulled back under the homomorphism $G \to G/N$ to give a chain of subgroups between N and G which, when augmented by a chain of subgroups from (1) to N, gives a chain of subgroups in G having the required properties. We leave the details as an exercise for the reader.   ∎

PROPOSITION 6.   A finite group G is solvable if and only if it contains a chain of subgroups $(1) = H_1 \leq H_2 \leq \cdots \leq H_n = G$ such that $H_1 \triangleleft H_{i+1}$ and H_{i+1}/H_i is cyclic for $1 \leq i \leq n - 1$.

Proof.   Any finite group having such a chain of subgroups is clearly solvable. Conversely, if G is solvable, then G contains such a chain of subgroups in which each quotient group is abelian. It follows easily by induction, however, that every abelian group contains a chain of normal subgroups in which successive quotient groups are cyclic. Therefore, G contains such a chain of subgroups. We leave the details as an exercise for the reader.   ∎

EXAMPLE 3. Solvability of the symmetric groups

Let us show that the symmetric group Sym(n) is solvable when $n = 1, 2, 3, 4$, but is not solvable for $n \geq 5$.

(A) Clearly, Sym(1) and Sym(2) are solvable since they are abelian, while for Sym(3) and Sym(4) we find that the chains

$$(1) \triangleleft \text{Alt}(3) = \{(1), (1\ 2\ 3), (1\ 3\ 2)\} \triangleleft \text{Sym}(3)$$

$$(1) \triangleleft \{(1), (1\ 2)(3\ 4), (1\ 3)(2\ 4), (1\ 4)(2\ 3)\} \triangleleft \text{Alt}(4) \triangleleft \text{Sym}(4)$$

satisfy the requirements for solvability.

(B) Now suppose that $n \geq 5$. To show that Sym(n) is not solvable, it is sufficient, in view of Proposition 4, to show that the alternating subgroup Alt(n) is not solvable. Suppose, to the contrary, that Alt(n) is solvable. Then Alt(n) contains a nontrivial normal subgroup N such that Alt(n)/N is abelian. Now, Alt(n) contains all 3-cycles, and is in fact generated by the 3-cycles; for if (ab) and (cd) are disjoint transpositions, then (ab)(cd) $= (abc)(bcd)$, while if (ab) and (bc) are transpositions having a letter in common, then (ab)(bc) $= (abc)$. Since every element in Alt(n) is the product of an even number of transpositions, it follows that every such element is the product of 3-cycles and hence that the 3-cycles generate Alt(n). Now, let (abc) be any 3-cycle. Since $n \geq 5$, there are letters d and e other than a, b or c. Let $x = (adc)$ and $y = (aeb)$. Then $x, y \in \text{Alt}(n)$ and $xyx^{-1}y^{-1}N = N$ since the quotient group Alt(n)/N is abelian. Therefore, $xyx^{-1}y^{-1} \in N$. But we find that $xyx^{-1}y^{-1} = (abc)$. Therefore, N contains every 3-cycle and hence $N = \text{Alt}(n)$, a contradiction. It follows that $\text{Alt}(n)$ is not solvable, and therefore $\text{Sym}(n)$ is not solvable when $n \geq 5$.

Let us now combine our results on solvable groups with our previous discussion on solvability by radicals to prove the main result of the chapter.

PROPOSITION 7.   (SOLVABILITY CRITERION) A polynomial is solvable by radicals if and only if its Galois group is solvable.

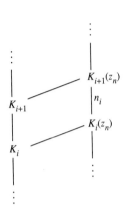

Proof.   Let $f(X)$ be a polynomial over a field k and let $G(K/k)$ be its Galois group, where K is a splitting field for $f(X)$ over k. If $f(X)$ is solvable by radicals, then, as we noted earlier, $G(K/k)$ is a quotient group of a solvable group and hence, by Proposition 4, is solvable. Now suppose, conversely, that $G(K/k)$ is solvable and let $(1) = H_1 \leq H_2 \leq \cdots \leq H_m = G(K/k)$ be a chain of subgroups such that $H_i \lhd H_{i+1}$ and H_{i+1}/H_i is cyclic. Let $K_i = K^{H_{m-i}}$ be the fixed subfield corresponding to the subgroup H_{m-i} for $0 \leq i \leq m - 1$. Then $k = K_1 \subseteq K_2 \subseteq \cdots \subseteq K_m = K$ is a tower of fields such that K_{i+1} is a normal extension of K_i, and $G(K_{i+1}/K_i)$ is cyclic since $G(K_{i+1}/K_i) \cong H_{i+1}/H_i$. Now let $[K:k] = n$ and consider the extension $K_i(z_n) \subseteq K_{i+1}(z_n)$, as illustrated at left, where z_n stands for a primitive nth root of unity. The field K_{i+1} is a normal extension of K_i and hence is a splitting field for some polynomial $g_i(X)$ over K_i. It follows that $K_{i+1}(z_n)$ is a normal extension of K_i since it is a splitting field for $g_i(X)(X^n - 1)$ over K_i. Therefore, $K_{i+1}(z_n)$ is a normal extension of $K_i(z_n)$ and every automorphism in $G(K_{i+1}(z_n)/K_i(z_n))$ restricts to an automorphism in $G(K_{i+1}/K_i)$. Thus, the restriction mapping $G(K_{i+1}(z_n)/K_i(z_n)) \to G(K_{i+1}/K_i)$ is a group-homomorphism and is, clearly, 1–1. It follows, therefore, that $G(K_{i+1}(z_n)/K_i(z_n))$ is cyclic since it is isomorphic to a subgroup of $G(K_{i+1}/K_i)$, which is cyclic. Finally, if $n_i = [K_{i+1}(z_n):K_i(z_n)]$, then n_i divides $[K_{i+1}:K_i]$ and hence n_i divides n. Let $d_i = n/n_i$. Then $z_n^{d_i}$ is a primitive n_ith root of unity in $K_i(z_n)$. It now follows from Proposition 2 that $K_{i+1}(z_n)$ is a radical extension of $K_i(z_n)$ since its automorphism group is cyclic and $K_i(z_n)$ contains an appropriate root of unity. Since $K(z_n)$ contains the splitting field K for $f(X)$ over k, we conclude that $k \subseteq k(z_n) = K_1(z_n) \subseteq \cdots \subseteq K_m(z_n) = K(z_n)$ is a tower of radicals for $f(X)$. Therefore, $f(X)$ is solvable by radicals and the proof is complete.   ∎

COROLLARY.   The general equation of degree n over $\mathbb{Q}$ is solvable by radicals when $n = 1, 2, 3$, and 4, but not when $n \geq 5$.

Proof.   The Galois group of the general polynomial of degree n over $\mathbb{Q}$ is the symmetric group $\text{Sym}(n)$, which is solvable if and only if $n \leq 4$. Therefore, the general polynomial is solvable by radicals when $n = 1, 2, 3, 4$, but not when $n \geq 5$.   ∎

COROLLARY.   Let $f(X)$ be an irreducible polynomial over $\mathbb{Q}$ of prime degree p, $p \geq 5$, and suppose that $f(X)$ has exactly two nonreal roots. Then $f(x)$ is not solvable by radicals.

Proof.   Let K be a splitting field for $f(X)$ over $\mathbb{Q}$, G its Galois group, and let $a_1, \ldots, a_p$ stand for the roots of $f(X)$ in K labeled so that a_1 and a_2 are nonreal. Then every automorphism in G permutes the roots of $f(X)$ and is uniquely determined by such a permutation, and hence we may regard G as a subgroup of $\text{Sym}(p)$. Now, complex conjugation interchanges a_1 and a_2 but leaves the remaining roots fixed; it therefore permutes the roots of $f(X)$ and hence defines an automorphism on K represented by the 2-cycle $(a_1 \ a_2)$. On the other hand, we claim that G contains a p-cycle. For $[k(a_1):k] = p$ since a_1 is a root of the irreducible polynomial $f(X)$, whose degree is p. Since $[k(a_1):k]$ divides $[K:k]$ and $[K:k] = |G|$, it follows that p divides $|G|$. Hence, G contains an element of order p. But the only elements in $\text{Sym}(p)$ that have order p are the p-cycles. Therefore, G contains a p-cycle and, by taking an appropriate power and relabeling the real roots if necessary, we may assume that $(a_1 \ a_2 \cdots a_p) \in G$. Thus, $(a_1 \ a_2)$ and $(a_1 \cdots a_p)$ are elements in G. We now claim that every 2-cycle lies in G. For clearly,

$$(a_1 \cdots a_p)(a_1 \ a_2)(a_1 \cdots a_p)^{-1} = (a_2 \ a_3) \in G$$

$$\vdots$$

$$(a_1 \cdots a_p)(a_{p-1} \ a_p)(a_1 \cdots a_p)^{-1} = (a_1 \ a_p) \in G,$$

and hence every 2-cycle of the form $(a_i \ a_{i+1})$ lies in G. Since

$$(a_{i-1} \ a_i)(a_1 \ a_{i-1})(a_{i-1} \ a_i) = (a_1 \ a_i)$$

$$(a_1 \ a_i)(a_1 \ a_j)(a_1 \ a_i) = (a_i \ a_j),$$

it follows that every 2-cycle $(a_i \ a_j)$ lies in G. But $\text{Sym}(p)$ is generated by the 2-cycles. Therefore, $G = \text{Sym}(p)$. It now follows that $f(X)$ is not solvable by radicals since its Galois group, $\text{Sym}(p)$, is not solvable when $p \geq 5$.   ∎

EXAMPLE 4.

Let p be a prime number, $p \geq 5$, and let

$$f_p(X) = (X^2 + 2p^3)(X - 2)(X - 4) \cdots (X - 2(p - 2)) - 2.$$

We claim that $f_p(X)$ is an irreducible polynomial over $\mathbb{Q}$ of degree p that is not solvable by radicals.

(A) Let us begin by showing that $f_p(X)$ has exactly two nonreal roots. Consider the polynomial

$$g_p(X) = (X^2 + 2p^3)(X - 2)(X - 4) \cdots (X - 2(p - 2)).$$

The roots of $g_p(X)$ are $\pm i\sqrt{2p^3}$ and $2, 4, \cdots, 2(p-2)$, and the figure here illustrates the graph of $y = g_p(X)$.

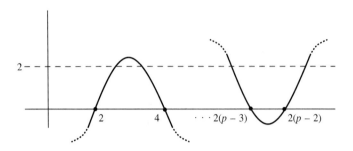

If $2k + 1$ represents the midpoint of the typical subinterval $[2k, 2k + 2]$ for $k = 1, \ldots, p - 3$, then we find that

$$g_p(2k + 1) = [(2k + 1)^2 + 2p^3](2k - 1)(2k - 3) \cdots (2k - (2p - 5)).$$

Therefore, $|g_p(2k + 1)| > 2$ and hence the graph of $y = f_p(X) = g_p(X) - 2$ cuts the axis in at least $p - 2$ points. It follows that $f_p(X)$ has at least $p - 2$ real roots. Now, if every root of $f_p(X)$ is real, then the sum of their squares must be a positive number. But if $a_1, \ldots, a_p$ stand for the roots of $f_p(X)$, then

$$a_1^2 + \cdots + a_p^2 = (a_1 + \cdots + a_p)^2 - 2(a_1 a_2 + \cdots + a_{p-1} a_p)$$
$$= s_1^2 - 2s_2,$$

where s_1 and s_2 are the symmetric polynomials in the roots. But s_1 and s_2 are the coefficients of X^{p-1} and X^{p-2} in $f_p(X)$, respectively, and these coefficients are the same as the corresponding coefficients in $g_p(X)$. Thus, $s_1^2 - 2s_2$ is the sum of the squares of the roots of $g_p(X)$ and hence

$$s_1^2 - 2s_2 = (i\sqrt{2p^3})^2 + (-i\sqrt{2p^3})^2 + 2^2 + 4^2 + \cdots + [2(p-2)]^2$$
$$= -2p^3 + 2^2[1^2 + 2^2 + \cdots + (p-2)^2]$$
$$= -2p^3 + 4\frac{(p-2)(p-1)(2p-3)}{6}$$
$$= \frac{4}{3}p^3\left[\left(1 - \frac{2}{p}\right)\left(1 - \frac{3}{2p}\right)\left(1 - \frac{1}{p}\right) - \frac{3}{2}\right],$$

which is clearly a negative number. Hence, at least one of the roots of $f_p(X)$ is nonreal and therefore so is its conjugate. Thus, $f_p(X)$ has exactly two nonreal roots.

(B) We now use Eisenstein's criterion to show that $f_p(X)$ is irreducible over $\mathbb{Q}$. Let

$$(X - 2)(X - 4) \cdots (X - 2(p - 2)) = X^{p-2} + A_1 X^{p-3} + \cdots + A_{p-2}.$$

Then the coefficients $A_1, \ldots, A_{p-2}$ are even integers since they are symmetric polynomials in the even numbers $2, 4, \ldots, 2(p-2)$. Now,

$$
\begin{aligned}
f_p(X) &= (X^2 + 2p^3)(X^{p-2} + A_1 X^{p-3} + \cdots + A_{p-2}) - 2 \\
&= X^p + A_1 X^{p-1} + (A_2 + 2p^3)X^{p-2} + \cdots + (A_j + 2p^3)X^{p-j} \\
&\quad + 2p^3 A_{p-2} - 2.
\end{aligned}
$$

The prime 2 divides each of these coefficients except that of X^p, and 2^2 does not divide the constant term $2(p^3 A_{p-2} - 1)$. Hence, by Eisenstein's criterion, $f_p(X)$ is irreducible over $\mathbb{Q}$.

(C) It follows from parts (A) and (B) that $f_p(X)$ is an irreducible polynomial over $\mathbb{Q}$ of prime degree $p \geq 5$ that has exactly two nonreal roots. Hence, by Corollary 2, $f_p(X)$ is not solvable by radicals. For example, we find that

$$
\begin{aligned}
f_5(X) &= (X^2 + 250)(X - 2)(X - 4)(X - 6) - 2 \\
&= X^5 - 12X^4 + 254X^3 - 3048X^2 + 11000X - 12002.
\end{aligned}
$$

EXAMPLE 5

Let n be a positive integer, $n \geq 5$. Then there is a specific polynomial of degree n over $\mathbb{Q}$ that is not solvable by radicals. For let $f_5(X)$ stand for the polynomial over $\mathbb{Q}$ constructed in Example 4 that is not solvable by radicals. Then $X^{n-5} f_5(X)$ is a polynomial of degree n over $\mathbb{Q}$ that is not solvable by radicals since not all of its roots, namely those of $f_5(X)$, may be expressed in terms of radicals.

Exercises

1. Let $f(X) = X^4 - 2$.
   (a) Find the roots of $f(X)$ and show that $f(X)$ is solvable by radicals.
   (b) Find the Galois group G of $f(X)$ and find the chain of subgroups in G satisfying the solvability criterion that corresponds to the radical tower in part (a).
2. Let $f(X) = X^4 + 2$.
   (a) Find the roots of $f(X)$ and show that $f(X)$ is solvable by radicals.
   (b) Find the Galois group G of $f(X)$ and find the chain of subgroups in G satisfying the solvability criterion that corresponds to the radical tower in part (a).
3. Find a specific polynomial of degree 6 over $\mathbb{Q}$ that is not solvable by radicals.
4. Suppose that $f(X)$ is an irreducible polynomial over $\mathbb{Q}$ having a root that is expressible in terms of addition, subtraction, multiplication and division, together with root extraction on the coefficients of $f(X)$. Show that $f(X)$ is solvable by radicals.

5. Let $f(X) = X^4 - 10X^2 + 1$.
   (a) Find the roots of $f(X)$, observing first that $f(X)$ is a quadratic equation in X^2.
   (b) Show that $f(X)$ is solvable by radicals and find a radical tower for $f(X)$.
   (b) Show that $f(X)$ is solvable by radicals and find a radical tower for $f(X)$.
   (c) Find the Galois group of $f(X)$ and the chain of subgroups satisfying the solvability criterion that correspond to the radical tower in part (b).

6. Let $f(X) = X^3 + 3X^2 + 6X + 5$. In Example 2 we used Cardan's formulas to find the roots of $f(X)$ and constructed a tower of radicals for $f(X)$ over $\mathbb{Q}$. Find the Galois group of $f(X)$ and the chain of subgroups corresponding to this tower of radicals.

7. Let $K = k(w)$ be a radical extension of a field k, where $w^s \in k$ for some integer s, and assume that k contains a primitive sth root of unity. Show that the automorphism group $G(K/k)$ is in fact a cyclic group.

8. Let $f(X)$ be a polynomial of degree n over a field k having n distinct roots $x_1, \ldots, x_n$ in a splitting field K, and let $\Delta = (x_1 - x_2) \cdots (x_1 - x_n) \cdots (x_{n-1} - x_n) \in K$. Then each automorphism in the Galois group $G(K/k)$ may be regarded as a permutation in $\mathrm{Sym}(n)$.
   (a) Show that $\sigma(\Delta) = \pm \Delta$ for every $\sigma \in G(K/k)$.
   (b) Show that $\sigma(\Delta) = \Delta$ if and only if σ is an even permutation of the roots of $f(X)$.
   (c) Show that the intermediate subfield of K/k corresponding to the subgroup $G(K/k) \cap \mathrm{Alt}(n)$ is $k(\Delta)$.
   (d) Show that $G(K/k) \leq \mathrm{Alt}(n)$ if and only if Δ^2 is the square of an element in k. Δ^2 is called the *discriminant* of $f(X)$.
   (e) If $f(X) = X^2 + bX + c$, show that $\Delta^2 = b^2 - 4c$.

9. **The general quartic polynomial.** Let $f(X) = X^4 + bX^3 + cX^2 + dX + e$ stand for the general fourth-degree polynomial over $\mathbb{Q}$. We find the roots of $f(X)$ as follows.
   (a) Let $X = Y - \dfrac{b}{4}$. Substitute this expression into the equation $f(X) = 0$ and show that $Y^4 + pY^2 + qY + r = 0$, where p, q and r are polynomials in $b, c, d,$ and e.
   (b) For the moment, let f be an undetermined constant. By adding $fY^2 + \dfrac{f^2}{4}$ to both sides of the equation in part (a), show that the equation may be written in the form

$$\left(Y^2 + \frac{f}{2}\right)^2 = (f - p)\left(Y - \frac{q}{2(f - p)}\right)^2 + \frac{R(f)}{4(f - p)},$$

where $R(f) = f^3 - pf^2 - 4rf - q^2 + 4rp$. The polynomial $R(f)$ is called the *resolvent cubic* of the original quartic equation.

(c) If we choose the constant f such that $R(f) = 0$, then the equation in part (b) is easily solved for Y. By doing so, show that the roots of the general quartic equation have the form

$$\frac{\sqrt{f-p} \pm \sqrt{(f-p) - 4\left(\dfrac{q}{2\sqrt{f-p}} + \dfrac{f}{2}\right)}}{2} - \frac{b}{4} \quad \text{and}$$

$$\frac{-\sqrt{f-p} \pm \sqrt{(f-p) - 4\left(\dfrac{q}{2\sqrt{f-p}} + \dfrac{f}{2}\right)}}{2} - \frac{b}{4}.$$

(d) Conclude that the general quartic polynomial over $\mathbb{Q}$ is solvable by radicals.

10. Using the method described in Exercise 9, find the roots of the equation $X^4 + 4X^2 + 32X - 48 = 0$. In this case, one root of the resolvent cubic is 8.

16

Euclidean Constructibility

In Book I of the *Elements*, Euclid lays out five postulates from which he begins the systematic development of geometry:

(1) To draw a straight line from any point to any point;

(2) To extend a straight line continuously in a straight line;

(3) To describe a circle with any given center and radius;

(4) That all right angles are equal;

(5) To draw a straight line parallel to a given straight line through a given point not on the line.

The devices used to carry out these constructions are, of course, the straight-edge and compass, with the proofs of the Euclidean propositions then reduced to finding an appropriate sequence of straightedge and compass constructions. For example, the reader is familiar with the standard Euclidean construction for bisecting an angle or for constructing the perpendicular to a given line through a given point. But to what extent is a line segment of any length constructible by Euclidean methods; that is, given a line segment of unit length, can we always find a sequence of straightedge and compass constructions that will produce a line segment having any prescribed length? It was not until the nineteenth century and the development of field theory that the algebraic framework for analyzing Euclidean constructibility became available and, with it, the realization that not every real number, or line segment of given length, is constructible by Euclidean means.

In this chapter our goal is to construct an algebraic model for Euclidean constructibility within the framework of field theory and obtain a criterion for when it is possible to construct a given point by means of Euclidean construction. We then use this criterion to discuss four well-known classical construction problems: doubling the cube, which asks for the construction of a cube having twice the volume of a given cube; trisecting the angle, which asks for a construction for trisecting any angle; squaring the circle, which asks for the construction of a square having area equal to that of a given circle; and finally, the cyclotomy problem, which asks for a construction dividing a given circle into any prescribed number of equal arcs. In each case we will see that these classical problems cannot be solved by Euclidean construction. Let us emphasize, however, that there are simple nonEuclidean constructions for solving some of these problems; for example, there is a simple nonEuclidean construction for trisecting any acute angle, as well as a method for trisecting an angle to within any desired degree of accuracy by Euclidean methods. The results in this chapter deal only with the possibility of carrying out constructions using straightedge and compass alone.

Let us begin by recalling two basic Euclidean constructions. First, if P is a point on a line L, then there is a Euclidean construction for the line through P perpendicular to L: simply construct a circle with center P and arbitrary radius cutting L at two points, Q and Q', as illustrated in Figure 1, then construct two circles with centers Q and Q', respectively. These circles intersect at two points, R and R', and the line determined by R and R' is then perpendicular to L and passes through P, as required. And second, if P is a point not on L, there is a Euclidean construction for the line through P parallel to L: simply construct the line L' through P perpendicular to L, as illustrated in Figure 2, then the line L'' through P perpendicular to L'. The line L'' is then parallel to the given line L.

Now, a Euclidean construction is a sequence of straightedge and compass constructions in which each new point is the intersection of two previously constructed lines, circles, or line and circle. To formalize this idea, let S be

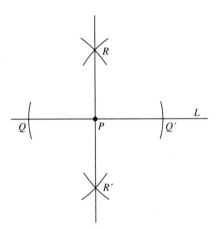

FIGURE 1.   Euclidean construction of a line perpendicular to a given line L through a given point P.

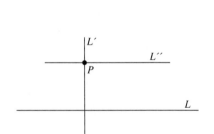

FIGURE 2.   Euclidean construction of a line parallel to a given line L through a given point P.

a finite set of points in the Euclidean plane containing at least two points. We say that a line is constructible from S if it is determined by two points in S, while a circle is constructible from S if its center lies in S and it passes through some point in S.

Definition

A point P in the Euclidean plane is *constructible from* S if there is a finite sequence of points $P_1, \ldots, P_n$ such that $P_n = P$ and where P_{i+1} is the intersection of two lines, two circles, or a line and circle each of which is constructible from the set $S \cup \{P_1, \ldots, P_i\}$ for $i = 1, \ldots, n-1$.

The points in S serve as initial data for the construction; they may represent the endpoints of a given line segment, for example, or the vertices of a cube. To say that a point P is constructible from S simply means there is a finite sequence of points beginning with those in S and terminating with P each of which is the intersection of lines or circles constructed from previous points. Any such sequence of points is called a *Euclidean construction* for P.

Let us now use the field of complex numbers to develop an algebraic model for constructibility. We begin by first choosing two points in S and labeling them $(0, 0)$, for the origin, and $(1, 0)$, for the unit distance. The line determined by these points is the x-axis. Now construct the line perpendicular to the x-axis through the origin and call this line the y-axis. Choose a point of intersection of the y-axis with the unit circle centered at the origin and label this point $(0, 1)$. We then have a Cartesian coordinate system for the plane. Observe that if (a, b) is a typical point in the plane, then (a, b) is constructible from S if and only if both of the points $(a, 0)$ and $(0, b)$ are constructible from S. Finally, associate to each point (a, b) in the plane the complex number $z = a + bi$. With these conventions, we now make the following definition for the constructibility of a complex number.

Definition

Let $z = a + bi$ be a complex number. Then z is *constructible from* S, or is a *constructible complex number*, if the point (a, b) is constructible from S. Let $\text{Const}_S(\mathbb{C})$ stand for the set of complex numbers constructible from S.

Clearly, a complex number $a + bi$ is constructible if and only if its real and imaginary parts a and b are constructible. In particular, a real number r is constructible if and only if a line segment of length $|r|$ is constructible. We now claim that the set $\text{Const}_S(\mathbb{C})$ of constructible complex numbers is a field under addition and multiplication of complex numbers and is closed under the taking of square roots.

PROPOSITION 1. The sum, difference, product and quotient of constructible complex numbers are constructible complex numbers.

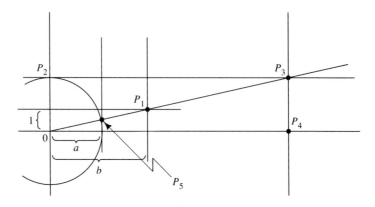

Proof. Let us first show that if a and b are constructible real numbers with $a, b > 0$, then $a + b$, $a - b$, ab, and a/b, for $b \neq 0$, are also constructible. For the sum and difference, it is easy to see that if we are given line segments of length a and b, then one may construct line segments of length $a + b$ and $|a - b|$. To construct the product ab and quotient a/b, first construct the points P_1, P_2, P_3, P_4, and P_5, as illustrated in the figure here. If we set $P_4 = (x, 0)$ and $P_5 = (a, y)$, then, by similar triangles, $a/x = y/a = 1/b$ and hence $x = ab$ and $y = a/b$. Thus, ab and a/b are constructible. It now follows that the sum, difference, product and quotient of constructible real numbers are constructible. Finally, let $z = a + bi$ and $z' = c + di$ be constructible complex numbers. Then

$$z + z' = (a + c) + (b + d)i,$$

$$z - z' = (a - c) + (b - d)i,$$

$$zz' = (ac - bd) + (ad + bc)i,$$

$$\frac{z}{z'} = \frac{ac - bd}{c^2 + d^2} + \frac{bc - ad}{c^2 + d^2}\, i.$$

Since a, b, c and d are constructible, the real and imaginary parts of these four complex numbers are also constructible and hence the numbers themselves are constructible. ∎

PROPOSITION 2. If z is a constructible complex number, then $\sqrt{z}$ is constructible.

Proof. Let $z = r(\cos \theta + i \sin \theta)$ be the trigonometric representation of z, where $r = |z|$ is the magnitude of z. Then $\sqrt{z} = \sqrt{r}(\cos \frac{1}{2}\theta + i \sin \frac{1}{2}\theta)$. Now, the number r is constructible since z is constructible and $r = \|z\|$. To construct $\sqrt{z}$, first construct the sequence of points P_1, P_2, P_3, P_4, and P_5, as illustrated in the figure (page 545), where P_4 is the midpoint of segment OP_3. If $P_5 = (r, y)$, then, by similar triangles, we have that $r/y = y/1$ and therefore $y = \sqrt{r}$. Hence,

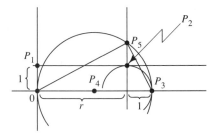

$\sqrt{r}$ is constructible. Finally, the complex number $\cos\frac{1}{2}\theta + i\sin\frac{1}{2}\theta$ is constructible since we may apply the Euclidean construction that bisects the angle θ. Therefore, $\sqrt{z}$ is constructible since it is the product of two constructible numbers and the proof is complete. ∎

It follows from Propositions 1 and 2 that the set $\mathrm{Const}_S(\mathbb{C})$ of constructible complex numbers is a field under addition and multiplication of complex numbers and is closed with respect to square roots. Clearly, $\mathbb{Q}$ is a subfield of $\mathrm{Const}_S(\mathbb{C})$ since every rational number may be constructed from the given line segment of unit length. Let $\mathbb{Q}_S$ stand for the subfield of $\mathrm{Const}_S(\mathbb{C})$ generated by $\mathbb{Q}$ and the coordinates of all points in S. Before we prove our main result characterizing constructible complex numbers, let us first make a few observations about these fields.

Let $P = (a, b)$ be a typical point in the plane and let $\mathbb{Q}_S(P) = \mathbb{Q}_S(a, b)$ stand for the extension field of $\mathbb{Q}_S$ generated by the coordinates of P. If $P' = (a', b')$ is any other point in the plane, we claim that the line through P and P' has an equation of the form $AX + BY = C$, where $A, B, C \in \mathbb{Q}_S(P, P')$; for if $a = a'$, the line through P and P' is vertical and hence has an equation of the form $X = a$, whose coefficients lie in $\mathbb{Q}_S(P, P')$, while if $a \neq a'$, then the slope of the line through P and P' is $(b' - b)/(a' - a)$, which is an element in $\mathbb{Q}_S(P, P')$, and hence the line has an equation of the form $Y - b = m(X - a)$, whose coefficients also lie in $\mathbb{Q}_S(P, P')$. Thus, the line through any two points P and P' has an equation whose coefficients lie in the field $\mathbb{Q}_S(P, P')$. Similarly, we leave the reader to verify that the circle with center P passing through P' has an equation of the form $X^2 + Y^2 + AX + BY = C$, where $A, B, C \in \mathbb{Q}_S(P, P')$.

Now, when we obtain new points by means of Euclidean construction, these points are intersection points of either two lines, two circles, or a line and circle. As such, we claim that the coordinates of these points extend the coefficient field of the equations by degree at most 2. For suppose that R is the intersection point of two lines, one through P_1 and P'_1, the other through P_2 and P'_2. Then, as we noted above, these lines have equations of the form

$$AX + BY = C$$

$$A'X + B'Y = C',$$

where $A, B, C, A', B', C' \in \mathbb{Q}_S(P_1, P_{1'}, P_2, P_{2'})$, and hence the simultaneous solution of the equations also lies in $\mathbb{Q}_S(P_1, P_1', P_2, P_2')$. Thus, the coordinates of R lie in the field $\mathbb{Q}_S(P_1, P_1', P_2, P_2')$ and therefore $[\mathbb{Q}_S(P_1, P_1', P_2, P_2')(R): \mathbb{Q}_S(P_1, P_1', P_2, P_2')] = 1$. On the other hand, if R is the intersection point of the line through P_1 and P_1' with the circle centered at P_2 and passing through P_2', then the line and circle have equations of the form

$$AX + BY = C$$

$$X^2 + Y^2 + A'X + B'Y = C',$$

where $A, B, C, A', B', C' \in \mathbb{Q}_S(P_1, P_1', P_2, P_2')$. In this case the simultaneous solution of the two equations has the form

$$x = \frac{D \pm E\sqrt{F}}{G}, \qquad y = \frac{D' \pm E'\sqrt{F}}{G},$$

and therefore $[\mathbb{Q}_S(P_1, P_1', P_2, P_2')(R):\mathbb{Q}_S(P_1, P_1', P_2, P_2')] = 1$ or 2. Similarly, if R is an intersection point of two circles, then the field $\mathbb{Q}_S(P_1, P_1', P_2, P_2')(R)$ has degree at most 2 over $\mathbb{Q}_S(P_1, P_1', P_2, P_2')$. Thus, when a new point is obtained by Euclidean construction, the field of coefficients is extended by degree at most 2. With these observations, let us now prove our main result.

PROPOSITION 3. (CONSTRUCTIBILITY CRITERION) A complex number z is constructible if and only if there is a tower of fields $\mathbb{Q}_S \subseteq K_1 \subseteq K_2 \subseteq \cdots \subseteq K_n$ such that $z \in K_n$ and where $[K_{i+1}:K_i] \le 2$ for $i = 1, \ldots, n - 1$.

Proof. Suppose first that $z = a + bi$ is a constructible complex number and let $P_1, \ldots, P_{n-1} = (a, b)$ be a Euclidean construction terminating with the point (a, b) corresponding to z. Let $P_i = (a_i, b_i)$ and $K_i = \mathbb{Q}_S(P_1, \ldots, P_i)$ for $i = 1, \ldots, n - 1$. Then for each i, P_{i+1} is the intersection of two lines, two circles, or a line and circle, each of which is constructible from $S \cup \{P_1, \ldots, P_i\}$, and hence, by our previous remarks, $[K_{i+1}:K_i] \le 2$. Finally, if we let $K_n = K_{n-1}(i)$, where $i = \sqrt{-1}$, then $[K_n:K_{n-1}] = 2$ and $z = a_{n-1} + b_{n-1}i \in K_n$. Therefore, $\mathbb{Q}_S = K_1 \subseteq K_2 \subseteq \cdots \subseteq K_{n-1} \subseteq K_n$ is a tower of fields having the required properties.

Conversely, suppose there is a tower of fields $\mathbb{Q}_S = K_1 \subseteq K_2 \subseteq \cdots \subseteq K_n$ such that $[K_{i+1}:K_i] \le 2$ for $i = 1, \ldots, n - 1$ and let $z \in K_n$. Then $K_{i+1} = K_i(\sqrt{u_i})$ for some element $u_i \in K_i$, $1 \le i \le n - 1$. Now, u_1 lies in $\mathbb{Q}_S$ and hence is constructible. Therefore, $\sqrt{u_1}$ is constructible and hence every element in the field $K_1(\sqrt{u_1})$ is constructible. Continuing in this manner, it follows that every element in K_n is constructible and hence, in particular, so is z. ∎

COROLLARY   If z is a constructible complex number, then $[\mathbb{Q}_S(z):\mathbb{Q}] = 2^s$ for some integer $s \ge 0$.

Proof.   Immediate from Proposition 3. ∎

Let us now use the constructibility criterion to discuss the classical construction problems mentioned in the introduction.

EXAMPLE 1. There is no Euclidean construction for doubling the cube

We choose for the points in S any two adjacent vertices of the given cube and label them $(0, 0)$ and $(1, 0)$. Then $\mathbb{Q}_S = \mathbb{Q}$. If there is a Euclidean construction for doubling the cube, the new cube has an edge length of $\sqrt[3]{2}$ units and hence the number $\sqrt[3]{2}$ is constructible. But $[\mathbb{Q}(\sqrt[3]{2}):\mathbb{Q}] = 3$, which is not a power of 2. Therefore, there is no Euclidean construction for doubling the cube.

EXAMPLE 2. There is no Euclidean construction for trisecting the angle

If there is a general Euclidean construction for trisecting the angle, then an angle of $60°$, as represented by the points $(0, 0)$, $(1, 0)$ and $(\cos 60°, \sin 60°)$, may, in particular, be trisected. But this is impossible in view of the constructibility criterion. For in this case $\mathbb{Q}_S = \mathbb{Q}(\sqrt{3})$ since $\cos 60° = \frac{1}{2}$ and $\sin 60° = \frac{1}{2}\sqrt{3}$. Hence, if there is a Euclidean construction for trisecting an angle of $60°$, then the point $(\cos 20°, \sin 20°)$ is constructible and hence so is the number $\cos 20°$. But, by DeMoivre's theorem,

$$(\cos 20° + i \sin 20°)^3 = \cos 60° + i \sin 60° = \tfrac{1}{2} + \tfrac{1}{2}i\sqrt{3}.$$

Expanding the left side, equating real parts and replacing $\sin^2 20°$ by $1 - \cos^2 20°$, we find that $8(\cos 20°)^3 - 6 \cos 20° - 1 = 0$. Therefore, $\cos 20°$ is a root of the polynomial $8X^3 - 6X - 1$, which we leave the reader to verify is irreducible over $\mathbb{Q}$. It follows, then, that $[\mathbb{Q}_S(\cos 20°):\mathbb{Q}_S] = 3$, which is not a power of 2, and therefore $\cos 20°$ is not constructible. Thus, there is no general Euclidean construction for trisecting the angle although, as we mentioned in the introduction, there are such constructions for certain angles. In the exercises we ask the reader to show that there is a Euclidean construction for trisecting an angle of θ radians if and only if the polynomial $4X^3 - 3X - \cos \theta$ is reducible over the field $\mathbb{Q}(\cos \theta)$; thus, an angle of $90°$, in particular, may be trisected by Euclidean construction. And finally, let us mention that there is a simple nonEuclidean construction for trisecting any angle that involves marking and transferring distances which we describe in more detail in the exercises.

EXAMPLE 3. There is no Euclidean construction for squaring the circle

We begin with a circle as given by its center, labeled $(0, 0)$, and a point on it, labeled $(1, 0)$. Then $\mathbb{Q}_S = \mathbb{Q}$. Now, the area of the circle is π square units and hence if there is a Euclidean construction for squaring the circle, it follows that the number $\sqrt{\pi}$ is constructible, in which case π is also constructible. But π is transcendental over $\mathbb{Q}$ and hence $[\mathbb{Q}(\pi):\mathbb{Q}]$ is infinite. Therefore, $\sqrt{\pi}$ is not constructible and it follows that there is no Euclidean construction for squaring the circle.

EXAMPLE 4. Cyclotomy: division of the circle into equal arcs

In the cyclotomy problem we begin with an arbitrary circle, as given by its center and a point through which it passes, and ask if there is a Euclidean construction that divides the circle into n equal arcs, where n is any positive integer. We claim that there is such a construction if and only if $n = 2^s p_1 \cdots p_m$, where $s \geq 0$ and where $p_1, \ldots, p_m$ are distinct prime numbers of the form $2^t + 1$.

(A) To show this, let $(0, 0)$ and $(1, 0)$ stand for the center of the circle and the given point through which it passes, respectively. Then $\mathbb{Q}_S = \mathbb{Q}$. Now, for any positive integer n, there is a Euclidean construction for dividing the circle into n equal arcs if and only if the complex number $z_n = \cos \dfrac{2\pi}{n} + i \sin \dfrac{2\pi}{n}$ is constructible. Recall that z_n is a primitive nth root of unity and that the field $\mathbb{Q}(z_n)$ is a cyclotomic extension of $\mathbb{Q}$ with $[\mathbb{Q}(z_n):\mathbb{Q}] = \varphi(n)$. If

$$n = 2^s p_1^{t_1} \cdots p_m^{t_m}$$

is the prime factorization of n, where $p_1, \ldots, p_m$ are odd primes, then

$$\varphi(n) = 2^{s-1} p_1^{t_1-1}(p_1 - 1) \cdots p_m^{t_m-1}(p_m - 1).$$

Therefore, $\varphi(n)$ is a power of 2 if and only if $t_i = 1$ and $p_i - 1 = 2^{s_i}$ for $i = 1, \ldots, m$, where $s_1, \ldots, s_m$ are nonnegative integers. Thus, if z_n is constructible, then $n = 2^s p_1 \cdots p_m$, where each prime p_i has the form $2^t + 1$, as required. Now suppose, conversely, that n has this form. Then the automorphism group $G(\mathbb{Q}(z_n)/\mathbb{Q})$ is a 2-group and hence contains a chain of subgroups $H_1 \leq H_2 \leq \cdots \leq H_n$, where $[H_{i+1}:H_i] = 2$ for $1 \leq i \leq n - 1$. Since $\mathbb{Q}(z_n)$ is a normal extension of $\mathbb{Q}$, it follows from the Galois correspondence that the extension $\mathbb{Q}(z_n)/\mathbb{Q}$ contains a tower of intermediate subfields $\mathbb{Q} = K_1 \subseteq K_2 \subseteq \cdots \subseteq K_n = \mathbb{Q}(z_n)$, where $[K_{i+1}:K_i] = 2$ for $1 \leq i \leq n - 1$, and hence, by the constructibility criterion, z_n is constructible. We conclude, therefore, that z_n is a constructible complex number if and only if n has the required form and consequently that the division of the circle into n equal arcs is possible by Euclidean construction if and only if n has this form.

(B) Any prime number of the form $2^t + 1$ is called a *Fermat prime*. It follows from part (A), therefore, that the division of the circle into n equal arcs is possible by Euclidean construction if and only if n is a power of 2 times a product of Fermat primes. Since dividing the circle into n equal arcs is equivalent to constructing a regular n-gon, it follows that a regular n-gon is constructible by straightedge and compass if and only if n is a power of 2 times a product of Fermat primes. For example, such constructions are possible for an equilateral triangle, a square, a pentagon, since $5 = 2^2 + 1$ is a Fermat prime, and a hexagon, since $6 = (2)(3)$, but not for a regular 7-gon.

This completes our discussion of Euclidean constructibility. As we have seen, Euclidean constructibility begins with an initial set S of points in the plane and adds new points through a sequence of straightedge and compass constructions. When such points are regarded as complex numbers, the collection of constructible complex numbers forms a subfield $\text{Const}_S(\mathbb{C})$ of the field of complex numbers. Not all complex numbers are constructible, only those satisfying the constructibility criterion in Proposition 3. Let us mention that if we restrict the tools of construction to only a straightedge, then the only constructible points are those in the ground field $\mathbb{Q}_S$. On the other hand, if we use only a compass, then the Italian geometer Lorenzo Mascheroni (1750–1800) showed that all Euclidean constructions are possible. If, instead of limiting our tools of construction we extend them, then many additional points may be constructed. The Greeks, for example, invented the trammel, a mechanical device for drawing a curve called a conchoid, as illustrated in Figure 3, and used this curve to trisect any angle.

Finally, let us conclude our discussion of Euclidean constructibility by looking at it from a more general point of view. Euclidean constructibility begins with an initial set of points in the plane and enlarges this set by means of two tools, the straightedge and compass. As we showed, not every point in the plane is so constructible. A computer operates in much the same way. A computer begins with a finite set of initial data as well as an algorithm, called the program, for calculating new objects. The question then arises: Is

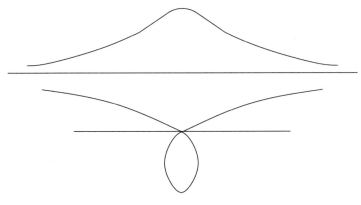

FIGURE 3.   The conchoid.

the computer, with its initial data and algorithms, capable of computing any desired function? This question, which is of great importance in computer science, deals with the theoretical computability of functions. Like Euclidean constructibility, it turns out that not all functions are computable. Finally, there is the matter of mathematical proof itself. When the concept of mathematical proof is formalized, we arrive at the axiomatic method. According to this method, we begin with a finite set S of initial data, which are the undefined terms and axioms of the system, and then apply a sequence of deductive reasoning, using laws of logic, to arrive at theorems. The question is: Can every mathematical statement that is true about S be proved from S? It is no exaggeration to say that the answer to this question was one of the most shattering developments in twentieth century mathematics: No, there are true statements about S that cannot be proved from S! This remarkable result was proved by the Austrian mathematician and logician Kurt Gödel in 1931 and led to a complete reformulation of axiomatic set theory and logic.

Exercises

1. Show that the polynomial $8X^3 - 6X - 1$ is irreducible over $\mathbb{Q}$.
2. Let $a > 0$ be a real number. Describe a Euclidean construction for $\sqrt{a}$.
3. Describe a Euclidean construction for dividing a given line segment into any finite number of equal parts.
4. Describe a Euclidean construction for a hexagon.
5. Describe a Euclidean construction for pentagon.
6. Let $\text{Const}_S(\mathbb{C})$ stand for the field of constructible complex numbers and $\mathbb{Q}_S$ the subfield generated by $\mathbb{Q}$ and the coordinates of points in S.
   (a) Show that $\text{Const}_S(\mathbb{C})$ is an algebraic extension of $\mathbb{Q}_S$.
   (b) Show that the degree $[\text{Const}_S(\mathbb{C}):\mathbb{Q}_S]$ is infinite.
7. Let θ be an arbitrary real number. Show that there is a Euclidean construction for trisecting an angle of θ radians if and only if the polynomial $4X^3 - 3X - \cos\theta$ is reducible over $\mathbb{Q}(\cos\theta)$.
8. **A Euclidean construction for trisecting a 90° angle.** We begin with a 90° degree angle determined by the points $A = (1, 0)$ and $B = (0, 1)$, as illustrated in the figure. Construct the circle with center 0 passing through A and the circle with center B passing through 0, and let P stand for the intersection point of the two circles. Show that $\angle POA = \frac{1}{3}\angle BOA = 30°$.
9. Using the construction in Exercise 8, describe a Euclidean construction for dividing a circle into 12 equal arcs.
10. **A Euclidean construction for trisecting an angle to within any degree of accuracy.** Let θ be a positive real number and let $\varepsilon > 0$ be an arbitrarily small real number.

    (a) Show that the infinite series $\sum_{1}^{\infty} \frac{\theta}{2^{2n}}$ converges to $\frac{\theta}{3}$.

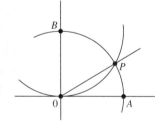

(b) Show that there are integers N and n such that $0 < \dfrac{\theta}{3} - N\,\dfrac{\theta}{2^{2n}} < \varepsilon.$

(c) Using the results in parts (a) and (b), describe a Euclidean construction that trisects an angle of θ radians with a maximum error of $\varepsilon.$

11. **A non-Euclidean construction for trisecting an acute angle.** Let θ be an acute angle given by three points 0, A, and B, as illustrated in the figure. Construct the circle with center at 0 passing through A and let r stand for the radius of the circle. Now, mark off two points on a straightedge at a distance of r units apart, and place the edge of the straightedge at point B in such a way that one of the points is at C on the circle, while the other point is at D on the line $A0$. Show that $\angle 0DB = \frac{1}{3}\theta.$

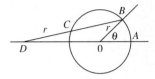

APPENDIX

A

Remarks on the History of Group Theory

Group theory, as we noted in Chapter 2, evolved out of early nineteenth-century attempts to solve algebraic equations by means of radicals. The first explicit use of the term *group* occurred in 1830 when Galois used it to describe a collection of permutations closed under multiplication. After Galois's death in 1832, the concept began to drift slowly out of the context of permutations and into its more axiomatic form that we know today; in 1872, for example, Felix Klein used the group concept as the basis for his Erlanger Program to reformulate geometry. In this appendix our goal is to present a brief historical survey of the individuals and ideas from which group theory evolved, beginning with the work of Lagrange on algebraic equations, and tracing it through Gauss, Cauchy, and Abel, until we arrive, finally, with Galois. We will then indicate some of the developments that occurred in group theory from the later half of the nineteenth century up to the present time.

Let us begin by recalling that if n is a positive integer, the *general algebraic equation of degree n* has the form $a_n X^n + a_{n-1} X^{n-1} + \cdots + a_1 X + a_0 = 0$, where the coefficients $a_0, \ldots, a_n$ stand for arbitrary complex numbers with $a_n \neq 0$. The equation is said to be *solvable by radicals* if each of its roots may be expressed by a formula that involves only the coefficients $a_0, \ldots, a_n$, together with the four arithmetic operations of addition, subtraction, multiplication, and division and including, possibly, root extraction. The general linear equation $a_1 X + a_0 = 0$, for example, is solvable by radicals since it has a unique solution X_1 expressed by the formula $X_1 = -a_0/a_1$. The general quadratic equation $a_2 X^2 + a_1 X + a_0 = 0$ is also solvable by radicals since its roots X_1 and X_2 are expressed by the well-known formulas

$$X_1 = \frac{-a_1 + \sqrt{a_1^2 - 4a_2a_0}}{2a_2}, \qquad X_2 = \frac{-a_1 - \sqrt{a_1^2 - 4a_2a_0}}{2a_2}.$$

The general cubic and general quartic equations are also solvable by radicals, but in these cases the formulas are more complicated than in the quadratic case.

Finding formulas for the solution of algebraic equations was an intriguing mathematical problem for centuries. The Greeks, for instance, developed geometric algorithms for constructing the roots of certain linear and quadratic equations. Later, the Renaissance mathematicians Tartaglia (1500?–1557) and Cardan (1501–1576), among others, obtained formulas for solving the general cubic and quartic equations; using clever algebraic manipulations, they transformed the original equation into another equation which was easily solved and whose roots could be transformed back to give the roots of the original equation. During the next few centuries there were many attempts to imitate these methods for higher degree equations, especially for the general quintic or fifth degree equation, in hope of finding formulas for their roots. These efforts were largely unsuccessful, however, except in special cases, and by the close of the eighteenth century there were still no formulas for the solution of the general quintic or higher degree equations—only a strong belief that such formulas existed and could be derived by a sufficient amount of ingenuity and algebraic manipulation. This belief was shattered in 1826 when the young Norwegian mathematician Niels Henrik Abel showed that the general quintic is, in fact, not solvable by radicals. Then in 1832 Galois answered the entire solvability question by creating a general theory for dealing with the solvability of any algebraic equation regardless of its degree. He first introduced the concept of a group and then associated a group with every algebraic equation. By studying the relationship between the solvability of the equation and the properties of its group, Galois showed that the general algebraic equation of degree n is not solvable by radicals whenever $n \geq 5$.

It is in the evolution of ideas leading to Galois's theory where we find the story of group theory; ideas that represent a change in thinking from the algebraic manipulation of equations that characterize the early approach, to a more subtle analysis of the nature of the solutions. The first person to approach the solvability question from this new direction was Lagrange.

Joseph-Louis Lagrange was born in Turin, Italy, on January 25, 1736. He spent the first 30 years of his life in his native Turin, teaching and carrying out his research on the calculus of variations and mechanics. Becoming disillusioned with the empty promises of the court of Turin, he accepted an invitation from the Prussian king, Frederick II, to fill the position vacated by Euler at Berlin, and left for Berlin in 1766. When the Prussian attitude toward the sciences changed following the death of Frederick II, he accepted an invitation from Louis XVI in 1787 to become a member of the French Academy of Sciences. In 1797 he was appointed professor of mathematics at the newly formed École Polytechnique in Paris, and held this position until his death on April 10, 1813. During his lifetime Lagrange received many awards, and was made a Count of the Empire by Napoleon. He made major contributions to virtually all areas of mathematics and is today regarded as one of the two most important mathematicians of the eighteenth century, the other one being Leonard Euler (1707–1783). His most important works are a treatise, *Traité de la résolution des équations numérique de tous degrés* (1767), dealing with numerical solutions of equations, and a book, *Mécanique analytique* (1788), which deals with the equations of motion of a dynamical system.

Lagrange's importance in the evolution of group theory stems from his paper *Réflexions sur la théorie algébriques des équations*, written during the years 1770–1771, in which he introduced several new ideas into the question of the solvability of algebraic equations by means of radicals. The paper begins with a detailed examination of the solutions to the general equations of degrees 1, 2, 3, and 4, with the hope of finding a general technique that may be applied to equations of higher degree. After examining the existing techniques and methods, Lagrange concludes that in each case the original equation may be transformed to a new equation, the *reduced equation*, which is easily solved and whose roots may then be transformed back to give the roots of the original equation. For example, to solve the quadratic $aX^2 + bX + c = 0$, we first complete the square to obtain an equation of the form $aY^2 = d$, where $Y = X - b/2a$, and then solve this equation and transform its roots back to find those of the original equation. Lagrange then applied the results of his analysis to the general quintic equation, but was unable to find a reduced equation whose roots would give the roots of the original equation.

After Lagrange, there are two developments that directly affected the solvability question for algebraic equations. The first came from Gauss, who showed that the particular equation $X^n - 1 = 0$ is solvable by radicals for all positive integers n, while the second came from Cauchy, who began to develop a formal theory of permutations.

Carl Friedrich Gauss was born in Brunswick, Germany, on April 30, 1777, and spent most of his adult life as professor of mathematics and director of the observatory at the University of Göttingen. In 1797, at the age of 20, he published a proof of the fundamental theorem of algebra, showing that every algebraic equation of positive degree n with arbitrary complex coefficients has exactly n complex roots, and which represents, for the first time, a completely satisfactory proof of this theorem. For his work Gauss was awarded the title *Doctor Philosophiae* by the University of Helmstädt. Virtually every area of nineteenth-century mathematics was influenced by Gauss, from pure and applied mathematics, to theoretical and experimental physics and astronomy, and he is generally regarded as the most important mathematician of that century. His first love, however, was number theory, and in 1801 he published, under the patronage of the Duke of Brunswick, his monumental masterpiece, the *Disquisitiones Arithmeticae*, which deals with his research in the theory of numbers. He died on February 23, 1855.

The fundamental theorem of algebra shows once and for all that an algebraic equation of degree n with complex coefficients has precisely n complex roots. The emphasis therefore shifts from the existence of solutions to how the solutions may be expressed; specifically, is every algebraic equation solvable by radicals? Gauss does not deal with this question in general but concentrates, instead, on the particular equation $X^n - 1 = 0$, whose solutions represent the division of a circle into n equal parts. To understand why this particular equation is important to him, we must turn to the *Disquisitiones Arithmeticae*. The first few sections of this book deal with the theory of congruences with applications to the theory of forms. But in the seventh section, "Equations Defining Sections of a Circle," Gauss discussed the classical geometric problem of dividing a circle into any finite number of equal arcs by means of straight edge and compass construction. He notes that the complex numbers $z_k = \cos(2\pi/n)k + i \sin(2\pi/n)k$, for $k = 1, \ldots, n$, correspond to the points $(\cos(2\pi/n)k, \sin(2\pi/n)k)$ of equidivision of a unit circle into n equal arcs, and that $z_1, \ldots, z_n$ are precisely the roots of the equation $X^n - 1 = 0$. Thus, the original geometric problem of subdividing the circle into n equal arcs now becomes a question of determining when the roots of the equation $X^n - 1 = 0$ are constructible by straight edge and compass alone. Using methods very similar to Lagrange's method of reduced equations, Gauss shows that the equation $X^n - 1 = 0$ is always solvable by radicals, but its roots cannot be constructed using straight edge and compass unless n is a product of primes of the form $2^s + 1$. Before we leave Gauss, let us mention that by applying his method to the case where $n = 17 = 2^4 + 1$, he was the first person to give an explicit straight edge and compass construction of a regular 17-gon.

Augustin-Louis Cauchy was born in Paris on August 21, 1789, at the height of the French Revolution and studied under Lagrange at the École Polytechnique. He first pursued a career in civil engineering, but gave this up at the urging of Lagrange and Laplace in favor of mathematics, returning to the École Polytechnique where he was elected to the chair of mechanics and to membership in the Academy of Sciences. But the unstable political climate in France around 1830, together with his own strongly held views, forced Cauchy into exile for 18 years, first to Turin and then Prague. He returned in 1848 to a professorship at the École Polytechnique and remained there until his death on May 22, 1857. Like his contemporaries Lagrange and Gauss, Cauchy made fundamental contributions to many areas of mathematics, most notably in analysis and the theory of functions of a complex variable, and is credited with introducing rigor into analysis. Many of the formal definitions and techniques in the modern-day calculus course are due directly to Cauchy.

Cauchy's main contributions to group theory are found in a series of papers written during the years 1815–1844 in which he studies the theory of permutations. In these papers he introduced the two-row matrix notation for permutations that we use today, defined the product of two permutations, the order of a permutation, cyclic permutations (cycles) and their powers, transpositions, even permutations and the alternating subgroup, and showed that every even permutation is a product of 3-cycles. It is in one of these articles that we find his theorem about the existence of elements of prime order in a group of permutations: if the order of a group of permutations is divisible by a prime p, then the group must must contain at least one permutation of order p.

With the terminology and basic properties of permutations in place, the stage was now set for the two individuals who finally put the solvability question for algebraic equations to rest: Abel and Galois.

Niels Henrik Abel was born on the island of Findöy, near Stravanger, Norway, on August 5, 1802 and studied mathematics at the University of Oslo. His mathematical ability was soon recognized by his teacher Bernt Michael Holmboe, and he was encouraged to read the works of Newton, Euler, Lagrange, and Gauss, especially the *Disquisitiones Arithmeticae*. In 1824, at the age of 22, Abel proved that the general quintic is not solvable by radicals. He also made important contributions to the theory of infinite series and elliptic functions. From 1824 until the time of his death, Abel traveled frequently to Paris and Berlin attempting to secure a teaching position, but poor heath and lack of finances made this difficult. When at last he did receive an appointment as professor of mathematics at Berlin, it was too late—he had died two days earlier, on April 6, 1829, from tuberculosis, at the age of 26. Today we honor Abel's memory with the terms *Abelian* group and *Abelian* function.

In his 1824 paper, *Mémoire sur les équations algébriques, ou l'on démontre l'impossibilité de la résolution de l'équation générale du cinquième degré*, Abel showed that the general quintic is not solvable by radicals. Building on the work of Lagrange, he first determined the possible forms for the reduced equation of the general quintic, and then showed, using some of Cauchy's results, that each possibility leads to a contradiction. Thus, the general quintic is not solvable by radicals.

In a second paper, published shortly before his death in 1829, Abel discussed the solvability of a particular class of algebraic equations. He showed that if the roots of an equation can be expressed as functions of a single root and if these functions commute, then the original equation is solvable by radicals. The significance of this result is that it provides, for the first time, a criterion for the solvability of a particular class of algebraic equations. It is from this work that we derive the term *Abelian* for commutative groups. Unfortunately, Abel died two months after its publication and it remained for Galois to bring about the final resolution of the solvability question.

Evariste Galois was born near Paris on October 25, 1811 and entered the College Louis-le-Grand in 1823. By all accounts he was a militant republican, caught up in the political turmoil surrounding Paris during the 1820s. He twice tried to enter the École Polytechnique, but each time was refused admission because he failed the entrance examinations. He was finally admitted to the École Normale in 1829, but was expelled in 1830 because of political activities associated with the July Revolution and spent the next two years in and out of jail. Although he wrote several mathematical papers ranging from number theory to elliptic functions, most of his work on the solvability question dates from the period 1829–1832. On the morning of May 30, 1832, he was wounded in a pistol duel, the result of a love affair gone awry, and died the next day, not yet 21 years of age.

In his papers dealing with the solvability of algebraic equations, Galois removes the problem from the computational schemes of Abel and Lagrange and places it within the context of permutations of the roots; that is, he concentrates on the functions that express one root in terms of the others. Galois is the first person to use the term *group*, by which he means a collection of permutations closed under multiplication. He then associates a group with every algebraic equation by considering those permutations that leave a rational function of the roots unchanged if and only if the function can be expressed rationally in terms of the roots. Today we call this group the *Galois group* of the equation.

Galois's fundamental paper on the solvability question was presented to the Academy on January 17, 1831. In it he reasons as follows. If an algebraic equation is solvable by radicals, then its roots may be expressed rationally by formulas that involve the four arithmetic operations of addition, subtraction, multiplication, and division applied to the coefficients, together with root extraction. It is then possible to build up such a formula by successively adjoining radicals, which correspond to roots of the reduced equations and which can be chosen to have prime degree. He then associates a subgroup of the Galois group with each of the reduced equations in such a way that the Galois group contains a chain of subgroups, each of prime index in its successor. In this way solvability of the equation corresponds to the existence of such a chain of subgroups in its Galois group.

By associating a group of permutations with an arbitrary algebraic equation, not just the general equation, and by using the properties of this group to study the solvability of the equation, Galois created a completely new and abstract approach to the solvability question of algebraic equations. His work not only gives a complete solution of the solvability question, but also contains many of the basic concepts of group theory: subgroup, cosets, normal subgroup and quotient group, although it would be several decades before these terms would actually be used to describe these concepts. Today, we say that a finite group is *solvable* if it has a descending chain of normal subgroups such that each consecutive pair of subgroups in the chain has a quotient group of prime order. The fundamental theorem of Galois theory then shows that an algebraic equation is solvable by radicals if and only if its Galois group is solvable. Since the Galois group of the general equation of degree n is the symmetric group Sym(n) and since this group is not solvable when $n \geq 5$, it follows that the general algebraic equation of degree n is not solvable by radicals when $n \geq 5$.

So what can be said of Galois, the young mathematician who, although not yet 21, completely answered the solvability question for algebraic equations, a question that had resisted the efforts of some of the most outstanding mathematicians for nearly 60 years? He solved the problem not by using the computational schemes of his predecessors, but by creating an abstract theory; by placing it, literally, in the realm of abstract thought. His approach was nothing short of a revolution in the way mathematics was done at that time. His reward: indifference. Indifference from a mathematical establishment that was, for the most part, unable and unwilling to understand his work. Some of his papers submitted to the Academy were lost, and when Poisson, who was assigned to report on Galois's fundamental paper dealing with solvability, finally got around to it on July 4, 1831, he reported that some of the results had already been obtained by Abel, while the rest were incomprehensible. Writing to his friend Guillaume-August Chevalier (1809–1868) on the night before his fatal dual, Galois, frustrated and perhaps sensing his fate, says ([22]):

> You will publicly beg Jacobi or Gauss to give their opinion not of the truth but of the importance of the theorems. After this, there will, I hope, be people who will find it to their advantage to decipher all this mess.

The next day he was mortally wounded in a duel and died the following day. In all, fate was cruel to him, for it blessed him with brilliance but robbed him of time. Thus was born group theory.

It was to be nearly two decades before Galois's ideas were clarified and given a formal exposition. The first textbook to deal with the subject was the third edition of *Cours d'Algebra superieure*, written in 1866 by **Joseph-Alfred Serret** (1819–1885). This was followed, in 1870, by the popular text *Traité des substitutions et des équations algébriques*, written by **Camille Jordan** (1838–1922). It is in Jordan's work that we find the term *simple group* used for the first time to describe a group that has no proper normal subgroup.

During the second half of the nineteenth century most of the basic concepts in group theory began to crystallize. The axioms defining a group, for example, were first formulated in 1854 by the English mathematician **Arthur Cayley** (1821–1895). Cayley was a lawyer by profession and mathematician by avocation until 1863, at which time he accepted an appointment to the newly created Sadlerian professorship at Cambridge University. He was a prolific writer who made many contributions to mathematics, especially in geometry and the theory of invariants, and is credited with introducing the concept of n-dimensional space into mathematics. Cayley is important in the development of group theory because of his abstract, axiomatic approach to the subject—an approach, incidentally, which was not popular at the time—but is best remembered for his 1878 paper in which he showed that every finite group is isomorphic to a group of permutations. Let us also mention that in 1872 the Norwegian mathematician **Ludvig Sylow** (1832–1918) published a 10-page paper in which he proved the theorems that now bear his name dealing with the existence of subgroups of prime power order in a finite group.

Research in group theory during the early years of the twentieth-century turned to the question of determining the structure of groups—how are groups built up from subgroups, and what, if any, are the basic building blocks? In 1889 **Otto Hölder** (1859–1937) showed that every finite group G admits a composition series $G = G_n \triangleright G_{n-1} \triangleright \cdots \triangleright G_1 = (1)$ in which the quotient groups G_{i+1}/G_i are simple groups, and are uniquely determined by G in the sense that two groups are isomorphic if and only if they have the same composition factors $G_n/G_{n-1}, \ldots, G_2/G_1$. This result, now called the Jordan-Hölder theorem, shows that the simple composition factors of a finite group completely determine the group up to isomorphism. The extension problem, on the other hand, deals with the problem of constructing a group from its composition factors: given a group G and a group N, determine all groups G^* such that N is a normal subgroup of G^* and $G^*/N \cong G$. Any such group G^* is called an *extension* of G by N. Extension theory thus provides a complete description of the extensions of a given group, and it is this theory that explains how groups are built up from normal subgroups. But more importantly extension theory, when taken together with the Jordan-Hölder theorem, identifies the basic building blocks of group theory as those groups that have no proper normal subgroups—that is, the simple groups.

Simple groups were not new to group theory. Galois had already identified the concept of a simple group and noted that the alternating group Alt(5) is simple. Jordan, in his *Traité* of 1870, not only extended this result by showing that the alternating groups Alt(n) are simple groups for all $n \geq 5$, but also constructed four other families of simple groups which today are called the classical linear groups: the projective special linear group, the symplectic group, the orthogonal group, and the unitary group, all of which are certain groups of matrices over finite fields. Let us also mention that in 1860 the French mathematician **Émile Mathieu** (1835–1890) discovered four transitive permutation groups, denoted by M_{11}, M_{12}, M_{23}, and M_{24}. The Mathieu groups were shown to be simple groups during the years 1895–1900 by **F. N. Cole** (1861–1927), after whom the Frank Nelson Cole Prize in Algebra awarded by the American Mathematical Society is named, and **G. A. Miller** (1863–1951).

In 1901 the American mathematician **Leonard Eugene Dickson** (1874–1954) wrote his classic work, *Linear Groups with an Exposition of the Galois Theory*, in which, among other things, he classified the known simple groups. After this, however, there was a rather lengthy hiatus in which little work was done on either finding or classifying the simple groups. But a new approach began to emerge by the mid-twentieth century. **Jean Dieudonné**, in 1948, succeeded in classifying all of the known simple groups, not by using matrices, which characterized Dickson's approach, but rather by interpreting these groups geometrically in terms of vector spaces and linear transformations. Then in 1955 **Claude Chevalley** approached the entire classification problem for simple groups from the point of view of Lie algebras, showing that most of the known simple groups were of Lie type, except for a few of exceptional type. Using Lie algebra techniques, Chevalley also found several new simple groups.

But in 1954 something happened that was to affect the search for simple groups: the American mathematician **Richard Brauer** (1901–1977) announced to the International Congress of Mathematicians meeting in Amsterdam that he and K. A. Fowler had proved a result showing that, to within isomorphism, there are only a finite number of simple groups having an involution whose centralizer is isomorphic to a given group. This result was to form the basis for a program leading to the eventual classification of all simple groups. Another major breakthrough came in 1963 when the American mathematicians **Walter Feit** and **John Thompson** announced that every finite group of odd order must be solvable, a result that brought them the Frank Nelson Cole Prize in Algebra from the American Mathematical Society in 1965. The proof of this result, known as the Feit-Thompson Odd Order Theorem, occupies 255 pages in the Pacific Journal of Mathematics. The Feit-Thompson theorem plays a crucial role in the study of simple groups for it shows that every non-abelian simple group must have even order and hence contain an element of order 2, that is, an involution, thus making Brauer's classification program a viable approach. As Brauer and Fowler showed, it is the properties of such an element that determine, to a large extent, the structure of the simple group.

The period from 1960 to 1980 saw an intense effort to complete the classification of the finite simple groups, much of it inspired and led by the American mathematician **Daniel Gorenstein** (1923–   ). Several new simple groups were discovered during this time that do not fit into any standard classification scheme and known, accordingly, as sporadic simple groups; groups with such esoteric names as Fischer's *Monster* and *Baby Monster*, discovered in 1974, and Conway's .1, .2, and .3 groups found in 1969 by John H. Conway and J. L. Thompson.

Although many more simple groups were discovered during the period 1960–1980, some of which fit into well-defined patterns, it was generally believed that they were too complicated and sporadic to ever be completely classified. But the end was in sight. In 1974, after seven years of work that occupies 407 pages spread out over six journal issues, Thompson announced that he had completed the classification of the minimal simple groups, that is, simple groups all of whose subgroups are solvable. Finally, in 1981 Gorenstein announced to the mathematical community what many believed to be impossible ([9]):

> In February, 1981, the classification of the finite simple groups was completed, representing one of the most remarkable achievements in the history of mathematics. Involving the combined efforts of several hundred mathematicians from around the world over a period of 30 years, the full proof covered something between 5,000 and 10,000 journal pages, spread over 300 to 500 individual papers.

With this monumental achievement we leave our story of group theory. Although the structure of every finite group is now known, at least theoretically, the work of understanding and interpreting these results and their consequences is just beginning. But let us leave the final word with Gorenstein who, perhaps more than any other individual, believed in the eventual success of the classification program and worked tirelessly to bring it about ([9]):

> During the last few years of the classification proof, the idea spread that its completion would somehow coincide with the end of the subject of finite group theory itself. The prevalence of this view was undoubtedly fostered by the unusually wide (for mathematics) press coverage of simple groups and many of the comments (including certainly my own) by finite group theorists. Indeed, the headline of an article in the *New York Times* Week in Review of June 22, 1980, read "A School of Theorists Works Itself Out of A Job. . . ." However, the first 'post-classification' conference quickly dispelled this gloomy prediction. Indeed group theory is 'alive and well': although the focus has shifted, its vitality continues unimpaired. . . . Thus the obituary for finite group theory has been totally premature.

References and Additional Readings

1. Aschbacher, Michael, "The Classification of the Finite Simple Groups," *Mathematical Intelligencer*, 3(2), 1981, 59–65.

2. Bell, E. T., *Men of Mathematics*, Simon and Schuster, New York, 1937.

3. Buhler, W. K., *Gauss, A Biographical Study*, Springer-Verlag, New York, 1981.

4. Eves, Howard, *An Introduction to the History of Mathematics*, Fifth Edition, CBS College Publishing, New York, 1983.

5. Freudenthal, Hans, "Cauchy, Augustin-Louis," *Dic. Sci. Bio.*, Vol. III, 131–148, ed. Charles Coulston Gillispie, Charles Scribner's Sons, New York, 1970.

6. Gallian, Joseph A., "The Search for Finite Simple Groups," *Mathematics Magazine* 49(4), 1976, 163–179.

7. Gardner, Martin, "The Capture of the Monster: A Mathematical Group with a Ridiculous Number of Elements," *Scientific American* 242(6), 1980, 20–32.

8. Gauss, C. F., *Disquisitiones Arithmeticae*, English translation by Arthur Clark, Yale University Press, New Haven, 1966.

9. Gorenstein, Daniel, *Finite Simple Groups, An Introduction to their Classification*, Plenum Press, New York, 1982.

10. Hamburg, Robin Rider, "The Theory of Equations in the 18th Century: The Work of Joseph Lagrange," *Arch. History Exact Sci.* 16(1), 1976, 17–36.

11. Infeld, Leopold, *Whom the Gods Love: The Story of Evariste Galois*, McGraw-Hill, New York, 1948.

12. Itard, Jean, "Lagrange, Joseph Louis, " *Dic. Sci. Bio.*, Vol. VII, 559–573, ed. Charles Coulston Gillispie, Charles Scribner's Sons, New York, 1970.

13. Kiernan, B. Melvin, "The Development of Galois Theory from Lagrange to Artin," *Arch. History Exact Sci.* 8, 1971–72, 40–154.

14. Kleiner, Israel, "The Evolution of Group Theory: A Brief Survey," *Mathematics Magazine* 59(4), 1986, 195–215.

15. May, Kenneth O., "Gauss, Carl Friedrich," *Dic. Sci. Bio.*, Vol. V, 298–315, ed. Charles Coulston Gillispie, Charles Scribner's Sons, New York, 1970.

16. Novy, Lubos, *Origins of modern algebra*, English translation by J. Tauer, Noordhoff International Publishing, The Netherlands, 1973.

17. Ore, Oystein, *Niels Henrik Abel, Mathematician Extraordinary*, Chelsea Publishing Company, New York, 1957.

18. Ore, Oystein, "Abel, Niels Henrik," *Dic. Sci. Bio.*, Vol. I, 12–17, ed. Charles Coulston Gillispie, Charles Scribner's Sons, New York, 1970.

19. Steen, Lynn Arthur, "A Monstrous Piece of Research," *Science News* 118, 1980, 204–206.

20. Taton, René, "Galois, Evariste," *Dic. Sci. Bio.*, Vol. V, 259–265, ed. Charles Coulston Gillispie, Charles Scribner's Sons, New York, 1970.

21. van der Waerden, B. L., *A History of Algebra*, Springer-Verlag, New York, 1985.

22. Wussing, Hans, *The Genesis of the Abstract Group Concept*, English translation by Abe Shenitzer, MIT Press, 1984.

APPENDIX

B

Remarks on the History of Ring Theory

Few centuries throughout the history of mathematics have had as profound an effect on the evolution of mathematical concepts as the nineteenth century. It was a time when fundamental concepts of group theory were emerging, as well as a more formal, axiomatic approach to the construction of the real number system and the subsequent introduction of rigor into analysis. It was a time when the foundations of arithmetic and algebra were being formalized and axiomatized in much the same way that Euclid axiomatized the basic ideas of geometry. At the beginning of the century the infinite was considered virtually taboo in mathematics, but by the end of the century Cantor and his theory of sets brought about an entirely new interpretation of the infinite. Put simply, the nineteenth century was a time when mathematics cast its critical eye upon itself.

The nineteenth century also saw the evolution of ring theory. The term *ring* was first used explicitly by Hilbert in 1897, although only in the context of real and complex number systems. It did not take on its modern, abstract meaning until 1914, when Abraham Fraenkel gave a general definition that is essentially the same one that we use today. The concept itself goes back to early 19th century attempts by the British school of algebraists to formulate a general axiomatic approach to algebra. The discovery of quaternions by Hamilton in 1843 and Cayley's introduction of matrix algebras in 1858 soon led to the general study of associative algebras, or hypercomplex number systems as they were called at the time, and their eventual classification, first over the real and complex numbers by Cartan, B. Peirce, and C. S. Peirce in the late nineteenth century, and then by Wedderburn in 1907, who developed the general structure theory for simple and semisimple algebras over arbitrary fields. At the same time, German mathematicians, notably Kummer, Kronecker, Dedekind, and Hilbert were beginning to develop a general theory of commutative number systems in an attempt to extend the ideas of unique factorization from the integers to the complex numbers.

The evolution of ring theory therefore stems from two distinct sources: first, the study of associative algebras and attempts to determine and classify all such algebras, which led to the evolution of noncommutative ring theory; and second, attempts to extend unique factorization from the integers to more general types of complex number systems—to save unique factorization, so to speak, for systems of hypercomplex numbers—which led to the evolution of commutative ring theory and the modern theory of ideals. Our goal in this chapter is to first discuss the evolution of noncommutative ring theory, beginning with Hamilton's work on quaternions, tracing it through Cayley's introduction of matrices and the Peirces' work on real and complex associative algebras, and concluding with Wedderburn's theory of simple and semisimple algebras, and then indicate how the basic ideas of commutative ring theory evolved under Kummer, Kronecker, and Dedekind into the modern theory of ideals. Finally, we indicate briefly how the geometry of algebraic curves and surfaces evolved, under Hilbert's use of commutative ideal theory, into the algebraic geometry of today. Taken as a whole, these ideas not only enriched algebra with new and important concepts such as rings and ideals, but in fact changed the very nature of algebra itself.

Up until the early 19th century algebra was regarded as an informal collection of rules for the symbolic manipulation of numbers. The first attempt to formalize, or axiomatize, the foundations of algebra came from the early 19th century British school of algebraists, including **George Boole** (1815–1864). **August DeMorgan** (1806–1871), and **George Peacock** (1791–1858). One of the earliest textbooks to approach algebra from this general point of view was Peacock's *A Treatise on Algebra*, published in 1830. Peacock's goal was to do for algebra what Euclid had done for geometry, namely to develop algebra from a formal, axiomatic point of view. In his book, he suggests what was, at the time, a revolutionary thought: that the symbols of algebra need not refer to numbers, but may represent more general mathematical objects. It was in this climate that Hamilton began his fundamental work on quaternions.

William Rowan Hamilton was born on August 4, 1805, in Dublin, Ireland. He was a precocious youth, educated by his uncle until the age of three, and who, by the age of five, had mastered Latin, Greek and Hebrew. As an undergraduate at Trinity College, Dublin, he read Newton's *Principia* and wrote several papers dealing with astronomy and optics, although he never completed his degree. He was appointed astronomer royal of the Dunsink Observatory in 1827 and remained in that position until his death on September 2, 1865. Hamilton was not a particularly effective practical astronomer, preferring, instead, to spend his time doing theoretical mathematics and pursuing his literary interests, especially poetry. He gained prominence as a member of the Royal Irish Academy and the British Association for the Advancement of Science, and was responsible for bringing the annual meeting of the Association to Dublin in 1835, at which time he was knighted by the lord lieutenant. Hamilton's most important scientific works include two papers, *Theory of Systems of Rays* and *On Caustics*, dealing with optics and dynamics, but he is best remembered today for his work on quaternions.

To understand how Hamilton was led to the creation of quaternions, we must go back to a letter from his friend, John Graves, dated 1828. Recall that complex numbers are expressions of the form $a + bi$, where $a, b \in \mathbb{R}$. Addition and multiplication are defined by means of the formulas

$$(a + bi) + (c + di) = (a + c) + (b + d)i$$

$$(a + bi)(c + di) = (ac - bd) + (ad + bc)i,$$

where $i^2 = -1$. Complex numbers may be represented geometrically by means of an Argand diagram, that is, the complex plane, in which $a + bi$ is represented as the ordered pair (a, b), although Hamilton was apparently unaware of this representation in 1828. Hamilton mentioned in a letter to Graves that the fundamental concepts of algebra needed to be put on solid ground. Graves sent him a book, John Warren's *Treatise on the Geometrical Representation of the Square Roots of Negative Quantities*, which described Argand's geometric approach to complex numbers. After reading it, Hamilton was led to ask the question: Is it possible to find a three-dimensional analogue of complex numbers, that is, a system of hypercomplex numbers, that are related to three-dimensional space in much the same way that complex numbers are related to two-dimensional space? It was this question that occupied most of his time for the next 30 years.

His goal was to start with an expression of the form $a + bi + cj$, where $a, b, c \in \mathbb{R}$, and define addition and multiplication in such a way that multiplication is associative, commutative, and distributes over addition. Addition is no problem since it may be defined componentwise:

$$(a + bi + cj) + (a' + b'i + c'j) = (a + a') + (b + b')i + (c + c')j.$$

But multiplication is different. By analogy with complex numbers, he set $i^2 = j^2 = -1$. Furthermore, to preserve the analogy with complex numbers, Hamilton insisted that the "modulus law" remain valid for multiplication of hypercomplex numbers. The number $a^2 + b^2 + c^2$ he called, by analogy with complex numbers, the modulus of $a + bi + cj$. For complex numbers, the modulus of $a + bi$ is $\sqrt{a^2 + b^2}$ and, if we set $(a + bi)(c + di) = e + fi$, then

$$\sqrt{a^2 + b^2} \sqrt{c^2 + d^2} = \sqrt{e^2 + f^2},$$

or, in terms of norms, which are the squares of the moduli,

$$\|a + bi\| \|c + di\| = (a^2 + b^2)(c^2 + d^2) = e^2 + f^2 = \|e + fi\|.$$

Thus, the modulus of a product is equal to the product of the moduli. The modulus of a complex number is the length of the vector represented by the complex number in the plane. Preserving the modulus therefore means that when two such vectors are multiplied by using their complex number representation, the length of the product vector is the product of their lengths. Hamilton absolutely insisted that this property should remain valid for hypercomplex numbers in three dimensions.

Now, for triples we find that

$$(a + bi + cj)^2 = (a + bi + cj)(a + bi + cj)$$
$$= (a^2 - b^2 - c^2) + i(2ab) + j(2ac) + bc(ij + ji).$$

Setting $ji = -ij$ therefore gives the required form for the norm of a triple. But we then find by calculation that

$$(a + bi + cj)(a' + b'i + c'j)$$
$$= (aa' - bb' - cc') + i(ab' + ba') + j(ac' + a'c) + ij(bc' - b'c),$$

whose modulus is

$$(aa' - bb' - cc')^2 + (ab' + ba')^2 + (ac' + a'c)^2 + (bc' - b'c)^2,$$

while the product of the moduli is

$$(a^2 + b^2 + c^2)(a'^2 + b'^2 + c'^2)$$
$$= (aa' - bb' - cc')^2 + (ab' + ba')^2 + (ac' + a'c)^2 + (bc' - b'c)^2.$$

This is exactly what it should be, except the product $(a + bi + cj)(a' + b'i + c'j)$ is not a triple but rather involves four terms!

For 13 years Hamilton tried to find computational schemes for the product of triples that would give him the results he wanted, but to no avail. Then on the morning of October 16, 1843, he and his wife were walking into Dublin to attend a meeting of the Royal Irish Academy when suddenly a thought burst upon him: if instead of dealing with triplets one used four terms, or quaternions, all of the arithmetic requirements would be satisfied. Hamilton immediately wrote down the appropriate formulas for the multiplication of quaternions. Setting $k = ij$ and

$$k^2 = j^2 = i^2 = -1, \quad ij = -ji = k, \quad jk = -kj = i, \quad ki = -ik = j,$$

he obtained the general formula for quaternion multiplication:

$$(a + bi + cj + dk)(a' + b'i + c'j + d'k)$$
$$= (aa' - bb' - cc' - dd') + i(ab' + a'b + cd' - c'd)$$
$$+ j(ac' - bd' + ca' + db') + k(ad' + bc' - cb' + a'd).$$

Quaternion multiplication is then associative, distributes over addition, and every nonzero quaternion has an inverse. Thus, except for commutativity, quaternions have all of the basic arithmetic properties of the real numbers. The lack of commutativity, however, represents a fundamental break with traditional algebra. In fact, when Hamilton told Graves of his success in creating the quaternions, Graves, echoing long-held sentiments about the nature of algebra, responded by saying that ([21]):

> There is still something in this system [of quaternions] which gravels me. I have not yet any clear view as to the extent to which we are at liberty arbitrarily to create imaginaries, and to endow them with supernatural properties.

By the end of the century, however, attitudes had changed. The French mathematician Henri Poincaré, writing in 1902, compares Hamilton's creation of the quaternions, a new and noncommutative number system, with the creation of non-Euclidean geometry by Lobachevsky, a geometry which, we recall, satisfies all the basic Euclidean axioms except the parallel postulate ([21]):

> Hamilton's quaternions give us an example of an operation which presents an almost perfect analogy with multiplication, which may be called multiplication, and yet is not commutative.... This presents a revolution in arithmetic which is entirely similar to the one which Lobachevsky effected in geometry.

After his discovery of quaternions in 1843, Hamilton spent the remaining 22 years of his life exploring properties and applications of quaternions. In 1853 he published his *Lectures on Quaternions.*

Following the creation of quaternions, new and more general types of hypercomplex number systems were discovered and identified. These number systems are examples of associative algebras, although the concept of an algebra was not formalized and named until 1870 by Benjamin Peirce. A finite-dimensional associative algebra over the real or complex number field is simply a ring containing elements $e_1, \ldots, e_n$ in which every element may be written uniquely in the form $a_1 e_1 + \cdots + a_n e_n$, where the coefficients $a_1, \ldots, a_n$ are real or complex; addition is defined componentwise, while multiplication is determined by means of the formulas $e_i e_j = \Sigma\, c_{ij}^k e_k$, where the c_{ij}^k are certain structural constants. Quaternions, for example, form a four-dimensional real associative algebra, while the "biquaternions," that is, quaternions with complex coefficients, form a four-dimensional complex associative algebra. Let us also mention that in 1844 John Graves discovered an eight-dimensional real algebra which he called octaves but which are known today as Cayley numbers. Cayley numbers form a nonassociative as well as noncommutative algebra, although neither Graves nor Cayley mentioned their nonassociativity.

In 1854 Cayley introduced a new concept into the mathematical vocabulary that would play a fundamental role in all future research on hypercomplex number systems—the idea of a matrix. At the time, however, Cayley was unaware of their relationship with hypercomplex numbers and was interested in them only as a convenient way of dealing with linear systems of equations. Nevertheless, he showed that they have all of the properties of an algebra.

Arthur Cayley was born on August 16, 1821, at Richmond, Surrey, England. He went to Cambridge University, studying mathematics and law, and in 1848 was admitted to the Bar, all the while continuing to pursue his interest in mathematics. In 1863 he was appointed to the newly created position of Sadlerian Professor of Pure Mathematics at Cambridge, and held this position until his retirement in 1892. He was frequently called upon by the university and various scientific societies for his legal opinion. Cayley was one of the most prolific writers on mathematics, writing more than 800 papers on topics ranging from symmetric functions, roots of polynomials, group theory, and elliptic functions, to the theoretical dynamics of planetary motion, and is credited with introducing the concept of n-dimensional space into mathematics. He received many awards and honors during his lifetime and died on January 26, 1895.

Cayley's contributions to the study of algebras are contained in three papers. In the first of these, published in 1854, we find for the first time a modern, abstract definition of a group. In the second paper, published in 1855, Cayley introduced the notion of a matrix, defined the inverse of a matrix and product of two matrices, and discussed the relationship of matrices to quadratic and bilinear forms. But there was no attempt to develop an arithmetic for matrices. It was not until his fundamental memoir of 1858, *A memoir on the theory of matrices*, that Cayley defined the sum and product of matrices, multiplication of a matrix by a scalar, and showed that the set of $n \times n$ matrices with real or complex entries is an associative algebra. He developed the basic algebraic properties of matrix operations, noting that matrix multiplication is not commutative, and proved what is today called the Hamilton–Cayley theorem: every $n \times n$ matrix M satisfies its characteristic polynomial, that is, is a root of the polynomial $\det(M - XI_n) = 0$.

Although Cayley introduced matrices and showed that they have all of the properties of an associative algebra, he apparently was unaware of the fact that every $n \times n$ matrix can be written in the form $\Sigma \ a_i e_i$ of hypercomplex numbers, although he did realize that this system of matrices contained Hamilton's quaternions. The fact that matrices and hypercomplex number systems are all examples of one and the same concept, namely associative algebras, is due in part to the efforts of two American mathematicians, Benjamin Peirce and his son, Charles Sanders Peirce, although at the time, 1870, neither of the Peirces were aware of Cayley's work on matrices.

Benjamin Peirce was born on April 4, 1809, in Salem, Massachusetts. He spent most of his adult life as professor of mathematics and astronomy at Harvard University, from which he received his M.A. in 1833. He was especially active in various professional organizations, such as the Smithsonian Institution, the U.S. Coast Survey, and the American Association for the Advancement of Science, and was one of the original fifty incorporators of the National Academy of Sciences. He wrote many textbooks on elementary mathematics, and, in 1870, at the urging of his son Charles, took up the study of algebras. He died on October 6, 1880 in Cambridge, Massachusetts.

Charles Sanders Peirce, second son of Benjamin and Sarah Peirce, was born on September 10, 1839, in Cambridge, Massachusetts, and attended private schools in the Boston area until enrolling at Harvard University in 1855, from which he received the M.A. degree in 1862. In the Peirce household he was constantly exposed to a variety of scientific and mathematical conversations; yet, despite his father's efforts, Charles preferred logic and methodology over a career in the sciences. From 1861 to 1891 he served in the U.S. Coast Survey, observing, cataloging, and investigating various natural phenomena ranging from eclipses to pendulums. He served as lecturer in mathematics at the Johns Hopkins University from 1879 to 1884, where he became acquainted with J. J. Sylvester, and where he first learned of the new mathematics being developed in Europe. He retired from the Coast Survey in 1891, but continued writing articles and reviews until his death on April 19, 1914. Throughout his life Peirce was a strong advocate of the scientific method in all aspects of knowledge, and wrote articles ranging from logic and philosophy, to mathematics and the philosophy of mathematics, probability, philology, criminology, telepathy, and optics.

Benjamin Peirce's fundamental contributions to the theory of associative algebras are found in his paper *Linear Associative Algebra*, which was read before the National Academy of Sciences in 1870. Here we find, for the first time, an explicit formulation of an associative finite-dimensional algebra. According to Peirce, an associative algebra consists of all expressions of the form $\Sigma\, a_i e_i$, where the e_i are a finite number of fixed elements of the algebra and the a_i are real or complex numbers. Addition is defined componentwise according to the formula $\Sigma\, a_i e_i + \Sigma\, b_i e_i = \Sigma\, (a_i + b_i) e_i$, and multiplication is defined by means of "structural constants" c_{ij}^k, where $e_i e_j = \Sigma c_{ij}^k e_k$.

In this paper, Peirce introduced the concept of nilpotent and idempotent elements of an algebra: a nilpotent element of an algebra A is any element $e \in A$ such that $e^n = 0$ for some integer n, while e is said to be idempotent if $e^2 = e$. He then showed that any algebra contains such elements. Moreover, if e is an idempotent of A, then every element $x \in A$ may be written as the sum $x = u + v = ex + (x - ex)$, where $u = ex$ has the property that $eu = u$ and $ev = 0$. This decomposition led Peirce to prove what is perhaps the single most important result of the paper, namely the Peirce decomposition of the algebra: if A is a finite-dimensional associative algebra, then $A = eAe \oplus eB_1 \oplus B_2 e \oplus B$, where $B_1 = \{x \in A \,|\, xe = 0\}$, $B_2 = \{x \in A \,|\, ex = 0\}$, and $B = B_1 \cap B_2$.

As we have already seen, the real numbers, complex numbers, and quaternions are examples of finite-dimensional associative algebras over the real numbers. In 1881 Peirce's son, Charles Sanders Peirce, showed that these three algebras are in fact the only finite-dimensional associative algebras over the real numbers. And it was C. S. Peirce, together with J. J. Sylvester at Johns Hopkins, who finally uncovered the relationship between the various hypercomplex number systems and the matrix algebras of Cayley.

Following the Peirces, research on hypercomplex number systems was concerned mainly with determining the structure of associative algebras over the real and complex number fields. Here the work of Cartan in France and Frobenius in Germany is most prominent. Finally, during the period from 1900 to 1907, a general abstract theory for the study and classification of all such hypercomplex number systems, regardless of the field of coefficients, was developed by Wedderburn, who proved the general classification theorem which today bears his name.

Joseph Henry Maclagan Wedderburn was born on February 26, 1882, in Forfar, Scotland, the tenth of fourteen children. He enrolled at the University of Edinburgh in 1898 and received his M.A. degree five years later with first-class honors in mathematics. He went first to Germany to study with Frobenius and Schur, and then to the United States as a Carnegie fellow at the University of Chicago. He returned to Scotland during the years 1905–1909, but was back in America in 1909 as a "preceptor" appointed under Woodrow Wilson at Princeton University. He enlisted in the British army at the outbreak of World War I, fought in France, and returned to Princeton, where he remained until his retirement in 1945 and death on October 9, 1948. His mathematical work is confined almost exclusively to his investigations on the structure of simple and semisimple algebras and division algebras.

Wedderburn's contributions to the theory of algebras are contained in his fundamental paper *On Hypercomplex Numbers* published in 1908. While Cartan and others had previously determined the structure of such algebras over the real and complex fields, Wedderburn's paper deals with a completely arbitrary ground field. If A is an associative algebra over an arbitrary field, he defined a *complex* as any linear subspace of A. A subcomplex B of A is said to be *invariant* if $AB \subseteq B$ and $BA \subseteq B$. Today we use the term *ideal* to describe such a subcomplex. Wedderburn's identification of such a structure marks the first time ideals were used in a general context. He also defined the notion of a nilpotent algebra, by which he meant an algebra A such that $A^n = 0$ for some integer n, and showed that every algebra has a unique maximal nilpotent invariant subalgebra that contains all invariant subalgebras of A. Although Wedderburn did not give a name to this maximal subalgebra, it is today called the radical of the algebra, a term introduced by Frobenius in his 1903 paper on hypercomplex numbers.

An algebra whose radical is trivial, that is, contains only the zero element, is called a *semi-simple algebra*, while an algebra in which every nonzero element is a unit is called *division algebra*. The main result of Wedderburn's paper is to give a complete classification of semi-simple algebras: every semisimple algebra is a direct sum of matrix rings $\text{Mat}_n(D)$, where D is a division algebra over some field.

Following Wedderburn's classification of simple and semisimple algebras, work turned to finding and classifying these algebras over specific fields such as the rational numbers and other algebraic number fields. During the 1920s and 1930s, for example, there was an intense amount of effort directed at determining the structure of such algebras, notably by A. A. Albert, Richard Brauer, Helmut Hasse, and Emmy Noether, and which culminated in the

Albert-Brauer-Hasse-Noether theorem. These structure theorems, when applied to the group ring of a finite group over a given field, for example, play a fundamental role in the theory of group representations: if G is a finite group and k is a field in which the order of G is not zero, then the group ring $k[G]$ is a semi-simple algebra and hence decomposes as a direct sum of simple algebras, each of which is isomorphic to a full matrix algebra over a division ring. Each component of this decomposition defines an irreducible, or simple, character of the group, and the group characters, which are functions and therefore easy to work with from an arithmetic point of view, turn out to describe completely all representations of the group. Today, character theory is an essential tool for the study of finite groups.

Let us now turn our attention to the development of commutative ring theory. As we noted earlier, Abraham Fraenkel gave the first abstract definition of a ring in 1914. According to Fraenkel, a ring was a system with two operations, which he called addition and multiplication, such that the system is a group under addition, and in which multiplication is associative, distributes over addition, and has an identity. He illustrated this new concept with examples such as the integers, integers modulo n, hypercomplex number systems and matrices, and proved a number of elementary properties that all rings have in common.

The evolution of commutative ring theory, and the theory of ideals in particular, did not begin with Fraenkel, however, but emerged during the period 1844–1897 from the work of Kummer, Kronecker, and Dedekind on algebraic numbers. An algebraic number is any complex number that is the root of a monic polynomial with rational coefficients. Numbers such as $\sqrt{2}$ and i, for example, are algebraic integers, while the ring $\mathbb{Z}[i]$ of Gaussian integers is a ring of algebraic integers.

The original motivation for studying algebraic numbers goes back to attempts to prove Fermat's Last Theorem, which asserts that if n is a positive integer, then the equation $x^n + y^n = z^n$ has no integer solutions when $n > 2$ other than the trivial ones. Mathematicians had attempted to prove this theorem ever since Pierre de Fermat (1601–1665), the French lawyer and mathematician, first wrote in the margin of his copy of Diophantus' book *The Arithmetica*, that he, Fermat, had found a truly wonderful proof, but unfortunately the margin was too small to write it down.

For $n = 2$ the equation is the familiar Pythagorean equation $x^2 + y^2 = z^2$, which has infinitely many integer solutions given parametrically by the formulas

$$x = s^2 - t^2$$

$$y = 2st$$

$$z = s^2 + t^2$$

For $n = 3, 4, 5$, and 6, Lagrange, Euler, Dirichlet, and others, including possibly Fermat himself, had shown that the equation has nontrivial solutions. Now, it is easy to see that if the equation $x^p + y^p = z^p$ has no solution whenever p is an odd prime, then none of the equations have any solution. Thus, one need only consider the equation for p a prime. It is at this point that Kummer began looking at the problem from the point of view of "ideal numbers" in an attempt to prove Fermat's theorem.

Ernst Eduard Kummer was born on January 29, 1810, in Sorau, Germany, and entered the University of Halle in 1828. At first he studied Protestant theology, but gave this up in favor of mathematics, receiving his doctorate in 1831. From 1832 to 1842 he taught at the Gymnasium in what is now Legnica, Poland, where one of this students was Leopold Kronecker, and in 1842, on the recommendation of Dirichlet, whose cousin Kummer had just married, was appointed full professor at the University of Breslau, a position which he held until 1855. When Gauss died in 1855, Dirichlet was appointed his successor at Göttingen and recommended that Kummer succeed him at Berlin. Kummer also arranged for Weierstrass to be appointed to Berlin, and together they formed the first seminar in pure mathematics in Germany. Kummer also served as dean of the University of Berlin from 1857 to 1858 and again from 1865 to 1866, and rector from 1865 to 1869. On February 23, 1882 he announced to the faculty that he noticed a weakening of his mental powers and his ability to carry out abstract, logical arguments, and therefore would request retirement from teaching. He died in Berlin on May 14, 1893.

Kummer and others had observed that if z_p is a primitive pth root of unity, p a prime, then the equation $x^p + y^p = z^p$ could be factored and written in the form

$$x^p + y^p = (x + y)(x + z_p y)(x + z_p^2 y) \cdots (x + z_p^{p-1} y) = z^p.$$

By considering prime divisors of these factors, Kummer and others hoped to show that the equation had no solution for odd primes p and hence that Fermat's theorem was true. But this first required an investigation into the nature of unique factorization for complex numbers, for he was aware of the fact that if one uses complex numbers such as z_p, factorization is not necessarily unique.

Kummer's main contributions stem from several papers written during the years 1844–1857 that deal with algebraic numbers. In these papers he attempted to find an appropriate definition of prime divisor that would extend unique factorization from the ordinary integers to algebraic integers and hence save unique factorization, so to speak, for systems of complex numbers. To do so, he introduced the concept of an "ideal prime divisor" of a complex number in the ring $\mathbb{Z}[z_p]$, where z_p is a primitive pth root of unity. Perhaps the single most important defect of his theory is that Kummer never says what an ideal prime divisor is; he only defines what it means for an integer to be divisible by such a factor. Nevertheless, with these results Kummer had essentially developed the theory of factorization in rings of cyclotomic integers generated by primitive nth roots of unity, for n a prime, and was then able to prove Fermat's theorem for certain primes.

The next stage in the development of the theory was to extend Kummer's results to arbitrary rings of cyclotomic integers; that is, develop a theory of factorization for the rings $\mathbb{Z}[z_n]$, where n is any positive integer. Kummer himself was apparently not interested in this problem; he left it, instead, to his friend and student Kronecker.

Leopold Kronecker was born on December 7, 1823 in Liegnitz, Germany, and received private tutoring at home until he entered the Liegnitz Gymnasium, where his first mathematics teacher was Kummer. He entered the University of Berlin in 1841, followed Kummer to Breslau in 1843, and returned to Berlin in 1844 to received his doctorate. His dissertation, *On Complex Units*, deals with the theory of units for arbitrary cyclotomic integers. Kronecker's father was a prosperous Liegnitz businessman, and shortly after receiving his degree, Kronecker returned to the family business. By 1855 he had become independently wealthy, gave up business, and returned to an academic life. He was appointed to the Berlin Academy in 1861 and thereupon began giving lectures, mainly on the theory of algebraic equations and the theory of numbers. He remained in this position until his death on December 28, 1891. Kronecker is best known today for his strict constructionist approach to mathematics.

Besides Kronecker's doctoral thesis on complex units, his main contribution to number theory is found in his 1882 paper *On the foundations of an arithmetic theory of algebraic numbers*. In this work, Kronecker extends Kummer's ideas by constructing a theory of prime divisors and unique factorization for arbitrary rings of cyclotomic integers, not through the use of Kummer's "ideal prime divisors," but rather through the use of greatest common divisors. Although Kronecker's conception of a prime divisor comes closer to the definition of an ideal than does Kummer's, he still never actually says what a divisor is. It remained for Dedekind to give what is essentially the modern definition of an ideal.

Richard Dedekind was born in Brunswick, Germany, on October 6, 1831 and attended the Gymnasium Martino-Catharineum from age 7 to 16, and then the Collegium Carolinum. In 1850 he enrolled at the University of Göttingen, where he did his doctoral work under Gauss on Eulerian integrals. In 1862 he accepted a teaching post at the Collegium Carolinum, recently renamed the Polytechnikum in Brunswick, and he remained in that position until his death on February 12, 1916. During the years 1872–1875 he was the director of the Polytechnikum. Dedekind received many awards and honors throughout his long life, and was also an accomplished pianist and cellist. Dedekind is perhaps best remembered today for his development of the real number system using *Dedekind cuts*, for his championing of Cantor's theory of sets, and for using sets as the basis for many of his theories, including his theory of ideals. He wrote several important mathematical works, including two short monographs on numbers: *Continuity and Irrational Numbers*, in which he develops the real number system and clarifies the concepts of limit and continuity, and *The Nature and Meaning of Numbers*, which deals with the construction of finite and transfinite numbers. His writing serves as a model of clarity and simplicity.

Dedekind was led to the study of algebraic numbers through his editing of Gauss's *Disquisitiones Arithmeticae* and Dirichlet's book, *Vorlesungen über Zahlentheorie*, on number theory. His work first appeared as Supplement X to the second edition of the *Zahlentheorie*, published in 1871. It is in this work that we find the concept of ideal and prime ideal expounded for the first time in its essentially modern formulation. If K is an algebraic number field, then ([13]):

> A subset A of the integers of K is called an *ideal* if it has the properties:
> I. If β, $\gamma \in A$, then $\beta \pm \gamma \in A$.
> II. If $\beta \in A$ and γ is an algebraic integer of K, then $\beta\gamma \in A$.
> An ideal is said to be *prime* if it is neither all integers nor 0 alone, and if it has the property:
> III. If $\beta\gamma \in A$, then either $\beta \in A$ or $\gamma \in A$.

Dedekind then defined the product of ideals and proved, as his main result, that every nonzero ideal is a product of prime ideals that are uniquely determined except for order. Thus, while the cyclotomic integers in $\mathbb{Z}[z_p]$ do not necessarily admit unique factorization, the ideals do.

Perhaps the most striking feature of Dedekind's approach is its simplicity. For unlike Kummer and Kronecker, it is no longer necessary for Dedekind to talk about a number by describing it in terms of all the numbers that it divides; for him, all numbers that are divisible by a given number form an ideal, and the given number is then replaced, or identified with, the ideal. It is ideals that we are working with, not numbers. This point of view, however, is completely alien to Kronecker and Kummer. At that time the idea of regarding an infinite set as a completed whole, as one, was taboo in mathematics for, among other reasons, Gauss's remark in 1831 in a letter to his friend Schumacher: "But concerning your proof, I protest above all against the use of an infinite quantity as a completed one, which in mathematics is never allowed. The infinite is only a *façon de parler*, where one would properly speak of limits." By introducing ideals as legitimate mathematical objects, Dedekind was one of the first mathematicians to use Cantor's theory of sets and thus to view an infinite set as a complete, single object.

As we leave Kummer, Kronecker, and Dedekind, let us emphasize that the evolution of the concept of an ideal represents a profound change in thinking on the part of mathematicians: infinite sets were finally being recognized as completed entities in themselves. What started out as an attempt to salvage unique factorization for cyclotomic integers ended with an entirely new approach to mathematics. Instead of dealing with cyclotomic integers, we now deal with ideals and unique factorization of ideas. Rather than construct a new theory of divisors that would save unique factorization, Dedekind chose instead to create new objects—ideals—out of the rings and develop a theory of prime divisors and unique factorization for the ideals.

By the close of the century, the single most important book on algebraic number theory was Hilbert's *Zahlbericht* written in 1897 by **David Hilbert** (1862–1943). Hilbert is generally regarded as the most important mathematician of his day. Speaking at the International Congress of Mathematicians held in Paris in 1900, he presented ten problems which he felt represented the most important mathematical questions at that time—he actually had 23 such problems, but decided to mention only ten of them in his talk. Many of his problems have been answered, while others, such as the Riemann hypothesis, are still open. In either case, the questions not only stimulated but in fact set the course for much of the mathematical research during the twentieth century.

At the 1893 meeting of the German Mathematical Society, Hilbert and Minkowski were charged with the responsibility of preparing a report, to be completed within two years, on the current state of knowledge in number theory. Minkowski shortly withdrew from the project and left Hilbert to complete it, which he did in 1897. But his work, the *Zahlbericht*, was much more than simply a report, for in it Hilbert undertook to reshape completely and to reorganize the basic concepts and fundamental results of algebraic number theory. Dedekind's theory of ideals played a central role in Hilbert's exposition, and it is in this work that we find the term "ring" used for the first time, albeit within the context of algebraic number fields. Writing in the *Dictionary of Scientific Biography*, Hans Freudenthal, mathematician and historian of mathematics, says that the *Zahlbericht* [(17)]

> ... is infinitely more than a report; it is one of the classics, a masterpiece of mathematical literature. For half a century it was the bible of all who learned algebraic number theory, and perhaps it is still. In it Hilbert collected all relevant knowledge on algebraic number theory, reorganized it under striking new unifying viewpoints, reshaped formulations and proofs, and laid the groundwork for the still growing edifice of class field theory. Few mathematical treatises can rival the *Zahlbericht* in lucidity and didactic care.

While the *Zahlbericht* represents the dawn of contemporary algebraic number theory, it was not the first time Hilbert used ideal theory to reformulate and solve classical problems. During the years 1894–1899 he used ideal theory to reformulate and solve some long-standing problems in the classical theory of invariants. The theory of invariants is concerned with finding polynomials that remain unchanged by a given group of transformations. One of the outstanding problems at the time was that of finding a basis for a given set of invariants, that is, a set of invariants by which all others could be expressed. Most of this work was very computationally oriented. Hilbert's approach, on the other hand, was more existential in nature. A given collection of invariants with complex coefficients form an ideal in the ring $\mathbb{C}[X_1, \ldots, X_n]$. Using ingenious arguments, Hilbert showed that every ideal I in this ring has a finite basis; that is, there is a finite collection of polynomials $f_1 = f_1(X_1, \ldots, X_n), \ldots, f_m = f_m(X_1, \ldots, X_n)$ in I such that every polynomial in I can be written in the form $g_1 f_1 + \cdots + g_m f_m$ for some polynomials $g_1, \ldots, g_m \in \mathbb{C}[X_1, \ldots, X_n]$. Thus, $I = (f_1, \ldots, f_m)$. In terms of invariants, this result, which is called the Hilbert Basis Theorem, shows that every invariant may be expressed in terms of only finitely many invariants $f_1, \ldots, f_m$. Hilbert's use of ideal theory to replace the complex computational schemes of his predecessors was so ingenious, in fact, that when Paul Gordan, the king of invariants, saw Hilbert's proof, he was led to exclaim, "this is not mathematics; this is theology!"

Through his study of polynomial rings and their ideals, Hilbert recognized that the study of algebraic curves and surfaces could also be reformulated within the context of ideal theory. If S is any nonempty set of points in complex n-space $\mathbb{C}^n$, let $I(S)$ stand for all polynomials in the ring $\mathbb{C}[X_1, \ldots, X_n]$ that vanish at every point in S. Then $I(S)$ is an ideal in the polynomial ring $\mathbb{C}[X_1, \ldots, X_n]$. On the other hand, if I is any ideal of $\mathbb{C}[X_1, \ldots, X_n]$, let $V(I)$ stand for the set of all points in $\mathbb{C}^n$ at which every polynomial in I vanishes; for example, the line $y = x$ in the complex plane is the set $V(Y - X)$. Any set of points in complex n-space of the form $V(I)$ is called an affine algebraic set. Since, by Hilbert's Basis Theorem, every ideal of $\mathbb{C}[X_1, \ldots, X_n]$ is generated by a finite number of polynomials, it follows that every affine algebraic set is defined by means of only finitely many polynomials, namely the common zeros of the basis polynomials $f_1, \ldots, f_m$. Consequently, all geometric properties of an algebraic set, whether a curve or surface, are contained in the algebraic properties of the polynomial ideals. Thus, the geometric study of algebraic curves and surfaces shifted from geometric arguments in complex n-space to algebraic arguments in the polynomial rings $\mathbb{C}[X_1, \ldots, X_m]$.

What remained in this algebraic approach to geometry was to determine precisely how to extract the relevant geometric information from the algebraic properties of the polynomials and their ideals. One fundamental result in this direction is Hilbert's Nullstellensatz, or zeros-theorem: if I is an ideal of polynomials having basis $f_1, \ldots, f_m$ and f is a polynomial that vanishes at all of the zeros common to the polynomials in I, then $f^s = g_1 f_1 + \cdots + g_m f_m$ for some polynomials $g_1, \ldots, g_m$ and some integer $s \geq 1$. With the Nullstellensatz, we now have a complete description of all polynomials vanishing on a given algebraic set.

Following Hilbert's use of ideal theory in polynomial rings to study geometric properties of curves and surfaces, the algebra of polynomial rings and their ideals was expanded and refined by Emmy Noether, W. Krull, O. Zariski, and others, so that today, modern algebraic geometry is essentially the algebraic study of polynomial rings.

With these remarks we come to the end of our discussion of ring theory. As we have seen, the evolution of hypercomplex number systems into associative algebras freed algebra from its traditional interpretation: it was no longer a symbolic system for dealing with real numbers, but had evolved into the abstract concept of a ring in which the laws of composition for combining elements are of fundamental importance, while the particular elements themselves are irrelevant. The evolution of the theory of ideals, on the other hand, which began as an attempt to extend unique factorization from the integers and thus save unique factorization for systems of complex numbers, not only solved the classical problem of unique factorization by shifting it out of the context of integers and into the realm of ideals, but also brought about a new sense of perspective: in number theory, problems that were originally discussed and solved by means of numerical calculations were being replaced with algebraic arguments about ideals, while in geometry, classical problems involving algebraic curves and surfaces were being reshaped and reformulated within the algebraic framework of polynomial rings and their ideals. The nineteenth century was indeed a time when number theory and geometry took on a new form. As we have seen, it was a time that not only changed the nature of algebra, but the nature of mathematics itself.

References and Additional Readings

1. Artin, Emil, "The influence of J. H. M. Wedderburn on the development of modern algebra," *Bull. Amer. Math. Soc.*, 56(1950), 65–72.

2. Bell, E. T., *Men of Mathematics*, Simon and Schuster, New York, 1937.

3. Bierman, Kurt-R., "Kronecker, Leopold," *Dic. Sci. Bio.*, Vol. VII, 505–509. ed. Charles Coulston Gillispie, Charles Scribner's Sons, New York, 1970.

4. Bierman, Kurt-R., "Kummer, Ernst Eduard," *Dic. Sci. Bio.*, Vol. VII, 521–524, ed. Charles Coulston Gillispie, Charles Scribner's Sons, New York, 1970.

5. Bierman, Kurt-R., "Dedekind, (Julius Wilhelm) Richard," *Dic. Sci. Bio.*, Vol. IV, 1–5, ed. Charles Coulston Gillispie, Charles Scribner's Sons, New York, 1970.

6. Bourbaki, Nicolas, *Commutative Algebra*, Addison-Wesley, Reading, 1972.

7. Boyer, Carl B., *A History of Mathematics*, John Wiley & Sons, Inc., New York, 1968.

8. Cayley, Arthur, "A Memoir on the Theory of Matrices," in *The treasury of mathematics*, ed. Henrietta O. Midonick, Philosophical Library, Inc., New York, 1965.

9. Curtis, Charles W., "The Four and Eight Square Problem and Division Algebras," in *Studies in Modern Algebra*, ed. A. A. Albert, The Mathematical Association of America, 1963.

10. Dedekind, Richard, "The Nature and Meaning of Number," in *Essays on the Theory of Numbers*, Dover Publications, New York, 1963.

11. Dick, Auguste, *Emmy Noether, 1882–1935*, Birkhäuser Boston, Boston, 1981.

12. Edwards, Harold M., "The Background of Kummer's proof of Fermat's Last Theorem for regular primes," *Arch. Hist. Exact Sci.*, 14(1975), 219–236.

13. Edwards, Harold M., "The Genesis of Ideal Theory," *Arch. Hist. Exact Sci.*, 23(1980), 321–378.

14. Edwards, Harold M., "Dedekind's Invention of Ideals," in *Studies in the History of Mathematics*, ed. Esther R. Phillips, The Mathematical Association of America, 1987.

15. Eisele, Carolyn, "Peirce, Benjamin," *Dic. Sci. Bio.*, Vol. X, 478–481, ed. Charles Coulston Gillispie, Charles Scribner's Sons, New York, 1970.

16. Eisele, Carolyn, "Peirce, Charles Sanders," *Dic. Sci. Bio.*, Vol. X, 482–488, ed. Charles Coulston Gillispie, Charles Scribner's Sons, New York, 1970.

17. Freudenthal, Hans, "Hilbert, David," *Dic. Sci. Bio.*, Vol. VI, 388–395, ed. Charles Coulston Gillispie, Charles Scribner's Sons, New York, 1970.

18. Hankins, Thomas L., *Sir William Rowan Hamilton*, The Johns Hopkins University Press, Baltimore, 1980.

19. Hankins, Thomas L., "Hamilton, William Rowan," *Dic. Sci. Bio.*, Vol. VI, 85–93, ed. Charles Coulston Gillispie, Charles Scribner's Sons, New York, 1970.

20. Hawkins, Thomas, "Hypercomplex Numbers, Lie Groups, and the Creation of Group Representation Theory," *Arch. Hist. Exact Sci.* 8(1972), 243–287.

21. Kleiner, Israel, "A Sketch of the Evolution of (Noncommutative) Ring Theory," *L'Ensignement Mathématique* 33(1987), 227–267.

22. Kramer, Edna, "Noether, Amalie Emmy," *Dic. Sci. Bio.*, Vol. X, 137–139, ed. Charles Coulston Gillispie, Charles Scribner's Sons, New York, 1970.

23. Nathan, Henry, "Wedderburn, Joseph Henry Maclagan," *Dic. Sci. Bio.*, Vol. XIV, 211–212, ed. Charles Coulston Gillispie, Charles Scribner's Sons, New York, 1970.

24. North, J. D., "Cayley, Arthur," *Dic. Sci. Bio.*, Vol. III, 162–170, ed. Charles Coulston Gillispie, Charles Scribner's Sons, New York, 1970.

25. Novy, Lubos, *Origins of modern algebra*, Noordhoff International Publishing, Leyden, 1973.

26. Peirce, Benjamin, "Linear Associative Algebra," in *The treasury of mathematics*, ed. Henrietta O. Midonick, Philosophical Library, Inc., New York, 1965.

27. Reid, Constance, *Hilbert*, Springer-Verlag, New York, 1970.

28. van der Waerden, B. L., "Hamilton's discovery of quaternions," *Mathematics Magazine* 49(1976), 227–234.

29. van der Waerden, B. L., *A History of Algebra*, Springer-Verlag, New York, 1985.

APPENDIX
C

Remarks on the History
of Field Theory

Throughout this book we have emphasized that the abstract algebraic structures that we call groups, rings, and fields did not arise spontaneously in the minds of mathematicians but, in most cases, evolved slowly out of a desire to unify existing mathematical concepts. While group theory evolved out of early nineteenth century attempts to solve algebraic equations by means of radicals, rings and fields arose out of a desire to unify the discussion of various hypercomplex number systems, free from the details of any one particular system.

As we have seen, the concept of a field provided the appropriate context in which to discuss the solution of algebraic equations. Indeed, field theory remained closely associated with the solution of algebraic equations and Galois theory for nearly a century. It was not until 1910 that Ernst Steinitz gave a more general and abstract definition of a field, the one that is essentially used today. After this, fields began to emerge as a distinct algebraic structure. Then during the years 1930–1948, Artin completely revised the classical approach to Galois theory, divorcing it from the solvability of equations and reformulating it in its modern, abstract form as a relationship between field extensions and their automorphism groups. In this brief chapter our goal is to indicate some of the early developments in field theory, beginning with the work of Kronecker, Dedekind and Steinitz, and conclude with a discussion of Artin's fundamental revision of Galois theory and its impact on field theory.

The first attempts to formulate the basic ideas of field theory came during the second half of the nineteenth century from Kronecker and Dedekind. As we indicated in Appendix B, both Kronecker and Dedekind were deeply involved with the development of ideals and algebraic number systems at that time. Indeed, it was through their study of such number systems that they were led to formulate some of the basic concepts of field theory. But their approach to creating mathematics represents two very different philosophical attitudes. Kronecker was a strict constructionist; for him, the existence of a mathematical object meant that one must be able to give a finite construction of the object, not merely assert its existence through some type of theoretical argument, especially an argument that relies on the infinite. Dedekind, on the other hand, embraced Cantor's set theory and used infinite sets as completed totalities in and of themselves. With Kronecker and Dedekind we have two mathematicians whose approach to doing mathematics is at opposite ends of the philosophical spectrum.

In 1882, as part of his work on cyclotomic numbers, Kronecker published a paper dealing with the arithmetic theory of algebraic numbers in which he used, for the first time, the term "domain of rationality." In his words ([2]):

> The domain of rationality $(R', R'', R''', \ldots)$ contains ... every one of those quantities which are rational functions of the quantities $R', R'', R''', \ldots$ with integer coefficients.

Domains of rationality were always understood to be complex number systems and hence extensions of the field $\mathbb{Q}$ of rational numbers. Thus, $(R', R'', R''', \ldots)$ refers to the extension field $\mathbb{Q}(R', R'', R''', \ldots)$. In this paper, Kronecker defined a number to be algebraic over a given domain of rationality if it is a root of some irreducible polynomial whose coefficients lie in the given domain. He then adjoined such a number to the given domain to form a larger domain of rationality. In order to form such an extension without assuming the existence of the algebraic number itself, Kronecker used the idea of quotient rings: if $p(X)$ is an irreducible polynomial over $\mathbb{Q}$, then the remainders of polynomials upon division by $p(X)$ is a domain of rationality containing a root of $p(X)$; that is, the quotient ring $\mathbb{Q}[X]/p(X)\mathbb{Q}[X]$ is a field and the element $x = X + p(X)\mathbb{Q}[X]$ is a root of $p(X)$. To Kronecker's way of thinking, there is an important philosophical difference between whether we assume the existence of the objects we are creating, or whether we create these objects from fundamental concepts such as polynomials and indeterminates. He would not accept the existence of $\sqrt{2}$, for example; to obtain such a number algebraically, he first constructed the field $K = \mathbb{Q}[X]/(X^2 - 2)\mathbb{Q}[X]$. If we let $x = X + (X^2 - 2)\mathbb{Q}[X]$, then $x^2 = 2$ in K and hence the element $x \in K$ has the required property.

Unlike Kronecker, who took a strict constructionist point of view in his mathematical work, Dedekind's approach to the creation of mathematical concepts was more general and abstract. Dedekind regarded mathematical objects from a more general point of view and used Cantor's ideas about infinite sets as the basis for many of his algebraic concepts, especially the view that an infinite totality can be regarded as a completed whole. As mentioned in Appendix B, this approach led him to his general definition of an ideal. In field theory, too, he used this more abstract approach. Most of

Dedekind's work on fields is contained in Supplement X to the second edition of Dirichlet's *Lectures on Algebraic Numbers*. Dedekind had edited Dirichlet's *Lectures* and included, as a supplement, many of his own ideas about algebraic number fields. Here we find, for example, Dedekind's definition of a field ([2]):

> Any system of real or complex numbers which satisfies the fundamental property of closure we will call a *number field* or simply a *field*.

But it was not until the 1894 edition of the *Lectures* that Dedekind sought to develop a general theory of fields. For example, he defined the concept of a subfield generated by a given set of elements as the intersection of all subfields containing the given elements. Note that in this definition there is no longer the constructionist point of view used by Kronecker; if S is a finite set of elements, Kronecker defined the subfield $\mathbb{Q}(S)$ by stating explicitly what elements lie in it, while Dedekind took a more existential approach by defining it as the intersection of all subfields of $\mathbb{C}$ that contain the elements in S. When S is finite, the two definitions are the same; but if S is infinite, Kronecker's definition fails. In this second supplement, Dedekind also defined the notion of field-isomorphism, and showed that the elements in a field that are fixed by a given isomorphism form a subfield containing $\mathbb{Q}$.

Another important concept that Dedekind introduced in the second supplement is the notion of a vector space. According to him, a collection of complex numbers $w_1, \ldots, w_n$ are "dependent" with respect to a field K if there are elements $a_1, \ldots, a_n \in K$, not all of which are zero, such that $a_1 w_1 + \cdots + a_n w_n = 0$; otherwise they are said to be "independent." If $w_1, \ldots, w_n$ are independent, then the collection Ω of all numbers dependent on $w_1, \ldots, w_n$ form a "family" having $w_1, \ldots, w_n$ as "basis" and is closed under addition and subtraction, closed under multiplication by elements in K, and any $n + 1$ elements are dependent. Today we refer to such a family as a vector space. If the family of elements Ω is itself a field, then he referred to n as the degree of the field over K, wrote $(\Omega, K) = n$, and observed that, in a tower of fields $k \subseteq K \subseteq K'$, $(K', K)(K, k) = (K', k)$. This result, that degree is multiplicative in a tower of fields, is fundamental in working with finite field extensions; ironically, Dedekind considered it to be so obvious that he did not bother proving it.

With Dedekind's work on field theory we begin to see the concept of a field drift away from the context of solvability of equations and into its more modern, abstract form, although fields were still primarily associated with hypercomplex number systems. It was not until the work of Steinitz in 1910 that the first abstract definition of a field was given, free from the context of complex numbers.

Ernst Steinitz (1871–1928) was a German mathematician who received his Ph.D. in mathematics from the University of Breslau in 1894, taught at the Technical College in Breslau until 1910, and then moved to the University of Kiel and remained there until his death in 1928. His most important contribution to field theory was his paper *Algebraische Theorie der Körper*, published in 1910. In it Steinitz gave what is essentially the modern definition of a field as a "system of elements with two operations (addition and multiplication) that satisfy associative and commutative laws (which are joined by the distributive law), the elements of which admit unlimited and unambiguous inversion up to division by zero" ([5]). He introduced the idea of prime field, separable elements, and the notion of the transcendence degree of an extension, which measures, basically, the number of algebraically independent indeterminates that generate the extension. But the main result of this paper is his proof of the existence and uniqueness of an algebraic closure for a given field: given a field k, there is an extension field K of k such that every polynomial over k splits completely into linear factors over K and hence has a complete set of roots in K. Moreover, he showed that the smallest such fields K are uniquely determined to within isomorphism. Steinitz used the axiom of choice to prove this result, although today it is usually proved by means of Zorn's lemma.

Up to this point in the evolution of field theory, the Galois theory of fields was still regarded as the basic mathematical theory for discussing the solvability of an algebraic equation by means of radicals. With Dedekind's and Steinitz's more general, abstract approach to fields, and especially with Dedekind's introduction of vector spaces, the stage was now set for a sweeping revision of Galois theory, a revision that would ultimately free it from the context of solvability of equations and turn it, instead, into an abstract relationship between the subfields of an extension field and the subgroups of its automorphism group. The individual who reformulated Galois theory from this abstract point of view was Emil Artin.

Emil Artin was born in Vienna, Austria, on March 3, 1898, but grew up in Reichenberg, Bohemia. Shortly after enrolling at the University of Vienna he was drafted into the army, served until the end of World War I, and then returned to the University, where he completed his Ph.D. in 1921. He taught at the University of Göttingen for a year, then at the University of Hamburg from 1923 to 1937. Artin, his wife, and two children then emigrated to the United States where he taught, successively, at the University of Notre Dame, Indiana University at Bloomington, and Princeton University. He returned to the University of Hamburg in 1958 and taught there until his death on December 20, 1962. Artin wrote many articles, monographs and textbooks that are classics of mathematical exposition because of their simple and lucid style; these include his *Galois Theory*, published in 1942, *Geometric Groups*, published in 1957, and *Class Field Theory*, coauthored with J. T. Tate in 1961. Although he made significant contributions to many areas of mathematics, most notably in algebraic number theory, perhaps the most lasting tribute to Emil Artin comes not from his mathematical work but from his style and approach to mathematics. His teaching style, in particular, and his ability to communicate abstract mathematical ideas clearly and simply is legendary. Artin's approach to doing mathematics and presenting mathematics had, and continues to have, a tremendous impact on generations of students.

There are few books in the mathematical literature that can rival the beauty and elegance of Artin's *Galois Theory*. Speaking of his fascination with Galois theory, Artin said in a lecture in 1950 ([2]):

> Since my mathematical youth I have been under the spell of the classical theory of Galois. This charm has forced me to return to it again and again, and to try to find new ways to prove its fundamental theorems.

During the 1930s, when Artin was beginning to reformulate Galois theory, Hans Zassenhaus, one of his colleagues at the University of Hamburg, recalls that ([6]):

> I was a witness how Artin gradually developed his best known simplification, his proof of the main theorem of Galois theory.
>
> The situation in the thirties was determined by the existence of an already well developed algebraic theory that was initiated by one of the most fiery spirits that ever invented mathematics, the spirit of E. Galois.
>
> But this state of affairs did not satisfy Artin. He took offense of the central role played by the theorem of the existence of a primitive element for finite separable extensions. This statement has no direct relation to the object of the theory which is to investigate the group of the equation, but it was needed at the time as a prerequisite for the proof of the main theorem.

Artin's approach to Galois theory was to reformulate the proofs using the language and techniques of linear algebra, in particular, systems of linear equations. This was especially true for those which, up to this time, had required the existence of a primitive element. But more importantly, he dissociated the theory from its classical origins in the solvability of algebraic equations and reformulated it as an abstract relationship between an extension of fields and its group of automorphisms. As we have already noted, Dedekind showed that a field-automorphism determines an intermediate subfield, namely its fixed field. Artin carried this idea through to its completion by showing that if H is any group of automorphisms of an extension of fields K/k, then K is a normal extension of the fixed field K^H and the automorphisms of K that leave K^H fixed are precisely the automorphisms in H. Let us note that up to this time the normality of an extension field meant that the extension was the same as all its conjugates; Artin replaced this with the equivalent but conceptually different definition of normality that we use today: K/k is normal if the ground field k is the fixed field of the automorphism group $G(K/k)$. With these results in place, and the fact that every intermediate subfield of a normal extension is in fact the fixed field of some group of automorphisms H, the main results of Galois theory could be stated and proved very simply.

By 1938 Artin had essentially completed his revision of Galois theory and published it as a set of lecture notes, *Foundations of Galois Theory*. Later, in 1942, it was published as a monograph titled simply *Galois Theory*, and it is this brief work of 82 pages that has inspired generations of students and which provides the modern interpretation of Galois theory. In it, Artin regards the solvability of algebraic equations, as well as the question of Euclidean constructibility, as applications of the Galois theory; the theory itself is a masterpiece of abstract mathematics, uncluttered by the specific details of either algebraic equations or Euclidean constructions. Gone are the arguments relying upon primitive elements and the calculations of his predecessors, replaced instead by abstract arguments dealing with vector spaces, linear dependence, systems of linear equations, and finite groups.

Let us close our discussion of the evolution of field theory and Galois theory with a quotation from Kiernan's extensive article on the subject ([3]):

> Artin seems to have come upon the scene at a time when the mathematical world was prepared for a new advance in theory. If the first phase of Galois Theory is computational, and the second is one of abstraction, then the third is one of generalization, even universalization.
>
> Artin's contribution involves seeing Galois Theory not as a separate entity, but as part of a larger picture. Indeed in much of his work, even apart from Galois Theory, Artin sought to break away from the confines of a particular problem and to treat it as part of a larger whole. In his development of Galois Theory it was seen that Artin could view the same set once as a field and another time as a group and still another time as a vector space, as the situation might demand. Indeed a parallel might be seen between his view of the fundamental theorem of Galois Theory, and the relation between a vector space and its dual space of linear functionals.
>
> Through the work of Artin, Galois Theory was being severed from all connections with its past. No longer was it viewed as having any necessary connection with the problem of solving equations algebraically. Galois Theory was a theory about the structure of fields and their automorphisms.

References and Additional Readings

1. Artin, Emil, *Galois Theory*, Notre Dame Mathematical Lectures, Number 2, Notre Dame, Ind., 1942.

2. Hadlock, Charles Robert, *Field Theory and Its Classical Problems*, Carus Mathematical Monographs, Number 19, The Mathematical Association of America, 1978.

3. Kiernan, B. Melvin, "The Development of Galois Theory from Lagrange to Artin," *Arch. History Exact Sci.* 8, 1971–72, 40–154.

4. Kleiner, Israel, "Thinking the Unthinkable: The Story of Complex Numbers (with a Moral)," *Mathematics Teacher*, October, 1988, 583–592.

5. Novy, Lubos, *Origins of modern algebra*, Noordhoff International Publishing, Leyden, 1973.

6. Schoeneberg, Bruno, "Artin, Emil," *Dic. Sci. Bio.*, Vol. I, 306–308, ed. Charles Coulston Gillispie, Charles Scribner's Sons, New York, 1970.

7. Schoeneberg, Bruno, "Steinitz, Ernst," *Dic. Sci. Bio.*, Vol. XIII, 23, ed. Charles Coulston Gillispie, Charles Scribner's Sons, New York, 1970.

8. Zassenhaus, Hans, "Emil Artin, His Life and His Work," *Notre Dame Journal of Formal Logic*, Vol. V(1), 1964, 1–9.

APPENDIX
D
Glossary of Basic Terms and Symbols

Algebraic element   If K/k is an extension of fields, an element $c \in K$ is *algebraic* over k if it is a root of a nonconstant polynomial over k; if not, c is *transcendental* over k. If c is algebraic over k, it is a root of a unique monic irreducible polynomial over k called its *irreducible polynomial*, denoted by $\mathrm{Irr}(c;k)$. The set $\mathrm{Alg}(K/k)$ of elements in K that are algebraic over k is an intermediate subfield of the extension K/k called the *algebraic closure* of k in K.

Algebraic extension   An extension of fields K/k is an *algebraic extension* if every element in K is algebraic over k. Every finite extension of fields is an algebraic extension. If K/k is an extension of fields and $c \in K$, the subfield $k(c)$ is an algebraic extension of k if and only if c is algebraic over k; in this case $k(c) \cong k[X]/\mathrm{Irr}(c;k)k[X]$, $[k(c):k] = n = \deg \mathrm{Irr}(c;k)$, and the elements $1, c, \ldots,$ c^{n-1} form a basis for $k(c)$ as a vector space over k.

Algebraically closed   A field K is *algebraically closed* if every nonconstant polynomial over K has a complete set of roots in the field, or equivalently, splits over the field. If k is any field, there is an extension field of k that is algebraically closed.

Automorphism   An *automorphism of a group* G is a 1–1 onto mapping $f : G \to G$ having the property that $f(xy) = f(x)f(y)$ for all $x, y \in G$. An *inner automorphism* of G is an automorphism of the form $f_g : G \to G$, where $f_g(x) = gxg^{-1}$ for all $x \in G$. The set $\mathrm{Aut}(G)$ of automorphisms of G is a group under function composition called the *automorphism group of G*; the inner automorphisms of G, denoted by $\mathrm{Inn}(G)$, form a normal subgroup $\mathrm{Inn}(G)$ of $\mathrm{Aut}(G)$. An *automorphism of a ring or field* R is a 1–1 onto mapping $f : R \to R$ such that $f(x + y) = f(x) + f(y)$ and $f(xy) = f(x)f(y)$ for all $x, y \in R$. If K/k is an extension of fields, a *k-automorphism of K* is any automorphism of K that fixes every element in k. The set $G(K/k)$ of k-automorphisms of K is a group under function composition called the *automorphism group of the extension K/k*. If H is any group of k-automorphisms of K/k, the set $K^H = \{x \in K \mid \sigma(x) = x \text{ for all } \sigma \in H\}$ of elements in K fixed by every automorphism in H is an intermediate subfield of the extension K/k called the *fixed field* of H; $[K^H:k] = [G:H]$.

Binary operation   A *binary operation* on a nonempty set X is a function that associates to each ordered pair (a, b) of elements in X a well-defined element $a * b \in X$. The operation is *associative* if $x * (y * z) = (x * y) * z$ for all $x, y, z \in X$; *commutative* if $x * y = y * x$ for all $x, y \in X$; has an *identity element* $e \in X$ if $x * e = e * x = x$ for all $x \in X$; and an element $x \in X$ has an *inverse* $y \in X$ if $x * y = y * x = e$. If Y is a nonempty subset of X, Y *inherits* the binary operation, or is *closed*, if $y * y' \in Y$ for all $y, y' \in Y$.

Cayley representation   The representation of a group G in which each element $x \in G$ is represented by the permutation $\pi_x : G \to G$, where $\pi_x(g) = xg$ for all $g \in G$. The correspondence $x \leftrightarrow \pi_x$ is a group-isomorphism between G and its representation by means of permutations; $\pi_x \pi_y = \pi_{xy}$ and $\pi_x^{-1} = \pi_{x^{-1}}$ for all $x, y \in G$. Also called the *left-regular representation* of G since π_x multiplies elements in G on the left by x.

Class equation   The equation $|G| = |Z(G)| + \Sigma^* [G : C(x)]$, where G is a finite group and Σ^* stands for the summation over one element x in each nontrivial conjugacy class of G.

Conjugation   Two elements x, y of a group G are *conjugate* if $y = gxg^{-1}$ for some $g \in G$. Notation: $x \sim y$. Conjugation of elements is an equivalence relation on G. The *conjugacy class of x* is the set $K(x) = \{gxg^{-1} | g \in G\}$ of elements conjugate to x; $|K(x)| = [G : C(x)]$, where $C(x)$ is the *centralizer* of x, that is, those elements that commute with x. Two subgroups H, K of G are *conjugate* if $K = gHg^{-1}$ for some element $g \in G$. Notation: $H \sim K$. Conjugation of subgroups is an equivalence relation on the subgroups of G. The *conjugacy class of H* is the set $K(H) = \{gHg^{-1} | g \in G\}$ of subgroups conjugate to H; $|K(H)| = [G : N(H)]$, where $N(H)$ is the *normalizer* of H, that is, all elements $x \in G$ such that $gHg^{-1} = H$. If K/k is an extension of fields, two elements c, $c' \in K$ are *conjugate* if there is a k-automorphism σ of K such that $c' = \sigma(c)$. In a normal extension of fields, two elements are conjugate if and only if they have the same irreducible polynomial over k; in this case the conjugates of c are the roots of its irreducible polynomial $\mathrm{Irr}(c; k)$.

Constructibility   A point P in the Euclidean plane is *constructible* from a set of points S if there is a finite sequence of points $P_1, \ldots, P_n$ such that $P_n = P$ and where each point P_{i+1} is the intersection of two lines, two circles, or a line and a circle each of which is constructible from points in the set $S \cup \{P_1, \ldots, P_i\}$; any such sequence of points is called a *Euclidean construction* for P. A complex number $z = a + bi$ is constructible from S if the point (a, b) is so constructible. The *constructibility criterion* states that a complex number z is constructible from a set of numbers S if and only if there is a tower of fields $\mathbb{Q}_S \subseteq K_1 \subseteq K_2 \subseteq \cdots \subseteq K_n$ such that $z \in K_n$ and $[K_{i+1} : K_i] \leq 2$ for $i = 1, \ldots, n - 1$.

Coset   If H is a subgroup of a group G, the *left cosets* of H in G are the sets $xH = \{xh \in G | h \in H\}$. Left cosets are either identical or disjoint; this partitioning of G into disjoint subsets is called the *left coset decomposition of G by H*. The set of left cosets of H in G is denoted by G/H; thus, $G/H = \{xH | x \in G\}$. The *index* of H in G, denoted by $[G : H]$, is the number of left cosets of H in G; $[G : H] = |G/H|$. *Lagrange's theorem* shows that $[G : H] = |G|/|H|$. The *right cosets* of H in G are the sets $Hx = \{hx \in G | h \in H\}$; right cosets partition G into disjoint subsets and are the same in number as the left cosets.

Cyclotomic extension   An extension of the field $\mathbb{Q}$ generated by a primitive nth root of unity. If $\mathbb{Q}(z_n)$ is the cyclotomic extension of $\mathbb{Q}$ generated by a primitive nth root of unity z_n, then $[\mathbb{Q}(z_n) : \mathbb{Q}] = \varphi(n)$, $\mathrm{Irr}(z_n : \mathbb{Q}) = \Phi_n(X)$, the nth *cyclotomic polynomial*, and the automorphism group $G(\mathbb{Q}(z_n)/\mathbb{Q}) \cong U(\mathbb{Z}_n)$, the multiplicative group of units of the ring $\mathbb{Z}_n$. The nth cyclotomic polynomial $\Phi_n(X)$ is the monic irreducible polynomial of degree $\varphi(n)$ over $\mathbb{Q}$, whose roots are the primitive nth roots of unity. Cyclotomic polynomials can be found recursively from the product formula $X^n - 1 = \Pi \Phi_d(X)$, where the product runs over all positive divisors d of n.

DeMoivre's theorem   If $z = r(\cos \theta + i \sin \theta)$ is the trigonometric representation of a complex number z, then $z^n = r^n(\cos n\theta + i \sin n\theta)$ for all integers n.

Division algorithm   If n and m are *integers*, with $m > 0$, there are unique integers q and r such that $n = qm + r$, where $0 \leq r < m$. If $f(X)$ and $g(X)$ are *polynomials* over a field k with $g(X) \neq 0$, there are unique polynomials $q(X)$ and $r(X)$ over k such that $f(X) = q(X)g(X) + r(X)$, where either $r(X) = 0$ or $\deg r(X) < \deg g(X)$.

Equivalence relation   A relation $\sim$ on a set X that is *reflexive*, meaning $x \sim x$ for all $x \in X$, *symmetric*, meaning that if $x \sim y$ for some $x, y \in X$ then $y \sim x$, and *transitive*, meaning that if $x \sim y$ and $y \sim z$ for some $x, y, z \in X$, then $x \sim z$. The *equivalence class* of an element $x \in X$ is the set $[x] = \{y \in X | y \sim x\}$ of elements that are equivalent to x. Equivalence classes are either identical or disjoint and hence partition X into disjoint subsets.

Euclidean domain   An integral domain R together with a *valuation function* φ that associates to every nonzero element $a \in R$ a nonnegative integer $\varphi(a)$ such that $\varphi(a) \leq \varphi(ab)$ for all nonzero $a, b \in R$, and for which there are elements $q, r \in R$ such that $b = qa + r$, where either $r = 0$

or $\varphi(r) < \varphi(a)$. Euclidean domains are principal ideal domains and unique factorization domains. An element $a \in R$ is a unit of R if and only if $\varphi(a) = \varphi(1)$.

Euclidean motion A function $f : V \to V$ mapping a Euclidean space V to itself such that f preserves the distance between all points in V; that is, $d(f(A), f(B)) = d(A, B)$ for all vectors $A, B \in V$. Euclidean motions, also called a *rigid motions*, include rotations, reflections, and translations—which are mappings of the form $T_A : V \to V$, where $T_A(X) = X + A$ for all $X \in V$, $A \in V$—and all compositions of these mappings. The set $E(V)$ of Euclidean motions of V is a group under function composition called the *Euclidean group* of V. The set $T(V)$ of translations is a normal subgroup of $E(V)$ called the *translation subgroup*; $E(V)/T(V) \cong O(V)$.

Euclidean space A finite-dimensional real vector space V together with an *inner product* $\langle\,,\,\rangle : V \times V \to V$ such that $\langle A, B \rangle = \langle B, A \rangle$, $\langle A + B, C \rangle = \langle A, C \rangle + \langle B, C \rangle$, $\langle cA, B \rangle = \langle A, cB \rangle + c \langle A, B \rangle$ and $\langle A, A \rangle \geq 0$ for all vectors $A, B, C \in V$ and all scalars $c \in \mathbb{R}$, with $\langle A, A \rangle = 0$ if and only if $A = 0$. The *norm* of a vector $A \in V$ is the scalar $\|A\| = \sqrt{\langle A, A \rangle}$. The *distance function* on V is the function $d : V \times V \to V$ defined by $d(A, B) = \|A - B\|$. Two vectors $A, B \in V$ are *orthogonal* if $\langle A, B \rangle = 0$. An *orthonormal basis* for V is any basis of unit vectors that are mutually orthogonal. *Euclidean n-space* $\mathbb{E}^n$ is the real vector space $\mathbb{R}^n$ together with the standard inner product defined by $\langle (a_1, \ldots, a_n), (b_1, \ldots, b_n) \rangle = a_1 b_1 + \cdots + a_n b_n$. The *standard orthonormal basis* for $\mathbb{E}^n$ is the set $\{e_1, \ldots, e_n\}$, where e_i is the n-tuple all of whose components are zero except the ith one which is 1. A mapping $\sigma : V \to V'$ from one Euclidean space to another is called an *isometry* if σ is a linear isomorphism that preserves the inner product on the spaces; that is, $\langle \sigma(A), \sigma(B) \rangle' = \langle A, B \rangle$ for all vectors $A, B \in V$. Every Euclidean space of dimension n is isometric to $\mathbb{E}^n$.

Euler phi function The function $\varphi : \{1, 2, \ldots\} \to \{1, 2, \ldots\}$, where $\varphi(n)$ is the number of positive integers less than n that are relatively prime to n for $n > 1$, and $\varphi(1) = 1$. Named after the Swiss mathematician Leonard Euler (1707–1783) who introduced it in 1763 to generalize Fermat's Little Theorem that $a^{p-1} \equiv 1$ (mod p) for positive integers a and primes p not dividing a, to $a^{\varphi(n)} \equiv 1$ (mod n) whenever a and n are relatively prime. $\varphi(n)$ is also the number of generators of a cyclic group of order n. The function φ has two important arith-

metic properties: first, it is multiplicative on relatively prime pairs, meaning that $\varphi(nm) = \varphi(n)\varphi(m)$ whenever n and m are relatively prime; and second, $\varphi(d_1) + \cdots + \varphi(d_s) = n$, where $d_1, \ldots, d_s$ are all positive divisors of n. If $n = p_1^{s_1} \cdots p_t^{s_t}$ is the prime factorization of n, $\varphi(n) = p_1^{s_1 - 1} \cdots p_t^{s_t - 1}(p_1 - 1) \cdots (p_t - 1)$.

Extension field A field K is an *extension field* of a field k if K contains k as a subfield. Notation: K/k. In this case K is a vector space over k whose dimension is called the *degree* of the extension K/k and denoted by $[K : k]$. K is a *finite extension* of k if $[K : k]$ is finite. An *intermediate subfield* of the extension K/k is any subfield k' such that $k \subseteq k' \subseteq K$; in this case *Dedekind's theorem* states that degree is multiplicative, that is, $[K : k] = [K : k'][k' : k]$. An extension K/k is an *algebraic extension* if every element $c \in K$ is algebraic over k, that is, c is a root of some polynomial over k; a *normal extension* if the only elements in K fixed by every k-automorphism are the elements in k; a *simple extension* if $K = k(c)$ for some element $c \in K$, in which case any such element c is called a *primitive element* for K/k; and a *radical extension* if K is generated by an nth root of some element in k, that is, $K = k(c)$, where $c^n \in k$ for some integer n.

Field A commutative ring with identity in which every nonzero element is a unit. First introduced into the mathematical vocabulary by the German mathematician Ernst Steinitz (1871–1928) in 1910. The *characteristic* of a field K, denoted by char(K), is the least number of times the identity element 1 added to itself gives zero; if no such integer exists, char(K) = 0. Char(K) is either 0 or a prime p. The *prime subfield* of K is the subfield generated by the identity 1; it is isomorphic to $\mathbb{Q}$ if char(K) = 0 and $\mathbb{F}_p$ if char(K) = p. A *finite field* is a field that contains a finite number of elements. Every finite field has nonzero prime characteristic p and contains p^n elements, where n is the degree of the field over its prime subfield. There is a finite field containing p^n elements for any prime power p^n, namely a splitting field for the polynomial $X^{p^n} - X$, and any two such fields are isomorphic. The multiplicative group of nonzero elements of a finite field is a cyclic group. The *Frobenius automorphism* of a finite field K is the mapping $\sigma_p : K \to K$ defined by $\sigma_p(x) = x^p$ for all $x \in K$.

Fixed point property of p-groups If G is a p-group acting as a permutation group on a finite set X and p does not divide $|X|$, then X contains some element fixed by every permutation in G.

Function   A rule $f:X \to Y$ that associates with each element $x \in X$ a uniquely determined element $f(x) \in Y$. X is called the *domain* of f, Y the *range*, and $f(x)$ the *image of x* under f. We write $x \mapsto f(x)$ and say that x maps to $f(x)$ under f. The *image of f* is the set Im $f = \{f(x) \in Y \,|\, x \in X\}$ of images of all elements in X. A function f is *1–1* if it maps distinct elements in X to distinct elements in Y—that is, if $f(x) = f(x')$ for some x, $x' \in X$, then $x = x'$— *onto* if every element in Y is the image of some element in X—that is, for every element $y \in Y$, there is some element $x \in X$ such that $y = f(x)$— and a *1–1 correspondence* if it is both 1–1 and onto. A function $g:Y \to X$ is an *inverse* for f if $g(f(x)) = x$ for all $x \in X$ and $f(g(y)) = y$ for all $y \in Y$; f has an inverse if and only if f is a 1–1 correspondence, in which case the inverse is unique. The identity function on a set X is the function $1_X:X \to X$, where $1_X(x) = x$ for all $x \in X$.

Fundamental theorem of algebra   Every nonconstant polynomial over the field $\mathbb{C}$ of complex numbers has a complex root. It follows that every such polynomial splits over $\mathbb{C}$, that is, can be written as a product of linear factors.

Fundamental theorem of arithmetic   Every positive integer is the product of a finite number of uniquely determined prime numbers.

Galois correspondence   The 1–1 order reversing correspondence between subgroups of the automorphism group $G(K/k)$ of a normal extension of fields K/k and the intermediate subfields of the extension.

$$
\begin{array}{ccc}
G(K/k) & & K \\
| \,\} & \diagdown \!\!\!\!\! \diagup & \{\, | \\
G(K/k') = H & \longleftrightarrow & K^H = k' \\
| \,\} & \diagup \!\!\!\!\! \diagdown & \{\, | \\
(1) & & k
\end{array}
$$

Under this correspondence, a subgroup H corresponds to its fixed field K^H; an intermediate subfield k' corresponds to the subgroup $G(K/k')$; $|H| = [K:K^H]$; $[k':k] = [G(K/k):G(K/k')]$; and $G(K/k') \lhd G(K/k)$ if and only if k'/k is a normal extension of fields, in which case $G(K/k)/G(K/k') \cong G(k'/k)$.

Galois group   The automorphism group $G(K/k)$ of k-automorphisms of a normal extension of fields K/k. If $f(x)$ is a polynomial over a field k and K is a splitting field for $f(x)$ over k, the Galois group $G(K/k)$ is also called the *Galois group of f(x)*.

Greatest common divisor   The greatest common divisor of two nonzero *integers* is the largest positive integer that divides both integers; it is unique. Notation: $\gcd(n, m)$. The greatest common divisor of two nonzero *polynomials over a field* is the unique monic polynomial of largest degree that divides both polynomials; notation: $\gcd(f(X)$, $g(X))$. In a unique factorization domain any two nonzero elements a and b have a greatest common divisor—which is any element d that divides both a and b and which is divisible by any element that divides both a and b— although it is unique only to within associates. Two elements of a unique factorization domain are *relatively prime* if their greatest common divisor is 1.

Group   A set together with a binary operation that is associative, has an identity element, and for which every element has an inverse. The concept was first introduced by the French mathematician Evariste Galois (1811–1832) in 1830, and later, in 1858, given its modern, abstract definition by the British mathematician Arthur Cayley (1821–1895). A group G is *abelian* (in honor of the Norwegian mathematician Neils Henrik Abel (1802–1829)), or *commutative*, if $xy = yx$ for all elements $x, y \in G$, and *finite* if it contains a finite number of elements, in which case the number of elements in G is denoted by $|G|$ and called the *order* of G. A finite group of prime power order p^n is called a *p-group*. A *subgroup* of G is a non-empty subset H closed under products and inverses. Notation: $H \leq G$. A *normal subgroup* is a subgroup N that contains all products of the form gng^{-1}, where $g \in G$, $n \in N$. Notation: $N \lhd G$. A group whose only normal subgroups are the identity subgroup and the group itself is called a *simple group*. The *order* of an element $x \in G$, denoted by $|x|$, is the smallest positive integer n such that $x^n = 1$, the identity element of G. The *cyclic subgroup* generated by x is the set $(x) = \{1, x^{\pm 1}, x^{\pm 2}, \ldots\}$ of integral powers of x; if $|x| = n$, then $(x) = \{1, x, x^2, \ldots, x^{n-1}\}$ and hence $|(x)| = n$. G is *cyclic* with generator x if $G = (x)$ for some $x \in G$. *Lagrange's theorem* states that if G is finite and $H \leq G$, then $|G| = |H|[G:H]$; it shows, in particular, that the order of every subgroup and every element divides $|G|$. *Cauchy's theorem* states that an abelian group of order n contains an element of order p for every prime p dividing n (cf. also the Sylow theorems for existence and properties of p-subgroups in a finite group). The *center* of G is the subgroup $Z(G) = \{x \in G \,|\, xy = yx$

for all $y \in G$} of elements in G that commute with every element in G.

Group-homomorphism A mapping $f : G \to G'$ mapping one group to another such that $f(xy) = f(x)f(y)$ for all elements $x, y \in G$. The *kernel* of f is the set Ker $f = \{x \in G | f(x) = 1\}$ of elements in G that map to the identity element under f; it is a normal subgroup of G. The *image* of f is the set Im $f = \{f(x) \in G' | x \in G\}$ of images of elements in G under f; it is a subgroup of G'. A group-homomorphism f is 1–1 if and only if Ker $f = \{1\}$. If f is both 1–1 and onto it is called a *group-isomorphism*; in this case G and G' are isomorphic and we write $G \cong G'$. The *fundamental theorem of group-homomorphism* states that the quotient group $G/\text{Ker } f \cong \text{Im } f$.

Ideal A non-empty subset of a ring R that is closed under addition, additive inverses, and multiplication by arbitrary elements in R; that is, $a + b$, $-a$, ax, $xa \in I$ whenever a, $b \in I$ and $x \in R$. The *trivial ideals* are $\{0\}$ and R. A *principal ideal* is an ideal generated by a single element. If R is a commutative ring with identity, a *prime ideal* of R is an ideal $P \neq R$ with the property that if $ab \in P$ for some $a, b \in R$, then $a \in P$ or $b \in P$; a *maximal ideal* of R is an ideal $M \neq R$ with the property that if $M \subseteq I \subseteq R$ for some ideal I, then either $I = M$ or $I = R$. P is a prime ideal of R if and only if R/P is an integral domain; M is maximal if and only if R/M is a field. Every maximal ideal is prime.

Integral domain A commutative ring with identity having the property that if $ab = 0$ for some elements $a, b \in R$, then $a = 0$ or $b = 0$.

Kronecker's theorem Every nonconstant polynomial of degree n over a field k has a root in some extension field of degree at most n over k.

Linear transformation A function $L : V \to V'$ mapping one vector space to another such that $L(A + B) = L(A) + L(B)$ and $L(cA) = cL(A)$ for all vectors $A, B \in V$ and all scalars c. The *kernel* of L is the set Ker $L = \{A \in V | L(A) = 0\}$ of vectors in V that map to the zero vector under L; it is a subspace of V. The *image* of L is the set Im $L = \{L(A) \in V' | A \in V\}$ of images of vectors in V; it is a subspace of V'. The *rank-nullity theorem* states that dim Ker L + dim Im L = dim V, and gives an important relationship between the dimension of these three subspaces. A linear transformation is 1–1 if and only if its

kernel contains only the zero vector. A linear transformation is *nonsingular* if it is 1–1, and a *linear isomorphism* if it is both 1–1 and onto. The set $GL(V)$ of linear isomorphisms of V is a group under function composition called the *general linear group* of V.

Matrix A rectangular array of numbers. The concept was first introduced by the British mathematician Arthur Cayley (1821–1895) in 1854 as a way of simplifying the discussion of linear systems of equations. A square matrix is *nonsingular* if and only if its determinant is nonzero; in this case the matrix has an *inverse*, or is *invertible*. The set $\text{Mat}_n(\mathbb{R})$ of $n \times n$ real matrices is a group under addition of matrices and a ring under addition and multiplication of matrices. The set $\text{Mat}_n^*(\mathbb{R})$ of $n \times n$ invertible real matrices is a group under multiplication of matrices.

Normal extension A finite extension of fields K/k for which the only elements in K fixed by every k-automorphism are the elements in the ground field k. An extension K/k with automorphism group G is normal if and only if either $|G| = [K:k]$, or the irreducible polynomial over k of every element in K has only simple roots, or if K is the splitting field of a *separable* polynomial over k, that is, a polynomial all of whose roots are simple. If K/k is a normal extension, the irreducible polynomial over k of any element $c \in K$ is the product of the linear factors $X - c_i$, where c_i runs over the distinct conjugates of c. Every normal extension of fields has a primitive element. The automorphism group of a normal extension of fields is also called the *Galois group* of the extension.

Orthogonal transformation A linear isomorphism σ of a Euclidean space V to itself that preserves the inner product on the space, that is, $\langle \sigma(A), \sigma(B) \rangle = \langle A, B \rangle$ for all vectors $A, B \in V$. A linear isomorphism is an orthogonal transformation if and only if it preserves the distance between points, the length of vectors, or the orthogonality of vectors. The determinant of an orthogonal transformation is ± 1; those of determinant $+1$ are called *rotations*, those of determinant -1, *reflections*. Reflections are further classified as either *symmetries* or *nonsymmetries* depending upon whether or not they reflect the space through a hyperplane. The set $O(V)$ of orthogonal transformations of V is a group under function composition called the *orthogonal group* of V. The set $O^+(V)$ of rotations is a normal subgroup of index 2 called the *rotation subgroup*. The matrix M of an orthogonal transformation relative to an orthonormal basis is an *orthogonal* *matrix*, meaning that $M^{-1} = M^T$, where M^T stands for the transpose of M.

Permutation A 1–1 correspondence of a set with itself. In matrix form,

$$\sigma = \begin{pmatrix} 1 & 2 & \cdots & n \\ i_1 & i_2 & \cdots & i_n \end{pmatrix}$$

is the permutation of the set $\{1, \ldots, n\}$ for which $\sigma(1) = i_1$, $\sigma(2) = i_2, \ldots, \sigma(n) = i_n$. The *orbits* of σ are the sets $O_\sigma(j) = \{j, \sigma(j), \sigma^2(j), \ldots\}$, $1 \le j \le n$. A *cycle* of length s is a permutation $(j_1 j_2 \cdots j_s)$ having one non-trivial orbit of length s, where $\sigma(j_1) = j_2$, $\sigma(j_2) = j_3, \ldots, \sigma(j_s) = j_1$, and where $\sigma(j) = j$ for all $j \notin \{j_1 \cdots j_s\}$. Every permutation is the product of its cycles: $\sigma = (j_{11} \cdots j_{1a}) \cdots (j_{s1} \cdots j_{sz})$. A *transposition* is a cycle of length two, and every permutation can be written as a product of transpositions; the transpositions are not unique, although the number of transpositions is always even, in which case the permutation is *even*, or always odd, in which case it is an *odd* permutation. The set Sym(n) of permutations of $\{1, \ldots, n\}$ is a group under function composition called the *symmetric group* on n letters. The subset Alt(n) of even permutations is a normal subgroup of index 2 of Sym(n) called the *alternating subgroup*.

Permutation representation A group-homomorphism $f: G \to \text{Sym}(X)$ that assigns to each element σ in a group G a permutation f_σ of a set X. G acts as a group of permutations on X by means of the permutations f_σ; f_σ permutes an element $x \in X$ to $f_\sigma(x)$. The representation is *faithful* if its kernel contains only the identity element; in this case distinct group elements define distinct permutations of X. If X is finite, the number of elements in X is the *degree* of the representation. The *orbit* of an element $x \in X$ is the set $O(x) = \{f_\sigma(x) | \sigma \in G\}$ of images of x under all permutations by elements in G. The representation is *transitive* if $O(x) = X$ for some element $x \in X$; this is the case if and only if, for any pair of elements $x, y \in X$, $y = f_\sigma(x)$ for some $\sigma \in G$. G acts transitively on every orbit in X; this is called the *transitive constituent* of f on the orbit. The *stabilizer* of an element $x \in X$ is the subgroup $G_x = \{\sigma \in G | f_\sigma(x) = x\}$ of elements in G that permute x to itself; $|O(x)| = [G:G_x]$.

Polynomial An expression of the form $a_0 + a_1 X + \cdots + a_n X^n$. The elements $a_0, a_1, \ldots, a_n$ are the *coefficients* of the polynomial and lie in a ring or field; a_0 is the *constant term*, a_n the *lead term*. If $a_n = 1$, the polynomial is

monic. The *degree* of the polynomial, denoted by deg, is the largest exponent n for which $a_n \ne 0$. The polynomial is *linear* if $n = 1$, *quadratic* if $n = 2$, *cubic* if $n = 3$, and *quartic* if $n = 4$. The set $R[X]$ of polynomials with coefficients in a ring R is a ring under addition and multiplication of polynomials called the *ring of polynomials in X over R*. $R[X]$ is an integral domain if R is an integral domain, and a unique factorization domain if R is a unique factorization domain. A polynomial over a field is *reducible* if it can be written as the product of two polynomials of smaller degree; otherwise it is *irreducible* over the field. A polynomial over a unique factorization domain is *reducible* if it has a proper divisor, which need not have smaller degree, and *irreducible* otherwise. Every nonzero polynomial over a unique factorization domain, including fields, can be factored into a product of irreducible polynomials which are unique to within order and constants or associates. *Eisenstein's irreducibility criterion* states that if $f(X)$ is a polynomial over a unique factorization domain R all of whose coefficients are divisible by some irreducible element whose square does not divide the constant term, then $f(X)$ is irreducible over the quotient field $QF(R)$. A *symmetric polynomial* is a polynomial in any finite number of variables over $\mathbb{Q}$ that remains unchanged by any permutation of its variables. Every symmetric polynomial can be expressed as a polynomial in the elementary symmetric polynomials.

Primitive element theorem If k is a field of characteristic zero, every finite extension K of k has a primitive element; that is, $K = k(c)$ for some element $c \in K$. More generally *Steinitz's theorem* states that a finite extension of fields, regardless of the characteristic, has a primitive element if and only if it has a finite number of intermediate subfields. Every normal extension of fields has a primitive element.

Principal ideal domain An integral domain in which every ideal is *principal*, that is, generated by a single element. Principal ideal domains are unique factorization domains; their nonzero prime ideals are generated by the irreducible elements of the ring and are the maximal ideals of the ring.

Principle of mathematical induction If a set S of natural numbers contains zero and contains the number $n + 1$ whenever it contains n, then S is the set of all natural numbers.

Principle of strong induction If a set S of natural numbers contains zero and contains a number n whenever it contains all numbers less than n, then S is the set of all natural numbers.

Real n-space The n-dimensional vector space $\mathbb{R}^n$ consisting of all ordered n-tuples $(a_1, \ldots, a_n)$ of real numbers $a_1, \ldots, a_n$. If $A = (a_1, \ldots, a_n)$ and $B = (b_1, \ldots, b_n)$ are vectors, *vector addition* and *scalar multiplication* are defined by $A + B = (a_1 + b_1, \ldots, a_n + b_n)$ and $cA = (ca_1, \ldots, ca_n)$, $c \in \mathbb{R}$. The *standard basis* for $\mathbb{R}^n$ is the set $\{e_1, \ldots, e_n\}$, where e_i is the n-tuple all of whose entries are zero except the one in position i which is 1.

Ring A set together with two binary operations called addition and multiplication—the sum and product of two elements x and y being written $x + y$ and xy, respectively—that is an abelian group under addition, and in which multiplication is associative and distributes over addition. The term was first used in its modern abstract form by Abraham Fraenkel in 1914. A *commutative ring* is a ring in which multiplication is commutative. A *ring with identity* is a ring having a multiplicative identity, denoted by 1. A *subring* is a nonempty subset of a ring that is closed under addition, additive inverses, and multiplication; for commutative rings with identity, subrings must also contain the identity element. In a ring with identity, an element having a multiplicative inverse is called a *unit*; if x is a unit, it has a unique multiplicative inverse denoted by x^{-1}. If R is a ring with identity, the set $U(R)$ of units of R is a group under multiplication called the *group of units* of R.

Ring-homomorphism A function $f : R \to S$ mapping one ring to another such that $f(x + y) = f(x) + f(y)$ and $f(xy) = f(x)f(y)$ for all $x, y \in R$. The *kernel* of f is the set $\operatorname{Ker} f = \{x \in R \mid f(x) = 0\}$ of elements in R that f maps to the zero element of S; it is an ideal of R. The *image* of f is the set $\operatorname{Im} f = \{f(x) \in S \mid x \in R\}$ of images of elements in R under f; it is a subring of S. If f is both 1–1 and onto it is called a *ring-isomorphism*; in this case R and S are *isomorphic* and we write $R \cong S$. The *fundamental theorem of ring-homomorphism* states that $R/\operatorname{Ker} f \cong \operatorname{Im} f$.

Root of a polynomial A *root* of a polynomial $f(X)$ over a field k is any element $c \in k$ such that $f(c) = 0$. *Kronecker's theorem* states that every nonconstant polynomial of degree n over a field k has a root in some extension field of degree at most n over k. The *multiplicity* of

a root c is the largest positive integer s such that $(X - c)^s$ divides $f(X)$ in a splitting field for $f(X)$ over k; c is a *simple root* if its multiplicity is 1, otherwise it is a *multiple root*. A root c is a multiple root of $f(X)$ if and only if c is a root of the *formal derivative* $f'(X)$. Irreducible polynomials over a field of characteristic zero cannot have multiple roots.

Root of unity A complex number z such that $z^n = 1$ for some positive integer n; in this case z is called an *nth root of unity*. The nth roots of unity have the form $z_n^k = \cos(2\pi/n)k + i\sin(2\pi/n)k$, $k = 1, \ldots, n$, and form a cyclic group of order n under multiplication of complex numbers. The generators of this group are called *primitive nth roots of unity* and have the form z_n^k, where k and n are relatively prime. There are $\varphi(n)$ primitive nth roots of unity, where φ is the Euler phi function.

Set Any well-defined collection of objects, which are called the *elements* of the set. The concept was first introduced by the nineteenth century German mathematician George Cantor (1845–1918). If S is a set, the *power set* of S is the set $\mathscr{P}(S)$ of all subsets of S. If A and B are subsets of S, the *union* of A and B is the set $A \cup B$ of elements in either A or B; the *intersection* is the set $A \cap B$ of elements in both A and B; the *complement* of A in S is the set $S - A$ of elements in S not in A; and the *Cartesian product* of A and B is the set $A \times B = \{(a, b) \mid a \in A, b \in B\}$ of ordered pairs whose first component is in A and second component is in B. Two sets are *disjoint* if they have no elements in common. The number of elements in a set S is called its *cardinality* and is denoted by $|S|$. Sets S and T have the same cardinality if and only if there is a 1–1 correspondence $f : S \to T$. A set S is *countable* if it is either finite or in 1–1 correspondence with the set of natural numbers; if S is finite and contains n elements, $|S| = n$; if S is infinite and countable, $|S| = \aleph_0$, aleph-zero. The cardinality of the set $\mathbb{R}$ of real numbers is called the *cardinality of the continuum*.

Solvability by radicals A polynomial $f(X)$ over a field k is *solvable by radicals* if there is a tower of fields $k = k_1 \subseteq k_2 \subseteq \cdots \subseteq k_n$ such that k_{i+1} is a radical extension of k_i for each i and $f(X)$ splits over k_n; in this case the roots of $f(X)$ can be obtained by adding, subtracting, multiplying, dividing and extracting roots of elements in the coefficient field k. The *solvability criterion* states that a polynomial over $\mathbb{Q}$ is solvable by radicals if and only if its Galois group is solvable. Combining this result with the fact that the Galois group of the general polynomial of degree n over $\mathbb{Q}$ is the symmetric group Sym(n), and the fact that Sym(n) is a solvable group if and only if $n \le 4$, it follows that the general algebraic equation of degree n is solvable by radicals if and only if $n \le 4$.

Solvable group A finite group G is *solvable* if it contains a chain of subgroups $(1) = H_1 \lhd H_2 \lhd \cdots \lhd H_n = G$ such that each quotient group H_{i+1}/H_i is abelian. The term derives from the fact that, under the Galois correspondence, a polynomial is solvable by radicals if and only if its Galois group contains such a chain of subgroups.

Splitting field A polynomial $f(X)$ *splits* over a field k if it factors into a product of linear factors over k. A *splitting field* for $f(X)$ over k is any extension field of k over which $f(X)$ splits but does not split over any intermediate subfield. Every polynomial over a field k has a splitting field, which is generated by k and the roots of the polynomial, and is unique to within isomorphism.

Symmetry A mapping of a Euclidean space through a hyperplane. The symmetry through the hyperplane $\langle A \rangle^\perp$ orthogonal to a nonzero vector A is denoted by S_A. If A is a unit vector, $S_A(X) = X - 2\langle X, A \rangle A$ for every vector X. Every symmetry is a reflection, although not all reflections are symmetries.

The Sylow theorems Three theorems named in honor of the nineteenth century Norwegian mathematician Ludwig Sylow (1832–1918) dealing with the existence and properties of p-subgroups of a finite group. The *first Sylow theorem* states that if G is a finite group and p^n is the largest power of a prime p that divides $|G|$, then G contains a subgroup of order p^n. Any such subgroup is called a *Sylow p-subgroup* of G. The *second Sylow theorem* states that all Sylow p-subgroups of G are conjugate and that every p-subgroup of G is contained in some Sylow p-subgroup. The *third Sylow theorem* states that the number of Sylow p-subgroups of G is congruent to 1 modulo p.

Unique factorization domain A integral domain in which every nonzero element other than a unit may be written as the product of a finite number of irreducible elements which are uniquely determined to within order and associates. The *associates* of an element x are all elements of the form ux, where u is any unit of the ring. A divisor of x other than an associate of x is called a *proper divisor*.

An element that has no proper divisors is said to be *irreducible*. In a unique factorization domain, every pair of nonzero elements has a greatest common divisor, although it is unique only to within associates.

Well-ordering principle Every nonempty set of natural numbers contains a least number.

$\mathbb{N}$	set of natural numbers, 1
$\mathbb{Z}$	set of integers, 1
$\mathbb{Q}$	set of rational numbers, 1
$\mathbb{R}$	set of real numbers, 1
$\mathbb{C}$	set of complex numbers, 1
$x \in S$	x is an element of S, 1
$x \notin S$	x is not an element of S, 1
$\varnothing$	the null set, 1
$A \subseteq B$	A is a subset of B, 2
$A \subsetneqq B$	A is a proper subset of B, 2
$\mathscr{P}(A)$	power set of A, 2
$A \cup B$	union of sets A and B, 2
$A \cap B$	intersection of sets A and B, 2
$S - A$	complement of A in S, 2
$|A|$	number of elements in A, 3
$A \times B$	Cartesian product of A and B, 3
$\bigcup_{\alpha \in I} S_\alpha$	union of sets S_α, 3, 4
$\bigcap_{\alpha \in I} S_\alpha$	intersection of sets S_α, 3, 4
$A \triangle B$	symmetric difference of sets A and B, 5
$f: X \to Y$	f is a function from X to Y, 5
dom f	domain of the function f, 5
1_X	identity function on the set X, 8
$x \mapsto f(x)$	x maps to $f(x)$, 5
Imf	image of f, 5
xRy	x is related to y in relation R, 9
$[x]$	equivalence class of x, 10
$X/\sim$	set of equivalence classes of $\sim$, 10
$a \mid b,\ a \nmid b$	integer a divides (does not divide) b, 16
$\mathbb{R}^n$	real n-dimensional space, 18
$F(X)$	set of functions from X to X, 33
$L(V)$	set of linear transformations on a vector space V, 33
$\text{Mat}_2(\mathbb{R})$	set of 2×2 matrices with real entries, 34
$\text{Mat}_n(\mathbb{R})$	set of $n \times n$ matrices with real entries, 277
$\text{Mat}_n^*(\mathbb{R})$	set of $n \times n$ invertible matrices with real entries, 47
$\|z\|$	norm of complex number z, 44

T	group of the triangle, 45
$GL(V)$	general linear group of vector space V, 48
$\lvert x \rvert$	order of a group element X, 51
D_4	group of the square, 53
$H \leq G$	H is a subgroup of G, 55
(X)	subgroup generated by X, 60
$Z(G)$	center of group G, 63
$SL_2(\mathbb{R})$	2-dimensional special linear group over $\mathbb{R}$, 63
$s \equiv t \pmod{n}$	integer s is congruent to integer t modulo n, 66
$\mathbb{Z}_n$	additive group of integers modulo n, 67; ring of integers modulo n, 276
$\mathrm{Rot}_n(\mathbb{R}^2)$	group of rotations of an n-gon, 70
ρ_n	rotation of the plane through an angle of $2\pi/n$ radians, 70
$\gcd(n, m)$	greatest common divisor of n and m, 71
$\mathrm{lcm}(n, m)$	least common multiple of n and m, 78
φ	Euler phi-function, 71; valuation on Euclidean domain, 368
$\cong$	is isomorphic to, 79
$\mathrm{Sym}(X)$	symmetric group on the set X, 90
$\mathrm{Sym}(n)$	symmetric group on n letters, 90
$O_\sigma(i)$	σ-orbit 94,
$(i_1\, i_2 \cdots i_s)$	s-cycle, 95
$\mathrm{Perm}(n)$,	set of $n \times n$ permutation matrices, 104
π_x	left multiplication by x, 105
G_L	Cayley representation of group G, 106
$x \equiv y \pmod{H}$	group element x is congruent to y modulo subgroup H, 111
xH	left coset of subgroup H determined by x, 111
Hx	right coset of H determined by x, 121
$[G:H]$	index of H in G, 111
$U(\mathbb{Z}_n)$	generators of $\mathbb{Z}_n$ as a cyclic group, 118; units of $\mathbb{Z}_n$ as a ring, 284
$x \equiv_r \pmod{H}$	group element x is right congruent to y modulo H, 121
G/H	quotient group of G modulo H, 123
$H \lhd G$	H is a normal subgroup of G, 122
$\mathrm{Ker}\, f$	kernel of f, 137
$T_{(a,b)}$	affine transformation, 148
$\mathrm{Aff}(\mathbb{R})$	set of affine transformations of the real line, 148
$\mathrm{Aut}(G)$	set of automorphisms of G, 150
$\mathrm{Inn}(G)$	set of inner automorphisms of G, 152
reg_H	left regular representation of G on G/H, 163
$O(x)$	orbit of x under a permutation representation, 165
G_x	stabilizer of x, 166
$F(\sigma)$	fixed-point set of σ, 175
$\mathrm{Color}(X;C)$	set of colorings of X by C, 177
$K(x)$	conjugacy class of an element x, 185
$C(x)$	centralizer of an element x, 186
$1^{c_1}\, 2^{c_2} \cdots n^{c_n}$	type of a permutation, 190
$K(H)$	conjugacy class of subgroup H, 198
$N(H)$	normalizer of subgroup H, 199
X^G	fixed-point set of G acting on X, 212
$\mathrm{Syl}_p(G)$	set of Sylow p-subgroups of G, 217
$\langle A, B \rangle$	inner product of vectors A and B, 226
$A \perp B$	A is orthogonal to B, 226
$U^\perp$	orthogonal complement of U, 226
$\mathbb{E}^n$	Euclidean n-space, 227
$\lVert A \rVert$	length of vector A, 226
$O(V)$	orthogonal group of V, 238
$O^+(V)$	rotation subgroup of $O(V)$, 238
S_A	symmetry through hyperplane orthogonal to vector A, 234
T_A	translation by vector A, 248
$E(V)$	Euclidean group of V, 250
$T(V)$	group of translations of V, 250
$\mathbb{E}(n)$	Euclidean group of Euclidean n-space, 250
G_S	symmetry group of S, 259
Tet	group of the tetrahedron, 260
$\mathrm{Sim}(\mathbb{E}^n)$	similarity transformations of $\mathbb{E}^n$, 269
$\mathbb{H}$	ring of quaternions, 278
$U(R)$	group of units of ring R, 283
$\mathrm{Hom}(G)$	endomorphism ring of an abelian group G, 303
$\mathbb{F}_p$	field of integers modulo p, 311
$QF(R)$	quotient field of an integral domain R, 316
$\mathbb{Z}[i]$	ring of Gaussian integers, 275
$\deg f(X)$	degree of a polynomial $f(X)$, 337
$R[X]$	ring of polynomials in X over R, 337
$R[c]$	ring of polynomials in c over R, 341
$R[X_1, \ldots, X_n]$	ring of polynomials in $X_1, \ldots, X_n$ over R, 342
$\gcd(f(X), g(X))$	greatest common divisor of $f(X)$ and $g(X)$, 346
$\gcd(a_1, \ldots, a_n)$	greatest common divisor of $a_1, \ldots, a_n$, 364

$PF(R)$	ring of polynomial functions on R, 357	$\sigma^*_{c,c'}$	field-isomorphism mapping c to c', 447
$R[[X]]$	ring of formal power series in X over R, 358	$G(K/k)$	group of k-automorphisms of K, 464
$C(f)$	content of the polynomial $f(X)$, 374	K^H	fixed field of H, 465
$k(X)$	field of rational functions in X over k, 388	$\Phi_n(X)$	nth cyclotomic polynomial over $\mathbb{Q}$, 486
$\dim_k V$	dimension of vector space V over a field k, 400	$Tr_{K/k}$	trace function of normal extension K/k, 493
$[K:k]$	degree of the field extension K/k, 401	$N_{K/k}$	norm function of a normal extension K/k, 493
$k(c)$	field of rational expressions in c over k, 409	$\Delta(A_1, \ldots, A_n)$	discriminant of a normal extension relative to basis $A_1, \ldots, A_n$, 520
$\text{Irr}(c;k)$	irreducible polynomial of c over k, 411	$\text{Const}_S(\mathbb{C})$	set of constructible complex numbers, 543
$k(S)$	subfield generated by k and S, 417		
$\text{Alg}(K/k)$	algebraic closure of k in K, 428		

Index

A

Abel, Niels Henrik (1802–1829), 523, 556
abelian extension, 576
abelian group, 41
   nth root of an element of, 121
action of a group, 160
adjunction of elements
   to a ring, 341
   to a field, 409
affine transformation(s), 148, 258
   group of, 148, 259
algebra of sets, 287, 304
algebraic
   closure, 428, 442
   extension, 421
algebraic element, 410
   irreducible polynomial of, 411
algebraically closed field, 442
alternating subgroup, 103
   simplicity of, 133
Artin, Emil (1898–1962), 471, 574, 582
associate, 361
automorphism
   canonical representation of, 172
   fixed-point free, 156
   fixed-point of, 156
   Frobenius, 469
   group, 150
   inner, 152
   of a field extension, 463
   of a field, 387
   of a finite field, 469, 477
   of a group, 149
   of real numbers, 476
   outer, 154

B

basis of a vector space, 400
   orthonormal, 226
   standard, 19
Basis theorem, Hilbert, 576
binary operation, 32
   associative, 34
   closure of a subset, 38
   commutative, 34
   general associativity, 37
   general commutativity, 37
   identity element for, 35
   inverse elements, 35
Boole, George (1815–1864), 564
Boolean ring, 280, 287
Brauer, Richard (1901–1977), 560

Burnside, William (1852–1927)
   counting theorem, 176

C

canonical homomorphism, 135
canonical representation
   of Aut(G), 172
   of Sym(n), 161
Cantor, George (1845–1918), 7
Cardan (1501–1576), 553
   formulas for roots of cubic, 526
cardinality of a set, 7
Cartan's theorem, 242
Cartesian product of sets, 3
Cauchy, Augustin-Louis (1789–1857), 555
   theorem for abelian groups, 216
Cayley, Arthur (1821–1895), 558, 567
   representation of a group, 106
   representation of the plane, 107
   table of a group, 46
center
   of a group, 63
   of a p-group, 208
   of a ring, 286
   of $Mat_n(\mathbb{R})$, 63
   of $SL_2(\mathbb{R})$, 63
   of Sym(n), 92, 197
centralizer
   of a subgroup, 205
   of an element, 186
characteristic
   function, 304
   of a field, 393
   of a ring, 286
   subgroup, 155
Chinese remainder theorem, 84, 86, 87
class equation, 195
class function, 197
Cole, F. N. (1861–1927), 559
colorings
   equivalent, 178
   number of, 178
   of a set, 177
commutative
   group, 41
   operation, 34
   ring, 275
commutator subgroup, 132
complex conjugation
   as a field-automorphism, 387
   as a group-automorphism, 155
   as a ring-isomorphism, 297
complex numbers
   addition of, 42